AF765238

Traditional Mexican Agriculture

Advances in Agroecology

Series Editors:
Stephen R. Gliessman,
University of California, Santa Cruz, California
Helda Morales,
El Colegio de la Frontera Sur (ECOSUR), Mexico

Energy in Agroecosystems:
A Tool for Assessing Sustainability,
by Gloria I. Guzmán Casado and Manuel González de Molina

Agroecology in China:
Science, Practice, and Sustainable Management,
edited by Luo Shiming and Stephen R. Gliessman

Climate Change and Crop Production:
Foundations for Agroecosystem Resilience,
edited by Noureddine Benkeblia

Environmental Resilience and Food Law:
Agrobiodiversity and Agroecology,
edited by Gabriela Steier and Alberto Giulio Cianci

Political Agroecology:
Advancing the Transition to Sustainable Food Systems, authored
by Manuel González de Molina, Paulo F. Petersen, Francisco Garrido Peña, and Francisco R. Caporal

Urban Agroecology:
Interdisciplinary Research and Future Directions,
edited by Monika Egerer and Hamutahl Cohen

Subtle Agroecologies:
Farming With the Hidden Half of Nature,
edited by Julia Wright and Nicholas Parrott

Food System Transparency:
Law, Science and Policy of Food and Agriculture,
edited by Gabriela Steier and Adam Friedlander

Traditional Mexican Agriculture:
A Basis for Sustainable Agroecological Systems,
authored by Alba González Jácome

For more information about this series, please visit: https://www.crcpress.com/Advances-in-Agroecology/book-series/CRCADVAGROECO

Traditional Mexican Agriculture

A Basis for Sustainable Agroecological Systems

Alba González Jácome

CRC Press
Taylor & Francis Group
Boca Raton London New York

CRC Press is an imprint of the
Taylor & Francis Group, an **informa** business

First edition published 2022
by CRC Press
6000 Broken Sound Parkway NW, Suite 300, Boca Raton, FL 33487-2742

and by CRC Press
2 Park Square, Milton Park, Abingdon, Oxon, OX14 4RN

© 2022 Taylor & Francis Group, LLC

CRC Press is an imprint of Taylor & Francis Group, LLC

Reasonable efforts have been made to publish reliable data and information, but the author and publisher cannot assume responsibility for the validity of all materials or the consequences of their use. The authors and publishers have attempted to trace the copyright holders of all material reproduced in this publication and apologize to copyright holders if permission to publish in this form has not been obtained. If any copyright material has not been acknowledged please write and let us know so we may rectify in any future reprint.

Except as permitted under U.S. Copyright Law, no part of this book may be reprinted, reproduced, transmitted, or utilized in any form by any electronic, mechanical, or other means, now known or hereafter invented, including photocopying, microfilming, and recording, or in any information storage or retrieval system, without written permission from the publishers.

For permission to photocopy or use material electronically from this work, access www.copyright.com or contact the Copyright Clearance Center, Inc. (CCC), 222 Rosewood Drive, Danvers, MA 01923, 978-750-8400. For works that are not available on CCC please contact mpkbookspermissions@tandf.co.uk

Trademark notice: Product or corporate names may be trademarks or registered trademarks and are used only for identification and explanation without intent to infringe.

ISBN: 978-0-367-46219-2 (hbk)
ISBN: 978-1-032-05714-9 (pbk)
ISBN: 978-1-003-19883-3 (ebk)

DOI: 10.1201/9781003198833

Typeset in Times
by codeMantra

Contents

Foreword xiii
Preface xvii
Acknowledgments xix
Author xxi

Chapter 1 Introduction to Traditional Mexican Agriculture 1

1.1 Introduction 1
1.2 Presentation 1
1.3 Background: A Historical Brief 1
1.4 The Anthropological Point of View 4
1.5 Shifting the Focus on Agriculture Studies in Mexico 5
1.6 Maize Studies in Agronomy 5
1.7 Considerations and Conclusions 6
Recommended Reading 7

Chapter 2 The Anthropological Concept of Traditional Agriculture 9

2.1 Introduction 9
2.2 The Anthropological Concept of Traditional Agriculture 9
2.3 Some Elements for the Conformation of the Concept 11
2.4 Mid-Twentieth Century Changes in Agriculture and the Third Stage of Agrarian Distribution 11
2.5 Great National Projects and Traditional Agriculture 12
2.6 Converting Traditional into Conventional Agriculture 14
2.7 Considerations and Conclusions 17
Recommended Reading 18

Chapter 3 Studies Conforming Traditional Mexican Agriculture 19

3.1 Introduction 19
3.2 From 1960s to 1980s, and New Thoughts, and Approaches 19
3.3 From 1980 to Date 21
3.4 Elements That Make Up the Concept of Traditional Agriculture 22
3.5 The First Steps 23
3.6 What We Call Traditional also Changes 24
3.7 Cultural Traditions, Food Systems, and Traditional Agriculture 25
3.8 Considerations and Conclusions 27
Recommended Reading 28

Chapter 4 Traditional Agriculture in Mexico (First Part) 29

4.1 Introduction 29
4.2 Classifications and Types of Traditional Mexican Agriculture 29
4.3 Chronology, Traditional Knowledge, and Environment in Traditional Agroecosystems 31
4.4 Traditional Mexican Agroecosystems 32

4.5 Traditional Agriculture in Regions Lacking Water ... 34
4.6 Traditional Agriculture in Regions with Water Excess ... 37
4.7 Local and Regional Development Projects ... 38
4.8 Water, Traditional Agriculture, Government Development Projects ... 40
4.9 Relevance of Irrigation Studies to Traditional Agriculture ... 40
4.10 Considerations and Conclusions ... 41
Recommended Reading ... 42

Chapter 5 Traditional Agriculture in Mexico (Second Part) ... 43

5.1 Introduction ... 43
5.2 Basic Aspects, Approaches, and Perspectives ... 43
5.3 Elements to Consider ... 43
5.4 Modernization, Urban Expansion, and Human Population ... 44
5.5 Agriculture and Health ... 44
5.6 Traditional Agriculture, Agrodiversity, and Markets ... 45
5.7 Traditional Agriculture and Diversity ... 48
5.8 Pests and Diseases in Ancient Agriculture ... 52
5.9 Agricultural Crises, Climate, and Disasters ... 53
5.10 Droughts, Deforestation, Climatological Events, and Volcanism ... 55
5.11 Recreation of Traditional Agroecosystems ... 56
5.12 Maize and the Emergence of Traditional Agriculture ... 58
5.13 Considerations and Conclusions ... 59
Recommended Reading ... 60

Chapter 6 How Traditional Agriculture Has Been Studied ... 63

6.1 Introduction ... 63
6.2 Centers of Agricultural Origin ... 63
6.3 Characteristics and Components of Traditional Mexican Agriculture ... 66
6.4 Diversity and the Combination of Components in Traditional Agriculture ... 67
6.5 Small-Scale Farmers and Solidarity ... 70
6.6 Traditional Agriculture: Diversity, Agrodiversity, and Cultural Diversity ... 72
6.7 Recent Discussions on Agrodiversity ... 73
6.8 Prejudices and Traditional Agriculture ... 76
6.9 External Elements to Consider ... 77
6.10 Considerations and Conclusions ... 78
Recommended Reading ... 79

Chapter 7 Studies of the Origins of Agriculture in Mexico ... 81

7.1 Introduction ... 81
7.2 Background ... 81
7.3 Vegetative Agriculture and Grain Agriculture ... 81
7.4 Mesoamerica, and the First Plants Domesticated ... 82
7.5 Domesticated Maize Diffusion in Mesoamerica ... 82
7.6 The Middle Basin of the Balsas River ... 84
7.7 Archaeological and Genomic Evidence ... 86
7.8 Domesticated Maize in Tropical Regions: Veracruz, Tabasco, Chiapas, and Honduras ... 87
7.9 Diffusion and Dispersion of Domesticated Maize ... 89

7.10 The Central Valleys of Oaxaca 89
7.11 The Tehuacán Valley 90
7.12 Maize Domestication in Tehuacán: According to the Genomics Studies (2009–2016) 95
7.13 Mesoamerican: Some Archaeological Remnants of Flora Evidencing Usage and Early Human Consumption 96
7.14 Considerations and Conclusions 97
Recommended Reading 97

Chapter 8 Domesticated Maize Dispersion 99

8.1 Introduction 99
8.2 Background 99
8.3 Routes to Southern Mesoamerica 100
8.4 Routes to Northern Mexico and Southern USA 100
8.5 Journey of Maize Outside the American Continent 104
8.6 Changes in the Cornfield and Food System 107
8.7 Cornfield Changes Outside Mexico 107
8.8 Pre-Hispanic Plants and Uses That Were Modified: The Conformation of the Mexican Model of Agriculture 108
8.8.1 Vines 108
8.8.2 Agave 110
8.9 Emergence of Traditional Mexican Agriculture 113
8.10 Considerations and Conclusions 115
Recommended Reading 116

Chapter 9 History and Discussions about Homegarden in Mexico 117

9.1 Introduction 117
9.2 Brief Historical Background 117
9.3 Homegarden Agroecosystem: Its Antiquity 118
9.4 Forest and the Homegarden as Hunting Places in Los Tuxtlas 119
9.5 Homegardens in the Tehuacán Valley 121
9.6 The Incidental Homegarden/Cornfield in the Tehuacán Valley 122
9.7 Emergence of the Mexican Homegarden, Orchard, and Gardens in the Sixteenth Century 124
9.8 Considerations and Conclusions 127
Recommended Reading 127

Chapter 10 The Central Highlands and the Homegarden 129

10.1 Introduction 129
10.2 The Nahua Homegarden, Orchards, and Gardens in the Sixteenth Century 129
10.3 The Novo-Hispanic Homegarden and Its Conformation 132
10.3.1 The Format ion of the Novo-Hispanic Homegarden 137
10.4 New Spain Types of Homegarden 140
10.5 Flower Gardens: In the Pre-Hispanic Time 144
10.6 Forests, Homegardens, and Gardens 145
10.7 The Indigenous Homegarden in Viceroyalty 146
10.8 Considerations and Conclusions 147
Recommended Reading 151

Chapter 11 The Ancient Mayan Homegarden and Its Transformations 153

11.1 Introduction 153
11.2 Background 153
11.3 The Homegarden and the Mayan *Solar* in Yucatan 154
11.4 The Homegarden and the Cornfield 156
11.5 The Mayan-Yucatecan Homegarden 156
11.6 The Ancient Maya-Yucatecan Homegarden 160
11.7 Domestic Fauna Cited in the Yucatan *Relations* 161
11.8 Working Tools and Changes after 1524 164
11.9 Changes in Homegarden and Cornfield Vegetation 164
11.10 Considerations and Conclusions 166
Recommended Reading 167

Chapter 12 History and Discussions about the Homegarden 169

12.1 Introduction 169
12.2 Background 169
12.3 Forest and the Homegarden as Hunting Places 170
12.4 Management of Domestic and Wild Fauna 171
12.5 Wild and Tolerated Edible Green Weeds in Homegardens 174
12.6 Is the Homegarden a Sustainable System? 174
12.7 Hunter-Gathered Bands, Their Sedentary Process, and Changes in the Food System 176
12.8 Deforestation and Ecological History 177
12.9 Homegarden Plant Diffusion and Domestication 179
12.10 Considerations and Conclusions 180
Recommended Reading 185

Chapter 13 History, Changes, and Diversity of the Homegarden 187

13.1 Introduction 187
13.2 Background 187
13.3 The Sixteenth-Century Homegarden 188
13.4 The Homegarden and Its Changes 191
13.5 Diversity and Change 193
13.6 Kitchens, the Homegarden, and Open Spaces 195
13.7 Considerations and Conclusions 198
Recommended Reading 202

Chapter 14 The Current Mexican Homegarden 203

14.1 Introduction 203
14.2 Changes in the Homegarden 203
14.3 Current Situation of Some Homegardens in Mexico 204
14.4 Homegarden Diversity 206
14.5 Homegardens and Orchards for Commercial Purposes 208
14.6 Considerations and Conclusions 213
Recommended Reading 213

Chapter 15 The Cornfield (*milpa*) in Mexico and the Food System 215
15.1 Introduction 215
15.2 What Is a Cornfield? 215
15.3 How the Cornfield Has Been Defined 218
15.4 Characteristics of Some Cornfields in Mexico 220
15.5 Emergence of the First Cornfields (*milpa*) in Mexico 224
15.6 The Cornfield in the Central Highlands in 1519 225
15.7 The Cornfield and Catastrophic Climatological Events, Epidemics, Diseases, and Famines 228
15.8 Old World Plants, Famines, and the Cornfield in New Spain 229
15.9 Changing the Native Production Systems 230
15.10 Considerations and Conclusions 231
Recommended Reading 231

Chapter 16 The Cornfield Evolution in Mexico 233
16.1 Introduction 233
16.2 Background 233
16.3 Early Mesoamerican Cornfields 233
16.4 The Cornfield in the Sixteenth Century and the Plants That Arrived from Spain 236
16.5 The Nineteenth-Century Cornfield 237
16.6 Marketing and Change in Agroecosystems: Agave and Pulque 238
16.7 The Cornfield: From the End of the Nineteenth to the First Quarter of the Twenty-First Century 240
16.8 Considerations and Conclusions 243
Recommended Readings 243

Chapter 17 Some Types of Cornfield in Contemporary Mexico 247
17.1 Introduction 247
17.2 Background 247
17.3 The Itinerant Cornfield in the Totonacapan 250
17.4 Marsh Agriculture: The Cornfield and March Maize (*Marceño*) in the Wetlands 254
17.5 El Palmar in Central Veracruz: The Cornfield and the Orchard 255
17.6 St. Isabel Xiloxoxtla, Tlaxcala: Combining Seasonal and Irrigated Cornfields 257
17.7 Traditional Agriculture and Industry 258
17.8 Agricultural Cycle 259
17.9 Small-Scale Irrigation in Traditional Agriculture 260
17.10 Commercial Agriculture 261
17.11 Traditional Agriculture and Migration 262
17.12 Daily Life and Cornfield Small-Scale Farmers 263
17.13 Steep Places and the Seasonal Cornfield: The *Tlacolol (e)*, the *Huamil*, the *Coamil*, and *Ecuaro* 263
17.14 Los Tuxtlas, Southern Veracruz: The Zoque-Popoluca Cornfield 267

17.15 St. Cruz Yagavila, Oaxaca, and the Altitude Cornfield Management, and Seed Selection 269
17.16 Considerations and Conclusions 274
Recommended Reading 285

Chapter 18 The Cornfield and the Food System 287

18.1 Introduction 287
18.2 Background 288
18.3 The Cornfield and the Food System in Its Early Stages 288
18.4 The Central Valleys of Oaxaca: Food System 290
18.5 The Tehuacán Valley Food System 291
18.6 Los Tuxtlas, Veracruz: Food System 292
18.7 Traveling South: the Food System in Southern Mesoamerica 294
18.8 The Basin of Mexico: Food System 297
18.9 The Tlatelolco Market (1536–1577) and the Food System 300
18.10 The Cornfield and the Nineteenth-Century Food System 303
18.11 The Contemporary Mexican Food System and the Cornfield 306
18.12 The Cornfield, Edible Green Weeds, and the Food System 310
18.13 The Mexican Cornfield and the Food System Today 312
18.14 Considerations and Conclusions 314
Recommended Reading 320

Chapter 19 Diversity, History, and Geography of Maize in Mexico 321

19.1 Introduction 321
19.2 Background 321
19.3 Some Changes in Agroecosystems and Weeds 321
19.4 Maize Travelled the World 322
19.5 Some Current Uses of Maize 324
19.6 Cultivated Maize for Forage 324
19.7 The Current Context of Maize in Mexico: Some Figures 325
19.8 Tolerated Edible Green Weeds in the Cornfield Agroecosystem 328
19.9 Considerations and Conclusions 330
Recommended Reading 336

Chapter 20 Soil, Water, and Erosion in Traditional Agricultural Systems 339

20.1 Introduction 339
20.2 Background 339
20.3 The Maya Lowlands 339
20.4 Soils in the Slash and Burn Cornfield and the Homegarden 341
20.5 Archaeological Evidence 342
20.6 Viceroyalty and Expansion toward the Hills: The Soil Erosion 344
20.7 Mexican Agave Fermented Mead Estate *(Hacienda Pulquera)*, and Soil Erosion Control 346
20.8 Management of Erosion in the Puebla–Tlaxcala Valley 348
20.9 Discussion on the Productivity and Knowledge of Soil in Mesoamerica 351
20.10 Soil Improvement in the Pre-Hispanic Era 352
20.11 Elements in the History of Soil in Mexico 356
20.12 Considerations and Conclusions 358
Recommended Reading 359

Chapter 21 Nature, Ideology, and Culture in Small-Scale Farmer Societies 361

21.1 Introduction 361
21.2 Background 361
21.3 Ideology, Cosmology, and Worldview in Traditional Agriculture 362
21.4 Ceremonies, Myths, Rites, and Beliefs Concerning Traditional Agriculture 363
21.5 Maize and the Worldview of Small-Scale Farmers 364
21.6 Natural Resources, Special Foods, Characters, Agricultural Festivals, and Celebrations 1980–2020 367
21.7 Xopilapa, Veracruz Central, and Conservation of the Environment 368
21.8 Los Tuxtlas in Southern Veracruz: Homshuk, Maize, and Seed Selection 369
21.9 El Naranjal, Quintana Roo, and Aid Against Hurricanes 369
21.10 St. Marta Chenaló, and the Gods of the Hills in the Highlands of Chiapas 370
21.11 The Matlalcueye, or Malinche Volcano Western Slopes 370
21.12 The Xinantecatl Volcano in the Toluca Valley in the State of Mexico 371
21.13 Considerations and Conclusions 371
Recommended Reading 374

Chapter 22 Some Final Considerations and Conclusions 377

22.1 Introduction 377
22.2 Background 378
22.3 Small-Scale Traditional Agricultural Societies: Foreign Impacts 379
22.4 Small-Scale Farmer Societies, Culture, and Agriculture 381
22.5 Relations with Other Disciplines 383
22.6 Some Reasons Why Traditional Agriculture Is Important 384
Recommended Reading 387

Glossary 389

Historical Sources 423

References 427

Index 477

Foreword

I met Alba Gonzalez Jácome for the first time in Cárdenas, Tabasco, Mexico, in July 1978. I was the organizer of a 4-week intensive course on tropical agroecology at a small college of tropical agriculture, the Colegio Superior de Agricultural Tropical (CSAT). The focus of the course was on building an ecological foundation for the development of alternatives to the large-scale Green Revolution monocultures of chemical-intensive crops beginning to dominate the landscape of this wet, humid tropical region. Alba enrolled in the course as a student, but I quickly learned that she was much more than a student. She was a new professor in the Anthropology Department at the Jesuit Iberoamericana University in Mexico City. When I asked her why an anthropologist was interested in taking an agroecology course, she explained that she viewed agriculture as the co-evolution between culture and the environment. She had studied the cultural side of this co-evolution, but not the environmental or ecological side. The course title had captured her eye. (At this time agroecology was a new academic term developed at CSAT.) She wanted to deepen her understanding of the ecological side of cultural ecology. So we agreed she could attend the course, but in the combined status as both an instructor and a student.

The CSAT was a small agricultural college located in the middle of a 90,000-hectare World Bank-funded green revolution project where tropical forest was removed and replaced with large-scale monocultures of "improved" high input varieties of corn, beans, sugar cane, rice, and pastures for intensive cattle production. The college was built to train young students from the tropics, in the tropics, producing a cadre of agronomists who would be capable of solving any problems that this new technology might encounter. I had come to the CSAT to teach ecology to these budding agronomists but soon found that natural ecosystem ecology was of no interest to them. They had been educated to increase production in farming systems. To get them to grasp ecological concepts, I had to begin teaching ecology in an agricultural context, and it was in this setting that *agroecología* emerged.

Those of us building this agroecological approach at the CSAT were able to observe the failings of large-scale industrial monocultures early on. These are common knowledge today with our deepened understanding of sustainability. These failings were exacerbated under the warm, humid conditions of the lowland tropics where we measured rainfall in meters, not centimeters. We also noticed that there was another agriculture in the small villages and roadsides of Tabasco – the traditional Maya agriculture covered so elegantly in Alba's book. We began doing ecological studies on farms, with farmers, using their seeds and practices, and began to get answers to questions such as how did their corn crop produce so much more when beans and squash were intercropped with the corn (the traditional three sisters!) as compared to corn planted by itself, how was it possible to grow 3–5 times more corn grain in seasonal wetlands than in the green revolution mechanized and mono-cropped fields, and how was it possible to integrate such high species diversity in the tropical home gardens around the dwellings in the traditional Maya communities? We found many ecological answers grounded in complexity, diversity, mutualisms, and reciprocity. It was Alba who pointed me toward the cultural side of these agroecosystems.

During the course, Alba stressed how important it was to respectfully engage traditional farmers in an exchange of knowledge, where "western" science equitably interacted with traditional or empirical knowledge. In a sense, what I was experiencing for myself was interculturality, where I was engaged in a process of co-creating knowledge that was grounded in both ecological science and cultural experience. Alba also insisted on using a participatory process and observation, where all voices were listened to and valued, especially in designing and carrying out field research. Finally, Alba stressed not only how much value traditional knowledge had, but how much of this knowledge was being ignored and lost, especially with the loss of the intergenerational transfer of knowledge in families and communities. She wanted to show that this knowledge was not backward or outdated, but that it had a long history of place-based change and evolution that meets the needs of local environments and cultures.

Ever since that early course in agroecology, I have deeply valued the journey Alba and I have shared, both professionally and personally. For the past four decades, we have maintained our trans-disciplinary dialogue – where the ecological and anthropological fields have contributed to the development of the field of agroecology. We have collaborated on research projects from upland Tlaxcala to lowland Tabasco. We have participated in writing projects, from research articles to book chapters. We have shared teaching responsibilities in agroecologically related courses from UC Santa Cruz, to the Iberoamericana University in Mexico City, to the International University of Andalusia in Spain. Many hours of fruitful discussion about agroecology and traditional knowledge were accompanied by home-cooked meals and sustainably produced red wine.

And now, you the readers of this book, will be able to partake of this unfolding knowledge-building process. Blending her historical anthropology and cultural ecology approaches with agroecology, Alba has amassed an amazing story of the traditional agriculture of Mexico. From pre-Hispanic time, to the Spanish Conquest, to the long road to Mexican independence, to challenging programs of land reform and rural development, to the current challenges of technological packages, industrialization, trade agreements and globalization, these agroecosystems have often persisted and some have even flourished. However, the loss of people from the rural regions of Mexico, a struggling economy, and the push toward an export agriculture put all of this knowledge at great risk. The detailed description of the incredible diversity of traditional knowledge mirrors the vast cultural diversity and ethnicity of Mexico. Having all of this knowledge and experience presented in one book is a monumental accomplishment and is the result of a life-long journey.

I invite you to take this journey in Alba's book. Travel from the diverse traditional corn fields (the *milpa*), to the multifunctional home gardens (the *huertos*), to other cropping systems. Explore agroecosystems adapted to steep slopes, to low areas and wetlands, from dry to humid environments, and from temperate to tropical parts of Mexico. Absorb the complexity of forms, structures, management, and uses that traditional agroecosystems have taken over the centuries. Cultural connections to food and agriculture form the underpinnings of this long history. Alba has devoted her life to capturing the stories of the traditional agriculture of Mexico. Similar studies are needed for other parts of the world, and Alba's book is a model for how to do this. The sooner this happens, not only will invaluable information and experience be captured and kept from being lost, but the combined ecological and cultural foundations for building meaningful sustainability can be gained.

Thinking back to that agroecology course in 1978, I also remember visiting a small Mayan village. It was the first time Alba and I had visited traditional agroecosystems together, and it began with a walk through the cluster of simple houses and their surrounding gardens. After we had passed through the village on our way to a farmer's *milpa*, she asked me if I had seen the bicycle leaning against one of the houses. I had seen it but thought nothing more than it being a bicycle. She asked me if I had seen the machete that was strapped to the back. No, I had not. What followed for me was a lesson in cultural observation. The machete had an entire story for Alba. Its length, width, shape, and sharpness told her that the farmer had a cornfield that he had to travel to, how far apart the corn was most likely planted, and that it was weeding time in the field. Alba had spent the time necessary, and in those days probably with a machete in her own hand, learning the intricacies of practices and knowledge that had evolved over many centuries. From that time on, I could never look at traditional farming systems the same way again.

There is so much to learn from traditional farming systems and livelihoods that the conventional agronomist too often does not value. It's no wonder that the top-down extension model of information transfer in agriculture has failed most of the rural world, especially in Mexico. With Alba's merging of agroecology and culture, the value of place-based and local knowledge is clear. It is a co-evolutionary process that is much more appropriate for developing the participatory partnerships that can lead to the co-generation of the knowledge needed for sustaining the food and farming systems of today and the future. We are all deeply indebted to Alba's contribution to agroecology and for showing us how the cultural foundations of sustainability must grow on the deep roots of the farming traditions in Mexico.

Steve Gliessman
Cuyama Valley, CA

Preface

The book does not intend to cover everything that has been published and discussed on traditional agricultural systems or the agroecosystems that make them up. Some texts were selected, based on the academic discussions we have had that were generated by the topic and the research done on it, and the progress made, allowing us to know the evolutionary, historical, and current situation of these agricultural systems in Mexico. There is no intent to include in this text all the research, academics, and people who have been involved in studies related to traditional agriculture, so I apologize to those whom I did not quote. This is a brief history of history, an environmental history, or an anthropological-ecological approach, focus in cultural ecology, for the studies in Mexican traditional agriculture.

This book begins with the studies and points of view of the cultural anthropology developed in Mexico. It continues with the advances in the studies in agriculture, agricultural systems, and agroecology realized in ecological anthropology, especially the studies on Mexican agriculture. It is clear that there are many other points of view, which is enriching, given the magnitude and importance of this field of study. The research results presented here were carried out by Mexican, American, and other specialists, whose studies were conducted in Mesoamerica and arose in this regard, generating new research, studies, questions, approaches, and new problems to be studied.

The Mexican Central Highlands are the principal focus in this book, including some comparative elements from other Mesoamerican regions in the north, northwest, and northeast of actual Mexico. The book includes historical data from New Spain and the nineteenth-century and contemporary Mexico. It covers studies from the fields of Social, Cultural, and Ecological Anthropology; Ethnology; Regional and Environmental History; and Agroecology. In addition, it takes into consideration several of the studies from other disciplines that have explored basic aspects of traditional agriculture in Mexico. Their findings on traditional agriculture have been widely disseminated, generating new questions, proposals, and discussions.

The knowledge arising from these investigations and their results is the basis, the background, and the information on which new lines of research are developed, to achieve a better knowledge of what is happening in the Mexican countryside. This strengthens the existing traditional agroecosystems and improves the living conditions of rural populations, particularly small-scale agricultural ones, showing the different types of work organization, and the roles of the families, kin, and close friends in these social and agricultural processes, and related ones, which arise from the interrelations among them.

This book is the result of a holistic study of Mexican agriculture. It offers the reader a perspective of traditional agriculture in Mexico from anthropology and its interdiscipline with the contributions of history and scientific disciplines to help obtain sustainable agroecology, where human societies obtain better ways of life, a healthy and nutritious food system, so the elements harmful to the health of both Man and the environments in which he lives and operates daily be reduced. The book also aims to recover ideas, management, and components of local knowledge of small-scale farmers at the different times when research on traditional agriculture in Mexico was carried out.

These investigations were coordinated by me and by graduate students in social anthropology at various educational institutions in Mexico, mainly at the Universidad Iberoamericana. In addition to the students, whose results are expressed in their theses and research that I constantly cite in this book, several professors and colleagues participated with their physical presence, financial, and academic support, and whom I thank for all the time they spent to discuss, give classes, and be part of the thesis committees of the then students, who are now doctors that continue working in the field of ecological anthropology and traditional agriculture.

Preface

Acknowledgments

Alba González-Jácome appreciates and gives thanks for the valuable opinions and contributions to this book by Drs Stephen R. Gliessman (Ecology and Agroecology), Artemio Cruz-León (Ethnobotany and Ethnoagronomy), Francisco J. Espinosa-García (Chemical Ecology), Francisco Gurri-García (Ecological Anthropology), William L. Crothers (English Grammar Revision), Jesús Sales-Colín (Ecological Engineering, and Social Anthropology), and Ernesto Cisneros-González. (Communication Sciences). I also appreciate the support given to me by the folks at codeMantra who have done careful editing and revising of the book.

Author

Alba González-Jácome has carried out studies in education, history, anthropology, and agroecology in different Mexican institutions of higher education. She also conducted studies in economic and demographic anthropology at the University of Texas at Austin. She has investigated various agricultural systems from the historical, social, cultural, and environmental points of view since 1969. To do this, she organized several research seminars and carried out fieldwork *in situ* in various parts of Mexico, directing undergraduate and graduate students in anthropology in various parts of Mexico. With the support of her undergraduate, graduate, and postgraduate students in Social Anthropology at the Iberoamericana University, she was able to carry out numerous research projects in various parts of Mexico. These resulted in several research projects and the direction of some 100 theses of the participants in said projects. Among these were the projects of Tlaxcala, Central and South Veracruz, Campeche, Sierra de Juárez in Oaxaca, Tabasco, Chiapas, the State of Mexico, and Sonora.

Alba González-Jácome performed various projects on the history of the ports of Veracruz, Alvarado, Acapulco, and San Blas. She has also studied the history of plants like coffee, vanilla, and cacao in Mexico, and the history of agroecosystems such as terraces and semi-terraces, homegardens, orchards, gardens, ridges, cornfields, and irrigation systems in various parts of the country. From her teachers Agustín Cué Cánovas, Ángel Palerm, Richard N. Adams, Bernardo García Martínez, and Alfred H. Siemens, she received the bases of interdisciplinary scientific knowledge and fieldwork.

With Dr. Stephen R. Gliessman, she carried out fieldwork in Tabasco and Tlaxcala. From Dr. Gliessman, she received academic encouragement to enter the interdisciplinary field of agriculture studies and training in the field of ecology and agroecology. She performed fieldwork with Dr. Alfred H. Siemens in central Veracruz, Marqués de Comillas in Chiapas, and in the Sierra Juárez of Oaxaca. With a Polish doctoral student, she did research in the peasant area of northern Poland. With a colleague from Paraguay, she undertook field research in the indigenous areas of the Chaco and in the east of that republic. With Dr. Juan Maestre Alfonso, a Spanish colleague from the University of Seville, she did a brief recent (2018), but very fruitful, research of the maize-producing areas in Galicia.

Throughout 50 years dedicated to teaching and researching, the author has received several awards for her work and contribution to the training of young researchers in the field of ecological anthropology (cultural ecology) and agriculture. Among them, the Miguel Guridi y Alcocer 2013 Medal for 40 years of research in Tlaxcala and the 2019 recognition of the Mexican Society of Agroecology for her contribution to social and cultural studies in agroecology. Currently, she belongs to the National System of Researchers (SNI), with level III, and to the Mexican Academy of Sciences. She is an emeritus professor at the Universidad Iberoamericana campus Mexico City and also a visiting professor at the Direction of Regional Centers at the Universidad Autónoma Chapingo in Texcoco, State of Mexico.

1 Introduction to Traditional Mexican Agriculture

1.1 INTRODUCTION

This book deals with some of the multiple aspects, project proposals, research, disciplinary, and interdisciplinary discussions that shaped traditional agriculture, along with advances in this branch of knowledge, and the results of recent discussions, hypotheses, and studies. These include elements of various disciplines such as Social, Cultural, and Ecological Anthropology (Cultural Ecology), Agroecology, and data related to agriculture in History (XVI–XIX centuries), interrelated with some data, and ideological questions in the evolution of the traditional Mexican agriculture (Prehistory, Archaeology, Regional, Environmental History, and Ecology).

A large number of traditional agroecosystems and their local and regional variants, which still exist in the country, made it difficult to register all of them; hence, there are only a few examples, which represented specific important advances, points of view, and problems to be discussed. The book deals with data on homegarden and cornfield (*milpa*) agroecosystems because in Mexico they are the oldest, and in them the domestication of plants began, starting thousands of years ago, and especially of maize, that has been the axis of the Mexican food system for thousands of years.

1.2 PRESENTATION

The agriculture in Mexico was modeled on the commercial agriculture of the United States. This situation drastically modified large areas of the country that became regions with monocultures for export exclusively (Duch 1992: 81–96). The scope, experiences, local knowledge, and considerations of studies in traditional agriculture are the basis for the creation of new fields of agricultural creation, approaches, proposals, and actions, particularly those that have developed in recent decades. Among them are agroecology, organic agriculture, permaculture, alternative agriculture, and other options, which emerged in the last century. This book aims to be the first step to make available to those who participate and promote new proposals, and options some of the results that traditional agriculture had achieved in the thousands of years since it arose, and in the little over one hundred years of academic life dedicated to it. The author hopes to provide elements that support some aspects that have been proposed or developed in the new options to achieve an agriculture that is ecologically and economically more sustainable and harmonious with the environment.

1.3 BACKGROUND: A HISTORICAL BRIEF

The great changes in traditional agriculture began with its invention, diffusion, evolution, the introduction of crops, and agricultural technology in pre-Hispanic Mexico, and the Viceroyalty. Since pre-Hispanic times, there were documents – such as the Mendocino Codex – made to control the taxes and sales taxes of the peoples for the benefit of the Mexica State. These documents were the basis for the construction of the formal economy in New Spain. Traditional agriculture was unimportant for local, regional, vice regal, or Crown authorities; both its study and its situation were left aside.

Travelers' chronicles written in those times related the functioning and descriptions of problems with the banditry of the time, on roads near zones where traditional agriculture was located, but the agricultural landscape was secondary and laterally mentioned (Carot and Hers 2011). Priests

DOI: 10.1201/9781003198833-1

addressed to their ecclesiastical authorities and public officials usually wrote the texts. These writings generally described – in passing – some components, characteristics, elements, technology, space and towns, and actions of human groups; and management, ceremonials, or cosmogony related to isolate aspects of traditional agriculture (Taylor 1990).

Some historical sources data were included, and they provide valuable information about the flora and fauna of New Spain as well as its introduction to the new landscapes in the sixteenth and seventeenth centuries. One example is the writings of the friars Alonso de Molina (1571), Alonso Ponce (1586), the Memorials of Alonso de la Mota y Escobar (1608 to 1612), the texts of Francisco de Burgoa (1672), and memoirs of Francisco de Ajofrín (1763). Including the *Geographical Relations* of 1579–1582, 1777–1778, 1789–1792, the writings of Juan de Navarro (1801), and many others.

In general, these sources do not provide much information about traditional agriculture, and its interrelationships with other activities, to which they refer in passing, but the data obtained from them is important. These little described aspects obtained from the historical sources on traditional agriculture helping to the reconstruction of past agroecosystems. There is little information on activities such as harvesting, capturing of wildlife, fishing, or hunting, which are very old and existed before crops but have accompanied agriculture and have supplemented the food systems of human groups so far. In general, the activities linked to traditional agriculture have been seen separately, without being related to each other, or with the food system until now (Naranjo-Piñera; Bello-Baltasar; Estrada-Lugo; Mariaca-Méndez; and Macario-Mendoza 2013).

The Rodríguez-Galicia study (Prologue in 2017; Chamomile 2017: 9) on Mesoamerican fisheries – written when he was dealing with archaeological excavations in Teopancazco, a site located southeast of Teotihuacan – identified and analyzed the coastal fauna that came to the table "[...] through a caravan system, with stops on allied sites, and through the multifaceted network of symbols, rituals, and communal banquets that a complex corporate group nestled in the Central Highland wove around the sea, and its inhabitants". The representations that were used to obtain information on fishing products in the aforementioned vicinity of Teotihuacan were glyphs, murals, codex, ceramics, and stone instruments. Information on the sources and the archaeological study were used.

Salting and smoking did the conservation of the fishing resource since it came from the coast of Veracruz. The preference for these fishery products left without explanation the resources obtained from the lagoon system of the basin of Mexico, much closer to Teotihuacan than the coast of the Gulf of Mexico. Consequently, this opens the question of the importance of long-distance trade. Fish, which was part of Moctezuma diet, was taken from the coast of Veracruz to Tenochtitlan (González-Jácome 2011: 142). These studies are more related to the food system, which will be discussed in Chapter 16.

There is scant information related to the ways of cooking food in pre-Hispanic times, although there are references to the forms of cooking when the Spanish arrived on the continent and in Mesoamerica. Very illustrative mentions of these activities appeared in historical sources occasionally, like that of Gonzalo Fernandez de Oviedo, who in his *Summary of the Natural History of the Indies* (2011 [1526]: 64–65) wrote about fishing, and hunting among the natives of Cuba, Hispaniola, and other islands of the Antilles, places where the natives roasted the fish on the same day they were caught.

> These Indians have their seats, some near the sea, and others near the river or stream, where there are fisheries, because commonly their main maintenance, and more ordinary is fish, and because they are very inclined to it, because more easily there may be plenty, better than pig, and deer savages, which they also kill, and eat. The way they fish is with nets, because they have them and they know how to make very good cotton, of which nature provided them at length, and there are many forests, and mountains full; but what they want to do more white, and better, heed it, and plant it in their seats next to their houses or places where they live.
>
> The deer and pigs arm them with traps, and other nets, where they fall, and sometimes they mount, and browse them, and with a quantity of people they arrest them, and reduce to place that they can with

FIGURE 1.1 Tlaxcala Market in the Sixteenth Century. Mural: Tlaxcala City, Governmental Palace.

arrows, and rods thrown kill; and after the dead as they have no knives for skinning cut them, and make them parts with stones, and flints, and pass them on sticks that they put, like grills or trivets, in a hole which they call barbecues, and the fire below, and in that same way they roast the fish.

Fernandez de Oviedo (1944, Vol. 8: 252–253) when describing the Tlaxcala market in 1519 narrated that it sold "[...] fish from the rivers [...] legumes, and things that the Indians eat, and more good fruits [...] a lot of wood, charcoal, and herbs to eat, and use for medicinal ends [...]". In 1609, Bishop Alonso de la Mota y Escobar (2013: 67–73), traveling along the Gulf coast from the north of Veracruz to Tampico, described the fisheries owned by Spaniards and Indians where white shrimp was smoked and roasted for its preservation and sale in regional markets. To date, the consumption of fish and sellfish in Mexico has been very important to the food system, and it is eaten fresh, smoked, or dried (Figure 1.1).

Historical sources allow researchers to understand the situation of traditional agriculture, transformed by the introduction of a series of components from the Old World that arrived in New Spain after 1519. During the New Spain Viceroyalty, there was little specific information about traditional agriculture systems. The studies mainly referred to conventional agriculture, aimed at the domestic market, and exportations, whose figures and components were linked to the income of the State through the payment of sales taxes (*alcabalas*).

Taxes in New Spain were charged by people related to the public accounting records in the Viceroyalty government, including: the Reports of the Viceroys to his Majesty, Sixteenth and Seventeenth Century *Geographical Relations*, the controls for the entry, and exit of goods in ports, the relationships on valuations, sales taxes (*alcabalas*), the lists of merchandise that was shipped in the ports, and other types of payments. The small-scale Indian farmers pay taxes with a combination between money, cacao beans, and domestic fauna.

The economic interests of the Crown and of Hispanic *encomenderos*, miners, merchants, and religious in New Spain were very different from those of traditional agriculture destined to feed indigenous, mestizo, and even Hispanic families, but their main objective was local, and regional-scale market, or for export to Spain, and its domains.

After 1821, with the emergence of independent Mexico, a series of travelers coming from the industrialized countries arrived in the newly established nation and described the Mexican economy. Those people were interested in commercial agriculture, and not in small-scale agriculture, which was the subject of some comment only briefly. Their reports referred to branches of the economy, and periodic publications appeared where the agricultural situation of the country was included. Reports were directed by the Ministry of Development, Colonization, Industry, and Commerce, which was reorganized in 1853 and restructured in 1891 when it lost some of its functions, and the secretariats of communications and finance were created.

Regional studies on crops (e.g., sugarcane, rice, coffee, Agave, rubber, and sisal hemp) important for the national economy and for export were done and published under the direction of the Ministry of Economy. In addition to reports, statistical tests, specialized studies, reports on economic activities, and geographical investigations in the national territory were included (Blanco-Martínez and Moncada 2011; Kaerger 1986). In the nineteenth century, Mexican landowners talked about modernizing agriculture, for which they acquired the machinery available at the time designed to introduce mechanized planting and harvesting systems, besides transportation to move their productions to national and international markets (Nickel 1996; Tutino 1990).

Despite the interest of some national entrepreneurs to modernize the countryside, and in general the economy of the country, some proposals took longer to develop than others. An example was the introduction of the railroad, which began in 1837, although the work progressed slowly, and it was only until 1872 when the 470,750km of the route running from the port of Veracruz to Mexico City was completed[1]. The estate holders (*hacendados*) of middle and large agrarian properties faced problems over their properties constantly with the small-scale farmers of the neighboring towns to obtain labor force in activities, like sowing and harvesting, their lands required. In the Central Highlands, these problems between landowners and neighboring towns had been detailed in information regarding the southwest of Tlaxcala and the state of Morelos (Nickel 1996; Rendón-Garcini 1994; Tutino 1990).

1.4 THE ANTHROPOLOGICAL POINT OF VIEW

In the Mexican case, anthropological research projects and studies on traditional agriculture –particularly the agriculture developed among indigenous peoples – are a hundred years old (1916–1919) since being developed and published (González-Jácome 1988: 55–189). Although the evolution and history of traditional agriculture are millenary, anthropological studies cover little more than a hundred years of their development. In 1916, *Suave Patria*, the book written by Manuel Gamio, promoted the study of agriculture in Mexico as an important and integrating factor for the development of the country. By 1917, Gamio created the Anthropology Direction in the Ministry of Agriculture and Development, from where he directed his project on Teotihuacan, whose results were published in 1922.

Some studies were carried out in the first third of the twentieth century with the purpose of describing and understanding the agriculture of Mexican small-scale farmers, especially the indigenous ones, to make proposals for its improvement. O.F. Cook's report (1919: 307–326), sent to the Smithsonian Institution of Washington DC, was published in 1921, in which he proposed that cornfield (*milpa*) agriculture with slash and burn be seen as a "primitive tropical system", which contributes to soil erosion.

This negative idea of Cook about slash and burn agricultural system was shared by many public officials in Mexico and helped to cement the idea that unconventional agriculture was moving backward, destroying soils, deforesting forests, using water poorly, and not being productive enough for its existence to be justified.

At the beginning of the twentieth century, the studies on indigenous ethnic groups and their agriculture systems made that public policies showed prejudices about the capacity of traditional farming systems – like slash and burn – to sustain and feed the human population of the country. This idea was sustained in later studies. The American anthropologist Ralph Beals (1942) affirmed that, without fear of being wrong, those prejudices continue to date, and consider that agricultural methods are "archaic" because only the maize and coffee produced with them are important crops. In addition, they should disappear or be transformed into other types of more productive systems aimed at obtaining profits (Nolasco-Armas 1985).

[1] García-Panes 2019; González-Jácome 1999: 157–207, 2009: 1–15, 2010: 503–526; Rendón-Garcini 1994, 1995; Tutino 1990; Villalobos-Nájera 2012, 2013: 79–85.

In the 1940s, anthropologists studying the agricultural societies of Mexico were focusing on ethnographic studies of the indigenous groups and of the need to change their agriculture and modernize it. In addition, to introduce soy so that, according to them, its consumption would improve the nutritional quality of native diets. Examples of works carried out in this direction are wide and include the institutions that at that time were dedicated to the study of peasant populations in Mexico, mainly indigenous, but did not exclude *mestizos*. Among these we can mention the ones written by Gamio (1945, 1948), Hinton (1945), the National Indian Institute (1943a, 1943b, 1945, 1946), and the Inter-American Indian Institute (1944). It is necessary to specify that in those times there were no nutritional studies in this regard (González-Jácome 1988: 55–189).

1.5 SHIFTING THE FOCUS ON AGRICULTURE STUDIES IN MEXICO

The introduction of cultural ecology in Mexico, in the 1950s and 1960s, modifies the situation, and the indigenous approach of Mexican anthropology described above, by expanding and introducing other proposals and studies on Mexican agriculture, and including peasants in them, a concept that – in addition to the indigenous – tacitly included small-scale *mestizo* farmers (González-Jácome 2007: 19–38; Toledo 2002: 511–522). Cultural ecology is a methodological theoretical approach to ecological anthropology that – in this case – is diachronic, synchronic, and interdisciplinary. It was initially proposed by the American anthropologist Julian H. Steward (1955, 2006: 5–9), who in 1937 was appointed advisor in anthropology of the Office of Indian Affairs in the USA, and who in April 1940 had participated in the meeting in Patzcuaro, Mexico, where the Inter-American Indian Institute (III) was created.

The theoretical-methodological approach of cultural ecology and ancient agricultural systems reached Mexico in the 1940s and 1950s, as the effect of a series of studies on the origins of agriculture and irrigation in ancient societies and pre-Hispanic societies in Mesoamerica. The studies focused on the *chinampa* agricultural system, and they were carried out by several researchers, among them: Armillas (1949, 1971), Palerm and Wolf (1956: 396–399), Calnek 1972, Coe 1964a, González-Pozo (cord.) 2010, 2016, Palerm (1954, 1955), Palerm and Wolf (1972), and West and Armillas (1950).

In addition, there were discussions on the increase of human population densities, the evolution and conformation of the State, and the emergence of urbanism in ancient societies (Palerm 1955a, 1955b, 2007). At the same time that occurred in Mexican Anthropology, similar interests arose in other disciplines too, as happened in Archaeology, History, Human Ecology, Cultural Ecology, Geography, and Agronomy. Of them, included in this book are several studies, reports, and proposals whose central focus is the research of ancient human populations, their histories, and their cultures (Cook and Simpson 1948; González-Jácome 1997: 167–188; Steward 1955).

1.6 MAIZE STUDIES IN AGRONOMY

In Agronomy, the interest to study traditional agriculture arose with botanical investigations on maize (Wellhausen, Roberts and Hernández 1952). In this direction, one of the most relevant figures has been the engineer Efraím Hernández X. (1988: 664–665). In an article published in 1988, Hernández X. elaborated a list with the characteristics and elements to consider for studies on Mexican traditional agriculture. This list gave an idea of agronomists' preoccupations with traditional agriculture in the 1980s and 1990s (Hernández X. 1988: 674–675). A summary is:

1. It was practiced on a 5 million hectare surface;
2. Involved a population of 1,806,703 people;
3. Its production was of national importance (7.5 million tons of maize) (CIMMYT 1988, 1993–1994);

4. It was urgent to study its ecological characteristics (geological, geographic, and edaphic), and also the technological, agro-ecological, and socioeconomic highlights;
5. Some of the technological characteristics were slash and burn (r-t-q), agriculture sowing in holes (Hernández X. 1980), predominance of the use of work animals and hand tools, use of indigenous seeds, preservation of germplasm, and smallholding;
6. Its socioeconomic characteristics were small agricultural area per production unit, limited access to market, production for self-consumption, shortage of credit and technical assistance, no information on the results of agricultural research, increase in the monetary needs of the population, constant migration of the most qualified workforce, and traditional agriculture was in a breaking process of its basic cultural elements.

The list, despite the advances in the interrelation between culture and agriculture, does not include some basic social and cultural elements that apart from migration and some economic components are shaping and allowing the functioning of traditional agroecosystems. Some of them are composition and organization of the workforce in these agricultural systems; nonpayments in money, but in solidarity systems (mutual aid) in services (mutual help); social and economic relations among the members of the farming communities; use, management, and exploitation of resources in the areas surrounding agricultural, communal, and natural ecosystems; the set of activities that small-scale farmers' families carry out to sustain the total economic system (hunting, fishing, gathering, work in industries, and in administrative, or tourism activities, migration and emigration, among others).

According to Espinosa-Pineda (1994: 17–18) and Heyden (1983), the waterfowl of the basin before 1519 had a "complex code" in which human destiny with the next rain, the destiny of the State, and that of the harvests were intertwined in a world where nature and divination were interrelated. A series of rites of ancient origin were performed – like the one that took place in the month of *ethzalcualiztli* – where the high priests attended the four smoke kitchens next to the water (*ayauhcalli*) and splashed their hands and feet in the water, imitating the shouting of waterfowl for four days, after which they were exhausted and slept on reed mats (*petates*) dreaming and moaning (Espinosa-Pineda 1994: 21)

1.7 CONSIDERATIONS AND CONCLUSIONS

The list of social and cultural components linked to traditional agroecosystems increases as there are issues to understand such as the food system, health, and disease; ideas about the world vision and cosmogony; popular religiosity, parties, celebrations, and religious events related to agriculture; the relationship between the members of the small-scale farmer family, nature, and the agricultural cycle; the social and ideological management of risks (e.g., economic, environmental, climatological, volcanic); and vulnerability of the human population to hazardous events (e.g., droughts, flooding, hurricanes, soil erosion, and pests). There are popular ideas about the uses and management of flora and fauna in agroecosystems as well as natural systems articulated to human communities (Espinosa-Pineda 1994: 17–18; Gurri-García and Molina-Rosales 2011: 71–98).

The modernization of agriculture in Mexico drastically modified large areas of the country that became regions with monocultures for export only. The effects of supporting the country's industrialization and agricultural modernization transformed traditional agriculture and maize, making them symbols of backwardness (Dutch 1992: 81–96). Among the numerous impacts there are the disappearance areas of land edges (*bordos, bordes*) with wild and cultivated plants on them, which protected the soil from erosion, and the crops from insects that damaged them. The land edges plants disappeared but also killed off wild pollinators that sheltered themselves in them and reduced diversity.

Another impact was the use and abuse of herbicides or pesticides that had several effects on agroecosystems such as the cornfield. Tolerated edible green weeds stopped growing in the cornfield and providing vegetal elements to the daily diet and herbalism of families. This reduction in

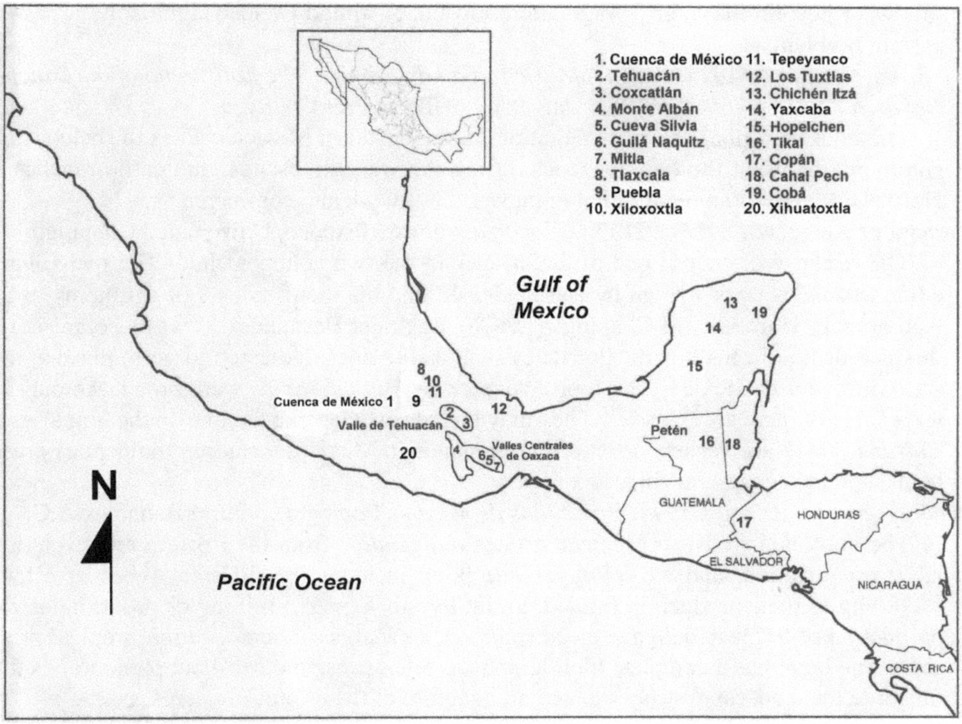

FIGURE 1.2 General Map. Mesoamerica: Places Cited in the Book.

agroecosystem agrodiversity occurred along with the reduction of labor force applied to the system, so favoring the departure of young people and adults to non-agricultural and salaried jobs elsewhere.

Finally, it is necessary to make a critical summary about the volume of research and studies in Agronomy, Ecological Anthropology (Cultural Ecology), Anthropology, and related disciplines, because articles, reports, and books appear currently in a large number of specialized scientific publications but are not always easily obtainable and are not well disseminated among non-specialists. This book includes only a small part of the accumulated existent knowledge, which it is hoped will be used to generate progress, new discussions, proposals, and further research in these nowadays interrelated cardinal areas (Figure 1.2).

RECOMMENDED READING

Cook, O.F. 1921. Milpa Agriculture, a Primitive Tropical System. In: *Annual Report*, Smithsonian Institution, Washington, DC.

The Report (1919: 307–326) was published in 1921, in which Cook proposes that cornfield (*milpa*) agriculture with slash and burn be seen as a "primitive tropical system", which contributes to soil erosion.

Espinosa-Pineda, G. 1996. *El embrujo del Lago. El Sistema Lacustre de la Cuenca de México en la Cosmovisión Mexica*. México, Universidad Nacional Autónoma de México (UNAM), Institute of Anthropological Research, and Institute of Historical Research.

The book is about ideas, beliefs, spells, rituals, representations, pleasant action related to the hunting and, or catching of birds in the basin of Mexico, including a guide of knowledge that the hunter should have in order not to take actions that would break the balance of the imagined cosmic order (Del Olmo and Roldán-Velasco. 2013 [2007]). All this went beyond the mere "pragmatic knowledge" aimed exclusively at solving a food system problem.

Harris III, Charles A. 1975 (2012). *A Mexican Family Empire. The Latifundio of the Sanchez Navarro Family*, 1765–1867. The University of Texas Press.

The book describes the work situation on the northern Mexico estates. It includes the commerce between the two countries during the war, the estates, and cattle ranches of Northern Mexico, and the internal organization of work and commerce.

Hernández Xolocotzi, Efraím. 2013. *Xolocotzia. Obras*. Texcoco, Universidad Chapingo.

The reader will choose one of the articles in the two volumes cited. The two-volume edition includes texts written by Hernández X. and his then students of agronomy at the Autonomous University of Chapingo (UACh). Engineer Hernández X. was a botanist, and this is reflected in his botanical articles such as the ones dedicated to germ plasma, and the Tripsacum in Mexico. The topics covered by Hernández X. were broad. Among the texts some of them are "classic". They may include articles like the one on the granaries in Tlaxcala; there are various articles on grasslands in Mexico, including indigenous grasslands, and the ecosystems in arid zones.

Rendón-Garcini, Ricardo. 1994. *Haciendas de México*. Fomento Cultural Banamex A.C.

The book is a study on Mexican estates (*haciendas*) from their origin and evolution, labor organization, and an epilogue. The book includes the different types of estates according to their production (mixed, grain, livestock, sugar, mining or profit, industrial production of fermented Agave mead (*pulque*), mezcal, sisal hemp, cotton, tropical products). The book has a complete bibliographical references, and excellent photographs that illustrate the landscape, structure, and architecture of the estates (*haciendas*).

2 The Anthropological Concept of Traditional Agriculture

2.1 INTRODUCTION

An important issue is whether traditional agriculture is backward, does not change, does not seek progress and, therefore, is not modernized. To enter the subject, the chapter begins with a synthesis of what has been called traditional agriculture in Mexican anthropology, including some relevant studies (González-Jácome 1988a: Vol. IV: 55–189, 2013). The study of traditional agroecosystems has differences in regard to the scientific studies, problems addressed by researchers, and the approaches of the different disciplines that have been used to do said studies.

There are important variants because the projects and studies have been developed with specific objectives to answer questions related to researchers' points of view and other external components. There are many common elements, questions, and approaches among them, especially with ecological anthropology and its branches of cultural ecology, political ecology, chemical ecology, the ecology of human food systems, and others.

Since 1982, the national public policy in Mexico has supported conventional agriculture whose productions are directed to international markets, unrestrictedly. At the same time, traditional agriculture suffered due to a lack of political and economic support. No one took into consideration that those who favored the industrialization of the country had caused several effects on the Mexican peasantry, who depended in large part on the existing conditions in the regions affected by industrialization and modernization.

2.2 THE ANTHROPOLOGICAL CONCEPT OF TRADITIONAL AGRICULTURE

The idea of modernizing Mexican agriculture has its roots at the end of the nineteenth and early twentieth centuries. Gamio (1945: 303–308; Hinton 1945: 180–182) and his postrevolutionary ideas written in the book *Forjando Patria* (1916) supported the introduction of "modern" agriculture, "[…] that improves the development conditions of indigenous groups […]", following the American conventional agriculture model. This involved buying maize from the USA, introducing soybeans to replace maize, using agricultural machinery, applying agrochemicals, and removing weeds ("bad herbs"), which prevents the transit of agricultural machinery and cargo trucks (Gamio 1945: 303–308; INI 1944, 1945; Quintana 1944: 129–132).

The "bad herbs" concept of Gamio had very distant connotations from that used by farmers in other parts of the country where there was no concept of "bad herbs", and that weeds, or other wild flora, could be very useful in understanding traditional Mexican agriculture. The "bad herbs" concept in Gamio (1945) had nothing to do with any other local study on plant classifications, and it was merely economic and functional.

This exemplifies how Gamio was based in pro of conventional agriculture, and that it should be against traditional Mexican agriculture. He was based on the idea of having open, clean fields, without trees or weeds that would hinder the passage of machinery and promote the indiscriminate use of agrochemicals needed for civilization development.

Years later, in the 1980s, a study on weeds ("good and bad herbs") was carried out by Chacón (1987) and Chacón and Gliessman (1982) in Tabasco. Small-scale farmers classified some plants as being "good or bad herbs"; to the latter are attributed characteristics such as heating the ground or burning the plants. Plants that are classified as "good forest" are considered fresh and medicinal.

DOI: 10.1201/9781003198833-2

Chacón (1978) showed how wild plants classified as poor forest have high allopathic potential and inhibit crops such as maize and beans.

On the contrary, those that are classified as good forest stimulate the radical growth of some crops. Among the "good weeds" were *Lagascea mollis*, *Guazuma ulmifolia*, *Cleome spinosa*, *Sida acuta*, and *Euphorbia* sp. The "bad weeds" were represented by *Bechum* sp., *Paspalum conjugatum*, *Melanthera aspera*, *Sida rhombifolia*, *Priva lapperlacese*, and *Panicum fasciculatum*. Chacón (1978) found weeds that could be classified as "good or bad" according to particular circumstances.

In the 1940s, in Gamio's (1945: 303–308) article on "Agricultural production, and the Industrialization of land endowed to a peasant (*ejido*)", he characterized Mexican agriculture as having: (1) immense temporary zones; (2) reduced and scarce irrigated land; (3) irregular, low, or nonexistent rainfall; (4) no effective application to fertilize crops; (5) excessive felling; (6) an imbalance of the crop regime; (7) outdated cultivated methods; (8) diseases that impeded the use of coastal land; (9) cornfield (*milpa*) that produced "the poorest agricultural crop"; (10) temporary small endowed parcels that did not allow integrated production; (11) peasants with no fund reserve, and thus were forced to emigrate or work for a low salary; (12) and peasants who receive limited official credit.

To mitigate Mexican agriculture conditions, Gamio (1945: 303–308) proposed: (1) DDT application in areas affected by malaria; (2) support of the synthetic manufacture of quinine; (3) colonization of the coasts in the future; (4) buying maize from the USA in return sowing and sending to the USA high-priced marketable agricultural products, such as tropical fruits; (5) connecting peasants with private capital; (6) industrializing peasants and combining agriculture with industry; (7) ensuring that machinery does not distort the "aesthetic" character of production; (8) facilitating the communication between villages and market centers; (9) and organizing industries according to the geographical location of raw materials.

Shortly after that, Gamio (1948) wrote in a 136-page special edition the series of elements of agricultural modernization that should be taken into account. They were that: (1) the elements of material and intellectual life are interdependent and inter-functional; (2) agriculture and food are a group´s cultural factors; and (3) studies and pilot programs are necessary. An experiment that lasted 2 years was conducted in an indigenous community, and the findings showed: (1) the introduction of the metal plow to replace the planter cane and the wooden plow; (2) the introduction of fertilizers; (3) the introduction of improved seeds; and (4) the introduction of dams or dykes to capture or to retain rainwater.

Despite unrestricted government support for modernization of the rural areas in the country, Gamio's project was rejected. However, shortly thereafter a series of projects based on these ideas were started and carried out. These were included in the proposal of the "Green revolution".

From 1943 to 1946, the promotion of agricultural products for the market was done in the *Bulletin* of the National Indigenous Institute (INI), (1943a, 1943b, 1945, 1946, 1946b). Some articles were about the economic importance and management of water coconuts, oranges, limes, soy, coffee, cabbages, and some legumes. At the same time, the INI spread the ideas of teaching "small farmers" how to build agricultural tools and women were given recipes for cooking with soybeans. Mexican and foreign anthropologists, such as Oscar Lewis (1944a: 281–314), wrote that: "[in Tepoztlán] the change of the *milpa* system to *tlacolol* is a return to more primitive horticulture, [which] solves immediate problems but it increases future problems [...]".

This opinion would conflict with the results of contemporary studies on this traditional agricultural system. Chávez-Segura and Cruz-León (2016: 201), after a thorough study on the tlacolole agroecosystem in the central-mountain region of Guerrero, consider that small-scale farmers see the system as "[...] a millenary knowledge structure in its long trajectory of resistance, transformations and adaptability [...]" that has several meanings related to agrodiversity, resistance, and agricultural invention space.

Oscar Lewis (1944a: 281–314), an American anthropologist who conducted research in Tepoztlán, Morelos, and Mexico City, considered that solutions would have to take into account problems such as population growth and the need to clear new lands that would cause erosion and wear, so the selection should generate: "[...] intensive agriculture, with fertilizers, cash crops, artificial irrigation, two annual crops, conversion of coffee and citrus into large-scale cash crops [...]". His proposals are related to the migration of inhabitants to places like Mexico City and the abandonment of agricultural activity (INEGI 1998, 2005).

2.3 SOME ELEMENTS FOR THE CONFORMATION OF THE CONCEPT

In Mexican cultural anthropology, the story begins in 1921, when Cook (1921: 307–326), published "Milpa Agriculture, a Primitive Tropical System", a report on the slash and burn (*roza, tumba, quema*) in the Mayan tropics of Guatemala, written in 1919 and published in 1922, for the Smithsonian Institute of Washington DC. The text includes the author's pejorative vision of the cornfield agricultural system that throughout the twentieth century was reinforced by similar opinions of other scholars of agriculture and rural societies. In a short time, these opinions generated negative and partial considerations about the cornfield, creating strong oppositions – which still exist – about its very existence and without comments on the importance that the system has in relation to the food system and economic self-sufficiency of farmers (González-Jácome 1988: 55–189).

The regions supporting large populations were considered by Cook (1919: 307) as scenarios of great activity that still at certain times were cultivated. His opinions were: primitive civilizations destroyed their basis of existence because nations can pass without leaving a history, yet leaving marks of devastation. Instead of resources of natural production remaining intact, now most of the tropical world is far from being a virgin forest, a fact that is often overlooked by tropical businesses. The woody vegetation of many tropical regions is "shrubby" or secondary growth, rather than the original virgin forest. It is possible to distinguish many reforestation stages by the very old shrubs from the original forest, and estimate approximately the period in which the land collapse occurred.

> The limitations of the cornfield system, and the resulting periodicity of agricultural populations tend to be more defined in districts occupied by larger, and more advanced communities, because denudation is more extensive, and complete. The large Indian cities in Central America are surrounded by belts of deforested grasslands, and many districts that were centers of ancient populations are now uninhabited.
>
> *Cook (1921: 326)*

Assuming that the ancient peoples lived in the same conditions, grew the same crops, and used the same farming system as the native populations of recent times, it is reasonable to infer that the consequences of deforestation and denudation were the same in the past as in the present, and that the same difficulties were encountered then to maintain the food supply in populous regions. The ancient cities and sculpted monuments of Central America show that there must have been relatively large and centralized communities, and the limitations of agriculture can clearly explain why the old populated centers were abandoned (Cook 1921: 326).

> The primary systems recognition of the limitations is not exempt from practical problems, pointing out the danger of excessive urban activity and congestion without the corresponding development of agriculture. The wider exploitation of natural resources by industrial nations is possible thanks to modern communication facilities, but the biological limitations of production must be recognized, just like the original system of the cornfield. While cultivation is more continuous when tillage methods are used, agriculture systems that do not maintain soil fertility are essentially nomadic and predatory. Population face problems like a more permanent agriculture and a more rational distribution. Agriculture is the root of civilization [...].

2.4 MID-TWENTIETH CENTURY CHANGES IN AGRICULTURE AND THE THIRD STAGE OF AGRARIAN DISTRIBUTION

In the 1960–1970s some ideas were established proposing that peasants were "traditional" and did not change easily. That is, the characteristic of opposing change of the small-scale farmers rendered them negatively recalcitrant to the ideas of modernization of agriculture. As a first step, the authorities set out to achieve the change, creation, arrangement, and modernization of roads and paths to connect those isolated populations with markets, and urban centers; to improve the education system through literacy campaigns that would include adults, also; and to create stores to supply rural areas with basic products, and to support the creation of coordinating centers that would break the dominance, and economic exploitation of indigenous people (Aguirre-Beltrán 1962).

The modernization ideas about agriculture had support among some anthropologists, that normally worked in government institutions and did applied anthropology work. But, there were other anthropologists, who usually worked in places independent of government policy, independent research centers, or universities, whose studies in the field showed contrasting conclusions, especially as to the visible advantages and contributions that maize, and the cornfield had given to people (Aguirre-Beltrán 1949; Cancian 1972).

However, the ideas that supported the public policies in favor of the modernization of agriculture following the American model won. In addition, those ideas and proposals, which were contrary to the existence of traditional Mexican agriculture, helped to swell the ranks of those who opposed its existence, and pursued conversion to modern commercial agricultural systems, and promote the extinction of traditional agricultural systems. So, in the eyes of the urban and "civilized" world, the cornfield became the enemy of progress, and maize a symbol of the country's backwardness.

The above ideas have not been modified, are current, still, and have been applied by politicians of all parties in the country. The list is long, and there are examples: the "march to the sea" (Revel-Mouroz 1980); the unrestricted use of agrochemicals; the abandonment of small-scale maize cultivation so now maize is purchased abroad (CIMMYT 1988, 1993–1994); the cultivation of agricultural products mainly fruits, aimed at the international market; the expansion of the crop of soybeans in the country, and government support for its export, and to marketing; the opening of roads and other modes to transport products to the ports, and to the northern border; the expansion of supermarkets, and convenience stores, where foreign agricultural products, and foreign industrialized foods replace national ones (Foster 1948: 153–162; González-Jácome 2008: 1–17).

These ideas exist in the mentality of public and private officials, and entrepreneurs, who support and favor the course of an agricultural modernization by adopting the US agricultural model, and NAFTA or USMCA (T-MEC), as a business model (Hussain 2010). This model directly and indirectly affects Mexican agriculture, and especially the traditional agriculture. The farmer receives low market prices for their produce; do not have access to markets, or very restricted access. Mexican politicians support the American agricultural model without any restrictions, or knowledge of the diverse physical, geographical, ecological, or climatological characteristics of Mexican regions.

Mexican politicians do not take into account the history, adaptation processes, and adjustments that Mexican peasants have made through time to local and regional agroecosystems. For all these reasons, it is important to record the evolution, and history of traditional agriculture in Mexico. The themes that have been addressed in relation to this agricultural model, and its local and regional adaptations are numerous, as will be seen below.

2.5 GREAT NATIONAL PROJECTS AND TRADITIONAL AGRICULTURE

The Papaloapan project was one of the best-known great Mexican National Projects inspired by the US Project of the Tennessee Valley Authority (TVA). The Papaloapan project focused on modernizing central Veracruz coastal agriculture, introducing sugarcane cultivated commercially for exportation. Since 1946, the first effects of the so-called "march to the sea" (Revel-Mouroz 1980: 19) can be observed. In this project, thousands of peasants from the North of the country, and from the Central Highlands were relocated to live in the tropical areas of Mexico, which caused a series of changes both in the vegetation (deforestation), and in the traditional agriculture of these regions as now cattle breeding became an activity of enormous importance, for which thousands of hectares were cleared to create grasslands.[1] Since the 1940s, anthropologists have expressed their opinions about that determination. Alfonso Villa Rojas (1948: 301–312) in an article for the magazine *Indigenous America* stated:

> The role of anthropology in Papaloapan works is based on the idea that the man-earth relationship, to be meaningful, must be exposed in connection with the cultural factors that condition its forms of life, work habits, ideas about property, cultivation, or irrigation, family organization.

Anthropologists have taken part in several national projects to create the country's hydraulic infrastructure, or to relocate the populations from their original places of residence that were affected by these projects, and their works. That happened with the Papaloapan Project, which reinforced the idea of promoting conventional agriculture aimed at foreign trade, in this case, by replacing maize with sugarcane, sown mechanically, and extensively.[2]

These projects were aimed at generating electricity for industry, and obtaining excessive profits for investors who enjoyed a flourishing capitalist economy. Commercial crops, especially those aimed at foreign trade (sugarcane), were favored by the economic support of government institutions, and their development policies.[3] In general, these megaprojects had displaced thousands of small-scale farmers, both indigenous and mestizo, from their original places of residence and created great distrust regarding the results of these policies with the forced relocation of human populations (Dominguez 2017).

The development projects carried out in the mid-twentieth century in several regions of the country had main purposes to reduce rural unemployment, improve road infrastructure, facilitate commercial transportation, boost the country's industrialization, and modernize agriculture. To do this, official authorities, sugarcane leaders, and civil authorities supported the expansion of cash crops (such as sugarcane) at the cost of cornfield crops. They committed themselves to the mobilization and mass transfers of populations, mainly indigenous people, who lived in areas that would be submerged by newly created dams (Acevedo 1980: 223–232; Aguirre Beltrán 1952; Blanco-Rosas 2011: 177–292; Goldsmith and Hildyard 1992.).

These projects continue to be carried out in the country, although currently they receive negative and even violent reactions from the town residents who would be affected by the displacements. The cultivation of small-scale maize, and its traditional agroecosystem the cornfield have been viewed with contempt, creating around it a propaganda that negatively affected its production on a local scale, and its forms of consumption. At the same time, social and cultural ideas that negatively affected human populations, and the environmental, agricultural, economic, social, and cultural consequences of their transfer to places alien to them, were ignored.[4]

Something similar to the aforementioned occurred at large-scale development projects destined to the construction of dams for hydroelectric purposes, because in the regions that were involved, and in its construction, development, organization, economic, and social effects in which several federal entities, and numerous state and local authorities, as well as entrepreneurs of different scales, local leaders, and the general population participated. These included the Papaloapan Project, Chontalpa, Tepalcatepec, La Angostura, El Zapotillo Projects, and many others from a long list (Barkin and Zavala 1978; Barkin and King 1972; Blanco-Rosas 2011: 177–202; Böege 1988).

---

1 Del Amo-Rodríguez 2007: 97–146; González-Jácome 1996, 2007, 2010: 141–158; González-Jácome and Velasco-Orozco 2008: 215–238; Molina-Rosales 2010; Palerm 1970, 1993, 2006; Rentería-Garita 2006; Revel-Mouroz 1980: 14–20.

2 Acevedo 1980; Blanco-Rosas 2011: 177–202; Ewel and Poleman 1980; González-Solano 2007; Palerm 1970, 1993; Poleman 1964: 60–61.

3 Barkin 1972, 1998; Barkin and King 1970; Barkin and Zavala 1978; Casas et al. 2008; Goldsmith and Hildyard 1992. González-Jácome 1996, 2008a, 2008b: 137–172, 2010c: 177–192, 2013d: 131–144; González-Jácome and Velasco-Orozco 2008e: 215–238; Palerm 1970, 1993; Poleman 1964.

4 Blanco-Rosas 1986, 2011: 177–202; Ewel and Poleman 1980; Goldsmith and Hildyard 1992; MacMahon 1973; Melesio 2002; Palerm 1970, 1993; Velasco-Toro 1990; Villa-Rojas 1955.

2.6 CONVERTING TRADITIONAL INTO CONVENTIONAL AGRICULTURE

The strong governmental supports and public policies aimed at modernization of the countryside and the cultivation of products for export are measures Mexico has shared with other countries. However, those countries do not have a traditional agriculture of such ancient origin as that of Mexico, for the most part. This Mexican agricultural model has shown its permanence with its capabilities and characteristics, some of which we will discuss in this book. At the end of the Mexican Revolution, in the first quarter of the twentieth century, national reconstruction was a priority. In the first three decades, the country's pacification was tied to the initial stages of Mexico's agrarian reform, with the creation of the National Irrigation Commission (CNI) and the National Agrarian Commission (CNA), both federal institutions.

Subsequent governments focused on controlling agricultural land tenure and irrigation, creating and directing several regional and supra-regional plans and programs. In the 1940s and 1950s, several models emerged to achieve development in the world. From 1947 to 1958, plans were made by the Mexican government to control the waters of the Papaloapan, Lerma-Santiago, Balsas, Grijalva, and Fuerte Rivers to boost regional growth with this model. Its objective was to "Execute the works that the Federal Executive considers indispensable for the economic recovery of the country, including a reasonable industrial development" (*Antología* 1985: 33).

The organization, planning, direction, and specific functions within the context were assigned to several ministries. For example, the Hydraulic Resources Ministry (SRH) had to take managed water flows to newly created areas of cultivation by irrigation, to improve existing irrigation works, the supply of drinking water, the defense of the lands against floods, and the drainage of land parcels, among other things (Aboites-Aguilar 1998; Antología 1985: 33; González-Jácome 1996, 2008a, 2008e). The Ministry of Agriculture and Livestock (SAG) was responsible for eliminating the national deficit of maize, wheat, beans, and sesame seeds for consumption, in addition to promoting rice exports. The Maize Commission was uncharged to handle the commercial distribution of improved seeds in areas of poor yield (*Antología* 1985: 33).

Since 1948, the National Colonization Commission has had to accommodate the surpluses of the rural population in densely populated regions toward others that were not developed agriculturally (*Antología* 1985: 33). The Papaloapan project, initiated by president Miguel Alemán in 1947, was the first to be carried out, but it had to correct issues caused by market forces, whose spontaneous action "[...] tends to concentrate economic growth in a few regions, leaving the rest behind [...]" (Ornelas 1993: 41). Executive commissions in each of the river basins organized the Papaloapan Project wherever the federal government decided to act (Poleman 1964; Tyrtania 1992b).

The actions undertaken to expand the agricultural frontier in the south of the country were conceived – in general – with the idea that the productive potential of the humid tropics would allow the conversion of those tropical regions into agro-productive enterprises whose highlands demographic pressures would be reduced through a colonization process of populating the humid tropics. At that time, the tropics were considered the new "granary of Mexico" (Palerm 1993, 2006: 98).

Numerous anthropologists participated in the projects carried out at that time that have been analyzed by other colleagues who not necessarily had taken part in them.[5]

The economic key of the basin, and – possibly – also of the Papaloapan project (1947), was its high possibilities to belong to a giant sugar complex, which would supply sugarcane to the state of Morelos sugar mills, but they were destroyed or semi-destroyed by the Emiliano Zapata agrarian combats and the Lázaro Cárdenas agrarian reform.

---

[5] Acevedo 1980; Agüero 1998; Aguilera 2004; Barabas and Bartolomé 1974, 1992; Bartolomé and Barabas 1990; Blanco-Rosas 1986: 2–12; Castro 1999; González-Jácome and Velasco-Orozco 2008: 215–238; Hernández 1999; MacMahon 1973, and many others.

Economic earnings justified the abandonment of traditional agriculture in Mexico and its conversion into a conventional commercial agriculture for exportation. Authorities gave unrestricted support to conventional agriculture, whose crops would be destined for the external market, and this is maintained and supported today. Currently (2019), commercial agriculture has grown overwhelmingly yielding national and foreign entrepreneurs in the country higher returns, which in the eyes of foreign trade economists and experts is a success and the reason for the existence of agriculture.

Several questions would have to be asked first. From 1940 to 1970, the economic governmental intention was to promote "industrialization at all costs" (Palerm 1962a, 1962b, 1964, 1965, 1966, 1993). As Ornelas (1993: 39) claimed, this intention and process started from identifying development as a result of industrialization, whose promotion was based on import substitution and the search "[...] for a high and sustained long-term GDP growth rate, through the concentration of capital resources, private savings stimulation, and internal investments, using public investment to boost the formation of private capital".

Several models have been used in Mexico to achieve development, both regionally, and nationally (Aboites-Aguilar 1998; González-Jácome 2006, 2008; Ornelas 1993). One of these was the "development by watersheds" model, of great importance in the 1970s (Vargas-Velásquez 2008). This model mainly was aimed at supporting the industrial development of the nation, and hydroelectric plants were built to support irrigated agriculture, which was aimed at exports. In the early 1970s, Palerm (1993: 299) noted and listed that macroeconomic indicators showed several important aspects that should be taken into consideration:

1. The disparity in the relations among the different sectors of the country (real product per capita of the labor force);
2. The unequal relations between Mexico and developed countries have economic effects;
3. The uneven development between the different sectors and national agriculture branches, which favored export crops stimulated by the national economic policy that put the basic crops destined for the consumption of the population and the domestic market, mainly, at a great disadvantage; for example, cotton, sugar, and coffee backed by the national economic policy earned 33% of all income obtained from exports from 1950 to 1964 (Bartra-Vergés, Cobo, and Paz-Paredes 2013).
4. Agricultural production aimed at the domestic market had two destinations. On the one hand, it focused on satisfying the selective demand of the urban population with high living standards; on the other hand, it was articulated to the subsistence of the peasant population and to the consumption of the urban population with low purchasing power – the upper urbanites had increased economic means;
5. Agricultural branches to promote export tended to be distributed in the north of the country, with the support of large investments for hydraulic infrastructure resulting in uneven development and regional imbalances that are two of the causes of unfavorable social, economic, and demographic effects;
6. The critical problem faced by agricultural productivity was that the sector producing for subsistence and for mass consumption is the one that occupied the lowest place on the scale of unit yields and of the real product per capita.

Wallerstein (2005) believes that the 1970s were influenced by the idea of achieving economic development as a tool to achieve rural development. The UN declared the 1970s the "decade of development". Yet this was a period of world economy stagnation. The oil-producing countries deposited large capitals in USA and German banks. These funds were used to offer loans to countries with economic difficulties, so in the 1980s, the payment balance became a serious problem for the indebted, who ended up creating the so-called "debt crisis".

For Wallerstein (2004), the opinion of developers in the USA was that development failure was attributable to the following factors: oversized bureaucracies; parastatals structures which were barriers to fruitful business achievements; and waste in "low priority" areas such as education and health. Due to the above, they proposed that State enterprises should be privatized at once, because they were inefficient and should be replaced by private companies, which represented maximum efficiency.

Sometimes the peasants lost the best lands. The water control by the Mexican State did not favor them either, as has been studied in several cases, such as in the Papaloapan (Aguirre-Beltrán 1952; Blanco-Rosas 2011: 117–202; González-Jácome 1996, 2002b: 167–188, 2011: 113–141; Palerm 1993, 2006).

The Mexican peasantry in the Central Highlands adapted to these situations by joining the industrial workforce, and, at the same time, maintaining traditional agriculture to produce at least a portion of the family food consumption. Other small-scale farmers went to the USA in search of a better future (González-Jácome 2003). This was the beginning of a race toward the monetization of their economy, the need to work in salaried activities, and the abandonment, modification, or deterioration of some traditional agricultural practices (e.g., reduction of the number of weedings from two to one, the increase of indiscriminate application of chemical fertilizers and herbicides). In fact, several areas where small-scale agriculture was practiced were in a frank process of deterioration and land abandonment, causing a decrease of population to zero.

Similar events occurred in the Central Highlands. In some states, such as Tlaxcala, small-scale traditional agriculture is in critical condition, currently. In 2014, it employed seasonal rain and only 11.7% irrigation.

The raised field agriculture was mainly located in the Atoyac and Zahuapan rivers basin. The former road construction and expansion, the destruction of the canals, and ditches of the entity's southwest changed the best agricultural region in the state. This region has been expanding the monoculture areas, roads, and urban areas drastically transforming this basin (Renard, Iriarte, Birk, Rostain, Glaser, and McKey 2012: 30–44). Its watercourses are contaminated, the water table has dropped, and the northern part of the basin has become of temporary use only (Crew 1985; Crew and Gliessman 1991; González-Jácome 1976, 1986: 201–220, 1988:1–73, 1989: 91–117, 1996, 2003, 2008, 2016: 349–350) (Figures 2.1 and 2.2).

The macro-region of the Atoyac and Zahuapan River basin (González-Jácome 1976, 1986: 201–220, 1988: 1–73, 1989: 91–117, 1996, 2002a, 2002b, 2008e, 2016) located in the southwest of Tlaxcala produced cold legumes for the national market that were distributed through the supply center in Mexico City. At that moment, the agricultural population focuses their efforts on economic activities to obtain monetary resources to support their families, since the once flourishing agriculture sector is no longer productive enough in agricultural or economic terms, as occurs in towns like

FIGURE 2.1 Tlaxcala: The Matlalcueye (Malitzin, Malinche) Volcano, North View (2012).

FIGURE 2.2 The Iztaccíhuatl Volcano, Northwest View (2016).

St. María Nativitas, St. Ana Nopalucan, S. Juan Huactzinco, and many other communities in the Zahuapan-Atoyac basin (Castañeda-Castillo 2014; González-Jácome 2003, 2016; Juárez-Choque 2013; Villalobos-Nájera 1998, 2008).

One aspect of the least social and economical elements considered in the 1980s when traditional agriculture was drastically modified in Tlaxcala had to be with the loss of traditional agricultural technology, including water management, and related local knowledge, about which there are only a few studies. However, the idea remains that, due to its backwardness, traditional agriculture technology must be changed for modern technology, without believing there is any need to be documented as to its efficacy and benefit (Cruz-León 1989, 1997, 2019; Cruz-León and Martínez-Saldaña 2001; Ramos-Galicia, Hernández and González-Osorno 1994).

2.7 CONSIDERATIONS AND CONCLUSIONS

In several cases, and in different parts of the country, profound changes occurred in traditional Mexican agriculture, mainly after the 1970s. These have been accentuated with the application of the neoliberal economic model, which contributed to highlight the economic and social distance among the different groups of the national society. However, despite the increase in poverty in rural areas, the agricultural and agrarian policy of the Mexican State has supported conventional and commercial agriculture destined for export over traditional agriculture.

For at least 200 years, Mexican public policy has avoided considering that traditional, small-scale agriculture is able to produce food and commercial materials for the national and international markets through a series of organic, biological, and specialized crops with high commercial value. At the same time, traditional agriculture has the ability to contribute to the achievement of adequate, balanced, nutritious, and healthful food for small-scale farmers, their families, and the market too. Traditional agriculture can reduce soil, water, and air pollution as well. Furthermore, by reducing pollution, traditional agriculture contributes to the sustainability, maintenance, and resilience of environmental and social systems.

The American agricultural model, in general, is characterized by large-scale monocultures, with extensive plains, deep soils, and irrigation; it is managed with machinery and inputs made up of agrochemicals and genetically modified plants. It is highly mechanized agriculture moved with energy derived from petroleum. This model has not been the most appropriate for the physiogeographic, altitudinal, and ecological characteristics of Mexico.

Due to its age, and adaptation to the physiographic, environmental, and climatic conditions of the country, the small-scale Mexican agricultural model has proven to be the most appropriate because

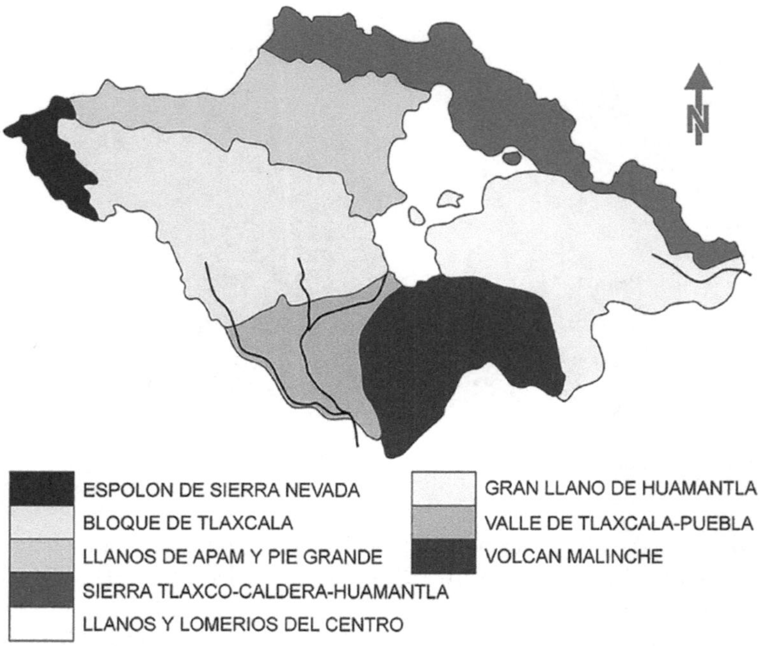

FIGURE 2.3 Map: Tlaxcala State: Macro Regions.

it adjusts to a rugged orography and soils that are generally thin, often lacking water for irrigation. The current and specific situation of wild, cultivated, and weed vegetation, wildlife, and human populations in all regions should be taken into account, including from the northwest, north, and northeast of Mexico to the south and southeast of the country.

RECOMMENDED READING

Palerm V., Ángel. 1967. "Agricultural Systems and Food Patterns". In: *Handbook of Middle American Indians*. Austin, The University of Texas Press, Vol. VI: 26–52.

Essay on the correlation between farming systems and feeding patterns.

Palerm V., Ángel. 1993. *Planificación regional y reforma agraria*. México, Ediciones Gernika, and Universidad Iberoamericana.

This book brings together nine writings on regional development and modernization of agriculture. Palerm analyzes the social effects of the applied models, and the importance of anthropology in the study of the impacts suffered by the peasant population. It includes cases studied in 1968 by the author in Mexico, United States, Israel, France, Italy, Yugoslavia, India, and Peru.

Poleman, Thomas T. 1964. *The Papaloapan Project. Agricultural Development in the Mexican Tropics*. Stanford University Press, California, The Food Research Institute.

This book deals with the governmental works of the Papaloapan Project in Mexico. It is a relation of the development plan of the Mexican government in the Mexican tropics dating from 1947. The book is based on data from published official sources and field research. Poleman tracks the characteristics, failures, and successes of the program, and in the final section summarizes a series of "guides for the future" related to the prospects of converting hitherto unexploited hot zones into new areas of agricultural development dedicated to sugar cane instead of the traditional cornfield agroecosystem.

3 Studies Conforming Traditional Mexican Agriculture

3.1 INTRODUCTION

At the end of the 1930s, and the beginning of the 1940s, research did not have the wonder of computers, the Internet, satellite mapping, electronic stereoscopes, GPS, drones, robots, or many other technological inventions that later facilitated and modernized research. In addition to anthropology, agronomy and several other disciplines and currents of thought were included in their studies plus the social question and especially the relationships between agriculture and small-scale farmers and their agricultural economics (Tourtellot and Hammond 1994: 119–124).

Ángel Palerm (1954, 1968a, 1968b, 1972a, 1972b, 1997, 2006, 2007) introduced the theory, practice, and perspective of Cultural Ecology in Mexican Social Anthropology. Palerm thaugh the proposals of Julian H. Steward (1949, 1955) related to the multilinear evolution of ancient societies. Shortly thereafter, in the decades of 1960 and 1970, Palerm's (2007) discussion with Karl Wittfogel (1945, 1966, 2007) about the characteristics and components of hydraulic societies was related to the population increase, development of agriculture, evolution of urbanism, and emergence of the State. At the same time, interest arose in understanding the role of Mexican agriculture, particularly with regard to the human population increase and its relationship to the emergence of Traditional Agriculture in Mesoamerica (Palerm 1955b, 1961, Vol. 11; Zizumbo 2017: 33–46).

The study of traditional agriculture changed due to questions arising as a result of the advances in their research. These include the participation of other disciplines with their distinct methodologies, which range from the evolution and history of traditional agricultural systems to irrigated agriculture. The current situation of agriculture and the traditional agriculture approach to study were incentives to innovate this research.

An important component is the study of interrelationships between humans and their social, cultural, and environmental structures (Men, Women and Nature, humans and culture, and ethnoecology as situated knowledge) as well as their interrelationships with other systems with which traditional agriculture is articulated (Descola and Pálsson 1996; Nazarea 2005: 34–39, 2006; Milton 1993: 1–17; Sahlins 1960; Tomé 2009: 7–22). These interrelations generated the need to carry out research where social actors are included. Human interactions where various academic interests and academic disciplines were combined with flora and fauna studies, and interrelations with social, economic, political, and cultural organization that generated and supported the emergence of the studies.

3.2 FROM 1960s TO 1980s, AND NEW THOUGHTS, AND APPROACHES

Some of the studies arose in the 1950s and continued until the 1980s. In 1954, Harold Conklin received his doctorate from Yale University with the thesis "The relationship of Hanunoo agriculture with the world of plants". He analyzes the interrelations between agriculture, settlers, edible plants, and medicinal plants. Other examples mention the approach of cultural ecology and ethnoscience, which Frake (1962: 53–59) brings to light his article "Cultural Ecology and Ethnography", published in *American Anthropologist.*

Sturtevant (1964: 99–113) writes an article on "Transcultural Studies in Ethnoscience", which was also published in the *American Anthropologist* magazine. In 1974, Berlin, Breedlove and Raven published the "Principles of Tzetzal Classification of Plants". Fowler (1977: 53–59) published

DOI: 10.1201/9781003198833-3

his article "Ethnoecology" in *Ecological Anthropology*, a book edited by Hardesty. Ford (1978: 13–49) published his article on "Ethnobotany, Historical Diversity and Synthesis", in the Museum of Anthropology at the University of Michigan.

The list includes authors and studies that try to understand the relationships between nature and culture, although this text is only an example of the beginnings. In addition, other historical, social, and cultural interests can be added, such as the study of the relationships of agriculture with other important components for human societies; for example, herbal medicine, allopathic and folk medicine, diets, health and disease, relations, classifications of animal categories with agriculture, and the culture of the peoples in addition to other factors (Ellen 1986, 1993, 1994, 1996a, 1996b).

In this regard, the statement made by Virginia Nazarea (2006: 38) on Ethnoecology, her proposal, scope, and perspectives as a research of the perception systems, knowledge, and use of natural environment that "[...] cannot continue to ignore the underlying history, and politics represented in the managerial aspects of culture, without moving away from distribution problems, access and power that shape the knowledge systems and resulting practices".

Within the field of Ethnobiology are the numerous writings of Roy Ellen (1986, 1993, 1996a, 1996b, 1999), that discusses the differences between food collecting, and food production. This flexible point of view also allows seeing how hunter-gatherers could temporarily do agriculture in smaller spaces, without strictly becoming cultivators. This idea can be applied very well to the understanding of how small-scale farmers carry out fishing activities without being fishermen, gather, without being gatherers, only, and there are more examples. According to Ellen (1999), agricultural systems are not uniform, and like in slash and burn, they can be set up in patches that are cultivated and abandoned according to a series of issues that include soil, water, family needs, and others.

Since the 1970s and 1980s, there have been discussions about interdisciplinary, multidisciplinary, and transdisciplinary theoretical models and practices, as viable possibilities to carry out these types of studies, that due to the nature of their problems to be investigated, require specialized and combined knowledge from various academic disciplines, their approaches, and proposals (Sturtevant 1964: 99–113).

One of the initiators of the interdisciplinary effort in the study of the origins of agriculture was Richard S. McNeish (1967a, 1967b, 1981, 1987, 1992, 1997), an American archaeologist of whom the first proposals on the aforementioned subject emerged. First, there were his studies on Tamaulipas, and then – the most famous – on Tehuacán, where he carried out a project involving a plethora of researchers, support staff, and others. McNeish had problems with financing and managing the project itself.

Several important contemporary scientific and practical currents emerged in the National Research Council in 1999 and developed at the beginning of the millennium as a response to the dangers for the planet originated by human action on the environment as the "sustainability sciences". Currently, there are various disciplines and approaches emerging such as human ecology in the 1860s and the Annals school by historians like Braudel (1949) or the cultural ecology by Steward (1955). The twentieth-century approaches add to the study of agricultural systems interdisciplinary scientific collaboration plus that of the "practiced" subjects (local knowledge).

The sustainable sciences are based on the notion that a common theme has been taken and studied by several sciences. Applying their knowledge in search of sustainable development for which the need arises to "co-produce" and achieve collaboration between scientists and practitioners (Gliessman 1998, 2015; Toledo 2003; Turner et al. 2003a: 8074–8079). The discussion of their similarities, differences, and scope will be on another occasion. It must be born in mind that the construction of any classifications systems involves in itself the intellectual and cultural management of those who propose them.

Ecological anthropology, and specifically cultural ecology, is theoretically and methodologically an approach enriched by the studies of other scientific disciplines and other ethnos – now so fashionable and expanding – where human activity is included in the framework of complex systems, in addition to the processes of vulnerability, sustainability, and resilience. For several decades, problems have been treated that are common, or answer questions that have also been focused on and studied by several of the above. These questions and their answers may be temporary, and its need to confirm hypotheses, but they all emerge from problems that affect human societies and are usable in understanding phenomena such as sustainable strategies of civilization or human responses to climate change in different moments in time.

The study of the past represents the generation of new ideas and possible solutions to the problems of the present. A good example of this are the answers that have been given over the years to the questions and issues related to the origins of agriculture, and the domestication and diffusion of plants; the interrelations between agriculture, the food system, and health; the interaction between agriculture, cultivated, weedy, and wild flora; or the elements of oral tradition related to issues about the environment, agriculture, culture, food system, the worldview of human populations, and many other environmental, climatological, geological, historical, social, and cultural aspects linked to this matter that provide the sustenance to enrich approaches such as those of agroecology and other alternative and sustainable agricultural models, types, and practices (Cruz-García 1992; Ford and Night 2010; Agroecología 1973; González-Jácome 1997: 167–188, 2001: 263–278; González 2008; Lentz 2000; Pope, Ocampo, Kingsland and Smith 1996; Turner II and Sabloff 2012: 13908–13914).

3.3 FROM 1980 TO DATE

Research in the 1980s and 1990s, in addition on the homegarden and the cornfield, focused on nutrition and social welfare (FAO 1992; Ford and Night 2010; Lentz 2000; Pope et al. 1996; Turner II and Sabloff 2012: 13908–13914). An example is the two investigations focused on the food system, of which some elements will be summarized. The places and times in which both investigations were located were very different. One corresponds to the pre-Hispanic period in the basin of Mexico, in the Mexican Central High Plateau, and the other one refers to the Mayan maize area of Yucatan at present. In both cases, the agricultural system that integrates the studies is the cornfield and the early production of food.

The archeological research of McClung and her colleagues Martinez et al. (2014: 97–121) of botanical materials from ten sites in the basin of Mexico, analyzed the early production of food in the basin of Mexico, for which they proposed the use of the term "traditional food", as one that:

> [...] conserves cultural patterns that have been transmitted from generation to generation (such as the forms of preparation, and consumption, as well as the type of food), although with the passage of time new products are incorporated making the food more varied.

Contemporary Mexican food systems are regionally diverse, and a traditional combination, as proposed by McClung et al. (2014: 98–99), is characterized:

> [...] There are various ingredients and flavors, many of them had pre-Hispanic origin, while others arrived in Mexico during or after the Conquest, such as cattle, cereals, wheat, and rice, legumes, peas or chickpeas, to name just a few. However, the variety of products, especially foods of vegetable origin, which have been used since pre-Hispanic times (for example, maize, beans or squash) continue to be the basis of the Mexican diet today, with all its regional variants.

McClung et al. (2014), in this archaeological study, analyzed macroscopic remains (seeds, wood, and fruits) and microscopic remains (pollen, phytoliths, and starch granules) found in archaeological excavations in the basin of Mexico. From north to south, they studied Xaltocan, Michpilco, Teotihuacán (housing units in Yayahuala, and Tetitla, Oztoyahualco, and Cuanalan), Otumba, Cuanalan, Tlatelolco, Iztapalapa, Terremote-Tlaltengo, Xico, and Temamatla (Map, McClung et al. 2014: 9). They incorporated comparative information from studies carried out by other archaeologists in different parts of the country. Information obtained from the Florentine Codex (1548 to 1585) is added to the research. The writings of Bernardino de Sahagún contain a rich font on material on the flora used by the inhabitants of the central Highland, including both the cultivated and the wild flora.

For this study, the authors divide the plants found, studied, and used by inhabitants in wild, cultivated, and domesticated. It must be taken into consideration that in those times, the generalized agricultural systems were the homegarden and the cornfield.

The periods established by the archaeological material recovered and analyzed – which were included in that investigation – corresponded: (1) from the end of the Pleistocene to the middle Holocene; (2) the Formative, or Preclassic; (3) Classic; (4) Epiclassic; (5) Post-classic; (6) and the sixteenth century. Specific information in this book it will be discuss in chapter XVIII on the food system. The results of this approach are a good basis for future research, where interdisciplinary studies will generate advances in the knowledge of traditional agriculture. An example of this, and within this theme, is the one mentioned below, which allows us to compare and establish differences with that of McClung et al. (2014).

A research carried out by Otero-Prevost et al. (2017) on the increase of the supply of industrialized foods in the diets of domestic units in rural Mayan communities found a relationship between the abandonment of the cornfield (*milpa*), as a basic agricultural system, and the increase of the supply of industrialized food consumption. Factors that may be responsible for the incorporation of industrialized foods into the diet of rural Mayan domestic units are analyzed, demonstrating how the amount of the supply of industrialized foods offered in their community of residence was lower than that consumed in community where they stopped growing in the cornfield, the food culture, and cultural values at that time (Toscano et al. 1998: 185–206).

In fact, studies of the pre-Hispanic past of agricultural systems and their relationship with the food system of human populations allow obtaining a better understanding of what happened in human agricultural societies with respect to basic aspects – as in these two studies. In both cases, the impact of new elements in the diet of human populations implies modifications to the original pattern. However, in the case of pre-Hispanic settlements, the cornfield does not disappear because new plants are introduced into it. In the case of contemporary maize societies with the introduction of industrialized food, maize consumption is dying out, just like with the other crops that were part of the cornfield and that actually seem to be in a critical situation (Linares and Bye 2016: 9–12). The impacts on soils and plants of pests, and diseases caused by the introduction of non-native plants or by the abandonment of the cornfield and its polyculture, are unknown. Nor do we have specific information related to the methods of cultivation or the impacts due to the introduction of technology (e.g., of the plow on the soils with the arrival of the Spaniards).

There is a paucity of detailed information about particular ecological aspects of the local agroecosystems. For example: how plants are managed (maize, beans, squash, chili) in situations with excess water and moisture, or with lack of water as a result of droughts; how plants relate to the biological control of insects; what the role of soil borders and plants on them is in relation to pollinating insects; and how plants are distributed on farmland and why (García-Espinosa 2007: 328–341). On the other hand, there is ample information on cultivated and wild medicinal plants, many of which fetch important income for families. These studies are by physicians, anthropologists, ethnobotanists, biologists, agronomists, and many others specialists.[1]

3.4 ELEMENTS THAT MAKE UP THE CONCEPT OF TRADITIONAL AGRICULTURE

It is essential to understand what happened when the labor force in the small-scale farmers' families was reduced in agroecosystems due to migration/emigration of some of its members, causing several effects on the traditional non-paid workforce (mutual aid) of agricultural systems such as the

[1] Bye and Linares 1987: 200–230; Estrada-Lugo 1989: 211–228; García de Alba et al. 2012: 29–44; García-Hernández et al. 2015: 12–30; Kumate (cord.) 1993; Lozoya 1987; Martínez 1987, 1996; Pérez-Ortega et al. 2016: 221–228; Robles-Cervantes 2008; Saynes-Vásquez et al. 2016; Vibrans 1999: 476–481, 2002: 558–561, 2016: 287–317; Viesca-Treviño 1992: 115–138, 1993; Vida and Carzola 2019: 1–15.

homegarden and the cornfield. In addition, the transmission of family systems of local agricultural knowledge from grandparents to parents, to their children was broken.

Using the "Vulnerability Framework Analysis" a model of the Sustainability Systems Program, Turner II et al. (2003: 8080–8085) propose that agriculture in the Calakmul Ecological Reserve (RBC) has undergone neoliberal political changes, causing agriculture and eco-archaeology tourism to compete for the use of land in tropical forests. The developers of the region in the 1960s and the established population deforested the forest significantly. By the 1980s, the Mexican government supported the economic growth through the program Maya World (*El Mundo Maya*, EMM) and the creation of the Mesoamerican Biological Corridor (CBM). Government plans were aimed at controlling land resources in the region, maintaining, and increasing agricultural food system reserves and farmers' incomes in addition to sustaining a mature forest with biotic diversity (Alayón and Gurri 2007: 243–260; Barrera 1980: 115–129).

The case occurred with migrant populations established in this region since the 1960s and 1970s, as a result of the lack of land in their places of origin (Molina-Rosales 2010; Vallejo-Nieto, Gurri-García and Molina-Rosales 2011: 71–98; Revel-Mouroz 1980). These new settlers were unaware of the tropics, their characteristics, and management; they introduce commercial agriculture in the region, based on products such as chili, with the inconveniences, and economic risks involved in their cultivation and commercialization. They also introduced cattle. In general, these new settlers did not have the local traditional knowledge necessary to implement traditional agricultural systems, such as the cornfield and the homegarden. For all this, they have been vulnerable to droughts, hurricanes, economic risks due to supply/demand, and market prices. Farmers who practice traditional agriculture, with their resilient and flexible systems, survive these events better and adapt to vulnerable situations (Vallejo-Nieto et al. 2011: 71–98).

Hence, new questions arose, such as: Do these aspects have to do with the increase in the consumption of industrialized foods? How does the change in diet relate to health and disease? To answer several questions, one of the methodological axes requires the realization of research projects and studies where several disciplines participate. In cultural ecology, the required theory and methodology starts from evolutionism, from the theory of complex systems, from the synchronic and diachronic study of the interrelations that occur between the environment (e.g. nature) and society and the agriculture of the human communities; without the interdisciplinary, transdisciplinary, or multidisciplinary approaches, it is very difficult to find solutions to the problems that afflict the contemporary world.

This interdisciplinary research is characterized by sharing a problem, one or more questions, or a common logic, which allows comparisons and places each scientific and social discipline, in possibilities of creating solutions – at least generate hypotheses – to questions that are – at the same time – specific for each scientific discipline, or science involved, or that are part of what happens in different places and times in a country, or a type of region on the planet. This approach also starts from scientific research, and specialized studies.

3.5 THE FIRST STEPS

From an evolutionary perspective of human societies, the diachronic approach is included in cultural ecology, which is carried out through studies of human evolution, plant genetics, palaeoecological, archaeological, historical, anthropological (ethnohistory), chemical ecological studies, and others that delve into the knowledge about the past of these societies. This approach is combined with the synchronous one, including cultural and ecological anthropology, ethnography, and ethnology that are required for obtaining and analyzing factual materials, which were obtained by performing fieldwork *in situ*, of the application of the ethnographic methodology; which are added to the results of the cartographic studies, of the reconstruction and change in agricultural landscapes, of the materials obtained in the corresponding historical, bibliographic, and hemerography sources.

The initial anthropological studies were carried out in the first half of the last century in simple societies, called "primitive" which were constituted by groups (bands) of hunters, hunters-gatherers, fishermen, and horticulturists. In Mexico, these initial studies were done in peasant societies – both indigenous and mestizo – who practiced traditional agriculture. The field of study of cultural ecology in Mexico quickly moved from Prehistory to the investigation of complex societies, with irrigation agriculture, high productivity, population increase, construction of the first permanent settlements, population centers, and cities which later became states. In addition to those evolutionary and historical processes, which explain its parallel emergence and diffusion in differenyt cultural areas and regions of the territory that would later be Mesoamerica (Niederberger 1996).

In ecological anthropology, studies of traditional and conventional agriculture have diversified into several branches of knowledge; these remain an important basis for the understanding of human societies in the rural world and their interrelations with those in the urban world.

Although human societies would increasingly like to move away from the rural world, which many despise, the reality is that we still need food for our bodies, which are not qualified to consume petroleum products. As we depend on food, the essential question is how we want to produce it, and how it is wanted to relate to the food system, health, nutrition, and social welfare. It is in this other sense, in which traditional agriculture and its study pave the way to obtain new economic perspectives and possibilities.

Mesoamerican studies led archaeologists and anthropologists to enter into the remote past, since the arrival of Man[2] in the Americas, which is long with ancient migrations, and the processes of obtaining food through hunting, fishing, and gathering, until Man's first steps in sedentary life, the discovery of agriculture, and its impacts on the life of human communities. The evolution of agricultural activity took several millennia in the history of humanity in America. The associated changes took hundreds of years along with the fact that agriculture became the most important axis of agricultural societies, including population growth, the emergence and development of urban planning, and the creation of the State and its cultures in Mesoamerica, which begins more or less during the Formative or Preclassic period in 2000 BC (Niederberger 1996; Plunket and Uruñuela 2016: 41–46).

That is, the agricultural life of human societies, the emergence of urbanism and the State, the differentiation of cultures, and the social and cultural adaptations of each one to the different types of geographies, environments, and climates are the result of relatively short processes, if we take into account the millions of years that the evolution of the planet has taken. However, human time is not very long if it is understood from the point of view of the entrance of human beings to the American Continent, more or less between 40 and 12,000 years ago, as the archaeologist Richard S. McNeish (1987: 66–67) affirmed. From this perspective, the agricultural history of Mesoamerican societies is a relatively recent phenomenon. The evolution and history of traditional agriculture are, therefore, a phenomenon and a relatively recent process.

3.6 WHAT WE CALL TRADITIONAL ALSO CHANGES

Defining agriculture as "traditional", as will be used in this book, does not mean conceiving it as late, nor does it mean that it is static and does not change, or that it is immutable over time. On the contrary, the concepts of evolution and change are immersed in the description and analysis of agroecosystems, which have been and are part of the studies on traditional agriculture in this case focused on Mexico.

The processes of change occur over time while changes may, or may not, be accepted and applied by all members of human society (bands, human groups, community). That is, only through time (diachronic methodology) that it will be known if they were incorporated – or not – into the social life of a population, and how they fit their needs.

[2] I use the word Man, or Men (capitalized), as a global and universal concept, which involves both women and men.

FIGURE 3.1 S. Juan Market in Mexico City, Foreign, and Gourmet Products.

The subsistence of traditional, small-scale agriculture exists for reasons that are important for human beings, such as those related to food system and health, the taste for foods that are not produced with agrochemicals, that are natural, considered healthful and, which are grown or developed by "traditional" methods. It is necessary to consider that a "necessary evil" has been the development of elites that can pay the costs of these productions, which have also been called "artisanal", and that generally include bread, vegetables, legumes, mushrooms, truffles, cheeses, sausages, meats, and fruits. This has stimulated product prices to allow farmers and small-scale farmers to make a profit.

In Europe, municipal markets are a good example of this and offer consumers high-quality food products, many of which are not offered in supermarkets or all convenience stores. These products are produced by hand and are usually considered "gourmet". They are sold at higher prices than those produced massively and industrially. In Mexico City, there are some markets of this type, such as that of S. Juan, located in the historic area of the city, where chefs and cooks of luxury restaurants are supplied with products (González-Jácome 2019). However, the sale profits obtained from their sales do not reach the small-scale farmers, whose products are not integrated into these types of markets (Adapon 2008) (Figures 3.1 and 3.2).

3.7 CULTURAL TRADITIONS, FOOD SYSTEMS, AND TRADITIONAL AGRICULTURE

In recent decades, studies on food system traditions and health have been important parts in the research and knowledge of traditional agriculture, agroecosystems such as homegarden and cornfield, and their advantages in the production of food for families. In addition to constant concern for international institutions such as FAO, where a number of programs support the study of the food systems and the formation of new homegardens and gardens including urban ones. Hydroponic systems for the production of legumes for family consumption and small-scale sale of surpluses have also been expanded. In several countries, there is also a taste to consume what the consumers themselves have produced (Figure 3.3).

The various types (biological, organic, permaculture, agroecological, alternative agriculture, traditional European agriculture), with their different proposals and models, have become producers of organic food or generator of gourmet products for the market, and they are costly. In the USA, the market for organic products has grown, and the number and variety of available produce also have increased and far exceeded those of non-organic foods. In these highly industrialized societies, buyers have the monetary resources to acquire – despite the prices – what this traditional

FIGURE 3.2 S. Juan Market in Mexico City: Woman Eating Scorpions.

FIGURE 3.3 O'Grove, Galicia, Spain, Municipal Market.

agriculture produces, including various types of small-scale production that utilize alternative agriculture models. The ideal would be that agroecological products be less costly in the market and could be acquired by a greater number of families (Figure 3.4).

In Mexico, despite the different situations, the number of organic markets has grown. In our country, these types of products are just being developed; although there are groups of buyers who want to buy healthy food, not contaminated with agrochemicals, as well as those called "exotic",

(a) (b) (c)

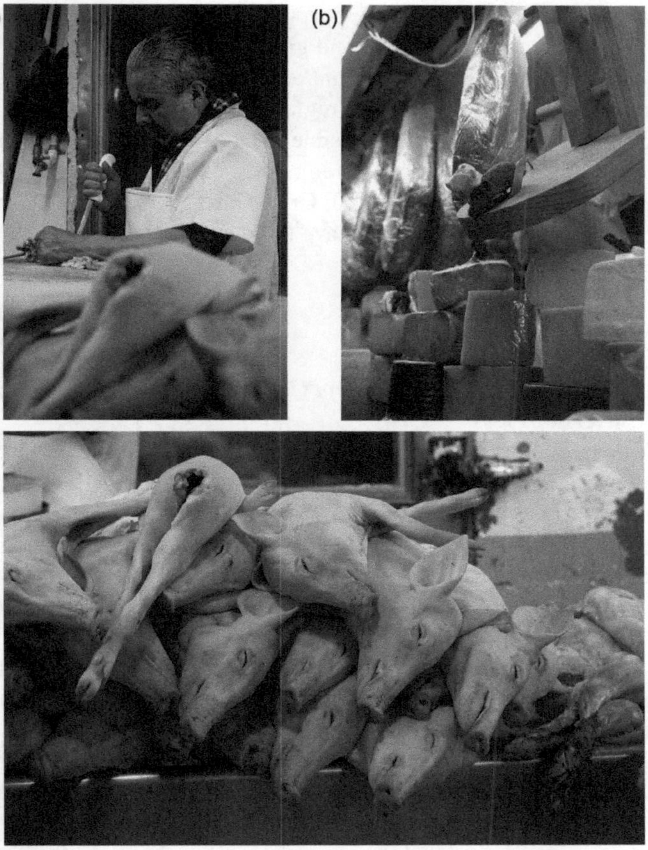

FIGURE 3.4 Piglet Sales Stall in San Juan Market in Mexico City (a, c). Stall with Delicatessen, selling Serrano ham in the San Juan Market, Mexico City (b).

which include vegetables from Asia and USA, and meat from wild animals that are brought usually from the USA (González-Jácome 2019). In these types of markets, agricultural products without agrochemicals are combined with others that possibly were grown using said chemicals, but that come from abroad. These markets are specialized in gourmet products offering articles that are not part of everyday consumption, and whose prices are not lower either. Currently, there are organic products markets throughout the country, but they are incipient, with little publicity that encourages buyers to look for them and buy their vegetables and fruits; besides their prices are not always within the reach of the entire population. There are studies done by educational institutions, and national and international research centers, public and private organizations, where relevance is given to the knowledge about the plants people is consuming daily and their origin; in addition to the ecological, nutritional, social, cultural, and agricultural importance they have.[3]

3.8 CONSIDERATIONS AND CONCLUSIONS

This book is a brief summary of the state of the art on the studies of traditional agriculture in Mexico, of the questions that have arisen, of those that have been answered, of those that remain pending, and of the situation in which these studies are in the scope of the relationship between

[3] Alayón and Gurri 2007: 243–260; Bye and Linares 2000; Cruz-León et al. 2012; McClung et al. 2014: 97–121; Entrena-Durán and Jiménez-Díaz 2013: 683–693; FAO 2006; González-Jácome 2018; Gurri-García 2007: 60–68, 2011: 65–84; Harper and Le Beau 2003; Villanueva-Verduzco 2017.

agri and culture. It has – among others – the purpose of giving way to new reflections, proposals, questions, and projects that continue to seek and give answers to what is traditional agriculture, as an activity that exceeds the subsistence and market economy of human societies. It must also be included in their interrelations with local and regional environments, and their adjustments to the environment and climate change; what it is possible to learn about traditional agriculture in Mexico, and the social, cultural, ideological, food system, and health aspects with which it is associated (Angulo and González-Álvarez 2008: 567–590; Cruz-García 1992; Lammel et al. 2008; Hodell et al. 2000: 13–38). In addition, other elements have to be articulated to achieve the welfare of small producers and their families.

RECOMMENDED READING

Golley, F.B. 1993. *A History of the Ecosystem Concept in Ecology. More than the Sum of the Parts*. Yale University Press.

The book deals with the early history of ecology, when basic concepts such as that of the ecosystem were being shaped, of which the author traces its development from the end of the nineteenth to the twentieth century, including its applications and creators in the United States and Europe. For Golley, the history of a discipline like ecology is his theory. The book is important because its comparative and historical information have the reader think about the basic concepts of a scientific discipline, how they were formed, and their deep meaning.

Kelly, Isabel, and Ángel Palerm. 1952. *The Tajín Totonac: History, Subsistence, Shelter and Technology*. The Smithsonian Institution and the Institute of Social Anthropology, Washington, D.C.

The book is based on the results of research carried out by two authors in Papantla, a region in the state of Veracruz. It includes, in addition to the social and cultural components of the ethnic group, a detailed description of their agriculture; the technological aspects related to this activity; and the interrelationships between the Totonac people with its environment, the cornfield, the homegarden, and the natural ecosystem that was the subtropical forest.

The Totonac agriculture was based on a combination of agroecosystems including the polyculture cornfield, the homegarden, and the use of resources from the forest and the exploitation of vanilla. The combination of these agroecosystems gave the small-scale Indian farmer and their family the necessary food as well as monetary resources to cover their physical and social needs. This is one of the anthropological studies that best integrate the social and cultural aspects with the ecological, geographic, and agricultural aspects.

Palerm V., Ángel. 1957. *Ecological Potential and Cultural Development in Mesoamerica. Studies in Human Ecology*. Pan-American Union, Social Science Monograph N° 1: 28–42, Washington D.C.

Palerm V., Ángel. 1968. The agricultural basis of Urban Civilization in Mesoamerica. *Man in Adaptation*; Yehudi A. Cohen (ed.). Chicago, Aldine Pub. Co.; pp. 348–361.

This book is based on the 1955 article and discussions on the importance of irrigation for the development of Mesoamerican civilization. Palerm classifies agricultural systems in: (1) slash and burn; (2) fallow; (3) and irrigated land for agricultural purposes. Palerm described them according to their environment, main agricultural characteristics, productivity, and the total population at that time, including the population density and the types of settlement where these people lived.

4 Traditional Agriculture in Mexico (First Part)

4.1 INTRODUCTION

This book seeks to provide the reader with a summary of complex agroecosystems and their main interrelated components. For this, a group of organized information elements is handled, which allow describing these agroecosystems, which are characterized by being holistic.

The numerous agricultural systems that still exist can be included in traditional agriculture since they are not static and are adjusted, abandoned, changed, reproduced, and reorganized over time according to the period, and the places where they are located, the human populations practicing them, and some of the external forces they can confront. There are combinations in which traditional agroecosystems work together with commercial and conventional agriculture. Something similar occurs with agricultural technologies applied to traditional agroecosystems.

One of the basic components in agricultural systems is water, including continental bodies, the rain that falls annually and, along with the humidity, and climate that occur in the various regions of the country. Since pre-Hispanic times, traditional Mexican agriculture has been rich, including a series of agroecosystems and their adjustments or local adaptations.

Among them are dry and irrigated homegarden, with areas ranging from a few square meters to several hectares; rainfed and irrigated cornfields; seasonally slash and burn agroecosystems in tropical lowlands; slash and burn in seasonal and marginal highlands (*tlacolol, tlacolole*); terraces (rainfed and with irrigation); semi-terracing (*bancales, metepantles*) (West 1970); cornfields in hills and mountains without terrace; coffee or fruit crops using interspersed areas within the forest such as *banquetera* (Servín-Segovia 2000, 2002), coffee forest, and coffee homegarden (Blanco-Rosas 2006; Bartra-Vergés et al. 2013; McClung de Tapia 2000: 251–280; Servín 2000, 2002; Moguel and Toledo 1999) for products with high market value.

Traditional agriculture in Mexico still requires studies and programs that are focused on developing improvements for the food system and health of small-scale farmers' families and their food system, including industrial food products. Also, it is necessary to know the mechanisms of articulation that occur between local productions and the market, and allow the recovery of their high exchange value in national food products and export products. In general, it can be said that the Mexican small-scale farmer considers his agricultural products as having little exchange value in the market.

Apparently, the many years of subordination and economic and cultural underdevelopment have left a difficult mark to erase, where the values of use, change, and cultural values of what is produced on a small scale are not appreciated. Perhaps that is why the ecological, social, cultural, and economic value that is implicit in the knowledge of traditional Mexican agriculture, and its applications to the future is not appreciated either.

4.2 CLASSIFICATIONS AND TYPES OF TRADITIONAL MEXICAN AGRICULTURE

Traditional agricultural systems can be grouped in different types depending on the particular interests involved, as of the studies and applications of them and how they are carried out. That is, the groupings researchers use in each case are constructs, which depend on the objectives of the projects, and of the investigators who are in charge of their study. Therefore, the classification can vary and contain numerous local or regional systems, which we can group as traditional because – almost

DOI: 10.1201/9781003198833-4

always – they are of ancient origin and have a long history of adaptation (cultural adjustments) to environmental, climatic, and social conditions.

These systems have several adjustments that include altitudinal, physiographic, ecological, climatological, historical, social, political, and cultural ones. They are generated, developed, and adapted by human populations over time, according to their physical, biological, social, and cultural needs (their food system culture, family's health, objective, senses, tastes, and worldview). They are adapted to regions and places with specific local characteristics where prehistoric human groups lived in a seasonally mobilized circuit or decided to settle permanently in places where they obtained their subsistence.

Traditional agricultural systems can be explained chronologically: Prehistory; Preclassic or Formative; Classic; Epiclassic or Postclassic; Late Postclassic or Historical; Conquest and the Spaniards' entry in Mesoamerica; the Viceroyalty; Independent Mexico in the nineteenth and twentieth centuries; and the beginning of the twenty-first century. Physiogeographic characteristics correspond to the spaces where these systems were developed (e.g., coastal plains, highlands, mountains, deserts, semi-deserts, mangrove, and others), and they can classify agroecosystem types.

The type of ecosystem where human agricultural societies live can be studied and organized: tropical rainforest, tropical deciduous forest, cold or temperate forest, semi-desert, desert, mangrove, and others. In addition to environmental components, the type of climate (tropical, semi-desert, cold, temperate, or the direct and indirect effects of weather events such as El Niño/La Niña, hurricanes, and tropical storms) that affects agriculture must be taken into consideration.

As a starting point, traditional agroecosystems can be classified by using the type of reproduction of the plants that were grown in the different ecosystems (i.e., fruits, tubers, grains, or cuttings), or by applying a selection of social and cultural traits that are specific to a human group (e.g. an ethnic or linguistic group). Similarly, their cultural areas and ethnic groups can be used to classify traditional agricultural systems (Nahua, Mayan, Mixtecs, Zapotecs, Totonac, Huaxtec, Zoque, Popoluca, Purhépecha, Yaqui, Otopame, Mayo, Pima, and others).

Anthropologists use this type of classification. Therefore, there are numerous studies where on ethnic group is related to a type of agriculture developed by it (e.g., coffee production among the Totonac of the sierra, slash and burn by the Maya, the cornfield by the Maya in Yucatan, the cornfield in Los Tuxtlas by the Zoque-popoluca, and many others).

The selection includes the area under cultivation and the methods used to work the land extensive, or intensive in surface area used for cultivation, or the labor force employed by the human group. It is basic to consider the characteristics of the components of the environment like thin or thick agricultural soils, and flat, rugged, or eroded soils; water (rainfall regime, regions with scarcity or abundance of water in natural deposits and rivers, droughts); irrigation in temporary dry and irrigated lands; flora; fauna; altitude; and agroforestry production (Willey 1983).

Available water resources and the irrigation in temporary, dry, and non-irrigated lands include the area under cultivation and the methods used to work the land (extensive, or intensive in surface area used for cultivation, or in the labor force employed by the human group). The type of climate is equally important: tropical, semi-desert, cold, temperate, and the direct or indirect effects of weather events that affect agriculture. The agricultural impacts caused by seasonal rains or irrigation on the land surface under cultivation by irrigation should be considered as well.

Classification can be based on the natural and geological processes with destructive impacts affecting human populations, their horizontal and altitudinal spatial mobility, their settlements, their environment, and agriculture. For example, soil erosion that prevented cultivation in Mesoamerican territories and deforestation that modified rainfall precipitation affecting the agricultural productivity of a human group or culture. The volcanism made human populations migrate to other places, like the eruptions of the St. Marta volcano in Los Tuxtlas and the Xitle volcano in the basin of Mexico.

The classifications are innumerable because their selection and construction depend on the participants' interest in each investigation, or the projects', or studies' objectives and purposes.

FIGURE 4.1 Mexican Blackcherry (*Prunus spp.*).

Traditional agricultural systems result from adjustments (ecological, political, economic, social, and cultural adaptations) to different environments, altitude settings, weather characteristics, and physiogeographic combinations of which there are several types with local and regional differences, according to the regions, their ecosystems, the amount of resources available, and the time involved.

Wilkerson (1987: 213–214) considers that Veracruz was a natural corridor, from near 7000 BC to 2400 BC, connecting from the center to the north and south of the Gulf Coast, and the numerous coastal ecosystems supported a considerable human population, which used various resources within close range. This situation was complicated by the regional volcanic activity in the Pleistocene, of which the most recent is the eruption of the San Martín volcano in Los Tuxtlas (1793). The corridor had its narrowest section in the central part of the current state of Veracruz (St. Ana, Palma Sola, St. Rosa, Laguna Verde, Villa Rica, and even near La Mancha lagoon), where archaeological studies traces of proboscides and camps, but no pottery were found.

These types of adjustments make them distinctive, with each one having specific characteristics. It is necessary to consider that a human group that lives in a place can handle several landscapes, ecosystems, and agroecosystems at the same time, and the combinations between them and with their ecological, and agroecological surroundings (Siemens 2002: 161–202; 2010: 183–203; 2009) (Figure 4.1).

4.3 CHRONOLOGY, TRADITIONAL KNOWLEDGE, AND ENVIRONMENT IN TRADITIONAL AGROECOSYSTEMS

This is a chronological summary review of the origins of horticulture and traditional agroecosystems of Prehistoric origin, located in semi-desert regions with bands of seasonal hunters/gatherers using rock shelters and caves as seasonal housing, where the domestication of maize began with the first cultivated plants (Xihuatoxtla). The incipient agriculture was born in the plains near rivers, streams, and lagoon systems. The food system was based on hunting, fishing, and capturing and collection of flora and fauna species. So far there is information based on archaeological research carried out: in the middle basin of Balsas River; in the central valleys of Oaxaca near Mitla; in the caves in the Tehuacán valley; in Zohapilco, in the basin of Mexico southern banks; in the caves in Tamaulipas; and in the remains of La Playa site, in Sonora.

It is in these places where the process of domestication of plants and continues to date. About the emergence of traditional agricultural systems in Prehistory (8700 BP–2000 AD), there is information on the location, seasonal movements of the inhabitants, and the regions where seasonal migration occurred.

This horizontal mobilization developed in regions with tropical and subtropical forests, or in semi-desert areas where the settler bands moved seasonally throughout the year. In nearby places, the emergence of the first garbage dumps and incidental homegarden developed, starting a process from the forest mixed with the incidental homegarden to domestication of the first plants (8700 BP–2000 AD) including natural forest and fruit trees. These plants were maize (*Zea mays*), Mexican hog plum (*Spondias mombin, S. purpurea*), Lead-tree (*Leucaena esculenta*), squash (*Cucurbita pepo* and *C. argyrosperma*), Sieva type beans (*Phaseolus vulgaris* and *P. lunatus*), Amaranthus and Chenopodium, Avocado (*Persea americana*), and chili pepper (*Capsicum annuum*) (Galindo-Tovar, Lee-Espinosa, Murguía-González, et al. 2013: 65–70; Weirsum 1997: 425–438).

With this domestication, a pre-agricultural phase began, until the forest ecosystem mixed with the homegarden and the cornfield arose, which over several centuries became a polyculture system. The first areas where prehistoric populations settled were located on the land of natural elevations, or on those that were artificially constructed in the plains near the lagoon and river systems. The initial settlements were rocky shelters and caves as well as near lands combined with areas for fishing and capturing and collecting fish, crustaceans, turtles, mollusks, and waterfowl. Among them were: S. Andrés, Tabasco; the Mayan zone; Soteapan, in Los Tuxtlas region in southern Veracruz, and S. Lorenzo, also in Los Tuxtlas; and the banks of the lagoon system in the basin of Mexico. For the clearing of the forests, fire was applied, giving rise to the slash and burn agroecosystem, which was not exclusive of the tropical regions.

From the prehistoric natural systems (At least from 13 to 12 thousand years BC–2300 BC to 2000 BC), the first traditional and incidental agricultural systems arose, whose evolution took about six millennia: (1) from forest to forest with incidental homegarden; (2) from forest to the incidental homegarden to the homegarden, and cornfield with the first cultivated plants, perhaps with variants depending on the particular region; (3) from the forest to the homegarden, and to the cornfield when both agricultural systems were independent; (4) from the extensive slash and burn cornfield; (5) from the process of the cornfield "travelling" to other places to later become the fixed cornfield with crop rotation, and rest periods; (6) to the cornfield with its own evolution processes that came about over time, with local differences related to the specific conditions in which the system developed in Mexico.

The themes and studies on traditional Mexican agriculture had been presented problems to be able to understand them. There are answers to various questions as to how it arose, how it works, how it has evolved, what elements characterize it, what changes it has undergone, and how the human population had responded to such changes. These answers come from various fields of scientific knowledge and are important concerning traditional agriculture. The information on the origins of agriculture in Mexico, or elsewhere in the world, is modified as new questions arise, along with new methods of collecting and analyzing information (Wheatherwax 1955: 1–16). In addition, new technologies and concepts give greater accuracy and depth to interpretations. Despite the progress made to date, there is clearly the need to conduct further studies to help to explain the evolution and history of this primary activity, and its interrelations with the environment, climate, human societies, and their cultures more extensively.

4.4 TRADITIONAL MEXICAN AGROECOSYSTEMS

The study of rural societies has generated a series of functions, interactions, and interrelations among the different components that explain the current situation in the different regions of the country. It is essential to study the relationships between the aforementioned factors and indigenous populations, particularly by those that live mainly from agricultural activity. It is here where the importance of the study of traditional agriculture sprang up and was highlighted, to understand what the country is, and what its needs are. Those interrelations go to different scales, from the local to the national and global.

This happened without considering the role of traditional agriculture in the various food systems, and in health, even though both issues are of the first order. Achieving food self-sufficiency is an essential objective in national and international institutions. This would have to be studied and analyzed from various points of view.

Traditional agriculture had and has to date a relevant role related to elements articulated to the maintenance of sustainability, biodiversity, cultural diversity, resilience (resistance) of social, cultural, and environmental systems in the occurrence of eventualities coming to them from abroad that impact them negatively. One of the elements that allow their successful maintenance over time is – without a doubt – the role that traditional local knowledge plays in agriculture, the forms of its transmission, and its relationship with the social forms of family organization.

Another element to consider is the relationship of traditional agriculture with the events or aspects that are currently essential for the conservation of human life on the planet, such as climate, an issue that requires attention from many points of view. All this is interrelated with aspects such as the conservation of diversity in systems and agroecosystems, and its sustainability. This concept – now in vogue – is also politicized, causing it to lose – or distort – its content and essential meaning. It is very useful for studies in agriculture, particularly traditional agriculture (Konrad 1985: 321–330, 1996: 98–130; Lipper et al. 2014: 1068–1072; Vasquez-Dávila (cord.) 1992).

The knowledge of traditional agriculture is closely linked with sustainability, the economy, politics, social organization, health, food system, leisure time, and popular religiosity (its traditions, ceremonies, and rites), in addition to other components, or features of the local and regional culture of human populations engaged in agricultural activity. These elements add resilience to local farming systems and their cultures to resist the onslaught of the larger society and its economic policies against traditional agriculture. It is important to remember what Efraím Hernández X. (1998: 673–678) wrote:

> The term traditional agriculture is derived from the way in which knowledge is disseminated, and it is distinguished by the reduced quantity, and quality of energy used in the agroecosystem; it predominates in the world's agricultural lands with favorable or marginal climates for production. [...] On the one hand, traditional agriculture, with a worldview that includes physical, and metaphysical aspects, on the other, a modern agriculture linked to a materialistic worldview. This dichotomy has been accentuated in Mexico due to uneven socioeconomic development, a cause of defeat, mass death, labor submission, and general marginalization of the indigenous population, as well as the domain of a reduced caste that monopolizes the natural means of agricultural production. The full penetration of capitalism accentuates this social, and technological inequality, giving the general judgment that traditional agriculture is carried out by indigenous farmers, poor, ignorant, clinging to their beliefs, little productive, which contrasts with modern agriculture, represented by productive agricultural systems, and capitalist, educated, and innovative people. These trials have discouraged studies on traditional agricultural technology in Mexico.

In 1941, Hernández X. (1998: 673–674) considered that one of the "Green Revolution" results was the governmental support of Mexican agriculture from the following points: (1) experimentation and application of innovations of the American agricultural sciences in ecological and social zones favorable for production; (2) generation of knowledge required for the specific situation of the country; (3) preparation of national professionals for the execution of the programs; (4) and analysis of the agricultural economic situation of the country and its causes. As the results of its application to more favorable regions for agriculture, credit facilities for small properties, and strong official support for technical services and dissemination, the results of the program were:

> [...] They were spectacular [...] the production of wheat, maize and beans grew rapidly; irrigation lands increased; agricultural export production increased; it was shown that modern technology is capable of solving national agricultural problems and there was full incorporation into the capitalist system dependent on foreign centers of power.
>
> *(Hernández X. 1998: 674)*

Consequently, Hernández X. (1998: 674) considered that the public policy to agriculture established the following bases of national agricultural development: (1) the massive introduction of modern technology, with its consequent capitalist dependence; (2) the establishment of an extensive network of disseminators who, together with the credit system, put pressure on producers to adopt agricultural innovations; (3) the provision of support to the fertilizer and industrial agricultural products; (4) and the strengthening of the specialized agricultural education system.

These bases were discussed by researchers in different areas. These studies revealed the limitations of the approach and established the points that should be strengthened. Hernández X (1998: 674) claimed that this situation originated and reinforced the need to conduct studies on traditional agricultural systems "[...] to which it was intended to modify the causes of their persistence, the social environment in which they developed, and their socioeconomic function".

4.5 TRADITIONAL AGRICULTURE IN REGIONS LACKING WATER

Ancient water storing systems in semi-desert regions was a vital factor in the survival of hunter/gatherer bands and early cultivators. In regions with a lack of water, or insufficient seasonal rains, early cultivators created systems to capture rainwater, retain it, and then convey it to homes and fields. Among the first systems were stonewalls crossing ravines from one side to the other; the accumulation of rainwater in natural hollows; open cisterns (*jagüeyes*, *sartenejas*); underground cisterns (*chultunes*); holes dug in the ground (locally called *jollas*, *ollas*, or *slat-board systems*); water boxes dug in the ground (*cajas de agua*) or on the banks of ravines, and terraces; foggaras or quanat (in-line-wells); different types of canals; and others.

In Mexico, these systems for collection, conservation, management, and distribution of rainwater are part of traditional agriculture. The open cisterns (*jagüeyes*) are receptacles for rainwater of a more or less circular shape, located in places where runoff can naturally and be captured in the rainy season, stored, conserved, and distributed during several months of the year. These systems to capture and direct water have pre-Hispanic origin, and are found in regions with climates ranging from cold to temperate to semi-desert like the Llanos de Apam (State of Mexico, Hidalgo, Tlaxcala, and Puebla), and the state of Morelos, also (Guzmán and Palerm-Viqueira 2005: 21–26) (Figure 4.2).

FIGURE 4.2 Tlaxcala: Aychqualco, Glyph of a Jagüey. Lienzo de Tlaxcala, Codex, 05/08/1520; Sheet, 23; p. 126.

These systems are distributed throughout the world, and there are them in large semi-desert regions of the Asian continent that follow what was the old silk route, also in Spain, Bolivia, and Peru. In Mexico, these systems for rainwater collection and distribution are part of traditional agriculture. They have pre-Hispanic origin and are found in regions with climates ranging from cold to temperate to semi-desert. They are ancestral systems for the collection and management of rainwater that are also known as "water traps" (Martínez de Azagra et al. 2006: 453–470).

For agroecosystems developed in semi-desert environments, the first thing that is necessary to know is how to capture and store rainwater, then how to distribute it along several months of the year. In the early days, dams were made with branches and stones, which crossed the serpentine paths and ravines to retain the water that ran through them during the rainy season (Sanders 1976: 85–150). These dams grew in size and importance until they formed the first pre-Hispanic dams, such as Purrón in Tehuacan that has been dated to 1050 +/–300 BC (Aiuvalasit; Neely; Bateman 2010: 1207–1213; Neely 2017; Rincón-Maurer 1999).

On the lower and eastern slopes of the Matlalcueye volcano, there are ducts made of stone or of hollowed tree trunks (*canoes*) that form canals to conduct water from the lakes running to the rivers that flow to the low areas and the plains destined for cultivation (Montero-García 2012). In the western slopes of the same volcano is in the subsoil a canal dug in the ground forming a foggara system (quanat, in-line-wells) (González-Jácome 2003). This system was developed in the slopes of hills and in regions with water shortage. The first issue to be solved by inhabitants is the capture, retention, and conservation of rainwater, and then its distribution. The canals dug in the ground are found in the mountains in the High Mixteca (Cruz-León 2019). The conservation of water in a healthy state has been a little-studied problem.

Another example is the one studied by archaeologists Sanders et al. (1979: 263–267) on the northern slope of Cerro Gordo, in Teotihuacan. There are the archaeological remains of the ancient canals that crossed through the canyon where the waters of natural currents go, crossing the zone there were open cisterns (*jagüeyes*). These systems have been –until now – difficult to date and have been considered as marginal, which maybe built before the large irrigation systems (René Millón 1957, 23: 160–166). Some of these water systems of pre-Hispanic origin are still in operation, although in recent decades many have been replaced by modern systems.

There are plains located on lake banks and lagoons, or in river basins where the villagers built drained and raised fields or ridges (*camellones*) (González-Jácome 2008; Renard et al. 2012: 30–44). In the Early Formative or Early Preclassic in Tlaxcala (1200 BC), open cisterns (*jagüeyes*) (Abascal and García-Cook 1972: 117–124; 1975: 199–210), artificial deposits like stirrups (*cajetes, estribos*), holes (*jollas*), and other local systems emerged. All of these are components of small, medium, or large-scale agro-hydraulic technology linked with traditional agriculture from early times.[1]

The conservation of water in a healthy state has been a little-studied problem. In addition to the development and role that water availability has had in traditional agriculture, especially in times of crises, such as that which occurred during the so-called "Mayan collapse", not to mention the current ones in urban, and rural populations.

Agriculture was practiced in the chinampas of the basin of Mexico or swampy lakeside areas since pre-Hispanic times. The *chinampa* was the most refined and productive agroecosystem developed in Mesoamerica, and maize was the principal *chinampa* crop (Dubmire 2004: 34–38; González 1992).

In tropical regions of south and southern Mexico, the drained or raised fields were built in central Veracruz, Campeche, and waterlogged regions; they were very important agroecosystems that were built in all parts of Mesoamerica, including the tropics and cold temperate lands (Darch 1983; Matheny and Gurr 1983; Puleston and Siemens 1972: 228–239; Renard et al. 2012: 30–44; Siemens 1983b, 1989a; Siemens et al. 1988).

[1] Böehm and Pereyra 1974; Cabrera 2005: 121–162; Doolittle 1990; Guzmán and Palerm-Viqueira 2005; Martínez-Saldaña et al. 2011; Neely 2017; Rincón-Maurer 1999; Palerm-Viqueira et al. 2008.

In addition, other traditional agricultural systems were generated in wetlands like *petenes* (Barrera 1982: 163–170), low lands or shoals, swamps, *rejolladas* (Kepecs and Boucher 1996: 69–91), Maize of March (*marceño, chococito* maize), and riverside summer crops agroecosystems (*tecallis, mawechis*).

In the Yucatan Peninsula with karstic soils, chultunes (*chulub tun*) are found. These artificial underground cisterns have a bottle or bell shape. They are built in areas where there are no sinkholes (*cenotes*), and seasonal rain is the main water source for residents of the peninsula and several archaeological sites, like those in Uxmal and Edzná in Campeche because they were a widespread system in the Maya Lowlands and the Puuc zone in Campeche (Matheny 1978: 185–210; Matheny and Gurr 1979: 441–449; 1983: Vol. 1, 1). To avoid filtration into the chultunes, they are covered with mortar or stones, like in Tenam Puente in Chiapas (Matheny and Gurr 1979).

The early farmers built canals to carry water to cover domestic and agricultural needs. Some chultunes were used as warehouses to store maize (Matheny 1983; Rojas-Rabiela 2009: 26–27). There are a few chultunes, which still are functioning in Uxmal and in the *Puuc* zone in Campeche (Matheny 1983) (Figure 4.3).

The conditions of climate, topography, physiography, altitude, and geological history modify the situation of the water supply at the local and regional scales, recording when it rains, how it rains, and what possibilities exist to capture and retain rainwater. An example of this in ancient times are the natural or artificial water deposits (*aguadas*), the sinkholes in the north of the Yucatan Peninsula, or the waterlogged lands in the wetlands, swamps, *petenes*, flowed lowlands (*bajos*), and *rejolladas* of the southern Maya lowlands (Barrera 1982: 163–170; Fedick 1996; Fedick and Ford 1990 22: 18–33; Kepecs and Boucher 1996: 69–91). Studies on the origin and management of water collection and control systems in semi-arid areas for agricultural and domestic use offer different chronologies and interpretations (Matheny and Gurr 1979; Palerm 1972; Whiteford and Enge 1989).

In pre-Hispanic times, agricultural landscapes in the Mayan lowlands included the use of a pattern composed of residential sites, intensive dryfield cultivation, and well-drained uplands that provided food on a regional scale (McAnany 2010). Well-drained cultivation used raised and drained fields in the periphery of the region where wetlands were found, riverine floodplains with karstic soil, and depressions where well-drained uplands were rare. Soil distribution was the instrument used by the Maya to manage pre-Hispanic settlement patterns and land use (Fedick and Ford 1990: 18–33; Luzzadder-Beach and Beach 2006: 91–117).

The information on ancient societies includes issues such as control over available water resources and their agro-hydraulic systems as well as the emergence of the State and the first civilizations

FIGURE 4.3 Yucatán, Cuzamá, X'togil Sinkhole (*cenote*).

(Palerm 1955, 1972, 1976, 1997; Palerm and Wolf 1961, 1968 [1955], 1972). In this sense, water, its quantity, and its quality – in addition to its management – are related to problems and discussions on the evolution of urban societies, the State, population increase, diseases caused by their management, and agricultural productivity (Palerm 1955, 1972, 1976, 1997, 2007; Palerm and Wolf 1961, 1968 [1955], 1972; Wittfogel 1956, 1966).

During the Formative period, the construction of different systems to retain and control water were built, including: the first dams made with stones and branches occurred. Stone walls (ringleras) were also built to retain moisture, located in the valleys between the hills (Spores 1969: 557-569); or dams on the river, such as the Purrón dam in the Tehuacán Valley, (Neely 2017; Rincón Maurer 1999). The ducts made of stone or with hollowed tree trunks (canoes) that form canals to conduct water from the lakes formed by thaws, the rivers that descend to the low areas, and the plains destined for cultivation. There were these types of water control systems in places such as the high Mixtecs Mountains or the eastern slopes of the Matlalcueye volcano (Montero-García 2012), among many others (Palerm-Viqueira and Martínez-Saldaña 2009).

In the twentieth century, water for agriculture had to deal with the processes of desiccation of lagoon areas, such as those of the Toluca Valley (Albores-Zárate 1992, 1995, 2018: 117–144; González-Jácome and Velasco-Orozco 2015), or the drainage of the basin of Mexico (Correa 2011: 203–225; Aréchiga 2011: 227–269) and its impacts in places such as Xochimilco-Chalco and chinampa agriculture in central Highlands of Mexico, or in economic activities and daily life of the affected populations (Canabal 1997; Espinosa 1996).

Moisture crops, located on the banks of rivers or lagoons, are distributed throughout the country. Del Amo et al. (1999: 77–101) studied in the Balsas River shore the moisture agricultural system called *tecallis*. In S. Mateo Atenco, State of Mexico, Albores-Zárate (2018: 117–143) studied the agriculture developed in the lakeshore of the Upper Lerma River. Robles-Linares (2018: 269–296) studied the agricultural systems called *Verano* and *Mawechi* that are found on the shore of the Mayo River in Sonora. The management of plank agroecosystem (*tablones*) to control the irrigation of crops through rectangular platforms delimited by wooden planks is found in plains as of Amatenango, Chiapas, and Guatemala (Wilken 1969, 1997).

Water for everyday use and for irrigation of crop fields represents problems and possibilities, depending on the region in question. Water for domestic and agricultural purposes has been one of the important axes in the description, understanding, and discussion of ancient and current societies. The water available in the different regions of the country and the way in which Mesoamerican societies obtained, captured, managed, control, and used it is the subject of controversies over social and urban evolution, and the emergence of the State.

If water for irrigation has great abundance, excesses, and annual floods occur; if it is scarce and seasonal, or soils do not retain it easily; both extremes are a problem for agricultural societies. The excess and the shortage of water place the farmer in stressful situations where he has to deal with several problems to retain it, conserve it, extract it, distribute it, or generate agricultural systems adjusted to those conditions. The farmer can modify plants (adapt them) so that they can resist the lack or excess of water.

4.6 TRADITIONAL AGRICULTURE IN REGIONS WITH WATER EXCESS

The interrelationship between available water and agricultural systems has been a controversial issue that encompasses water history and human uses. For agriculture, ecological, social, and cultural adjustment to the situations in which the different regions of the country are found is essential. The conditions of excess or lack of water during several months of the year explain – at least in part – the characteristics of the agricultural systems, the crops that are produced in them, and the adjustments or adaptations that the farmers made in this regard (Palerm 1955, 1972, 1976, 2007; Palerm and Wolf 1961, 1968, 1972; Wittfogel 1956, 1966).

As water is a fundamental element for agricultural activity, its relationship with culture and agriculture has been the subject of numerous studies; the same as its forms of sociopolitical organization and its relationships with the diversity of systems developed by mankind had modeled the results depending on the lack or excess of it. There are evident differences in the human adaptations to water occurrence and management macro-regions, from the north, center, south, and southeast of present-day Mexico. From these different conditions, we can divide the systems for the collection, storage, and distribution of water in several ways.

Water management in different ecosystems and regions has to be understood according to aspects like: (1) rainfall in the different regions, the monthly and annual average of the rainfall and its location; (2) the size of surface of underground currents, their location, characteristics, and geological history; (3) the way in which the amount of water running down the rivers, their tributaries, and through the ravines occurs; and (4) the availability of water in the different seasons.

Traditional agricultural systems in tropical regions are challenging to farmers because their soils are permanently or temporarily flooded for several months of the year. In these regions, the soils are not permeable and do not allow easy water outflow for several months, especially when heavy rains and downpours fall. Thus the aquatic surface of wetlands, swamps, petenes (*akalché*), flooded lowlands, and *rejolladas* increases gradually with the first rains, which occur in the spring and summer, from the end of May or beginning of June, lasting until autumn.

In this category is "swamp agriculture", which includes the March Maize agroecosystem (*marceño*) described in Chapter 17. 3. It is possible to find the Maize of March in the temperate-cold regions, with a continuous excess of water. This agricultural system is found in flooded lowlands regions with the confluence of rivers that increases their level and flows in the annual rainy season. In these systems, other types of swamp agriculture are found such as the raised and drained fields (ridges, *calal*, *camellones*) and the furrowed fields with canals, as happens in the Atoyac-Zahuapan river basin in southwestern Tlaxcala (González-Jácome 2008), in the old basin of Mexico (Rojas-Rabiela 1974, 2009: 9–26), and in the today extinct Lerma river basin in the Toluca Valley (Albores-Zárate 2018: 117–144; Molina-Hampshire 2009). These agricultural systems are more sophisticated and require larger populations, as they require the participation of the inhabitants in their construction and maintenance.

4.7 LOCAL AND REGIONAL DEVELOPMENT PROJECTS

There were scientific development projects (Barkin and King 1970; Barkin and Zavala 1978; MacMahon 1973; Molina 1976; Palerm 1970, 1993). An experimental program directed and organized by the ecologist Stephen R. Gliessman was little known, unfortunately, but funded and developed in the mid-1970s, by the Higher College of Tropical Agriculture (CSAT), located in Cárdenas, Tabasco. Gliessman, his undergraduate, and graduate students[2] in tropical ecology, with the support of Ángel Ramos – then director of the College – initiated a local development project based on what they called the "ecological bases of traditional agricultural systems".

Studies were organized, and fieldwork was carried out in the Emiliano Zapata and Tacotalpa endowed lands (*ejidos*) in Tabasco. At the same time, the project included field experiments, classroom classes, and the use of laboratories and facilities of the College, where the ecological components of local agriculture were studied. Experiments (*modules*) were organized with the information obtained. In addition, discussions were generated, and the opinions of several specialists were heard at the school's facilities, where courses were organized to create a postgraduate course in Tropical Ecology and disseminate the importance of such studies. One of them was Efraím Hernández X., who took part in the graduate courses (Rosado-May 2018; Mariaca-Méndez 2018).

This agroecological project aspired to generate in the residents the interest to produce their food, using modules seeded with the plants that were of their ordinary consumption, without the application

[2] Among them Moisés Alarcón, Ángel Martínez-Becerra, Francisco Rosado-May, Octavio Ruiz-Rosado, Elizabeth Lacedelli, Mario Alejos-Peraza, Miguel Ángel Hernández-Bautista, Martín Sánchez-Libreros, Ramón Mariaca-Méndez, Juan Carlos Chacón, Orlando Solano-Vidal, Sergio Alavez, Tomás González Estrada, and Tomás Chan Tzuc.

of agrochemicals, and following the elements of traditional agriculture, but studied and reorganized. In spite of all the efforts, the project was not successful in its initial phase because the small-scale farmers were more interested in abandoning agriculture and developing pastures for the introduction of cattle rising, which left more money than agriculture. However, years later with the CSAT, the project's team and the Gliessman experiments disappeared, but land agricultural workers (*ejidatarios*) redesigned and reorganized the modules relying on memories (local oral tradition) of those people who had been involved in the modules and their conformation (scientific knowledge).

Dr. Gliessman had supported a subproject of a sociological type to evaluate the operation, organization, human, and technical problems of the modules. For the information obtained, there are elements to help understand what happened. Gliessman hired Luis F. Leriche (1983), a Mexican student who was pursuing a graduate degree in the Department of Latin American Studies at the University of Texas in Austin. For several months Leriche did fieldwork in the project-endowed lands (*ejidos*), where modules had been settled. With observations *in situ* and to interview information, he wrote his thesis and graduated. Apparently, the module project had failed. However, sometime later the villagers – when they needed it – collected what they remembered of it and rebuilt the system according to their needs.

The result of the High College of Tropical Agriculture (CESAT) project in the Lázaro Cárdenas and Tacotalpa endowed ejidal lands is important, although initially it was not encouraging nevertheless sometime later, without institutional support, the peasants reincorporated their agriculture and livestock, raising their modus vivendi by using several elements that they took from the original project, and the learned scientific knowledge (Leriche 1983).

A series of development projects have been carried out by biologists and agroecologists; some worked, and others did not. The causes of the success or failure of the programs must be explained individually, locally, and within specific contexts, since they are due to very different causes, many of them local and related to the situations in which communities live and develop (Del Amo and Gómez-Pompa 1994; Gómez-Pompa and Del Amo 2001; Gómez-Pompa et al. 2003).

In the 1990s, some private-capital projects emerged that were aimed at the creation of non-governmental nature reserves, where scientific studies for their conservation and use by the settlers in them were carried out. A good example was El Eden, in Quintana Roo. After 10 years of operation, the participants in the project published the results of their efforts (Gómez-Pompa et al. 2003). These were uneven. The biological section is detailed, innovative, and important. This is logical since an important part of its founding members was made up of biologists. The archaeological section has given important results, which clarify the role of the Mayan forest in the evolution of society, especially in the lowlands (Fedick 2010, 2012; Gómez-Pompa et al. 2003). However, the social part of the project is practically non-existent.

Some development projects worked in the short term but failed in the medium and long term, perhaps as a result of local political and administrative situations, internal divisions that occurred in the communities, and problems among the people involved them.[3] On other occasions, the people working in the projects modified the original conditions, to adapt them to their particular needs, as was the case with the *tropical chinampas* of Nacajuca, Tabasco (Pérez-Sánchez 2007). In these cases, the peasants involved showed their capacity to adjust to new environmental, political, and social situations and conditions. They were a living example of the resilience of the system.

Hernández X. (1998: 364) considered two worldviews in farmers. One includes physical and metaphysical aspects, the other a materialistic worldview. This dichotomy in the country generated unequal economic and social development, labor submission, and the marginalization of the indigenous population. At the same time, it has put in the hands of a few the natural means of agricultural production to hoard and has extended the idea that peasants are very poor, ignorant, clinging to their beliefs, and little productive.

---

[3] Aguilar et al. 2014; del Amo and Gómez-Pompa 1982: 8–18; del Amo and Rorive 2003: 551–559; del Amo and Vergara 2007, 3: 31–45, 2009, 4 (2): 27–33; del Amo et al. 2010: 141–156.

These characteristics are opposite to modern agriculture, represented by productive agricultural systems, and capitalist, educated, and innovative people. These widespread judgments have discouraged studies on traditional agricultural technology in Mexico, whereby the recovery of the concept and practice of traditional agriculture is taken seriously proving that is hugely crucial.

4.8 WATER, TRADITIONAL AGRICULTURE, GOVERNMENT DEVELOPMENT PROJECTS

Water uses, management, and their relations with agriculture began to be studied in anthropology in the 1940s, with the Papaloapan project, the construction of a basin, and the Miguel Alemán Dam (Tyrtania 1992b; Poleman 1964). Anthropologists participated in the mobilization of the Mazatec indigenous population out of the basin, but few of them questioned the role, the results, and impacts of the State policy related to the agricultural modernization of regional agriculture (Poleman Vol. 1, 1977: IX).

The basin included 14 municipalities in Oaxaca, Veracruz, and Puebla states. It had 656,774 hectares for sugar cane cultivation, that in 1950 covered 31.8 thousand hectares in the lower basin, including the Tuxtlas highlands, compared to 67.5 hectares for maize cultivation. In the upper basin, excluding the Orizaba-Córdoba-Tehuacan area, there were 8.5 thousand hectares of sugar cane cultivation, compared with 113.9 hectares of maize cultivation. Despite this imbalance in agricultural areas, in the lower Papaloapan basin, the sugar cane production returns were 54.2 million pesos, compared with 24.5 million pesos of returns from maize. In the upper basin, they amounted to 12.5 million pesos for sugarcane and 32.5 million for maize. Sugarcane covered 11% of the total cultivated area but represented 25% of the total value of regional agricultural production and 25% of sugar production throughout the country (Palerm 1993, 2006: 98).

There were 17 sugar cane mills, which in 1957–1958 had begun to concentrate under the auspices of the Papaloapan project. Five mills then handled 72% of the harvest, and one of them, the S. Cristóbal, handled 43% of it, thus placing itself in first place in the country, and in one of the first places in the world. About commercial agriculture "[...] the most benefited by a regional development project were, not the settlers, or the alleged colonizers, but the elements of national, and foreign private capitalism" (González-Jácome and Velasco-Orozco 2008: 215–238; Palerm 2006: 102; Vargas-Velásquez 2008; Vargas-Velásquez and Mollard n/d). The *Papaloapan Executive Commission* (CODELPA) was created, in 1946, and functioned until 1984. "To plan to design and build the works required for the integral development of the extension of the country that constitutes the Papaloapan river basin".

> Granting this commission, according to Article 3 [...] the broadest powers for planning, project, and construction of all the river defense works, those of irrigation use, energy development, and sanitary engineering, those of communication roads comprising waterways, ports, roads, railways, telegraphs, and the related creation, and expansion of villages, will have the power to dictate all measures, and provisions in industrial, agricultural, and colonization matters, as long as all of the above refer to the integral development of the Papaloapan basin.

4.9 RELEVANCE OF IRRIGATION STUDIES TO TRADITIONAL AGRICULTURE

The irrigation related or not to the termed hydraulic societies in Mexico has been a central element in the studies of traditional Mexican agriculture. These include academic discussions between Ángel Palerm (2007) and Karl Wittfogel (1956, 1966, 2007). The approach of political ecology deals with issues and discussions related to water uses, management, and scales over time. Water is such an important element in agriculture that studies in this regard currently constitute a whole branch of ecological anthropology that is called Political Ecology. However, in this book, it will only give some basic elements, without trying to cover what has been studied and analyzed within by Ecological Anthropology.

Current research on the situation of traditional agriculture, its evolution, and the consequences of having – or not – irrigation is most important. It all depends on the characteristics of the region where a human group is located, on the permanent or seasonal water available, and on the location of the farmland in relation to canals, dams, wells, water reservoirs, rivers, and water conduction courses to the land, including the local or external systems. It considers annual rainfall, its distribution throughout the year, and its impacts on the soils. These droughts occur mainly in the years affecting both soils and crops. Particularly, it is necessary to consider the frequency and the duration of the drought season – if it is extreme or continues for several years – about which several studies and hypotheses have arisen concerning the need of irrigation for agricultural activity.

Among Mexican agriculture researchers, the discussion about the social and political role of water itself in rural and urban societies has been fundamental in the theoretical discussions that began in the 1950s and continue to date (Rutsch and González 2011; Sánchez-Izquierdo 2000; Vargas-Velázquez 2008; Vargas-Velázquez and Mollard n/d; Vallejo-Nieto et al. 2011: 71–98). When these traditional agricultural systems have irrigation, they become intensive agriculture and are capable of withstanding high population densities. They also direct a part of their production to the market and, at the same time, meet the needs of farmers and their families (González-Jácome 2011; Vargas-Velázquez 2008; Vargas-Velázquez and Mollard n/d; Vallejo-Nieto et al. 2011: 71–98).

Researchers included in the discussion the traditional irrigated agricultural systems, which have high productivity. Irrigated agricultural systems have been the engines in increasing agricultural production, in the total population, and population density. These systems were fundamental in the conformation and emergence of cities (urban planning) and the State. Furthermore, including the production of agricultural surpluses, the complexity of the work organization, the separation of social groups from the direct food system production, plus the birth and expansion of the domestic market, and long-distance trade was necessary (Palerm 1955, 1970, 1972, 2007; Palerm and Wolf 1972).

4.10 CONSIDERATIONS AND CONCLUSIONS

A characteristic of the traditional agricultural systems of Mesoamerican origin is the combination by which human populations include the consumption of cultivated plants along with tolerated edible green weeds and wild flora. Also, they increase the flora and fauna of the ecosystems and agroecosystems where they are found too. This has been a way to increase the species diversity in natural and agricultural ecosystems. They also modify the characteristics and uses of soils and vegetation. Diversity was increased in the sixteenth and seventeenth centuries by contact with the Spanish when the changes in Mesoamerican agriculture and its agroecosystems were greater.

Currently, the increase or decrease in diversity is related to small-scale farmers' interests and necessities, and also of the large agri-food companies, and in general the economic interests of the national State programs, and transnational companies inversions whose focus is on the rural towns' agricultural land, forests, water resources, and the use of nearby ecosystems because they are not interested in the environment conservation but in earning money. Traditional agricultural systems include the cultivation on low and high hills, and sloping hills, and are managed by peasants through the construction of terraces, semi-terraces, and deposits to store rainwater (Martínez-Menes et al. 1999).

There are different types of canals to retain, divert, and provide water to residents for domestic purposes and irrigate the crops. The types of water management systems includes the construction of canals, water boxes (*cajas de agua*), the cultivation in pots or jollas in the terraces hilltops in the Highlands of the Mixteca of Oaxaca, the water boxes in Guanajuato (García-Zamacona 1999: 249–266), or the storage in pots or "jollas" in the High Mixtec of Oaxaca (Guevara et al. 2008: 6–16). There are also areas with rows of artificial rectangular shape rainwater deposits like stirrups (*cajetes, estribos*). These rainwater deposits are used for the storing of water and also for the softening of soils in the creation of soils in laterized areas, like with hardened soils in the central and southwest region of Tlaxcala (González-Jácome 2013a: 69–82, 2016: 111–145; Werner 2012).

The construction use problems, agriculture, livestock, and the natural resources affected, including human activities at local and regional scales,[4] have uprooted the native population in the region. The role that cultural anthropology played in those times and the situation of the populations that were mobilized by the floods caused by the dams were important topics of discussion concerning the impacts on agriculture and on the economy and the daily life of the settlers of these regions.[5]

The development studies and the programs supported by the Mexican State were generally supraregional and focused on the mobilization of populations as a result of the construction of dams and hydroelectric plants. Traditional agriculture expressed in the cornfield and the production of maize was strongly suppressed in favor of the cultivation of sugarcane for export. Livestock was also privileged, which was used to sell milk for Nestlé and meat to supply the meat trails of Mexico City. Sugarcane became the center of the economy, of the reorganization of the daily life of the ejidatarios who, turned into sugarcane growers, remained under the control of the sugar mills and of the economic policies of the federal government and its institutions.

RECOMMENDED READING

Barkin, D. and T. King 1970. *Regional Economic Development. The River Basin Approach in Mexico.* Great Britain, Cambridge University Press.

The book deals with the development governmental program of the Tepalcatepec River basin and Fuerte Projects in Michoacán, Mexico. Irrigation development had an impressive effect on local agriculture output and incomes. The investment in roads was very high, but less than that on health and education. It was not a coordinated planning of public investment, and a large number of decentralized spending agencies exist.

Poleman Thomas, T. 1964. *The Papaloapan Project. Agricultural Development in the Mexican Tropics.* California, Food Research Institute, Stanford University Press.

The Papaloapan project was started by Miguel Alemán government (1947–1952) to develop the basin of the Papaloapan River in the southeast of the country. One of its goals was to modernize agriculture, changing from polyculture cornfield to monoculture of sugar cane cultivation.

Eliminating the polyculture cornfield, supporting the extensive monoculture of sugarcane, and introducing fertilizers, and other agrochemicals to increase crop yields carried out the modernization of agriculture. The construction of roads was one of the best-achieved objectives in the Papaloapan project. Its impacts contributed to the modernization of the settlements located along roads and highways.

[4] Aguirre-Beltrán 1946, 1952, 1958; Barkin 1972; Barkin and Zavala 1978; Basauri 1940; Blanco-Rosas 2011:177–202; González-Jácome and Velasco-Orozco 2008: 215–238; González et al. 2018; Molina 1976; Ángel Palerm 1969, 1970; Poleman 1964; Rosenblat 1945; Rutsch and González 2011; Villa Rojas 1948, 1955.

[5] Aguirre-Beltrán 1946, 1952, 1958; Barkin 1972; Barkin and Zavala 1978; Basauri 1940; Blanco-Rosas 2011:177–202; González-Jácome and Velasco-Orozco 2008: 215–238; González et al. 2018; Molina 1976; Ángel Palerm 1969, 1970; Poleman 1964; Rosenblat 1945; Rutsch and González-Jácome 2011; Villa Rojas 1948, 1955.

5 Traditional Agriculture in Mexico (Second Part)

5.1 INTRODUCTION

The generalized poverty, sometimes object, that connects human populations with hunger and malnutrition on the planet, including Mexico, goes hand in hand with a growing global and national inequality of wellness and wealth (Gurri-García 2007: 60–68). Immanuel Wallerstein (2005) considers that these economic and social problems have increased since the last century, especially with the rise of neoliberalism, because monetary resources, profits, and benefits do not reach all the social and economic groups of the nation, as well as this system has contributed to the impoverishment of numerous groups on the planet.

It seems to be that neoliberalism has worsened poverty by making it object, now being a growing phenomenon for countries in recent decades, particularly those that "are developing". On the contrary, the human groups formed by those who were rich, already, are now much wealthier, a situation that occurs in Mexico also. Neoliberal's entrepreneurs and politicians justify this situation by saying that in order to distribute wealth, one must first create and have it. The problem seems to be that the cash never trickles down. In addition, those who obtained great profits and wealth become greedy. In this context, the role that agricultural activity has played, and still plays, is fundamental.

5.2 BASIC ASPECTS, APPROACHES, AND PERSPECTIVES

Small-scale agriculture is based on the traditional one and is on a feasible path for the conformation and development of new models and new types of agriculture that contribute to the reduction of negative impacts generated by human societies in relation to the environment (water, soil, flora, and fauna), physiography, and climate. These proposals and innovations are vulnerable to catastrophic events, including the economy, and the low formal education that many people have, health, food, work organization, justice, and forms of social, and economic organization.

It is hoped that the new models and proposals, such as the sustainability sciences, agroecology, alternative agriculture, permaculture, and many other approaches to the problem of environmental drastic changes in the planet, are searching for a better comprehension of vulnerability of person, places, ecosystems, and others that will result in studies and proposals directed to a better sustainable condition of material and non-material life for the present and future human generations (Bettencourt and Kaur 2011: 19540–19545, 2011; Clark 2007, 104(6), pp. 1737–1738; Turner et al. 2003: 8074–8079).

5.3 ELEMENTS TO CONSIDER

The accumulation and distribution of wealth, together with population growth, modernization processes in rural areas, women's spending several hours daily doing salaried work outside their homes, and other factors such as mass media propaganda generate an increase in the consumption of industrialized foods. This causes pressure and negative impacts on small-scale agriculture in Mexico because of the increased in the prices of processed food. Indirectly, the production, distribution, and circulation of industrialized food have effects on small-scale agricultural production, especially on what is grown through traditional agricultural systems such as the homegarden, orchard, cornfield, or terraces. These agroecosystems are unknown by urban populations and even by some

DOI: 10.1201/9781003198833-5

of the people that propose and/or establish public policies related to agriculture (Blanco-Rosas 2011: 177–202; Hernández-Matamoros 2016).

The effects that industrialized food production and its myriad of plastic and aluminum packaging have on the environment and climate should be considered and compared, without forgetting the economic, social, and cultural consequences affecting urban and rural populations. Some of them are well known; for example, the application of agrochemicals to large-scale crops or to food for its preservation, plus the unfortunate effects of packaging and distribution methods employed on products going long distances. The negative effect of the excessive use of plastics, which unintentionally end up in the seas, is well known and has generated controversies about the contamination of air, soils, and waters.

5.4 MODERNIZATION, URBAN EXPANSION, AND HUMAN POPULATION

It is important to study the interrelations that rural populations have with other towns and cities or the market, and the problems arising from supply-demand, the crop prices paid to small producers, the circulation routes, and the forms of control for the produce (Palerm 1968a, 1978b, 1978, 1980). These products go to rural and urban markets, and outdoor markets (*tianguis)*, but there are other forms of marketing and exchange between small producers and villagers (Cámara-Barbachano 1966: 273–280; González-Jácome 2019). The high cost of industrialized food and the impact derived from its production, circulation, distribution, and consumption affect the family economy (Palerm 1980).

The salaried jobs in non-agricultural work, such as employees in stores, supermarkets, coffee shops, restaurants, and shopping centers, have separated men and women of younger generations from agricultural activity and thus incorporated them into the consumption of industrial products, including food. There is a strong road construction and large-scale commercial agriculture expansion over rural areas (Palerm 1980), accelerating changes in land tenure, in the structure, and organization of small-scale agricultural families, and in their economic activities. They are articulated with demographic processes like horizontal mobility (migration, emigration, and return migration) in its different types (González-Jácome 2010a; INEGI 1998, 2005).

The departure of a member of the family labor force not only reduces the non-salaried activities necessary in small-scale agriculture, but also increases the need to obtain more monetary resources to pay for work activities, such as planting, weeding, and harvesting, that previously were exclusive to family members and their kindred. In addition, it influences intergenerational communication, and with it, the loss of the transmission and permanence of rural and local knowledge, traditional ecological knowledge (TEK), and traditional agricultural knowledge (TAK). Added to this, there are the local and regional demographic impacts that affect families such as the lack of available labor force for small-scale agriculture, and the impacts of modernization processes, plus the loss of peasant knowledge, of traditional agriculture practices, and biological, and cultural diversity.

5.5 AGRICULTURE AND HEALTH

Salaried work in non-agricultural activities has subtracted the labor force applied to small-scale agriculture. Food system distribution mechanisms through large corporations, supermarket chains, and convenience stores have contributed to the disappearance, impoverishment, or extinction of agricultural products directly or indirectly. Food buyers in these stores do not select the products due to the agricultural characteristics related to the production itself, the quantity of agrochemicals, or the cost of transportation over long distances. They do this by taking into account its shape, flavor, color, state of ripeness, and price. In the last decades, organic products have become more important on the tables of buyers, who consider their health as an important element to consider.

The expansion of these mechanisms has modified the number and variety of plants that are grown and distributed for consumption through local markets, tianguis, or grocery stores. This happens because diets have been homogenized, and the acquisition of industrialized food and beverages has increased in an accelerated manner occurring in both cities and the countryside.[1]

The need to conduct studies on diets and their health impacts must be added. Some effects are negative, such as obesity, and there has been an increase in the number of people affected by diseases like diabetes, hypertension, and heart problems (Gurri-García 2011: 65–84; Tecontero 2005). Studies that interrelate food and health have become important in recent decades. These studies have not been extensively studied, but their harmful effects are important indications noted through some research carried out by anthropologists, physicians, and nutritionists.[2] From this perspective, interest continues in the study of the relationships that exist between diseases and the consumption of plants, or with folk medicine as well as social and health aspects developed in human societies. It would be necessary to include local or regional concepts and classifications related to food system, consumption, and cultural ideas on food and health, like cold and hot foods, their consumption, and their relation to health or diseases.[3]

The role of traditional agriculture in relation to food, health, and economy of the rural and urban families is an important research topic. It includes the types of grains consumed, such as maize, and the plants, and fauna associated with traditional agroecosystems and cultural systems, like the cornfield or the homegarden (Linares and Bye 2016: 9–12). Between the 1940s and 1950s, demographers, historians, and anthropologists initiated these studies and projects, but they were abandoned for a few decades. Other studies were done like medical anthropology, health history, psychosocial diseases, historical demography, and traditional or folk medicine, also known as "mestizo medicine".[4]

5.6 TRADITIONAL AGRICULTURE, AGRODIVERSITY, AND MARKETS

The national economy has negatively impacted the agrodiversity of local and regional small-scale agroecosystems, directly or indirectly. In the last four decades, both the number and the variety of crops targeted for self-supply and the market have been reduced. In turn, the expansion of industrialized food has diminished the agrodiversity of cultivated plants and weeds in traditional agroecosystems, as in the case of the cornfield, which in some places has become a monoculture. In fact, some farmers call a single maize plant (cornfield, *milpa).* The indiscriminate application of agrochemicals, mainly herbicides, has contributed significantly to this agrodiversity reduction.

In some ways, convenience stores and supermarkets homogenize the agricultural products they sell and the taste of buyers. National and international commercial agro-enterprises contribute to the elimination of local crops varieties, which are difficult to commercialize due to factors like their fragility, duration between the time of their harvest, their long distance from the place of the plots, and the distant markets. Many of the traditional agroecosystems' productions are unavailable throughout the year; therefore, they cannot supply commercial companies throughout the year. In addition, many of their products do not meet the characteristics required by the market, which reduces their access to it. For example, this occurs in avocados known as Creoles locally, like the strong avocado, pahua, black or thin-skinned green; in several sapodillas locally called Sunday, yellow, green or black, and white; in tropical fruits like the yellow native papaya that in recent years

---

1 Blanco-Rosas 2006; Cahuich 2012; Dewey 1981: 151–187; Entrena-Durán and Jiménez-Diaz 2013: 683–693; González-Jácome 2019; Gurri-García 2007: 60–68; Linares and Bye 2014: 151–178; Otero-Prevost et al. 2017; Poleman and Freebairn 1974; Tecontero 2005.

2 Alayón and Gurri 2007: 243–260; Daltabuit and Ríos 1992: 23–33; Dewey 1981: 151–187; Gurri 2007: 60–68, 2011: 65–84; Otero-Prevost et al. 2017.

3 Cahuich 2012; Chevalier and Sánchez 2003; McClung de Tapia et al. 2014: 97–121; Tecontero 2005; Vargas and Bourges 2012: 99–123; Vargas and Casillas 2010:87–115.

4 Aguirre-Beltrán 1943, 1947, 1952, 1963, 1965, 1986; Chevalier and Sánchez 2003; Cook and Simpson 1948; Cook and Borah 1960; del Amo and Malvido 1990: 345–381; Florescano and Malvido 1982; Malvido 2003: 65–78; Márquez 1994; Reff 1990; Rosenblat 1967; Toledo and Burlingame 2006; Viesca 1982, 2010: 47–54; Zolla 2008.

has been replaced by Maradol, which is more resistant to movement; and local varieties of bananas, tomatoes, or pumpkins, among other crops.

The local varieties of which are sold in markets and supermarkets have been reduced. For example, tomatoes sold in Mexico City are of three types: green; round red for salads; and elongated reds (guajillo or saladeti) to boil or roast, and thereby prepare the broth for soups, marinades, stews, and so on. In the urban markets and tianguis of the cities and villages, vendors sell wild and cultivated green small tomatoes (tomatillo), wild mushrooms, and fruits, like the Mexican hog plum (*Spondias mombin*) that are seasonal products, which according to their flowering, are sold in different seasons. Local and regional varieties of bananas are almost unknown in modern supermarkets. Normally only Tabasco, Macho, and Dominico bananas are sold; occasionally, they have the Purple or Roatan (SAGARPA, SIAT no/date).

A similar agro-economic phenomenon occurs in other countries and favors the production, handling, and long-distance distribution of industrialized food. The European Union (EU) tries to control the situation by imposing rules and generating new responses from local agricultural producers. In Spain, it is forbidden to sell agricultural products directly from the farmer to the consumer. This includes legumes, vegetables, and fruit that were ground in gardens, homegardens, or the adjacent land to homes. To end these bans on retail sales, farmers distribute and sell their produce in markets, on roadsides, and highways, at gas stations, or out places that are not controlled by intermediaries, and large hoarders of fruits and vegetables. Furthermore, food companies do not acquire farmers' products, but if they do, only pay what they consider appropriate, not the fitting price.

Such European Union (EU) prohibition is based on the idea that the quality and hygiene of the products must be controlled to avoid diseases and/or epidemics. This situation affects food processing too. For example, in Spain, the Spanish omelette has to be made with eggs that are industrially packaged and that were previously pasteurized. Small egg producers cannot sell eggs directly to consumers, so they are regulated, with quality and controlled prices by large hoarders. This regulation has something positive, since it eliminates diseases caused by the improper handling of eggs (e.g., salmonellosis). However, a negative effect is that this regulation has taken small producers out of the market because they cannot compete with companies (Maestre-Alfonso 2018).

In the municipal market in O'Grove, on the low estuaries of the Spanish Galician coasts, it is still feasible to find legumes and vegetables grown without agrochemicals on a small scale in nearby gardens of the region. They are not found throughout the year, since many of them are seasonal: tomatoes, pumpkins, peppers and Padrón peppers, lettuce, mushrooms collected in the forests, and table grapes, fresh chicken, cheeses, and processed sausages which are seasonally produced. The sales included handcrafted baskets filled with agricultural products controlled by hoarders, by large supply centers, or by importers who handle these products, but get them from abroad (Figures 5.1 and 5.2).

Several crops are sold in the harvest season like Padrón peppers, grown in Herbón, a town close to the town of Padrón located in La Coruña municipality, just over 34 km southwest of Santiago de Compostela. Franciscan friars took these peppers to the region in the seventeenth century (Maestre Alfonso 2018) (Figures 5.3 and 5.4).

Supermarkets in Mexico encourage the consumption of foreign fruits and vegetables. Examples of this are cherries and apples brought from the USA; peaches, pears, and plums from Chile; and table grapes from Chile and the United States that according to the harvest season in those countries alternate for sale with Mexican table grapes (Table grape producer, Sonora 2010). Although they are cultivated in Mexico, foreign products are sold in Mexican supermarkets, tianguis, and mayor markets: kiwis from New Zeeland and Chile; berries from the USA; lettuces and fruits of Asian varieties, which arrived from China; white, yellow, and green cauliflowers come from the USA; artichokes and white and green asparagus come from Spain; varieties of pumpkins called "fantasy" no longer grown in Mexico but have been brought from the USA and Spain to be sold on the festivities dedicated to the "dead saints" and Christmas.

FIGURE 5.1 O'Grove Municipal Market: Tomatoes and Vegetables from Nearby Gardens.

FIGURE 5.2 O'Grove Municipal Market: Shrimp, Mussels, and Oysters from the Galician Coast.

FIGURE 5.3 O'Grove Municipal Market: Organic Tomatoes.

FIGURE 5.4 O'Grove Municipal Market: Padrón Peppers.

The prices of these imported agricultural products are higher than those produced in Mexico and cover a market for people with high monetary resources.

In Mexico City, the S. Juan market located in the historic center of the city has several sections dedicated to the sale of foreign products, gourmet foods, beverages, vegetables, fish, seafood, meats, and national and foreign sausages. They sell meat from wild animals (lion, buffalo, crocodile, ostrich, deer, and others) that come directly from Texas. There are the sale of edible insects, scorpions, snakes, and all kinds of "exotic" fauna.

The different markets and their types, organization structures, size, national and international, from street markets, small grocery stores, to transnational chains, have prepared food to eat there or to take home; "fast food"; bottled water; industrialized drinks; drinks prepared with water or milk; tortillas; and bread (baked, handmade, or industrialized of different brands). Markets contribute positively to reduce the working day for women, facilitating their household chores and leaving them more leisure time. Industrialized products with their processes like pasteurization reduce gastrointestinal diseases.

But, at the same time, this entails several elements and behaviors that affect the family organization, the pocketbook, the food system of families, the children's education, and the family's health. The aforementioned, indirectly, diversity in everyday food has been reduced, also, not to mention the quality of food. The feeding industry creates foods with characteristics that fit the needs of the modern world. It announces that products have fewer chemical preservatives; are gluten free, with no added sugar or no fat; are lactose and salt free; are organic; and so on. As the industries' interest in retaining customers is fierce, other forms of food production and food packing augment, contributing to reduced diversity in agriculture and increased contamination, which occurs due to the paperboard, aluminum, and plastic used for food packing.

Research of the food industry must consider other ways to preserve food and to pack food products, with non-polluting materials for waste they generate. There is some progress in this regard with the production of natural-material bags that replace plastics and that are biodegradable (Harper and Le Beaus 2003; Long and Vargas 2005).

5.7 TRADITIONAL AGRICULTURE AND DIVERSITY

In recent years in studies on traditional agriculture, an interest has emerged in "[...] the richness of food species as a measure of biodiversity [...]" (Lachat et al. 2918: 127–132). These researches used Simpson's diversity and functional diversity index in the study of species richness (SR) resulting in stronger associations and better diagnostic properties with micronutrient adequacy. One of the ideas applied in this study is that research has to link biodiversity with human diets: "The richness of species in the diet, or an account of the different species consumed per day, assesses the nutritional adequacy, and dietary biodiversity of diets for women, and children in rural areas". This positive association of food species richness with the quality of the diet was observed in the study of the wet and dry seasons in ancient Mesoamerican societies (Williams et al. 2017: 269–287).

According to this approach, in high biodiversity areas, food diversity is interrelated with human population vulnerability and the quality of these people's diet. For which it is necessary to know the number of species consumed during the evaluation of the diet. This can be considered as an opportunity for cross information on two components of sustainable development: human health and environmental health, which complements the knowledge of existing indicators for healthy and sustainable diets.

Various traditional agroecosystems can coexist in the same region. In general, they are important because they are compatible with the characteristics of the environment, they are combining resources from other agroecosystems, or they obtain other economic returns. These agroecosystems allow families to be self-sufficient to a high degree food-wise. For this reason, they are protective shields, kinds of "nest eggs", against the economic risks that are produced by the market, eventualities, and climatic fluctuations.

These agroecosystems are generally adapted to the local environmental, and regional and local climatic conditions as well as altitude, relief, land gradient, type of soil, flooding, annual rainfall, and the humidity of each area where they are located (Blanco-Rosas 2006; Cahuich 2012; González-Jácome 2010: 141–158; González-Jácome et al. 2007; Palerm 1955; Palerm-Viqueira and Martínez-Saldaña 2009). Annual and cyclic phenomena are included, such as cyclones, hurricanes, rains (with their scarcity, or abundance throughout the year occurring within non-catastrophic margins), prolonged droughts, frosts, and hailstorms. Regional and local flora and fauna play a basic role in hunting, fishing, and/or gathering activities of the inhabitants for their daily diet.

Traditional agricultural systems can be self-sufficient, and because of their ecological basis, farmers maintain –or seek to maintain – a more or less "friendly" relationship with the environment (Gliessman 1998, 1990, 1999: 25–31). Consideration should be given to discussions about these systems' intensity, their required workforce, and the time needed to build them, in addition to their relationship with human work and productivity (Harrison and Turner 1978). Also, that soil conditions, humidity, climate, altitude, flora, and fauna are locally modified constantly.

Therefore, there are adjustments that are local or regional, and the knowledge that human populations have in this regard is heeded (Cruz-García 1992). It is important how over time this population has created agroecosystems that share certain characteristics, but whose local components have to be studied and explained in detail. Is this capacity to adjust to change that traditional agricultural systems have had over time that has made them sustainable (International Council for Science 2002).

Resilience in these systems arose from the local knowledge of the small-scale farmer, the importance of solidarity (mutual aid), and people's ability to adjust to external situations. The contributions that small-scale farmers impart of their knowledge of environmental and climatic characteristics on the micro-scale are fundamental to understand the operation and management of traditional agroecosystems. The characteristics are associated with several micro-scale local issues, such as soil features, gradations, depth and type, relief, consistency, proximity to water sources, moisture retention, capacity, and distribution of the cultivation land.

Adaptation to a specific environment, management of diversity, and local knowledge are articulated with traditional agroecosystem economy and environmental risks. For example, the place that flora occupies in agroecosystems like homegarden, cornfield, terraces, or other Mexican agroecosystems, which often are located in small areas (Robles-Linares 2018: 269–302).

The direction of the prevailing winds throughout the seasons, its relationship with the construction of natural barriers (tree curtains) or artificial barriers (edges of land, or stone walls) on the farmland, and the placement of plants grown in specific spaces on the microscale (cultivation area, characteristics of the plot, land soils, and available water) are inseparable here. Places on the plot destined for housing, cooking, bathing, and toilet use must also be chosen. Areas used for crop storage (Hirth 2012: 178–186; Sánchez-Muciño and Velasco-Orozco 2018: 235–267) and areas for domestic animals must be delineated, too. In some places, where the steam bath (*temascal, temazcal*) is still accustomed, its location has to do specifically with the direction of the wind and where the barn has been built (Romero-Contreras 1998). The old barns were located in the patio or on plot, and treated with lime to counteract the harmful actions of insects and pests (Hernández X. 1949: 153–211; 1985, Vol. 1). Although crop storage is a fundamental element in agriculture, its study has been scant in Mexico (Bortot et al. 2012; Hirth 2012: 178–186; Hernández X. 1949: 153–211) (Figures 5.5 and 5.6).

The country's economy – directly or indirectly – has negatively impacted the agricultural diversity of local and regional small-scale systems. In the last four decades, both the number and variety of crops targeted at self-supply and the market have been reduced. In turn, the expansion of industrialized food has diminished the agrodiversity of cultivated plants and weeds in traditional agricultural systems, as is the case of the cornfield, which in some places has become monoculture. In fact, some farmers, as we pointed out before, call a single maize plant *milpa*. The use of agrochemicals – mainly herbicides – has also contributed significantly to diversity reduction (Terán 2010; Terán et al. 1998).

Agrodiversity reduction in traditional agroecosystems has favored, or generated changes in local and regional food systems. In Mexico, it is necessary to have more information to know them in depth. In recent decades, studies on food cooking are very popular in Mexico, and the same is

FIGURE 5.5 Cacaxtla, Tlaxcala: Prehispanic Ovid Type Barn (*Cuexcomate*).

FIGURE 5.6 S. Tadeo Huiloapan, Tlaxcala (1981): Contemporary Barn (*Cuexcomate*).

happening in other countries of the world (Goody 1982). Possibly, the UN/FAO and national and international research centers, which wish to improve the negative impacts of agri-food policies that are associated with poverty and malnutrition, have contributed to this concern.[5] The FAO (1992, 2001, 2009) developed and/or supported projects for the introduction of family homegardens in various parts of the world.

In some ways, convenience stores and supermarkets homogenize the agricultural products they sell and the taste of buyers. They contribute to the elimination of local varieties, which are difficult to commercialize, due to factors such as their fragility, duration between the time of their harvest and circulation distribution in the market, the ability to transfer over long distances, or availability throughout the year. These drawbacks eliminate the consumption of local varieties. This phenomenon occurs daily in other countries and favors the production, handling, and industrial distribution of food.

White asparagus, artichokes, and Padrón peppers from Spain are beginning to be introduced in Mexican supermarkets, in addition to other fruits and vegetables that are displacing national crops. The wines of the low estuaries of Galicia, known as Albariño, now are offered in Mexico City supermarkets and restaurants. Since the sixteenth century, spices, rice for paella, wines, olive oils, capers, saffron, Cayenne pepper, Christmas sweets, Spanish hams, sausages, and sheep cheeses have been imported and sold in urban markets. This phenomenon is historical and not exclusive in Mexico. Actually, the same occurs in other countries and is part of the globalized economy and the growth of trade controlled by transnational food companies, but the effects on traditional agriculture and its diversity in Mexico have not been studied sufficiently (Gurri-García 2011: 65–84; Gurri-García and Moran 2002: 41–47; Guzmán-Sebastian et al. 2016: 871–883) (Figure 5.7).

In a study conducted by Lachat et al. (2018: 127–132), researchers recommend to link biodiversity with human diets: "The richness of species in the diet, or a count of the different species consumed per day, assesses the nutritional adequacy, and dietary biodiversity of diets for women, and children in rural areas".

This positive association of food species richness with the quality of the diet was observed in the study of the wet and dry seasons. According to this approach, in high biodiversity areas food diversity is interrelated with human population vulnerability, and the quality of these people's diets. For which it is necessary to know the number of species consumed during the evaluation of the diet. This can be considered as an opportunity for cross information on two components of sustainable

[5] Beals and Haycher 1943: 295–304; Blanco-Rosas 2006; Cahuich 2012; González-Jácome et al. 2007; Long and Vargas 2005; Tecontero 2005; Vargas and Casillas 2010: 87–115; Vargas and Bourges 2012: 99–123; Vargas and Casillas 2010: 87–115.

FIGURE 5.7 Fancy Pumpkins Imported from Spain for Sale in Mexican Markets.

development: human health, and environmental health, which complements the knowledge of existing indicators for healthy, and sustainable diets.

5.8 PESTS AND DISEASES IN ANCIENT AGRICULTURE

In the 1990s, anthropological and archaeological studies resurface on the subject of the impact of pests and diseases related to ancient agriculture, and that related to health in ancient, nineteenth-century, and contemporary Mexico. Ethnohistory takes this basic theme as an axis for the studies on agriculture,[6] and the same occurs in ecological studies. There were numerous studies on agricultural pests and diseases developed by specialized personnel working in government agencies and research institutions, mainly investigating the technical and biological aspects to deal with them (García-Rodríguez et al. 2012: 217–251; Ruiz-Corral et al. 2013).

There is little diffusion about this specialized research and its results. García-Rodríguez et al. (2012: 217–252) deal with phytochemical diversity as well as how wild plants defend themselves against pests and diseases through the diversity of secondary metabolites in plant tissues. The origin of phytochemical diversity and its maintenance came from the coevolution between a plant and its consumers. Knowledge of phytochemical variation is related to biological control and the search for alternative plant biopesticides to use instead of synthetic pesticides. Studying the phytochemical diversity applied to the biological control of species or varieties of volatile emitting plants attractive to natural enemies of herbivores is essential to reduce environmental pollutants (García-Rodríguez et al. 2012: 244).

Chemical ecology applied to the study of agriculture is one of the foundations of scientific knowledge when interrelated with traditional agricultural practices. Among these practices, there is the rotation of maize crops with *Tagetes* spp. In tropical areas, polyculture includes fruit trees such as Soursop or Sweetsop (*Annona squamosa*); in the Totonac region, there is the *Nicotiana tabacum*. In the middle basin of the Mayo River, in Sonora, the wild tobacco plant (*Nicotiana glauca*) locally known as *tabaquillo*, *giganton*, *corneton*, *maraquiana*, or aspen is allowed to grow next to the crops. The study cited in the following paragraph corresponds to plants that are not Mesoamerican, and it gives an idea of their use as plant bioinsecticides.

[6] Altieri et al. 1989 (12): 82–107; Báez et al. 2013; Barrera 2002: 81–84; Campos 2012; García-Quintanilla 2005; Malvido 1982; Márquez 1994; Quezada 1995; Viesca 2010: 47–57.

> The most used families as plant insecticides are Meliaceae, rutaceae, asteraceae, annonaceae, labiatae and canellaceae. Recently, plant insecticides are being developed from seeds of Soursop (*Annona muricata*), Florida custard apple (*Annona triloba*), *Melia volkensii* [Kenyan plant commonly known as Melia, boba] and *Nicotiana gossei*. In addition, very promising results have been obtained from root extracts of *Tagetes* spp., foliar extracts of *Ginkgo biloba*, seeds of *Vitis vinifera*, and *Lupinus* spp.
>
> *(Isman 1999, in García Rodríguez 2012: 243)*

Plant dependence on pollen and seed dispersion is based on agent dispersion and attraction of pollinators where terpenes are the main components of volatiles in flowers (Gershenzon and Dudareva 2007, in García Rodríguez et al. 2012: 226). The same compounds are transmitted by defensive and attractive activities. For example, there are cases where anthocyanin and mono terpenes in flowers attract insects, and at the same time have an insecticidal and antimicrobial activity (Wink 2003, in García-Rodríguez et al. 2012: 226).

In traditional Mexican agriculture, the wild vegetation and weeds that are found on the earth banks, borders, edges, or fringes that surround the cultivation areas remain, as is the case in the Atoyac-Zahuapan river basin. The use of land edges, huts, hedges, boundaries, riverside groves, or ridges of land planted with trees that mix with wild vegetation is common in several agroecosystems in the world. They are related to the diversity of agroecosystems. In Tlaxcala and other parts of the country, farmers' families drink herbal or medicinal infusions, and they used weeds and some wildflowers taken from the land edges (Vibrans 1997: 21–67).

Examples of the importance of flora in the land edges are the hedges in the northwest of Holland (Friesland), where trees and weeds are mixed on the loins of earth, and the neighbors collected the blackberries that grow in them; hedges are pruned each year in the fall (van der Ploeg 2008: 286, 299, 302–303, Figure 7.2).

In the Cantabrian mountain range in northern Spain, hedges of trees, and shrubs are preserved as boundaries between farms, land for livestock use, cultivated land edges, margins, banks, unevenness between fields of work, or terraces. They also serve as windbreaks for crops. In the southeast of Spain, hedges are used as windbreaks in streams and riverbanks. In these southern regions, the hedges had almost disappeared because they were settled on the best agricultural lands and river valleys.

In Porcón Alto, Cajamarca, Peru, the hedgerows of *Polyiepis* spp. are associated with mutiyo or saliwa (*Cassia* sp.), nettle (*Urtica* sp.), suro (*Chusquea* sp.), cocho or tarwi (*Lupinus* sp.), and malva (*Malvas* sp.) (Padilla-Mendoza 1995: 29–30).

A study that was carried out in Tepoztlán and Texcoco want to know how many weeds the locals used as fodder. The percentage of the total number of weeds that had been used for fodder gave the results of a lower than 1.5% of the weed list, and the maximum had reached 38% of weeds used as fodder, thus indicating the importance of these weeds in crops (Espinosa-García and Díaz Pérez 1996a).

5.9 AGRICULTURAL CRISES, CLIMATE, AND DISASTERS

Research has been important, especially in the study of the causes of the Classic Mayan collapse. The discussions began in the 1940s and 1950s and the theme was about feeding large human populations only with the agricultural production obtained from the slash and burn agroecosystem. The other idea was that the quality of the soils in the Mayan area was deficient, and along with erosion did not allow the realization of more than one annual harvest, which did not cover the needs of the population. Other reasons related to the weather helped to establish this idea.

The emphasis that cultural geographers, archaeologists, and Mayan specialists put on the water supply for domestic and irrigation purposes is important, particularly in relation to the study of parasitic diseases and the quality of drinking water.

The Classic Maya faced diseases and epidemics. The remedies they used to solve the problems continue to be studied. Those that have been carried out to date included the relationship with water, its types of storage, structure, and the forms of social organization associated with the conservation and distribution of the liquid. They have taken into consideration its population settlements, urban centers, and communal reservoirs for the collection and storage of the vital liquid. In addition, the use of plants to purify the water accumulated and conserved in wetlands, reservoirs, and communal deposits was considered, as well as the sociopolitical forms of organization to manage it all have been central. The capture and storage of water have been investigated to explain the Mayan collapse that occurred from 800 AC to 900 AC, causing depopulation of cities (Argivo 1994: 187–188; Conrad 1905; Dunning et al. 2006; Lucero 2017; Lundell 1937; Luzzadder-Beach; Beach 2006: 91–117; Siemens 1978: 117–143).

In the 1980s, interest resurfaced in the study of agricultural crises in New Spain's agriculture (Malvido 1982; Florescano 1982; Hernández-Rodríguez 1982; Quezada 1995; Ocaranza 1995). A relevant aspect in traditional agriculture has been studies on the history of medicine, with emphasis on human diseases and epidemics that drastically impacted the Novo-Hispanic populations in the sixteenth and seventeenth centuries, causing high mortality rates, the abandonment of agricultural fields, and consequent famines. Primarily physicians, medical anthropologists, and historians wrote these texts. They deal with health, disease, or food in human populations who practice traditional agriculture.[7]

The studies on traditional agriculture have emphasized the interrelationships between climate events and unforeseen catastrophic events. These include volcanisms, hurricanes (Konrad 1985: 321–330, 1996: 98–130), winds, storms, tropical storms, tornadoes, vortexes, frosts, hailstorms, El Niño/La Niña, annual floods, and others. All of them generate social reactions, because of the urgency to find preventive measures and an ideology on how to control them, especially in relation to nature's violence and urban planning. In current times, researchers from the Southern Frontier College (ECOSUR) have studied small-scale farmers' vulnerability in the face of hurricanes and droughts in the Calakmul communities in Campeche (Alayón and Gurri 2007: 243–260). Known and important studies on droughts, climate, and the Mayan collapse of the Terminal Classic have been published.[8]

In Calakmul, Campeche, a comparison of vulnerability to hurricanes and droughts in six units with traditional agriculture and four with commercial agriculture found how local adaptations of traditional agriculture were more resilient and flexible in the face of catastrophic events than commercial agriculture. Units with commercial agriculture took more risks, based on the consideration that they had sufficient economic capacity to rescue many products and start the agricultural cycle anew. This would be subject to the amount that would be saved and intervals with low market prices (Vallejo-Nieto et al. 2011: 71–98).

Climatic events and their relations with the environment (especially soil and vegetation[9]) and agricultural activity have been the object of historical and anthropological studies, especially after they occur in various parts of the world. For example, the cases of volcanism, fires, tremors, earthquakes, tsunamis, prolonged droughts, and floods have drastically impacted both urban and rural human populations.[10] An example of these, and the impacts that such events may have, is the Mayan collapse. In this regard, cultural geographers and archaeologists have carried out a large part of the numerous studies and discussions. They have concluded that this cannot be explained without

[7] Aguirre Beltrán 1986: 151–178; Argivo 1994: 187–188; Ashburn 1981: 33–44; Hernández-Rodríguez 1982; Lucero et al. 2015; Mandujano-Sánchez et al. 2003; Ocaranza 1995; Rodríguez 1982: 139–153; Turner II 1990: 178–211; Vargas and Bourges 2012: 99–123; Viesca 1982:157–165.

[8] Angulo and González-Álvarez 2008: 567–590; Bose et al. 2003; Chase and Chase 2014; Cooper and Duncan 2016; Cuevas and Navarrete 2005: 39–49; Gill 2000; Lammel et al. 2008; Lewis Parry and Carter 1989: 95–116; Ianone 2012; Lucero 2017; Scheffran et al. 2012; Siebe 2009; Siemens 2004, 2009; Turner II and Sabloff 2012:13908–13914; Yaeger and Hodell 2008.

[9] Bellón et al. 2009: 355–382; Sohn et al. 1999: 947–958.

[10] Albores and Broda 1988; Angulo and González-Álvarez 2008: 567–590; Campos 2012; Cuevas and Navarrete 2005; Dahlin 2002; Fedick and Ford 1990; Ford and Nigh 2016; González-Jácome 1988: 467–501; Konrad 1985: 321–330, 1996: 98–130; Lammel et al. 2013 [2008]; Márquez-Mireles 2000; Ortiz 1947; Pagania et al. 2015: 5607–5612; Yasper and Hodell 2008.

the competition of several causes that include environmental causes, like the impacts of climate on agricultural activity, especially in the long periods of droughts, and sociopolitical causes, like the role that political and social organization has in human societies.[11]

5.10 DROUGHTS, DEFORESTATION, CLIMATOLOGICAL EVENTS, AND VOLCANISM

One of the events that have been more and better studied by specialists from various disciplines involved in the study of traditional agriculture in the Mayan area is the abandonment of the cities in the Terminal Classic (Mayan collapse), which takes place from 800 AD to 1000 AD, causing the reconfiguration of the entire urban system and agriculture of the Maya then. This event was gradual and has been explained through several interrelated events that occurred during several agricultural cycles. These included deforestation of the forest, continuous droughts, soil erosion, insufficient water for domestic and agricultural purposes, reduction in production for human consumption, loss of power of the ruling classes, and social and demographic impacts on the Maya culture,[12] affecting its economy and cultural history.

Deforestation remains a serious problem in the northern center of the Yucatan Peninsula (Sohn et al. 1999: 947–958), as is the retaining of healthy soils. In traditional agriculture, farmers have learned how to grow maize through various practices that relate to altitude (Abasolo 2006; Perales, Brush and Qualset 2003: 7–20), management of microenvironments and microniches (Latournerie et al. 2006: 1771–1775; Liebman 1999: 191–203), crop rotation (which breaks disease cycles and keeps soil healthy), polyculture, non-tillage practices, and the selection of seeds that are tolerant of pests, droughts, and adjustments to weather events (e.g., windbreak curtains, tree planting, shrubbery protecting crops, and others). This is combined with agroforestry, hunting, fishing, and gathering practices, which provide food for families and the external salaried work of some of its members.[13]

The study of the effects of climatic events and their interrelation with the emigration of human populations affected by these phenomena must be understood. The same occurs with the association of the results on crops during a prolonged drought for periods of time that excessed the annual ones, or the results on crops resulting of excessive, or scarce rain, or the insufficient weathering of the soils. These have been key to understanding what happened, particularly to agriculture, and the management that populations have adopted to survive these events that usually have catastrophic results (Buendía-Rodríguez et al. 2008; Espinosa 1996; Rodríguez-Tapia et al. 2004: 11–21).

In the Mayan culture in the past, the climatological, environmental, and geological elements, as well as other factors of a political and social nature, combine to cause the loss of prestige of the ruling elites, affecting their social, political, and ideological control. Apparently, despite their efforts, the rulers were unable to mobilize water deities to produce the indispensable rains to reduce the impact of droughts. These events in turn were associated with phenomena such as mortality in human populations.[14] It would be necessary to add the importance that the investigations and

---

[11] Acuña-Soto et al. 2005; Andrews et al. 2003; Argivo 1994; Barrett and Scherer 2005; Chase and Chase 2014; Conrad 1905; Dahlin 2002; Demarest et al. 2004; Dunning et al. 2006; Gyles 2014; Kepecs 2007; Kepecs and Masson 2003; Killion 1990; Krupinsky et al. 2002: 198–209; MacAnany 2012; Marcus 1993; Masson and Peraza n/d; Pagania et al. 2015: 5607–5612; Pope et al. 1996; Rice 1985; Siemens 1978: 117–143; Turner II and Sabloff 2012: 13908–13914; Sabloff and Rathje 1975.

[12] Andrews et al. 2003; Butzer 2012; Butzer and Enfield 2012; Dahlin 2002; Demarest et al. 2004; Douglas et al. 2015: 5607–5612; Hodell, Breener and Curtis 2000: 13–38, 2006: 215–240; Kepecs and Mason 2003; Lentz 2000; Marcus 1993; Santley et al. 1986: 123–159; Shaw 2003; Turner II 1990; Turner II and Sabloff 2012: 13908–13914; Yaeger and Hodell 2008.

[13] Abasolo 2006; Atran et al. 1993: 633–700; Gurri-García et al. 2001: 590–602; Toledo 1992: 197–218; Velasco-Orozco 2007: 331–335; Ziervogel et al. 2006: 294–305.

[14] Aimers and Hodell 2011; Bose et al. 2003; Giel 2000; Hodell et al. 2006: 215–240; Ianone 2012; Lentz 2000; Lucero 2008, 104: 814–826, 2017; Yaeger and Hodell 2008.

studies that different specialists have done concerning processes such as the abandonment of the Mesoamerican cities at the end of the Classic.

Currently, deforestation in Mexican regions with temperate-cold climates and rainforests is one of the most serious problems, and it occurs in other parts of the planet, such as the Amazon. This issue is closely related to important issues such as climate change, global warming, rain reduction, and its effects including droughts or torrential rains causing floods in large regions of the world (Del Amo-Rodríguez 2007: 97–146).

Some governmental plans and programs have been organized to counteract the effects of deforestation. An example occurred in Tlaxcala, Mexico from 1970 to 1975, when two governors of the entity, Ignacio Bonilla Vázquez (1969–1970) and Luciano Huerta Sánchez (1970–1975), decided to reforest the slopes of the Malitzin volcano by planting fruit trees, thinking that thus the population settled in that region would stop hewing the pines to make charcoal to sell to the inhabitants of the region. Their idea was that the former deforesters now could sell the fruits of trees used in reforestation.

The program provided fruit trees and training for their planting to the peasants. The project was unsuccessful, but the facilities of the nursery in charge of carrying out the program remained until the 1980s. The program failed because those in charge of the program were unaware of the history of the lower slopes of the volcano villages, whose inhabitants had sold charcoal to the neighboring Puebla textile industries since the nineteenth century (Robichaux 1991: 43–68). That is, the production and sale of charcoal had a safe market organized many years ago before the reforestation program with fruit trees was implemented (Robichaux 1991: 43–68).

A few years later, governor Tulio Hernández-Gómez (1981–1987) created a program to use the zone and forests of the volcano as an area for the practice of athletics. From this program, the facilities of the center for the training of athletes were built but currently are used as a restaurant, hotel, and place for family recreation; however, the deforestation continues there.

5.11 RECREATION OF TRADITIONAL AGROECOSYSTEMS

The relationship between agriculture and the food system has been an important proposal for the understanding and development of traditional agriculture as it has been for the development and implementation of projects and proposals of international institutions like FAO (2006b, 2015), national research centers, and universities. Said institutions seek ecological solutions to current industrialized and globalized systems of food production that impact the environment and climate of the planet negatively because these systems lack sustainability.

The study of traditional Mexican agricultural systems is reinforced in the late 1960s and 1970s, trying to recreate pre-Hispanic agricultural systems. This was because these agroecosystems have biodiversity, high production yields, and the capability to feed populations of high densities. These agricultural systems contribute to cultural diversity in addition to being sustainable over long periods and managing agricultural spaces combined with natural ecosystems.[15]

The agricultural model that served to recreate these traditional agricultural systems was that of the *chinampa* in the basin of Mexico. The results of studies of the chinampa and groups of small-scale agriculturalists that taught locals how to manage the system were taken in regard. As an example are the programs for the recreation of "tropical chinampas" in Nacajuca, Tabasco, in the southeast of the country (Pérez-Sánchez 2007) and the chinampas in the Toluca valley (Ramírez-González 2004).

The chinampas in the Valley of Toluca were of a little-known local development program that emerged after the desiccation of the lagoon system of the valley as a possibility of recovery and agricultural use of a small part of the land flooded in the area. These areas of the old basin of the Toluca

[15] Barkin 1998; Barkin and Zavala 1978; Barkin and King 1970; Gómez-Pompa et al. 1978: 221–232; Guzmán and Palerm-Viqueira 2005: 21–26; Leriche 1983; Palerm 1967: 26–52, 1969, 1970, 1993.

Valley continue to be flooded every year during the rainy season, generating serious problems in towns such as S. Mateo Atenco. Currently, the chinampas are virtually abandoned.

At least in the Viceroyalty, the Matlatzinco people built stone and land edges (*bordos*), as well as artificial dams, on the fringes of the rivers that flowed in the rainy season. At the same time, there were earth banks that stopped and then controlled the rain runoff from the Xinantecatl volcano (Nevado de Toluca) toward the lower parts of the valley, to direct them to the Tejalpa River. These earth banks were built near the city of Toluca and in Zinacantepec.

There is also news of the construction of this type of land edges in Amealco, Querétaro. In the artificial dams, there were aquatic and terrestrial fauna and flora, forming artificial ecotones. The flora consisted of oaks, willows, mesquites, sweet acacia or huisache (*Vachellia farnesiana*), Willow ragwort (*Senecio salignus* DC.), and Southern cattail (*Typha domingensis* Pers.). The fauna included herons, ducks, turtles, snakes, two-banded sea bream, axolotl (*Ambystoma mexicanum*), salamanders, and freshwater shrimp (*acocil*) (*Cambarellus montezumae*).

The tropical chinampas were made using heavy machinery to excavate canals and form platforms in the Nacajuca River, which is connected to the Mecoacán lagoon system in Tabasco. The weight of the machinery compacted the soils, but the canals were very high and wide, phenomena that affected the possibility of infiltration irrigation in the cultivation platforms. In spite of everything, farmers were able to plant several plants that had utility for them and that could be sold in the nearby markets.

The idea of achieving local development through the recreation of pre-Hispanic agricultural systems persists to this day and is related to issues of ideology and politics. In the 1970s and 1980s, researchers from various fields of knowledge, mainly economists, agronomists, biologists, ecologists, anthropologists, and sociologists, participated in several regional development projects and organized – or were involved – in local development programs. From the results of the projects, there are evaluations showing which ones had failed and why.[16] In fact, historical information on ancient traditional agricultural systems has not been used in conjunction with the environmental characteristics of antiquity vis-a-vis current ones.

Officials and personnel involved in these programs organized by the government of Mexico City try to "recreate wetlands" in the existing space on the upper strip of the Viaduct tunnel. The waters of the old rivers of La Piedad, Tacubaya, Becerra, and Churubusco flow through their tunnel. Recalling the achievements of the past remains an important aspect in these types of plans, programs, and their nationalist and identity ideology. Apparently, failures do not discourage the conformation of these projects, and several of which have been applied to the reconstruction of the chinampas in the former lakes of Xochimilco and Chalco. This has not been possible entirely, due to factors such as water pollution that prevents the cultivation of vegetables and legumes, but allows the successful cultivation of a variety of flowers (Aréchiga-Córdoba 2011; Correa-Ortiz 2011; González-Pozo et al. 2010; Quiñonez 2005; Olivares 2007).

These projects last six years, usually the future will tell which projects will emerge in this six-term period, and which will have positive results in the medium and long term. In general, there are several problems that prevent their success: the existing pollution in air, soil, and water; the environmental, social, and cultural context in which the projects are initiated; the market for cultivated products; the varieties used in the systems introduced; the local knowledge that farmers have about the old agricultural systems; the environment and the climate of the places where they want to rebuild; the techniques and management required by crops; and factors such as urban expansion, the processes of modernization of the countryside in Mexico, and many other aspects of local and regional character.

[16] Andrade et al. 2011: 309–328; Barkin and King 1970; Barkin and Zavala 1978; del Amo 2000a, 2000b; del Amo and Flores 2001; del Amo and Moctezuma 2008, 5: 1–15; García-Acosta 2000; Gómez-Pompa et al. 1995a, 1995b; MacMahon 1973; Molina 1976; Palerm 1970, 1993.

From the 1980s to date, critical studies have been added as to development projects and their social outcomes, plus the discussion on sustainability, and how to achieve it. From the 1930s to date, changes in small-scale agriculture continue to be proposed to modernize it following the American model of conventional agriculture, whose production is directed to foreign markets. However, the current characteristics of small-scale production in traditional agricultural systems are not compatible with the structuring, characteristics, and circulation of goods like those produced on larger scales under conventional systems (A. Palerm 1968, 1998). Projects sponsored by FAO and IDB are based on the idea of reducing poverty, improving food in the world, giving rural populations sovereignty, and providing food security to give them access to better standards of living. On the contrary, conventional agriculture is not compatible with sustainable environments or healthy food systems for human or domestic animal populations (FAO 2006b, 2015; International Council for Science 2002).

From the latter part of the last century to the present, FAO (2006a) and IDB are interested in the reconstruction of parcels, school homegardens, and gardens. The programs have had different types of impacts in Mexico and in many other countries on the planet. These programs are generally implemented, complemented, or supported with international monetary resources because their success is associated with improving the quality of life of human populations. Their promoters believe that small-scale farmers can reach high percentages of agricultural production for human consumption. However, in the medium and long term, the programs do not seem to have had the expected success.

Historians, agronomists, cultural geographers, archaeologists, ethnobotanists, ecologists, agroecologists, and anthropologists have studied homegardens in Mexico for more than half a century. However, applications to promote the expansion of them in plots and land for school gardens have been made by agricultural extensionists, educators, public officials, and local staff hired by social development institutions. Apparently, the results in the medium and long term are not always satisfactory (Barkin 1972, 1998; Fernández 2017), at least in Mexico. Programs have been implemented with relative success in several states of the country, including Tabasco, Puebla, and Tlaxcala. In the state of Tlaxcala, programs to promote family homegardens and gardens have had little success, mostly because products that are not part of the daily consumption of the inhabitants were planted.

5.12 MAIZE AND THE EMERGENCE OF TRADITIONAL AGRICULTURE

The first Mexican anthropological studies on traditional agriculture are related to maize as well as its domestication and dissemination processes. This grain is the axis of the human food system in Mexico. According to FAO (1999), Barkin (1972, 1998), Fernández (2017), the use of maize as human food is essential in several parts of the world. In the case of Mexico, there are numerous ways for consumption, distribution, handling, and use in general (CIMMYT 1988, 1993–1994; FAO 2010; González-Jácome 2018a; González-Torres 2007: 45–80, 2008: 179–219). The greatest variation in maize uses is found in Mexico (Paliwal 2009).

Other countries in Latin America and Sub-Saharan Africa where this grain remains a staple for humans follow with the greatest variation in uses. Maize use as fodder is important in Spain, France, and other European countries.

An issue that has interested researchers, academics, scholars, politicians, and the general population has been knowledge of practices and management related to maize. This plant is a source of national food systems, whose traditional cultivation in a small-scale way has had support or rejection over time, especially in the nineteenth and twentieth centuries, of ideas that favored or rejected its cults, rituals, ceremonies, and representations of different types. Like happened with the rituals organized in Otomi, Totonac, Huaxtec, Zoque-Popolocas, and Maya towns where maize cultivation is done in small-scale production units, such as the polyculture cornfield.[17]

[17] Acle-Tomasini 2000; Blanco-Rosas 2006; Broda 1983: 145–165, 2004a: 61–81, 2004b: 245–261, 2004c; Broda and Good 2004; Caamal and del Amo 1987, 37(1): 195–210; Del Amo and Moctezuma-Pérez 2008: 1–15; Hoppe et al. 1975: 489–505; Martínez 2006: 77–98; Wake 2015: 155–175.

Since the first third of the twentieth century, a series of social and political proposals by researchers were published in specialized magazines and reports that defend the idea that maize is not a food nutritious enough for human populations. They believe it should be substituted, or at least published, and complemented with other crops, like soybeans, in spite of the economic, nutritional, social, and cultural importance of maize, and its wide distribution on the national scale to date. This proposal exists with some success. Currently, soy consumption is attained by several means. One of them is to mix soy with maize used to make tortillas; another is to process it to produce drinks that replace cow's milk, putting an end to its consumption (Hinton 1945: 180–182; INI 1944: 93–96, 1945: 254–259).

A soybean drink per liter is expensive for the popular classes, although State-owned low-price stores (*Liconsa*) sell a drink called soymilk for low-income families. The approach of introducing milk derived from products such as soybeans, rice, or almonds is related to the old idea of modernizing agro-livestock production and technology, and its handling and destination in the market, advocating the idea that these products are better than milk, more nutritious, and have added vitamins.

This tactic includes the disappearance of traditional agriculture in Mexico, which has been encouraged and sponsored, for its replacement of more conventional, modern, and commercial agriculture. The small-scale farmer can introduce agrochemicals (fertilizers, herbicides, and so on), electric pumps to irrigate his land, and hybrid and genetically modified maize varieties. But generally this is at their own expense, which requires having the capital or borrowing from local lenders.

Small-scale farmers have some degree of access to the market through direct trade in the tianguis, local, and regional markets, or the direct sale of their products when they are in season like boiled or roasted maize, and boiled maize kernels (*esquites*) at roadside and street stands, likewise the sale of *quesadillas* filled with cornfield plants, cooked pork, cooked chicken, and others (Diskin 1969). Farmers sell seasonal fruits in baskets, bags, by the piece, or by volume.

At the same time that traditional agriculture has been assailed, there were some positive results of the agricultural distribution of land, and the legal form that were applied in Mexico through the control of the State over the endowment lands (*ejido*) and water (implanting irrigation districts). In the Central Highlands, authorities generated small property (5 ha) and small-scale agriculture that without major economic resources followed in the manner of traditional temporary agriculture, with diverse agricultural systems adapted to specific environmental conditions, both local and regional, to their economic and social interrelations with other farmers, to the market, to self-consumption, and to family work. In the north of Mexico in the cattle regions, the surface delivered by means of endowments was greater.

5.13 CONSIDERATIONS AND CONCLUSIONS

After 1970–1980, regarding the cultivation of maize and production in the cornfield, it is basic to consider the traditional social and cultural forms related to the food system distribution, reciprocity, and mutual aid applied to agricultural work. The forms of mutual aid between nuclear family members, kin, and friends are bartering, macoa, yetza or quetza, tetlamacá, or exchange without money involved. Daily while they are being carried out, and when they finish the unpaid agricultural work (mutual aid), the participants receive food that was cooked by the women of the family that receives this support in unpaid work.

The tasks corresponding to the mutual help is aimed at work carried out for the benefit of the community included canals, communal wells, and open cisterns (*jagüeyes*) cleaning; cutting of the ruderal vegetation on the roadsides; public building repairs; administrative and legal procedures for the introduction of services in the villages; religious building repairs; and other communal activities.[18]

---

[18] Ignacio-Felipe 2007: 24; Juan-Pérez 2003; González-Jácome 2018b; González-Jácome et al. 2007; Harper and Le Beau 2003; Rivas et al. 2008: 11; Romero et al. 1986.

In these times, without traditional agriculture, the food system and health of many small-scale farmers will suffer major mishaps because of market risks (supply–demand and the uncontrolled increase of availability and prices for products), and environmental and climatic crises. It seems that poverty and malnutrition would continue to increase, not to mention the surge in negative social behaviors articulated with crime and gangs. In addition, the abandoning of their places of origin of migrant young people to work in remote regions or other countries seeking to raise their living standards. The countryside remains in the hands of the elderly due to those family members who migrate or emigrate (seasonally, temporarily, or permanently).

Given the loss of local knowledge and traditional agricultural knowledge (TAK), it is important to reconsider the contributions and benefits that have remained over time, despite the plight. This has been possible due to the ability to achieve sustainability and to have resilience, especially in the face of catastrophic natural phenomena, and external economic, social, and political changes (especially drastic ones), which usually occur and affect the human societies that live in rural areas and their environment at the local and regional scales.

RECOMMENDED READING

Gurri, F.D. and E.F. Moran. 2002. Who is interested in Commercial Agriculture? Subsistence Agriculture and Salaried Work in the City Among Yucatec Maya from the State of Yucatán. *Culture and Agriculture* 24(1): 41–47.

The slash-and-burn agroecosystem, and the uneven distribution of water resources in the state of Yucatan have been limiting factors for the total population growth that could earn a living on the land surrounding the town. The pressure of excess population, and the concentration of land, or sinkholes in the hands of a few would eventually lead to the rupture of the community. People who having lack of sufficient land and water, would migrate from the area in search of new areas, unclaimed sinkholes with enough land when it limited the periods of rest of the soil. The agrarian reform did not take into account the over-exploitation of the land when it limited the periods of rest of the soil. The Yucatec Maya in order to improve their living conditions have to work in salaried activities away from their communities, generally.

Murra, John V. 1989. High altitude Andean societies and their economies. In: Genovese, E.D., and L. Hochberg (eds.). *Geographic Perspectives in History*; B. Blackwell, pp. 205–214.

For the historian, geography is part of history. For Murra, altitude in the Andean case explains the economy, what is produced, how it is produced, and where it is produced. This element interrelates and explains the production and circulation of agricultural products and the mobility of people. This context appears related to Andean political control closely. All events that are studied have causes and consequences, and these depend upon where the action occurs, in part. Altitude is the causation of where, why, how, and what are interconnected events in the economies of Andean societies.

Thomas, Julian, 2010. Understanding past landscapes: experience, memory and materiality. In: E. Ortiz Díaz (ed.), *VI Coloquio Pedro Bosh Gimpera. Lugar, espacio y paisaje en Arqueología: Mesoamérica y otras áreas culturales*; UNAM, Instituto de Investigaciones Antropológicas (IIA), pp. 27–46.

This article deals with the concept of landscape in archaeology that emerged during the twentieth century. This concept attempted to peel apart the sequence of fields, walls, buildings, and roads that cover the land. Landscape archaeology emerged in Britain is concerned with landscapes that have been in continuous use over several millennia. As a subdiscipline has tended to be multiperiod: far more so than forms of archaeology that concentrate exclusively on excavation. Archaeologists tend to concern themselves with all of the transformations that have overtaken a given landscape, and to talk of the land as a palimpsest of traces of human activity, layered upon one another.

Landscape archaeology is in some degree overlapping with environmental archaeology. Although a series of different kinds of archaeology claim to be studying the landscape, in reality they are doing a variety of different things, and the past landscapes that they claim to be reconstructing are actually very different kinds of entities. Those different kinds of landscape may be incompatible and mutually contradictory.

The vision of landscape as an accumulated record of continuity and tradition tends to play down the human element in the land. Historically, the landscapes that can be seen today are not just the outcome of the slow continuity of agriculture and rural life. The notion of landscape can serve to conceal inequality and historical upheavals. The notion of landscape can be ideological. Landscape can mean the topography and land-forms of a given region, or a terrain within which people dwell, or a fragment of the land which can be overseen from a single vantage point and represented as such. In the modern west, our predominant conception of landscape is a visual one; the author thinks of landscape as something to be looked at, perhaps from a distance.

6 How Traditional Agriculture Has Been Studied

6.1 INTRODUCTION

In anthropology, studies on traditional agriculture began in the early twentieth century, related to the work of cultural anthropologists, historians, and geographers on the indigenous in areas of Mexico, their food system, and economic activities (Aguirre-Beltrán 1952, 1958: Armillas 1949, 1971; Beals 1943; Beadle 1932, 1939, 1972, 1978, 1980; Böehm and Pereyra 1974; Bruman 1944; Coe M.D. 1964a, 1964b; Coe 1994, 2004; Sauer 1952). Historical and archaeological studies carried out in the Old World, especially those of Gordon V. Childe (1995), Karl Wittfogel (1956, 1966), and Alexander V. Chayanov (1966), lead scholars to investigate the agricultural activity in communities studied and to ask questions related to various issues, including that of the origins of cultivated plants and their diffusion.

In agriculture, one of the forerunners of this type of study is Nicolai Ivanovich Vavilov (1992 [1926a]; 1926b, 1927, 1931a, 1994 [1931b]: 15–33). He was a Russian botanist, geneticist, and agronomist who from 1924 to 1940 conducted numerous studies on the origins of agriculture, cultivated plants, and regional agrological studies. To do this, Vavilov proposed several centers of origin of world agriculture. He carried out with the support of 98 investigations, including fieldwork *in situ* and collection of materials for laboratory analyses.

These studies were looking for obtaining information on agricultural techniques and were done in Mediterranean countries of Morocco, Algeria, Tunisia, Egypt, Portugal, Spain, Italy, Greece, Asia Minor, Syria, Palestine, Sicily, Sardinia, Crete, Cyprus, and Rhodes islands. Also investigated were Abyssinia, Eritrea, Persia, Afghanistan, western China, agricultural Mongolia, Japan, Korea, Formosa, India, the former agricultural regions of Transcaucasia, and Turkestan. Vavilov and his team studied in the New World Mexico (Yucatán), Guatemala, Colombia, Peru, Bolivia, and Chile (Vavilov 1994 [1931b]: 15–33).

6.2 CENTERS OF AGRICULTURAL ORIGIN

Vavilov (1926, 1931) considered that the center of origin of cultivated plants and domesticated animals was associated with time and space. To obtain information about the project the researchers asked themselves: Where should we look for the origins of agriculture? Are there independent origins, on different regions, or on different continents? How should the geographical location of agriculture be explained? What were the first cultivated plants? What were the first domesticated animals, and where were they domesticated? Where can we find the sources of cultivated plants? How do domesticated animals and current cultivated plants connected with their wild holotypes? How did the evolution of cultivated plants and animals take place? How did the first agricultural civilizations communicate? What tools did primitive farmers in different regions use?

Vavilov (1926a, 1994 [1931b]: 15–33) developed his theory about centers of origin of cultivated plants throughout his botanical-agronomic expeditions *in situ*, including where he collected cultivated plants in the regions studied. The centers of origin of most of the current cultivated plants were in the botanical areas or regions where powerful type formation processes were active. The research of these plants in the different experimental stations continued. The Leningrad Cultivated Plants Industry Institute (former Office of Applied Botany) was responsible for organizing the collection of genetic material of plant origin.

DOI: 10.1201/9781003198833-6

These regions included an important plethora of forms, endemic characteristics, and concentrate of genetically related species or wild relatives. Also, they were rich in plant species. Human populations since ancient times had inhabited these areas. In these regions, knowledge accumulated about such diversity was carried out. It is now thought that variability was a characteristic of the knowledge, management, and use of species in Mesoamerican cultures (Casas and Caballero 1995: 2).

Studies classified the global geographic distribution of species and varieties. In addition, several species were discovered that were unknown to botanists, growers, and agronomists. Their often-possessed valuable "practical qualities". New species of wheat, hundreds of new varieties of small grains, and wild and cultivated plants were discovered. Vavilov and his collaborators of the *Plant Growth Institute* considered that the American continent was an independent center of origin of cultivated plants because the cultivated species and their genera were unknown in the Old World. Two cultivated plant centers of origin were established, one in Central America (The Mayan area), and the other in the Andes. In Peru and Bolivia, 12 new species of potatoes were discovered. Only one (*Solanum tuberosum* L.) had been known outside of these Andean countries before.

Vavilov (1992 [1926a], 1926b, 1927, 1931) proposed eight cultivated plant centers of origin to be the basis of ancient agriculture in the world: (1) China; (2) India (with the Indo-Malaya Region); (3) Central Asia (including Pakistan, Punjab, and Kashmir, Afghanistan, and Turkestan); (4) the Near East; (5) the Mediterranean; (6) Ethiopia; (7) the Southern Mexico and Central America; and (8) South American countries like Ecuador, Peru, Bolivia, the Chiloe zone, and Brazil-Paraguay. Botanists Casas and Caballero (1995: 36) made some modifications to Vavilov's list to include all of Mesoamerica, the Andean zone (Peru, Ecuador, and Chile), Equatorial Africa, the Mediterranean region, Southeast Asia, and Northern China (Chávez and Thompson 2006; Grobman et al. 1961).

Nicolai Vavilov's (1931: 16–17) proposal on "Mexico, and Central America as the Basic Center of Origin of New World Cultivated Plants" stated that American Continent agriculture arose independently from that in the Old World,

> [...] its own wild floristic diversity, and in different material conditions [...] Without having agricultural domestic animals [...]. Nor did they know the iron instruments or the plow, and [...] only occasionally used some copper tools, as they made their agricultural tools of wood, bone, and stone.
>
> *(Agricultural Geography Nº 20, 1994: 15–34)*

Vavilov's (1994:18–22) list of cultivated plants that were domesticated in Mexico comprised maize, teosinte, cotton (*Gossypium hirsutum*), common beans, Scarlet runner bean (*Phaseolus coccineus*), Lima beans, Tepari beans, Black seed squash (*Cucurbita ficifolia*), Vegetable Pear, Merliton, Christophine (*chayote*) (*Sechium edule*), annual chili, perennial chili, Mexican yam bean (*Pachyrhizus erosus*), *Agave atrovirens* (wild and cultivated), sisal (*A. ixtlero*), sisal (*A. fourcroydes*), *Agave americana* wild and cultivated, *A. lechuguilla*, *Opuntia* (wild, and cultivated), species of the genus Cereus, *Opuntia* to host Mealy bug (*Dactylopius coccus*), huautli (*Amaranthus cruentus*), Goosefoot (*huauzontle*) (*Chenopodium nuttalliae*), Jaguar flower (*oceloxochitl, cacomite*) (*Trigidia pavonia*), *Physalis aequata* (ruderal and cultivated), tomato (*Solanum lycopersicum*, ruderal, and cultivated), chia (*Salvia hispanica*), cacao (*Theobroma cacao*), and Annatto (*Bixa orellana*). There were about 30 wild species of Mexican potatoes, within that are the current boundaries of Mexico and Central America. Wild potatoes were collected and consumed by Mexican natives, but no evidence was found about cultivation.

Among the fruit trees were Custard apple, Medlar; sapota fruit (*Manilkara sapota*), White sapota (*Casimiroa edulis*), Mammee Apple (*Pouteria sapota*), Graft (*Calocarpum viride*), Sansapote, Yellow sapota (*Pouteria campechiana*), papaya (*Carica papaya*), Avocado (*Persea americana*), Guava (*Inga edulis*), Mexican Hog Plum, Red Mombin, Yellow Mombin (*Spondias* spp. Wild, and cultivated), Mexican Hawthorn (*Crataegus mexicana*), Mexican Black cherry (*Prunus* sp.), Hicoria pecan (*Carya illinoiensis*), Monkey pod, and Blackbeard (*guamuchil*) (*Pithecellobium dulce*).

Other trees and bushes were Ceiba-silk-cotton tree (*Ceiba pentandra*), Wax tree (*Mexican Myrica*) (wild and cultivated), and the Rubber plant or Guayule bush (*Parthenium argentatum*) found in the mountainous regions of northern Mexico.

Ornamental plants included eight species of Dahlia and 15 species in Mexican and Guatemalan highlands; at least four species of *Cosmos*; four species of *Tagetes*; three species of *Zinnia*; *Brugmansia candida*; six species of *Ipomoea*; *Lantana camera*; *Philadelphus mexicanus*; two species of *Poinsettia*; *Mirabilis longiflora*; *Bomarea acutifolius*; *Polyanthus tuberose*; two species of *Yucca* in Mexico and Guatemala; *Bouvardia ternifolia* (one wild species that was used as medicinal and another cultivated one used for decoration). Several species of orchids and cultivated bromeliads were found.

Vavilov and his team (1931: 22) considered a clear evidence of wild links that corresponded to many of the cultivated plants in Mexico and Central America [Mesoamerica]. This was found in at least half of the cultivated plants. The transition from wild to cultivated species was associated with agricultural practices, as in the case of the forest, where the farmer left the Mexican hog plum (*Spondias mombin*) and the guavas (*Psidium guajava*) uncut, so the wild Scarlet runner bean (*ayocote*) became entangled in the bushes. There is little difference between species of cultivated Agaves (*A. atrovirens* and *A. sisal*) and wild Agaves, and whether they reached the fields by "random effect" is not known. Among the wild plants are the *Physalis ixocarpa* weeds, a tomato species (*Lycopersicum esculentum*), and wild cacao.

Vavilov, and his research team (1931: 22–23) observed the transition from wild to cultivated plants in papaya and avocado. There were cactus species that were both wild and cultivated. The team considered the cultivated *Agave atrovirens* to be native of the southern Mexico mountainous region that corresponded, according to the map (1931: 27), to Mesoamerican territory.

In the 1950s, Sauer (1952) the American geographer, promoter of the School of Cultural Geography in Berkeley, was studying the origins of agriculture in America, and he proposed a division between two types of agriculture: (1) a vegetative one, based on cuttings, stolon, auxiliary shoots, basal shoots, or bulbs, which are typical of tropical regions where vegetation is naturally reproduced in this system, so these clonal spots formed allow to humans to propagate desirable phenotypes (Gentry 1982; Colunga et al. 2017; Torres et al. 2015b); (2) and the agriculture based on grain cultivation, the grains capable of being stored for longer times and being transported to regions far from their center of cultivation. Sauer (1952) considered that grain agriculture arose in the humid tropics of Central America, in South America, and later in the American dry and semi-arid regions.

Based on these two ideas, the study of the origins of agriculture in Mexico has had several versions and several hypotheses. About Mexico, the questions began with the hypothesis about whether agriculture arose in tropical or semi-desert areas, but with the presence of rivers, lagoons, or natural water deposits. The study of traditional agriculture had two lines of development that followed: (1) one which comes from anthropologists' interests to study culture, including the agriculture of the country's indigenous peoples; and (2) another which arose from the studies of other specialists in the genetic characteristics, composition, forms of propagation, and diffusion of cultivated plants.

This second line was part of the studies on the origins and domestication of maize, and the diffusion of studies carried out in places of the American continent (Index of domestication).

At the outset, several centers were proposed. Studies of genetics combined with archeological, palaeoecological, pollen, and phytolithic analyses concluded that the origins were found in the middle basin of the Balsas River, in Mexico.

It is necessary to consider that the information can vary easily, since it depends on the new investigations in situ that are carried out all the time, updating the information. So what is written down at this time is not necessarily true permanently. Hence, traditional agriculture is understood from various fields of scientific knowledge, on-site studies, and interest in understanding the evolutionary and historical processes through which the different agricultural systems that were formed at the beginning had passed. Traditional agriculture research follows various lines to study regarding it. Some come from anthropology (ethnology), and others from agronomy, geography, history, economics, ecology, ethno-agriculture, ethnobotany, ethnopedology, ethnoecology, and agroecology (Hernández X. 1987b; Rosado-May 2015, Vol. 10(2): 9–17).

In general, traditional agriculture has been defined as one where, generally, agroecosystems are of ancient origin[1] and due to the minimal use of inputs, lack continuous disturbances, and exhibit complex interactions between factors such as soils, crops, flora, fauna, and others, all optimal scenarios in the evaluation of properties such as sustainability and resilience. Its study allows obtaining criteria for the design and management of alternative agroecosystems (Clark and Dickson 2003: 8059–8061; Gliessman et al. 1981).

This traditional agroecosystems had sought and often found solutions to the problems and risks they had faced, such as the phenomena generated by alterations to the environment (climate, volcanism, prolonged droughts, fires in tropical forests, and temperate-cold forests) should be included, the same as the social and cultural problems faced by the societies that managed traditional agroecosystems is key. The local knowledge, its features, and its numerous contributions to traditional agriculture cases added to traditional agricultural knowledge (TAK) and its interrelations with other disciplines have been of chief importance. Efraím Hernández-Xolocotzi (Hernández X.) and his students have carefully studied these agricultural systems, as will be seen below.

6.3 CHARACTERISTICS AND COMPONENTS OF TRADITIONAL MEXICAN AGRICULTURE

Hernández X. (1981, 1998: 673–674, 2013) stated first that in traditional agriculture (a system) energy consumption is lower than that required by conventional agriculture systems. Currently, small-scale farmers combine inputs of agrochemicals, with natural fertilizers. In zones of Tlaxcala in the 1970s and 1980s, small-scale farmers applied chemical fertilizers for three or four years in a row, and then used natural fertilizers, acquiring at least one truck of compost to apply to the land. Next, they tilled the ground and flattened it to let it rest. It was tilled anew previous to sowing (González-Jácome 2003; Hernández Rojas 2014).

Traditional systems use little agricultural machinery, or they combine machines with draft animals and farm implements intended for agricultural work. Thus they mix the components and technology of conventional agricultural systems with traditional agricultural systems (Cruz-León 1989; Cruz-León and Martínez-Saldaña 2001).

Small-scale farmers have not rejected the components of modern agriculture but have selected them according to their needs, adjusting or adapting, and combining them with traditional components. These combinations are born of needs like performing the work in less time and with less economic cost without losing control of the essential components of the system (soil, water, flora, and fauna), mainly it has been historians and ethno historians who had studied the variety selection of cultivated plants. An example refers to the changes in pre-Hispanic agriculture with the arrival of plants, draft, and load animals, and agricultural metal tools, and the plow, in the sixteenth century.

An example of the advances in Mesoamerican agriculture is found in the refined knowledge of herbalism. Medicinal plant studies have been of great interest to anthropologists, physicians, biologists, and ethno-botanists. Examples of this are the writings of Argueta (2016: 38–46); Bye and Linares (1987, 2010, 2016), and Robles-Cervantes (2008) that highlight the importance of medicinal plants, their main systems of production (e.g., homegarden), and of gathering on the earthen borders with wild plants or of systems like Agave rows (*metepantle*). There are wild weeds and cultivated herbs that are grown specifically for delivery in large national markets, such as Sonora in Mexico City, or to regional open-air markets (tianguis).

[1] Altieri 1991; Armillas 1949: 85–113, 1971: 653–661; Cox and Atkins 1979: 111–138; Flannery 1982; Gliessman 1999: 25–31; Gliessman et al. 1981; González-Jácome 1985: 521–538; Harrison and Turner 1978; Hernández 1981, 1987a, 1987b; 1998: 673–678; Palerm 1955, 1972; Sanders et al. 1979; Siemens 1989a: 233–250.

Machuca (2013), using 45 *Geographic Relations* (1580) as sources, built an important proposal regarding the arrival of plants to the Balsas region in Jalisco. For example, cotton crop in the Michoacan region that went from Sinagua to Tepalcatepec was displaced with the introduction and cultivation of sugar cane and plantain. Other articles made with the *Saccharum officinarum* cane replaced the products manufactured with reed, previously.

Caballero and Casas (1995: 36–45), in their article on the domestication of plants and the origin of agriculture, presented archaeological and ethnobiological examples of past and present forms of plant management that Mesoamerican Indians developed. These two scholars' opinion on traditional Mexican agriculture does not oversimplify the matter but ignores the research in Mesoamerica that is based on the social and cultural knowledge of interdisciplinary and multidisciplinary work, where studies involved a huge number of researchers. These social and cultural studies have been carried out over a century, at least, and have been seen and will be seen as very important contributions to the knowledge of traditional agriculture.

6.4 DIVERSITY AND THE COMBINATION OF COMPONENTS IN TRADITIONAL AGRICULTURE

Traditional agriculture is characterized by combining components of the system itself, with others of the natural environment. These include insects as well as wild, cultivated and tolerated plants, which appear on farmlands seasonally, edges of land or borders areas riverbanks, lagoons, roadsides, divisions between plots, weedy strips, wooded areas, and galleries (González-Jácome 1986: 201–220; Vela 2019; Wilken 1969, 1987). Traditional agriculture systems include the biological control of insects that affected crops by means of soil borders where wild vegetation grows and ruderal plants; many of which are used by herbalists and are insect attractors. It is important to include the pollinators that dwell in these patches of vegetation.

Consequently, other important input to agroecological traditional studies came from chemical ecology, which has touched on the related aspects between the knowledge of plant domestication and artificial selection that produced changes in the human population genetic and phenotypic frequency characteristics, causing divergences with their wild ancestors (Bautista-Lozada et al. 2012: 253–267; Krupinsky et al. 2002: 198–209).

Ramos-Elorduy (2009, 2019: 11–14) referred that in the Florentine Codex (+/–1521) Bernardino de Sahagún listed 96 species of edible insects of various types in the valley of Mexico and its surroundings. In addition to zones with forest remains, where wildlife was mixed with domestic wildlife – as is the case of pigs in the southeast of the country insects, birds, and other local fauna make up an important part of the food system.

The remains of the wild fauna consumed by the population had several uses. For example, Götz and Emery (2013) in their introduction on Mesoamerican fauna deal with the subject of the remains found in the dumps and the fillings of the excavation structures. The remnants of wildlife that had been consumed were associated with the homes of the elites in five pre-Hispanic Mayan sites in the northern Maya lowlands: Champotón, Chichén Itzá, Dzibilchaltum, Sihó, and Xcambó. Dating from the Classic and Postclassical times (from 200 to 1500 AD), all wildlife materials found were "cultural accumulations" (Ramos-Elorduy 1999: 11–14).

In the Mexican central highlands, duck eggs and *Opuntia* mucilage were added to mixtures, and mortars were applied to buildings and their coatings (a type of waterproofing). Lagoon systems, such as the basin of Mexico, were important places to obtain fish, edible birds, eggs, and feathers. Skeletal remains and fish bones were found there (Ramos-Elorduy 1999: 11–14). The identification and comparison of wildlife remains of the ancient Maya indicate various subsistence adaptations and strategies for collecting and hunting vertebrate animals (Harrison 1982: 119–130).

Different exploitation patterns are revealed from coastal to inland sites, as they adapted to each unique local environment. Elite residents at inland sites appeared to had used well-adapted strategies to modify the environment of secondary forests and agricultural fields. The food system of the elite settlers in the coastal sites was mainly based on marine fauna and completed with small quantities of terrestrial vertebrates some of which were possibly obtained through nearby communities or long-distance exchange.

It must be considered that small-scale farmers managed the soil to prevent erosion. They used marine fauna mixed with terrestrial flora, and fauna remains as fertilizers. Among the methods for this activity were the conformation and obtaining of water from canals and rainwater tanks (trenches, artificial rainwater deposits like stirrups (*cajetes, estribos*) or canals with ground divisions along) conforming a mixture to the finely crushed soil from the canal system and the microscopic remains of local flora fauna, and sediments which were trapped in these catchment basins. In addition, peasants applied green manures to the soil. These practices prevent salinization of the land and control the effects on the soil of diseases that attack plants (Espinosa-García 1994, 2007; González-Jácome 2008; Lumsden et al. 1989: 83–103).

Diversity is fundamental and applies to both components of the traditional agricultural systems and combinations related to other economic activities, which not necessarily are agricultural (Acle-Tomasini 2000; Acosta-Naranjo 2002; González-Jácome 2007: 15–40). Thus, working in the field is combined with the care of domestic animals, hunting, fishing, gathering, temporary jobs, artisanal production, or work in industries or urban services.

Some family members, especially young people, work as public employees, factory workers, office workers, or shop employees, or in tourism or other jobs.[2] In tourist areas, such as Cancun, and Playa del Carmen work is temporary and is associated with the arrival of important tourist groups. Although in several cases, tourism has become an area of permanent employment for people in the more or less nearby villages (González-Jácome 2010: 141–158).

Small-scale trade is important to consider, as is the retail sale of agricultural produces from the cornfield the homegarden, or the terraces, the collection of cultivated and wild green weeds (*quelites*) with commercial value, dry chili, or other regional issues, such as the sale of fresh, and dried fish, river shrimp (*acamayas*)· dry and freshwater shrimp, Mexican orange dwarf crayfish (*Cambarellus*) (*acocil*), and others. It is important to include the elaboration, distribution, and sale of various types of bread, quesadillas (Castañeda-Castillo 2014), maize tortillas, or wheat flour tortillas (Figures 6.1 and 6.2).

FIGURE 6.1 Sonora. The Process for Making Wheat Flour Tortillas (*Sobaqueras*).

2 Castañeda-Castillo 2014; González-Jácome 1989: 86–107, 2008b: 9–24, 2007a: 15–38, 2007b: 59–95; Juárez-Choque 2015, 2016: 9–12; Thome-Ortiz 2015: 1373–1386; Villalobos-Nájera 1998, 2008.

FIGURE 6.2 Sonora. Iron plate to make wheat flour tortillas (*sobaqueras*).

In all the country open-air markets (*tianguis*) and markets, or in the nearby streets and parks, there are stalls to sell in small-scale prepared and cooked foods like *mixiote*; tacos; tamales; and *tlacoyos* ready to be heated and seasoned by the consumer; jellies; breads (Juárez-Choque 2015), sandwiches; hot dogs; hamburgers; hot coffee with cinnamon and brown sugar; atole of various types and flavors; some type of cacao drink (*Cacao, cacahuatole also called agua de barranca, posol, posol agrio, chorote, tescalate, tejate, popo, puntal, chocolate con pimienta, atol de mujer, tanchuca, champurrado*). Local drinks that are sold in streets, parks, and markets such as tuba in Colima and Agave mead in Zacatecas; locally made sweets, flowers, and other things. Flower selling was also a large activity for farmers (Figures 6.3–6.5).

FIGURE 6.3 Sonora. Strings with Dry California Chile, and Garlic in Small-Scale Trade, 2013.

FIGURE 6.4 Sonora. Strings with Dry California Chile and Garlic String, 2013.

FIGURE 6.5 Sonora. Garlic Strings and Pumpkins in Small-Scale Trade, 2013.

6.5 SMALL-SCALE FARMERS AND SOLIDARITY

There is great solidarity among the small-scale farmer families who combine their resources in unremunerated collective labor and remunerated activity selling agricultural or working in salaried jobs. The support and solidarity (mutual aid) of the nuclear family, friends, kin, acquaintances, and members of the community where they live is a basic social element, which allows agricultural work to be carried out regarding no monetary expenses.

This unpaid work is reciprocated when it is necessary for activities such as planting, weeding, or harvesting. Solidarity is so important that it applies to the transport of commercial agricultural produce to its destination, as occurs in places like Progreso Hidalgo, a town in the State of Mexico, where strawberries and flowers are cultivated, and later sold at the Market Central Supply in Mexico City. Cornfield production is destined solely for self-consumption (Juan-Pérez 2003).

Agricultural landscapes are interrelated with demographic aspects: the horizontal mobility of the total population; the densities, increases, age, and sex of the family members; formal education;

and the number of people conforming the families, their lifestyle, and how they organize their work in the family.[3]

At least one of the family members is part of the daily, weekly, seasonal migration, national or of international emigration to complete monetary family income to pay for industrialized food; the cost of public transport (or gas) to go to other towns or places; the expense of illnesses; medicines; debt payments for communal social events in their home community; important holiday celebrations for the family; expenses of stewardships; and other expenses that arose from events related to group members like payments for the educational cycle, payment of transportation costs for children to study in places other than the town, purchase of appliances, or exceptional medical expenses due to accidents.[4]

The family labor force applied to cultivation in traditional agricultural systems is the main work force but receive no economic payment (salary). The emigration of male members modifies the size of cultivable area and the volume of labor force applied to activities such as planting, weeding, or harvesting. Production is aimed at consumption and the market. However, the balance between both elements is organized throughout the year, depending on Family needs, including social and cultural ones, which are of great importance for the group and the community networks. A part of feeding is derived from the traditional and commercial agricultural systems that families have. Food system varies in the year, at times due to changes in the daily diets, which adjust to family needs, tastes, and resources (Blanco-Rosas 2006; Cahuich 2012; González-Jácome 2003, 2009: 40–65; Vargas-Guadarrama 2007a, 2007b).

In traditional agricultural systems, the non-remunerated labor force is very important. It includes: tight nuclear or extensive family members, the kin, and "*compadres*", close friends, and acquaintances that want to participate. The families of small-scale farmers articulate themselves with religious and civil festivities and ceremonies, where the community participates and integrates into a worldview shared by the social group. In fact, unpaid work in a peasant society has two aspects: one directed to communal works that included the construction or repair of roads, cleaning communal water tanks, arrangements in the church, and public buildings, plus other works for common benefit like the introduction of electricity or sweet water systems in towns (mutual help).

The other aspect of mutual aid focuses directly on agricultural activity (especially planting, weeding, and harvesting), the construction or reconstruction of homes, the collaboration in the payment of fees to cover the expenses of the stewardships, and other family activities requiring greater amounts of money than the families' income. Social and religious expenses included weddings, baptisms, confirmations, and completion of studies of any of the children in the family; all of these constitute a support in work, in money and/or in kind for nuclear, extended kinship families, *compadres*, and close friends (mutual aid).

In the 1980s, in Tlaxcala, a wedding cost a minimum of 70–90 thousand pesos, so couples who did not have parents and godparents who could face the expense, preferred to "run away", and later marry in a civil ceremony only or participate in a mass communal religious wedding. In fact, the parents knew about the escape and did not become angry. The sponsorship system was reinforced, to where – in addition to the baptismal godparents – friends were invited to be sponsors for the weeding and groom paid for different things: the wedding rings; wedding attire; the bride's make-up and coiffure; the bride's bouquet, flowers to decorate the church; music; alcoholic beverages (beer, brandy, mezcal, tequila); payment of the mass; the wedding cake; money for the honeymoon; and anything that was required in each case.

For smaller expenses, the sale of poultry and eggs could help them out of the pinch. Some families saved to meet these expenses; this could be through domestic animals, which cared to be sold

---

3 Del Amo 2000a; Castañeda-Castillo 2014; González-Jácome 1989: 86–107, 2007: 15–38, 59–95, 2008b: 9–24; Ignacio-Felipe 2007: 24–25; Juan-Pérez 2003; Juárez-Choque 2015; Villalobos-Nájera 1998, 2008.

4 Caloca-Rivas 1999; Castañeda-Castillo 2014; González-Jácome 2003, 2009: 71–111, 2010: 141–158, 2011b: 71–111; González-Jácome et al. 2007; Mariaca-Méndez 1997; McC Netting 1993; Moctezuma-Pérez 2013a, 2013b; Juárez-Choque 2015.

at any given time (cattle, sheep, and pigs). These small-scale agricultural systems can be highly productive (Gliessman 1998; McC Netting 1993), but they do not necessarily allow them to face such high expenses.

6.6 TRADITIONAL AGRICULTURE: DIVERSITY, AGRODIVERSITY, AND CULTURAL DIVERSITY

In 1999, the United Nations Organization (UN/FAO) defined the concept of biological diversity as the variety and variability of animals, plants, and microorganisms that are used directly or indirectly for food and agriculture, including crops, livestock, forestry, and fishing. It included the diversity of genetic resources (varieties, races) and species used as food, fodder, fiber, fuel, and pharmaceutical products. Also added in the diversity list the unharvest species that support, or harm crops and their production. These should include soil microorganisms, predators, pests, pollinators, and those in the wider environment that supported agricultural, pastoral, forest, and aquatic ecosystems, as well as their varieties.

De Avila (2008) stated as important the study of the relationship between ecological diversity, agriculture, and cultural diversity. This point had been relevant to the domestication process likened to the areas or zones with greater biological and biogeographic diversity, together with the traditional local knowledge. Different human groups have this knowledge about nature and its "codification" in the lexicon through the languages spoken by them. From ethnobiological studies, De Avila (2008: 497–556) listed 68 groups in Mexico that belong to 11 indigenous linguistic families, and two others that spoke Indo-European languages. Linguistics has been a useful tool for the studies on the emergence and diffusion of cultivated plants in the Southwest of the United States (Huckell 2006: 97–107).

Agricultural diversity is considered a subcomponent of biodiversity that participates directly in livelihoods and agricultural productivity, regardless of geographic context and resource availability (Blanco-Rosas 2006; Durán and Méndez. 2010; Smale 2006; Hunter and Fanzo 2012). According to Jarvis et al. (2011), agro biodiversity includes the biological diversity components, which are related to the varieties and variability of plants, edible mushrooms (Briones and Moreno 2019, N° 87), insects (Jarvis et al. 2011), worms, and other fauna. Microorganisms must be included on the scale of genes, species, and ecosystems because they are necessary to maintain the functions, structures, and key processes of agroecosystems (Jarvis et al. 2011; Robichaux and Yetman 2016).

About cultural diversity's components should be added. This includes the role that the family organization of family non-paid work (mutual aid), the size of the family farm, the demographic factor on farmer families, the role of kin, and friends in mutual aid, the available community workforce for agricultural purposes, and the role they all play in traditional agriculture. These components must include trade, market, tianguis, agricultural festivals and ceremonies, factors directly – or indirectly – intervening in crop selection, and varieties of all these components, like the food system. In addition, there are a number of external political, economic, and social issues that are related to land tenure. The farmers living in specific regions, the surrounding areas, and the access to natural resources and water for irrigation are also important.

The agrobiodiversity is associated with sustainable agricultural models, including agroecological (Gliessman 1984, 1999, 2007, 2012, 2013: 19–26, 2015), alternative, and biological, ecological, organic, permaculture, and others. These are based on the sustainability of the agroecosystems and on the local knowledge.

The rescue and conservation of native species contribute to traditional agricultural knowledge (TAK) and traditional environmental knowledge (TEK). Both are included as parts of traditional agriculture. In Mexico, traditional agriculture is destined for family consumption mainly and must be included in the family income diversification, in the economic, and the food security of families. In Mexico traditional agriculture is detined for family consumtion mainly, and must be included the formal education of family members and their non-agricultural occupations articulated to the family members' horizontal and vertical mobility, the age of family members who belong to different generations, and their subcultures which should all be taken into account, also (Blanco-Rosas 2006; Smale 2006; Hunter and Fanzo 2012; Wilk 1997).

FAO (1999) considers that agrodiversity is the result of the interaction between the environment, genetic resources, management systems, technology, and practices used by peoples of diverse cultures. Small-scale farmers have local and regional knowledge that was obtained through direct experience, trial/error, through the vicarious observation of others (Bandura 1977; Moctezuma-Pérez 2016: 79–94). Therefore, knowledge of the environment, agriculture, and local culture are integral parts of agrobiodiversity because it is the human activity of agriculture that shapes and preserves this diversity, which is biological and cultural. Some agroecologists, economists, ecologists, ecological anthropologists, rural sociologists, and ethno agronomists (Cruz-León 2016, 2019) have addressed these research problems.

6.7 RECENT DISCUSSIONS ON AGRODIVERSITY

For ecological anthropology, agrobiodiversity is a product of Man's intervention in the ecosystem, of his inventiveness and creativity, and his interaction with the natural environment. The practices of management, cultivation, and selection of species developed by farmers over thousands of years generated the diversity of cultivated plants and agroecosystems. Therefore, in addition to environmental, climatological, and physiographical factors, agrobiodiversity cannot be separated from the contexts, processes, and the cultural and socioeconomic practices that explain and define it. The effects of the economic policy and development plans of the Mexican State should be included (Cruz-León 2016; González-Jácome et al. 2007; Hernández-Matamoros 2016).

Agrodiversity results from an interaction process among Man, agriculture, and nature in several dimensions, including obtaining resources in natural systems. These natural systems have a strong human intervention-causing flora and fauna species to alter their production cycles, ergo adjust them to human interests. Calaza (2010) considers that in this sense biotechnology, biodiversity, and agrobiodiversity are constitutive parts of the same process in which Man's interests transform animal and plant species. Something similar occurs with the production of the food obtained from agricultural activity, plus other activities like fishing, hunting, and the capture of species destined for the family house and the food market.

From the origins of agriculture to date, a total of 300 thousand plant species has been estimated by FAO (2010), but only ten thousand have been used for human nutrition. Of these, 150–200 have been commercially cultivated, and four species are the basis of world food systems (rice, wheat, maize, and potatoes), which meet 50% of the energy needs for a human. Some 90% of the world's caloric intake comes from 30 crops, mainly. Of 15 thousand estimated species of mammals and birds, only 30–40 have been domesticated for the production of food, and less than 14 species – including cattle, goats, sheep, and buffaloes – represent 90% of the world production. In recent decades, these species have undergone genetic erosion, and some are at risk of disappearing or their status is unknown (Del Olmo and Roldán-Velasco 2013 [2007]; FAO 1997, 2015; Hunter and Fanzo 2012).

The diversity of indigenous peoples' agroecosystems generates a rich source of nutrients, mainly proteins, and vitamins (Becerril 2013; Gurri-García 2007: 60–68, 2011: 65–84; Otero et al. 2017). However, conventional farming practices are reducing agrobiodiversity. This, together with the increase in the consumption of industrialized food in homes, reduces agrodiversity and promotes components with negative impacts on nutrition (Santillani 2009). This loss of agrodiversity is caused by the substitution of local and traditional varieties, which are characterized by their wide genetic variability (Alarcón 2013; Gliessman 1988, 2015). Long-distance homogenization is closely related to the market and its need to transport agricultural production great distances, without suffering great losses in quality or quantity.

In recent years, literature has emerged on the relationship between the importance of agrobiodiversity and the availability of food for human populations. It has been highlighted that most of the food produced on the planet is obtained from small-scale agriculture (Alarcón 2013; Gliessman 1988, 2015). In this context, FAO (2009) questions the planet's ability to feed the 9,100 million people that are estimated to inhabit it by 2050. Agroecology in Mexico could become a possible solution to reach a sustainable and equitable development, since it provides the scientific, methodological, and technological bases to obtain a new "agrarian revolution" (Gliessman 1999; Hodgkin et al. 2010; Altieri et al. 2011).

Another answer to the problem of agrodiversity reduction and its impacts on the interrelations with flora and fauna has been suggested. The development of agroecology, with its contribution to the design and redesign of combined systems that, in addition to crops of flora and bevies of fauna, include cultural elements that contribute to sustainability. These are based on ancestral concepts and practices which have contributed to, and continue to contribute to, the sustainability of ancient cultures – both agricultural, and hunter-gatherer groups (Mollison and Holmgren 1978). For this proposal, the role of traditional agriculture and its history are fundamental.

On the other hand, agrodiversity or agricultural diversity is the axis of the food chain, and its use is linked to the nutritional security and health of human populations. It can be understood as a mechanism to combat hunger and a source of nutrients to have a basic diet with diversity and quality, strengthening local food systems, and environmental sustainability. For Hodgkin et al. (2009) and Santillani (2009), diet plays an important role in income generation and the development of environmental services.

The presence of a diversity of cultivated, tolerated, and wild plants, together with domestic animals in the cornfield system, generates the capacity of social and cultural adaptation to the adverse conditions of the environment (contributes to the system's resilience), served to cover specific family needs and provided farmers with a chance to achieve a successful survival in areas that are subject to environmental stress. Agrodiversity contributes to the sustainability of the system. The inhabitants' need for food producers and the market's need are not necessarily the same, making interrelations between the participants difficult.

Santilli (2009), in his "Agrodiversity and Farmers' Rights", established that agrodiversity is directly related to the sustainable production of food and its quality. A diversified diet is balanced in proteins, vitamins, minerals, and other nutrients. That is why nutritionists and health specialists recommended it. It can be can concluded that agrobiodiversity contributes to the provision of food, the diversity of diet, the obtaining of fibers, the purchasing of ingredients for medicines, and fuels. This generates of monetary income, also. Frison (2006) considered of importance the role-played by agrodiversity in mitigating the effects caused by micronutrient deficiency, mainly in children and women in developing countries (Cahuich 2012; Frison et al. 2004; Toledo and Burlingame 2006; Burlingame et al. 2009; Johns et al. 2009).

Toledo and Burlingame (2006: 477–483) believe that nutrition and agriculture are interrelated on the road to food security, along with the goal of achieving the "Millennium Development Goals". In their article on biodiversity and nutrition, these authors considered that biodiversity has three levels: ecosystems; the species they contain; and genetic diversity within species (Toledo and Burlingame 2006: 477–483). That is why diversity can contribute to achieving food security and better nutrition.

The Food and Agriculture Organization of the United Nations (FAO) and the International Plant Genetic Resources Institute (IPGRI) proposed the "Convention on Biological Diversity", to promote the sustainable use of biodiversity in programs that contribute to food security and human nutrition, awareness of the importance, and the achievement of sustainable development (International Council for Science 2002). Toledo and Burlingame (2006) said that more research is required to fill the gaps of present knowledge with better inventories, and more information on the composition and consumption of cultivated plants. In addition, a systematic analysis of nutrients, the dissemination of various edible plant species, diversity and knowledge of interspecies in relation to edible plant varieties, and subspecies of edible animals (food system), including domesticated and wild birds, along with aquatic fauna.

The World Declaration and Plan of Action for Nutrition adopted in 1992 by the International Conference on Nutrition, recommended promoting diet diversity, and the use of local, indigenous, and traditional foods as a strategy against food insecurity, and malnutrition. Hence, that nutrition, and biodiversity seek food with nutritional security, thus achieving sustainable development (International Council for Science 2002). The role of wild and intraspecific species in global food security made it possible to manage public policies that guarantee the use of local food (Becerril 2014; Bharucha and Pretty 2010).

The proposals to obtain food without deteriorating the environments where it is produced are diverse. One of the most important for Latin American countries is that proposed by agroecology. Peasant agriculture, based on the principles of agroecology, is characterized by: (1) maintaining high levels of biodiversity in farmland (plots); (2) using ingenious management systems and technologies in landscape production; (3) contributing to local food security; (4) increasing the resilience of agroecosystems; (5) applying peasant knowledge and technological skills; (6) and maintaining the cultural values of the population (Barrera Bassols and Floriani 2018; Fernández et al. 2002; Gliessman 1988, 2015; Koohafkan and Altieri 2010; Toledo and Barrera Bassols 2008).

Agrodiversity includes both biological and cultural diversities. It encourages the appropriate management of new technologies so that they do not impact the environment drastically but reduce the negative effects of the climate. Moreover, it respects peasant ideologies and their role in the culture of these agricultural societies (Gliessman 1988, 2015; Koohafkan and Altieri 2010; Toledo and Barrera Bassols 2008).

The continuity and permanence of diversity and cultivation systems of several ethnic groups are directly linked to their cultural foundations; production processes transmitted generationally based on beliefs about life and religious ideology; and direct biological maintenance through life in food system production (Ferraz de Rocha 2009; Woodley 2002; Zazur et al. 2003).

Then, the role that agricultural diversity has – among others factors – is acting together with traditional ecological knowledge (TEK), the local cultural knowledge of small-scale farmers, and traditional agricultural knowledge (TAK). It can be proposed and discussed that the relationship of Men with nature when he observed, collected, experimented, domesticated, and learned to cultivate and adapt plants and animals to his necessities, interests and tastes have been shaped over time.

There is also the proposal of Sarker (2017: 47–52) to form a new branch in cultural anthropology, termed by him as *Agricultural Anthropology*, with the aim of using "[...] anthropological methodology especially to study the conservation of the agrobiodiversity, participatory, and collaborative research with farmers, technology adoption behavior of farmers, and the historical study of some dangerous diseases, and insect pests" (Del Amo 2000b). For Sarker (2017: 47–52), "Agricultural Anthropology addresses a techno-economic combination that explained the sociocultural, and ideological components, and their interaction with the natural environment".

This proposal reflects the need for studies that include both ecological and biological factors, as well as social and cultural factors that are articulated in the study of agriculture. The proposals are numerous, and some of them show studies that make them promising in relation to the objective of studying complex phenomena and processes in systems, which are of great complexity, like agriculture. There are new proposals based on the need to study traditional agriculture including human actions, management, and interventions in it, plus the forms of social, economic, political, and cultural organization that are related to this activity.

In conclusion, there is a close relationship between agrodiversity with local, popular, or folk knowledge; and generational local knowledge of small-scale farmers with the conformation of traditional agricultural knowledge, which has been generated from the coexistence between human communities, nature, culture, and agriculture. There is agrobiodiversity that generates relevant knowledge, culturally appropriate to use and manage the environment in ways that are more in line with nature, consequently contributing to well-being, whether in the food system, medicine, social, or cultural satisfiers, and others, like a "good life" and "living well".

In small-scale farmers' logic a "good life", "living well", "quality of life", and other local definitions are broad concepts with variations from one town to another, from one village to another, or from one region to another. The concept used by farmers can include several aspects that make the individual and his nuclear and extended family feel in harmony with kin, close friends, society, nature, and other inhabitants with whom the family lives, including such aspects like sports, music, ceremonies, rituals, festivities, and other activities, beliefs, and ideas about the world with which he and his nuclear family coexist daily (van der Ploeg 2008).

This possibility of creating, recreating, and transmitting knowledge depends on the continuity of the relationship those human communities – especially agricultural communities maintain with nature (Dehouve 1995). Traditional agricultural knowledge (TAK), associated with biodiversity and cultural diversity, results from the interaction between communities contributing to the development of complex, diverse, and adapted (adjusted) agricultural systems, and local or regional environmental, climatological, social, and cultural conditions in each territory, along with the knowledge of small-scale farmers. It is, in fact, a relationship between management scales, where one of the central ideas is the maintenance of the environment so that it continues to feed human populations, with their societies and cultures.

6.8 PREJUDICES AND TRADITIONAL AGRICULTURE

Some misconceptions about understanding small-scale agriculture are well known. One of the most widespread is that this type of agriculture is not profitable in economic terms and does not yield high profits. On the first issue, it must be analyzed the cases where small-scale production focuses on products that include family food, along with others that target certain types of markets, for example organic. In relation to productivity, research and studies of agroecologists, such as Gliessman (1998), have shown how small-scale agricultural systems can obtain high yields per unit area that are higher and far outweigh those resulting from large commercial holdings.

An example of academic prejudice is related to cultures and their agriculture, as it was developed and found in the Mayan case where Fedick (2015: 47) said that the first European studies on agriculture in tropical areas considered these environments to be hostile to this activity, that the soils were used agriculturally by slash and burn, only, that allowed a very limited number of crops (maize-beans-squash). Also, the soils were poor in nutrients because of their shallow depth, and their being of limestone. Therefore, the Mayan lowlands were characterized as very limited for agricultural production because the lowlands Mayan landscape was poorly suited for plowing (Hernández X. et al. 1987: 338–340; Hernández X. and Levy 1995).

After the 1970s, archaeological, geographical, ethnobotanical, anthropological, and ethno agronomical studies found that: (1) the total human population in the region was greater than the carrying capacity that a slash and burn agriculture would sustain; (2) food system production went beyond the initially considered three crops ("Mesoamerican triad"); (3) there were alternative crops, such as tubers and *Brosimum alicastrum* (Ramón), with some flavor like cacao was used to make flour; (4) other intensive farming practices were found, like terraces on the hillsides or the conversion of wetlands into raised and drained fields for various crops; and (5) there was genetic adaptation of March Maize (*marceño*, an early maize maturation), which was cultivated in wetlands (Gliessman 1999; Mariaca-Méndez 1992:69–76; Orozco 1999: 111–122; Siemens 1983a, 1983b; Siemens and Puleston 1972).

Soil studies conducted by specialists showed that in these Mayan lowlands were some of the best soils, with great agricultural potential, and that there was great local and regional variability in the soil-mosaic of lands with different agricultural potentials (Fedick 1996, 2015: 48; Gómez-Pompa et al. 2003).

Aspects related to the organization of Mayan agricultural production are still unknown: if there were specialists who produced food for the elites; which groups did not participate in agricultural production; what were the differences between agricultural production for larger cities, such as

Tikal and Calakmul, and that of rural areas; what were the differences between agricultural producers and consumers; and what environmental, demographic, and political changes impacted agriculture requiring a setup of it (Alayón and Gurri 2007: 243–260; Beach et al. 2003; Dunning and Beach 2000; Fedick 2015: 47–48; Gill 2000).

6.9 EXTERNAL ELEMENTS TO CONSIDER

Production systems are discussed together with other components of the global system and their derivations. These include agro-livestock, agroforestry, permaculture, polyculture, and others (Liebman 1999: 191–203; Marzal 2007; Mollison and Holmgren 1978). It is important to consider the time and the scale these combinations had in relation to agricultural activity (González-Jácome 1988b, 1988c: 253–262, 1994: 1–12, 2008c: 1–17, 2010a: 141–158, 2012: 487–520, 2016c: 233–257; Martínez-Saldaña 1997; van der Ploeg 2008).

Also, there are important social and demographic aspects to regard such as the workforce available to families and the seasonal or periodic times to return to help them, as well as the emigration of some family members (Caloca-Rivas 1999). Another element to consider is whether internal (national) or external (international) migration affects the environment, and the small-scale agriculture in different degrees and forms, or if there are differences among these processes, small-scale farmers' families, and cultivation (González-Jácome 2010a: 141–158, 2011b: 71–111). The answer to a large extent will depend on the relationships among migrants, their families, and the migrants' hometowns; the monetary contributions that migrants send throughout the year (remittances); and their uses and applications (Cortés Villavicencio 2012).

There are diverse cases in this regard. For example, small and medium-scale farmers' families from towns in Guanajuato migrated in steps to Fort Worth, in Dallas, Texas (Andrade-Torres 2004). Migrants send money to those who remain in the village, and thus Mexican families live without working in any activity, because their ejido and private lands were rented to the transnational company Green Giant. So their children spend their childhood in Mexico and went to school there. When they reach adolescence, they were transferred to Fort Worth where they learn English usually and spend the rest of their lives in the USA.

Remittance from migrants to USA does not reach agriculture in the St. Barbara region, although originally the migrants from there were small-scale farmers living and working on granted ejido lands. The St. Barbara, Guanajuato people periodically return to town to visit relatives and friends, in addition to participating in important familial and town parties and celebrations (Andrade-Torres 2004, 2008).

There are other cases of migrants in the USA who send remittances to families still living in the town. A part of the money of the remittances that the migrants send to their families in Tonatico, a town to the south of the State of Mexico, is applied to the houses construction or reconstruction of the family home. Another part is for the family's subsistence in the hometown. Annually, these migrants return to attend the feasts of September 27, the celebrations and the fair organized for the festivities of the patron saint (Candelaria virgin), or February 2, for those of the shrine of "Our Lady of Tonatico".

The February 2nd celebration originally was an agricultural one, dedicated to the Candelaria virgin. The original feast was to ask for a good harvest and bless the seeds that would be sown. These ideas and customs have already been lost, because the town had ceased to be agricultural primarily and currently the neighbors are mainly engaged in tourism. Tonatico has hot springs, a spa, and the waterfall of Zumpantitlan, and is called Sun's Park, from where it is possible to see a waterfall and the suspension bridges (Cortés-Villavicencio 2012).

In the USA, there is an organizing committee for the festivities in Tonatico, including the equestrian events (*jaripeos*), cockfights, and folk-dance presentations. There are stalls for selling typical Mexican food, footwear, clothes, candies, and candied fruits. Salesmen come from various states of the Mexican Republic to sell handicrafts like wooden pieces, basketry, jewelry, and saddler (Cortés-Villavicencio 2012). Migrants live in the USA but have continuous communication with their families in Tonatico. They had a network using the Internet where everything related to the town, parties, and families are transmitted. They pay fees that cover most of the annual hometown party expenses. The events are videotaped and circulate among those who contributed money for the celebrations and who, due to family or personal circumstances, were unable to attend (Cortés-Villavicencio 2012).

There are other cases – such as the one in S. Lucas Tecopilco, Tlaxcala – where the money from remittances earned by legal seasonal agricultural work in Canada was applied by migrants to purchase agricultural machinery and assist in the modernization of local agriculture in their town of origin. Migrants learned to use agricultural machinery and grow wheat and barley in Alberta, which has huge areas with monocultures. They worked near Montreal and Ottawa, also. Agricultural workers are hired every year for a few months to work in the cultivation, care, and harvest of cereals or vegetables. There are given temporary seasonal work visas, but they cannot stay in Canada to live and must return to their villages (Binford et al. 2004; Caloca-Rivas 1999).

This legal seasonal migration has existed for several decades, and many workers have been retired already, but their children still have a working relationship with the Canadians who own the agricultural land. The money made in this seasonal migration is taken to Mexico and was used for mutual help in several activities: fixing the floors and walls of the town church, paving the streets, putting electricity and drinking water in the houses, conditioning their homes to store agricultural machinery and vehicles, buying trucks and freight transport, breaking the boards that separated the system of terraces and semi terraces, and compacting soil to level the water canals. With the purchase and help of machinery, agriculture was mechanized, so now new migrants sow barley and wheat as monocultures.

6.10 CONSIDERATIONS AND CONCLUSIONS

There is a long list of cases and their uses of remittance money.[5] In several of them, seasonal work paid for goods and different activities linking them to nature, including agricultural activities. A part of the cause of the bankruptcy of small and medium-scale Mexican farmers is farming itself, but not always because of a lack of productivity. It seems that they had a better income from other activities than that of small-scale farming, thus allowing them to have a better standard of living in Mexico. Traditional agriculture has not given farmers the support to get loans for the improvement of their *ejido* lands or enough monetary resources to have a better standard of living (González-Jácome 1988c, 2003b, 2007a, 2010a; van der Ploeg 2008).

Traditional agriculture is characterized by components which are contrasted with those of conventional, mechanized, commercial, export agriculture, which is economically dominant, even if in the long term is not ecologically sustainable. This happens because the objectives of conventional agriculture are to maintain productivity and to improve cultivated species genetically by utilizing cutting-edge technology to increase earnings, having utter disregard for small-scale agriculture, with its respect and concern for the habitat (Almeida 1999). In some historical moments in Mexico, state policy has been openly opposed to the existence of small-scale agriculture, which has lacked economic support and has been fought against in favor of exportable monocultures, which are a source of immense earnings for producers (Anthology 1985; Hernández-Matamoros 2016; González-Jácome 2007a).

Another issue to consider is that the money obtained by migrants and sent to peasant families in Mexico, does not always reach agricultural activity, and when it does is not applied necessarily to strengthening traditional agriculture. In fact, that is happen is the turning agroecosystems such as the cornfield or the homegarden into land for monoculture of barley and/or commercial crops. In the Tlaxcala Block, terraces have become land for the cultivation of commercial fruit trees, abandoning the cornfield and the cultivation of maize. In Sonora, soybean cultivation has become more and more widespread in areas that were previously for polyculture crops. In

[5] This year (July 2020) despite the Covid-19, migrants have sent more than 4 billion dollars to their families in Mexico.

Michoacán, the expansion of the avocado orchards had been carried out on what were forested areas, which had been deforested.

As the prices paid to small-scale farmers for their traditional crops are so low, they will abandon traditional agroecosystems. Add to this the fact that young generations do not want to work as farmers and are engaged in different economic activities, away from agricultural land because they have an urban mentality and have other economic expectations and ways of life (González-Jácome 2016: 347–350).

Consequently, the abandonment of traditional agriculture seems to be inevitable. Another factor influences this situation and is related to migration to the USA, and the wage differences between the USA and Mexico. Some authors considered that a process of peasants disappearance occurs, which organization and functions have other conditions and forms of organization of production like occurred in the 1990s with the so-called urban and peri-urban agriculture with values shared by a community that generates higher yields, including maize and beans (van der Ploeg 2008).

As long as the differences between the dollar and the peso are so great, and the prices paid to domestic agricultural products grown on a small and medium scale are so paltry, or they would not add other commercial characteristics that increase their exchange value, traditional agriculture will not lead to a life of well-being ("good life") for a very high percentage of the rural population in this country that still is practicing it.

Furthermore, the information from various small-scale farmers groups and organizations in various states of Mexico like the Teenek people of S. Luis Potosí; the Totonac Nahuatl Indigenous Union from the North Sierra de Puebla; the Huichol community El Roble in Nayarit; and 20 towns in the state of Guerrero, including Chilapa, Ahuacuotzingo, and Zitlala (García-Martínez 1987). Alvarez Buylla-Roces et al. (2011: 6) considered that the "producers" and their families' experiences must be important for achieving and developing a comprehensive agroecological approach that protects native seeds.

RECOMMENDED READING

Kirchhoff, Paul. 1943. Mesoamérica. Sus límites geográficos, composición étnica y caracteres culturales. *Acta Americana* 1: 92–107.

According to the proposals of the American anthropologist Alfred Kroeber, the German-Mexican historian, and anthropologist Paul Kirchhoff conducted research on the cultural area he called Mesoamerica using a large number of historical sources and archaeological studies available in the 1930s and early 1940s. He defined Mesoamerica through various components: those cultural features that are exclusive to this cultural area, those that it shares with other cultural areas, and those that were not Mesoamerican, but were integrated into this cultural area. The results of Kirchhoff' research were published in 1943.

Palerm, Ángel, and Eric Wolf. 1955. The agricultural basis of urban civilization in Mesoamerica. In: Julian H. Steward (ed.), *Irrigation Civilizations: A Comparative Study*. Pan American Union, Science Monograph No 1, pp. 28–42.

Text based on discussions on the importance of irrigation for the development of Mesoamerican civilization. Palerm classified the agricultural systems in slash and burn, fallow, and irrigated. After, he compared and related agricultural systems to population density and types of settlement.

Sanders, William T., and Price, Barbara J. 1968. *Mesoamerica: The Evolution of a Civilization*. Random House.

This book is about a theory of civilization, which included the growth processes that shaped the cultural history of Mexico and Guatemala long before the arrival of the Spanish in these lands. The authors used the concept of cultural evolution to express the exact meaning of the development of a civilization and its culture. Mesoamerican civilization was a high culture, a society as complex as those of Sumer, Egypt, China, and the Indus

Valley in the Old World. The book showed how the rose of a civilization could be seen as a process of ecological adaptation, a process independent that separated it from simple societies through highly productive irrigation agriculture.

The book is divided into three parts and ten chapters: Part 1 deals with civilization as an ecological system; Part 2 is about the evolution of Mesoamerican civilization; Part 3 includes the summary and conclusions.

Chapter IV The Dynamic of the Ecosystem (pp. 74–97) in Part 1 of the book is recommended, where the authors discussed three basic processes that played a central role in the evolution of New World civilizations: (1) Population and Civilization; (2) Population and Agriculture (including the nutritional value of various foods per 100 grains of weight); and (3) Competition and Cooperation.

7 Studies of the Origins of Agriculture in Mexico

7.1 INTRODUCTION

There are several theories about the origin and spatial mobility of the first human groups (bands) arriving in the Americas about 35 or 40 thousand years ago (MacNeish 2007: 57–67). The weather, geo-environmental conditions, flora, and fauna influenced the formation of several routes and corridors from where the first hunter-gatherers moved from Asia and Beringia to the American Continent. Dikov (1987: 47–57), a Russian archaeologist who studied Beringia described the elements that the first settlers brought with them from Asia: the knowledge of fire, domesticated dogs, Clovis-type spearheads, and some technological stone elements associated with hunting such as scrapers.

One question is what is, or not, the concept of resource, it does change over time and from one place to another. This is more relevant regarding ancient times, when mobility and rapid communication of human populations were hindered. Vicarious observation of others was their mean of acquiring folk knowledge, through personal experience and the transmission of what was learned by trial and error. The processes of vicarious observation, invention, experimentation, dissemination, and parallelism generated a food system based on hunting, fishing, and the collection of flora and fauna species. The invention of agriculture in the Americas enriched the cultural features, daily life, and food system of the first migrants to the continent.

7.2 BACKGROUND

Scholars discussed the origins of agricultural activity and its relationship with the domestication of plants as beginning with vegetative reproduction or by cutting. This has been very important in tropical regions, probably, where the branches and fruits of numerous plants fall reproducing naturally as a result of wind and heavy rains. Grain-based agriculture experience required other circumstances, like observation, trial and error (initial experimentation), and the accumulated folk or local knowledge. In Mexico, it was found that it developed in dry and semi-desert tropics or in regions, places, and areas near natural watercourses and reservoirs. Grain agriculture arose in Mesoamerican prehistory around 8990-8610 BP in the middle basin of the Balsas River (Hastorf 2009: 4957–4958; McClung de Tapia 2000: 251–280; Piperno et al. 2009: 5019–5024).

7.3 VEGETATIVE AGRICULTURE AND GRAIN AGRICULTURE

The vegetative reproduction or cuttings (banana, yam, and taro) of agricultural origins seemed to be older than that of grains, as happened in Papua, New Guinea (from near 7000 BC to 10000 BC). Denham et al. (2003) multidisciplinary studies placed its origins from 6950 BC to 6440 BC, having less antiquity than those found for maize domestication in the Americas. Lenfer and Boyd (2018) traced the antiquity of banana cultivation on Watom Island, in Papua, New Guinea, in 2300 BC. Piperno and Pearsall (1998) calculated a similar chronology for the New World (Lentfer and Boyd 2018).

For the first settlers, vegetative reproduction process observation of the flora was relatively easy to put into practice, using those that were selected for their characteristics (e.g., size, taste, and smell). There are other factors to consider, such as their growth, their duration before spoiling, maturation time, and the ease of collecting and transporting. Plus other characteristics related to issues such as scarcity or abundance in the areas where the collection was done as happened with

DOI: 10.1201/9781003198833-7

the ethnic group of Kung! San, which lives in the Kalahari Desert (Lee 1979). There is also the proposal that both agricultural systems (cutting and grain) arose more or less at the same time; however, more information is necessary to solve it.

The study of the origins of Mexican traditional agriculture includes several points of view. For example, on agroecosystems and the origins of agriculture, Harris (1969: 3–15) considered that the domestication of crops with vegetative reproduction and with starchy roots preceded the domestication of seed crops, which took place in the tropical lowlands, where the dry season covered from 2.5 to 2.7 months year. For the aforementioned author, the probable areas for the emergence of agriculture were the Orinoco basin, the South American Caribbean coasts; Guinea in Africa; the Ganges lowlands; and much of the Indochina peninsula in Asia.

Harris (1969: 3–15) used as evidence the fact that about 1000 BC maize in Venezuela and Colombia replaced root crops. He also pointed out that seed crops are likely to spread faster than root crops. Although they present the problem of depleting the soil to a greater extent, they provide a more balanced diet. The origins of agriculture are explained from the study of the evolution and history of the first agricultural systems, the processes of invention, domestication, and dissemination of domesticated plants (Archaeology Lab n/d).

7.4 MESOAMERICA, AND THE FIRST PLANTS DOMESTICATED

The origins of agriculture in Mexico started with the domestication of teosinte grass (*Zea mays* subsp. parviglumis) that was the initial cultivated plant at the beginning of a new food system that after some millenniums will be the basic crop of human populations in several countries of the world (Staller 2010; USDA 2013). The cultivation of grains, together with the agriculture of cuttings, was the base of traditional agriculture in Mexico and other countries of the continent. In the tropics and semi-desert areas of current Mexico, the domestication of teosinte occurred almost parallel with other plants.

As a matter of fact, squash was the first cultivated plant in places like Oaxaca. The cornfield agroecosystem had its starting with teosinte, and some trees like Mexican hog plum (*Spondias*), lead tree (*Leucaena*), and other cultivated plants corresponding to both grain production systems and root crops of cuttings or vegetative reproduction.

According to Iltis (1972) and Doebley (1980, 1990), hybrids between maize and several teosinte taxa are easy to do because the pairing of the chromosomes goes from good to excellent. The evolution between *Zea mays* subsp. parviglumis produced *Mexican Zea* (Schrader) Kuntze Raza Balsas. Wilkes (2004: 10) stated that of the 49 races of maize found in Mexico, 7 had counterparts in Guatemala, 6 in Colombia, 5 in Peru, and 2 in Brazil. Of those 49 races, 30 are endemic in Mexico, 6 are included in the Guatemalan Mayan zone, and another 12 races increased the sum. Other specialists, including Linares and Bye (2011: 9–12), gave the figure of 60, as the number of races of maize from Mexico, which is shared by other researchers[1].

7.5 DOMESTICATED MAIZE DIFFUSION IN MESOAMERICA

There are – at least – two hypotheses about the origin in America of cultivated plants, especially maize. Wilkes (2004: 5) proposed that maize was domesticated in several places on the continent because there are more than 20 native species of the *Tripsacum* genus. Wilkes (1967, 2004: 5) also argued that *Zea* and *Tripsacum* hybrids are feasible because there is gene exchange but that F_1 generations may be marginally fertile. Sarukhan-Kermes (2009: 8) considered that maize was domesticated in Mexico, and the process "[...] occurred simultaneously in various regions, and spread practically throughout the national territory, inhabited by dozens of indigenous groups [...]".

---

[1] Grobman, Salhauana and Sevilla 1961; Sánchez and Goodman 1992; Stross 2006: 577–598; Grobman, Bonaviab, Dillehayc, Piperno, Iriarte, and Holst 2012; Grobman 2013; Sánchez et al. 2000; Wilkes 2004: 10.

Wellhausen et al. (1952, 1957) proposed that the racial diversity of maize in Mexico was due to four factors: (1) the preservation of primitive taxa; (2) the influx of exotic races from the south of the country; (3) the hybridization with the teosinte; and (4) the varied habitats and insulating factors in the Mexican geography that led to a rapid evolution. Since the 1930s, Vavilov (1931), Mangelsdorf (1974, 1981), Beadle (1932, 1939, 1972, 1980), and others affirmed the Mexican origin of the domestication of maize.

The *Palomero Toluqueño* and *Chapalote-Nal-Tel* races are descendants of the lineage founders. Mangelsdorf (1974) considered that due to their geographical dispersion, it is feasible that these initial lineages were domesticated in several places. Randolph (1976) suggested that more than one species of wild maize were involved in the domestication origins.

In this second hypothesis, Wilkes (2004: 10) considered that teosinte was domesticated and dispersed from Mexico to the rest of America, following several routes: (1) Oaxaca, Chiapas, coast in the Pacific; (2) Migration from the Central Highlands, through places with higher altitudes of the types belonging – mainly – to the Conical group. They had an extension to the north, to Chihuahua, with different races. (3) Migration of the Pepitilla and broad maize race, which goes from the north of Morelos and Guerrero to the Pacific coast, crossed the Isthmus of Tehuantepec, and the Tuxpeño race reached the coast of the Gulf of Mexico. According to Hernández X. (in Wilkes 2004: 12), Pepitilla is a "key" race to understand maize in Mexico; (4) and from the highlands of Guatemala, where migration was divided into two routes: one to the north of Mexico, and another to South America.

Linares and Bye (2011: 9–12) proposed other routes like that from Guilá Naquitz in Oaxaca, which went to Tehuacán in Puebla, and from there to Ocampo cave, in Tamaulipas. This same route started in Xihuatoxtla, Guerrero, from where it reached Guilá Naquitz, in Oaxaca, and continued to Tehuacán and Tamaulipas. There was another route to Central and South America that passed through Belize, Honduras, and Costa Rica. There were at least two or three to reached the north of the continent, to the United States and Canada: (1) by Jalisco, Sinaloa, Sonora, Arizona, and New Mexico; (2) by the Central Highlands to Chihuahua; (3) and by Tehuacán to Tamaulipas (Carot y Hers 2011) (Figure 7.1).

There is enough information to propose an initial diagram, which is modifiable to the extent that more archeological, archaeobotanical, phytolithic, and pollen studies arose in addition to more accurate chronologies.

Before maize was a fundamental element for human populations, the food system was based on hunting, fishing, and capture of species in the forests, areas with vegetation, wetlands, lowlands, swamps, mangroves, marshes, lakeshores, lagoons, rivers, and other aquatic ecosystems.

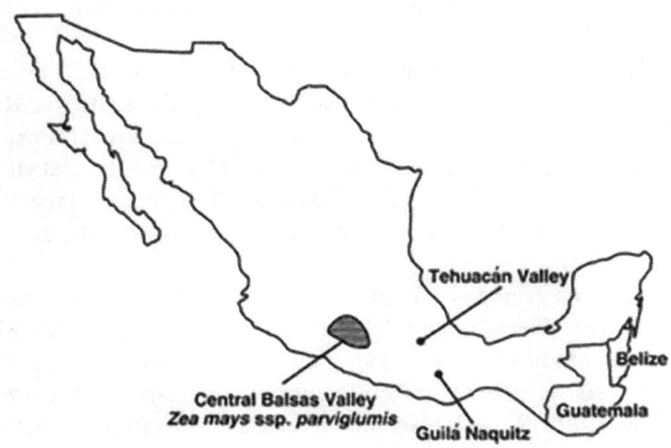

FIGURE 7.1 Map: Places of the Origins of Maize Domestication in Mesoamerica.

Human survival was based on the consumption of fish, crabs, river shrimp, oysters, octopus, crustaceans, turtles, mollusks, waterfowl, and other species was vital for humans (Cyphers 2004: 21–40; Rodríguez-Galicia 2017).

This activity remained important to date and still occurs in towns such as in villages of the Atasta peninsula, or Arena Island, in Campeche (Gómez-Nava 2005; Méndez-Guevara 2004), in the coastal towns of the Yucatan Peninsula, in the S. Luis Potosí, in the Huastec, or in towns such as Palenque in Chiapas. Fishing has been an important activity, along with the capture of aquatic species (Gómez-Nava 2005; Méndez-Guevara 2004), and had been fundamental since immemorial times (González-Jácome 2002: 261–280, 2010: 141–158; Ochoa and González 2009a: 145–168). In the basin of Mexico, along with the hunting and gathering of aquatic fauna, it existed until the relatively recent lagoon system desiccation (Espinosa-Pineda 1996).

The basin of Mexico lake system undergoes continuous drying processes, but it subsists until the first half of the twentieth century (Aréchiga-Córdoba 2011: 2267–269; Correa-Ortiz 2011: 203–225; Rodríguez-Galicia 2017). Although very damaged, the relics of the old lakes in Xochimilco-Míxquic remain. It is necessary to include the fish that arrived from the coast of the Gulf of Mexico to the central Highlands through trade. This has been studied since Teotihuacán, and it was described for the Tlatelolco market in 1540 (Sahagún 1830). Vegetable species and Spirulina algae were collected in the basin of Mexico. This activity allowed fisherman and agriculturalist to learn about edible species, how they reproduced, at what seasons of the year when, and how they were collected (Francisco Hernández 1958 [1763]).

Traditional Mexican agriculture began with the domestication and dispersion of plants found southwest of the Mexican Central Highlands, in Xihuatoxtla, Guerrero, and a semi-arid zone of the dry tropics in the middle basin of the Balsas River. The human groups were organized in bands living in rock shelters where Piperno et al. (2009: 5019–5024) collected phytoliths, pollen, teosinte starch, Pipian squash (*Cucurbita argyrosperma*), native Mexican hog plum (*Spondias* sp.), and Lead tree (*Leucaena esculenta*).

7.6 THE MIDDLE BASIN OF THE BALSAS RIVER

This dry tropics region in the middle basin of the Balsas River has a climate ranging from semi-dry very warm to dry very warm, with summer rains. The aridity condition is accentuated to the east of the basin. The mountainous systems with their altitudes produce that the Balsas along its route passes through an ecological mosaic, characterized by a great variety of altitudes, climates, soils, flora, and fauna (Fernández-Nava et al. 1998 (9): 1–151). Grain farming arose in the 9000 cal BP, in the early Holocene, when there is information on the maize and squash domestication (8990 BP–8610 BP) (Piperno 2011; Piperno et al. 2009: 5019–5024; Smith 1997: 932–934, 2001b: 1324–1326, 2006: 12223–12228) (Figure 7.2).

The Balsas basin runs through a depression descending from the Tlaxco Sierra, north of Tlaxcala (Zahuapan River), to join the south of Puebla state, with the Atoyac River; from where it descends transversely, in an east/southwest direction to Oaxaca, crossing through the Tributaries Rivers of Tlaxiaco, Mixtepec, Juxtlahuaca, and Mixteco; Morelos; the Cutzamala tributary river in the State of Mexico; Guerrero; Michoacán (Tepaltepec, Cutzamala, Tacámbaro or Carácuaro, Tributary Rivers, and Jalisco) until it flows into the Pacific Ocean, with the name of Zacatula River (Rodríguez-Galicia 2017: 18).

The average altitude of the average basin of the Balsas is 1,000 masl. The basin crosses two different physiographic regions (Sierra Madre del Sur and the Neo-volcanic Axis Province), together with the altitudinal slope; produces climatic variations; and has several types of vegetation: alpine grasslands, pine-oak forests, xerophyte thickets, and tropical forests (Rodríguez-Galicia 2017: 18).

Fernández Nava et al. (1998 (9): 2–3) research on the Balsas river middle basin, with a river total route of 771 km, described the tropical deciduous vegetation 50 km north of Iguala. This is combined with savannas and scattered trees of *Bursera*, *Ipomoea*, and *Acacia*, which extended along the

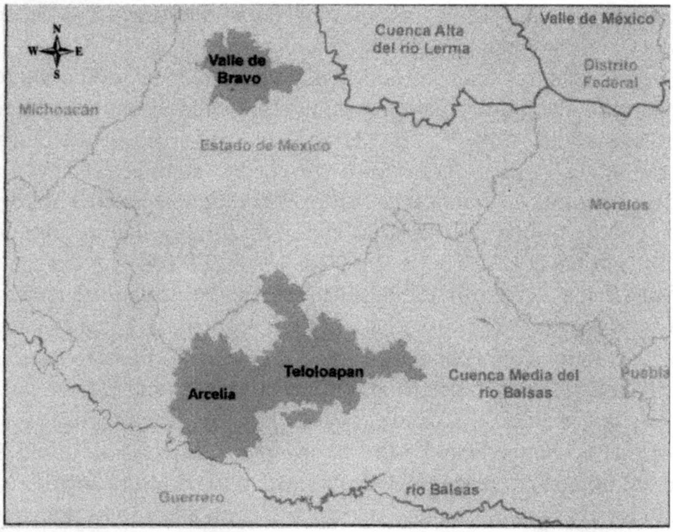

FIGURE 7.2 Map: The Middle Basin of the Balsas River, and the Teosinte Parviglumis Region.

slopes of the hills and stream banks. It has two tree strata the lower one with 8–15 m high, and the upper one from 15 to 35 m high, with an average 25 m. The main species of upper tree stratum are *Enterolobium cyclocarpum, Andira inermis, Hura polyandra, Ficus maxima, F. pertusa, Bursera simaruba, Platymiscium lasiocarpum, Lonchocarpus eriocarinalis*, and *Cedrela odorata*. On the lower slope of the Southern Sierra Madre, the most frequent species are *Ficus goldmanii, F. maxima, Andira inermis, Licania arborea, Aphananthe monoica, Brosimum alicastrum, Thouinidium decandrum, Inga vera, Lysiloma divaricatum, Cordia elaeagnoides, Pterocarpus orbiculatus*, and *Dalbergia granadillo*.

Much of the basin has a semi-dry warm climate; in the east, there is greater aridity, with xerophyte vegetation predominance. Fernández Nava et al. (1998 (9): 2–3) found that the tree main species in Guerrero are *Lysiloma divaricatum, Cordia elaeagnoides, Cyrtocarpa procera, Tabebuia impetiginosa, Haematoxylon Brasiltto, Amphipterygium adstringens, Bursera copallifera, B. coyucensis, B. discolor, B. fagaroides, B. hintonii, B. kerberi, B. trimera, Ruprechtia fusca, Vitex pyramidata, Pterocarpus orbiculatus, Cochlospermum vitifolium*, and *Ceiba aesculifolia*.

The secondary vegetation main species are *Acacia angustissima, A. cochliacantha, A. farnesiana, A. macilenta, A. pennatula, Haematoxylon brasiletto, Gliricidia sepium, Cordia curassavica, C. elaeagnoides, Piptadenia flava, Mimosa polyanthus, Senna skinneri, Caesalpinia eriostachys, C. platyloba, C. pulcherrima, Pithecolobium acatlense, Mimosa egregious, M. guatemalensis, M. langlassei*, and *Guazuma ulmifolia*. The researchers affirm:

> According to this floristic list, there are 202 families, 1,246 genera, and 4,442 species. The best-represented families are Compositae with 136 genera, and 573 species, Leguminosae with 90 genera, and 486 species, Gramineous with 84 genera, and 253 species, and Orchidaceous with 51 genera, and 197 species. The states best represented to date are Guerrero, Michoacán, and Morelos. Their biodiversity, and high degree of endemism make them undoubtedly one of the regions of the country of greatest botanical interest, little known today, and that needs to be studied.

Some 8–10 thousand years ago, the Prehistoric human population was organized in itinerant bands, which were made up of six to ten people, moving seasonally through the territory and living on hunting, fishing, and gathering. These human groups sucked or chewed the young teosinte reeds to obtain their sugary juices, or drank those obtained from their fermentation. They consumed the edible parts of the grass (Benz and Staller 2006: 29; Blake 2006: 68–69; Iltis 2006: 25; Smalley and Blake 2003: 675–689). In the early nineteenth century, Alexander von Humboldt (1822) on his trip to Mexico wrote that maize sugar was an important product consumed in Mexico, at least until the end of the Viceroyalty and the beginning of the nineteenth century.

Studies on the Balsas middle basin showed that maize domestication commenced from parviglumis (Balsas' teosinte). The hunter-gatherer bands that carried out this domestication lived in the mountainous reliefs south of the Central Highlands and north of the central Rafts between Teloloapan and Arcelia. This region goes from the current state of Guerrero to Valle de Bravo in the State of Mexico, and its altitude ranges from 900 to 1,400 masl. The region had large and dense populations endemic to the *Zea mays* L. spp. parviglumis that developed over millions of years (Iltis 2006: 25; Vargas 2007a, 2007b, Vol. 30, No. 3: 97–104).

To the south of the Balsas River middle basin in the current state of Guerrero, the teosinte was domesticated, and its dispersion began with the hunter-gatherers. Linares and Bye (2011: 9–12) considered that changes in teosinte began when the grass was modified to have a single main stem instead of several branches. Several small cobs were modified, and a few cobs appeared but are larger. The previously wrapped grains were exposed. The initial number of two rows increases to four. The number of grains changed from 6 to 300 or more. The teosinte ceased disperses itself and becomes dependent on humans to disperse it, and they had to trim it to sow it.

Seven to eight thousand years ago, a mega genetic mutation allowed its domestication by modifying the external glume, reducing its size, and softening it, therefore making possible its human consumption (Doebley 2004: 37–59; Doebley and Steec 1991; Dorwelier, Steec, Kermicle and Doebley 1993: 233–235; Iltis 2006: 29).

For Kaplan et al. (1992), teosinte (or the domesticated maize) traveled along with other plants and their domesticators to the north and south of the continent. According to Buckler et al. (1998: 151–164), the Mesoamerica on the southeast route began with the Proto maize in Guilá Naquitz cave, and Silvia rock shelter, near Mitla, in the Oaxaca central valley (6250 BC).

According to Iltis (2006: 29), the softening of the external glumen in teosinte architecture (TGA) was the outcome mega genetic mutation. It made the grain accessible to be harvested by flattening the maize dome, and tilting the raquis and kernels so being naked this facilitated their domestication, removal, and harvesting. This process removed the ability of maize to reproduce by itself; this was left to hunter-gatherers who discovered it. Doebley et al. (1993: 233–235) considered that the gradual selection of teosinte improved its size, number of rows with kernels, qualities, and nutritional value; and that occurred from a previous maizoid after TGA modifications.

For Buckler and Holtsford (1996: 459–465), the fossil evidence found to date showed that domesticated teosinte was already in a common form in 6250 BC. Apparently, from the middle basin of the Balsas River, teosinte traveled south to reach Tabasco, crossing the central valleys of Oaxaca and the Tehuantepec Isthmus. It would be necessary to study whether teosinte traveled from its place of origin, to the east to the Toluca valley, and to the lagoon system in the basin of Mexico.

7.7 ARCHAEOLOGICAL AND GENOMIC EVIDENCE

Plant geneticists indicate that the first races of maize arose from teosinte (*teocinte*, *teocintle*) or *Zea mays*, or a mixture of both. In Xihuatoxtla, a rocky refuge that is located in a tropical deciduous forest, Piperno et al. (2009: 5019–5024) found remains of pollen and phytoliths of the teosinte parviglumis (Balsas' teosinte) in addition to pollen from domesticated maize. Currently, no teosinte grows there, near Xihuatoxtla. It grows a little farther north in the same middle basin of the Balsas.

Piperno et al. (2009: 5019–5024) considered that both teosinte pollen and maize pollen are not distinguished, precisely. *Zea* pollen grains date from the late Pleistocene, and they were recovered in Lake Ictaxyola (20 km west of Xihuatoxtla), so this suggested that during Prehistory, about 8700 BP, teosinte occurred in this region. The phytoliths sample indicated that in the Xihuatoxtla refuge, teosinte was not consumed as a grain or as a stem. The researchers argued that the lack of evidence on the use of teosinte in other places in Mexico, and the fact that the hard glumen tightly enclosed its grains made wild maize unsuitable for consumption as food, by hunters/gatherers and the first growers. They proposed that teosinte was exploited and cultivated for the sugary sap inside the stem, and, or for its inflorescence and tender grains.

To give strength to this argument, Iltis (2006: 25), Blake (2006: 68–69), and Smalley and Blake (2003: 675–689) considered that the magnet that led teosinte to have acceptance and subsequent diffusion by the human population, and its subsequent consumption and domestication was the possibility of obtaining sweet sap from the stem and using it to made a fermented beverage (a kind of beer), a product that was important in the ceremonial life of human communities. This hypothesis seems feasible given the evidence found. However, it still needs proof. A similar process occurred with other plants, such as cacao, whose process began when it was used to make a fermented beverage. Chocolate from cacao beans happened much later (Henderson et al. 2007: 18937–18940).

Henderson et al. (2007: 18937–18940), based on chemical analysis data of extracted residues of ceramic vessels collected in Puerto Escondido, Honduras, demonstrated that the first cacao drinks were made with the sweet mucilage of the cacao cob that houses the seeds. They found that, before 1000 BC, and at least until 500 BC, before the cacao beans were extracted from the mucilage, cleaned, dried, roasted, and ground; the beverage was made with the fermented in water mucilage. It was later when the rainforest inhabitants began using cacao beans to make chocolate. Several fertilized drinks made with plants – or their fruits – were made since ancient times, although there is no precise information about their antiquity.

7.8 DOMESTICATED MAIZE IN TROPICAL REGIONS: VERACRUZ, TABASCO, CHIAPAS, AND HONDURAS

Archaeological and palaeobotanical studies of cores and excavated places in S. Andrés and Tabasco provided evidence that maize pollen, phytoliths, starch granules, and Sunflower seeds were cultivated since 7300 BP (Pohl, Piperno, Pope, and Jones 2007: 6870–6875). Piperno and Pearsall (1998: 314) announced an age for maize pollen as 7100 BP. They found what had been considered as the earliest Mesoamerican glyph where maize is shown (Pope et al. 2001; Pohl et al. 2002). Piperno and Pearshall (1998: 314) announced a date for maize pollen as 7100 BP. That pollen was found in a place near the Veracruz port (Sluyter and Dominguez 2006: 1147–1151). They argued then, "[...] humid, low, tropical lands were the main places for the origin, and development of agricultural systems [...]" (Sluyter 1994: 557–584).

Several cores and excavations near La Venta, in Veracruz, showed evidence of maize pollen, the first cultivated sunflower seeds, and the aforementioned maize glyph, considered being the oldest of those early times (Pope et al. 2001; Pohl et al. 2002). Although Piperno and Pearsall (1998: 314) announced 7100 BP as the date of maize pollen found near Veracruz and maize pollen in S. Andrés, Tabasco dated 7300 BP, they argued, "[...] wet, low, tropical lands were the main places for the origin, and development of agricultural systems [...]".

This idea is contradicted by the discoveries in Xihuatoxtla; therefore to elucidate the question it is necessary to wait for more studies. According to Pohl, Piperno, Pope, and Jones (2007: 6870–6875), it is feasible that the pollen found in Tabasco is the same as the Balsas' teosinte, and that it got there by a route that passed through Oaxaca and the Tehuantepec Isthmus. That could be done following the natural corridor of the Gulf coast (Wilkerson 1987: 212–214).

In the St. Marta Ocozocuautla cave in Chiapas, in the Mayan zone, macro and micro botanical evidence of the age of domesticated maize was found (Acosta-Ochoa 2008, 2010: 1–4, 2011: 227–235). This gave information on the transition initiated with the collection of wild plants in a broad spectrum until an increase in the number of domesticated plants was reached. The study reported the presence of *Zea* pollen, with a date ranging from the late Pleistocene (10,456±50 years ago, to 10,050±90 years ago), and *Zea* starch grains associated with grinding stones, with a date of 9800±50 years BC (Early Holocene) (Acosta-Ochoa 2008, 2010: 1–4, 2011: 227–235). Chistom and Blake (2006) found evidence of maize grown in Soconusco, Chiapas, as 4400 BC.

Kennett et al. (2017: 9026–9031) gave data on the Gigante rocky refuge in the Honduran Mayan area beginning with the molecular and micro botanical study of 10 thousand macrofossils of maize and 37 well-preserved rachis. The most primitive kernels (4340 BP to 4020 BP) had 10–14 rows, suggesting a strong selection to increased yield. The last ears (2350 BP to 980 BP) were comparable with the first ones but showed the appearance of various other features, like the increase in the width of the cob, the length of the segment that corresponded to the rachis, and the width of the dome containing the maize kernels.

These research data indicated that the domesticated native maize varieties were productive enough to be a basic grain for human populations, and already existed in Central America in 4300 BP. There is a lack of information on this matter to provide materials and analyzes going a clearer idea of what happened in these regions, especially in the Mayan area, in relation to the origins of agriculture in Mesoamerica, and the domestication of maize there. The earliest maize macrofossils in South America were dated on 6775 BP to 6500 cal BP). On the domestication, and diversification of maize, Kennett et al. (2017: 9026–9031) conclude:

> [Domestication] had resulted in local varieties with remarkable genetic, and morphological diversity adapted to a range of geographical restrictions, and climatic conditions. A number of 59 creole varieties of maize are documented, and there is morphological variability in other parts of North, Central, and South America. The existing local breeds known represent only a fraction of the diversity that existed during the evolutionary history of this globally important grain. However, this diversity is currently threatened by the proliferation of genetically modified (GM), and improved hybrid varieties, as well as by the associated loss of traditional agricultural knowledge, culture, and practices [...] In the context of global climate change, germplasm in the seed banks, and the macrophiles archived in the archaeological deposits provide an important source of genetic information in the face of growing environmental instability [...]. Less is known about domestication, hybridization, local adaptation, and diversification of maize in the Americas after 4950 BP, as well as the pace, and character of subsequent changes as Mesoamerican farmers switched to food production maize-based in the context of state-level partnerships between ~3500 and 1000 BP.

Smalley and Blake (2003) and Blake (2006) studies in Tlacuachero, a site in Acapetagua municipality in southern Chiapas, settled at 30 masl, indicated that there was already maize, avocado, two varieties of beans, and Opuntia cultivated by 2800 BC. However, the food system remained strongly composed of terrestrial fauna: squirrel, armadillo, rabbit, wild boar, dog, opossum, white-tailed deer iguana, and snakes. The food system was completed with fish, hunted or caught in the estuaries; crocodiles; *masacuate* snakes; turtles; crabs; shrimp; clams; catfish; crappie; lizard, and numerous fishes that had not been identified.

Besides maize domestication, there was one that occurred with other plants, some of which were domesticated before grain and other plants, which accompanied it, including those grew along with wild weeds or tolerated weeds (Altieri 1992: 262–281). It should be taken into account that fruit trees, Agaves, and cacti in general were an important part of the diet of the first settlers, especially in dry and semi-desert regions of the tropics like the middle basin of the Balsas River, the Tehuacán valley, or the present-day Northwest Mexico.

For the consumption of the Agave, the first settlers created open ovens dug in the earth, where the aforementioned plants were cut into pieces, roasted or toasted, and cooked in the ovens, to be consumed in the form of sweet, sugars, honey, and fermented drinks (Evans 1990: 117–132; González-Jácome 2018: 21–74). In the cases of Agave and Opuntia, both domesticated and wild specimens were found in the explored archaeological sites, located in several regions of the country (Sonora, Oaxaca, Tehuacán, and Tlaxcala). The villagers used both; Agave to produce honey and sugars, and to make fermented beverages, Opuntia, for human consumption, was used both for its fruits, as well as for the cladodes, which were eaten as vegetables.

This combination between domesticated plants with weeds, and wild existing in Mexican agricultural systems continues to date (Vibrans 1997: 21–67). Actually they are in the abandonment process caused by factors such as the application of herbicides in agroecosystems. Examples are the wild weeds, the cultivated herbs and the seasonal green edible weeds (*quelites*), the small wild green tomato, some varieties of wild chili (e.g., the wild Piquín chili in Sonora, which is collected in the Alamos municipality), and the domesticated and wild native Mexican hog plum (*jobo*, *hobo*), which are still collected and consumed.

7.9 DIFFUSION AND DISPERSION OF DOMESTICATED MAIZE

There are several hypotheses about the diffusion of maize to the Mexican south and southeast. One of them supported the feasible idea of a multiple origin (parallel invention) of domesticated and cultivated maize. Although theoretically it is feasible, to date there is a lack of evidence to confirm it. The second hypothesis is the idea of teosinte dispersion from the middle basin of the Balsas River. This had a greater number of studies and scientific evidence than the first, but the possibilities of obtaining more information – with better samples and chronologies for Central and South America – are not ruled out. For now, archaeological, palaeoecological, palaeobotanical, and plant genetic materials showed an origin for the maize domestication and dispersion in Mexico, older than in other parts of the continent, considering that the data obtained for earlier maize found in El Gigante rock shelter in Central America (4340 and 4020 cal BP) is less older than calculated data for the Balsas River teosinte (Kennett et al. (2017, 114 (34): 9026–9031).

Kennett et al. (2017, 114 (34): 9026–9031), summarizing the work of more than 55 researchers, specialized in genetics, archaeology, geography, studies on maize, phytoliths, pollen, and agricultural origins, conclude that (1) genetic data indicated that maize was initially domesticated from the annual herb known as teosinte (*Zea mays* ssp. parviglumis), which currently grows from 400 to 1,800 masl in the Balsas region, southwest of Mexico; (2) tests of phytoliths and starch of archaeological sites in this region indicated the early use of maize in at least ~ 8700 BP cal; (3) in addition to a widespread dispersion of this domesticated grain, through the Neotropics in the Mayan lowlands, which occurred shortly after this time; (4) a consistent observation is that the earliest maize macrofossils in South America occurred from 6775 BP to 6500 BP cal. (Bird 1982; Kennett et al. 2017: 9026–9031; Lippi 1984; Pearsall 1978: 177–178).

7.10 THE CENTRAL VALLEYS OF OAXACA

Archaeological evidence on the domestication of maize after Xihuatoxtla (8990 BP to 8610 BP) was found in S. Andrés, Tabasco (7100 BP), and in the central valley of Oaxaca near Mitla, where the Guilá Naquitz cave and the Silvia rock shelter are located. In the cave, three ancient inflorescences of *Zea mays* were found; for Guilá Naquitz, the selection of domesticated teosinte was dated in 6250 BC and 5400 BC. Two specimens, one with two and other with four-grain rows in the cob concurred. In addition to other morphological features in the ears, which supported the hypothesis, based on molecular analysis and quantitative analysis, that maize evolved from teosinte, and the domestication of the wild ancestor occurred before the end of the fifth millennium BC (Benz 2001a: 2104, 2001b, 98: 1324–1326; Doebley 1990: 6–27, 2004: 37–59).

After Xihuatoxtla (8990 BP to 8610 BP), a rachis was found in Panama (7800 BP), where maize could reach along the coast from Belize and Costa Rica. From the central valleys of Oaxaca, the teosinte (parviglumis) already domesticated took two paths: one traveled in the direction of the Tehuacán Valley, and the other toward the south to the Mayan zone. From there to Central America, on routes that took the current Belize, crossing the Corozal swamp (5575 BC to 5341 BC) and following the coast to the St. Rosa swamp (ca. 5000 BC), where cassava and maize pollen were found. Passing through Guatemala, and Honduras (El Gigante rock shelter (4340 AP to 4020 AP), maize was able to reach South America.

Grobman (2013) affirm that maize was domesticated first in Peru, where the oldest macrofossil of maize was found in two sites of the Chicama Valley, located on the Peruvian coast. It dates from Late and Medium Pre-ceramic Period, ca 6000 to 3000 calibrated years BP, and the maize remains included maize cobs, husks, stalks, and tassels[2]. In Paredones and Huaca Prieta, maize was domesticated, and there is evidence of its previous origin in the Sierra. Data confirmed on maize domesticated are: Huaca Prieta from 7200 AP to 2300 calibrated AP, and Paredones from 6700 to 4000 calibrated AP (Grobman et al. 2012: 1755–1759, 2013).

It is feasible that the squash (*Cucurbita argyrosperma*) accompanied maize on this trip. However, new questions arose since in Guilá Naquitz remains of another domesticated squash (*Cucurbita pepo*) were also found, which are dated at 6980 BC; that is to say, that its domestication is older than that of maize, at least in the central valleys of Oaxaca (6250 BC) (Benz 2001b, 98:1324–1326; Bruce 1997; Buckler et al. 1998, 39(1): 151–164). This had been published as a result of other studies, among others, those of Bruce D. Smith (1997 (276): 932–934), where the researcher considered an antiquity for the domestication of squash (*Cucurbita pepo*), about 10 thousand years.

Archaeological information indicated that before the Spaniards' arrival in the continent, *Cucurbita pepo* L. was widely distributed throughout Mesoamerica, northern Mexico, and the southwestern United States 7000 years ago (Huckell 2006: 97–107). Historical information indicated that it was grown in the center and east of the USA (Villanueva 2017: 15). Studies by Villanueva (2017) indicated that in Mexico there are 26 cultivated and wild species of *Cucurbita* genus. Bear in mind that the pumpkin was sown more for its seeds than for its fruits, whose taste is bitter. The nutritional value of the seeds is high as that they are rich in protein (40%) and oil (40%). Villanueva (2017: 15) stated that squash is distributed according to altitude: *Cucurbita argyrosperma* and *C. moschata* in warm places from 0 to 1,800 masl; *Cucurbita pepo* at altitudes from 1,000 to 2,200 masl; *Cucurbita ficifolia* in places whose altitude varies from 1,300 to 2,400 masl.

The results of the Guilá Naquitz cave and the Silvia rock shelter research showed that by 6000 BC, the human population food system included: maize, squash, domesticated chili peppers along with wild chilies, wild beans, avocados, Agave, Opuntia prickly pears, tropical sapodilla, wild onions, and berries (Kaplan and Kaplan 1992; Perry and Flannery 2007: Vol. 104, No. 29: 11905–11909). Evidence was found that the first tortillas appeared until the Formative or Preclassic from 500 BC to 300 BC. In the Formative or Preclassic, the villagers' food system included avocado (*Persea americana*), squash (*Cucurbita* spp.), beans (*Phaseolus vulgaris*), and maize (*Zea mays*). All of this was strongly complemented with wildlife resources obtained in the forests, rivers, estuaries, wetlands, lowlands, swamps, surrounding land areas, and even coastal meanders (Chisholm and Blake 2006: 161–172).

Iltis (2006: 29) found that Guilá Naquitz maize appeared in the Coxcatlán cave in Tehuacán (5310 BC–4700 BC±110) about a thousand years after its arrival in the Oaxaca central valleys. Then it began its adaptive and evolutionary stage in the Tehuacán Valley and southeast of the current state of Puebla. From there it spread to the rest of the Mexican Central Highlands, going north of the basin of Mexico and arriving later at the Ocampo cave in Tamaulipas. It is not clear what route teosinte took from the middle basin of the Balsas River to the Basin of Mexico, but it could have reached Valle de Bravo (north of the middle basin of the Balsas River), the Toluca Valley, and the Basin of Mexico (Zohapilco 3090 BC±115). Human bands could have it arriving from Tehuacán, Tlaxcala, and Texcoco, or they could have used both routes.

7.11 THE TEHUACÁN VALLEY

The Tehuacán Valley is located 150 km southeast of Mexico City. It extends to Cuicatlán, with a portion adjacent to northern Oaxaca. The studies and descriptions of Whiteford and Enge (1989), and MacNeish (1992, 1997: 81) established that the arid valley has a dry climate, and temperatures ranging from 18 °C to 22 °C, that its rainfall varies annually from 250 to 500 mm, concentrated

[2] Grobman, Salhauana and Sevilla 1961; Sánchez and Goodman 1992; Grobman, Bonaviab, Dillehayc, Piperno, Iriarte, and Holst 2012; Grobman 2013; Sánchez et al. 2000; Wilkes 2004: 10.

from May to October, with the most intense rains occurring from June to September. For Rzedowski (1978) and MacNeish (1997: 81), the Tehuacán Valley is one of the few arid zones located below the 20° parallel, and its main vegetation consists of xerophytes (Mora-Jarvio 2010) (Figure 7.3).

According to Rzedowski (1978), the types of vegetation in the valley are deciduous tropical forest, xerophytic scrub, thorny forest, grassland, Quercus forest, coniferous forest, mountain mesophyll forest, sub-deciduous forest, and aquatic and underwater vegetation. In addition, Rzedowski (1978) included other types such as palm grove and gallery forest (Tehuacán-Cuicatlán Biosphere Reserve, Management Plan 2013).

The valley limits are to the northeast and east with the mountain range of Zongolica, to the southeast with the Mazatec mountain range (2,600 masl), to the south and west with the Zapotitlán mountain range, and in the Highland Mixtec or mountain range of the Mixtec. About 10,000 years ago, the valley had an arid and semi-arid zone, with a semi-warm semi-dry climate.

Rzedowski (1978) considered that this region is the center of evolution of the flora, Agave species and the columnar cactus, which reaches 15 m. in height. Earl Smith Jr. (1967: 133–142) found that 29.1% of the 253 species collected in the valley are endemic, with at least three genera of phanerogams that are exclusive to the zone.

Fernández-Nava et al. (1998) described the xerophyte scrub of the region, a dense thicket that is located on the slopes of the hills, formed by bushes of *Castela tortuosa*, *Schaefferia stenophylla*, *Gochnatia obtusa*, and *Fouquieria Formosa*. Shrubs of *Cercidium praecox* are interspersed; the associations of *Escontria chiotilla* with large extensions of spiny thickets of *Celtis pallida*, *Randia armata*, and *Schaefferia pilosa* are common.

Aquatic vegetation has tulars and reeds consisting of *Typha latifolia* and *Scirpus californicus*. There are submerged aquatic plants such as *Myriophyllum hippuroides* and *Potamogeton illinoensis*. The floating vegetation is *Lemna gibba*, *Azolla mexicana*, and *Eichhornia crassi*pes, who live in fresh, moderately brackish waters in calm places or in places with currents adapting to them. According to Fernández Nava et al. (1998, 9: 2–3), the most representative aquatic plants in the riverine basin in the Tehuacán valley are *Salvinia auriculata*, *Pistia stratiotes*, *Heteranthera limosa*, *Nympha odorata*, *Callitriche heterophylla*, *Ludwigia erecta*, *Marsilea mollis*, *Neptunia natans*, *Polygonum amphibium*, *Ranunculus hydrocharoides*, *Utricularia gibba*, and *Najas marina*.

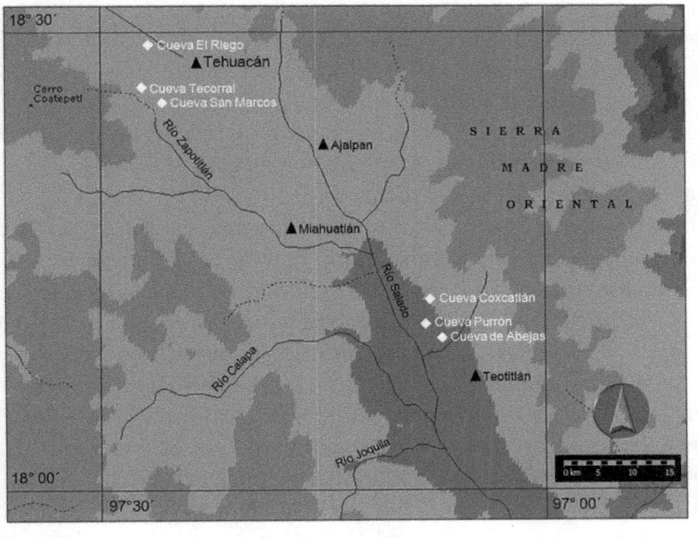

FIGURE 7.3 Map: The Tehuacán Valley, Puebla.

The Prehistory of the valley had a sequence of occupation that spans about 12 thousand years. During this time, the newcomers got to know the environment and how to manage it, exploring the different plants, especially the edibles, and developing their food preparations. These settlers developed lithic artifacts and created new technologies; elements of material culture, artifacts whose vestiges facilitated the reconstruction of how they lived, and the use they made of natural resources. Those first phases of occupation showed disturbances in some areas, due to extractable activities to regional flora and fauna.

About 11 thousand years ago, the inhabitants began selection, and with it the domestication of some plants, among them the Agave, whose stems, leaf bases, and peduncles of the inflorescences were baked, roasted, or tamed to feed them (Smith Jr. 1967, 1986; MacNeish 1967; Flannery 1986).

The human population was scarce and seasonal. MacNeish (1997: 81) and the researchers of his interdisciplinary team studied the valley since 1961 and considered that the regional environment was able to sustain a limited population, whose nomadism was conditioned by the seasons. From 7000 BC to 5000 BC, during the optimal climate, the vegetation of Tehuacán was a tropical deciduous mountain forest, semi-arid, with mesquite and cacti, lasting until 1,500 masl. Lauer (1979: 37, 39) said that from 2500 BC to 2000 BC, the vegetation in the higher altitude areas was mixed forest, with *Pinus* and *Quercus*.

Apparently in the last 10 thousand years, there was no drastic climate change, but geological evidence indicates that a gradual decrease of the water table occurred throughout that period. At that time, the nipple cactus (*Coryphantha pallida*) and lead tree (*Leucaena esculenta*) were widely used by the villagers. According to Brewbaker and Sorensson (1990: 2) and Earl Smith Jr. (1967: 220), this legume – native to the Neotropical regions – was found from what is now Texas to Peru, and its pods were an important part of the food system for local populations in Prehistoric times.

MacNeish (1997: 82) evidence showed that from 6800 BC to 6500 BC bands formed by four to eight people lived in caves and changed their camps from one place to another about three or four times a year. They were bands of hunters and trappers of smaller animals (hares, moles, rats, snakes, turtles, birds, lizards, and small mammals). They collected plants, and in Coxcatlán, they began to domesticate some, such as avocado (6500 BC), thus beginning the selection process (Mhammed et al. 1997: 279–286; Millar 1996: 71–122; Schroeder 1968: 29–34). Anderson (1967: 93) wrote that the instruments used by hunters were made of lacquered stone (knives, and foliated projectile points, scrapers, cutters, and razors). They had mortars to crush their food. They did not use polished stone objects.

Seasonal mobility prevented the production of food from agriculture, so the plant selection process was based on the taste of the inhabitants for the consumption of certain fruits and parts of some plants. Accidentally people formed areas creating garbage dumps, with remains of plants and with the waste they tossed on the ground. This is the most remote antecedent about the beginnings of agriculture in the valley (incidental domestication). For Richard MacNeish (1992, 1997: 82), it must be considered that the weather was colder than today.

Shortly after than onset, there is evidence of several plants that were used (Smith Jr. 1967, table 26: 232): *Setaria* cf. *macrostachya*, an unidentified grass seed, which served as food; the *huautli* so-called by the Nahuas (*Amaranthus* spp.); avocado (*Persea americana*); legume web (*Prosopis juliflora*), whose roasted pods and seeds were eaten; American mastic or American pepper seeds (*Cyrtocarpa procera*); Prickly pear cactus (*Opuntia* spp.) from whose cladodes and prickly pears were consumed; and a palm (*Brahea dulcis*) used to obtain fiber. There were two other unidentified pastures, which had uses that have not yet been established (Early 1992: 17–199). With the increase in a sedentary lifestyle, several phases or stages can be established and organized according to the materials found in several caves and sites studied in the MacNeish project.

In Phase 1 "Coxcatlán" (5137±32 years BC), the oldest human occupation in the valley occurred. The human remains found in the Purrón and Coxcatlán caves suggested that in the dry season, the inhabitants' food system subsisted on what they hunted and in the rainy season on what they collected (Anderson 1967: 93). In this stage, the first incidental homegarden arose, combining plants in different stages of domestication, wild and induced; several of them were for fibers. The irrigation cave was occupied from 6500 BC to 5000 BC. The remains of several plants used by the population were found in it and identified: mesquite tree gum (*Prosopis juliflora*), gourds (*Lagenaria siceraria*), prickly pears (*Opuntia* spp.), and Hechtia (Smith Jr. 1967: 133–142).

The transition from hunting/gathering to agriculture began in this phase. Semi-sedentary human groups inhabited the valley. Fifty to ninety people composed a band, which was divided into smaller groups. From 5300 BC to 5000 BC, maize was in its initial domestication processes, already having a rachis and a central stem that differentiated it from teosinte, but it maintained similar characteristics to teosinte – such as having a limited number of seeds –, and genotypes like its ancestor. This was a transition phase to domestication, so maize was partially domesticated there (CINVESTAV 2016).

The food system of the inhabitants of El Riego cave included: avocado; amaranth; sweet acacia pods; Mexican hog plum (*Spondias mombin*); *Opuntia* (prickly pears, and cladodes); two types of cacti; Agave; seeds of Baboso organ cactus (*Lemaireocereus hollianus*), whose seeds were eaten; Spanish physic nut tree (*Jatropha*); star apple fruits (*Sideroxylon* ef. tempisque Pitt-Sapotaceae); wild squash (*Cucurbita pepo*); cultivated squash (*Cucurbita mixed*); and chili (*Capsicum annuum*). There is evidence of the consumption of the seeds of foxtail (*Setaria geniculata*) grass seeds (Earl Smith Jr. 1967: 133–142).

There were remains of lead tree (*Leucaena esculenta*) and native cotton, or *pochote* (*Ceiba parvifolia*), whose roasted seeds were eaten. Recent studies on gourds show the spontaneous hybridization of legumes in incidental homegardens by a process favored or precipitated by human action that preceded domestication. The pods were collected and consumed in two seasons of the year (fresh and dried). Several of the seeds collected were roasted, and could be ground to form flour, as was done in northwestern Mexico, even until the eighteen century (Nentvig 1764).

In Phase 2 "Bees" (4177±47 BC), the population continued to be hunter-gatherers, but a lifestyle increased and populations began to locate in small camps and terraces on the valley banks, allowing them to have an agricultural production that maintained a population of 500–900 people. By then, maize had a limited structure, and the grains were partially covered by their glumes. In 4000 BC the White lead tree (*Leucaena esculenta*) was pre-domesticated, cultivated, and used as human food from the center to the south of the current Mexican territory (Castello-Yturbide 1986). This plant went through numerous independent transitions from its wild state to its domesticated state for cultivation.

Hughes et al. (2007: 14389–14394) stated that from lead tree (*Leucaena esculenta*) two to six local species and a total of 13 resulted. Its hybridization was extensive and complex as well as involved nuclei encoded in DNA sequences and diversity in the hybrids resulting from the juxtaposition of cultivated species. Collected plants were used as food; others were to obtain fibers, branches, their dry stems, and bushes to make fire; and gourds were used as containers.

In Tehuacán, the earliest evidence of domesticated amaranths came from this stage, when the hunter-gatherer bands have been consuming them for some 5500 years. A change from hunting and gathering made amaranth, together with pumpkin and chili, integrated 6% of the population's diet. In the next millennium, it reached 14% of the total food system. *Amaranthus cruentus* was grown in Mesoamerica, while *Amaranthus hypochondriacus* was cultivated in the Andes. The Spanish took the plant to Europe where it was used for human food and fodder (Quer and Martínez 1762).

Archaeological evidence showed that at the late stage of the Bees Phase (about 2300 BC), at the end of the Late Prehistoric Period (late Paleo-Indian or Early Archaic), shortly before the Formative or Preclassic began, in the Tehuacan Valley the construction of a hydraulic structure started. The "Purrón" dam was built, followed by the S. Marcos well, and afterward a series of "fossilized" line-canals appeared. These constructions facilitated agriculture, increasing their annual production cycles and food stability, like in the next phase.

In Phase 3 "Palo Blanco" (1560±249 BC), a radical change in agriculture arose with the appearance of irrigation ("systematic irrigation"). Villages with "ceremonial centers" were created. The population of the Tehuacán Valley was from 50 to 90 thousand inhabitants. As of that phase, the maize of the valley was similar to the current races, with a structured and massive rachis and bare grains. Some of the rachis that have been studied of that stage have been described and referred to as belonging to the "Chalqueño" breed, one of the oldest maize races in the Mexican Central Highlands (Linares, and Bye 2011; Hernández X. and Alanís 1970: 3–30; Wellhausen et al. 1952).

These irrigation works were the beginning of the management of hydraulic systems in Mesoamerica, where people adapted to the semi-arid environments in which they were built. The water management and agricultural intensification systems in the Tehuacán Valley are the oldest and studied in Mesoamerica to date.

The recent Neely et al. (2017) archaeological fieldwork proved that agricultural intensification occurred much earlier than previously was thought. Neely et al. (2017) dated the S. Marcos Necoxtla well, located in a small village in the central part of the Tehuacán Valley, from 7000 BC to 4000 BC, as the oldest well for water management, and the "head" or the starting place for the fossilized in-line canal system. Neely et al. (2017: 142–143) affirm:

> [. . .] Were skillful adaptations that allowed early water management and agricultural intensification well adapted to the topography, hydrology and arid surroundings in Tehuacán [...] In conclusion, it is clear that these three forms of water management [Purrón dam, wells, in-line-canal] were ingenious in their design and engineering, especially at the time of their construction and were skillful adaptations that allowed early access to water and promoted agricultural intensification in the variety topographic and hydrological of the arid environment in the Tehuacán Valley.

According to the evidence obtained by Neely and his colleagues (2017: 142–143), the use of new technologies and research methods applied to structures and components related to water management in the Tehuacán Valley consolidated an early date for its domestic use and agricultural intensification. Evidence suggested that agricultural intensification might extend up to 1000 years, earlier than currently indicated. The Purrón dam was a complex with small sites, where the houses were associated with corporate groups that cooperated and undertook large water management projects during the Formative Period (Aiuvalasit et al. 2010: 1207–1213).

These results had implications for understanding the causal links between political complexity and agricultural intensification. They supported recent ethnographic and archaeological research in Mesoamerica, as well as questioning the argument that only a growing social complexity leads to the construction of large-scale systems for water management. This idea was originated by the discussions of Ángel Palerm (1973, 2007) with Karl Wittfogel, and these discussions were very important to understand the role of water, the magnitude of the works, and their political control of them in relation to the type of urban society and sociopolitical system developed in it.

Phase 4 "Salt Sale" (419±130 BC), the Tehuacán Valley had a total population of 250 thousand people. Agriculture was well developed within a political, economic, technological, and religious structure. At this stage, the emergence of multiple races of maize began, such as the "Conic". The Tehuacán Valley maize was identical to the current breeds, showing a considerable increase in the number of grains, as well as in the size of the rachis, which meant an increase in the amount of available food. From the total selection of domestication characters and the implementation of new technologies, possibly this transformation resulted.

Phase 5. According to the authors of the CINVESTAV Project (2016) report, this phase corresponded to the current rachis. They consider it the only one for which studies had evidence of the influence of the processes that gave rise to hybrids, based on Shull's works that were done in the late nineteenth century. However, Shull did not describe those processes in detail. The report ignored that the processes with respect to maize developed in the Classic, Post Classic, Epiclassic, and Historical periods in Mesoamerica.

7.12 MAIZE DOMESTICATION IN TEHUACÁN: ACCORDING TO THE GENOMICS STUDIES (2009–2016)

The research group of the National Laboratory of Genomics for Biodiversity (LANGEBIO), of the CINVESTAV research institute located in Irapuato[3], in a symposium on ancient DNA (2009) presented a work titled "The Origin Of Maize: A Genomic Perspective". In it, they announced that some progress had been made in the project they were developing and affirmed:

1. Maize was domesticated from its wild progenitor, the teosinte (*Zea mays* ssp. parviglumis, or teosinte del Balsas), a grass native to Mexico and Central America that differs from maize in its importance as a plant, and a reproduction of unique traits.
2. In accordance with archaeological studies, the genotyping of multiple microsatellite loci indicated that a single domestication event occurred approximately 9,000 years ago in central Mexico, giving rise to a wide variety of native races through a constant diversified selection.
3. From a combination of complete genome sequencing by "shotgun" and sequencing filtered by gene enrichment, we sequenced the approximately 2 Gb of the genome of the Mexican breed *Palomero Toluqueño* and compared its characteristics with those of the improved variety B73. This comparison reveals differences that confirm its ancient origin, including small genome size, retrotransposon content, and a high frequency of repetitions of low complexity.
4. The assembly and annotation of the *Palomero* gene space resulted in the identification of large chromosomal regions of low nucleotide variability enriched with genes involved in plant responses to the environment. Our recent results open new perspectives for understanding the environmental changes that shaped the transformation of teosinte into maize.

In 2012, the Mexican archaeologist Ángel García-Cook joined the CINVESTAV research group. In 1962, García-Cook had conducted research in the Tehuacán Valley, along with Richard MacNeish and his team. They explored the Coxcatlán, Purrón, El Riego, Tecorral, and S. Marcos caves, discovering more than 24,100 maize specimens. In the CINVESTAV laboratories, three samples of the ancestral maize genome were studied, and their chronology ranges from 5000 BC to 5300 BC. This maize is considered the second oldest in Mexico. In November 2016, CINVESTAV published a report with the project results. The study was also published in the Proceedings of the National Academy of Sciences (PNAS).

The Ramos-Madrigal et al. (2016: 1) paleogenetic studies, based on a maize rachis found in Tehuacán, which is estimated to be 5310 BC, affirm that: (1) the complex evolution of maize from the genomic level to modern races and wild teosinte which has been documented by archaeological studies suggested that the maize domestication occurred from 6,250 to 10,000 years ago, in southern Mexico; (2) under human selection, maize reached "[...] remarkable phenotypic transformations, as well as adaptations to varied environments"; (3) many questions about the domestication process remain unanswered, "[...] because modern specimens do not represent the full range of past diversity due to abandonment of unproductive lineages, genetic drift, ongoing natural selection, and recent reproductive activity".

Through the use of D statistics, model-based clustering algorithms, and multidimensional scaling analysis, the palegenomic study compared this ancient sample of maize, with a reference panel of local breeds and modern teosinte. It was found that the studied specimen came from the same modern maize population origin. The researchers concluded that in 5310 BC, maize in the Tehuacán

[3] The investigators in the project were Jean-Philippe Vielle-Calzada, Octavio Martinez de la Vega, Gustavo Hernández-Guzman, Enrique Ibarra-Laclette, César Álvarez-Mejía, Julio Vega-Arreguín, Beatriz Jiménez-Moraila, Araceli Fernández-Cortés, Guillermo Corona-Armenta, Luis Herrera-Estrella, and Alfredo Herrera-Estrella.

Valley was – in general – genetically closer to modern maize than to its wild counterpart. But, many genes associated with key domestication traits existed in ancient maize, "[...] which contrasts with the ubiquity of alleles derived from native varieties". The genomic characterization findings suggested, "[...] that during domestication much of the evolution may have been gradual, and encouraging" process, which occurred from 10000 BC to 6250 BC in southern Mexico; but more palegenomic research is required. It represents a baseline lineage of teosinte, related to all modern races.

For Ortega-Paczka and Chamorro-Cobaxin (2017: 232), the current situation of maize is based on the proposal that both in Mexico and in Guatemala there is the center of origin of the domestication and diversity of maize, which explains the current existence of its 59 or 64 Mexican races. Recent studies show the origin of such domestication in the middle basin of the Balsas River, from where it spread to several parts of Mexico, including the central valleys of Oaxaca, the Isthmus of Tehuantepec, Tehuacán, and the Central Mexico Basin. Ortega-Paczka and Chamorro-Cobaxin (2017: 232) considered that this diversity had been maintained for 70 years, because:

> Seed selection to maintain the identity of maize populations and improve their characteristics plays a fundamental *in situ* conservation and evolution role. Although between local native populations crossing with exotic introductions are also very relevant. On the other hand, although it is not well studied, it does not seem that the gene flow of the teosinte, close relatives towards maize has been an outstanding evolutionary factor in the last century. In some warm low-lying areas, as well as in some of intermediate heights, improved maize has partly replaced native diversity, especially white-maize used for tortillas, without matching them in food quality. On the other hand, they have hardly replaced native maize for traditional special uses, nor those grown in large regions with soil limitations or low temperatures. On the other hand, in advanced regions, advanced generations of improved maize combined with native maize are part of the bicultural heritage of the communities of farmers, peasants, and indigenous people who manage these populations as they do with native maize.

The current inventory of managed species and forms of management of plants registered in the flora in the Tehuacán-Cuicatlan valley, according to Casas et al. (2017: 81) are: (1) total species 2,621; (2) useful species used by regional cultures 1,608 (about 54% of the regional flora); and (3) 39% of some use in the species in the register regional flora. The Tehuacán Valley had an important diversity in relation to the flora known and managed by its early occupying populations (Mora-Jarvio 2010). In addition, it must consider the cultural elements participating in the domesticated flora selection.

7.13 MESOAMERICAN: SOME ARCHAEOLOGICAL REMNANTS OF FLORA EVIDENCING USAGE AND EARLY HUMAN CONSUMPTION

In Guilá Naquitz, Oaxaca early levels (8000 BC to 6000 BC) were domesticated: acorns, edible almez berries (*Celtis* sp.), mesquite (*Prosopis juliflora*) dahlia, *Opuntia* sp., and *Cucurbita pepo*. In more recent levels, lead tree (*Lagenaria* sp.), avocado (*Persea americana*), and bean (*Phaseolus vulgaris*) (Flannery 1986). In the Brazilian and Peruvian Amazon, the first hunters/gatherers domesticated the *Cucurbita ecuadorensis* and cassava from 6000 BC to 5000 BC (Clement 2017: 27–42; Piperno, and Pearsall 1998). Before the Spanish entry, 52 exotic species were domesticated, 41 are semi-domesticated, and even there is an incipient domestication in 45 species.

In the Tehuacán Valley, from 3500 BC to 1500 BC, these domesticated species were *Setaria* cf. *macrostachya*, *Capsicum annuum*, *Amaranthus*, *Prosopis juliflora*, *Persea americana*, *Spondias Mombin*, *Opuntia*, *Arachis hypogaca*, *Canavalia*, pig bean, sword bean or machete bean (*Canavalia* sp.), green beans or big beans (*Phaseolus coccineus*), bean or white bean (*Phaseolus lunatus*), common beans, guava (*Psidium guajava*), cotton (*Gossypium hirsutum*), *Cucurbita pepo*, *C. mixed*, *C. moschata*, and *Zea mays* (Smith 1967; Long et al. 1989).

To the south of the basin of Mexico, in Zohapilco, the archeologist Niederberger (1979: 132–142) found the following domesticated plant species: goose foot (*Chenopodium* sp.), *Amaranthus* sp., green tomato (*Physalis* sp.), and teosinte (*Zea mays* ssp. mexican). These plant species were domesticated from 6000 BC to 2200 BC.

In S. Lorenzo, in Los Tuxtlas, Veracruz, from 1200 BC to 800 BC, Cyphers (2004) found: cultivated maize, tubers, coyol palm (*Acrocomia aculeata*), palms and other fibers used in rope making, gourds used as containers (*Crescentia cujete*), and numerous plants that settlers collected for food consumption. In S. Lorenzo site, the information for that time shows the greater importance of the fishing, collection, and capture of aquatic fauna, and the one coming from the marsh.

In northern Mexico, B.D. Smith (1997a) and MacNeish (1992) found some domesticated plants in the Ocampo caves in Tamaulipas like lead tree (*Lagenaria* sp), *Cucurbita pepo*, avocado, acorns, common beans, green beans, foxtail grass (*Setaria* cf. *macrostachya*), chili, Sunflower (*Helianthus annuus*) amaranth, and maize (Whitaker; Cutler; MacNeish 1957: 352–358; Whitaker 1986 [1965]: 344–349).

In the USA, Lentz (2000: 89–120) collected plants that were the background of domestication. In the site of Hester from 9000 BC to 6000 BC: hard nut hickory (*Carya* pp.), acorns (*Quercus* spp.), walnuts (*Juglans nigra*), hackberry (*Celtis* sp.), and *Prunus americana*. In Gainesville site: acorns and the American walnut nuts (*Diospyros virginiana*). In the Rodgers shelter: American walnut nut, walnuts (*Juglans nigra*), and almez (*Celtis occidentalis*). On the Koster site: squash, marshaled (*Iva annua*), American walnut, acorns, nuts (*Juglans regia*), walnut (*Carya illinoensis*), grapes (*Vitis* sp.), a perennial aquatic plant "smartweed or grass ready" (*Polygonum* spp.), quinoa (*Chenopodium* spp.), wild bean (*Strophostyles* sp.), foxtail grass (*Alopecurus* sp.), almez, black cherry (*Prunus* spp.), *Phytolaca americana*, hawthorn (*Crataegus* spp.), green heather (*Smilax* sp.), *Viburnium* sp., and Solomon seal (*Polygonatum* sp.).

7.14 CONSIDERATIONS AND CONCLUSIONS

Some archaeological plant remnants that had been found in present-day Mexico and the southern United States showed an early human use of local and regional flora. The list can be continued to the extent that more information emerged in this regard and is only intended to provide the reader with a general idea of this ethnobotanical wealth. For some specialists in the maize studies, there are doubts regarding its domestication origins and dissemination throughout Mesoamerica and the Americas.

An element in the understanding of the domestication and diffusion of maize processes and other cultivated species can – in some way – start from the flora diversity existing in the regions involved in the discussion and various studies carried out in this regard.

The group of the National Laboratory of Genomics for Biodiversity (LANGEBIO) made a historical review of maize origins in the Tehuacán Valley. They do not consider what happened during the Viceroyalty and the nineteenth century. Its report provided a good idea of the elements that arose from the maize domestication, which were discovered by these investigations in the Tehuacán Valley. It is important concluding that the Tehuacán maize came from the teosinte, from the middle basin of the Balsas River which arrived, and was domesticated in Tehuacán, from where it was spread to other parts of the Mesoamerican territory, and reached places so far as the Ocampo cave, in Tamaulipas, at the north of the country (MacNeish 1947, 1958, 1964, 1967a, 1967c, 1981: 31–47, 1992, 1997: 80–93; MacNeish, Whitaker, and Cutler 1958; Mangelsdorf 1958; 1967b: 178–200).

RECOMMENDED READING

McClung de Tapia, E.; D. Martínez-Yrízar, E. Ibarra Morales, and C.C. Adriano-Morán. 2013. Los Orígenes Prehispánicos de una Tradición Alimentaria en la Cuenca de México. In: *Anales de Antropología*, 48-I (2014): 97–121. México, UNAM.

For the human population, food is a fundamental part of their subsistence that can be defined as the resources that human beings need to satisfy their needs. Food is considered traditional when it preserves cultural patterns that have been passed down from generation to generation (forms of preparation, and consumption). Although new products have been incorporated making, the food system more varied (Castello-Yturbide 1986).

The study of plants that were used as food by ancient populations was done through botanical materials recovered from sediments from archaeological excavations. These materials are macroscopic seeds, fruits, and charred wood; and microscopic pollen, phytoliths, and starch granules. This work deals with some hypotheses regarding the pre-Hispanic use of food resources in the basin of Mexico through the comparison of archaeological data with data from sixteenth-century documents.

The authors found a recurrent set of plants in the domestic contexts of different periods including Postclassic: Otumba, Xaltocan, Tlatelolco, and Iztapalapa; Epiclassic: Michpilco, and Xico; Classic: Tetitla, Yayahuala, and Oztoyahualco; and Formative: Terremote-Tlatenco, Cuanalan, and Temamatla, suggesting a tradition of use over time.

Benz, Bruce F. 2001. Archaeological evidence of teosinte from Guilá Naquitz, Oaxaca. In *Proc. Natl. Acad. Sci. U.S.A.* (PNAS), February 13, Vol. 98, No. 4: 2104–2106.

The article analyzed three fragments of inflorescence from *Zea mays* that are the oldest found in the cave of Guilá Naquitz, Oaxaca, and Mexico. The sample was not naturally disarticulated, indicating that domesticated teosinte agricultural selection was underway in 5400 AP (14C). That means about 4,200 years dendrocalibrated BC.

The coexistence of two-rank, two-row, and four-row grain specimens and several additional morphological characteristics of these specimens support hypotheses based on molecular and quantitative genetic analysis that maize evolved from teosinte. The domestication of the wild teosinte occurred before the end of the fifth millennium BC. The morphological characteristics of three maize ears of Guilá Naquitz compared to the first three maize specimens from the S. Marcos cave in the Tehuacán valley (4750 BP 14C) showed that the two populations were not distinguishable for statistical reasons. Therefore, for 5400 BP (14C), and over the next 700 years, human selection focused on stabilizing the diffuse, non-disarticulating, bare-grain phenotype, and increasing the number of grain-containing spikelets per node from one to two. Guilá Naquitz and the first specimens from Tehuacán suggested that the human population sought to maintain or increase grain productivity.

8 Domesticated Maize Dispersion

8.1 INTRODUCTION

From early times to its domestication in the middle basin of the Balsas River, maize was dispersed in the north and south of the territory that would later form Mesoamerica. This dispersion started 8,700 years ago, continued over several millennia and centuries. Upon the arrival of the Spanish in the New World, shortly after 1492, Christopher Columbus took maize to Spain, from where it began to extent to various countries in the Old World. From Europe, maize dispersed by Asia and Africa becoming an important food product. That is, it became part of various food systems in the known world of the sixteenth and seventeenth centuries.

The consumption of maize as food generated various food systems, including those in the American continent. Today, maize has become an industrial product used as a substitute fuel for oil, as the basis for the industrialization of plastics of vegetal origin, and as the fundamental element of new products derived from this grass, its grains, leaves, reeds, roots, and other parts. The food industry in particular had created a series of products that supply family food systems, and that in some way have an impact on agroecosystems.

The organic input to the soils is an important part of traditional agricultural systems. When the Agave began to be used for fermented Agave mead extraction (*pulque*), the heart was scraped, and the waste was thrown to the ground, so the person in charge of the task added this biomass to the soil. Similar effects occurred in all agroecosystems, with specific areas that the farmer develops to manage the crop fields, such as the edges of land and areas with shore boards next to canals and surrounding lands with secondary vegetation, or with tolerated plants where organic input to the agroecosystem has beneficial effects on plants and soils. In soils, the input produces biological changes that contributed to the suppression of pathogens, making the soil and its microbiome a system that assigns important roles to biological components (Álvarez-Ríos, Figueredo-Urbina, and Casas 2020: 1–17; Vida, de Vicente, Carzola 2019: 1–15).

8.2 BACKGROUND

Confirmed genetic data indicated that maize was domesticated from the annual teosinte grass (*Zea mays* ssp. parviglumis), which grows today from 400 to 1,800 masl, in the middle basin of the Balsas River, in the southwest of Mexico (Van Heerwaarden et al. 2011; Matsuoka et al. 2002; Doebley 2004). The tests of phytoliths and starch of the archaeological sites in this region indicated the early use and domestication of maize at least in 8700 BP (Piperno et al. 2009; Ranere et al. 2009).

The widespread dispersion of this domesticated teosinte through the Neotropics and toward the Maya lowlands took place simultaneously (Piperno 2011; Pope et al. 2001; Pohl et al. 2011; Kennett et al. 2010). For Grobman et al. (2012), this observation is consistent since the earliest South American maize macrofossils corresponded to later dates (6775 BP–6500 BP). Also, the middle basin of the Balsas River is the site where the dispersion of teosinte in several directions began.

According to Hernández X. (1977), Ortega Paczka (2003: 126), and Wellhausen et al. (1952), it should be considered that this native maize diversity arose from the adaptations of the grass to different physiogeographic, environmental, and agronomic conditions. Altitude and weather factors on the list must be added. On its long road to the center, south, and north of the continent, maize adapted to the specific conditions it encountered in its path, forms of consumption, and the taste of humans to consume it. Along its many routes and roads, new races, and varieties were emerging.

DOI: 10.1201/9781003198833-8

Summarizing the history of maize dispersal, and based on the Kennett et al. (2017) conclusions, in the last 10 thousand years, the domestication and diversification of maize (*Zea mays* ssp. mays) have resulted in local varieties with notable genetic and morphological diversity adapted to a range of geographical restrictions and climatic conditions (Corral et al. 2008; Kato et al. 2009; Piperno et al. 2009; Ranere et al. 2009; Vielle-Calzada et al. 2009). From 48 to 59 native maize races are documented in Mexico (Vielle-Calzada et al. 2009), with morphological variability in other parts of the continent (da Fonseca et al. 2015; Grobman et al. 2012; Merrill et al. 2009). The existing known local breeds are only a fraction of the diversity of breeds that existed during the evolutionary history of the grain and what is of global importance now.

8.3 ROUTES TO SOUTHERN MESOAMERICA

There are several proposals about the various routes traveled by maize dispersion across the continent. For those that went south, there is macro botanical evidence of teosinte in S. Andrés, Tabasco (7100 BP). Also, a domesticated rachis fossil was found in S. Carlos, a place located in Soconusco, Chiapas. According to Clark (1994), the fossil rachis dated 3335±55 BC indicated the route taken to the Mayan lowlands. The two crops of maize and squash joined the growing number of plants that were placed under cultivation and domestication in both Mexico and South America from 9000 BP to 7500 BP, almost at the same time that agriculture emerged in the Old World (Blake 2006: 55–72; Clark 1994; Jones 1996; Kennett et al. 9026–9031; Lippi et al. 1984: 118–124; MacNeish 1960; Pearsall 1978; Pohl et al. 1996) (Figures 8.1 and 8.2).

8.4 ROUTES TO NORTHERN MEXICO AND SOUTHERN USA

There is a list of these routes that includes just over 30 sites; among them (1) Old Maize on the Colorado River highlands, located west of New Mexico. They're the archaeologists Sánchez-Miranda and Carpenter (2018) who dug wells finding maize storage and roasting places (2460 BC–2060 BC). As happened at La Playa site, north of Sonora, where about 80 ovens dug in the ground were found. (2) The MacEuen Cave in the Gila Mountains east of Arizona (2270 BC–1940 BC). (3) The Clearwater site on the Tucson River banks south of Arizona (2200 BC–1960 BC). (4). The riparian site Las Capas in the Colorado highlands in southern Arizona (2200 BC–1940 BC) (Merrill et al. 2009: 21019–21026). (5) The Three-Fir Shelter located in the Colorado highlands northeast of Arizona (2470 BC–1540 BC) (Merrill et al. 2009: 21019–21026; Staller et al. 2006, MAP: 99).

FIGURE 8.1 Sonora: Squashes in the Basin of the River Mayo, Sonora.

Maize appeared in sites located both in highlands and in lowlands as researchers suggested. This means that ancient native varieties were tolerant of a wide range of environments. Studies on the maize appearance in Tamaulipas and the southwestern United States, as well as the genetic differences found in early maize recovered in the two areas, suggesting that the grain was not spread to the southwest through northeastern Mexico (Jaenike-Després, and Smith 2006: 83–95). Benz (1999) believes that the route followed passed along the eastern and western flanks of the Sierra Madre Occidental on the Mexican Pacific coast. To identify other places of maize origins in the early southwest, additional information is lacking.

Another possible route for maize arrival in the USA goes from Tehuacán to Tamaulipas and passes north from Texas. In these two routes, the studies showed the dispersion of teosinte and different races of maize toward the north of the continent. One route took teosinte from Tehuacán, and the other one took the pre-Chapalote first, and then teosinte. Perhaps Mesoamerican farmer migration was motivated by weather factors from 4300 BC to 2100 BC, when maize agriculture would have been spreading northward. The climate of the central highlands of Mexico was characterized by an effective humidity decrease. Ely et al. (1993) stated that after 3900 BC, the southwestern United States and northwestern Mexico experienced an increase in effective humidity.

For Metcalfe et al. (1997), farmers who lived in the ecotone or transition zone between these two regions were forced to migrate north at that time, where the effective available humidity was relatively higher. Time elapsed between the process of change from the first settler's semi-sedentary and seasonal life, to the discovery of agriculture and the domestication of plants –especially maize – in its first phase. The inhabitants learned how to know plants and became aware of their diverse uses (especially as food), the times of the year when they were fruitful, and how to consume them. This was a long period that apparently lasted from 6–7 thousand years. The places where the studies were conducted did not coincide necessarily with the places where there was information regarding domestication.

Lentz (2000) believed that the greatest impact on the environment where these bands moved was the result of human interaction with organisms and the domestication of them (mainly of plants). Thus when they were incorporated into trophic networks, they became interconnected with human beings as primary consumers. Rindos (1984: 122) defined agroecology as a system that included humans, their domesticated plants, and associated ruderal weeds. It contains large groups of other organisms that are close and essentially involved with agroecological systems, including the autotrophic and heterotrophic ones that are out of the question, generally.

To understand a series of aspects related to the chronology of domestication and dispersal of maize from Mexico to the north, including this country, and the current southwest of the United States, it must consider the sequence that is possible reconstruct, organized according to the chronology established by archeobotanical, and archaeological discoveries, which were dated with Carbon 14 (14C) and had been located in several places.

Archaeologists Sánchez-Miranda and Carpenter-Slavens (INAH, 13, 2018) proposed a route that took maize from the northwest of present-day Mexico to the southwest of the USA, based on the archaeological discoveries at the La Playa site, in northwest Sonora. This site has an approximate area of 12 km^2 and a human occupation of about 10,000 years. The site was excavated with the support of E. Villalpando and J. Watson, discovering traces of extinct Pleistocene fauna as of 13 thousand years ago, traces of Paleo-Indian groups using Clovis-type spearheads 10 thousand years ago, and hunter-gatherer bands dated from 7500 BC to 1500 BC. Hence they propose that the route that reached the southwest of the USA traveled:

> [...] From the north of Jalisco to Sinaloa, Sonora, and even Arizona, at least five thousand years ago, when a great relevant cultural corridor was opened facilitating the exchange of ideas, and inputs among the pre-Hispanic Yuto-Aztec-speaking groups, of whom the current Pápagos, Pimas, Tepehuan, Tarahumares, Guarijíos, Mayos, Yaquis, Coras, and Huichol descend in the Sonoran, or Mexican division, and the Hopi, and Comanche among others, in the USA Yuta division.

FIGURE 8.2 Chihuahua City Market: Dehydrated Squash Slices.

Sánchez and Carpenter (2018) argued that despite the importance of maize in Mesoamerica, the grass found in northwestern Mexico was not paramount but allowed desert groups to secure food in the winter because dried maize cobs could be preserved for 3 or 4 months. Dehydrating plant was an important method of conservation. In parts of the southern United States, and in Chihuahua, Mexico, squash is still sliced; placed in strips of wood, which are put on four posts; and dried in the sun, to consume time after rehydration. Dehydrated squash slices and pumpkin seeds are still sold in the Chihuahua city public market. Dehydration of plants should be considered for conservation and subsequent consumption as a common technique in these semi-desert areas (Figure 8.2).

However, the main food was gotten by hunting and collecting of wild plants in the desert that were the basic diet of these villages. In 80 kilns, the plants found there were seeds, calcined remains of mesquite (*Prosopis glandulosa*), and some cacti such as the Toothpick weed (*Ammi visnaga*).

Also found in the La Playa site, cultivation areas, irrigation canals, lithic tools, burial sites, workshops for manufacturing shell articles, and more than 1300 earth ovens in 140 of them the archaeologies found botanical remains of roasted mesquite, toothpick weed, and Agave. Seventy percent of the kilns had traces of maize. In La Ciénaga phase from 800 BC to 200 BC, a settlement appeared with large populations with greater dependence on cultigens and associated with storage wells, residential structures, increased location, a complexity of grinding stones, the emergence of irrigation canals, and "raw" ceramics (Gregory 2001; Mabry 1998) (Figure 8.3).

The information shows several routes maize took and its diffusion to northern Mexico from the Central Highlands. The Sierra Madre Occidental plain was the first route. Another route was from east Tamaulipas to Texas. Perhaps, a third route was from Mesoamerican Chichimec region to Chihuahua. The commercial exchange route proposed by the archaeologist Beatriz Braniff (2009: 27–29) can serve as a key point between Mesoamerica and the Great Chichimeca, including the Southwest of the current United States and the northwest of Mexico. For Carot and Hers (2011: 139–179), this route followed the plains running parallel to the Sierra Madre Occidental, passing through Guanajuato, Michoacán, Jalisco, Nayarit, and the Aztatlán corridor.

FIGURE 8.3 Northern Sonora Metate Without Legs on a Stone Base, 2013.

The other two routes to the northwest were the coastal and the eastern-central possibly, that passed through Zacatecas and Durango. There is found evidence affirming that at least from 600 AD to the tenth century AD shells, metalwork, copper rattles, cacao beans, and macaws were moved through them, in addition to ideological cultural figures, such as idols, iconographic elements for instances, the cipactli, snakes, eagles devouring snakes, and ceramic decorations that were incorporated into northern cultures such as the Hohokam. For Braniff (2009: 27–29), the things found in archaeological investigations were part of trade and gave an idea of the routes products were transported.

The pre-Hispanic route followed by maize and other cultivars that came to northwestern Mexico is unknown. For Braniff (2009: 27–29), the Great Chichimec included the northwest of present-day Mexico and the southwest of the United States, a large semi-desert region, inhabited by diverse groups of horticulturists and nomadic or semi-nomadic bands; some very bellicose. The south-central part adjoined Chichimec Mesoamerica, which in the east reached the Gulf of Mexico.

This huge region connected the center-south with the west (from the Fuerte River in the north to the Balsas River to the south). The south was bordered by Mesoamerica, a mobile frontier area, whose limits moved from north to south, and *vice versa* at different times. In the sixteenth century, with the Spaniards' arrival, Chichimec Mesoamerica was populated by hunter-gatherer bands (e.g., Guamares, Guachichiles, and Zacatecs) that migrated seasonally through this vast region, consuming plants they collected like *Opuntia* and prickly pears from the area known as "the great prickly-pear zone" (*El Gran Tunal*).

During the Novo Spain Viceroyalty, whose government included current northern Mexico, California, Arizona, and New Mexico, plants arrived from Asia by way of commerce with the "Manila Galleon", which went from the Philippines to Acapulco where a very important commercial fair was organized. From this port, the products were moved by land to Mexico City, the Veracruz port, and the cities in central and northern New Spain (González-Jácome 2019; Siemens 1990). Plants arrived by ships to the north through the S. Blas, S. Diego, and Los Angeles ports (Del Río 1990; Schurz 1939).

Among the plants that arrived from the Philippines were palms (*Cocos nucifera*), tamarind (*Tamarindus indica*), rice (*Oryza sativa*), ginger (*Zingiber officinale* L.), vines (*Calamus* spp.), cane grass (*Cassia fistula* L.), cinnamon (*Cinnamomum zeylanicum* Blume), nutmeg (*Myristica fragrans* Houtt.), onion (*Allium cepa* L.), poppy (*Papaver somniferum* L.), almond (*Terminalia catappa* L.), and others (Bruman 1944: 220–243, 1945 (25) No. 2: 212–223, 1947 (27) No. 3: 572–573; Machuca 2013 (34) No. 136; Zizumbo and Quero 1988 (52) No. 1: 68–77).

Pacheco-Olvera (2010: 593–607) mentions that conservation techniques to protect plants during the long trip were diverse. Fleshy species of seeds crops were preserved in ground sugar, while root crops – such as ginger – were preserved in dry sand, cuttings to reproduced plants were wrapped in moss and before sowing they were soaked, and onions were preserved and propagated with bulbs.

8.5 JOURNEY OF MAIZE OUTSIDE THE AMERICAN CONTINENT

In the sixteenth century, maize spread outside the Americas. Baron Alexander von Humboldt (1822, Book IV, T. 2: 256–264) wrote that when Christopher Columbus arrived in Cuba in 1492, maize was distributed throughout America. It was grown from southern Chile to Pennsylvania. Apparently, from 1492 to 1493, Columbus took several hard maize seeds to Spain. Humboldt (1822, Book IV, T. 2: 256–264) stated that since 1525, the grass was planted in Andalusia, already, and its cultivation was carried out near the Atocha chapel, in Madrid. Via the Atlantic, maize arrived in Galicia in the 1600s, where it was distributed along the Cantabrian Cornice. There are some places where – on minimal scale – maize is cultivated in a traditional way. Currently, most of the maize is modified genetically and planted for forage purposes as a monoculture.

From northern Spain, maize spread to other European countries (France, Germany, Hungary) to areas where the average temperature did not reach 7 °C or 8 °C and dropped in the warmest regions of the Torrid Zone, where wheat, barley, and rye did not grew. Humboldt (1822, Book IV, T. 2: 256–264) explained that this grain was known as "Turkish", because it was thought to be native from Turkey. Maize in these European countries had different functions. Initially, it was used as human food but later as fodder. Friar Motolinía (1541: 401) wrote that in New Spain freshly ripened maize and a reed that was always a green weed was obtained from the lake and served as fodder for the horses of Mexico City.

Maize cultivation developed in the same two original forms of its evolution. That is, as a system interspersed with trees (forest-homegarden-cornfield) or cultivated with a combination with maize, beans, and legumes (the old Hispanic garden model). In 1498, Christopher Columbus (Anonymous 2017) wrote that maize was grown in Castile. By 1525, its cultivation had expanded through Andalusia, Castile, and Catalonia. Maize arrived in Portugal in 1523, and in Italy in 1530. In 1526, Gonzalo Fernández de Oviedo (2011) wrote that there were maize crops in areas near Madrid. The Baron of Humboldt (1822, Book IV, T. II: 257–258) wrote that in the late 1500s, maize was extensively cultivated in Spain, Italy, southern France, and Hungary a country where a variety of maize with a short cycle was introduced.

At the end of 1604, Gonzalo Méndez de Casio (1541–1622) sailing from America arrived with the first maize shipment at the port of Tapia de Casariego in Asturias, carrying the "maize ark". The first harvest of the grain was in May 1606, at the Río Barba estate, and its effects in the short and medium-term affected the economy, gastronomy, landscape, folklore, and festivals. Maize was low cost to grow but rendered high productivity. A subsistence crisis, which occurred after 1630, contributed to its rapid spread in rural areas where cornbread (*boroña*) became common food among the poor (Rolmarta 2017). Fernández de Oviedo (2011: 29–30 [1526]) wrote:

> [...] The Indians on the mainland have another use for this bread, and it is in this way: they grind it especially in a concave stone, with another round one that they use with the force of their arms. They usually do it the same way painters grind colors, little by little pouring in a little water, thus grind it mixing it with the maize, making dough. They take a little of that, and wrap it in a grain leaf which they already have for this, or on a cane leaf of the maize itself or one similar. They throw it on the coals, where it roasts, hardens, and turns into a white bread. The latter forms crust but the inside of this bun is somewhat tenderer than the crust. They eat it hot, because when cold it neither tastes as good nor is it so easy to chew, because it is drier, and harder. These buns are also cooked, but they don't have a good taste. This bread, after cooked or roasted lasts very few days, and after four or five days, it molds, and is not eaten.

In the sixteenth century, a strong emigration to the American continent left from Guipúzcoa (Gipuzkoa). Seasonal agriculture and several years with a lack of water did not allow wheat or barley to produce enough for food. The Basques took maize to the Cantabrian Cornice, beginning its food consumption in the form of *talos*. The *talo* is a large, hard maize *tortilla*, brittle, with a sandy texture, and eaten as bread. The leftovers are eaten for dinner, prepared as a kind of soup, or dipped in milk. Talo is prepared mixing cornmeal with water to obtain dough that is grilled; wheat flour and salt can be added to consume it with different fillings. Talo is a typical food in the Spanish and French Basque countries as well as the north of Navarra. It is eaten on holidays, such as December 21st.

From the Basque country, maize passed to southern France by 1826. It was well established there, and the use of figs as food for geese and ducks was abandoned, being replaced by maize (Balmaceda 2014: 8–9). In northern Spain, maize was planted in May along with beans, and it was harvested from October to November. Later, farmers planted ray grass (vallico, ballico) (*Lolium rigidum*) and turnips. The maize cobs were hung in the barns (*hórreos*) to dry, after which the farmers organized a party called *esfoyaza*. This involved, on the one hand, removing the leftover husks (*panoya*), leaving only about three or four leaves that were used to make the bundles, and hanging them on the roofs of the barns. This prevented maize from being attacked and eaten by insects and rats.

An *hórreo* is a granary for storing maize and preserving food away from both moisture and pests. These granaries are of medieval origin and originally were to preserve millet, but they were redesigned in the sixteenth century to save and preserve maize. Currently, there are granaries in the north of the Iberian Peninsula, in Asturias, Galicia, some places on the Cantabrian coast, and in northern Portugal. Their uses have changed, and now they serve as warehouses or use as ornaments for tourist areas, like in Combarro, a coastal village in the Galicia low estuaries. Although hórreos current use is no longer to store maize, they are still preserved and kept up. They are a reminder of traditional agriculture that in those regions of the Hispanic world have become conventional nowadays, and whose production is aimed at the market of forages. Today, maize is a monoculture for fodder. Squash, peppers, and tomatoes are grown in the gardens attached to the houses in the towns (Figures 8.4 and 8.5).

In the past, the maize harvest was done with mutual aid in which family members and neighbors participated. The *panoyas* were woven into braids or strings, four or five rods long, called pine nuts. At night, the *esfoyaza* became a party, which lasted until one or two in the morning, where people sang, laughed, and told stories. In addition, tortillas with sardines or ham with boroña were served. In some places, each participant was given a half-pound muffin, "ties", cheese, fresh pears, and apples. Gaspar de Jovellanos (1744–1811), in *Jovellanos en La Rioja* (1993, 2017), said that he was born in Gijon, Asturias, and in his Travel Journal (1808) wrote:

> The operation of the *esfoyaza* is done in shifts in the houses of the farmers, the young men of the round attending them: the women unwrap the leaves, discovering the corncobs, separating the unfit leaves, and leaving three or four good ones, and the men weave these leaves together forming four [five] long ropes, which they call pine nuts when they are shorter. This operation is very joyful; they sing a lot; the *panoyas* are thrown at each other; the people frolic, and snack on sardines or ham with hot burgundy, cheese, and pears or apples cooked with the same *boroña*. Elsewhere, instead of a snack, each person is given a half-pound roll, and another *garulla*, tied up in a pear, and raw apples inside. This *esfoyaza* is always at night, and they end at one or two, also. Then the gallants accompany the girls to their homes, which are usually distant, and at dawn they are back at work.

In the early 1500s, Portuguese navigators took maize from Europe to Africa. Miracle (1966), a student of maize in the African continent, thinks that maize was introduced in several tropical African places at the same time. Linguistic evidence suggests that many areas received maize through the Sahara, probably from Arab merchants. By 1526, maize had reached the Coptic Christian regions of Ethiopia, and Eritrea (Anonymous 2017). In 1561, it was cultivated in Mozambique and was the

FIGURE 8.4 Galician Historical Barn (*Hórreo*) Seventeenth Century, Poio, Spain, 2018.

FIGURE 8.5 Combarro, Spain, *Hórreos* Located along the Coast, 2018.

staple food of the Portuguese who lived there. By 1573, maize had reached Iraq via trade routes from Tripoli to Aleppo (Syria) and Jerusalem. Using the crescent routes, maize arrived in Baghdad (Iraq). In 1605, maize was grown in Angola. The Dutch merchant Pieter van den Broecke (1634) reported on his trip to Cape Verde, Guinea, and Angola, that it was a "[…] fertile land with all kinds of sustenance, such as the so-called millet masse, or Turkish wheat".

Brandolini (1970) stated that in the early 1500s, maize arrived in South Asia, procured in Zanzibar by Portuguese and Arab merchants. Dowswell, Paliwal, and Cantrell (1966) proposed that Silk Road merchants had introduced maize to the northwest of the Himalayas from where it spread to neighboring regions. From 1615 to 1627, maize reached the Kathmandu valley in Nepal, where its introduction occurred during the reign of Jagatjyotir Malla. By 1687, the grain was sown in Thailand, as a French king's envoy reported. From 1687 to 1688, the Siamese were growing maize as a vegetable garden for domestic use (Anonymous 2017).

Marszewski (1978) argues that in pre-Columbian times there were contacts between human populations in the New World and the Old World, and Asia, and that was when primitive forms of maize appeared in Asia. Anderson (1945) and Stonor and Anderson (1949) argue that the Himalayas were a center of origin of maize domestication because in Sikkim and Bhutan there are some types of maize that do not exist elsewhere in the world, such as the primitive Sikkim maize. In 1778, maize and wheat were introduced into the British colonies of Australia to supply ships. It should be remembered that maize has a great capacity for adjustment or adaptation to different physiogeographic, environmental, and climatic conditions.

Mangelsdorf (1974, 1986) considered that there was insufficient evidence to support the origin of maize domestication in India or Asia in pre-Columbian times. Ho Ping-ti (1956) stated that maize was introduced in China at the beginning of the sixteenth century on sea and land routes. Suto and Yoshida (1956) stated that Portuguese navigators introduced the grain in Japan in the 1580s. Maize spread as a food crop in South Asia around 1550, and by 1650, it was an important crop in Indonesia, the Philippines, and Thailand. In 1750, maize cultivation was widespread in Fukien, Hunan, and Szechuan provinces in southern China. Dowswell, Paliwal, and Cantrell (1996), and Paliwal (2001) stated that in less than 300 years, maize traveled around the planet to become a food crop in many countries, or it was fodder in others.

8.6 CHANGES IN THE CORNFIELD AND FOOD SYSTEM

Cornfield (*milpa*) diversity can be expressed through a continuum, which goes from its first interrelation with the forest, the homegarden, to the polyculture cornfield. A mixture of forest/homegarden/cornfield appeared before the system was instituted near human settlements. Then the number of plants that grew in the system increased, and later it began to decrease, becoming monoculture, and occasionally, back to being polyculture (Mariaca-Méndez 2003; Terán; Rasmussen; and May Cauich 1998).

As happened in the past in some cases, the cornfield is complemented by the homegarden to cover a wide range of basic needs. There are cornfields (*milpas*) where trees are interspersed, as in the popoluca where *chijol* or *kuauitl* (*Piscidia grandifolia*), palms, and fruit trees are part of the agroecosystem (Blanco-Rosas 2006; Foster 1975; Linares, and Bye 2015; and 2016: 9–12). Perhaps there is a relic of the homegarden/cornfield, and they stem from the utility that these trees have for the small-scale cornfield farmers (*milperos*) and their families.

8.7 CORNFIELD CHANGES OUTSIDE MEXICO

As happened in Spain, some plants arrived from the Mexican cornfield, like maize, beans, squash, chilies, and tomatoes. These were mixed with other crops that had existed in Europe, already, many of them from long before. They were cultivated in gardens. In September, squash (*Cucurbita pepo*) was harvested. Quer (1762, T.5, pp. 73–75) wrote that the pumpkin seed "[…] is a cold food of the nature of melons [...]. It is used in the preparation of tiger nut milk (*horchata*), and in cooking, snacks, pectorals, and refrigerants [...]. Usually, it is mixed with aromatic herbs [...], such as parsley, oregano, mustard, or pepper, or with sugar, and honey to make sweets".

An interesting case about cornfield changes in Galicia, Spain, at present, where maize, squash, chili, beans, and tomatoes arrived in the seventeenth century. The last four were sown together with maize to create what Mexicans call *milpa*, where the *Vicia faba* replaced the bean, and other plants originating in the Old World were incorporated. However, the Galician cornfield, associated with a series of factors (e.g., the granaries for storing maize, the harvest celebrations, and the food made with maize), had several crises, the most important was that which occurred in the twentieth century and persists to date, when maize was replaced by transgenic maize seeds and sown as a monoculture for forage production (Esteva and Marielle 2003).

From New Spain, several components of the system entered into the Spanish homegarden (*huerto*), like the avocado. The garden (*huerta*) included legumes and vegetables, which originated in the New World, that supply the family pantry and the market, now. Currently, the cornfield is a monoculture, separated from the other components that passed to the garden. Remember the difference between homegarden (trees and fruit trees) and garden (legumes and vegetables), but that is not applicable in Mexico, where the terms can be used as synonyms. Then there are the lands surrounding the villages (*aldeas*) that are vineyards, where the vines protect the grapes from being close to the ground so they do not rot from excess water and humidity.

Presently, garden produce is taken to the market and sold, despite the rules of the European Union (EU) to avoid the direct sale of producer to consumer of agricultural products cultivated by families on their lands. It is traded and is part of a diet that combines lettuce salads with tomato and *Padrón* peppers fried in olive oil, seasoned with salt, with freshly caught seafood from estuaries or the sea. This is accompanied by white wine (Albariño) produced by small-scale grape-growing families. The wine is artisanal of low production, and despite its fame is not found throughout Spain all year long because its production is familial and not standardized.[1]

To exemplify some issues to be included in traditional agriculture, consider the *Agave* and the vine (*Vitis vinifera*), an unknown plant in Mesoamerica as a crop. The *Agave* was domesticated since ancient times, but it had very strong changes in its agricultural and commercial destination. In the cases of the *Agave* and the vine, both plants are characterized by the series of transformations that, as a consequence of their expansion and management, modify the landscapes of vast regions in New Spain, and integrate into the capitalist economy with the arrival of the Spaniards to American lands.

8.8 PRE-HISPANIC PLANTS AND USES THAT WERE MODIFIED: THE CONFORMATION OF THE MEXICAN MODEL OF AGRICULTURE

8.8.1 Vines

Between 1519 and the end of the Viceroyalty in New Spain, several elements emerged that would modify the landscapes and the pre-Hispanic agroecosystems. The flora, fauna, crops, and technologies that arrived from the Old World had strong impacts on Mesoamerican cultures and agricultures, modifying agroecosystems in practice prior to the sixteenth century. An example of these technological changes in Mesoamerican agriculture was the result of the arrival of the Spaniards, including the grafting of fruit trees like Mexican Hawthorne (*Crataegus mexicana*) used for grafting pear, apple, and plum.

[1] In recent months, products from Galicia have begun to arrive in supermarkets such as the City Market, of the Mexican Commercial enterprise, where you can now find Padrón peppers, Spanish squash, Albariño, and porcelain tableware.

The native wild grape (*Vitis* spp.) was the growing plant for *Vitis vinifera*. Grape was a crop to produce an alcoholic beverage, a practice nonexistent in Mesoamerican lands before 1519. The small and middle size vineyards appeared in various regions of New Spain, including south of Puebla (1531), S. Juan del Río, Querétaro, Fresnillo, Zacatecas, Durango, St. María de las Parras (Siglo XVII), Parral, and California (Corona Páez 2002). It began with the introduction of vines brought from Spain, especially for the production of wine. In 1522, the vines brought from Spain (*Vitis vinifera*) were grafted onto wild vines (*Vitis rupestris*, *Vitis labrusca*, and *Vitis berlandieri*) because the native ones were too acidic to produce wine (Ibarra-Crespo 1992: 239–267).

Hernán Cortés (1524) established that on the native wild vines they should, "[…] blend the strains of the land with those of Spain". These "vines" were very important because the expansion of the wine culture in the Viceroyalty forged a path from the central highlands towards the north of New Spain. Toribio de Benavente, Motolinía (1541: 431) wrote about the native varieties that:

> Some types of wild vines were used rudimentarily [...]. There are in many parts of these mountains very thick grape vines, without knowing who planted them, that have very long stems, and are loaded with many clusters of green grapes that are to be eaten; and some Spaniards make vinegar from them, and some have made wine, although it has been very little.

In March of 1524, the Ordinances, dictated by Hernán Cortés, provided that the neighbors who had forced labor Indians (*Indios de repartimiento*) would have to plant a thousand vine shoots every year "[…] even if they are from imported plants, until they reach the number of 5,000 of these strains". There were vineyards already established and in production at the beginning of the Viceroyalty in spite of the problems that these Ordinances would cause. At the same time, the missionaries planted vines into their convents to produce their own wines, which were used at masses on their tables, and for the sick, but these wines were not sold. The leftover wine was distributed to the missionaries' benefactors, or it was exchanged for provisions (Ibarra-Crespo 1992: 239–267).

The Spaniards used the native strains to graft the vine-shoots brought from Spain on them. Eustaquio Ibarra Crespo (1992: 240) says that were by 1594 there were vineyards in Michoacán, and in 1595, the friars were cultivating vines in Parras, Dolores, Celaya, S. Luis de la Paz, and the Sierra Tetela del Río. The friars went to California together with the Jesuits who founded missions there, which have been there since October 1697, on land with wild vines. With the expulsion of the Jesuits from New Spain, the missions passed to the Franciscans who extended the wine culture to the north. By 1769, their grapevine crops had reached S. Diego, California, from where they were taken north, where wine is produced to date.

There is a lack of information on the impacts that the introduction of *Vitis vinifera* had on these semi-arid regions where wild vines existed in natural systems. Nor do we know if it had impacts on the expansion of Mesoamerican agriculture, because except for a few exceptions, there was no previous complex agricultural base in these places. In the northern regions of New Spain, the native cultures were mainly of hunters-gatherers. Only some indigenous groups lived in fixed settlements, having a horticultural agriculture combined with hunting and fishing (Nentvig 1764: 106–107). The impacts on landscapes, soils, vegetation, the mobility of ethnic groups, and their cultures, including sedentary groups settled on the ancient borders of Mesoamerica with the Great Chichimec, and the Mesoamerican Chichimec are unknown.

The friars mentioned some cultural impacts. The Jesuits, like Juan Nentvig (1764: 106–107), complained about the consumption of intoxicating beverages locally produced by the natives, because its originated disorders. These mentions refer to fermented beverages, produced from maize, prickly pears ("figs of the Indies"), or elder, because their effects lasted several days. There was no mention that those intoxicating beverages were derived from wild vines. So wine was a new product in these regions. There is no knowledge if grape-wine cultivation and the production of wine had some type of diffusion among native Indian groups of Sonora as if they were in the hands of the Spaniards, exclusively. The northern landscapes included new settlements and vineyards within their regional ambits.

8.8.2 Agave

An important impact on the indigenous plants in Mexico was the reduction of pre-Hispanic multiple uses, and the redirection or reinforcement of their use as happened with the *Agave*, and the production of intoxicating drinks, both fermented and distilled. The uses of some other Agave species were modified (Serra-Puche 2017). In Oaxaca, Kent V. Flannery (1986) discovered in the Guilá Naquitz cave that Agave fibers were obtained and interwoven to produce ropes, and that occurred in 10800. It was between 9000 BC and 7000 BC when the first products made with *Agave* and *Yucca* fiber appeared, including ropes, nets, and sandals.

Flannery (1986), MacNeish (1967), and Earl Smith Jr. (1967) consider that the first settlers in Mexico began using Agave 11 thousand years ago, when it was a food for hunter-gatherer bands. To consume it, they cut it into pieces and burned it (*tatemar*) on a clay plate, and baked – or roasted – the stems, leaf bases, and peduncles of the inflorescent plant. Parsons et al. (1990) established that the Agave mead began to be used in 3500 BC, apparently.

According to King (1986), since 9 thousand years ago there have been 48 species of fiber plants, of which 12 were domesticated, and dedicated to the production of sacks, ropes, twines, and clothing (Colunga et al. 2917; Crane and Griffin 1958; Gentry 1982; Schery 1972; Smith 1967; and Smith and Kerry 1968). Several researchers found and studied textiles, and clothing made with Agave fiber from 1300 BC to 800 BC (MacClung de Tapia et al. 1986; MacNeish 1992; Niederberger 1978; and Parsons et al. 1990). In the archaeological zone of Tecaxic Calixtlahuaca, Matlatzincas planted Agaves in their agricultural land to retain soil and moisture, and the archaeologist García-Payón (1979) files of Agave (*metepantles*) served to support soil for crop protection (Vida, Vicente and Carzola 2019: 1–15) (Figures 8.6 and 8.7).

This leads us to consider the agricultural and ecological uses of *Agave* in different regions of ancient and current Mexico: (1) live fences; (2) reinforcement for the edges of terraces and semi-terraces, together with Opuntia, medium-size trees, and thickets, plus several tolerated plants; (3) control the access of larger mammals, to protect crops (e.g. deer, and after 1518–1519, from major and minor livestock); (4) retainment of soil and residual moisture, particularly in areas with slopes (e.g., the *metepantles* on the Matlalcueye or Malinche volcano slopes in Tlaxcala); (5) help to cover the ground with waste from the process of scraping the heart of the Agave, plus the fallen leaves, fruits, and residues of plants, and animals.

FIGURE 8.6 Agave Palma, Due to Its Long Leaves, It Was Used to Obtain Fibers.

FIGURE 8.7 Agave "Palma", Panotla, Tlaxcala.

In the eleventh century BP, the natives began to elaborate fermented Agave mead (*pulque*), but its consumption was restricted and controlled. Disobedience to established norms implied very strong punishments. Francisco Hernández (1574, Book 7, Chapter LXXI) made a long description of the uses that were still in practice in relation to this plant in the last quarter of the sixteenth century:

> Almost countless are the uses of this plant. The whole plant serves as firewood, and to fence the fields; to its empty stems as wood; to apply to cover the ceilings, such as tiles; as plates or boules; to make papyrus; to make thread to fashion footwear; to make fabrics, and all kinds of dresses that among us are usually made of linen, hemp, cotton, or other similar materials. From the tips they make nails, and spikes, which the Indians used to pierce their ears to mortify the body when they worshiped demons. They also made pins, needles, war pins, and hard brushes to card with. The juice is distilled in the middle cavity by cutting the interior sprouts or softer leaves with *iztli* knives (sometimes a single plant produces fifty amphorae). They make wines, honey, vinegar, and sugar from it too. Said juice provokes menses, softens the belly, causes urination, causes the menstruation, and cleanses the kidneys, dissolved kidney or gallbladders, and washes the urinary tract. Of the fiber they make very strong, and useful ropes for many things also. The thickest parts of the leaves, as well as the trunk, are cooked underground (cooking mode that the Chichimec call barbecue), are savorous, and taste to eat like citron seasoned with sugar. Agave also closes recent wounds admirably, because its juice, cold and sappy,

> becomes glutinous when roasted. The roasted leaves are applied to cure seizures, and soothe pains, even if they come from the Indian plague, especially if the same hot juice is drunk; they dull sensitivity, and produce drowsiness. By distillation the juice becomes sweeter, and by cooking sweetest and the thickest until it is condensed into sugar. [...] The Agave juice in which the roots of *piltzintecxóchitl*, and *matlalxóchitl* have been cooked lowers high fevers.

Also, Francisco Hernández (1574, Book 7: 349) explains that Agave was cultivated with sprouts, which spring around the plant in any type of soil, but mainly fertile and cold. For the protomedic, the Agave alone "[...] could provide everything necessary for a frugal, and simple life since it is not damaged by storms or harsh climate, nor doe it wither from drought". In the Viceroyalty, Agave goes from being a multifunctional plant to become the first basic element for the elaboration of *pulque* and then of distilled and fermented drinks, along with *quapatli* and others. Agave reaches its peak in the second half of the nineteenth century with the rise of estates dedicated to *pulque* production, the construction of railways, and their utility for the transportation and commercialization of pulque in the central Highlands (Figures 8.8 and 8.9).

The use of several plants known by the ancient Mesoamerican human population was modified after 1519, as new uses and applications were developed. The list included the manufacture of rubber soils for sandals, for offerings, for the makeup of deities, for the people who represented them, or for medicinal purposes. In pre-Hispanic times (probably around 1700 BC), the rubber latex was used in the preparation of balls, kneepads, and solid rubber mittens for the ball game of that period. These products were destined for trade. According to Martinez-Cortés et al. (1986: 4–8), the Olmec extended the knowledge and use of rubber to the south as far as Costa Rica, and to northwestern Mesoamerica, reaching Michoacán.

FIGURE 8.8 Agaves Described by Francisco Hernández (1574) in the *History of Plants*, Book 7, and Chapters LXX–LXXXII).

FIGURE 8.9 Agave rows (*metepantle*) in South Tlaxcala.

8.9 EMERGENCE OF TRADITIONAL MEXICAN AGRICULTURE

Some agroecosystems have similar components like homegardens intermixed with forest and fruit trees, but their differences are significant (e.g., in those Spanish homegardens, the fruit tree production was for the market mainly, although it was for self-sufficiency). Some are (1) the home gardeners practiced the fruit tree grafting technique, which did not exist in pre-Hispanic homegarden; (2) the major and minor livestock waste were used to fertilize the soil; (3) the soil breakage in lines marked by the plow to sow in holes at a similar distance vs. holes made with a planter cane distributed in the land irregularly, to take advantage of the soil moisture and the specific characteristics of the plot; (4) grain cultivation agricultural systems with monocultures vs. polyculture agroecosystems like the cornfield; (5) the combined management in pre-Hispanic agroecosystems of grains, legumes, and wild and tolerated weeds.

Agroecosystems have common elements: (1) the use of tree curtains to counteract wind impacts, to control the light on planted plants, and negative impacts on crops from heavy rains; (2) the combination of green manure, and animal fertilizers uses; (3) the combination of cultivated areas with forest, and zones with natural vegetation, like ravines, riverbanks, canal borders, river, and marsh margins, slopes and so on; (4) the storage of rainwater, its conduction, and its distribution to crop fields through reservoirs, canals, and dike systems; (5) intermixing of crops; (6) and the collection of mushrooms, and edible or medicinal wild plants that complement the food system and the family economy (Figure 8.10).

With the Spanish arrival in Mesoamerica, pre-Hispanic agriculture had to be adjusted to new environmental conditions: (1) a move from irrigated lands to marginal ones because their irrigated and flat areas were appropriated by the Hispanics; (2) new crops with new management and agricultural technologies were adapted by Indigenous farmers; (3) different relations with the local and regional economy, and the market; (4) new forms of social and labor organization, and land tenure; (5) new agroecosystems arising from these changes were different from the old ones (Vines, Agave); (6) changes of several plants changed of their pre-Hispanic uses for others related to the needs of the Spaniards, such as Agave, rubber, or Mexican hawthorn; and (7) disappeared of other plants as part of the agroecosystem and the food system were replaced by others, as happened with native onions and some edible green weeds.

Hence my proposal that change in time and space characterizes traditional Mexican agriculture. This has given it its resilience to external situations that were difficult and not controllable by

FIGURE 8.10 Wind Impacts and Tree Curtain to Counteract It.

the agroecosystem itself, but the small-scale farmers and their local agricultural knowledge, food system, and culture readapted to them. Traditional Mexican agriculture was born after 1519 (Long 1997, 2003, 2009; Long and Vargas 2005).

The list of modifications and changes that arose as a result of the mixture of Mesoamerican agriculture with Old World agriculture is long. The multiple impacts those matter on natural ecosystems, agroecosystems, and agricultural landscapes.

Should be emphasized owing to the introduction of the plow, metal-tipped agricultural, instruments, draft, and load animals, plants such as wheat, barley, beans, rye, oats, and others modified native landscapes to form new agroecosystems and landscapes as did the emergence of a new agricultural model, which Ángel Palerm (1968a) called "Mexican agriculture".

That was a new phase or new stage of traditional agriculture in Mexico. Some of the most important transformations in the conformation of this new agricultural model were: (1) the introduction of farrows forming straight lines, or following contour lines made with a plow pulled by oxen, that had greater depth than the holes produced with human strength and the cane planter; (2) the increase in the number and height of the stone walls that were placed on the boundaries or borders between the land, and the cultivation plots located in stony areas where the plow could not enter.

The appearance of draft and load animals opened a new agricultural technology in traditional agriculture. For example, soil compacting, the introduction of animal fertilizers, forage cultivation, and new diseases and crop pests. This without considering the problems caused by the herds of cattle, flocks of sheep, herds of horses, and pack animals because of their size and seasonal invasions into the cornfields and cultivated plots of the native Indians.

A series of new crops modified the structure of agroecosystems like the cornfield (e.g., fava bean instead of beans, which are interspersed with maize), and there was an increase in medicinal plants that were sown in the plots, plus pests, and diseases that entered along with those plants. Meat from domestic animals was added to human diets and modified the food systems, and the waste of those animals entered into agricultural soils. Then there was a huge need to produce fodder to feed the cattle. In addition, it is important to consider the impacts sparked by the introduction of new techniques and tools for agricultural and domestic use for food production.

Agroecosystems that changed or emerged in the nineteenth and twentieth centuries that were thought to be modern at the time were compatible with traditional agriculture.

8.10 CONSIDERATIONS AND CONCLUSIONS

From 1519, to the end of the Viceroyalty in New Spain, several elements emerged that would modify landscapes and pre-Hispanic agricultural systems. The flora, fauna, crops, and technologies that arrived from the Old World had strong effects on Mesoamerican cultures, their agroecosystems modifying agricultural systems in practice prior to the sixteenth century. An example of these technological changes in Mesoamerica with the arrival of the Spaniards is the grafting of fruit trees using as a growing plant native trees, and, in the case of vines, the wild grapes were used as growing plant for grafting the Spanish grapes.

Another change was the introduction of new agricultural systems aimed primarily at the capitalist market, primarily. The demand for European and Hispanic products will give rise to a transformation of pre-Hispanic agricultural landscapes. Also, there was the formation of new units of land tenure and production, or commercial extraction like the grana or cochineal, the Campeche stick, quinine, rubber, chocolate, and vanilla that would reach its great development, and boom in the nineteenth century.

The wheat production estates (*haciendas trigueras*), the sisal production estates (*haciendas henequeneras*) making ropes and derivatives, the estates, and ranches of agricultural products and livestock, which were destined for the nineteenth-century domestic and foreign market will be added to this model (Corona-Páez 2002; Rendón-Garcini 1994: 40–45, 307–309).

The introduction of all Old World grains and garden vegetables, added to the medicinal and ornamental cultivated flora of the native agricultural systems are considered to have increased the floristic and faunal diversity of the Mesoamerican systems. As the new elements and agricultural systems were developed before the diffusion of agrochemicals and the use of fossil fuels, they incorporated existing agricultural elements, and knowledge of that time, hence without a doubt, they can be classified within traditional agriculture.

However, new agroecosystems emerged in the nineteenth century. Among them, there are wine northern regions created in areas where there were native vines, and the climate was adequate; and semi-terraces (*bancales*) where *Agave salmiana* was planted on the boards and intercropped with cereals such as maize, wheat, or barley in flat crop strips, as occurred in the Apam Plains in the states of Hidalgo and Tlaxcala (González-Jácome 2016: 63–64; Hernández 2014).

All this information leads to rethinking the role that diversity played in relation to the origins and development of agriculture in Mesoamerica. That may explain the large number of traditional agricultural systems, which exist still, each one adapted to specific conditions of the place where it is located. That is the physiogeographic, environmental, climatological, social, and cultural conditions of specific human groups and their particular habitats. The correlation between diversity and the local, has been made possible and carried out through the domestication, diffusion, and expansion of domesticated maize from the teosinte parviglumis.

During the sixteenth and seventeenth centuries, Mesoamerican societies were modified. With regard to agriculture, there were a number of changes: forms of land tenure; the introduction of plants, animals, and technologies from the Old World; the abandonment or substitution of other plant resources; and the emergence of a new form of traditional agriculture. The new model of "Mexican agriculture" was described by Ángel Palerm (1968) as the conformation of a mixture of native and foreign knowledge that modified the old agroecosystems and formed new cultivation systems adapted to the specific local and natural conditions.

Traditional Mexican agroecosystems are changing and adapting all the time to new economic and political necessities. The Mexican model of agriculture is traditional, with historical, ecological, economic, and political bases to which it adjusts according to specific circumstances. It is not a static model, and it is this capacity for change that makes it resilient to some degree.

One of its characteristics is the ability to survive despite the constant attacks of the major society, which prefers the application of mechanized models, based on the use of energy derived from oil. It is necessary to maintain its capacity for change and adaptation to conditions that are not only agricultural but have an important influence on its sustainability.

Finally, the great changes that agroecosystems in Mexico have had since their origins become evident after 1519, especially in the nineteenth century with the emergence and accelerated growth of the industrialization processes which included agriculture. Some forage plants grew in importance (e.g., grasses, turnips), as did the industrialization strived with the development of monocultures and their rotation to control soil erosion. The emergence of and demand for cotton-made underwear (taken to England from Egypt and India) that required a textile industry with new machinery (Braudel 1984a, 1984b, 1984c; Hobsbawn 1975a, 1987 [1975b], 2003 [1971]: 41–47), and a need grew for agricultural machines moved by gasoline and other inventions moved by that fuel, too, that were conceived since the industrial revolution.

RECOMMENDED READING

Ellen, R.F. 1999. Models of Subsistence and Ethnobiological Knowledge: Between extraction and cultivation in Southern Asia. In: D.L. Medin and S. Atran (eds.). *Folkbiology*. A Bradford Book, The MIT Press; Cambridge, Massachusetts, London, England; pp. 91–117.

When there is enough food available among hunter-gatherers, there is a degree of sedentary lifestyle that is feasible without these groups ever needing crops. That is why there are human groups that could include part-time horticulture in their economic structure and food system.

Agricultural systems are not uniform; they are flexible and can change their spaces, or be cultivated or abandoned according to the subsistence needs of people, as occurs in the Itinerant slash and burn where the system can be patched and then abandoned. This idea can be included in the explanation of agroecosystems such as the itinerant cornfield, which is compatible with semi-sedentary human populations with low population densities.

Gliessman, S.R. 2015. *Agroecology. The Ecology of Sustainable Food Systems*. CRC Press, Taylor and Francis Group. Section VI Bringing about a Sustainable World Food System, Chapters 24, 25, and 26; pp. 301–344.

The author wonders why human societies prefer industrial agriculture, despite knowing the negative impacts on the environment, or why they prefer to increase productivity in large-scale monocultures, even if the result is a lack of food and hunger for numerous groups in society. The agro-business model is a "system organized to meet the demands of capital and little else" (Gliessman 2015: 315). It is important to consider that the urbanization of agricultural societies is part of the distance between growing and eating. People do not see how the quantity and quality of food they consume affects their diets and health.

The industrial food system has adopted other food systems, like that of organic food, and has adopted the language of movements that promote healthy food, like slow and local food. This raises questions as to whether agroecology has the capacity to shape a sustainable food system. Also, the question is how to organize a sustainable healthful food production and distribution system. If the problem of negative impacts on the health of many people is added to the reflections discussed in this text, this chapter makes the reader meditate on where the population is moving within an apparent irrationality.

In reality, it is all in relation to different types of economic rationality. From the capitalist economy, the axis of rationality is the capacity of the system to enrich the owners of large industries, agro-business, and companies. In the rationality of the small-scale farmer, the well-being of the family and the group (the social and political community) is the main objective. However, in the last 50 years, this rationality has been changing and has been increasingly adapted to the idea of earning, earning, earning more money.

9 History and Discussions about Homegarden in Mexico[1]

9.1 INTRODUCTION

This chapter starts from the disseminated and accepted idea that homegardens are sustainable systems, which have the capacity to maintain themselves over time (sustainability and resilience). Homegardens had attracted the attention of scientists because of their ability to mitigate environmental problems, such as the loss of biodiversity or the elevation of CO_2 levels in the atmosphere (González-Jácome 2016c: 233–257, 2018a: 58–150; International Council for Science 2002; Kumar and Nair (eds.) 2006; Mariaca-Méndez (comp.) 2012; Ordoñez-Díaz (comp.) 2018).

Homegardens are agroforestry systems, and some of the oldest and most widespread in the world. Their antiquity is explained by their structure and characteristics, which are similar to those of nature. Their emergence had been casual, accidental, or incidental. Homegardens are distributed throughout the planet, and they exist in cold, semi-desert, temperate, or warm climates, although it is in the tropics where they have a greater extension and use, with their multilayers that characterize them. Homegardens have high biodiversity and cultural diversity; they provide food, products, and economic resources to families, who participate in the production and consumption of their fruits throughout the year (González-Jácome 2001, 2003; Katz 2008). These products cover various needs in the daily life of their growers and can become highly profitable monetarily.

9.2 BRIEF HISTORICAL BACKGROUND

From 14000 BC to 12000 BC, in the early Holocene, the first inhabitants reached the territory of what later would be called Mesoamerica (Alexandrovskiy et al. 004: 9–17). These initial semi-nomadic settlers were organized in bands of 10–30 individuals who moved seasonally through the regions where they found favorable conditions to survive. One was the existence of fauna for hunting, fishing, and capture, plus plants to consume in the places where they were temporarily established. These settlers brought with them the knowledge of fire, domesticated dogs, Clovis-type spearheads, and some stone implements associated with hunting by herding animals to fall off cliffs, and thus die. Also, they caught sight of large animals, cornered them in swampy areas, and killed them, like in the Mexico basin (González-Jácome 1987).

The first homegardens arose in forests or semi-desert areas with vegetation and nearby water sources. They were the result of the observation of wild edible plants, which healed them of their wounds or diseases, or provided elements for their daily lives (e.g., fibers to make ropes and nets, sticks to hold spearheads, or wood to make fire). Hunting and gathering was based on experience, trial and error. In the collection and consumption of fruits, the first garbage dumps were accidentally formed, where they threw the waste of what they consumed (eg, seeds, peels, or fruit seeds, in addition to the bones of hunted and consumed fauna). These garbage dumps served as hosts for the plants that were most liked and eaten by the bands temporarily settled in the region to grow, and as a result, the first homegardens incidentally emerged. In these forests, or wooded areas, wild plants were mixed with those that began to be domesticated. As Argueta et al. (2011) wrote the local knowledge of prehistoric farmers and the accumulated wisdom of these first farmers' experience in soil and water management using seemingly simple tools merged, also (Argueta; Corona and Hersch 2011).

[1] This section of the chapter summarizes several writings of the author, published on homegarden in 1985, 2003, 2007, 2010, 2011a, 2011b, 2012, 2016c, and 2018. They are noted in the references at the end of the book.

DOI: 10.1201/9781003198833-9

Despite their importance in the feeding of the families that own and manage them, the anthropomorphic characteristics that make homegardens sustainable have been ignored, and, their social and cultural components have not been analyzed completely. These components are articulated to the economy, and their ability to generate food contribute to the maintenance of health, obtain monetary resources, and other social and cultural uses that are important for families, articulated to their system, and everyday life.

9.3 HOMEGARDEN AGROECOSYSTEM: ITS ANTIQUITY

Investigations provide data related to the archaeological chronology, archaeobotanical, and genetic discoveries, and some related to the antiquity, of homegardens in Mexico. Not all the results of the researches coincide in their chronology. Reasons are explained in each case, and according to the specific conditions of their development, so it is more or less clear concerning what it is known about homegardens to date.

In Mexico, the first homegardens emerged accidentally and slowly in pre-Historic times (about 12–14 thousand years ago) when the first settlers arrived in forests or in areas with vegetation and near water sources in the semi-desert regions, which have edible fruit and wildlife. The forest/incidental homegarden is older, from 6 to 7 thousand years old. According to Piperno and Smith (2012: 152), human management of early homegardens involved hybridization as happened between two species of domesticated Lead tree (*Leucaena* spp.).

Human consumption formed "trap places" (garbage dumps), and they were preferred by wildlife. Flora and fauna were abundant and selected through consumption and taste of the human bands who obtained their food system in these places. The first forest/incidental homegarden emerged and formed in the Holocene, from 10000 BC to 9000 BC, and up to 4000 BC.

Several millennia elapsed during time the villagers built the first homegardens that arose mixed with natural vegetation. These included several ecosystems, but in all of the cases aquifer resources (lakes, lagoons, or rivers) existed nearby. Bands lived in caves, rock shelters, or nearby places where they settled temporarily. Homegarden in pre-Hispanic Mexico seems to have an age that corresponded to the Formative Period (2500 BC–200 AD). To understand the technology of the pre-Hispanic era, it is necessary to imagine the sixteenth century without certain agricultural practices, such as grafting and the use of metal tools.

The earliest domesticated, tolerated, and cultivated plants initiated the development of the original traditional agricultural systems. Apparently, those that began to be domesticated were different in each of the regions where the inhabitants seasonally moved (Piperno and Stothert 2003). Also, the sequence of their domestication varied from one place to another; for example, 12 thousand BP cassava and *Cucurbita ecuadorensis* were domesticated in the Amazon. In studies, such as those of Smith (1997, 2001: 1–43), it is shown that squash (*Cucurbita pepo*) was domesticated in Oaxaca from 9000 BP to 7000 BP, before maize was subject to the human management that led to its domestication (Piperno and Smith 2012: 154). Silver-seeded squash (*Cucurbita pepo*) was domesticated in the cave of Guilá Naquitz, Oaxaca before that maize (Basurto-Peña; Castro-Lara; Ovando and Juárez-Castro 2015: 43–53). Silver-seeded squash (*Cucurbita pepo*) found in the Romero cave in Tamaulipas was domesticated in 5085 BP. The possible date for cultivated *Cucurbita argyrosperma* in Xihuatoxtla is 8700 BP (Piperno and Smith 2012: 154).

The diversity of the ecosystems in which people moved and settled explains – at least temporarily – the plants that the inhabitants began to domesticate. The homegarden/cornfield (*huerto – milpa*) disturbed the natural ecosystem, but the inhabitants worked with the forest in an integral way giving rise to a sustainable system. For some researchers, the current Mayan forest is the consequence of human activities in the virgin forest during Prehistory, the Viceroyalty, and recent times. The forest/homegarden/cornfield was an initial agro-food stage, which was the outcome of the sedentary lifestyle of the first permanent settlers. The whole process took from 7,000 to 6,000 years (Long, and Vargas 2005).

Actually, at least a part of the production of the homegarden, orchard, and vegetable gardens are directed to the market, but the agroecosystem still contributes to the family food system, and the family social and labor organization, as well. This is an old characteristic, which remains to this day.

A good example of the economy of the avocado homegarden is that of S. Francisco Tepeyanco, and the neighboring towns of the southwest Tlaxcala. At the end of the nineteenth century (1895), the small Tepeyanco avocado had an annual profit of 350 pesos locally. The avocado was sold in the region in an annual circle. From June to September it was sold in Atlixco, Puebla, giving a profit of 415 pesos per year. North of Tlaxcala, the avocado load in Zacatlán was sold for 100 pesos per year; In Zacapoaxtla, the avocado had a profit of 261 pesos per year. In total, the avocado producing towns received $ 1,401.00 pesos per year for their sale (by load, percent or per piece) (Arvizu-Barrón 2013; Carpology 1895: 17–33.

9.4 FOREST AND THE HOMEGARDEN AS HUNTING PLACES IN LOS TUXTLAS

The first incidental homegardens of which there is news are La Joya and Bezuapan, two sites located in the Sierra de Los Tuxtlas that emerged along with the first domesticated plants. Vanderwarker (2006) pinpoints their evolution during the Formative or Preclassic (2500 BC to AD 200) in the forests near the seasonal settlements where the inhabitants obtained food by hunting large fauna, like deer, and small ones, like rats, rabbits, and others that along with food collection complemented the settlers' diet.

On the Los Tuxtlas coastal plain, Cyphers (2004: 29) found that from 1200 BC to 800 BC, homegardens were located in higher places, such as Tenochtitlan and Loma del Zapote (but not in S. Lorenzo, a site of major importance), where there was little farmland, and most of it was subject to waterlogging that lasted several months. Therefore, food system was based on what was fished, hunted, and collected of species obtained in the marsh and the lagoon systems.

Vanderwarker (2006: 79–80) states that in the Formative agriculture of the Tuxtlas were found in the two places: (1) *Zea mays*, beans (*Phaseolus vulgaris*, *P. coccineus*, and *P. acutifolius*), avocado (*Persea americana*), coyol palm (*Acrocomia mexicana*), and sapodilla (*Pouteria sapote*). The wild fruits and plants which completed the diet of the inhabitants in the two places were maize, beans, avocados, coyol palm (*Acrocomia mexicana*), and several sapodilla varieties; (2) were found only in Bezuapan: Beans (*Phaseolus vulgaris* and *P. coccineus*), *Opuntia* Prickly pears (*Opuntia* sp.), guavas (*Psidium guajava*), walnut (*Juglans*), grape (*Vitis*), Candlewood (*Cupania glabra*), and morning glories (*Convolvulaceae*); an unidentified tuber and acorns of *Quercus* sp.; (3) were found in La Joya, only: Horse purslane (*Trianthema* sp.) and Annatto (*Bixa orellana*).

The role of the homegarden in these processes had the following sequence, probably: (1) from forest to forest hunting-gathering; (2) from forest hunting-gathering to incidental homegarden; (3) from the incidental homegarden to the homegarden-cornfield; (4) and from the homegarden-cornfield to (5) the homegarden and the cornfield all as individual agroecosystems. Several proposals had been generated, as these variants formed the first traditional agricultural system, and were invented by mankind. There were trees and domesticated plants mixed with wild plants and tolerated weeds leading to ways that improved food by obtaining several resources, including wild flora and fished, hunted, caught, or gathered fauna.

The combination of the forest with the incidental homegarden and the homegarden-cornfield is one of the pre-agricultural systems, which are explained from the history of maize, its domestication, and initial dispersion from 9000 BP to 8000 BP. Other flora species were domesticated and spread little by little (Blanco-Rosas 2006; Killion 1990; Lentz 2000; Vanderwarker 2006).

The first incidental homegarden was structured and functioned in a similar model as the natural ecosystem. It is not clearly distinguished from the temperate forest, the rainforest, or the semi-arid regions, except for the greater number of fruit trees and edible plant remains found in it, and for what was thrown in the "garbage dumps". The creation of a homegarden near the housing and seasonal settlement derived from human preferences for certain collected fruits, berries, and plants. The people in turn facilitated the care of the fruits from pests and predatory fauna because fruits served as a wild animal attractor, as a result making the homegarden a handy "hunting garden".

It is feasible that the first homegardens were formed by a combination of wild plants and plants that were domesticated from the initial stages but were not clearly distinguished from those in the forest. The abundance of archaeobotanical remains of bones belonging to the hunted and consumed fauna, together with the remains of edible fruits and plants that were domesticated from the initial stages is what allowed researchers to talk about anthropomorphically modified areas, which were inhabited or seasonally used by the first settlers. In the spaces used as habitation by the bands of hunter-gatherers, including caves, rock shelters, garbage dumps, or temporary housing areas, including the surrounding areas, the phytoliths, pollen and remains of the consumed fauna show the taste of these inhabitants for their consumption

By the Middle Formative (1000 BC–400 BC), the inhabitants preferred hunting in disturbed environments, which abounded in deer. There was an increase in the use of the homegarden as a hunting place. McCormack (2002) argued that the political center of Los Tuxtlas was not yet established and everyday life took place in independent and self-sufficient villages. In the Late Formative (400 BC–100 AC), the first regional center (Chuniapan de Abajo) arose but was never as important as the lowlands Olmec settlements.

The settlers planted maize and collected avocados and macaw palm (*Acrocomia aculeata*). Ceramic articles were diverse, and the range of their forms and size depended on what was cooked in them. McCormack (2002) and Vanderwarker (2006: 195) considered that the manufacture and use of specialized grinding instruments and their abundance are indicators of the expansion of production and consumption of maize leading to a greater focus on maize production and its processes (such as grinding) associated later with what it is called maize hominy (nixtamalized maize) in Mexico.

Archaeobotanical information showed that maize and tree fruits were the main staples in the food system. Maize grew in areas close to residential areas, and at the same time, the corncob underwent changes in the rachis and in the spokes of the corncob dome. Faunal materials suggested a greater disturbance in terrestrial biomes, indicating an increase in hunting in the homegarden, where cultivation was a less risky subsistence strategy. Some researchers (Santley et al. 1997; Santley and Arnold 1996, in Vanderwarker 2006: 196) claimed that there appeared to be some correspondence between agricultural intensification and increased sociopolitical complexity.

Los Tuxtlas region was abandoned at the end of the Formative (100 BC–300 BC) because of a volcanic eruption, an event that influenced population emigration. Those who remained continued to grow maize. Data on stable carbon isotopes and nitrogen indicated that the grain was basic in the diet of the population. The consumption of avocado, sapodilla, and macaw palm (*Acrocomia aculeata*) increased and drained field systems were built for cultivation. Villagers diversified hunting sites, with a wider range than in the previous stages. Hunting in the homegarden was less selective, indicating an increase in the subsistence risk. Vanderwarker (2006: 197) said that despite everything, cultivation was the most important strategy for feeding the population in the region.

In the ancient times in Los Tuxtlas, the development of the systems were: (1) forest-incidental homegarden/hunting; and (2) forest/homegarden/cornfield/hunting. Both systems persist today. Studies by George Foster (1940, 1942, 1967, 1975) and Blanco-Rosas (2003, 2006) showed that the Zoque-Popoluca settlers of Soteapan, in Los Tuxtlas, in southern Veracruz, caught edible birds (partridges, and pigeons) in cages, in addition to small mammals that they hunted in homegardens and cornfields. That meat provided the daily and annual food system of the settlers with the amount of animal protein needed and available. Meat could also be sold in nearby markets, affording small-scale farmers the possibility of obtaining monetary income.

9.5 HOMEGARDENS IN THE TEHUACÁN VALLEY

In the Tehuacán Valley, studies by archaeologist James Neely (2017: 136) and his team in the project carried out in the Lencho Diego Canyon and the Purrón Dam showed evidence of homegardens found in the aforementioned canyon and was dated from 1350 BC to 1050 BC. These homegardens had irrigation coming from the stream that ran through the ravine and corresponded to the Formative or Preclassic.

According to Hughes et al. (2007: 14389–14395), the processes of domestication of maize were parallel to those of other domesticated perennials, such as Agave and Opuntia. Hybridization in backyard homegardens played a central role in the domestication of plants, being a trigger for the process. In Tehuacán, bands settled in camps near the creek and water sources. In the lowland seasonal forest of the Tehuacan Valley, by 7000 BP, the Mexican hog plum (*Spondias mombin*) was probably domesticated (Piperno and Smith 2012: 154–155).

Backyard homegardens contributed to the processes of domestication of plants. MacNeish (1964) and Smith (2001) when discussing this process said that Avocado (*Persea americana*) began to be planted in places near homes. Trees with larger avocados that had an anise smell were selected from the forest and taken to more favorable habitats. This increased the adaptation of the tree for more specific uses and is considered as the second phase in domestication processes, where already domesticated avocado was grown in regions whose altitude was higher than 2000 masl.

The archaeological information of MacNeish (1997) and his colleagues showed the Tehuacán Valley as a fundamental place in maize domestication and the evolution of the homegarden and the cornfield, along with the emergence and development of the first irrigation systems. Anthropomorphic changes in the first homegarden were formed by a combination of fruit trees and trees for shade, or for obtaining construction wood, charcoal, and firewood. In addition to plants domesticated incidentally, those that were domesticated agriculturally, wild flora, weeds, epiphytes, vines, and fungi were found, also. See Map: The Tehuacán Valley (See Chapter 7, Figure 7.1).

Over time, especially in arid or semi-arid areas, homegardens became irrigated. MacNeish (1997) and Neely (2017) found that irrigated agriculture was probably practiced in Tehuacán from 1050 BC to 150 BC. As humans altered the wild vegetation by introducing new plants from outside the area, clearing other forest areas, and using them for planting, they created new habitats and established pioneer weeds, which grew in open places attracting insects and their predators. However, agriculture based on irrigation was developed later. Archaeological materials indicated that irrigation was practiced in Tehuacán at least as of 1050 BC.

The Tehuacán Valley, with a semi-desert climate, had xerophyte vegetation, and the grassland abounded in mesquite (MacNeish 1981, 1997: 82). Subsistence depended on the collection of wild plants and hunting. Studies showed that humans preferred to consume certain fruits, whose remains were thrown on the ground ("garbage dumps"), where they piled up, seeded, grew, and consequently created the first incidental homegardens. The introduction or the constant elimination of plants (selection) allowed satisfying human tastes and needs. The men threw the remains of vines they had used to manufacture utensils, containers, ropes, and other objects into the dumps.

Studies by Bruce D. Smith (2005: 9438–9445), in the Coxcatlán cave, in the Tehuacán Valley, indicated that, until later times than originally established, the following three plants were domesticated and consumed: squash (*Cucurbita pepo*) in 7920 BP, gourd (*Lagenaria siceraria*) in 7200 BP, and *Cucurbita argyrosperma* in 2065 BP. Date still to be confirmed, but in different places, is the domestication of the Butternut squash (*Cucurbita moschata*), which was not present in Tehuacán for this phase but occurred in others places.

The transition from hunter/gathered to a new way of food production in Mesoamerica was found in S. Marcos and Coxcatlán caves in Tehuacán, Puebla, a center of agricultural origin, and have resulted from the reanalysis of early domesticated plants that were recovered in the 1950s and 1960s from caves Romero and Valenzuela in Tamaulipas; and some years later in Guilá Naquitz, in

Oaxaca. Piperno and Smith (2012: 154–155) considered that Silver-seed squash (*Cucurbita argyrosperma*) was domesticated by 8700 BP in Xihuatoxtla, Guerrero, and by 5085 BP in other places.

These caves had shaped the understanding of the early history of plant domestication and food production economies in Mesoamerica. Disturbed areas abounded with natural vegetation and contained wild plants, that were used for human consumption. These were mesquite (*Prosopis* spp.), American mastic or American pepper seeds (*Cyrtocarpa procera*) (*chupandilla*), prickly pear (*Opuntia*), nance (*Byrsonima crassifolia*), avocado, Mexican hog plum (*Spondias mombin*), edible cacti, and gourds (*Lagenaria siceraria*). Piperno and Smith (2012: 154–155) included among the cultivated trees in the Tehuacán Valley: tree gourd (*Crescentia cujete*), 3000 BP; avocado (*Persea americana*) 3000 BP; lead tree (*Leucaena esculenta*) from 6000 BP to 3000 BP; white lead tree (*Leucaena leucocephala*) 3000 BP; Mexican hog plum, Yellow Mombin, and Red Mombin (*Spondias mombin*) 7000 BP.

Anderson (1949) proposed that incidental homegardens were disturbed habitats, and diverse places suitable for spontaneous hybridization processes. The disturbed habitats favored spontaneous hybridization and preceded domestication. Among the first plants subjected to these processes were *Lagenaria*, *Agave*, and *Opuntia*. These homegardens were suitable places for the hybridization, selection, domestication, and experimentation of plants, giving rise to the first crops.

The first homegardens depended on rainwater and without receiving greater care occupied these territories seasonally. Casually the bands added plants whose taste and uses satisfied them. After thousands of years of seasonal occupation, zones were created with the first incidentally domesticated plants. These pre-Historic homegardens were located near permanent streams or seasonal rivers; later, land edges and walls appeared to protect them.

The oldest indications of the uses of fruit trees in forests are based on the existent remains of the fruits that were consumed by the hunter-gatherer bands during the Formative, both in the Tuxtlas region and in the Tehuacán Valley. An example of this is found in Coxcatlán, in the Tehuacán Valley, where the inhabitants consumed avocados, which they began to select by size and smell by 6500 BC (Storey et al. 1986), being totally domesticated by 900 BC (MacNeish 1967; Smith Jr. 1967: 232, Table 26).

9.6 THE INCIDENTAL HOMEGARDEN/CORNFIELD IN THE TEHUACÁN VALLEY

Rzedowski (1978: 47) described the Tehuacán Valley as one of the few arid zones located below the 20° parallel with xerophyte vegetation (MacNeish 1997: 81). The boundaries of the valley are: on the northeast and east with the Zongolica mountain range, in Oaxaca on the southeast with the Mazatec mountain range (2600 masl), to the south and west with the Zapotitlán mountain range in the Mixteca mountain range. About 10,000 years ago the valley had an arid and semi-arid zone as well as a semi-warm and semi-dry climate (Alavez-Chávez 1992). For Rzedowski (1978), some areas were in environments where the humidity conditions were lower (Zapotitlán Valley) or higher (south of Caltepec). Both areas were covered by xerophyte scrub. The region is the center of the evolution of this type of flora, exemplified by the Agave species, or with the columnar cactus that reaches 15 m in height.

Earl Smith (1967: 133–142) considered that 29.1% of 253 species collected in the valley are endemic, and at least three genera of phanerogams are exclusive to this arid zone. From 7000 BC to 5000 BC, during the optimal climate, the Tehuacán vegetation was of tropical deciduous mountain forest, semi-arid, with mesquite and cacti that occurred until 1500 masl. Lauer (1979: 37, 39) indicated that from 2500 BC to 2000 BC, the vegetation in the highest zones was of mixed forest of *Pinus* and *Quercus*. The villagers appreciated the biznaga (*Coryphanta pallida*) and lead tree (*Leucaena esculenta*) highly. Lead tree is native to the neotropical regions and was found from Texas to Peru, and the young pods were an important food for local populations (Brewbaker and Sorensson 1990: 2; Smith Jr. 1967: 220).

Apparently, there was no drastic climate change in those last 10 thousand years, but geological evidence indicates a gradual decrease of the water table over that period. The sequence of occupation of the Tehuacán Valley covered about 12 thousand years. At that time settlers had several approaches to the environment, different food preparations, lithic artifacts, and technological systems. The material culture includes a series of vestiges which facilitated the reconstruction of a part of the daily life of the population, and the use that they made of natural resources.

The first phases of occupation show disturbances in some areas due to extractive activities of regional flora and fauna. There was a paucity of human population, and bands were seasonal. MacNeish (1997: 81) and his team studied the valley from 1961, considering that the regional environment was able to maintain a limited population, with nomadism conditioned by the characteristics of the seasons. Only the emergence and development of irrigated agriculture allowed the growth of a larger population.

MacNeish (1997: 82) stated that from 6800 BC to 6500 BC, evidence indicated the presence of hunter-trapper bands, living in caves and changing their camps three or four times a year. Bands were small-animal hunters formed by four to eight people, who were also plant gatherers. They trapped small wild animals like hares, moles, rats, vipers, turtles, birds, lizards, and additional small mammals. Anderson (1967: 93) said that their instruments were crude and made of lacquered stone (knives and foliated projectile points, scrapers, cutters, rakes, burls, and razors). They had mortars to crush their food and did not use polished stone objects.

The most remote antecedent on the beginnings of agriculture in the valley was the dependence of people on their seasonal mobility. This dependence hardly allowed them to obtain their food system from agriculture; so the process began with the selection of plants of the taste of the first settlers, and the accidental action of throwing the remains in places that allowed the formation of areas with the preferred plants. MacNeish (1997: 82) stated that the weather of those times was colder than today. The bands were settled in camps near the creek and water sources in the valley.

The pioneer of archaeobotanical studies, Earl C. Smith Jr. (1967: 232, Table 26), found evidence that, shortly after 6500 BC, seasonal settlers used several plants for diverse purposes. Among these plants were the *Setaria* cf. *macrostachya*, an unidentified grass seed, which served as food; the amaranth (*Amaranthus* spp.); avocado (*Persea americana*); mesquite (*Prosopis juliflora*), whose pods, and seeds were roasted and eaten; American mastic, or American pepper seeds (*Cyrtocarpa procera*); *Opuntia* spp., from which the cladodes and prickly pears were consumed; a palm (*Brahea dulcis*) used to obtain fiber. There were two other unidentified grasses, which had uses that are still unknown.

These bands consumed amaranth, which by 5500 BC was domesticated by them, along with pumpkin and chili, and made up 6% of the population's food system. In the next millennium, this combination makes up 14% of the total food consumption (Early 1992: 17–19). Anderson (1967: 93) based on the human remains of the Purrón and Coxcatlán caves proposed that in the dry season the bands subsisted by hunting and in the wet season by gathering.

The incidental homegarden combined domesticated, wild, and induced plants, and several of them were used to obtain fibers. In the El Riego cave, occupied from 6500 BC to 5000 BC, remains of mesquite (*Prosopis juliflora*), gourds (*Lagenaria siceraria*), prickly pear (*Opuntia* spp.), and *Hechtia* were identified. The people ate avocado; amaranth; acacia pods; Mexican hog plum (*Spondias mombin*); prickly pear (fruits, and cladodes); two types of cacti; Agave; seeds of *Lemaireocereus hollianus*; seeds of *Jatropha curcas* were edible, and the latex of the plant is medicinal, curing infections caused by bacteria, fungi, bee stings, and wasps; the fruit of Cossack (*Sideroxylon* ef. *tempisque* Pitt-Sapotaceae); wild squash (*Cucurbita pepo*); cultivated squash (*Cucurbita mixed*), and chili (*Capsicum annuum*). There is evidence that foxtail grass seeds (*Setaria geniculata*) were consumed too (Smith Jr. 1967: 133–142).

Not all the plants used and consumed by the settlers were scattered by them. Some, like the avocado, which is a tree native to the Sierra Nevada, in California, arrived in Tehuacán 15 thousand years ago, before the first settlers arrived, who when they arrived began to consume avocados, thus initiating the selection and domestication processes (Schroeder 1968: 29 -38). Avocado was already cultivated in the homegardens of Coxcatlán since 6500 BC (Storey et al. 1986). They also found remains of the white lead tree (*Leucaena esculenta*), whose seeds were roasted and consumed, forming part of the food system, and they also found seeds of the Silk cotton tree (*Ceiba parvifolia*), which were also eaten roasted. That predominant plant was used from the center to the south of what is the current Mexican territory. It has had numerous independent transitions from its wild state to its cultivated state. It was an established crop by around 4000 BC.

Hughes et al. (2007: 14389–14394) considered that from *Leucaena* resulted in two to six local species and a total of 13 species. They considered that two to six local species resulted in *Leucaena*. They considered that the hybridization of *Leucaena* was extensive and complex; it involved nuclei encoded in DNA sequences and diversity in the hybrids resulting from the juxtaposition of species under cultivation. In general, collected plants were important food. Fibers were obtained from the branches, dried stems of the plants, and shrubs that served to make fire, and the gourds functioned as containers for common use. Hughes et al. (2007: 14389–14395) proposed that these processes were parallel to those of other domesticated perennials –such as Agave and Opuntia – and that the hybridization in backyard homegardens played a central role in the domestication of plants, being a trigger for the processes.

In the Central Highlands, plants domesticated in early times were avocado, the Mexican hog plum, and cold and temperate White sapota. Lentz (2000: 93) and Smartt (1988: 149) indicated that it is unknown whether maize and beans coevolved as part of the domestication process. However, it had been proved that they have a long history and tradition as intercrop or mixed with successive crop rotation. In this case, the cultivated bean was of the twining type and used maize stalks to sustain itself during its growth, thereby reducing the chances of being affected by pests and diseases. The management of teosinte genetically by the first settlers transformed maize, with the outcome that the cornfield was born as part of the incidental homegarden, which was located in the vicinity of the first and temporary human settlements.

At the same time, this combination favored the ability of the soil to maintain nitrogen through a symbiosis (Farrell 1990: 169–183; Guiller 2001: 17). Nitrogen fixation by microorganisms colonizes the roots of trees, improves the mineral nutrition of plants, and minimizes the use of fertilizers in soils; furthermore of include the root diseases effects on agricultural crops (García Espinosa 2007, 2010; Gliessman 2002: 231–232; Wood and Bollinger 1990: 184–204).

9.7 EMERGENCE OF THE MEXICAN HOMEGARDEN, ORCHARD, AND GARDENS IN THE SIXTEENTH CENTURY

In general, it is easy to imagine that pre-Hispanic homegarden was an agroecosystem similar to those known now, and local variations were the result of climate and gradient characteristics. Native Mesoamerican homegardens could be seasonal or irrigated, unlike Hispanic ones that were irrigated always. The entrance in Mesoamerica of the Spanish homegarden through the convents and monasteries introduced new fruit trees, new crops, and new technologies. In order to understand the changes in agroecosystems like the homegarden and the cornfield, it is necessary to have in mind the impacts on pre-Hispanic agriculture after 1519, caused by the introduction of working, and pack animals. This included the use of manure for fertilization of soils, the new technologies with iron instruments, and the long list of plants that were introduced in New Spain, and their use as green fertilizers, among others elements.

The Spaniards distinguished the homegarden with different types of trees (polyculture) and orchards, with the dominance of fruit trees. In the sixteenth century, there were indigenous orchards with monocultures of fruit trees (avocado, Mexican hog plum, or Mexican hawthorn). Spanish homegardens had fruit for selling or their derivative products such as jellies, jams, sweets, or spirits that were traded or redistributed. The Indian and mestizo homegardens had horizontal and vertical strata (multi-strata), for the cultivation of trees, grains (beans), aromatic herbs, medicinal herbs, ornamental plants, edible green weeds, legumes, vegetables, Agave, and Opuntia.

The result was the emergence of Mexican agriculture and agroecosystems such as the polyculture homegarden, which combined plants and technologies from pre-Hispanic and Hispanic agroecosystems. This homegarden modified the structure of flora strata, including trees for different uses (fruit, shade, wind curtains, household objects, firewood, domestic wood to manufactured household goods, wood, domestic charcoal, canals, gutters, logs hollowed and used for honeycombs, agricultural tools, and others), and altered domestic fauna, equally. This combination made diversity increase. Trees remained the key to the agroecosystems' continuing its long road in the agricultural history of Mexico (Figures 9.1 and 9.2).

On the homegarden/cornfield agroecosystem in the Mayan zone, Ford and Nigh (2009, 2010, 2016) considered that at the beginning this was a diverse agroforestry system with cornfield a polyculture and that it endures in the Mayan homegarden today. The Peten Mayan forest in Belize was studied by Ford (1990, 2002), Ford and Fedick (2009), Ford and Night (2009, 2016), and Nigh (2014) when dealing with the impact of droughts and climate change in the Mayan area, sustainable cultivation, and the Maya settlements. Lundell (1978) considered that the negative impacts of drought were lessened by the adaptive changes made by the Maya (Ford and Night 2010, 2016). Researchers proposed that the population required the invention and adaptation of more sophisticated agroforestry practices (Hurtado and Román 2012). At the same time, the pre-Hispanic homegarden was the pivot in the creation and maintenance of the forest, as it is known (Figures 9.3–9.5)

FIGURE 9.1 Dresden Codex: Honeycombs.

FIGURE 9.2 Yaxcaba: Honeycombs in Hollowed Logs.

FIGURE 9.3 Yaxcaba, Yucatán: Honeycombs on Pierced Tree Trunks.

FIGURE 9.4 Yaxcaba: Interior Zone of a Honeycomb.

FIGURE 9.5 Yaxcaba: Honeycombs in Wooden Boxes.

9.8 CONSIDERATIONS AND CONCLUSIONS

In the majority of the cases, traditional agriculture had ancient origins; hence, some aspects of the evolution and history of the homegarden and the cornfield are narrated in this book. Antiquity is important because the age of agricultural systems is closely related to their sustainability, which is proved by their permanence over time. The initial selection processes preceded the incidental domestication of fruit trees and other plants important for human consumption, such as teosinte. As is known, one of the characteristics that are recognized about the homegarden is its diversity, which among other components apart of naturals, it was a product of the human action, such as to bring new plants to the forest, as it is done with the homegarden, currently. The first agroecosystems like the homegarden and the cornfield were intermingled with natural ecosystems.

This close interrelation continued over time, being interrupted by situations such as human population growth, urban expansion on agricultural land, and regions that have resources that are used and exploited by industry or mining for the production of large-scale goods, the creation or expansion of roads, or the construction of large-scale hydraulic works for hydroelectric power.

The homegarden increased the diversity of forests by introducing new plants into them. The traditional agroecosystems in a tropical or subtropical region began with the homegarden/cornfield, integrating to it little by little plants that were domesticated, together with tolerated weeds and wild weeds that were in the surrounding natural ecosystem. In the case of the cornfield, an ecotone was formed around the polyculture crops, and due to its uses and management, it can be considered a subsystem of the agroecological system.

This ecotone can be an edge of land with wild and tolerated plants on it, or a row of Agaves (*metepantle*), or ruderales, or trenches, a ditch or canal to distribute water on the plot, or a hedge next to the canals, and the plot in whose soil aromatic and medicinal herbs were planted to sell in nearby markets, like happened in Tepeyanco homegardens before the 1980s.

RECOMMENDED READING

Dunmire, William W. 2004. *Gardens of New Spain. How Mediterranean Plants and Foods Changed America*. Texas University Press.

The book is divided into eleven chapters, of which it is recommended to read Chapters 1 and 2. Chapter 1 deals with pre-Columbian Spain. Chapter 2 is about Mexico before Columbus. On pages XIII, XIV, and XV of the Preface, there is a figure on the migration of food plants to the New World.

The chart included: grains, leafy vegetables, stem vegetables, root and bulb vegetables, legumes, stone fruits, citrus fruits, other orchard fruits, other fruits, nuts, culinary herbs, culinary spices, and miscellaneous. The places in the New World where these food plants arrived and were distributed into the Caribbean Islands, Mexico, New Mexico, El Paso, Sonora, Arizona, Inland of Texas, and Alta California. A series of marks in boxes established the arrival dates of the food plants from 1492 to 1848. A letter included symbols to establish if the place had an inhospitable climate for plant cultivation. The data about plants was checked with experts who reviewed the information.

Chapter 2 is dedicated to pre-Columbian Mexico. It deals with the diversity of the country, including numerous ecosystems ranging from the desert to the rainforest and alpine mountain areas. For Dunmire (2004: 31), the country is very similar to Spain. He treats in detail what he calls swamp gardening in the chinampa, cultivation in mountainous regions, Aztec crops in rain-fed lands, festivals, pleasure gardens, cultivation on slopes, northwestern Mexico, the mountains of the Tarahumares, Casas Grandes before Colón, and the north and northeast of Mexico (Chamoux 1997).

Aghajanian, Alfred. 2018. *Chinampas. The Role in Aztec Empire-Building and Expansion*. (An Academic Research). Copyright 2018. Indo-European Publishing.

The book is an excellent summary of the chinampa agriculture in the basin of Mexico. It has six chapters that included an introduction, the historical background of chinampas agriculture, the historical reference of the Aztecs and the Valley of Mexico, the role of the chinampas in the formation of the Aztec Empire, the role of the chinampas in the expansion of the Aztec Empire, and conclusions. References are well chosen and extensive.

Chinampas are defined as a network of man-made ridges, such as low islands built into lakes, marshes, and shallow lakes. Nahuas worked with seedlings, willow trees planted on the edges of the platforms that prevent erosion and missed ground. Those trees also provided shade and firewood restrained the crop-damaging pests. The chinampa farmers used the muck they extracted from the canals as compost. Chinampas were garden plots in which crops were planted. They were from 15–30 feet and 300 feet long; they were only a few feet above the water level.

The hypothesis of the research was that the Aztecs could remain as a nomadic tribe like they were before settling on the islet of Lake Texcoco surrounded by marshlands and swamps; however, the chinampa agroecosystem gave Aztecs the opportunity to adopt it and from there grow to become an empire. This could have happened because the chinampa production allowed them to create a superstructure that provided them with food and tributes in addition to allowing them to expand and dominate beyond the Valley of Mexico. In addition, the war permitted them to obtain tax subjects.

10 The Central Highlands and the Homegarden

10.1 INTRODUCTION

Upon the arrival of the Spanish in the Central Highlands in 1518–1519, the homegarden agroecosystem was of various types. Among them, Nahuas included the polyculture homegarden with trees: the Mexican black cherry (*Capulín*), avocado, the North American native nut (*Carya illinoinensis* (Wangenh.) K. Koch), white sapota (*Casimiroa edulis*), Poinsettia (*Cuetlaxóchitl*) (*Euphorbia pulcherrima*), and the American coral tree (*Erythrina americana*). Homegardens had Agave, Opuntia, maize, beans, native onion (*xonacatl*), and native legumes (squash, cultivated, and wild green edible weeds). Also, there were orchards where fruit trees such as avocado, the Mexican Black cherry, and Mexican hawthorn were planted, in addition to vegetable gardens and spaces for the cultivation of squash, red and green tomatoes, edible green seasonal weeds (*bledos*, *quelites*), some of which were cultivated, and others were wild tolerated herbs.

Nahuas had homegardens dedicated for pleasure exclusively, where fountains and singing birds brought together for the enjoyment of personages like Moctezuma. They had forested areas where the sovereign and eminent people (*principals*) hunted wild fauna. In summary, the types of homegardens included those that were producers of food, raw materials, enjoyment, and pleasure. Pre-Hispanic homegardens very soon incorporated flora and fauna that came from the Old World (Figures 10.1–10.3).

10.2 THE NAHUA HOMEGARDEN, ORCHARDS, AND GARDENS IN THE SIXTEENTH CENTURY

In the Central Highlands after the native plants were mixed with trees and plants from the homegarden, orchards, and gardens brought by the Spaniards to New Spain (Dunmire 2004, González-Jácome 2011), the native agroecosystem had a structure made up of trees, wild and cultivated plants, wild fauna, water sources, cultural management, and agricultural instruments. The local knowledge of

FIGURE 10.1 Tepantitla Mural, representation of a Teotihuacan Garden dedicated to Tlaloc, full of water, with a maize plant, a buttertfly, and happy people.

DOI: 10.1201/9781003198833-10

FIGURE 10.2 Pontevedra, Spain: Trellis for Vines, and Entrance to the Orchard in San Xoán de Poio Monastery. Poio, Spain: Monastery, Trellis for Vines, and Entrance to the Orchard.

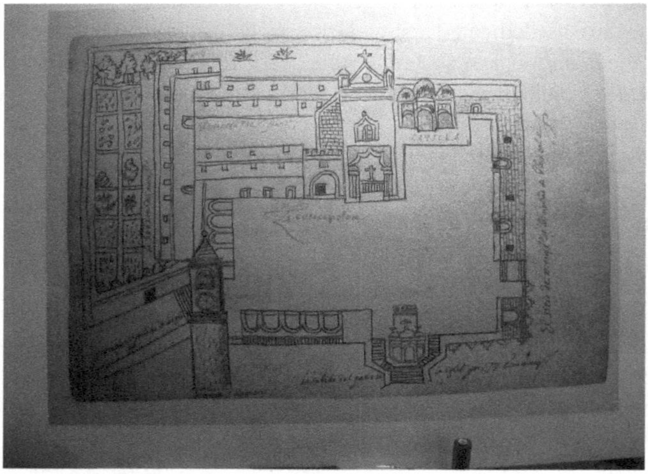

FIGURE 10.3 Tlaxcala City: S. Francisco Monastery with Homegarden and Garden on the Left Upper Side of the Plane.

pre-Hispanic farmers and the accumulated wisdom of farmers' experience in soil and water management using seemingly simple tools existed (Dunmire 2004). Before 1519, the agricultural technology lacked metal tools, plow, and working and pack animals (González-Jácome 2011: figures 10, 16, 30, and 31, pp. 437, 450, 494–495).

There are examples of the sixteenth-century homegarden in the Highlands of Central Mexico, and its flora like the Mexican hawthorn tree (*Crataegus mexicana*), the pear, the apple tree, grafted fruit trees, and Agave. Other fruit trees that were included were the Avocado (*Persea americana*), the Mexican Black cherry (*Prunus* sp.), the White sapodilla (*Casimiroa edulis*), and the "creole" walnut (*Carya illinoensis* Wangenh K.) (Dunmire 2004: 31–145). In Tlaxcala, the homegarden included Opuntia whose cladodes were used to house *Dactylopius coccus* from which mealybug (cochineal) dye was extracted. Unlike in Spain, where homegardens were irrigated, indigenous Mexican homegardens could be irrigated or not.

The pre-Hispanic central Highlands homegardens had different types of trees mixed with other plants, such as maize, beans, and legumes. An important part of the vegetables for the food system of pre-Hispanic societies came from cultivated and wild weeds, as was the case with wild amaranths, watercress, roseroots (*Suaeda* sp.), pápalo (*Porophyllum ruderale*), and a large number of edible foliage seasonal green herbs (quelites). The crops were organized in areas within the land according to the local characteristics of the soil, like the hardness and depth of the agricultural soil, humidity, and the gradient of the land (Figures 10.4 and 10.5).

Homegarden functions were aimed at supplying fruits and other edible or medicinal plants, obtaining firewood, elaborating ropes and nets, providing shade, obtaining elements for ceremonial and ritual purposes, or being planted for ornamentation purposes (Albores-Zárate 2016). Upon the Spaniards' arrival, there were several types of orchards, which were mentioned by Alonso de Molina (1571: 9, 12v, 112v). Sahagún (1830, 1956; Vol. 2: 144) described the orchards of Mexican black cherry (*capulquauhtla*) (*Prunus* sp.), avocado (*auacamilli*) (*Persea americana*), and Mexican hawthorn (*texocoquauhtla*) (*Crataegus mexicana*).

FIGURE 10.4 Tlaxcala City: Main Entrance to S. Francisco Monastery.

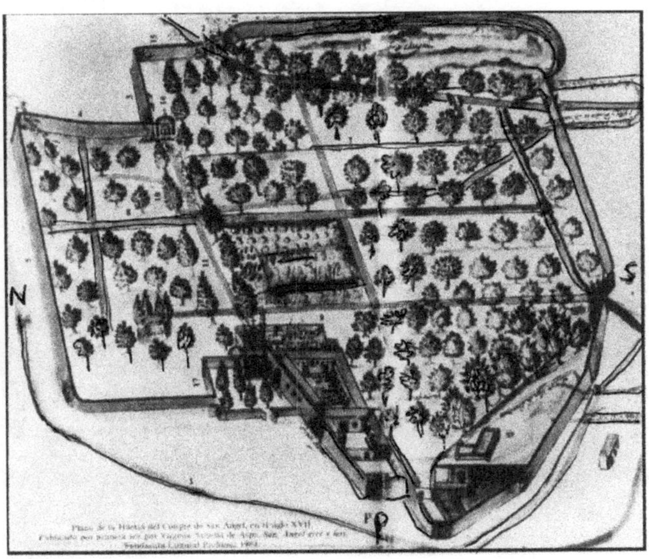

FIGURE 10.5 Mexico City: San Ángel: Plane of The Carmelite Homegarden, and Vegetable Garden.

Several homegarden trees or those that were planted on the edges of land surrounding the plots and raised earth edges were a source of edible flowers, such as those last century in Tlaxcala, with the flowers of *Erythrina* sp., *Agave*, and *Yucca elephantipes*. There was a wide range of edible flowers that to date has been studied little in relation to their role and importance in the preparation of food and beverages (Albores-Zárate 2016), as with aromatic flowers to put in chocolate, like some orchids. There were other flowers, such as those of the tree called *teonacaztli*, that were ground with cacao, "most aromatic, beautiful, very yellow, and with a strong smell", *Myrodia funebris* (*Quararibea funebris*) (*Cacahuaxóchitl*), flowering ear (*Hueynacaztli*), and string flower (*Piper auritum*) (*Mecaxoxitl*) (Appendix 1, *Revista Mexicana de Biodiversidad* 2005, Vol. 1 (76): 11–35).

These flowers scented the chocolate and were sold in the markets, like Tlatelolco, as were other edible flowers like squash flowers, Scarlet runner bean flowers (*Phaseolus coccineus*), and Garambulla cactus flowers (*Myrtillocactus geometrizans*). The list could be expanded regionally (Sahagún 1956, Vol. III: 289).

The homegarden in the basin of Mexico was next to the houses; this was not true necessarily for all the regions of Mexico. The types of homegardens that are found today had undergone major transformations, especially those that were organized in areas where land tenure, modernization, and urbanization processes reduced the area of arable land of the native population. There is a lack of information about the role that children and the elderly played in the work applied to the agroecosystem in pre-Hispanic times (González-Jácome 2011: 101–103) (Figures 10.6–10.8).

10.3 THE NOVO-HISPANIC HOMEGARDEN AND ITS CONFORMATION

From 1518 to 1519, with the arrival of the Spaniards to the Central Highlands, and after 1524, in the Yucatan Peninsula, the native homegarden began to undergo important transformations with the introduction of citrus trees and fruit trees, such as pear, quince, fig, apple, Castile walnut, olive, and Old World plum, plus grape vines (González-Jácome 2011; Mariaca-Méndez et al. 2010). Some native plants were displaced and/or replaced by others, as happened with the Mexican hog plum (*Spondias* spp.) or the native onion. There are plants that were grown in pre-Hispanic homegardens that are unknown or in disuse, currently. Others have lost the importance that one day they had, like Mexican Black cherry (*Prunus* spp.), Mexican hawthorn (*Crataegus* sp.), white sapota (*Casimiroa edulis*), paper nut or pecan (*Carya illinoiensis*), and coral bean tree (*Erythrina americana*) (*Zompantle, colorín*).

FIGURE 10.6 S. Felipe Ixtacuixtla Convent: Main Entrance and Homegarden.

FIGURE 10.7 S. Francisco Tepeyanco, Monastery and Homegarden.

FIGURE 10.8 Entrance to the Convent of S. Francisco, Tlaxcala City.

The fruits of the pear, quince, fig, and apple are still used in the preparation of sweets and jellies. Mexican Black cherry (*Prunus* sp.) called "the cherry of the earth" by the Spanish, was used in the elaboration of *atole*, *tamales*, spirits, and sweets, and continues to be used today. Its branches and leaves – like those of White sapota – served to wrap bread made with kneaded flour for the called *Pulque* bread (*pan de pulque*) and Holiday bread (*pan de fiesta*), giving them freshness and fragrance (González-Jácome 1985, 2003b, 2016; Villalobos-Nájera 1998, 2008, 2009). Roasted and spiced Mexican Black cherry seeds with chili were sold retail, as were peanuts, chickpeas, and roasted pumpkin seeds, all of which were consumed as snacks. They have now been partially replaced by junk food and other industrialized products.

Currently, the white sapota tree is in the process of abandonment. Its fruit is still used as food to fatten pigs, although the importance that its young branches and leaves had to wrap bread and keep it fresh has been reduced since the appearance of plastics in 1861 and their diffusion in Mexico since the 1950s (González-Jácome 2003b, 2016; Villalobos-Nájera 1998, 2008, 2009).

Pulque bread and Holiday bread are made still, and they are bought and consumed by the attendees of the festivities in various rural and urban towns of the Mexican Central Highlands. They reach as far away as the borderlands. Currently, stalls set up to sell these breads during the festivities of S. Jude Thaddeus on the 28th of each month. Said festivities are organized in the temples, such as the former convent of St. Hippolytus (known as S. Jude) or the S. Jude temple in the north of Mexico City. Holliday bread and *Pulque* bread have lost the importance they once had, so although their producers and sellers continue to market it, they are replacing the bread much more with pizzas they prepare in portable ovens (Castañeda-Castillo 2014).

In the Tlaxcallan homegarden, paper walnut or thin-shelled pecan (*Carya illinoiensis*) was cultivated but was partially replaced by walnuts (*Juglans regia*) brought from the Old World (called Castile walnut or European nut) before 1580. Currently, the nut crop is being abandoned in homegardens. It is still produced and collected for commercialization in the town Amaxac de Guerrero.

Among the trees that in 1574 were not mentioned as part of the agroecosystem is the Mexican hawthorn called "earth apples" by Spaniards (*tejocote*, *Manzanita de la tierra*), which after the conquest was used as a base to graft pears and apples. The Mexican hawthorn was an important tree in the New Spain homegarden or on raised borders (agroecotones, *metepantles*, earth edges, stone walls). It was grown between cultivated areas, also (González Jácome 2006 (3): 1946, 2016c: 233–257, 2018[a]: 58–150).

In Tlaxcala, Mexican hawthorn cultivation is in a state of extinction. Some Mexican hawthorns remain on the land edges of the Atoyac and Zahuapan rivers basin where they were planted on the sides of the canals, or between the plots as boundaries. The consumption of their berries as fruit has been reduced to the holidays for the preparation of sweets, jellies, and punches. However, along with peanuts and sugarcane, hawthorns berries have been replaced by industrialized sweets wrapped in cellophane, chewing gum, confetti, flour, and plastic toys, which are used to fill the *piñatas* for holiday parties, and children's birthday parties now (González-Jácome 2003b, 2011, 2018d).

In the southwest of Tlaxcala, until the 1970s, the Coral bean tree (*Erythrina americana*) had several important uses: its flowers were eaten, covered with stiffly beaten eggs whites, and fried; masks and various items were made from its wood; the tree was planted on the shore, next to the fences or walls between the houses, on the cultivation plots, and their edges as protection against evil spirits.

There are manufacturers of wooden masks in an old neighborhood in S. Pablo Apetatitlan that argued that the current size and thickness of the Coral bean tree do not allow the elaboration of masks made of a single piece, so they been replaced with the wood of other trees. However, they consider that those that are made of *Erythrina* are the authentic ones and have the highest quality.

Francisco Hernández (1570–1571) gave information that the homegarden in the central Highlands provided an important part of the diet and health for families. Flowers were cultivated for their aroma and religious purposes to make or perfume drinks as well as to make some dishes. Wild aromatic plants, such as orchids, were collected and sold in the market. Several flowers to aromatize chocolate were sold at the Tlatelolco market, along with cacao beans, and some sweetener like Agave sugar, Agave stirrup, maize sweet, or wild bee (Melipona) honey (González-Jácome 2011: 96–103).

Paulina Machuca (2013) published the results of a study based on the *Geographic Relations* of the sixteenth century, about the arrival of plants in the Balsas region, in Jalisco. These plants came from both the Old World through the Old World and through the Philippines. Among the conclusions of the study expressed as a reflection, Machuca considered that the consequences of the introduction of these plants in Mesoamerica included: the expansion of cereals, vegetables, and tools; and they were cultivated "[...] in the European way". González-Jácome (2011: 212) wrote:

> Until now we have little information on specific ways that tell what happened when packing, and moving the first live plants from the other side of the Atlantic Ocean. Perhaps the initial and most documented case is that of the introduction of the orange by Bernal Díaz, occurring in the year 1518-when

> Díaz participated in the Juan de Grijalva expedition- and he was in the lands located between Tonalá, and Coatzacoalcos, where he planted some orange seeds next to a temple [...].
>
> *(Dunmire 2004; López-Piñero, and Pardo-Tomás 1994; Romero-Contreras et al. 2004: 237–245)*

The entry into New Spain of plants, animals, tools, and agricultural technologies for crop management introduced elements for the conformation of a new agricultural model, which would take several centuries to become combined with the varied models of traditional Mesoamerican agriculture, at least with those of Central Highlands, mountains, central valleys of Oaxaca, and those located in the highlands and lowlands of the Mayan area. These models were the result of sociocultural adjustments or adaptations to local conditions. Also, they would impact local and regional societies (regardless of ethnic groups), their languages, and their cultures (González-Jácome 2001b; Terán and Rasmussen 1994).

Cultivated plants that arrived in New Spain from the Old World were pears, figs, quinces, apples, lemons, oranges, peaches, apricots, walnuts, grapevines, and olive trees (Dunmire 2004: 83–145). The information means that homegarden was dedicated not only for self-consumption but also for needs of the market in order to provide some monetary income to indigenous families (González-Jácome 2011: 247–249). The fruit trees of the Old World spread rapidly through New Spain and Guatemala, where by 1586, the Franciscan Friar Alonso Ponce (1586) on his visit found them grown and flourishing around Lake Atitlan and neighboring regions. The small-scale farmers' and their families' social needs generated new ways of obtaining money, which were associated with seasonal migration, interchange, and commerce.

Cruz-León (1989: 24) defined the agricultural technology after 1519, as composed of pulleys and hand tools for agricultural work that included wooden and metal containers used throughout the work process, along with equipment proper for the activity, such as harnesses and work instruments used in relation to animal traction. With the entry of the Spaniards to the Central Highlands, the homegarden underwent several transformations, beginning with the introduction of fruit tree species from the Old World. The friars had a lot to do with it, especially the Franciscans, who were the first to enter these regions. Motolinía (1541: 435) wrote:

> In New Spain the same friars have planted almost all the fruit trees, and persuaded the other Spaniards to plant them too; and they taught many to graft, which has been the cause that today there are many, and very good homegardens, and there must be many more; because the Spaniards have seen that the land produces one hundred and one of what they plant, given much to plant, and ingested good fruits, and trees of esteem.

In the native homegarden, the Spaniards introduced: citruses, pears, apples, quinces, figs, and cherries (Dunmire 2004: 83–115; González-Jácome 2011, figures 13, 14, 15). The homegarden in the Franciscan convents was separated from the vegetable garden. Legumes and vegetables constituted the daily food of the friars (González-Jácome 2011: 268–295, 513). The number of monasteries and convents was of importance for the entry and diffusion of fruit trees, grains, and legumes.

Motolinía (1996: 296) calculated that by 1524, in Mexico City there were 12 monasteries, about 10 churches, and a secular parish (Santiago Tenochtitlan). Peter Gerhard's (1986: 187) records show that each of the four Indian neighborhoods was converted into a Franciscan visit (stopovers), and some of the visits were later entrusted to the Augustinian and Carmelite friars.

Alonso Ponce (1976, Vol. I: XXVIII, CLII-CLIII, 38, 52–54, 74–77) described the Franciscan convent in Tlaxcala City as founded in 1525 and rebuilt and expanded before 1548. Its facilities were a community house with two bedrooms and another bedroom on the floor above. It had an infirmary for the friars of the Archbishopric of Mexico and S. José de los Naturales, and for those of one attached chapel. It was the head and main house of the Holy Gospel province, where theology was taught. In the convent the offerings of the Inidians were keep, and there was a friar living there. By 1579, the monastery has two sources of water, a ditch, and its arm, from which water was put inside the house.

By 1580, the convent homegarden was located at the northeast of the grown, where Castile walnut, quince, prisco peach, apple, pomegranate, cherry, plum, pear, orange, lemon, and other trees, plus grape vines were sown. The garden had vegetables, legumes, flax, and hemp from Spain. Another one occupied a space with smaller dimensions, located to the north of the convent.

In the sixteenth century, the 13 Franciscan convents of the Tlaxcala Province were S. Francisco Tlaxcala, Our Lady of the Assumption, S. Juan Totolac, S. Francisco Tepeyanco, St. Ana Chiautempan, St. Mary Atlihuetzía, St. Mary Nativitas, S. Luis Huamantla, S. Juan Atlangatepec, S. Felipe Tequemeca, S. Ildefonso Hueyotlipan, S. Felipe Ixtacuixtla, and Calpulalpan. Their homegardens and gardens had Castile walnut, chestnut, peach, pear, quince, apple, olive, prisco peach, fig trees, and grapevines. *Opuntia*, with prickly pears, was added to the list but was the only native plant that was included in the Hueyotlipan convent homegarden. In addition, there were Castilian roses, lilies, asparagus, and other vegetables. Flowers were very important to place on altars (González-Jácome 2011: figure 44, p. 513).

With information from 1574, by Francisco Hernández, proto-medical of Felipe II, it was possible to reconstruct the composition of the flora and fauna of the homegarden in the basin of Mexico. This included trees, shrubs, maize, other grasses, legumes (beans), vegetables, and flowers. The preparation of food, chocolate, atole, and herbal drinks was executed using grains, herbs, and spices for flavorings. Trees for wood or firewood were grown there, too. The flora with medicinal uses obtained from homegarden plants, but mostly from forests, nearby ecosystems, and ecotones.

The homegarden fauna included birds and fish that were caught or fished in the fountains or in the canals (*apantles*), whose waters ran through the agroecosystem. In the houses, or next to them, were cages with various types of birds, which beautified the sites with their plumages and singing. Some of the birds were for consumption González-Jácome (2011: 137–148, 2018: 43–85). Both in the Old and New World, food was linked to diet and health, but this did not mean that the population ignored the senses when taking pleasure in dishes that were time-consuming in their preparation, as is noted in the texts of Bernal Diaz del Castillo (1960) and Francisco Hernández (1984 [1574]).

In Mexico City, during the second half of the sixteenth century, women cultivated homegardens and many of the medicinal plants were focused on them and their needs. In 1574, Francisco Hernández, Philip II's proto-medical, obtained the information to rebuild the plants cultivated in the homegardens in the Basin of Mexico. Mexican flora was collected when the effects of the deaths of young men and adult males who participated in the war of the conquest that occurred 50 years before suffered and emphasized the female role of working in homegardens. This can be an indication that women were responsible for maintaining the agroecosystem, and this idea is reinforced because a large number of homegardens were located next to the houses.

In Spain, since 1501, a Royal Decree established that with the commercialization of fruits and vegetables the owners of homegardens should pay tithes to the Church. In 1624, the Royal Council of the Indies demanded of the religious orders the payment of tithes to the cathedrals from the earning obtained from the products of their labor and livestock that they had sold. Laws related to this were issued in 1655, 1657, and 1664, when orders were condemned to the payment of tithes. In New Spain, by 1677, the Carmelites began a lawsuit to avoid paying any tithe (Abundis-Canales 2007, Vol. 1: 460). This litigation allows researchers to know some figures related to the production and commercialization of products from the New Spain homegardens that were created and managed by religious orders.

The Carmelite order had homegardens in Tenanitla, Chimalistac, S. Ángel, or St. Ana, and they are good examples of their functions and economic earnings. The convent in San Angel and its homegarden became very large in extension, its fruit production was directed to the market, and it was a very important source of income for the order (Abundis Canales 2007, Vol. I: 460). The S. Angel homegarden was fenced, and the Carmelites founded in 1613, the new facilities of the Valladolid school with the name of S. Angel Martyr (Ramos 1992: 64).

In 1624, the Royal Council of the Indies in Seville demanded of the religious orders the payment of tithes to the cathedrals from the earnings obtained from the products of their labor, and livestock that they had sold. Laws related to this were issued in 1655, 1657, and 1664, when religious orders

(Jesuits, Dominicans, Augustinians, Mercedarians, and Carmelites) were condemned to the payment of tithes. The Carmelites began litigation (1677) to avoid paying any tithes (Abundis-Canales 2007, Vol. 1: 460).

The Hondo River, the Chico stream, their very large ditches, and three ponds irrigated the S. Angel Carmelite homegarden and vegetable garden. The homegarden had 30 thousand fruit trees – the majority pear and peach – which were used for internal consumption and for sale. It had an area where legumes and vegetables were grown for the daily consumption of the religious community. Some fruit was sold on the grounds of the convents, but most of the production was sent to the Mexico City, Coyoacan, Pachuca, and Puebla markets. From 1671 to 1672, the sale of fruit produced 9,000.00 pesos to the S. Angel homegarden (Abundis-Canales 2007, Vol. 1: 460).

10.3.1 The Formation of the Novo-Hispanic Homegarden

From the Old World, the following fruits, vegetables, and legumes were integrated into the indigenous homegarden and the cornfield: the melon and turnip in 1526; carrot, lettuce, and garlic by 1529; and the cabbage in 1538. Later, but before 1580, onion, borage, parsley, and watermelon arrived in New Spain (Dunmire 2004). Several native plants were replaced by other Hispanic equivalents, altering the native agroecosystem. For example, lettuce and cabbage began to be grown in the chinampas of the basin of Mexico. The leaves that were removed from the plants were used to make the new green fertilizers, and feces, and remains of turkey feathers and bones, with the mud from the canals were used to maintain soil fertility. The time the plants were growing in the chinampas soil was shortened at least 1 month with the use of seedlings (chapines), and the agroecosystem allowed 3 harvests/year, then. The impact of new crops on the agroecosystem is not known, but the leftover leaves were larger than weeds and plant residues that were used before 1519 (Dunmire 2004: 34–38).

Information began with the plants brought by the Spaniards to the New World, saying that some native species disappeared very soon after that from the list of Mesoamerican crops, or that others replaced them. Thus happened that the native onion (*xonacatl*) was replaced with the onion from the Old World, despite the fact that Spaniards like Francisco Hernández argued that they liked the milder flavor of the native onion more. The same happened with the edible green seasonal weeds (quelites, *bledos)* (*Amaranthus hypochondriacus*), which only in recent decades, and mainly as a result of vegan and vegetarian food fads have regained their importance in the food system.

McClung de Tapia et al. (2014: 106–107), based on the sixteenth century information provided by Bernardino de Sahagún and Francisco Hernández, indicated that food plants found at different sites in the basin of Mexico[1] were at that time: *Amaranthus*, *Spondias mombin*, *Asteraceae*, *Helianthus annuus*, *Opuntia*, *Mammilaria*, *Myrtillocactus geometrizans*, *Cucurbita*, *Chenopodium*, *Dysphania ambrosioides*, *Mexican Suaeda*, *Crotalaria*, *Phaseolus*, *Prosopis*, *Salvia*, *Persea americana*, *Oxalis corniculata*, *Zea Mays; Portulaca oleracea*, *Crataegus Mexicana*, *Prunus serotina*, and *Solanum*.

Not all the plants that were brought to New Spain bore fruit in the places where they were transplanted initially, as the vines in Yucatan. In three Ordinances of Good Governance, signed by Cortés on March 20, 1524 (Hernán Cortés 1960: 172; Martínez 1990, Vol. 1: 277–283), the following was established:

> Item: [...] Any neighbor who has Indians of distribution is forced to grow in every year, per every one hundred Indians a thousand plants, choosing the best they could find; it is understood that neighbor plant them quickly, and well cured, so that they can fruit; which said plants, can be grow in the part that seems to him, not harming another; and that he plant them in every year, in the times that should be planted, until he arrived of every one hundred Indians to five thousand strains; under penalty that for the first year that he did not put them or cultivate, he pays half a gold frame applied as he is said; and for the second, the bent penalty; and for the third, he lose the Indians of distribution he has.

[1] Xaltocan, Iztapalapa, Tlatelolco, Otumba, Xico, Michpilco, Terremote Tlatenco, Cuanalan, Temamatla, and Teotihuacán housing units.

> Item: [...] that having vines plants of those of Spain in the land in quantity that can be made, they are forced to graft the strains that they had of the plant of the earth, or of new planted of again, just with those penalties.
>
> Item: [...] that having other tree plants in Spain, or wheat or barley, or any other legumes, they are also obliged to plant, and sow in the villages of the Indians that they have, under the aforementioned penalties.

At the end of the Ordinance, Hernán Cortés (4th Relation, 10/15/1524) asked at the emperor:

> Plants of all places will be brought to this land and, because there are all kinds of agriculture in her, because so far nothing was provided to this land. Begging your majesty, because will be very useful, order to the House of Procurement of Seville so that each ship brings a certain amount of plants, and that it cannot came without them, because it will be very good to the population [...]. I have also informed your Caesarian Majesty about the need for plants to be brought, because there are of all kinds of agriculture in this land [...].

Because of its importance in Mexican life and economy, indigenous agriculture from the sixteenth to nineteenth centuries has been treated in several historical studies. Although the native homegarden and the cornfield suffered strong impacts from the introduction of new plants, animals, agricultural equipment, management, and techniques from the Old World, they were reorganized as a new traditional agricultural system model (Mexican agriculture). Native agriculture adapted and incorporated European farming systems and technology to the New Spain environmental conditions. This led to new ways: doing agriculture; creating other forms of land tenure; working it; and including technology, tools, new crops, and ways of adjusting to environmental, climatic, and economic market conditions.

Several scholars of Mexican agriculture discussed the results of the introduction of new technologies. Some argued that they increased deforestation, soil erosion, or dried up lake areas, as happened with the basin of Mexico. Cruz-León (1989, 1997) considered that one of the strongest impacts on the soils was caused by the introduction of the plow and the draft, and load animals used in agriculture (Cruz-León and Martínez-Saldaña 2001). However, the opinions and studies underlying these proposals are varied.

A conclusion of what happened can be drawn from the entry of new concepts, plants, animals, and technologies for Mesoamerican societies, their economies, social, cultural, and labor organization forms, legislation, food systems, and daily life. These concepts proposed agricultural practices that ignored the knowledge that Mesoamerican pre-Hispanic farmers had developed along several millenniums.

The Indian farmers in New Spain were influenced by Hispanic farmers on agricultural methods, management of soils, technologies, social organization of work, cultures, worldview, religion, and associated ceremonies. Undoubtedly, in the Viceroyalty, the Novo-Hispanic small-scale farmers received numerous influences from the Old World, which they adapted to the existing ones to form a new model of agriculture. Sahagún (1830 [1567], T. III, L. X., Chapter XII, pp. 28–29) wrote about the agricultural tasks of farmers:

> The good farmer is strong, diligent, and careful, he gets up very early for not losing his farm, and for increasing it he stops eating, and sleeping, works hard in his trade, agrees to know, in digging the ground, weeding, in time of dry, dismantling, paving the dug, making ridges, put soft the earth well [*mollir*], plowing it in its time, making boundaries, and fences, and also breaking ground in time of rain water, knowing how to choose the good one to make it, make holes to pour the seed, and water it in dry weather; sow spilling the seed, drown the earth to sow the beans, blind the holes where the maize is sown, or cushion or gather the earth, to what is born: remove the grass [*vallico, ballico*], entrench the reeds by breaking them, and remove the cobs, and remove the stalks, because what is born grows well, entrench the green maize cobs in their time: at the time of harvest, break the reeds by picking them up, and take the maize when it is well seasoned: removing the husks from the maize on the cobs, and tying one with the other, adding the nightgowns with each other, and making corncobs by tying with each other, and carrying home cooked and silage it: breaking the reeds

> that nothing they have by pounding them, threshing, cleaning, throwing, raising the threshing wind. The bad farmer is very negligent, lazy, and all work is made serious, and annoying; in his work he is coarse, gross, rude, villain, who eats a lot, scarce, enemy of giving, and friend of taking.

The Viceroyalty was a period of relearning to do things. The changes and adjustments in agriculture were not as easy or as fast as one might suppose. The innovation processes suffered several disasters, probably. The environmental and economic risks could not be controlled as they were before the Spanish entry to Mesoamerican lands. In this sense, the role played by homegardens and gardens in convents was very important, which is quite well known (Gibson 1952). The friars taught the Indians how to sow and care for newly arrived plants, and educated them on the technology associated with them, especially during the sixteenth and early seventeenth centuries (Galeote 1997: 119–161). The role of the gardeners, many of the friars in the convents, was fundamental.

Bernardino de Sahagún (1830 [1567], Vol. III, Book X, Chapter XII, pp. 28–29), when numbering different offices of the population, defined the gardener: "[A man who] has the office of sowing seeds, plants, and trees, caring for the grounds, digging, and softening the earth well. The good gardener is usually discreet, careful, prudent, of good judgment, and has a ledger with the month, and the year". The gardener of the homegardens in the convents specialized in the cultivation of trees and flowers, which were to decorate altars. Many convents had gardens, where the cultivation of vegetables satisfied the friars' basic needs. In general, the New Spain convents followed the model of the Spanish convents and monasteries.

The indigenous homegarden, in addition to having native and introduced plants, included *Agave* and *Opuntia*, which helped the economy of the native families. The *Agave* is associated with soil conservation when grown on terraces, and in rows on land borders (*metepantles*). In addition, *Agave* offered a wide variety of products for everyday life that ranged from food, drink, construction materials, to others for religious uses. A document dated 1789 (AGN Tierras, Vol. 552, Exp. 2, fol. 47v.), cited by Trautmann (1981: 112), deals with the use of Agave cultivated on eroded land surfaces, hardened by wind, rain, and loss of vegetation (*tepetate*), located in S. Esteban Tizatlán. This settlement sits on a surface that includes hillside lands descending to a flat area.

Trautmann (1981: 112) found that at the end of the eighteenth century Agaves were cultivated for tax purposes in the Tlaxcala and Chiautempan jurisdictions, but by then those lands were very eroded (Trautmann 1981: 112. AGN, Tithes, Vol. 20). In fact, these pre-Hispanic functions were not focused on fermented-Agave-mead (*pulque*) production, a phenomenon that occurred later with the configuration of *pulque* estates (*Haciendas pulqueras*). Apparently, they began in 1590 and grew when pulque production became a lucrative business, reaching its peak in the nineteenth century.[2]

In the sixteenth century and the first half of the seventeenth century, the Tlaxcallan indigenous homegarden was cultivated with *Opuntia*, mainly because the plant was used as a cochineal host and cochineal dye was an important trade between New Spain and Spain. The common indigenous people and the members of the Indigenous Council (*Cabildo Indígena*) had problems with the production of grana, or cochineal, as a consequence of the enrichment of the former. The constant complaints of the indigenous authorities of the town hall were for the lack of obedience of the Tlaxcallans enriched by the grana trade and the abandonment of their obligations of agricultural activities that were not as lucrative for them as the production of cochineal paste and its commercialization.

The Tlaxcallan cochineal trade was done with Hispanic merchants in Puebla City, but there were problems due to quality, because the Indians mixed the paste with sand, to increase its weight (Gibson 1952; Trautmann 1981). During the Viceroyalty, *Opuntia* was cultivated in various places,

[2] Gibson 1967: 326; González-Jácome 2010: 521–554; Hernández-Palomo 2004: 239–257; Ramos-Galicia (cord.) 2014; Rendón-Garcini 1995, 1996; Vela et al. 2014.

like Oaxaca, because the insect dwelled in the *Opuntia* cladodes. During the Viceroyalty, dye production created great richness, but in the second half of the nineteenth century natural dyes were displaced by chemical dyes. Gonzalo Gómez de Cervantes (1944 [1599]: 138) wrote:

> New Spain produces four genres of things that each one of them is a treasure, of all it is done every year it is taken out to Spain, of all they are taken, and navigated to the kingdoms of Castile, with which those kingdoms are enriched, and they are ennobled. [This things] are: silver, indigo, grana, and leathers [...].

Very soon, the indigenous population of the Central Highlands that had homegardens increased their diversity, including a mixture of native fruit trees with those brought from the Old World. The native trees were Avocado (*Persea americana*), Mexican black cherry (cherries of the earth, *Prunus* sp.), Mexican hawthorn ("apples of the earth": *Crataegus* sp.), pecan walnut (shelled pecans, *Carya illinoensis*), and White sapota (*Casimiroa edulis*). Trees and vines from the Old World that were introduced to the homegarden were pears, figs, quinces, apples, lemons, oranges, peaches, apricots, European nuts (*Juglans regia*), grapevines, and olive trees, all of which had spread from the Central Highlands to distant regions in Guatemala, already by the end of the sixteenth century (Alonso de la Mota y Escobar 2012).

Based on information of the Bishop of Tlaxcala, Alonso de la Mota y Escobar *Memorials*, Trautmann (1981: 112) considered that by 1614, there was a large fruit crop in the province, and district elders documented its surface cultivation in Topoyanco (Tepeyanco). This extensive fruit cultivation can be considered as an antecedent of the fruit growing of current homegardens. It was initiated by the dissemination actions of the Franciscans who founded the convent of Tlaxcala, "[…] who were the bearers of innovations even in the rest of Mexico".

10.4 NEW SPAIN TYPES OF HOMEGARDEN

There are differences due to climate affecting trees, cultivated flora, and the wild herbs of the homegardens. In the tropical and temperate-cold regions, the homegarden agroecosystem introduced the citrus fruits that arrived in Mexico very soon like occurred in the basin of Mexico and in Tlaxcala-Puebla Valley. In the tropical and subtropical regions, there were other cultivated trees, such as cacao, Mexican hog plum, gourds, and vines. These trees and vines were grown, and their fruits were marketed inside and outside their regions because they were important objects of trade. Farmers used and consumed their fruits along with those of their wild relatives.

On the edges of land and nearby places, there were wild varieties of cacao and gourds (*Crescentia alata* Kunth) (*tecomate*, *comate*, *cuestecomate*). The latter were tolerated flora sponsored by homegarden owners. Their flowers were used in herbalism, and their fruits were made into drinking utensils known as *xicalli* (*jícaros*, *jícaras*). Gourds were found as far as Michoacán and Guatemala. Friar Alonso de Ponce (1586: 7) described them:

> In that town there were many gourd trees, which are medium, with small leaves that cover the branches poorly, the fruit they carry is like medium-sized pumpkins, very round, and stuck by the nipple of the same trunk, and thick branches. Like cacao pods, these are cured, and sawn in half, as an orange is ordinarily cut, they cut each one into two, which serve as bowls, casseroles, and cups, and glasses in which they drink chocolate. This is the common service of the Indians, blacks, and even poor Spaniards, called in the Mexican language xicalli, a corruption of the word *jícara*, they make some very curious, frayed, and painted them, which have a lot in Mexico. They are also seasoned without splitting them, as jars, with their mouth, and they served to pour water, wine, vinegar, and other liquors. These they are made in New Spain […] in Michoacán, and other very large gourds, as sources, and large plates, which are not from trees, but some sort of very large pumpkins, which taken from their bushes, and they are cut in half, and cured they are varnished, and paint them, and sell very expensive, and take them to Mexico, and other parts of New Spain.

Homegarden fruits were for self-consumption and sale in the markets. Other parts of trees were destined for herbalism (leaves, flowers, fruits, branches, and sections of bark). Everyone provided some monetary income for the small and medium-scale farmer families. The Mexican homegarden agroecosystem was the result of a combination of the elements of the homegardens and gardens of Hispanic origin, with the Mexican pre-Hispanic ones, resulting in greater biological and cultural diversity. There are several types and sizes of homegardens, which depend on the region where they are located, and their altitude, gradient, physiographic, environmental, and diverse characteristics.

These homegarden types can be placed in a continuum: (1) with greater or lesser biodiversity (Toledo 2010 [2015]); (2) with a range from polyculture to monoculture; (3) with rainfall, being irrigated or having seasonal rains; (4) with spaces minimal or with large extensions; (5) with use or not of industrialized inputs and machinery; (6) with production for self-consumption or directed to the market; (7) with the sole use of family labor force or mixed with wage labor.

The food system and indigenous herbalism of the sixteenth and seventeenth centuries included cultivated plants, weeds, and wild tolerated plants, which were not necessarily the result of a crop, but were preserved and collected in ecotones, natural vegetation strips, galleries, forest relics, and in nearby woods. Those areas included natural and artificial edges of agroecosystems, the sides of ravines, river canals, riverbanks or lagoon areas, marshes or territorial boundaries.

A large number of plants, fruits, mushrooms, wild fauna, and some minerals were part of the life and economy of farmers. They were collected in the neighboring natural environments. Sahagún's information (1830) shows the importance of forests and ecotone zones for the collection of wild plants destined for herbalism. In fact, agricultural systems, such as homegarden and cornfield, had more food and herbal plants because they were stocked with wild plants collected in natural areas.

Mesoamerican agroecosystems had many changes after 1519. Several native plants were replaced by other Hispanic equivalents, altering the native agroecosystem. For example, lettuce and cabbage began to be grown in the chinampas of the basin of Mexico. The leaves that were removed from the plants were used to make new green fertilizers, and turkey remains and the water mud from the canals were used to maintain soil fertility. The time the plants were growing in the chinampas soil was shortened at least 1 month with the use of seedlings (chapines), and the agroecosystem allowed three harvests per year. The impact of new crops on the agroecosystem is not known, but the leftover leaves were larger than weeds and plant residues that were used before 1519 (Dunmire 2004: 34–38).

Information began with the plants brought by the Spaniards to the New World, saying that some native species disappeared very soon after that from the list of Mesoamerican crops or that others replaced them. Thus happened that the native onion (*xonacatl*) was replaced with the onion from the Old World, despite the fact that Spaniards like Francisco Hernández argued that they liked the milder flavor of the native onion more.

Not all the plants that were brought to New Spain bore fruit in the places where they were transplanted initially, as the vines in Yucatan. In three Ordinances of Good Governance, signed by Cortés on March 20, 1524 (Hernán Cortés 1960: 172; Martínez Vol.1 1990: 277–283), the following was established:

> Item: [...] Any neighbor who has Indians of distribution is forced to grow in every year, per every one hundred Indians a thousand plants, choosing the best they could find; it is understood that neighbor plant them quickly, and well cured, so that they can fruit; which said plants, can be grow in the part that seems to him, not harming another; and that he plant them in every year, in the times that should be planted, until he arrived of every one hundred Indians to five thousand strains; under penalty that for the first year that he did not put them or cultivate, he pays half a gold frame applied as he is said; and for the second, the bent penalty; and for the third, he lose the Indians of distribution he has.

> Item: [...] that having vines plants of those of Spain in the land in quantity that can be made, they are forced to graft the strains that they had of the plant of the earth, or of new planted of again, just with those penalties.
>
> Item: [...] that having other tree plants in Spain, or wheat or barley, or any other legumes, they are also obliged to plant, and sow in the villages of the Indians that they have, under the aforementioned penalties.

At the end of the Ordinance, Hernán Cortés (4th Relation, 10/15/1524) asked at the emperor:

> Plants of all places will be brought to this land and, because there are all kinds of agriculture in her, because so far nothing was provided to this land. Begging your majesty, because will be very useful, order to the House of Procurement of Seville so that each ship brings a certain amount of plants, and that it cannot came without them, because it will be very good to the population [...]. I have also informed your Caesarian Majesty about the need for plants to be brought, because there are of all kinds of agriculture in this land [...].

Because of its importance in Mexican life and economy, indigenous agriculture from the sixteenth to nineteenth centuries has been treated in several historical studies. Although native homegarden and cornfield suffered strong impacts from the introduction of new plants, animals, agricultural equipment, management, and techniques from the Old World, they were reorganized as a new traditional agricultural system model (Mexican agriculture). Native agriculture adapted and incorporated European farming systems and technology to the New Spain environmental conditions. This led to new ways: doing agriculture; creating other forms of land tenure; working it; and off including technology, tools, new crops, and ways of adjusting to environmental, climatic, and economic market conditions.

Several scholars of Mexican agriculture discussed the results of the introduction of new technologies. Some argued that they increased deforestation, soil erosion, or dried up lake areas, as happened with the basin of Mexico. Cruz-León (1989, 1997) considered that one of the strongest impacts on the soils was caused by the introduction of the plow, and the draft, and load animals used in agriculture (Cruz-León and Martínez-Saldaña 2001). But the opinions, and studies underlying these proposals are varied.

A conclusion of what happened can be drawn from the entry of new concepts, plants, animals, and technologies for Mesoamerican societies, their economies, social, cultural, and labor organization forms, legislation, food systems, and daily life. These concepts proposed agricultural practices that ignored the knowledge that Mesoamerican pre-Hispanic farmers had developed along several millenniums.

The Indian farmers in New Spain were influenced by Hispanic farmers on agricultural methods, management of soils, technologies, social organization of work, cultures, worldview, religion, and associated ceremonies. Undoubtedly, in the Viceroyalty, the Novo-Hispanic small-scale farmers received numerous influences from the Old World, that they adapted to the existing ones to form a new model of agriculture.

The Viceroyalty was a period of relearning to do things. The changes, and adjustments in agriculture were not as easy, or as fast as one might suppose. The innovation processes suffered several disasters, probably. The environmental, and economic risks could not be controlled as they were before the Spanish entry to Mesoamerican lands. In this sense, the role-played by homegardens, and gardens in the convents was very important, which is quite well known (Gibson 1952). The friars taught the Indians how to sow, and care for newly arrived plants, and educated them on the technology associated with them, especially during the 16th, and early 17th centuries.

Bernardino de Sahagún (1830 [1567], Vol. III, Book X, Chapter XII, pp. 28–29), when numbering different offices of the population, defined the gardener: "[A man who] has the office of sowing seeds, plants, and trees, caring for the grounds, digging, and softening the earth well. The good gardener is usually discreet, careful, prudent, of good judgment, and has a ledger with the month, and the year". The gardener of the homegardens in the convents specialized in the cultivation of trees, and flowers, which were to decorate altars. The role of the gardeners, many of them friars in the convents, was fundamental. Many convents had gardens, also, where the cultivation of vegetables satisfied the friars' basic needs. In general, the New Spain convents followed the model of the Spanish convents, and monasteries.

The indigenous homegarden, in addition to having native, and introduced plants, included *Agave* and *Opuntia*, too, that helped the economy of the native families. The *Agave* is associated with soil conservation when grown on terraces, and in rows on land edges (*metepantles*). In addition, *Agave* offered a wide variety of products for everyday life that ranged from food, drink, construction materials, to others for religious uses.

Trautmann (1981: 112) found that at the end of the eighteenth century Agaves were cultivated for tax purposes in the Tlaxcala, and Chiautempan jurisdictions, but by then those lands were very eroded (Trautmann 1981: 112. AGN, Tithes, Vol. 20). In fact, these pre-Hispanic functions were not focused on fermented-Agave-mead (*pulque*) production, a phenomenon that occurred later with the configuration of *pulque* estates (*haciendas pulqueras*). Apparently they began in 1590, and grew when pulque production became a lucrative business, reaching its peak in the nineteenth century.[3]

In the sixteenth century, and the first half of the seventeenth century, the Tlaxcallan indigenous homegarden was cultivated with *Opuntia*, mainly, because the plant was used as a cochineal host, and cochineal dye was an important trade between New Spain, and Spain. The common indigenous people, and the members of the Indigenous Council (*Cabildo Indígena*) had problems with the production of grana, or cochineal, as a consequence of the enrichment of the former. The constant complaints of the indigenous authorities of the town hall were for the lack of obedience of the Tlaxcallan enriched by the grana trade, and the abandonment of their obligations of agricultural activities that were not as lucrative for them as the production of cochineal paste, and its commercialization.

The Tlaxcallan cochineal trade was done with Hispanic merchants in Puebla City, but there were problems due to quality, because the Indians mixed the paste with sand, to increase its weight (Gibson 1952; Trautmann 1981). During the Viceroyalty *Opuntia* was cultivated in various places, like Oaxaca, because the insect dwelled in the *Opuntia* cladodes. During the Viceroyalty, dye production created great richness, but in the second half of the XIX century natural dyes were displaced by chemical dyes. Gonzalo Gómez de Cervantes (1944 [1599]: 138) wrote:

> New Spain produces four genres of things that each one of them is a treasure, of all it is done every year it is taken out to Spain, of all they are taken, and navigated to the kingdoms of Castile, with which those kingdoms are enriched, and they are ennobled. [This things] are: silver, indigo, grana, and leathers [...].

Very soon, the indigenous population of the Central Highlands that had homegardens increased their diversity, including a mixture of native fruit trees with those brought from the Old World. The native trees were: Avocado (*Persea americana*); Mexican Black Cherry (cherries of the earth, *Prunus* sp.); Mexican Hawthorn ("apples of the earth": *Crataegus* sp.); pecan walnut (shelled pecans, *Carya illinoensis*); and "Mexican Apple" (White sapota) (*Casimiroa edulis*). Trees and vines from the Old World that were introduced to the homegarden were: pears; figs; quinces; apples; lemons; oranges; peaches; apricots; European nuts (*Juglans regia*); grapevines; and olive trees, all of which had spread from the Central Highlands to distant regions in Guatemala, already by the end of the sixteenth century (Alonso de la Mota y Escobar 2012).

Based on information of the Bishop of Tlaxcala, Alonso de la Mota y Escobar *Memorials*, Trautmann (1981: 112) considered that by 1614, there was a large fruit crop in the province, and district elders documented its surface cultivation in Topoyanco (Tepeyanco). That extensive fruit cultivation can be considered as an antecedent of the fruit growing of current homegardens. It was initiated by the dissemination actions of the Franciscans who founded the convent of Tlaxcala, "[...] who were the bearers of innovations even in the rest of Mexico".

3 Gibson 1967: 326; González-Jácome 2010: 521–554; Hernández-Palomo 2004: 239–257; Ramos-Galicia (cord.) 2014; Rendón-Garcini 1995, 1996; Vela et al. 2014.

10.5 FLOWER GARDENS: IN THE PRE-HISPANIC TIME

Zelia Nuttall, in *The Gardens of Ancient Mexico* (1923) and in *Mexican Archeology* (Vol. X, No. 57, 2002: 16–17), noted that there were pre-Hispanic gardens dedicated to flower cultivation. There were a number of types: (1) *Xochitla*, place of flowers, that was a general term; (2) *Xoxochitla*, place of many flowers; (3) *Xochitepanyo*, or walled garden; (4) *Xochiteipancalli*, or palace of flowers, where the royal gardens for the pleasure of the ruling classes were; (5) *Xochichinancali*, site of flowers surrounded by a fence made of reeds or branches, that belonged to the common Indians (*macehual*); (6) *Tlalocan* in *Tepantitla* and *Teotihuacán* mythical gardens; (7) private flower gardens, such as those of Cuitláhuac, the emperor in Iztapalapa, and those of the rulers of Oaxtepec, and Texcoco, where fragrant, brightly colored flowers were mixed with fruit trees and horticultural products.

Morales-Folguera (2004: 355–357) included other two types to the list: (8) the gardens of the rich Tenochcas; and (9) the pre-Hispanic religious gardens, which continued to exist after 1518. The friars in the convents continued the custom of cultivating these because their flowers were destined to adorn the altars of the Catholic temples. The ideological role of flowers in the indigenous cosmogony, and the interrelations between water birds and flowers present in the gardens should be considered (Albores-Zárate et al. 2016; Espinosa-Pineda 1994: 17–22; Heyden 1983, 1995, 2002).

Pre-Hispanic gardens have their antecedents in Teotihuacán. One of the first representations of a garden is in the Tlalocan mural in Tepantitla: "[…] an earthly place considered mythical, flowery, and fragrant. It is an underground place full of wealth, and abundant food". Those chosen by Tlaloc "[…] go to this place" (Morales-Folguera 2004: 352). The Tlalocan Mural was dated from 550 AC to 650 AC. It is thought to be a water reservoir, where rains and rivers were born, which connected with hills, caves, and the sea. It had a forest, with lush and abundant vegetation. In the center was a large tree that served as inspiration for the little trees that were planted in the courtyards of the temples (Morales-Folguera 2004: 353). The Florentine Codex (1979, III, pp. 182–183) described the mural:

> The second place to which [the chosen ones of Tlaloc go] is Tlalocan. And in Tlalocan there is a lot of well being, there is a lot of wealth. They never suffer. The maize, the squash, the pumpkin flower, the huauzontle, the *chilchote*, the tomato, the green beans, the cempoalxochitl are never lacking. […] And there go the Tlaloque, similar to priests, to those who leave, similar to the offers of fire. […] And there are those who have been struck by lightning, the drowned, those who died in the water and they, those who have the divine disease, the bubo, the tumorous, the white spots in the skin (*jiote*), and the one who has rot, the paralytic. And they take to the full of swelling, to the one who dies hydroponically. [...] And these, when they die, were not burned, they just buried them. They painted their foreheads blue, and put paper on the back of their neck. In front of them put mountain figures. And they were dressed in paper. And they put a wooden stick in their hands. […] And they say that in Tlalocan the plants are always green, the plants are always sprouting, and it is always the rainy season that remains.

Morales-Folguera (2004: 353) proposed that to a greater or lesser extent, all the homegardens of Pre-Hispanic Mexico had their antecedents in this model, which, due to their importance in culture and religion, was described by Bernardino de Sahagún (1990, Book III, Chapters 2 and 3, pp. 237–238):

> The other part where said that the souls of the dead are going is the earthly paradise, named Tlalocan, in which there are many joys, and refreshments, without any penalty. Never before are green maize cobs, and squashes, and sprigs of green weed [*Amaranthus*], and green chili, and tomatoes, and green beans in pods, and flowers. And there live some gods who call themselves Tlaloque, who look like the idol ministers who bring long hair [...]. The other part where the souls of the deceased live is the sky, […] and in it there is a grove and forest of various tree genera. And the offerings that they were given in this world the living went to his presence, and there they received them. And after four years passed, the souls of these deaths become in various genera of birds of rich feather, and color, and they are sucking all the flowers, and crave in the sky as in this world, as the Pinzon bird do.

Salvador Toscano (1970: 127) called it "The original Paradise, Tlalocan-Tamoanchan". In the lower part of the mural there are a double trunk tree which [...] is loaded with all kinds of flowers, from which the nectar is spilled, and on whose branches birds are seen with the singing virgule, spiders, and insects. This tree has been linked to hallucinogenic plants, and "[…] a series of edible, and medicinal plants" such as maize, the cacao tree, and the Jointed flat sedge (*Cyperus articulatus* L.), along with others, such as the Angel's trumpet [*Brugmansia arborea*] used to cure rheumatism, and soothe pain, associated with the god of rain".

The Mexican gardens, like those of the European Renaissance, were for pleasure, which was to enjoy the colors and aroma of the flowers. These gardens gave prestige, dignity, and greatness to sovereigns, aristocrats, and the nobility (*pillis*). Among the Nahuas, the members of the royal family, including the monarch, were responsible for the conservation and management of the gardens. With the Spanish entered of the basin of Mexico, most of these gardens, "[…] conceived as ideal, and paradisiacal gardens, in which a small-scale recreation of all the wonders of nature was sought", disappeared in the first half of the sixteenth century (Morales-Folguera 2004: 351).

10.6 FORESTS, HOMEGARDENS, AND GARDENS

Morales-Folguera (2004: 360–361) included the forests that surrounded Tenochtitlan in his homegarden typology. This homegarden was planted and tended for the Mexican emperor for his entertainment in Tetzcotzinco, a few kilometers from Texcoco, with ponds carved out of rock, now known as "baths". The forest that surrounded the homegarden was on hill terraces. Cervantes de Salazar (1971, Book IV, Chapter IX, Vol. 1: 317–325) wrote on the Chapultepec forest that it was a recreational place for the Mexican kings and a sacred space, where Tlaloc and Chalchiuhtlicue, the protective gods of water and nature lived and had their temples carved into the rock. Also, he said that Montezuma had:

> Large circuit forests, and water fences, so that the vermin, and the hunt were safe. Within these forests were fountains, rivers, pools with fish, rabbit hutches, snakes, cliffs, and rocks in which deer, roe deer, hares, foxes, wolves, and other similar animals homed, in whose spaces the Mexican lords hunted frequently.

Morales-Folguera (2004: 357) established the existence of five types of gardens and two types of homegarden, which had been described in the sources, for the pre-Hispanic populations of Mexico: (1) recreational gardens; (2) botanical gardens; (3) homegardens with fruit and vegetables that belonged to the tlatoani, and wealthy people in Tenochtitlan; (4) cornfield and vegetable gardens in the chinampas that were articulated to natural aquatic resources (fountains, rivers, and freshwater lagoons, or artificial fountains and canals). These types of vegetable garden include the construction of aqueducts, a central channel, and derivative canals, which transported water from fountainhead to valleys, plains, and urban settlements; and (5) forest/homegarden that had trees and vegetables. This type is distinguished from botanical gardens because they were spaces where trees, medicinal plants, and flowers were concentrated (Heyden 1983, 1995, 2002).

Cervantes de Salazar (1971, Book IV, Chapter IX, pp. 317–325) said that gardens dedicated to recreation were exclusive to Moctezuma's and did not contain plants for food consumption. However, they did have medicinal plants, fragrant flowers, trees with irrigation, and pathways. Moctezuma said "[…] it is not proper for kings to have farms, or profits from places for their delights, and that homegardens were for slaves, or merchants, although he did have homegardens with fruit trees, but far away, and where he rarely went [they were] farms or profit" in places destined for pleasure (Cervantes de Salazar 1971, Book IV, Chapter IX, p. 294).

The gardens were built in places that could be contemplated (from houses, palaces, gazebos, and high corridors). The flowers were ornamental, fragrant, and medicinal. They were classified by plot, planted in blocks, and had pathways inside reticular spaces, formed by demarcated platforms with cane walls of interwoven reeds covered with aromatic herbs. The trees were evergreen cedars, cypresses, willows, and Montezuma Bald Cypresses (*Taxodium mucronatum*).

Most of the gardens had one or several ponds. Moctezuma had two gardens: one in Chapultepec, and the other in the Peñón islet where he had a forest for hunting. On the vegetation in Mexico City, Motolinía (1996, XLVI: 297) wrote "And then there are groves of cedars, cypresses, and willows; together with floral trees. The Indian lords do not go there for fruit trees because their vassals bring them their fruit. The lords go there to pick roses, breed birds to enjoy their singing, or to shoot them with blowguns, with which they are great shooters".

Antonio de Solís (1970: 195), in his *History of the Conquest of Mexico*, noted, "[...] they had herbs for all diseases, and pains, whose juices, and applications made up their remedies, and achieved admirable effects [...]. Medical herbs were "distributed from the gardens of the king were all the herbs that doctors prescribed or asked of the mourners".

Cervantes de Salazar (1971, Vol. 1, Book IV Chapter IX, pp. 317–325) stated that Moctezuma's gardens had more than 10 ponds, some with brackish water, and others with fresh water. They were used for the breeding of fish and birds. These were kept with great care, and then their feathers were plucked, to make blankets, tapestries, shield adornments, masks, and various other objects in which gold and silver were combined with the feathers.

Hernán Cortés (1993: 244–245) wrote that Moctezuma had a garden and a house with corridors and lookouts located on ponds to be able to see the multicolored spectacle of the thousands of birds that populated them. Cervantes de Salazar (Book IV, Chapter IX, pp. 289–290) wrote:

> Moctezuma had near the palace a very beautiful house with many good rooms, with large corridors in block, raised on rich jasper pillars, all in one piece. There were other corridors more colorful, and rich than these, that sloped down to a very large garden in which there were ten or more ponds. Some of salt water for the sea birds, others of fresh water for those of river and lagoon, which were emptied, and filled many times for the cleaning of the feather.
>
> So many birds walked in these ponds, that did not fit inside or outside; they were in such diverse ways, feathers and workmanship, that they admired ours the first time they saw them, which, with being from different lands of Spain, where there is, as of other things, great diversity of birds, they missed so much these, that very few said they looked like ours. The others, which were of many genres and species, did not know, because never until then did they speak or see them.
>
> Some birds were given flies and other vermin, which was their food. There were three hundred birds for the service and more; some cleaned the ponds, others fished, others fed them, others spouted them, others kept the eggs, others threw them when they were brooding; others healed them, getting sick; others, in hot weather, removed the thinnest feather, so that so much expense, and diligence was done. They made rich blankets, tapestries, and roundups, plumages, mosquito nets, and many other things with gold, and interwoven silver, a very colorful, and very strange work for ours.

10.7 THE INDIGENOUS HOMEGARDEN IN VICEROYALTY

The combination of the pre-Hispanic indigenous homegarden, the Spanish homegarden, and floral gardens resulted in the formation of the Mexican homegarden. It was a space for the cultivation of fruit trees, vegetables, medicinal herbs, and aromatic flowers. It was a recreational space for its owners. According to its surface, a forest for hunting could be integrated into it. The most famous homegardens were in Tenochtitlan, Iztapalapa, Texcoco, Oaxtepec, Huaquechula, Puebla City, and Atlixco.

Cervantes de Salazar (1971, Book IV, Chapter IX, p. 151) wrote that the Oaxtepec homegarden had fragrant flowers, medicinal herbs, rooms, beautiful views, stairs, monumental fountains, and irrigation. It was both recreational, a utilitarian, being a forest, it was a refuge for animals like deer, but which were hunted as well. Moctezuma used to visit it, and Cortes (1993: 353) considered it is the "[...] largest and most lush and fresh garden I have ever seen". On the house, and homegarden of the Iztapalapa ruler, Solis (1971, Book III; p. 270) wrote:

> It had large patios, high, and low beautiful rooms, many very fresh gardens; all stonewalls, and very well carved wood. The rooms are many, and very spacious, with cotton curtains, very rich in their own way. There was a homegarden with lots of fruit, and vegetables on one side. The terraces of the garden

> were made of reeds, covered with roses, and very fragrant flowers. There were freshwater ponds with lots of fish. The garden was so large that there was a pool of lime, and smooth slanted borders four hundred steps in a frame, and six hundred millimeters around, with steps to the water, and even to the ground in many places. It had lots of fish; many egrets, Great Crested Grebes [*Podiceps cristatus*], seagulls, and other birds came to them, which often covered the water, a real sight to see [...]. Iztapalapa has up to ten thousand houses; half of it is founded on the salty lagoon, and the other half on the mainland; it has a fountain on the road to Mexico, surrounded by tall trees with lots of clear, and fresh water.

Toribio de Motolinía (2001: 209) described the homegarden in Texcoco. The aqueduct built during the reign of Nezahualcoyotl supplied water for it. That homegarden had gardens, houses for relaxed pleasure, and a plantation with "[...] more than a thousand very large and beautiful cedars". It is possible that under the generic designation of cedars, Motolinía (2001: 209) was referring to the Montezuma Bald Cypress (*Taxodium mucronatum*). Fernando de Alva Ixtlilxóchitl (1965 [1610–1640], XVI: 264–265) described how Nezahualcoyotl had sumptuous palaces, and that his son Nezahualpiltzintli built his own with "[...] labyrinths, gardens, baths, fountains, ponds, lagoons, which ran under the earth, which could be seen communicated with the large lagoon, to go through them when he wanted to go to his gardens, and recreational hideaways he had in Acatelco, and Tepetzinco, and to go to Mexico City". The lagoon, located to the east of the houses, had [...] a diversity of poultry birds" (Martínez-Saldaña 2015: 95–138).

For Motolinía (2001: 277–278), the best homegardens in New Spain were in the valleys of Atlixco and Huaquechula. In his opinion, the Atlixco homegarden exceeded those of Granada and Orihuela. This homegarden was located in an extensive valley, with numerous rivers, streams, and fountains that collected the waters of the perpetual snows of the Popocatepetl volcano, which then formed a great river that watered the very wide, long, and fertile valley.

There were gardens on the chinampas in the Xochimilco, Chalco, Zumpango, and Xaltocan lakes, where the fresh waters allowed the cultivation of maize, fruit, and vegetable crops. The chinampas are used for the cultivation of flowers or have become the homes of the old peasants, now dedicated to other occupations (González-Pozo (coord.) 2010).

The Jesuit Francisco Javier Clavijero (1971: 229–230) said that gardens on the chinampas sprang from the first seedlings that the Mexicans had after the founding of Mexico. They were constituted in plots surrounded by canals and their owners moved the productions from one place to another with their flat-bottomed canoes (*trajineras*). The *chinampas* survived until the middle of the twentieth century, when the desiccation of the lakes of the basin caused their abandonment as an agricultural system. However, men working on the chinampas plant flowers on them to date.[4]

10.8 CONSIDERATIONS AND CONCLUSIONS

In summary, the types of homegardens in the central Highlands described and classified are an antecedent in the understanding of the evolutionary history of the agroecosystem (Morales-Folguera 2004). The pre-Hispanic homegarden can be seen as the immediate antecedent of the Viceroyalty homegarden, and then homegardens histories are the result of historical adjustments to the Man-nature-agriculture interrelation. The homegarden and vegetable gardens in the chinampas resulted from the agricultural pre-Hispanic knowledge, including the management of water, soil, and vegetation for maize and vegetable production.

Furthermore, if forest is included in human activities and its relationships with the types of homegarden, they are continuation over time of the anthropomorphic relationships and adaptation processes that can be described as a process: (1) forest-hunting; (2) forest-hunting-incidental homegarden; (3) forest-incidental homegarden-initial domestication of plants and the beginnings of agriculture; (4) homegarden-vegetable garden-cornfield; and (5) homegarden and cornfield as separated agroecosystems (Table 10.1).

[4] Armillas 1949, 1971; Dunmire 2004; Jiménez-Osornio 1995b; Jiménez-Osornio and del Amo-Rodríguez 1988; Jiménez-Osornio et al. 1995a; Quiñonez 2005; Rojas Rabiela 1983; West and Armillas 1950.

TABLE 10.1
Tolerated Weeds, and Wild Plants Used for Medicinal Purposes (Sahagún, 1830, III)

No	Name	Uses	Raw	Cooked	Roasted	Total
1	Nanacatl	These edible mushrooms are grown in the fields. However, they cause vomiting or gases if they are undercooked or eaten raw, and they can be deadly. This effect is counteracted with enema.		X		1
2	Tzontecomanatl	Large round mushrooms that grow together on one stalk, some tall, and others low.		X		1
3	n/d	Wide, round mushrooms used in dishes.		X		1
4	n/d	Round white mushrooms easily cooked or roasted on clay plates. They are very tasty.		X	X	2
5	Cimatl	A wild weed with a root called *quavecoc*, *cimatl*, or *eimatl*. It produces large beans similar to broad beans. They are boiled for 2 days.		X		1
6	Huauhquilitl	A very green annual plant similar to amaranth with thin branches and broad leaves. The stems and seeds are called den (Huautli). It is cooked with salt, has an ashy taste, and is like Spanish varieties. When it is cooked, it is squeezed and used to make tamales (*quiltamalli*), and tortillas.		X		1
7	Quilonilli	It has broad leaves and is edible when small. When it is large, it is cooked with the *tequixquitli* and then squeezed before using. Its black seed is called *pitzitl*.		X		1
8	Itzmiquilitl	Brown weed with large branches, round, and flat leaves.		X		1
9	Aioxochquilitl	Flowers, lumps, or limbs of pumpkin branches. The skin is removed from above them before cooking. Note of Sahagún (1830 Vol. III: 245). "In Oaxaca they are eaten together with the guides and green pumpkin stems, mix some balls of maize dough with salt, and some cooked green chilies. It is the food of poor peasants and very delightful, because it is dominated by openness, simplicity, and the good humor of the countryside".		X		1
10	Axoxoco	Weed with long, wide leaves. It is tasty and sour.		X		1
11	Mizquilitl	It is attic, very green, with bladed leaves. It is tasty to eat.		X		1
12	Acuitlapali	Long and brown weed grows at the water's edge. It is good to eat.		X		1
13	Tziunquilitl	It grows by the water's edge. It has blue and harp leaves. It is good to eat.		X		1
14	Conalquilitl	It grows in the mountains; it is ash colored; its root (*tacanalli*) is eaten.		X	X	2
15	Mamaxtl	It grows at the water's edge; it is tasty.		X		1
16	n/name	It is like a nettle; it is edible weed.		X		1
17	Ueiquauhquilitl	Wild smoothies that before cooking are bitter; they are tasty.		X		1
18	Etenquilitl	Herb of beans, which spill when they are picked. It is eaten.		X		1
19	Tlalaioquilitl	They are wild pumpkins.		X		1
20	Xaltomaquilitl	It is eaten cooked.		X		1
21	Tzitziquilitl	It is dark green, with flowers and seeds, very good to eat.	X			1
22	Eloquilitl	It is very green and tender; it has flowers; it is very tasty.	X			1

(*Continued*)

TABLE 10.1 (*Continued*)
Tolerated Weeds, and Wild Plants Used for Medicinal Purposes (Sahagún, 1830, III)

No	Name	Uses	Raw	Cooked	Roasted	Total
23	Quauheloquilitl	It is wild; it is born in the Opuntia *tunales*. It is very tender and good to eat.	X			1
24	Mozoquilitl	It is green, fluffy, and very tender. It is tasty.	X			1
25	Tzaianalquilitl	It is made in water; its branches are hollow and harp; it is good to eat.	X			1
26	Achochoquilitl	It is light green; it is born near the water; it's good to eat. "They say of this grass that if the boys or girls eat it they become powerless to beget; but after they grow up everyone eats it ". Sahagún, Volume III, p. 247.	X			1
27	Tzonquilitl	It is light green; it is born near the water; it is good to eat.	X			1
28	Iztacquilitl	It is low and flat; it has a salt flavor.	X	X		2
29	Tepicquilitl	With long and pointed leaves. If you eat a lot of this herb, it gives cameras.	X			1
30	Ecoquilitl	They are the leaves and branches of beans. They are a little rough and fluffy. They provoke regurgitate.	X			1
31	Vitzquilitl	Thistles of the earth have thorns; the leaves below are ashy and the ones above are green. Inside they have lint, like the thistles of Castile, make them at the water's edge. It is also a garden weed. They are good to eat.		X		1
32	Qaavitquilitl	Wild thistles are born in the mountains. They are prickly and itch those who eat them.		X		1
33	Chichicaquilitl	Edible grass that grows near water and in sweet, tilled land. It has white roots; it is very tender and somewhat bitter.		X		1
34	Coiocuexi	Herb that is not prickly, it is carve and flourishes, is bitter; it makes the voice hoarse in children. Girls don't eat it.		X		1
35	Popoiauh	It is a black and green bush; it is kneaded with maize; and tortillas are made from it.	X	X		2
36	Exixi	It has tiny leaves; its seed is yellow; it is chewy; and if you eat a lot, it produces blisters; and it is too hot. It is used to make tortillas and tamales and atole or porridge (for those with chambers of matter and blood, since it purifies the intestines).	X	X		2
37	Xoxocoiolli	Perennial plant with an acidic flavor that is used as a condiment. It is sour.	X	X		2
38	Xuxocoiopapa-tla	Herb with tall, slender feet, broad, rounds leave. The leaves at the tip of the foot are tasty to eat cooked.		X		1
39	Xoxocoiolcue-cuepoe	Weed with fat round feet, rare leaves. It is born when the rains begin. They are tasty like tomatoes.	X			1
40	Xolxococolol-vivila	It is a stopped herb with small and round leaves; it is tasty.		X		1
41	Miccaxoxocoili	Herb with a thick, hairy foot; the leaves are wide and very acidic and cause an unpleasant sensation that is experienced in the teeth and gums when eating (*dentera*).		X		1
42	Quauhxoxocoili	They are soft to eat.		X		1

(*Continued*)

TABLE 10.1 (*Continued*)
Tolerated Weeds, and Wild Plants Used for Medicinal Purposes (Sahagún, 1830, III)

No	Name	Uses	Raw	Cooked	Roasted	Total
43	Quananaca-quilitl	Herb that the chickens of Spain eat. They are the *serrajas* (*Sonchus oleraceus*) de Castilla. Indians say there were none in these lands before the Spanish came. Now there are so many that the whole earth is filled with them, because the seed flies and they have multiplied everywhere.	X			1
44	Cultivated xonacatl	Little onions from this land. They are planted, Indians are horticulturalists, and they eat the onions of Spain.		X		1
45	Wild xonacatl	Wild onions are found in these fields and burn a lot.		X		2
46	Maxten	Wild onions that are born many, and together, they are faint and flourish are bland. Indians eat the root, or the onion head.		X		1
47	Papaloquilitl	Round-leaved, fragrant, and tasty herb. It is found in hot lands.	X			1
48	Aiauhtonv	Bush wild and low. It is found in the fields and mountains. Its flowers are fragrant.		X		1
49	no/name	It is edible; Sahagún named it as "the sweet potato".		X		1
50	Xícama	Indians also eat its leaves.	X			1
51	Tolcimoquilitl	Herb with very beautiful and very delicate flowers. The root is edible.		X		1
52	Xaltomatl, xaltotomatl	Fruit from the plant called *xaltomaxilhuitl*. There are black and white, very juicy, sweet, and round. The root is edible.	X	X	X	3
53	Coiototomatl	Fruit like that of the small tomatoes called *miltomatl*. It has yellow coverage, they are sweet, and they slightly lock the throat. If you drink a little of the root of this herb, it is medicinal and cleanses the intestines. Women who breastfeed drink it because it purifies their milk.	X	X	X	3 It is medicinal.
54	Atlitiliatl	It has a sweet, blackish strawberry that declines to bittersweet. The leaves are good for the sick steam bath (*temascal*) to add to the water.	X			1
55	Tlalilotl	Edible weed.	X			1
56	Tlalaiotl	Edible weed.	X			1
57	Iacanalxivitl	Medicinal herb that takes away the fever. It is ground with maize and drunk with water.	X			1
58	Coanenepille	Herb to purge. After drinking it, you eat bird broth or an atole (*yolatolli*).		X		1
59	Teonanacatl	Cures gout and fever		X		1
60	Tlatlalayotli	Its leaves are edible; the root is medicinal.	X	X		2
Total	60					

Source: Frair Bernardino de Sahagún, *Historia de las Cosas de la Nueva España*, Tomo III, pp. 244–249, 257.

RECOMMENDED READING

Bye, Robert A. Jr. and Edelmira Linares. 1990. Mexican market plants of the 16th century- I. Plants recorded in Historia Natural de Nueva España. *Journal of Ethnobiology* 10(2): 151–168.

This article by Robert Bye and Edelmira Linares was based on the information that Francisco Hernández provided in his *Historia de las Plantas* (1574). It is about some plants that were sold in the Tlatelolco market in the sixteenth century, like the plants used to flavor chocolate (*Hueinacaztli*, *Tlilxochitl*, and *Mecaxochitl*). The ear flower (*Huanacaxtle*) (*Enterolobium cyclocarpum*) is a tree from tropical forests. In the sixteenth century, its dried flowers were sold in the Tlatelolco market (Bye, and Linares 164). It was medicinal (Bye, and Linares 161). The flavor of the Mexica ear flower has been compared to that of black pepper, nutmeg, allspice, and cinnamon. Currently, these flowers have almost disappeared from commerce in central Mexico, but can be found in Guatemala, where it has continued to be used as a medicinal-infused drink to aid digestion (Bye and Linares 164).

Dunmire, William W. 2004. *Gardens of New Spain. How Mediterranean Plants and Foods Changed America*, University of Texas Press.

In Chapter 3, Dunmire (2004: 59–82) deals with early cultivators in the American Southwest (Hohokan, Odham, Hopi, Zuni, Pueblo, and Nomadic tribes). Those settlers including Northwest Mexico were horticulturalists that depended on soil nurturing skills, agricultural techniques, and dry farming of maize, beans, squashes, cotton, and gourds. The chapter opens a door to the history of agriculture in the Southwestern United States and northwestern Mexico (Fish 2000). The ethnic groups are well established through the two Maps that illustrate the routes followed by crops and food in their pre-Columbian spread from Mesoamerica.

11 The Ancient Mayan Homegarden and Its Transformations[1]

11.1 INTRODUCTION

The swamp agriculture among the pre-Hispanic Maya (at least since the Late Classic) included several agroecosystems: drained or raised fields, canal irrigation systems, wetland cultivation (e.g. March Maize), cultivation in wet soils after flooding, and terraces. These systems were connected to the homegarden (obtaining, hauling, applying, and deliberate handling of the periphyton to improve soils and increase agricultural productivity) and maize crops. Those Mayan communities were self-sufficient and planted in areas far from urban centers.

In Yucatan, the ancient homegarden had no stonewalls or fences limiting its space. The agroecosystem was interspersed with areas of forests and natural vegetation. The distinction between the forest vegetation and the homegarden was made by the increase in the number of fruit trees and plant remains ("garbage dumps) that were left by hunter-gather bands. Archaeologist Barba-Pingarrón (1987: 71–86) information doubts this statement, at least for Cobá, and believes it can be resolved only with specific studies on cultivated plants in these archaeological sites, which would provide more information about it (Benavides-Castillo and Manzanilla 1987: 11–24).

On the antiquity of the homegarden, agroecosystem in the Mayan Area, Annabel Ford, and Ronald Nigh (2016) considered the Mayan cornfield as a diverse agroforestry system, which arose from polyculture in prehistoric Mayan homegardens that had their climax during the Classic (500 AD to 900 AD), while lasted until the Spaniards' arrival in 1524.

11.2 BACKGROUND

Acosta's studies (2010: 1–4, 2011: 227–235) in St. Marta, Chiapas, place the origins of incidental homegardens, referred by this author as "scattered", at the end of the Pleistocene, and the beginning of the Holocene (10,460 BC±50–10,050 BC±90). His studies associated the presence of cacao pollen (*Theobroma* sp.), with tomato seeds (*Physalis* sp.), nance seeds (*Byrsonima crassifolia*), figs (*Ficus* cooki), and forest species, which differed according to their location in the mesophilic forest, the humid tropical, or the low deciduous forest. For Acosta (2010: 1–4, 2011: 227–235), vegetation in St. Marta was associated with various gathering activities (e.g., freshwater jute snails *Pachychilus* sp.), and the hunting of small and medium-sized fauna like deer (*Odocoileus,* and *Mazama*), peccary (*Dicotyles*), and rabbit (*Sylvilagus*), as well as snakes (*Crotalus*), iguanas (*Iguana*), and turtles (*Kinosternon*).

Acosta (2010, 2011) considered that grinding stones and botanical samples indicated an incipient horticulture existence, which began in Chiapas at the end of the Pleistocene. The chronology proposed by this researcher deemed this region as one of the oldest in Mesoamerica and linked it to the emergence of the first incidental homegarden. The dates are very far from those provided by other studies, such as Xihuatoxtla (8990 BP–8610 BP), S. Andrés, Tabasco (7110 BP), Oaxaca (6250 BC), and Tehuacán (5310 BC).

[1] Some data use in this chapter was obtained during research conducted by Ramón Mariaca-Méndez, Alba González-Jácome, and Manuel Arias Reyes in Yucatán. Their results were published in the book *El Huerto Maya Yucateco in the 16th Century*, published in 2010. The other two authors gave me permission to use the information from this book.

DOI: 10.1201/9781003198833-11

Acosta (2010, 2011) left unanswered the inquiring about the dating of the domestication and cultivation of plants such as maize. This generated several problems regarding the interpretation of the origins of Mesoamerican agriculture. One premise is that this case is related to the emergence of vegetative agriculture, and not that of grains, that would coincide with dates obtained in other parts of the world, such as South America and Oceania.

11.3 THE HOMEGARDEN AND THE MAYAN *SOLAR* IN YUCATAN

The homegarden and the Mayan *solar* in the Yucatan Peninsula had prehistoric origins. According to Colunga and Zizumbo (2004), the avocado arrived in the Mayan area before the first settlers.[2] In 3400 BC, the Mayas consumed avocado, along with Mexican hog plum and sapodilla. Archaeological remains were found with what appear to be homegardens, which are in the Mayan area of Cobá, now Quintana Roo. They were dated in the Classic period (400 AD–900 AD) by Barba-Pingarrón (1987: 71–86), who excavated two areas of a surface about 1300 sqm delimited by a stonewall (*albarradas*, *tecorral*, *ringleras*). He found these zones, which he called "activity areas", and a depression, with an organic layer of soil, which is contemporary with pre-Hispanic occupation.

The Barba-Pingarrón discovery found that stonewalls existed there long before the arrival of the Spaniards in the region. This proposal should be explored more deeply about other places, and what their archaeological and paleoecological bases are. From my point of view, pre-Hispanic agriculture did not need to remove stones from soils that were going to be cultivated, since the coa or spear cane plants did not require it, and it is laborious to strip stones from terrain and to sweep it, when the ownership of the land was communal and functional. Stonewalls have functions related to soil, to prevent the land from being windswept, to maintain humidity, and to control water, also.

The homegarden or the Mayan plots fenced with stones (*solar*) could have Pre-Hispanic origins. One of the oldest evidences about the first induced fruit trees correspond to the conformation of the first homegarden found in the Los Tuxtlas region and the Tehuacán valley. Later dates of similar archaeological remains appear to be homegardens found in Cobá.

The archaeological sites registered in the Yucatan Peninsula (1615) suggest that the population density was quite high as of the Middle Formative (300 BC–50 BC). The sites were built and inhabited during the Middle Formative and continued until shortly before the arrival of the Spaniards (1517–1550) in the peninsula. Before the conquest of the peninsula, the common Mayan peasantry lived scattered in hamlets close to sinkholes (*cenotes*), and these settlements were established in areas near the cities and distributed in the form of patches in the fertile areas. The Hispanics applied the congregation laws for the use of the territory, changing its ownership, and places to resettle the population in towns close to sinkholes (*cenotes*). These new settlements concentrated the Mayan population and were implemented for the religious, social, political, and economic population control in the territory during the sixteenth and seventeenth centuries (Dunning 1998; Fedick 1996; Rice 1993; Sanders 1973) (Figures 11.1 and 11.2).

Mayan peasants managed a combination of agroecosystems, which included family homegardens, fields with short fallow systems, and fields with long fallow periods that were worked with slash and burn system. In these last two, people planted maize, beans, and squash (Flannery 1982; Harrison and Turner 1978; Killion 1990).

In the lowlands south of the peninsula, intensive agricultural systems were drained fields, terraces, and irrigated lands with dams and canals (Dunning 1997), which were not under the control of the ruling elite (Demarest 1992; Tourtellot 1993; Marcus 1982). Recent archaeological studies showed that the Maya of the Classic used periphyton obtained from wetlands, along with shells of amphibious snails to fertilize homegarden soils (Lohse 2020: 372–397; Tavera and Novelo 2011: 42–43).

[2] Recent studies in the Yucatan Peninsula have found, and dated human remains in the sinkhole Hoyo Negro as those of a young woman named Naia, dated in 12 thousand BC. There are other important discoveries of human remains older than 12 thousand years located in the Chan Hol and Las Calaveras cenotes.

FIGURE 11.1 Chan Kom, Yucatan: Sinkhole (*Cenote*).

FIGURE 11.2 Xtogil, Yucatan: Sinkhole (*Cenote*).

The north of the peninsula had a relative lack of thick soils for agriculture and a lack of surface water. Therefore, the slash and burn was the agroforestry agroecosystem, which is associated with the emergence of local cultures. After the Spanish expansion, pre-Hispanic borders changed, and minor indigenous populations survived as tax, political, or territorial divisions, or they emigrated.

Peter Gerhard (1991) said that the southeastern provinces (Uaymil and Chactemal) disappeared before 1582 that the interior of the peninsula was sparsely populated so it was not conquered, and that the territorial limits of the Yucatan Viceroy decreased in the sixteenth, and seventeenth centuries. With the exception of the Itzaes, the interior of the Yucatan Peninsula was conformed of dispersed peasant settlements, who lived from a simple slash and burn agriculture that had long periods of ground rest (Jakemann 1952; Peter Gerhard 1991: 45, 52).

For Velasquez-Morlet (2004) and Nalda (2005: 31), the Maya of the peninsula combined the homegarden and the cornfield with commerce and war. From the Preclassic until the end of the pre-Hispanic era, the seaside communities of the peninsula lived on long-distance coastal trade, which reached central and South America. In 1502, Christopher Columbus encountered a Mayan ship on the coast of Honduras and did one of the first exchanges with Mayas (Andrews 1998: 16–23; Flores 2013, Vol. 14, No. 6). Mayan merchants transported salt, clothing, cacao, slaves, jade beads, and products from the interior regions leading to trade in the ports of Holbox, Ecab, El Meco, Paamul, Tulum, and Xelhá (Drenan 1988; Ford 1990; Lucero 2001, 2002 Vol. 104: 814–820; Santley 1990).

11.4 THE HOMEGARDEN AND THE CORNFIELD

For Yucatán, the homegarden-cornfield and the solar were described in the Maya Cordemex Dictionary where Barrera et al. (1980: 773) recorded the terms *tankab* (from the Motul Dictionary of the sixteenth century) as "[…] the cornfield or inheritance within the town or next to it […]", and *tankabal* as "[…] the earthly corral or belonging that the Indians sow, that is next to and around the house […]". Recent writings mention it as *tankabal* or *tankabalil ich nah*, unoccupied place, as *solar* (Dictionary of Vienna Spanish-Maya, seventeenth century), "[...] earthly, what the Indians sow behind their homes [...]" (Juan Pío Pérez, mid-19th century). Friar Pedro Beltrán de Santa Rosa (1746) defined it as "[…] the fenced house […]". That is, the families combined the homegarden with the cornfield on land surface circumscribed by stonewalls (*albarradas*) (Mariaca-Méndez et al. 2010: 2).

At the end of the Formative period in Los Tuxtlas, Vanderwarker (2006: 160–161) found that hunting in the homegarden gave the inhabitants of La Joya meat of two types of squirrel, two species of rabbit, collar peccary, two types of field rat, white-tailed deer, and other animals that furnished sufficient food, favoring the creation of anthropogenic habitats through the clearing of the forest and hunting. Recent studies showed that one of the functions of the homegarden was to attract wild fauna that was easily hunted to contribute to the food system, including daily consumption of meat of villagers (Blanco-Rosas 2006; Foster 1948, 1975; Naranjo-Piñera y Rangel-Salazar 2012: 1–18; Naranjo-Piñera et al. 2013: 87–118).

This continues today. The studies of Blanco-Rosas (2006) and Foster (1948 (13): 153–162, 1975: 448–477), in Soteapan and neighboring places, show the importance of the hunting and capturing of species in the homegarden and in the cornfield. Hunting, in addition to providing meat to the families, also supplied products for local and regional trade, giving families extra money to purchase necessary products throughout the year.

11.5 THE MAYAN-YUCATECAN HOMEGARDEN

Roys (1957) and Scholes and Adams (1959: 199) found that from 1550 to 1560 a large number of the Yucatecan Mayan Indians abandoned their settlements, to settle in what the Hispanics called "ordained towns, or formed towns". These spaces were created around a square, with a reticular trace where the church, town hall, and public buildings were. The reticular trace of New Spain cities was different from the trace of Spanish cities, where not all public buildings were built around a square.

The larger communities absorbed the smaller ones, which lost their identity. To be supplied with water, the new towns were founded near the ancient Mayan cities and close to sinkholes (*cenotes*).

Resettlement in new towns was not always successful. Gerhard (1991: 53) found that in 1761, the governor ordered, "[...] that all the peasants who lived more than three leagues (About 12 km) from a town be transferred to it [...]". The Maya opposed it, and the order was suspended.

With the congregation of towns and their peoples' policy, the "solar" was born, legally as land tenure, and a form of organization of human populations. Thus, the Indian towns were units of distribution and reorganization of the land. A plot measured about 50 by 50 m, with four lots in a block of 100 by 100 m, except for the main houses, which were 100 by 100 m. The first regulations on settlement and creation of settlements sought that each neighbor has legal spaces for agriculture that produce for self-consumption and the creation of homegardens. Doña Juana, on June 15, 1510, in Monzón, Spain, established the communal property law by issuing Law 8th ordering: "Let the Fruit Mountains be Common" of what it says in this regard, "Our will is to be done, and hereby we make the common mountains of wild fruit, and that each one can take it, and take the plants to put in their estates to keep, and take advantage of them as a common thing".

With respect to the Spaniards who settled in New Spain, the 1st Law says: "That the new settlers be given land, and plots, and be entrusted with Indians; and that is peony, and a cavalry". This was signed by Fernando V, in Valladolid, on June 18 and August 10, 1513, Carlos I, in Toledo, on June 26, 1523, and May 19, 1525. Felipe II, in Toledo, on May 25, 1596, mandated:

> Our vassals are encouraged to discover, and populate the Indies, and can live with the comfort, and convenience, that we want. It is our will that houses, plots, lands, cavalries can be distributed, and peonies to all who are going to populate new lands in the Villages, and Places, which by the Governor of the new population are indicated, making distinction between squires, and pawns, and those that are of lesser degree, and merit, and increase them, and better attentive the quality of their services, so that they take care of the tillage, and upbringing.

On the organization of the solar land, the Hispanic legislation brought to America established:

> And because it could happen, that when he distributed the lands he would move doubt in the measures, we declare, that a peony is a plot of fifty-five feet wide, and one hundred in length, one hundred bushels of working land, of wheat, or barley, ten of maize, two for a vegetable garden, and eight for plants of other trees from dry areas, pasture land for ten belly pigs, twenty cows, and five mares, one hundred sheep, and twenty goats. A cavalry is solar one hundred feet wide, and two hundred long; and of everything else, such as five peonies, which will be five hundred bushels of work for wheat bread, or barley, fifty-five maize, ten of land for vegetable gardens, forty for plants of other trees from dry areas, grazing land for fifty-five pigs belly, one hundred cows, twenty mares, five hundred sheep, and one hundred goats. And we order that the distribution be done in a way, that everyone participates in the good and the medium, and in what will not be such, in the part that each one should be pointed out that a friar Hernando de Guevara went to all these villages and brought them to Tinum and Temozón [...].
>
> *(Relationship of Tinum and Temozón: TII-158)*

The towns' congregation and their people had several consequences including environmental and economic, in addition to the effects of the political and social organization of the population. This separated the homegarden from its physical relationship with the forest. The documents indicated that in Yucatan the mandate's application transformed the urban layout of human settlements, plus the physical structure of the homegarden – now converted into a solar – whose form was regular and was surrounded by the stonewall (*tecorral, albarrada*) and fences that prevented the passage of a large part of the wildland fauna increasing the need to introduce domestic animals instead:

> The towns that are now populated with Indians have no form, they do not have streets, because the houses are made of straw-covered wood, [...] and so they are and seem in sight, a congregation of cabins (Relationship from the city of Mérida: TI-71) [...]. These villages in plain, but stony they have no streets drawn but all scrambled, without any concert.
>
> *(Titzal and Tixtual Relations: TI-237; Jakemann 1952)*

> This town of Chauac-Ha was a thousand neighbors at the time that Captain Francisco de Montejo populated there, who were in town in some permanent, and formed way, with their straw-covered masonry houses, where they made their congregations, and markets, although the streets are not compared [...].
>
> *(Relations of the Villa de Valladolid: TII-32)*

The documents of that time show some of the important changes in the size, distribution, and lifestyle of the population. The sharp increase in the mortality of the natives left land free and reduced the labor force for agricultural activities or directed it to the construction of the convents. The mortality of natives worsened, in large part due to the war and diseases or epidemics that ravaged the region. This situation is clear in the following paragraphs:

> Their total destruction has been to move them from their old seats, reducing them from many villages in one, and this with too much and barbaric rigor [...]. When fray Francisco Aparicio arrived in the town Temaza, with great uproar, he ordered all the houses to be set on fire, which were more than one hundred and seventy, [...] likewise, he sent fire to all the fruit trees, that Mayans had in front of their houses in said town [...] seeing the principals, and chieftains without their houses, which had been very sumptuous although made of wood, taken from their old seat. In eight days the chiefs and most of the principals died, so that the changes have been the main part of the population impairment, beyond that the friars of this governorate have built so sumptuously that it has also been a lot of help undermining the Indian sayings, and it will be, because the four friars, [...] there is a house for more than one hundred [...]. It was also the cause why these Indians have come to such a decline, having left their villages and old seats, where they used to be settled and located. Which they leave because Thomas Lopez, his Majesty's Hearer (*Oidor*) was urging them to move, and leave their houses, fruit trees, crops, and seedlings [...].
>
> *(Relationship of Chahuac-ha, Chichimila and Chancenote: T II: 247)*

> And also they were taken away from them because they did not to make a wine, which they used, that they say that it was healthy for them. They call it balché, and were made of water and honey, and threw a root that is the name balcheétree itself. And this they threw in large recipients, which made fifty more arrobas of water, and cooked and weed there for two days. [This beverage] was a very strong thing and it smelled very bad, and in their dances and songs, walking dancing and singing, they gave each of those who danced and sang to drink for a small glass.
>
> *(Relationship of Kanpocolché and Chocholá: T II: 325–326)*

> Also at that time they had the women who loved each one and left them when it seemed to them, and they felt like it. And now Justice for one end, the religious on the other urged them to be Christians, and to leave their bad habits, idolatries, rites, and past and bad uses [...]. And because they have taken away everything that has been said above, they say they have been saddened and given them such grief, that they let themselves die [...], and so on. I say today there are not, twenty parts of Indians that used to be, now twenty years fifteen and this is the cause they give because they die, and it seems to me that it may be because freedom is a big deal.
>
> *(Relationship of Kanpocolché and Chocholá: T II: 325–326).*

The temporary disappearance and the loss of fruit trees and plants in the new settlements occurred for some time – sometimes years – reducing the role of the homegarden provider probably related to the increase of famine. To this situation were added the negative impacts of the occurrence of weather events – some of them catastrophic for the population – such as the very strong north winds (*nortes*), hurricanes, cyclones, and droughts:

> And if good years happen they always have an abundance of food, although some years are sterile of water, they suffer from hunger, and some years come in August and September some storms of very strong winds that bring down trees and the seedlings, and when this happens we all need the Indians and we the Spaniards of food [...];
>
> *(Relations of Dzonot: TII-88)*

The spatial structure of the homegarden changed; first, by mobilizing populations outside their original places and by being assigned to reticular spaces in the new settlements. The family homegarden underwent changes in its composition, since they had to be surrounded by stonewalls (*tecorral, albarradas*). It is known that stone fences help maintain residual moisture in the systems where they are used. This situation in regions such as Yucatecan, where there are no superficial rivers could contribute to its horizontal mobility of population. These had to reduce the wild animals that previously entered and left freely through the grove with the exception of birds and insects. There are references about homegarden at the time of contact with the Spanish population in a veiled manner and with little descriptive mentions, but the fact that the phrase "[...] trees grown around houses [...]" is mentioned may be evidence of their existence:

> And that this Tomás López ordered the depopulation of the said towns to bring them together with the monasteries as I have said, and the towns that did not want to depopulate their seats sent them to set fire to the houses, and since leaving these lands he left power to the religious friars of San Francisco [...] and they burned their houses and cut the fruit trees they had [...].
>
> *(Relations of Dzonot: TII-85–86)*

> They were urged by a Tomás López, Hearer (Oidor) of his Majesty [...] and he urged them to move and leave their houses, fruit trees, crops and seedlings, to reach the monasteries [...] and those who did not want to move willingly, moved them by force and against their will, burning their houses and cutting their trees and plants that they had in their homes, which they had planted and sown themselves [...].
>
> *(Relations of Kanpocolché and Chocholá: T II: 325)*

> Their total destruction has been moving them from their old seats, reducing them from many villages in one, and this with too much and barbaric rigor [...] fray Francisco Aparicio. And when he arrived at the town (Temaza) with great uproar, he ordered all the houses to be set on fire, which were more than one hundred, and seventy [...] likewise, he sent fire to all the fruit trees they had before their houses in the said village [...];
>
> *(Relations of Chahuac-ha, Chichimila, and Chancenote: T II: 247)*

The mention, without being conclusive, allows us to intuit that if they were trees of collective property, and they were planted around the houses. Areas of collective ownership could also be treated with trees of individual and family ownership. In those years, homegardens were constituted mainly by native fruit trees, and the cultivated ones were mixed with the wild ones, as is seen in the sources: "There are in this town of Chauac-Ha many fruit trees of the earth [...]"; (Relations of Chahuac-ha, Chichimila, and Chancenote: T II: 248). [...] They have many fruits and trees from them, planted by hand as well as wild ones [...] (Relation of Dzonot: TII-90).

In 1544, Friar Tomás de la Torre (in Cabrera-Bernat 2011 (1): 168–170), when describing Campeche, in the Province of Yucatán, wrote that they ate maize (*tauli*) called in Spain "wheat of the Indias", plus sweet potatoes, Mexican yam bean, or Mexican turnip (*Pachyrhizus erosus*) (*Jícama*), and chicle tree fruit (*Manilkara sapota*) (*chicozapote*). There were very good grapes and figs. There was an abundance of wax and honey that were used to pay the tribute to the Spaniards, who in turn sold them to merchants who took them to Mexico, Chiapas, and all over. There was a lot of yellow and red cotton from which they made blankets and used to pay tribute; with the money they have left over, they buy chickens, which they also pay Hispanics. Tomás de la Torre (1544, in Cabrera Bernal 2011 (1): 168–169) wrote:

> They make the bread [tortillas] in this way, if it is to be very good, they show each grain in itself and then cook a pot of it with lime and after they have done it well washed. The women grind it on some stones and this is continuous trade and work of the women of the Indians. Then make dough pellets and put on a little fire a large clay griddle that they call *comalli* and on the bake each bread in itself as a Spanish omelette, and so the Spaniards call these breads tortillas. If you eat it hot, this bread is

> very tasty and you could eat it by frying the pan. But if it is hard is like a shoe sole and tasteless and so they do not begin to do so until they sit at the table. The Indians however they eat it and this is common bread of all the Indias.

11.6 THE ANCIENT MAYA-YUCATECAN HOMEGARDEN

According to archaeological studies, the pattern of housing distribution seems to have been common in the Maya area of the Classic period. Gordon Willey (1983: 445–462) mentions the existence of three types of residential settlements: minimum units, informal group units with a central patio, and groupings (formed by several group units). The first two had the main function of building homes for nuclear and large families, respectively. These were distributed in quadrangle, almost always, with a central patio with trees, whose canopy created a roof that reduced the impact of the solar rays and the high temperatures typical of the region (Vogt and Leventhal 1983).

The Maya grew large numbers of plants. Some were sown in the homegarden, and others in the cornfield. Then, when opening a space that would be used for the homegarden, the first thing that they created was the cornfield around the house that was developed for 1 or 2 years. Solar plants would eventually include maize, beans, squash, cotton, some tubers, and roots. Landa (1983 [1566]: 141) mentions the domestic fauna, which was raised around the houses: "It is much praise for the [...] crowd of trees that in this [...] and of all the Indians and even the Spaniards have their services and benefits". In the Yucatan *Relations*, plants and animals introduced by the Spanish population during the sixteenth century were recorded. The plants that were sown in the homegarden around the houses were 50, not including their varieties (Mariaca-Méndez et al. 2010: 43–45).

The *Historical-geographical Relations of the Government of Yucatán* cite more than 74 useful plants, in addition to medicinal plants, which are mentioned generically. Most medicinal plants are "[...] found in the forest, as are the cultivated ones [...]". The information allows us to consider that in the Postclassical period, the peninsular Maya maintained a process of plant domestication, planting around the houses an important number of plants, which existed in the wild forest normally. This coincides with the proposal of Ford and Nigh (2016) that polyculture in Mayan homegardens was part of sustainable strategies in agriculture. Polyculture was largely the result of human action and the interest in having in agroecosystems the necessary plants to meet the Maya requirements.

The registered plants are: (1) *kopte', chak' k'oopte', k'aan koptte'* [*siricote*] (*Cordia dodecandra* D.C.); (2) *Kumche*' (*Pileus mexicana* (D.C.) Johnston); (3) *Chich puut, ch'i'ch'puut, puut, puut ch'iich*' (*Carica papaya* L.); (4) *Theobroma cacao* L.; (5) *choluul, chuluul, kik che', kirikche', sak tintaj che', xkik´che´, xulul* (*Apoplanesia paniculata* Presl); (6) *Chi', nan che*' (*Byrsonima crassifolia* (L.) HBK); (7) *Kuj'che', k'ul che'* (*Mexican Cedrela* Roem); (8) *Pichi', julu'* (*Psidium guajava* L.); (9) *Nukuch pool* (*Dioic pepper* (L.) Merr.); (10) *Julok xa'an* (*Sabal yapa* (Wrigth.) Stand.); (11) *Tsilbuul, ixijum* [soap] (*Sapindus saponaria* L.); (12) *Chi'keejil* (*Chrysophylum cainito* L.); (13) *Choh* [drunken sapodilla?] (*Lucuma hypoglauca* Stand); (14) *K'uts, sak k'uuts, ya'ax k'uuts* (*Nicotiana tabacum* L.); (15) *Ixluch* (?); and (16) *Kup bez* (?).

The historical sources mention that the Maya had an important herbal knowledge. However, there was no specific mention of the flora they were using. However, those herbal medicinal plants were very important in the treatment of diseases and that was narrated. Apparently, those Mayan herbal plants were wild, mostly, and were collected. However, some were cultivated, as was the case in the basin of Mexico (Sahagún 1830, Book III). Now, in a large part of the solar, a specific site for the medicinal plants is destined, and in the case of the disappearance of the homegarden, some plants and seeds are being conserved in pots (González-Jácome 2012: 487–521; Juan- Pérez 2005; Stuart 1993: 53–61).

> There are a lot of medicinal herbs on this land [...] because there is no disease for natural Indians to not apply herbs [...] but asking them the reason for their property they don't know how to give anything other than being cold or hot [...] there are very poisonous and deadly (*Relation* of Merida city: 78). [...] Regarding the herbs and plants with which the Indians are cured, there are many roots and very

> healthful herbs, there are large numbers that give a lot of health to the natives; there are also others Indians have, for being cured, are very good and many, that if they knew them it herbs with which people kill [...] (Relation of Tekit: 288). [...] The herbs would be very important thing (*Relation* of Cacalchén, Yaxa and Sihunchen: 339). [...] Most of the Indians do not get cured in their illnesses if it is not with herbs and roots, and they bleed in the parts that hurt them.
>
> *(Relation of Dzidzantún: 414)*

The *Government of Yucatán Historical-Geographic Relations* considered eleven species as curative and medicinal plants, which were grown in the solar: (1) *Kantubub* (*Sanvitalia procumbens* L., *Melampodium gracile, Pinnaropapus roseus*?); (2) *Poom, poom te'* (*Protium copal* Schl. et Cham.) Engl.); (3) *Balché, xul, xu'ul* (*Lonchocarpus yucatanensi* Pitt.); (4) *Ix kabal jaw, ix kaanbal jaw, ix kaanpa jaaw, kaanbal jaaw, xkaanpal jaw* (*Dorstenia contrahierba* L.), *maculan, mak'ulan, xmak-ulan, xmak'olan, momo* (*Piper auritum* HBK); (5) *Xiw kin* (*Plantago major* L.); (6) *Cacaoche', cacaw che', jazz che', xkakaw che', manzanillo, papelillo, tabaquillo* (*Alseis yucatanensis* Stand.); (7) *K'uts, sak k'uuts, ya'ax k'uuts* (*Nicotiana tabacum* L.); (8) *Iik* (*Capsicum annum* L.); (9) *Xtu'ja'abil, yerbamora* (*Solanum americanum* Miller); (10) *Xtuu ch'ub, ka'an xiw, cruz xiw, ibin xiw, tal che', xpaay che', pak'un pak, sayun tsay, tsayentsay, xpak'al pak', xtak' al tak', verbena* (*Stachytarpheta cayennensis* (L. Rich.) Val., *S. jamaicensis*, and (11) *Priva lappulacea* (L.) Pers.?), and it was mentioned only "Plants to kill" (?).

The Yucatán *Relations* indicated that, with these 11 plants, 19 health problems were treated. The use of some of them should be added because they were causing damage, and they were poison, but they are not specified. Through *Relations*, it is possible to be aware of the uses Mayan-plants had to heal the bite of poisonous spiders, to restrict the blood of any wound and heal it in a very short time, to refresh the liver, to purge and worm, to lower the fever, to calm hemorrhoids, to disinflame, to counteract evil eye, to cure headache, to heal sores, to open swelling, to grow long hair, to grow hair where there is none, to help in labor, and to employ as an antidote.

11.7 DOMESTIC FAUNA CITED IN THE YUCATAN *RELATIONS*

There are some mentions of the domestic fauna of the solar. Some were raised in homes or their surrounding areas, either throughout their life cycle or only in some period. This information is being understood as a broad process of animal domestication in different stages. A total of 39 species were mentioned, cited with 21 generic names: 2 insects, 3 mammals, 2 reptiles, and 32 birds.

The Pre-Hispanic Yucatecan domesticated fauna include (Mariaca-Méndez et al. 2010: 163–167): (1) *ko'lel kab, yilk'il kab* (*Melipona beecheii*) (Stingless bee); (2) Mealy bug (Cochineal), *Muk'ay* (*Xeroderma pigmentosum*); (3) *Pek, alak pek* (Meek dog rose at home), *k'iik'bil* (Landless dog); *tsom* (*Canis familiaris*) (Dog from this land with very shorthaired); (4) *Keh, wakna', yuk'* (*Odocoileus virginianus* and *Mazama mazama*) (Deer); (5) *Chu* (*Nasua nasua*) (Coati); (6) Iguana (*Iguana iguana, Ctenosaura* sp.); (7) *Kut'sa, kutsil ximbal, kuts ha'* (*Cairina moschata*) (Cairina, Indian, and Perulero ducks); (8) Tree ducks types but there is not a description of them; (9) *maxix* (Small duck); (10) *Pixixeh, pihihi, pijije* (*Dendrocygna bicolor*); (11) *Tso', ulum* (*Meleagris gallopavo*) (Domestic turkey); (12) *Kuts* (*Agriocharis ocellata*) (Wild or bush turkey); (13) *K'ambul* (*Crax rubra*) (Pheasant); (14) *Nom* (*Tinamus major percatus* Van Tine (Partridge of Yucatan according to Pearse, mount hen), *Crypturellus cinnamomeus goldmani* Nelson, *C. boucard, C. soui*); (15) (ah) *ki 'tsutsuy,* (*ix*) *kukutkib, ukum, akpakal, tsutsuy, k'ankab* (*Columba flavirostris* Wagler (Pigeon wild red-billed), *Columba speciosa* (Royal pigeon, chukib), *Leptolila verreauxi fulviventris* (Lawrence) (Wild pigeon), *Oreopelia montana Columba passerina*? *Claravis pretiota*?); (16) *Kota', (ix) k'an, moo', op, t'ut ', x-op'* (*Ara militaris, Amazona xantholora, A. albifrons, A. ochrocephala*) (Macaw Parrot, or Yucatecan parrot (yellow-headed or white-headed); (17) *Cox* (*Penelope nigra, P. purpurascens*) (Little pad); (18) *Bach, (ah) koba', tso bach*' (*Ortalis vetula*) (*chachalaca*); (19) *Bech*', (*ah*) *sul* (*Colinus virginianus, C. nigrogularis* (Gould) (Black-necked and short-tailed

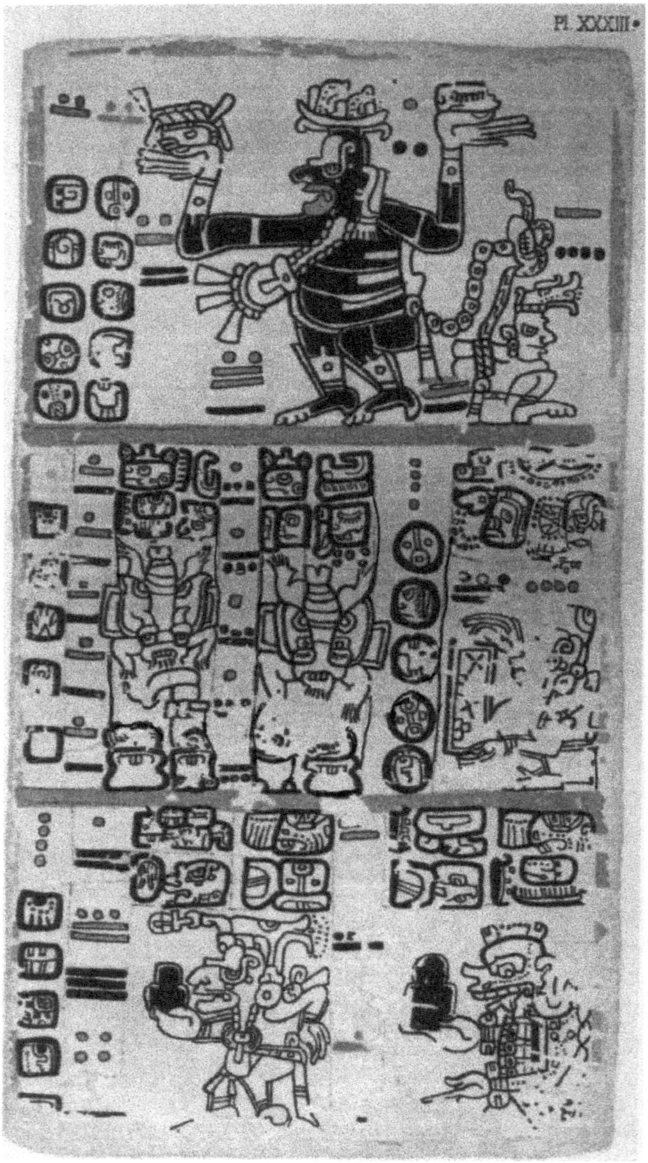

FIGURE 11.3 Madrid Codex: Plate XXXIII, Honeycombs, and Melipona Bees.

quail); *Dactylortyx thoracicus Sharpei*, *Odontophorus guttatus*); (20) *lxyalchamil* (?); and (21) *K'uk*, (*ah*) *k'uk'um*, quetzal (*Pharomachrus mocinno*) (Figure 11.3).

According to Schlesinger (2001: 249) in the Madrid Codex, they are bees in a honeycomb (*jobon*) with their cells. The circular symbol represents the *Cauac* day.

In the Viceroyalty, the breeding of the Mesoamerican dog for food and ceremonial purposes disappeared. Possibly pigs were bred, whose skin, meat, and fat were accepted very soon by the natives, who had the practice of consuming peccary that was hunted in the forest with the use of traps (Dresden and Trocortesiano, Codexes, Plates 32 No. 4, and 33 No. 4, 6). As the natives could not have cattle, they focused on the care of swine, sheep, and goats. Bishop Diego de Landa (1983 [1566]: 66) provides information on work in the Mayan homegarden, apparently under the charge of women:

> [Women] are great workers [...] they do the oldest and most jobs to support their homes and education of their children, pay their taxes, and if necessary, they sometimes carry a greater load working, and sowing its food. Women are wonderful farmers, watching at night the time that serve in their houses, going to the markets to buy and sell their things. They raise birds of their own and those of Castile to sell and eat. They raise birds for obtaining feathers, which they make their well-decorated clothes. The women raise other domestic animals like roe deer that they breastfeed, and they raise them so meek that they never know how to go to the mountain, even if they carry them, and take them through the mountains [...].

This reference provides information on fauna species that were not mentioned explicitly but were raised in the solar apparently, and was cited in the sixteenth-century chronicles. The plants brought from the Old World were tested on the Spanish properties in the convent homegardens and gardens where the indigenous population may have learned to handle them. With the the fauna introduced in the sixteenth-century, small-scale farmers received a great impact on the agriculture, economy, food system, and their daily life. With the Spanish, pigs (*Sus scrofa*), chickens (*Gallus gallus*), dogs, other than the Mesoamerican ones (*Canis familiaris*), cats (*Felis catus*), and pigeons (*Columba* sp.) entered the cornfield and the homegarden. Also, other species impacted agricultural landscapes, such as horses (*Equus caballus*), donkeys (*Equus asinus*), bulls, oxen, and cows (*Bos taurus*, *B. indicus*).

There were species, like sheep (*Ovis aries*), and goats (*Capra aegagrus hircus*), that did not adapt to warm climates and humidity, were taken to other regions (Mariaca-Méndez et al. 2010). The sheep in Yucatán had several problems of adaptation to the climate, which is why they did not prosper. But their introduction was successful in temperate-cold places, such as Tlaxcala, Texcoco, and Querétaro (González-Jácome 1989b: 86–107, 1991, 2008b: 9–24).

The historical sources have little information about the domestic management of animals in Yucatan. The Spanish dog was accepted to the extent that it became a means of defense of the family unit and was trainable for hunting. The Spanish cat was similar to the local felines, but unlike them, it was domesticated and offered protection against rats and vermin (Mariaca-Méndez 2017, Vol. 2: 475–508). Hens, other birds, and pigs were adopted quickly and became part of the cornfield and the homegarden. The draft and load animals arrived, also (Mariaca-Méndez et al. 2010: 162).

> There is also lots of light, and smooth horses, cows, rams [...] pigs, cats, dogs and everything is raised very well [...] (Relation of Merida city: 81). [...] Animals are given well on this land, and it can be raise cattle, and pigs, mares, and goats [...]; there are pastures for them on the seacoast, sixteen leagues from this town [...] (Relation of Valladolid Villa: TII-43). [...] Although sheep, due to the roughness of the land, are not as good as the others, and the lack of water is the major cause of not being much the young animals [...] (Relation of Mérida city: 81). [...] Sheep cattle are not so good [...].
>
> *(Relation of Valladolid Villa: TII-43)*

The population soon became fond of consuming the meat of animals of Castile (birds, cows, pigs, sheep, and goats). Sahagún (1956 [1569], T. III, XXII: 138, 144–145) tells how the meat of these animals was sold raw, cooked, roasted, and cooked in underground kilns (barbecue type) in the Tlatelolco market. The meat of wild animals was also marketed. For example, geese, ducks, birds quails, eagles, gray four-eyed Opossum (*Philander opossum*), wild beasts; and fish. The native diets combined pre-Hispanic products with those of the Old World (e.g., lard, olive oil, olives, capers, garlic, onion, cane sugar, and seasonings). In Yucatan, the Maya raised turkeys, mountain chickens, pheasants, and cambul as well as productive and easy-to-handle birds, just like pigeons brought from the Old World, considering that the Mayans raised native pigeons and doves (Mariaca-Méndez et al. 2010: 162).

> They have domestic birds that breed in the houses, such as chickens and roosters in large numbers, although they are hard to rise. Have been born chickens and they breed wonderfully, and at all times of the year there are chickens of them. They breed some tame pigeons, and they multiply them a lot.
>
> *(Landa 1983 [1566]: 51)*

The food system, with the consumption of peccary and dogs in pre-Hispanic times, was the forerunner for the breeding of domestic fauna by the native population. Dog meat had a similar taste to pork, and the Mayans ate wild board meat having stronger flavor. Accidentally, the Mayans let the domestic pigs run free, and they bred with the wild boards in the forest next to the houses. The native population managed the local bees (Melipona), whose wax and honey were subject to trade and taxation. The introduction of the European bee was not rejected became it was more productive than the native one (Mariaca-Méndez et al. 2010: 162; Mariaca-Méndez 2013: 29–43, 2017: 475–508; Vázquez-Dávila and Lope-Alzina 2012).

11.8 WORKING TOOLS AND CHANGES AFTER 1524

In addition to working with iron instruments that were different from the implements used in America, the following two references illustrate the initial adoption process of the Spanish techniques and tools by the local population. The first refers to the order given by Governor Tomás López in his Ordinances (1566: 232), where he asks the chiefs (*caciques*), main Indian elders (*principales*), and ordinary farm workers (*macehuales*) (Ordinances 1552–1553. Appendix, Document No. 8):

> That farming and cattle rising be introduced among the Indians. That they should teach them the necessary mechanical trades in the republics, to single young men, and that knowingly they would return to their villages, where they would be compelled to use them, and teach them to others.
>
> *(Diego de Landa 1983)*

Regarding the new tools, the *Muxuppipp Relations* (p. 377) say that: "[the natives] are great thieves of all the things that Spaniards have, particularly steel, as they are very anxious for iron because with this they till their soil more easily than with sticks [. . .]".

11.9 CHANGES IN HOMEGARDEN AND CORNFIELD VEGETATION

On the knowledge of how to manage plants, the *Relation* of Merida city (p. 77) said that among the fruit trees that Spaniards brought to New Spain from the Española Island "[. . .] there are: bananas, coconuts and mamey sapota (*Mammea americana*) that taste like peaches [...]". There were eight registered and mentioned fruit trees brought by the Spaniards in the sixteenth century. In addition, the Mexican Highland was the area that served for the adaptation and reproduction for some plants:

> The seeds of radish, and lettuce, cabbage, and all vegetables, which come from Mexico, are purely irrigated, because they do not take advantage of this place, and each year they must bring some again (Cacalchén, Yaxa, and Sihunchen Relations: 339). [...] The vegetables are good, but their seeds have to be brought from foreign lands because these seeds that were brought to this earth are not good [...]. There have been quince, and peach but they have not borne fruit, although they have made flowers; there are no more trees of them (*Oxkutcab Relations*: 356). [. . .] There are vegetables from Spain that the Spanish neighbors eat, and the seed with which it is sown comes from New Spain, because if any seeds from which they are sown came from New Spain; because the seeds from here are no good as the land is lazy [. . .].
>
> *(Muxuppipp Relation: 381)*

Twenty-six plants were introduced in the Mayan solar that soon adapted to the soil and peninsular climate conditions. This was one of the first places where the orange grew. Bernal Díaz del Castillo introduced orange in 1518 when he traveled with Juan de Grijalva to the Tabasco coasts (Martínez 1990: 124).

> [...] Since in that river there were many mosquitoes ten soldiers and I, went to sleep in an idols high house, and next to that house I sowed the orange seeds which I had brought from Cuba, because it was fame that we came to populate, and they were born very well, because the Priests of those idols benefited and watered and cleaned them, and they saw that they were plants different from their own; from there the whole province was made of orange trees.

Spanish cuisine required, and herbalists promoted, the importation of grains, vegetables, fruit trees, and European medicinal plants. Through trial and error, they learned to adapt the plants from the Old to the New World, a task that was difficult in the Yucatan Peninsula. The gardeners, many of them friars in the convents (González-Jácome 2006: 19–46, 2011: 269–274), did this work. Most of the plants had been adapted to tropical and subtropical climates, previously, because they were originally grown in Cape Verde or the Caribbean islands, from where they arrived in the Americas.

The *Mani* and *Kantemo Relations* (p. 113) said that: "Spain has brought orange, lime, lemon, citrus, large date, and fig trees, although they do not grow very well, because I have sown them, and many plants dried up [...], and coconuts from St. Domingo came, also". The Mérida Relation (pp. 77–78) reports in this regard: "[...] the vegetables of Spain: radishes, lettuce, cabbage, turnips, parsley, coriander, peppermint, onions, carrots, borage, and spinach grow very well, but in dry weather they must be watered by hand, and with the rain the plants are full of worms, and plants are lost [...]. There are also very good melons, and cucumbers".

The plants introduced in Yucatán by the Spaniards that adapted to local conditions were: (1) sweet orange (*Citrus sinensis*), (2) sour orange (*Citrus aurantium* var. amara), (3) lime (Ceuti: *Citrus aurantifolia*), (4) lemon (*Citrus limon* L.), (5) cidra or grapefruit (*Citrus medica*), (6) pomegranate (*Punica granatum*), (7) date (*Phoenix dactylifera*), (8) banana (*Musa paradisiaca*), (9) radish (*Raphanus sativus*), (10) lettuce (*Lactuca sativa*), (11) cabbage (*Brassica oleracea* var.), (12) cabbage (*Brassica oleraceae* var. Viridis), (13) turnip (*Brassica rapa*), (14) parsley (*Petroselinum crispum*), (15) coriander (*Coriandrum sativum*), (16) peppermint (*Mentha spicata*); (17) onion (*Allium cepa*); (18) carrot (*Daucus carota*); (19) borage (*Borago officinalis*), (20) spinach (*Spinacia oleracea*), (21) melon (*Cucumis Melo*), (22) cucumber (*Cucumis sativus*), (23) cucumber (*Cucumis* sp.), (24) garlic (*Allium sativum*), and (25) mustard (*Brassica* spp.).

Not all the plants brought by the Spaniards to Yucatán adapted to the environmental and climatic conditions. It is argued that it was due to the lack of water. Among the plants that did not adapt to the Yucatecan climate were grape (*Vitis vinifera*), mulberry (*Morus alba*), quince (*Cydonia oblonga*), peach (*Prunus persica*), and fig (*Ficus carica*). Apparently, the fig fruited in some places. In 1544, Friar Tomás de la Torre (Diario, in Cabrera Bernat 2011 (II): 169), during a few days that he was in Campeche, said: "[...] We saw in the Mayor's house a five-month large fig tree graft and having small figs".

The lack of adaptation of plants and the lack of water are explained by climatic conditions (seasons, humidity, hurricanes, and strong winds) that hit the peninsula (Hemond and Goloubinoff 2008: 133–169; Lammel et al. 2008; Mariaca-Méndez et al. 2010: 158–160).

Wheat, barley, beans, and other grains were better adapted to temperate-cold plateaus. However, oranges, limes, and lemons as Friar Tomás de La Torre noted in 1544, dispersed rapidly through tropical regions (de La Torre, *Diario*, in: Cabrera Bernat 2011 (1): 192; Ochoa-Salas and González-Jácome 2009: 145–168). About citrus fruits mixed with native fruits, de La Torre wrote about the town of Xicalanco that there were "[...] many oranges, limes, lemons, and the fruitful trees of the earth in great abundance". The town square was full of these trees, which provided shade under which he and his traveling companions rested.

Information by Dunmire (2004), Riley (1973), Cline (1966), and some plant geography scholars, provided the data obtained from historical sources like Bernal Díaz del Castillo and Alonso de Molina (1571) for these plants arrival to New Spain (González-Jácome 2011: 212). In the sixteenth-century plants that entered the homegarden were sweet orange in 1518; grape in 1529; pomegranate and fig in 1530; citron, lime, lemon, and sour orange were presented in 1531; quince in 1536; and fig and peach between 1530 and 1540 (Dunmire 2004).

There is little information about the pack and transfer of plants to New Spain. Many arrived as part of the ballast and traveled in the wine casks of the ships. Salinity affected the seeds and rodents ate them, so they ended up in barrels and arks. The quadrangular wooden crate known as "Spanish ark", or boxes larger served to bring the grass seeds to New Spain (González-Jácome 1987, 2011: 212, 222). In 1520, live plants were sent as seedlings and stakes, including almond trees, pomegranate, quince, and olive trees (Romero-Contreras et al. 2004, 11 (3): 237–245). The ways of moving them – both by sea and by land – were established in the eighteenth century, by Casimiro Gómez Ortega who developed the *Instructions* in 1779 (AGI, Gómez Ortega, Casimiro, Instrucciones 1779).

The interest in bringing plants from Spain to the New World was reflected in requests, such as those of Hernán Cortés (4th Relation 10/15/1524), who in his *Relation Letters* (1960 [1524]: 172) showed his desire to introduce plants to New Spain from Cuba, Hispaniola, S. Juan de Puerto Rico, and Jamaica. These plants arrived adapted to warm climate. Among those plants introduced to New Spain from the Caribbean islands were sugar cane (1530) and pear. In the *Good Government Ordinances*, signed on March 20, 1524, the conqueror established:

> Item: that any neighbor who has Indians of distribution is forced to put with them in every year: with every one hundred distributed Indians a thousand branches, even if they are from the native place, and choosing the best one could find. It is understood that he puts plants, and has them quickly, and well cured, so that they can fruit. The branches can be grown in the part that seems to the neighbor, not harming another, and to put them in every year. As it was said plants should be cultivated in time, until it arrives in quantity with every one hundred Indians, five thousand strains. Penalties: in the first time pity it is said; and for the second the bent penalty, and for the third lose the Indians who the neighbor had [...]. The first year that I did not put them or cultivate, pay half a gold frame applied as:
>
> Item: that having vines of those from Spain in quantity that can be made, they are forced to graft the strains that they had on the native plant or to grown again, so suffer the said penalties.
>
> Item: that having other tree plants in Spain, or wheat or barley or any other legumes, they are also obliged to plant, and sow in the villages of the Indians that they have, or they received the aforementioned penalties.

At the end of the 4th Relationship (1524), Hernán Cortés asked the emperor:

> I have also informed your Caesarian Majesty about the need for plants [of all kinds] to be brought to this land, [...], and because of the rigging that there are all kinds of plots in this land, and agriculture, and because so far nothing has been provided, I begging to your majesty, because it will be very useful, send your provision to the House of Recruitment of Seville so that each ship brings a certain amount of plants, and that you can not leave without them, because it will be much cause for the population and perpetuation of it.

11.10 CONSIDERATIONS AND CONCLUSIONS

The proposal that original homegardens were places of attraction for fauna that was hunted seems not to adapt to all homegardens in the past, which was explained by the specific environmental, climatological, and physiogeographic conditions, and the history of each agroecosystem. The domestication of animals brought from the Old to the New World was successful when they adapted to environmental and climatic conditions appropriate to their physiological characteristics, and if their ability to adapt to the climate of the area was resilient (Hemond and Goloubinoff 2008: 133–169; Lammel et al. 2008; Mariaca-Méndez et al. 2010: 158–160; Mariaca-Méndez et al. 2014: 323–359). Native domestic fauna combined with newcomers to be part of the Mexican agroecosystems.

The dissemination of Old World plants was initially carried out by the members of the convents, particularly with the knowledge and work of the missionary friars. Very shortly after the entry of the Spaniards to Mesoamerica (1519), fruit trees, vegetables, medicinal plants, and condiments brought to the New World spread.

The Mexican homegarden was born in the sixteenth and seventeenth centuries after the entry of the Spaniards into ancient Mesoamerica. It is a combination of the pre-Hispanic and Spanish homegarden as well as orchard and vegetable garden. The Mesoamerican had avocado, Mexican black cherries, and Mexican hawthorn orchards. In places close to forests and natural areas, the Mexican homegarden included orchards and vegetable gardens that were used for hunting and gathering. When the available land and the characteristics of the soil for agricultural purposes allowed it, some crops of the cornfield were added to the homegarden.

This combination resulted in greater biological and cultural diversity, beginning with the knowledge of wild flora, cultivated plants, tolerated weeds, as well as of herbal and edible plants. Food processing is modified, with the entry of products such as lard and olive oil, resulting in the cooking model of frying (Katz et al. 1974: 765–773). Stews increased the number of plant elements and meats in them. Methods of barbecuing and roasting were applied to the meat of newly arrived domestic fauna. The kitchen was one of the places in which this combination and integration among plants, animals, and agroecosystems was done. The market was another important component in the dispersion of these new combinations, and the increase in biological and cultural diversity created between the Old and New World agricultures.

RECOMMENDED READING

Warwick, Bray. 1983. "Landscape and Figures: Settlement Patterns, Locational Models, and Politics in Mesoamerica". En: Vogt, E.Z. and R.M. Leventhal (eds.); *Prehistoric Settlement Patterns. Essays in Honor of Gordon R. Wiley.* University of New Mexico Press, Peabody Museum, Harvard University; pp. 167–193.

The main objective of the chapter is to provoke discussion about the relationships that exist between archaeological settlement patterns and political institutions. The information was obtained from a survey conducted in regional settlements, population, size, size ranking, and spatial distribution of the site. Several models are discussed, including Alfred Sauvy's on-site population optimality, the Central Place Model general principles, Hexagonal Patterns, Primate Centers, and Dendritic Marketing Patterns. Also included are Boundary Problems and Levels of Analysis. The cities studied are Teotihuacán and the Mayan centers of the Classic.

Gibson, Charles. 1964. *The Aztecs under Spanish Rule. A History of the Indians of the Valley of Mexico.* 1519–1810. Stanford University Press, California.

Charles Gibson discusses how the Spanish organized economic production, religion, land use, tribute, agriculture, and other things in the Mexican lands after the conquest, and until the war for independence from 1810 to 1821. The emphasis is on weighing the relative importance of different Spanish institutions and practices during the viceregal era, and contrasting them using statistics when they are available and indirect estimates when they are not accessible. This book has 14 chapters, one dedicated to work (IX, pp. 220–256), other to land (X, pp. 257–259), and another to agriculture (XI, pp. 300–334). Reading these three chapters is recommended. The research is meticulous; the final notes and the bibliography have almost 200 pages.

12 History and Discussions about the Homegarden

12.1 INTRODUCTION

Throughout the world, the homegarden has attracted the attention of scientists because of its ability to mitigate environmental problems, such as the loss of biodiversity or the elevation of CO_2 levels in the atmosphere (Kumar and Nair 2006; Mariaca et al. 2012; Toledo 2010 [2015]; Toledo and Böege 2010 [2015]: 160–192). The historical, social, and anthropomorphic characteristics that make it sustainable have been ignored. In Mexico, the homegarden seems to have an age that corresponds to the Formative or Preclassic period (2500 BC–200 AD). However its antecedents of forest, to hunting and gathering as the background of the incidental homegarden, are older, between seven and six thousand years.

The first cornfields seem to have arisen incidentally in prehistoric times, from the initial selection processes, which preceded the domestication – also incidental – of fruit trees and other plants important for human consumption, such as teosinte, which people sucked to consume his sap, beginning with its selection and domestication in the middle basin of the Balsas River, about eight thousand years ago (Piperno et al. 2009: 5019–5024).

Antiquity is important because the age of agroecosystems is closely related to their sustainability, which is proved by their permanence over time. It is important to substantiate the proposal that in the majority of the cases, traditional agroecosystems have ancient origins, so some aspects of the evolution and history of the homegarden and the cornfield are narrated in the book. It is emphasized that the major transformations suffered by the homegarden through time reach to the present.

The changes of the homegarden agroecosystem begin with the conversion of the forest, used to hunt the available fauna, and the gathering of fruits and edible flora into the forest-incidental homegarden, from that to – the incidental homegarden–cornfield that included teosinte and other plants, to then move to the homegarden, and the beginning of the domestication, and the cornfield agroecosystems.

Another change was the conversion of the homegarden into orchards and to recreation gardens. These changes included the conformation of the Mexican homegarden, which combined pre-Hispanic and Hispanic homegardens, with vegetable gardens. That is, they became productive agroecosystems, like orchards with monocultures, or they were converted into ornamental gardens (Espinosa-Pineda 1994; González-Jácome 2012: 487–520; Heiden 1983, 1995, 2002: 18–23). To date, there are some homegardens that disappeared and were converted into residential areas or tourist areas.

12.2 BACKGROUND

These ancient and modern processes occur despite the homegarden characteristics as a permanent source of food and additional resources including monetary ones for families as well as characteristics that remain important. Its conversion nowadays into ornamental gardens or in residential areas for family members who are getting married or who migrate to work seasonally or permanently dates from the sixteenth century when sovereigns like Nezahualcoyotl or Montezuma II enjoyed these types of gardens. In these areas, trees, flowers, water, and birds completed the enjoyment of their holders. However, it has been proposed that they were ritual spaces and their main use was related to the official state religion and the cult to deities (Cervantes de Salazar 1971 [1557–1564]; Díaz del Castillo 1978; Espinosa-Pineda 1994: 20; Heiden 1983, 1995, 2002: 18–23; Nuttall 1923).

DOI: 10.1201/9781003198833-12

> Let us not forget the gardens of flowers and fragrant trees, and the many genres that they had and the concert and walks in them and their pools and freshwater ponds. How the water comes from one end and goes to another and from the swimming pools that they had inside. And from the diversity of little birds that grew in the trees and what medicinal herbs and benefits they had in them was a thing to see. And for all this many gardeners, and all carved of stonework and very whitewashed, as well as bathrooms, walkways, and other toilets, and apartments with walkways, and also where they danced, and sang. And there was so much to look at in these gardens or in everything else, that we weren't tired of seeing their great power [...].
>
> *(Díaz del Castillo 1998: 170)*

Gabriel Espinosa-Pineda (1994: 20) – based on his own studies on the birds of the basin of Mexico – and Doris Heiden (1983, 1995, 2002: 18–23) – based on her publications on the mythology and symbolism of the pre-Hispanic flora of Mexico – when dealing with the ornamental gardens of the sovereigns, especially those of Moctezuma II and Nezahualcoyotl, explain that in them there were ritual and medicinal plants, next to which there were also numerous waterfowl, whose function was ritual and divinatory:

> [...] The water birds concentrated in the gardens must have had a divinatory, and religious function of importance. They were not "zoos" or capricious and curious collections of specimens that were adored and replenished every time they die. But they were carefully planned centers, with specialists who fed and took care of the birds, curing them when they got sick, and certainly using various healing methods, drugs, and surgery.
>
> *(Espinosa-Pineda 1994: 20)*

12.3 FOREST AND THE HOMEGARDEN AS HUNTING PLACES

Amber Vanderwarker (2006: 148) considers that the hunting of animals in the forest and the incidental homegarden in Los Tuxtlas initiated an anthropogenic process generating an animal protein reserve, available quickly and easily for the human population. The predatory cycle was established in disturbed habitats, where it was qualitatively and quantitatively different from that of the primary and undisturbed environments, shortening the time devoted by man to hunting and fishing, thus reorganizing the major subsistence systems.

The homegarden-hunting-gathering agroecosystem persisted through time; its efficiency led it to continue through thousands of years, until at least half a century ago (Blanco-Rosas 2006: 172–173; Foster 1940, 1942, 1967).

The homegarden-hunting model proposed by Vanderwarker (2206: 149) for the Early Formative and the Early Classic periods (1400 BC–300 AD) at La Joya and Bezuapan sites was based on Emslie's (1981: 306), Linares' (1976: 331), and Neusius' (1996: 276) studies. She proposed that much of the fauna were crop pests, and their hunting in induced homegardens had the dual purpose of providing the inhabitants with animal protein and protecting their crops. Vanderwarker (2006: 151, 181) considers the model feasible using examples of hunters in the Amazon, archeological studies in the southwest of the United States, and her own analysis of zoo archaeological materials. These analyzes found an increase in the subsistence risk related to volcanic activity and human factors.

These cases represent two options chosen by the population in similar circumstances. The feasible options may have affected the operation and self-sufficiency of the settlements. The environmental conditions reduced the options available for the inhabitants of La Joya and Bezuapan in the early and middle Formative periods (1400 BC–300 AD) in the Tuxtlas. The people made the necessary changes to sustain their economy in a self-sufficient way. The analysis of stable isotopes in carbon and nitrogen of human, dog, and deer bones allowed examining the nutritional contribution of terrestrial and aquatic animals to the food system of the daily diet of the dwellers.

From the information provided by these analyses, combined with archaeobotanical, zoo archaeological, and palaeoecological data, some changes in subsistence were known (Gorman 1992). This proposal does not apply necessarily to places such as the basin of Mexico, where the abundance and permanence of the vital liquid in the lagoon system made fishing, hunting, and capture of fish, birds, batrachians, algae, aquatic flora, and others very important activities from ancient times until the extinction of the lagoons and navigation in the canals and lagoons of the basin. This last process ends until the first half of the twentieth century (Aréchiga-Córdoba 2011: 227–270; Correa-Ortiz 2011: 203–226; Espinosa-Pineda 1994: 17–22, 1996; Tortolero-Villasor 2001).

In the 1950s and 1960s, George Foster (1940, 1942, 1967) observed and recorded that the Zoque-Popolucas hunted: whitetail deer (*Odocoileus virginianus*), deer (*Mazama americana*), common tapir (*Tapirus bairdii*) (*tapir, anteburro*), wild boar (*Sus scrofa*), Paca (*Cuniculus paca*) (*tepescuintle*), Armadillo (*Dasypus novemcinctus*), Badger (*Meles meles*), Guatin (*Dasyprocta mexicana*) (*cerete, zerete prieto*), rabbit, and squirrels.

In Ocotal Chico, a town in the Tuxtlas, Blanco-Rosas (2006: 172–173) showed how the Zoque-Popolucas, inhabitants of the region, obtained meat from animals hunted in the cornfield, the homegarden, and the forest. This gave the daily diet of their food system important animal protein throughout the year. The meat they got was on average 941–1,000 kilos/year, which was distributed among families for consumption. Each of the families had an average of 10 members, and they consumed from 90 to 100 kilos per person, annually. Hunting was the most important source of animal protein for families, although some hunters preferred to sell the hunted, and gathered animals for money.

In that same period, the Ocotal Chico Zoque-Popolucas hunted foul: pheasants, doves, plain chachalaca (*Ortalis vetula*), partridges, toucans, *chuco* birds, quails, and parakeets. They provided them with about two kilos of meat a day or about 710 kg of meat a year. The capture of snails and river shrimp gave Popoluca an average of 460 kilos/year, and fishing about 460 kilos of fish too. The hunt was done using firearms and trap cages to capture. The information obtained by Blanco-Rosas (2006: 174) confirmed the data of Foster's study (1940, 1942, 1967) on the Popoluca municipality of Soteapan that since 1940 was more deforested than the Ocotal Chico region.

12.4 MANAGEMENT OF DOMESTIC AND WILD FAUNA

The prehistoric groups of semi-nomadic hunters-gatherers, fishermen, and bands of people with a shell culture (*concheros*) had wildlife as the main source for their food system. In addition, they made clothing, some tools, and objects with the wildlife bones and pelts (e.g. needles, hooks, flutes, fur clothing) (González-Jácome 2002c: 261–280; López-Austin and López-Luján 2001). Today, the wild flora and fauna are basic components in the food system and for the survival of small-scale farmers, both indigenous and mestizo (Mariaca 2017: 475, 486–508).

Mariaca et al. (2011) wrote that wildlife introduced into the sixteenth-century Mayan site included: chickens (*Gallus gallus*), dogs, other than the Mesoamericans (*Canis lupus familiaris*), cats (*Felis silvestris catus*), and pigeons (*Columba* sp.). Other species that were not destined for the indigenous site but which were incorporated over time were horses (*Equus caballus*), donkeys (*Equus africanus asinus*), and cattle (*Bos taurus*, *B. indicus*). The European bee was more productive than the native bee, so Mayan beekeepers soon learned to manage the former's aggressiveness.

This situation continues to date in places where there is abundance of aquatic resources or where people live near jungle or forest relics. Regarding the management of domestic and wild fauna in the Viceroyalty, there are sources, such as the narrations of friar Alonso de Ponce (1586: 10), in his text about the capture of wild pigeons in the S. Miguel Guard, then located behind the Xiriualtique settlement, in the Guatemalan Mayan area:

> There is a large sum of pigeons like those in Spain, which in the summer they eat, after fed they will sit in the trees The Indians go there in the blazing force of the sun, and go flipping, and frightening them, and they flee little by little, from tree to tree. They reach the savanna where there are no trees, and they fall there exhausted on the grass, where the Indians take them alive: hunting this way is very valuable, and profitable [...].

In the Lacandon land, domestic animals were left free in the homegarden and the forest, as was the case with pigs, which joined wild animals in the herds. In the southwest of Tlaxcala, wooded natural areas and waterlogged areas were important for hunting, fishing, and gathering. The Spaniards used the waterlogged areas of the Zahuapan and Atoyac river basin to confine the cattle in the dry season. The natives kept the herds in the swamps during the dry season. This has not been studied profoundly enough in relation to ancient times (Trautmann 1987). In the Viceroyalty, lard was used in the manufacture of soap for washing wool in the Tlaxcallan textile mills (Giordano-Sánchez 2000, 2002: 1–21). Pork, bacon, and lard were incorporated into the food system of the indigenous and mestizo families (González-Jácome 2009c: 250–281). Pigs were traded in markets live or slaughtered (Trautmann 1981: 113) (Figures 12.1 and 12.2).

Since 1521, the Spanish Crown promoted the introduction of domesticated animal species among the natives. The raising of chickens provided them with a safe and economical food base, which required little labor investment, and it was an important innovation for the people's economy. Gradually other domesticated animal species were introduced. In 1609, the fisheries on the north coast of the Gulf of Mexico, located in Nautla, Raudal, Tecolutla, the Cazones River, the Tuxpan

FIGURE 12.1 Yucatán: Turkeys and Pigs in Homegardens.

FIGURE 12.2 Chan Kom, Yucatan: Homegarden and Chicken Coops.

River (Tancupo), and the Tamiagua Lagoon (Tampamachoco), were in the hands of Spaniards and some indigenous people. These supplied fresh or smoked and roasted fish and shrimp to the inhabitants of the neighboring towns and regions (Mota y Escobar 2013: 55, 65, 68, 70–73).

In 1736, the Tlaxcallan farmer's families owned three to four oxen or mules for agricultural work in addition to pigs, which were mainly for marketing and eventually for consumption. Domestic animals were fed with carob bean, beans, and barley (Trautmann 1981: 130). For that same year, in Tlaxcala, Puebla, Cholula, and Querétaro, the wool of the sheep was of great importance for textile works. The flock of sheep was handled through transhumance, but there is little information for Tlaxcala (Castañeda-Castillo 2014; Melville 1994, 1999; Miranda-González 1972; Mora 2013). Trautmann (1981: 113) considers that in this Province herding sheep was almost completely abandoned. In S. Francisco Topoyanco (Tepeyanco), St. Inés Zacatelco, and S. Pablo del Monte, carded cotton was imported into clews, which were delivered to the weavings of the city of Tlaxcala, and it was sold in local stores (Trautmann 1981: 113; AGN, Indios, Vol. 43, Files 167, 169; FMMN, Serie Tlaxcala, Roll 30, No. 111; Roll 29, No. 63) (Figures 12.3 and 12.4).

FIGURE 12.3 Cuaxonacayo, Tlaxcala Bloch: Domestic Fauna.

FIGURE 12.4 Sonora: Homegarden and Caged Birds.

Since its introduction in the Province of Tlaxcala, the cattle herd caused problems and lawsuits between the natives and the Spaniards, which were the cattle owners. In theory, the flocks paid the land after the harvests, but dates and limits where animals could graze were not always respected (González-Jácome 1991, 2008: 1–17, 2009c: 250–281). This theme deserves more attention in its relationship with agriculture. Old World domestic animals modified the environment, the economy, labor, agricultural systems, and diet of the native populations.

12.5 WILD AND TOLERATED EDIBLE GREEN WEEDS IN HOMEGARDENS

In traditional Mexican agricultural systems, including the Mexican homegarden and the cornfield, the management of wild and tolerated edible weeds, as much as cultivated, tolerated, and harvested wild plants, is an ancient ecological and cultural element of these agroecosystems.

There is information on edible weeds and cultivated herbs through Bernardino de Sahagún' texts that were the result of accusative and comparative research in towns of the Mexican Central Highlands carried out over three decades. A team of assistants from the Tlatelolco College was supporting the Franciscan friar from 1530 to 1577–1578.[1] Tolerated weeds (*arvenses comestibles*, *quelites*, *bledos*) and cultivated herbs could be eaten raw or cooked, mixed with food, or alone as a salad.

The list elaborated in this book with that type of information provides a total of 60 edible weeds, in addition to the cultivated and wild herbs that the Franciscan classified as medicinal (Sahagún 1830, Vol. III, Chapter VI, pp. 229–240; Chapter VII, pp. 241–281) (Figures 12.5 and 12.6; Table 12.1).

12.6 IS THE HOMEGARDEN A SUSTAINABLE SYSTEM?

Homegarden has been described as sustainable agroecosystem that was studied by geographers since the 1960s (Kimber 1966) and, subsequently, by other specialists who have been interested in homegarden for its ability to maintain itself as an agroecosystem over time (González-Jácome 2016: 233–257, 2018: 43–85; Kumar and Nair 2006; Mariaca-Méndez et al. 2007: 119–138, 2012a; Mariaca-Méndez 2012b). On the homegarden, several aspects have been analyzed, among their social and cultural components, those that are articulated to their economy and their ability to produced food, how they contribute to the health of the families, and also how from them the families can obtain money and other important resources. The homegarden has attracted the attention of scientists for their ability to mitigate environmental problems (Mariaca-Méndez et al. 2007: 119–138; Kumar and Nair 2006) in addition to its great resilience, which is one of the factors that explain its survival through thousands of years.

Regardless of the characteristics attributed to this agroforestry system, one of them is to be one of the oldest and most dispersed on the planet, due to its similarity with nature, and because its emergence was incidentally. This agroecosystem is found in tropical, cold, semi-desert, and temperate climates, with lower or higher masl, in areas with complex orographic relief or in plains. Tropical homegardens now have a greater extension (Mariaca-Méndez 2012a, 2012b). In Mexico, they are better preserved than cold-temperate homegardens or those in semi-desert regions.

A characteristic that makes homegarden sustainable is diversity in their multilayers (herbs, shrubs, trees, and dosal). In addition, they may also have mushrooms, moss, epiphytes, and others components of the agroecosystem depending on the specific areas where they are found. Homegarden provides families with food, various products, and economic resources in a consumption process that is distributed over several months of the year. They also cover needs in the daily life of farmers (shade, wind protection, wood, firewood, and fibers), including the possibilities of obtaining monetary resources (González-Jácome 2016: 233–257, 2018: 58–150).

[1] Among them are known: Antonio Valeriano from Azcapotzalco, Martín Jacobita, and Andrés Leonardo from Tlatelolco, and Alonso Bejarano from Cuautitlán.

FIGURE 12.5 Relict of Lerma lagoon: Water Potato (*Papa de Agua*), 2008.

FIGURE 12.6 Relict of Lerma Lagoon: Wild Green Weeds (*Bledos*), 2008.

In Mexico, homegarden and cornfields agroecosystems seem to have arisen almost parallel in prehistory, from the incidental domestication of fruit trees in the forest and other plants – such as maize – that have been and are important to the inhabitants. Anthropomorphic action in forests increased their diversity because humans introduced several species, which were originating from other places.

The first settler's bands in the Petén forest are established before the Holocene thermal peak, and they were scattered throughout the tropical forest, using its resources (Alexandrovskiy et al. 2004: 9–17). But about four thousand years ago, a drought begins, which seems to consolidate the management of resources centered on the forest-cornfield (Nigh 2008). The climatic changes that impacted the Mayan area may have been reduced by human action, by establishing the system of forest-incidental homegarden, which was a diverse, constructed landscape, whose importance increased with the establishment of fixed settlements and the emergency of urban centers in the Formative or Preclassic (Ford and Nigh 2009, 2016; Turner and Sabloff 2012: 13908–13914). Ford and Night (2009) consider that the Mayan adaptations inherited the biodiversity and current complexity of the Mayan forest.

For Neff et al. (2006a, 2006b), the radical climatic changes in the environment were reduced with the homegarden agroecosystem and with the human investment in the landscape construction, which was increased with the emergence of the centers in the Formative or Preclassic. In the Mayan area, 2,000–1,000 years ago, there was a dry and stable climate, without the extremes of the previous years. Ford and Nigh (2009) consider that the dry years occurred in the Terminal Classic, but their intensity and frequency were minimal, compared to the Preclassic. At this time, the management of resources and the variability of the homegarden made by the Maya were the mitigating factors (Burroughs 2005: 43–47; Leyden 2002: 88–90: Steele et al. 1998).

The Mayan collapse in the Terminal Classic is explained by the climatic conditions, droughts, and recurrent famines, and by the abandonment of the civic obligations of the population (Ford and Nigh 2009; Hurtado and Román 2012; Neff et al. 2006a, 2006b; Scarborough 2003; Webster 2002). Scarborough (2003) considers that after the Post-Classic period the Maya could not maintain the system, because the work they invested had to be applied to the landscape construction (Andrews et al. 2003: 151–156).

The Spanish changed the pre-Hispanic structure of the homegarden with the introduction of a new system of land tenure. The indigenous populations also suffered the spatial effects of the congregation process, and the construction of fences (*tecorrales*, *albarradas*), as dividing boundaries of the land where the new homegardens were installed (Mariaca-Méndez et al. 2010). The congregations were located near the sinkholes to have access to water (Mariaca-Méndez et al. 2010: 36–38). The homegarden in the Viceroyalty combined the flora brought from the Old World with those native to Mesoamerica undergoing changes. Among the characteristics, they acquire are floristic diversity and the management of various strata in their structure.

12.7 HUNTER-GATHERED BANDS, THEIR SEDENTARY PROCESS, AND CHANGES IN THE FOOD SYSTEM

Vanderwarker (2006: 194–195) proposes that in the Early Formative (1400 BC–1000 BC), human groups organized in bands moved seasonally and annually in larger areas. At the end of the period, they begin their establishment in more permanent settlements, which is associated with the volcanic explosion in the area. They grew maize by slash and burn, and the crops were scattered throughout the adjacent areas to that ones to obtain food. Maize cultivation was complemented by the collection of fruits from wild and domesticated trees (incidental homegarden) in addition to hunting local wildlife and fishing. The abundance and distribution of wild flora could have negative impacts with the eruption of the volcano, which limited the territory for collection and cultivation favoring its domestication.

In the Middle Formative (1000 BC–400 BC), the settlers in the ranges of the Tuxtlas modified their strategies for obtaining terrestrial fauna. They preferred hunting in disturbed environments, where deer abounded. People also continued planting maize and collecting avocados and macaw palm (*Acrocomia aculeata*). The pottery was diversified, and the range of containers for cooking food was expanded. McCormack (2002) and Vanderwarker (2006: 195) believe that the manufacture and use of grinding instruments specialized and its abundance indicate the expansion of maize production and consumption. At this stage, materials show an increase in hunting in homegarden, which suggests that there was also a greater focus on the maize production and its processes (such as grinding).

These are associated with the invention of the process of cooking maize grains with lime, before washing them, grinding, and converting in soft dough (maize hominy process). This process is characteristic of maize consumption in Mexico. McCormack (2002) considers that at this time the political centers in the Tuxtlas had not been established, and life was done in independent and self-sufficient villages. Ciphers (2004: 23) find that in the Tuxtlas' plains from 1200 BC to 800 BC the region had its apogee, and the population reached 13 thousand inhabitants. The settlements were built, on artificially made headlands, as in S. Lorenzo. Much of their food came to S. Lorenzo through trade, although its inhabitants planted maize.

In the Late Formative (400 BC–100 BC), the first regional centers arose, such as Chuniapan de Abajo, which were never as important as the Olmec centers of the lowlands. Archaeobotanical information shows maize and fruit trees as the main axis of the food system. Maize consumption stabilized and was grown in areas close to housing, while the cob underwent changes in the pot and the radii of the dome. Fauna materials suggest a greater disturbance in terrestrial biomes, which indicate the increase in hunting in the homegarden, and cultivation became a less risky subsistence strategy. Santley et al. (1997) and Santley and Arnold (1996; cited in Vanderwarker 2006: 196) consider that there was a correspondence between agricultural intensification, and the growth of sociopolitical complexity.

In the Terminal Formative (100 BC–300 BC), Vanderwarker (2006: 197) found a decrease in the regional population caused by the volcanic eruption that influenced the abandonment of the region. Those who remained in the region continued to grow maize, which according to the data on stable isotopes of carbon and nitrogen was basic in their diet. There was an increase in the consumption of avocados, sapodilla, and macaw palm fruits. Cultivation remained the most important food system strategy, and agricultural ridges were built. Villagers diversify hunting sites, with a wider range than in the previous stages; hunting in the homegarden becomes less selective, which seems to indicate an increase in the subsistence risk.

A current description of the Tuxtlas landscape, made by Siemens (2004: 41–58, 2009), explained the biological diversity of the region and the volcanism observable impacts in the landscape, including difficulties in the settlement process and human adaptations to the environment. Following a geographical sequence, we would go from the homegarden in Los Tuxtlas, to those in Tehuacán, with irrigation, which according to Mac Neish (1997) and Smith (1995), was practiced from 850 BC to 150 BC and that, with later studies Neely takes it to at least 1050 BC. However, due to the importance of the studies of the homegarden in Tehuacán, and their agricultural evolution will be discussed in another section of the book.

12.8 DEFORESTATION AND ECOLOGICAL HISTORY

Ecological history provides evidence on the forest deforestation in the Maya area, until the arrival of the Spaniards and their model of agriculture, including grains monoculture, plow, draft and load animals, metal agro technology, and flora and domestic fauna. In the Viceroyalty and until the nineteenth century, the exploitation of fine woods and dye stick began deforestation in Campeche, as happened on the banks of the Candelaria, Hopelchén, and the Chenes regions. In Campeche the chewing gum was extracted from the sapota chicle tree (*chicozapote*) in the Papaloapan basin in Veracruz and other tropical areas.

Bishop Alonso de la Mota y Escobar described in his *Memorials* (1609: 45–467) that the wood that was cut down by the loggers, who extracted from the forests in the Papaloapan basin, was thrown and sent downstream until it reached Tlacotalpan and Alvarado, from there to the Port of Veracruz, or it could also be taken outside of New Spain. In 1609, when he was passing through Tlacotalpan on his way to Alvarado, Veracruz noted in his *Memorials* (1609: 45–467) that 12 poor Spaniards lived there, and they were fishermen. Those with the greatest resources:

> They also deal with some cedar wood, which they release upstream, which they call wood from Cozamaloapa, and this wood is sold in Veracruz, for its buildings, made boards [...] entertain yourself in cutting with the help of these Indians many woods of rich cedars, which are in these mount very thick and closed (*arcabuco*), which they bring down to Alvarado river, and there they saw them and make boards to take to sell to Veracruz. Boats are also made of boards in this bay of Alvarado and they are very extreme for this purpose.

It is considered generally that the greatest deforestation of the forest occurs in the twentieth century and that it is related to the expansion of intensive pastures for cattle. In addition, impact of soil erosion was caused by the introduction of the plow for the fields and by the construction of roads. Road construction was a public policy that caused the reduction of agricultural land and forest leading to deforestation as happened in the second half of the twentieth century in the region.[2]

Secondary forest patches (*acahuales*) are important in the regeneration of the forest processes. In Tabasco and Veracruz, farmers leave these relics of secondary forest to offer shade for the cattle (Wilkerson 1983). They have been considered as forest relics, which at any given time can induce their regeneration. The discussions related to whether or not it is possible to reconstitute a tropical forest that has been deforested have some answers from history, which opens up new questions and the possibility of understanding the natural process of forest reforestation.

Bernardo García-Martínez (1999: 55–74) found historical information about the deforestation process, and the natural regeneration of the Mount Mixtlán tropical forest, which in the eighteenth century had cedar trees (*Cedrela odorata*). This forest is located on the coastal slope of the Gulf of Mexico, near Tuxtepec, a region of the Bishopric of Oaxaca. García-Martínez' text deals with the links between the human population, and its advance or retreat in these humid forested areas of the former Huaspaltepec or Guaspaltepec that was abandoned around 1730. Without the presence of people, the tropical vegetation of the region originally covered by a "[…] thick, and impenetrable forest […] and by swampy areas […]" was first deforested and abandoned for about four or five generations. However, without human presence or action, the forest recovered itself alone.

From 1766 to 1786, the population of Tuxtepec was in dispute with the lumberjacks (*monteros*) who had arrangements ("right of floor") with the towns of Suchiapa and Latani to cut down the trees. The workers floated the logs downstream to what is the current Vicente Beach. The dispute resulted in the establishment of a boundary in 1778, as Suchiapa and Tesechoacán passed in 1776 to the Cosamaloapan jurisdiction. The lawsuit among towns, landowners, and loggers for the property and the use of the mountains in New Spain was constant (Ayala 1999: 75–92). In pre-Hispanic times, the use of the mountains was of common property, as in Spain, where the Castilian administration established it, but in New Spain the use of the mountain forest was established by the Viceroyalty administration.

The Indian towns constantly protested the lack of compliance of the laws on the patrimonial domain of the land and the common use of mountains, pastures, and waters established in Castile and brought to New Spain, but which were not fulfilled. Hernán Cortés was the first to discard them, who "[…] banned every person from entering the mountains and pastures of Cuernavaca indicated at his mercy. He also banned the cutting of firewood, and wood without his license" (Ayala 1999: 76–77). In the first two centuries of Viceroyalty, the issuance of laws and ordinances to control the exploitation of forests was not very successful. At the end of the eighteenth century, new ideas appeared to protect forests and species in them. Reforestation arises as a possibility to "[...] help nature in its reproduction" (Ayala 1999: 91).

Deforestation in New Spain had several impacts. In Independent Mexico, it generated changes in agricultural systems and their environment. It must add the processes of population decline, as a consequence of the war against the Spaniards and their allies. Since 1518, New Spain suffered the sequels left by diseases, epidemics, famines, and depopulation. The diseases brought by the conquerors affected the New Spain population during the entire Viceroyalty period but were especially destructive in the sixteenth and seventeenth centuries (Moreno 1987).

Land tenure changes are added to these phenomena and have to be considered as biological, social, and cultural elements and processes, modifying pre-Hispanic agroecosystems and forming part of the reconfiguration in the Mexican agriculture of the Viceroyalty. These occurred as a result of epidemics, famines, and depopulation. In addition: the granting of land to the Hispanic conquerors as payment for their participation in the conquest enterprise; the Royal Grant (*Merced Real*), the sale of land to the Spaniards by the natives (as happened in Tlaxcala); the expansion of cattle on indigenous lands; the abandonment and emigration of the natives to other more remote regions·

[2] Altieri 2002; Buendía-Rodríguez et al. 2008: 1–9; Dunning et al. 2014; Griffin et al. 2014; Molina 2010; O´Mansky 2014; Parsons 1976; Raimowitz 1996; Rentería-Garita 2006; Schwartz 1987.

Juan Carlos López-Acosta et al. (2018), in an article on the recovery of floristic diversity in citrus plantations in Uxpanapa, Veracruz that had been abandoned in the 1970s found that 80% of the vegetation in the region had been deforested in these years. Due to its low profitability, the orange plantations were converted into areas with secondary vegetation. The result of the study was that there were similar effects to what happened and was described in the historical case narrated:

> Of plant species was considered. 0.1 ha were sampled in each of 6 plantations; with to analyze the dynamics of regeneration in these systems, the composition, diversity, and turnover of plant species were characterized in a chronosequence of abandoned orange groves. In addition, the establishment of strategies, life forms and dispersal strategies different abandonment times (3, 8, 17, 26, 36, and 48 months). In each transect two groups of plants were distinguished: adults, and seedlings. In total, 191 species (57 families and 122 plants genera) were recorded. Floristic wealth increased after the time of abandonment. Trees with zoochory dispersion dominated all plots. Abandoned orange groves are attractive to vertebrate dispersers, and their proximity to sources of propagules favors an accelerated regeneration process.

The two cases that occurred at different times, but in places in the same region, show how both in the historical case and in the current one, the permanence of secondary vegetation in warm humid regions is linked to the population abandonment of these areas in several decades. This has allowed the floristic restoration of the ecosystem in a natural way, as the outcome of the forms of dispersion of the propagules. Other cases studied and documented on the same type of events were in Veracruz, Puerto Rico, Costa Rica, and other neotropical areas.[3] The role of propagation of propagules by animal agents is fundamental in the recovery of this type of system (Fedick 1996, 2012: 953–954).

12.9 HOMEGARDEN PLANT DIFFUSION AND DOMESTICATION

The domestication process of trees can be explained with a model in four stages: (1) that humans consumed the fruits of the trees that existed in a region; (2) that they selected the ones they liked, which gave bigger and better fruits; (3) that they took them to more favorable habitats close to their homes, thus accelerating their adaptation processes, and where the avocado with anise aroma gives larger fruits; and (4) that they dispersed the trees, generating new species. In the case of the avocado, there are two varieties: the one adapted to altitudes greater than 2,000 masl; and the Guatemalan avocado, which was adapted to altitudes from 1,000 to 2,000 masl due to its phenotypic characteristic of tolerance to cold, became its oil content permitted its adaptation to the altitude.

By genetic analysis, these two varieties are considered to have their common origin in California (Mhammed et al. 1997; Schroeder 1968: 29–34), having arrived there before human populations reached the Central highlands. From California, they continued south, via the Western Sierra Madre. Scott Fedick (1995) stated that avocado (*Persea americana*) reached the highland forests in Chiapas and Guatemala before the first settlers and grew there naturally.

It is the abundance of palaeobotanical and archaeological remains of edible plants, which allows us to talk about areas. These areas were used annually and routinely by the first itinerant and seasonal bands, and show what their taste in plant consumption was. The remains of animals that were cooked and eaten in the settlers' living spaces or the surrounding areas forming what can be called the first local "garbage dumps".

The domestication of plants is studied from genetics, as in the cases of the cultivated Mexican Hog Plum (*Spondias purpurea*), which is one of the first domesticated trees. It distinguishes itself from wild trees (*Spondias mombin*), which reproduce by seed, as it is of the cultivated ones that reproduce from branch cuttings. Wild *Spondias* reproduced by seeds is derived from native populations, which have a structured age and contain a variety of young and ripe fruits.

The pre-Hispanic domesticated Mexican hog plum, found and classified by specialists, is *Spondias mombin*, which is wild, currently, although there are areas where it is grown. This exemplifies the

[3] Aide et al. 1996; Bedoya 2009; Gómez-Pompa and Yáñez 1985: 1–25; Gómez-Pompa and del Amo 1985.

different modifications that with respect to domestication of plants occur over time. These modification and transformations occur depending on the characteristics of the plants as food and the taste of the inhabitants for their consumption. According to Miller and Schaal (2005: 12801–12806), in the case of Mexican hog plum, there was a process of domestication of multiple varieties in the same place, therefore showing the importance of the homegarden as a reserve with genetic variability.

Cruz-León et al. (2012) considered that the contrast in morphology and the method of reproduction between wild and cultivated Mexican hog plum indicated that *Spondias purpurea* became a genetically altered species throughout the course of its domestication. During the domestication process, humans selected trees with large, fleshy, sweet fruits. These trees could be reproduced easily with cuttings. Presently, wild populations of *Spondias* are found in the dry tropical forests of Mexico and Central America. Their natural habitat is restricted, and it is estimated that they are found in less than 2% of the Mesoamerican dry tropical forests (Tovar-Soto et al. 2012: 11–29; Cruz-León and Gutiérrez 2012: 32–62; Ruenes-Morales 2012: 63–75).

On the other hand, the study by Fedick (2012) in Copán, Honduras, established the existence of several problems related to the identification of plants that were domesticated by human action, or those that were pollinated by insects or by wind. On the latter, in relation to the plants that were managed by farmers, citing McNeil et al., Fedick (2010), wrote that it is important:

> Recognize that the increase in forest cover, as indicated in the pollen of a sediment core, may represent intentional manipulation of the forest in the form of forest agriculture, and the cultivation of trees in homegardens. The interpretations offered by McNeil et al. make an important contribution, also, to the growing recognition that the increase and decrease of maize pollen in sediment records is not necessarily the only or the best measure of plant cultivation for the Maya.
>
> A review of the literature on Mayan ethnobotany reveals that these, in recent times, make use of more than 500 species of food plants that are indigenous to the Mayan lowlands, complemented by many other species that have been introduced, some more in Northern Mesoamerica, or South America, most likely in pre-Columbian times. These food plants vary among 92 different taxonomic families.

12.10 CONSIDERATIONS AND CONCLUSIONS

Recent research and its interrelation with the social, cultural, and ideological ways of life of the Mayan populations associated with these phenomena and processes and studies on homegarden in the Mayan area gave important results in understanding many relationships. Among them are the environment, natural catastrophes, and the economic, social, political, and cultural effects that explain phenomena such as the so-called "Mayan collapse" of the Terminal Classic, or the migrations of human populations caused by volcanisms, fires, hurricanes, floods, earthquakes, and other natural phenomena.[4]

The first incidental homegarden is distinguished from the forest by its greater number of fruit trees, and edible plants whose remains have been found in it. The anthropomorphic impacts of man on the forest environment began with the hunting, gathering, and fishing of wild animals, complemented with the collection of fruits and their consumption. They continued with the tossing of the remnants of animals, fruits, and plants consumed on the ground, thus forming areas, incidentally, with the plants that consumers liked. It can be said that these were the first prehistoric dumps. The knowledge acquired by the residents as a result of this was the necessary tool for the selection and domestication of plants.

The domestication of plants in the original homegarden is being studied from the perspective of various disciplines and fields of knowledge (genetics, archaeology, ethnobotany, and agronomy). This is important for the understanding of selection processes, as in the case of the wild or cultivated Mexican hog plum (*Spondias purpurea*). The Mexican hog plum is distinguished

[4] Andrews et al. 2003: 151–153; Fedick 1996; Ford 2002, 2010; Ford and Nigh 2009, 2015, 2016; Killion 1990; Nigh 2008; Turner and Sabloff 2012; Vanderwarker 2006; Webster 2002.

because wild trees (*Spondias mombin*) reproduce by seed, while the cultivated ones do it by the cuttings. That is, it leads us back to the question posed in the 1930s of whether cuttings were first or grain farming was first.

Miller and Schaal (2005, Vol. 102, No. 36: 12801–12806) demonstrated the importance of the homegarden as a reserve of genetic variability. Without the conjunction of several disciplines and their results, it would be impossible, almost, to understand what happened and answer the questions which arose. The answer seems to be that – at least in the incipient and incidental Mesoamerican homegarden – the grain agriculture (maize) and that of cuttings originated almost at the same time, practically.

The agriculture of grains took much longer to evolve because it took about five to six thousand years to modify maize genetically from the original teosinte to be able to feed larger human populations. These modifications must be considered as the prime vegetation of the regions where hunter-gatherers passed before arriving at the sites where they were established seasonally and temporarily. These settlers carried the plants that they liked as food or that were useful (e.g. hard-shell pumpkins, gourds, or water containers) from various places, which contributed to increasing diversity of the forest.

The paleobotanical and archaeological remains, which include the bones of wild fauna as well as the residue of edible fruits and other plants, allow researchers to talk about areas that were anthropomorphically modified when used temporarily, seasonally, or annually by the first settlers. They show that the domestication of plants, such as maize, resulted from the selection and consumption of several grasses of propagated vegetative plants and animals that were trapped or hunted in the forest to be roasted or cooked, and consumed by the people in their residential spaces or in surrounding areas.

There are differences in the chronology of archaeologists and paleobotanists regarding the emergence of the first incidental homegarden. This is probably the result of insufficient factual information.

Acosta's studies (2010: 1–4; 2011: 227–235) in St. Marta, Chiapas, seem to put the origins of the incidental homegarden, referred to by this author as "scattered", in the Pleistocene period (10460 BC±50–10050 BC±90). All of the studies those researchers associate the presence of cacao pollen (*Theobroma* sp.) with tomato seeds (*Physalis* sp.), golden-spoon or nance seeds (*Byrsonima crassifolia* (L.) Kunth), and figs (*Ficus cooki*), together with forest species, which differ according to their location in the mesophilic forest, in the humid tropical, or in the low deciduous forest.

This vegetation is associated with the collection of freshwater jute snails (*Pachychilus* sp.), and the hunting of small and medium-sized animals, such as deer (*Odocoileus* and *Mazama*), peccaries (*Dicotyles*), rabbits (*Silvylagus*), snakes (*Crotalus*), iguana (*Iguana*), and turtles (*Kinosternon*). The activities of hunting, fishing, gathering, and trapping were dominant in the daily life of the first settlers, as shown by studies such as those of Vanderwarker (2006) and Ciphers (2004) in the region of Los Tuxtlas, in Veracruz.

For Acosta (2010, 2011), in St. Marta, Chiapas, both grinding stones and botanical samples indicate the existence of an incipient horticulture initiated at the end of the Pleistocene period. This date is very far from those given in other studies, such as in the Tuxtlas (3400 BP), or those of Tehuacán (more than 6,000 years ago), hence leaving the discussion about the dating of plants, such as maize, unanswered. That is why several problems arose in the interpretation of the origins of agriculture in the Mexican Central Highlands and in the south since there was a lack of information on vegetative agriculture and the domestication of maize.

Fedick's study (2012) in Copan, Honduras, showed some problems in the identification of plants that were pollinated by insects or by wind, in relation to those managed by man. Fedick (2010) proposed that for studies on plants it was important to know the origin of their pollination in places like the Mayan area, where they did not have anthropomorphic origins, always. Almost all the pollen was represented in the sediment cores that were carried by the wind, while most of the food plants used by the Maya were pollinated by animals. Among the pollinators were bats, beetles, bees, moths, and other insects. Ford's (2010) study of 18 homegardens in Belize shows that, in 20 dominant species forests that were cultivated in the homegardens, only *Brosimum alicastrum* (*ramón*) was pollinated by wind. This phenomenon was consistent with what happened in the floristic recovery of Uxpanapa (Ewell and Poleman 1980; López-Acosta et al. 2018: 1212–1233).

It is necessary to reconsider the information on the origins of the homegarden, where trees and plants of vegetative reproduction, and trees combined with grasses were found, all of which were genetically modified, like maize. The incidental creation of a homegarden near the settlement of the hunter-gatherer bands was an event resulting from the taste of these ancient settlers for eating certain fruits and dumping their waste on the ground. This creation facilitated the care of fruits from pests and predatory fauna. Vanderwarker (2006: 23, 148–181) stated that in the Tuxtlas the homegarden served as a trap to attract wild animals, which were food, thus the agroecosystem was a hunting area. It should be included that the homegarden attracted insects, which served in the dispersion of the flora in the system.

Humans altered the natural ecosystems by introducing plants from outside the forest, managing deforesting, or reforesting according to the necessity of obtaining resources like firewood and wood to cover them. They cleared the forest and planted other plants. After 1518, they increased this by planting those that came from the Old World, creating new habitats and introducing a wide range of pioneer weeds. These plants grew in abundance in open places attracting insects and their predators. The pre-Hispanic habitats included wild, tolerated, and cultivated plants, favoring diversity and influencing the density of small animals; larger animals – like deer – preferred disturbed environments. The homegarden agroecosystem changes occurred in several stages of their history, producing a series of differences, which were expressed in the history of the agroecosystem and were found in its components and elements that appeared to be similar.

TABLE 12.1
Edible Weeds and Herbs Cultivated and Collected from 1530 to 1578. Source Bernardino de Sahagún 1830, Vol. III

Nº	Name	Uses	Raw	Cooked	Roasted	Total
1	Nanacatl	These mushrooms are born in the fields they are good to eat. But they cause vomiting or chambers if they are raw or undercooked. They can be deadly. Its effect is counteracted with enema.		X		1
2	Tzontecomanatl	Large round mushrooms that grow together on one foot, some tall and others low.		X		1
3	no data	Wide and round mushrooms as dishes.		X		1
4	no data	Round white mushrooms that are not hard to cook and are also roasted on clay plates, they are very tasty.		X	X	2
5	cimatl	Wild weed with a root (*quavecoc, cimatl*, or *eimatl*); produces some large beans, like broad beans. They are boiled for 2 days.		X		1
6	Huauhquilitl	Very green smoothies with thin and tall branches, broad leaves. The stems and seeds are called Huautli. It is cooked with salt, tastes of ashes and is like those of Spain. It is squeezed and they make tamales with it (*quiltamalli*) and tortillas.		X		1
7	quilonilli	It is of broad leaves, when it is small it is edible. When it is large it is cooked with *tequixquitli* and the water is squeezed. Its black seed is called *pitzitl*.		X		1
8	itzmiquilitl	Brown weed with large branches, round and flat leaves.		X		1

(*Continued*)

TABLE 12.1 (*Continued*)
Edible Weeds and Herbs Cultivated and Collected from 1530 to 1578. Source Bernardino de Sahagún 1830, Vol. III

Nº	Name	Uses	Raw	Cooked	Roasted	Total
9	aioxochquilitl	Flowers, lumps, or limbs of pumpkin branches. The skin is removed from above them before cooking. Note of Sahagún (1830 Vol. III: 245). "In Oaxaca they are eaten together with the guides and green pumpkin stems, mix some balls of maize dough with salt, and some cooked green chilies. It is the food of poor peasants and very delightful, because it is dominated by openness, simplicity, and the good humor of the countryside".		X		1
10	axoxoco	Weed with long, wide leaves. It is tasty and sour.		X		1
11	mizquilitl	It is attic, very green, with bladed leaves. It is tasty to eat.		X		1
12	acuitlapali	Long, brown weed, grows at the water's edge. It is good to eat.		X		1
13	tziunquilitl	It grows by the water's edge. It has blue and harp leaves. It is good to eat.		X		1
14	conalquilitl	It grows in the mountains; it is ash colored; its root (*tacanalli*) is eaten.		X	X	2
15	mamaxtl	It grows at the water's edge; it is tasty.		X		1
16	n/name	It is like a nettle, it is edible weed.		X		1
17	ueiquauhquilitl	Wild smoothies that before cooking are bitter, They are tasty.		X		1
18	etenquilitl	Herb of beans, which spill when they are picked. It is eaten.		X		1
19	tlalaioquilitl	They are wild pumpkins.		X		1
20	xaltomaquilitl	It is eaten cooked.		X		1
21	tzitziquilitl	It is dark green, with flowers and seeds, very good to eat.	X			1
22	eloquilitl	It is very green and tender, it has flowers, it is very tasty.	X			1
23	quauheloquilitl	It is wild; it is born in the Opuntia *tunales*. It is very tender and good to eat.	X			1
24	mozoquilitl	It is green, fluffy, and very tender. it is tasty.	X			1
25	tzaianalquilitl	It is made in water, its branches are hollow and harp; it is good to eat.	X			1
26	achochoquilitl	It is light green, it is born near the water; it's good to eat "They say of this grass that if the boys or girls eat it they become powerless to beget; but after they grow up everyone eats it ". Sahagún, Volume III, p. 247.	X			1
27	tzonquilitl	It is light green, it is born near the water; it's good to eat.	X			1
28	iztacquilitl	It is low and flat, it has a salt flavor.	X	X		2
29	tepicquilitl	With long and pointed leaves. If you eat a lot of this herb, it gives cameras.	X			1
30	ecoquilitl	They are the leaves and branches of beans. They are a little rough and fluffy. They provoke regoldar.	X			1

(*Continued*)

TABLE 12.1 (*Continued*)
Edible Weeds and Herbs Cultivated and Collected from 1530 to 1578. Source Bernardino de Sahagún 1830, Vol. III

Nº	Name	Uses	Raw	Cooked	Roasted	Total
31	vitzquilitl	Thistles of the earth, they have thorns; the leaves below are ashy and the ones above are green. Inside they have lint, like the thistles of Castile, make them at the water's edge. It is also a garden weed. They are good to eat.		X		1
32	qaavitquilitl	Wild thistles that are born in the mountains. They are prickly and itch those who eat them.		X		1
33	chichicaquilitl	Edible grass that grows near water and in sweet, tilled land. It has white roots; it is very tender and somewhat bitter.		X		1
34	coiocuexi	Herb that is not prickly carves and flourishes, is bitter, it makes the voice hoarse in children. Girls don't eat it.		X		1
35	popoiauh	It is a black and green bush, it is kneaded with maize and tortillas are made from it.	X	X		2
36	exixi	It has tiny leaves, its seed is yellow, it is chewy and if you eat a lot it produces blisters and it is too hot. It is used to make tortillas and tamales and atole or porridge (for those with chambers of matter and blood, since it purifies the intestines).	X	X		2
37	xoxocoiolli	Perennial plant with an acidic flavor that is used as a condiment. It is sour.	X	X		2
38	xuxocoiopapa-tla	Herb with tall, slender feet, broad, rounds leave. The leaves at the tip of the foot are tasty to eat cooked.		X		1
39	xoxocoiolcue-cuepoe	Weed with fat round feet, rare leaves. It is born when the rains begin. They are tasty like tomatoes.	X			1
40	xolxococolol-vivila	It is a stopped herb with small and round leaves it is tasty.		X		1
41	miccaxoxocoili	Herb with a thick, hairy foot; the leaves are wide, very acidic and cause an unpleasant sensation that is experienced in the teeth and gums when eating *(dentera)*.		X		1
42	quauhxoxocoili	They are soft to eat.		X		1
43	quananaca-quilitl	Herb that the chickens of Spain eat. They are the *serrajas* (*Sonchus oleraceus*) de Castilla. Indians say there were none in these lands before the Spanish came. Now there are so many that the whole earth is filled with them; because the seed flies and they have multiplied everywhere.	X			1
44	cultivated xonacatl	Little onions from this land. They are planted, Indians are horticulturalists, and they eat the onions of Spain.		X		1
45	wild xonacatl	Wild onions that are found by these fields and burn a lot.		X		2
46	maxten	Wild onions that are born many, together, faint and flourish are bland. Indians eat the root or the head.		X		1
47	papaloquilitl	Round-leaved, fragrant and tasty herb. It is found in hot lands.	X			1

(*Continued*)

TABLE 12.1 (*Continued*)
Edible Weeds and Herbs Cultivated and Collected from 1530 to 1578. Source Bernardino de Sahagún 1830, Vol. III

Nº	Name	Uses	Raw	Cooked	Roasted	Total
48	aiauhtonv	Bush wild and low. It is found in the fields and mountains. Its flowers are fragrant.		X		1
49	no/name	It is edible; Sahagún named it as "the sweet potato".		X		1
50	xícama	Indians also eat its leaves.	X			1
51	tolcimoquilitl	Herb with very beautiful and very delicate flowers. The root is edible.		X		1
52	xaltomatl, xaltotomatl	Fruit from the plant called *xaltomaxilhuitl*. There are black and white, very juicy, sweet and round. The root is edible.	X	X	X	3
53	coiototomatl	Fruit like that of the small tomatoes called *miltomatl*. It has yellow coverage, they are sweet, and they slightly lock the throat. If you drink a little of the root of this herb it is medicinal, it cleanses the intestines. Women who breastfeed drink it because it purifies their milk.	X	X	X	3 Medicinal, also.
54	atlitiliatl	It has a sweet, blackish strawberry that declines to bittersweet. The leaves are good for the sick steam bath (*temascal*) to add to the water.	X			1
55	tlalilotl	Edible green weed.	X			1
56	tlalaiotl	Edible green weed.	X			1
57	iacanalxivitl	Medicinal herb that takes away the fever. It is ground with maize and drunk with water.	X			1
58	coanenepille	Herb to purge. After drinking it, you eat bird broth or an atole (*yolatolli*).		X		1
59	teonanacatl	Cures gout and fever		X		1
60	tlatlalayotli	Its leaves are edible; the root is medicinal.	X	X		2
Total	60		24	43	4	71

Source: Frair Bernardino de Sahagún, *Historia de las Cosas de la Nueva España*, 1830, Vol. III, pp. 244-249, 257.

RECOMMENDED READING

Benz, F. Bruce. 2006. Maize in the Americas. In: *Histories of Maize. Multidisciplinary Approaches to the Prehistory, Linguistics, Biogeography, Domestication, and Evolution of Maize*. J. Staller, R. Tykot, and B. Benz (eds.). Academic Press; pp. 9–20.

This chapter reviewed the archaeological evidence of plant domestication and provides critical evidence of human manipulations of genetics, absolute dates, and the cultural context of the origins of maize cultivation and its evolution. There is little archaeological interest in the documentation of sustained agriculture, selection, and continuous domestication after recording the initial steps humans took in the selection of traits that allowed distinguishing between domesticated species and their wild ancestors.

The author showed little interest in maize production processes and manipulations by human populations. Instead, archaeologists abound in studies related to domestication processes, the morphology of inflorescences, the size of pollen, and the abundance and shape of phytoliths. The author highlights the efforts made by biological evidence on this topic, and he considered that on the historical-cultural processes the archaeological evidence only leaves inferences.

Turner II, B.L., Jeremy A. Sabloff. 2012. Classic period collapse of the Central Maya Lowlands: insights about human-environment relationships for sustainability. *Proceedings of the National Academy of Sciences of the United States of America* 109(35): 13908–13914.

The collapse and abandonment of the central Mayan lowlands of the Yucatan peninsula in the ninth century resulted from complex interactions among the small-scale farmers, the environment, and the Mayan rulers. Increasing demands on ecosystem resources and services produced highly stressful environmental conditions, which increased with climate aridity. In addition, trade shifted from overland transit through the peninsula to coastal maritime transit. The environmental, social, economic, and political conditions created social conflicts, reduced the political control of the Mayan elite, and produced the emigration of the farmer population to other regions on the peninsula, to not maintain the high costs of the existing human-environmental systems.

Dunmire, William M. 2004. *Gardens of New Spain. How Mediterranean Plants and Foods Changed America*. Austin, University of Texas Press.

This book is a history of the plants that arrived on the American continent from Spain since the sixteenth century. It begins with a chapter on pre-Columbian Spain, agriculture of Mexico, and the American Southwest before the arrival of Christopher Columbus on the continent (Huckell 1996). The book included a chapter on trade, technology, and livestock in Spain. Several chapters are dedicated to describing the Mediterranean-type homegarden in New Mexico, Sonora, Arizona, the corridor to Texas, and the Mediterranean connections to Florida and California. The book is a study on the migration of plants to the New World and their arrival to New Spain.

13 History, Changes, and Diversity of the Homegarden

13.1 INTRODUCTION

The Hispanic homegarden, orchards, and gardens with their flora, domestic fauna, and agricultural instruments arrived in Mexico through the convents. From them, the Spanish homegarden and gardens spread to the villages to mix with the native homegarden and orchards, generating new combinations of agroecosystems and crops nevertheless their structure continued to be based on trees and their combinations with other trees strata (González-Jácome 2011: 177–193, 243–294).

The homegarden has undergone a series of changes over time, but its trees and human management have given the system its resilience and sustainability over time. That is why the homegarden agroecosystem has been and continues to be an important source of information for all those dedicated to the study of agriculture. The homegarden differs according to environmental, physiographic, climatic, economic, social, or cultural factors, as well as local and regional factors of the places where it is located. Currently, the small-scale and polyculture homegarden includes several strata: herbaceous, shrubby, and arboreal including the canopy.

The homegarden functions are multiple, varied, and characterize, and explain their diversity (Cahuich 2012; Leyva 2017; Mariaca-Méndez 2012; Ordoñez-Díaz 2018). These characteristics depend on the local conditions, specific to each of the regions where they are found, and their environmental, climatological, and demographic characteristics.

13.2 BACKGROUND

The information of the results of the pre-Hispanic archaeobotanical analyses indicated that there were differences in the zones caused by the climate, topography, chronology, and cultural context. The botanical remains recovered and identified in the INAH investigations included archaeological and contemporary items. According to Xelhuantzi-López (2015: 109), they correspond to those found in caves and rock shelters, but they were not specifically studied in regard to the pre-Hispanic homegarden agroecosystem.

The archaeological remains at INAH included: corncobs, maize kernels, corn husks, and teosinte kernels; fruits, seeds, peduncles of chili; squash seeds, stalks, and bark; gourd peduncles and seeds (*Lagenaria siceraria*); leaf, bagasse, and fiber remains (basketry and cordage); Agave remains; chayote pericarp; Mexican hog plum fruits and seeds (*Spondias purpurea*); and American mastic or American pepper seeds (*Cyrtocarpa procera*); palm fibers (*Brahea* sp.); bean seeds (*Phaseolus vulgaris*) and beans (*Canavalia* sp.); remains of White sapota seeds (*Casimiroa* sp.); and remains of gourd fruits (*Crescentia alata* and *C. cujete*).

It is always a temptation to imagine that the pre-Hispanic homegarden was an agroecosystem similar to the current one. Also, local variations were caused by characteristics only related to climate, altitude, and direction of the homegarden products toward self-consumption, interchange, or the market. The pre-Hispanic and Viceroyalty homegardens were for the cultivation of edible fruits and food plants, medicinal plants, obtaining firewood, materials for making ropes and nets, providing shade, ornament, and items for ritual purposes.

The proposal that in ancient times homegardens were articulated with forests and the cornfield, and were places of attraction for game animals should be studied in more depth and in tandem with their regions. These characteristics do not seem to suit all homegardens in the past, which were

DOI: 10.1201/9781003198833-13

defined by local conditions, specific to each of the regions where they were found, and its environmental, climatic, and demographic characteristics. Before the arrival of the Spaniards in the central Highlands, it was established very clear that there were several types of homegarden with zones for legume and maize cultivation, monoculture orchards of avocado, Mexican hog plum, Mexican black cherry, and recreational flower gardens (*jardines floridos*).

In the Mexican Central Highlands, the homegarden agroecosystem formed an important assemblage along with the fauna, agricultural instruments, local knowledge, soil and water management, and their interrelationships with surrounding natural ecosystems. After 1519, the Mexican homegarden began to mix its components with those of flora and fauna in the convents, creating a new way to use the space, both in the vertical, and horizontal structures, increasing biodiversity, and the food system. The new structure was a mixed one, with the native plants and those brought by the Spaniards which were introduced to the homegardens, orchards, and gardens.

13.3 THE SIXTEENTH-CENTURY HOMEGARDEN

The convents with friars, new religion, the Hispanic homegarden, orchards, gardens, and all their components mixed then with the native homegarden and avocado, Mexican hog plum, and Mexican black cherries orchards, thus combining crops and agroecosystems. The Mexican homegarden structure continued to be based on trees, but the Indians integrated the Old World fruits, medicinal herbs, flowers, and vegetables to their pre-Hispanic homegarden. This new agroecosystem included Opuntia and Agave (González-Jácome 2011: 177–193, 243–294).

There is little archaeological and historical information about the flora found in pre-Hispanic times in agroecosystems, such as the homegarden, orchards, and gardens. However, it can be established that upon the arrival of the Spaniards in the central Mexican Highlands there were several types of homegarden, orchards, and recreational gardens already. The recreational homegardens were intended for the cultivation of trees and colorful flowers, whose fragrance was an important part of their presence in the system.

There were homegardens with trees and cultivated herbs, like the ones described by Friar Juan de Navarro in his *Historia de las Plantas de Nueva España* (Vol. II, Chapter XXXV; p. 527, Fig. 116; Vol. II, Chapter CIX: 594–598, Fig. 14). This book was written in the last quarter of the sixteenth century; and at that time, the list included native and foreign flora, including one tree from South America.

The flora included: *mizquixochicopatli* (*Ipomoea jalapa* Pursh), (*Cacamotic Tlaloquiloni*, or *Batata purgante*), *holquahuitl*, *huitzxochitl*, *quauhconex*, *maja balsam* or *maxipenela*, *balsamotohi*, Peruvian peppertree (*Schinus molle*) (*pirul*), copal tecomaca (*Bursera tecomaca*), sweet gum tree (*Liquidambar styraciflua*), various kinds of *Opuntia* (*nopalli*), various types of palms, *cagua*, *amacastic*, *cuauhxiotl*, *coacamachalli*, *tlatzcan*, *bitonco*, *xumerl*, papaya (*Carica papaya*), a kind of sempervivum called *coapatli* (*Euphorbia prostrata*), guyabaza (*Ligustrum* sp.), *aguaparti*, *chichiltic tlapa tezquiahuitl*, Montezuma bald cypress (*Taxodium mucronatum*), soap beads, *macanillo*, *xahuali*, *higuexo* (*Ficus carica*), *tlalzilacayotli* (*Cucurbita ficifolia*), vegetable pear (*chayotl*, *chayote*) (*Sechium edule*), *tiplacaca*, Chinese pomegranate (*Punica granatum*) that was brought from Spain, *tlatlarizpatlia*, *tomatzitzitzicaztli*, *iztaccihuatl*, and Dragos'blood (*tlapal ezpatli*, *Sangregado*) (*Croton draco*).

Regarding the tree called *tlapal ezpatli* (*Croton draco*) in Nahua (Reko 1947: 29), the medical information is that the bark of this tree was sold in street markets (*tianguis*) and was used for pectoral diseases. It was cooked in water, with other herbs, and mixed with pips and chili (Sahagún 1990: 738). A bread roll called *ezpatli* was made with *Croton draco* by mixing with water and ground (Sahagún 1990: 740).

Primary historical sources (Bernardino de Sahagún 1830, 1956; Francisco Hernández 1959, Vol. 2) showed that at the beginning of the sixteenth century the homegarden in the basin of Mexico included fruit trees, and depending on the physiographic and climatic characteristics, it had aromatic trees such as copal (*Protium copal* (Schltdl. & Cham.) Engl.). The homegarden contained shrubs, herbs, vines, plants for obtaining grains (beans), and vegetables. The plants were used for food, drink, herbalism, condiments, and flavorings; and the trees were used for wood, firewood, and shade. The branches and bark of some trees were used for medicinal purposes or for ceremonial drinks such as balche. Some vines, in addition to being used in the manufacture of ropes, nets, and basketry, were used for medicinal purposes.

The flora for medicinal uses made up the majority of plants in the homegarden. It is important to consider that food and health went hand in hand in ancient societies.

Based on information from Sahagún (1956), González-Jácome (1985 (2): 137–148, 2018: 43–85, 521–537) said that something similar was happening in the European societies of those times. In Spain, the homegarden was irrigated (Maestre-Alfonso 2011), unlike the homegarden in ancient Mexico that was irrigated by systems like canals, ditches, or manual watering (*zoquimaitl*, *ollas*), or watered with seasonal rains. Anyway, the Mesoamerican kitchen was articulated to dietary aspects and health, but this did not mean that Mesoamerican people ignored the delight of taste, sight, and smell through a series of dishes that in many cases were time consuming in their preparation (Beals and Hatcher 1943; González-Jácome 2011: 137–148; Lachat et al. 2018: 127–132; Méndez-Mercado 1993: 32–40) (Figures 13.1 and 13.2).

FIGURE 13.1 Tlaxcala, Atoyac-Zahuapan Rivers Basin: Ditch between Drained Fields.

FIGURE 13.2 Tlaxcala, Atoyac Zahuapan Drained Field: Earth Ditch.

In the second half of the sixteenth century, women were in charge of cultivating and maintaining the homegarden agroecosystem. Many medicinal plants in it were focused on women and their needs. The information compiled in 1577–1578 by the proto-medic of Felipe II, Francisco Hernández, about the effects on young males who had participated in the war of the conquest 50 years before, and in which many had died, may have intensified the female role of work in the homegarden. Furthermore, the location of the agroecosystem next to the house facilitated female participation in its management.

There is a lack of information about the role of children and the elderly in the work applied to the agroecosystem and their activities, like removing diseased leaves from trees, sweeping the surface of the homegarden, picking mature fruits from trees manually, and after 1518, grafting trees and doing other related activities (González-Jácome 2011: 101–103, 2016: 233–257, 2018: 58–150).

The native Mexican Central Highlands homegarden underwent major transformations after 1518. Citrus and other fruit trees were introduced, like pear, quince, fig, apple, walnut, and Old World plum (Dunmire 2004: 125–130), as can be read in Juan de Navarro (1801). Some native plants were displaced and replaced by the ones brought from the Old World, such as the Mexican hog plum (*Spondias* sp.). Several native plants, which were grown in homegardens, are unknown, currently. Other trees lost the importance they had before: the Mexican black cherry (*Prunus serotina* var. capuli (Cav.) McVaugh, the Mexican hawthorn (*Crataegus mexicana*), the White sapota (*Casimiroa edulis*), and the Coral bean tree (*Erythrina americana*).

The Mexican hawthorn and the Mexican black cherry were grown aligned on the raised and drained fields and edges of land on the limits of the plots where their leaves and fruits fell to the ground contributing to the maintenance of the soil fertility and offering food for birds and humans. Erythrina trees and edible flowers were planted for religious and protective purposes (González-Jácome 2004, 2006: 19–46, 2011, 2016, 2018: 58–150; Wilken 1969).

Some of the native plants were displaced by those brought from the Old World. One of them was the native onion (*xonacatl*), with a sweeter flavor than the one that arrived in the sixteenth century. Francisco Hernández (1574) wrote that the native onion had milder taste and smell than that brought from the Old World. The native onion seemed to him tastier. The same thing happened with wild edible green seasonal weeds and tolerated weeds (*quelites* or *bledos*), which were replaced by lettuce (1526) in salads, or green cabbage (*berzas*) and cabbage used in cooked foods, cauliflower (1526), leek (pre-1580), beet (pre-1580), carrot (1526), horseradish (1526), radish (pre-1580), turnip (1526), cucumber (1530s); eggplant (1530), and parsnip (pre-1580) (*Pastinaca sativa*) (Dunmire 2004: 126–127).

Old World vegetables were integrated into both the homegarden and the cornfield, although they entered the cornfield in greater quantity, giving this agroecosystem a greater diversity. Also integrated into the homegarden were citron (1531), lemon (1531), lime (1531), sweet orange (1518), sour orange or Sevilla (1531), shaddock (pre-1580), apple (1536), fig (1530s), jujube (mid-1500s), medlar (mid-1500s), olive (1530s), pear (1536), pomegranate (1530s), quince (1536), banana (1520s), grape (1529), melon (1530), and watermelon (pre-1580) (Dunmire 2004: 126–127).

The sixteenth-century food system in the basin of Mexico included new elements like the broth made with vegetables, poultry or fish, stews, and roasts with meats of different domestic animals, like ducks that were a very important part of the food system. In addition, the remains of vegetables brought from the Old World, such as lettuce and cabbage, that had larger remains than those of native plants were given to poultry and pigs as food. These remains were added to the soil as green manure.

From Francisco Hernández' data (1959 [1577]), it can be considered that in the sixteenth century the Mexican Central Highlands homegarden was diverse and included different types of trees, medicinal herbs, green cultivated and tolerated weeds, condiments, grains, and vegetables. The rich diversity of this homegarden was important for the food system and health of families. Some flowers were grown for their scent, for decoration, and for religious or ceremonial purposes. Others were used to make drinks or to perfume the environment. Some wild plants, such as orchids, were collected and added to chocolate to flavor it, but not all aromatic plants added to chocolate were orchids (González-Jácome 2011: 96–103; Sahagún 1830, Vol. III).

Aromatic flowers to be added to cacao were sold in the Tlatelolco market, along with cacao beans, different types of sweets such as Agave syrup, maize sugar, honey from wild bees, or honey of domesticated bees. The flowers were the Flower ear (*Hueynacaztli*, *Teonacaztli*, *Xochinacaztli*) (*Cymbopetalum penduliflorum*), Black flower of the Aztecs, or Vanilla orchid (*Tlilxochitl*), Rope flower (*Mecaxochitl, Macazuchil*) (*Piper amalago* L.), and Mexican blood trumpet or Scarlet trumpet vine (*Tamacaxochitl*) (*Phaedranthus buccinatoria*) (Chamoux 1997; González-Jácome 2011: 96–103; Sahagún 1830, Vol. III).

13.4 THE HOMEGARDEN AND ITS CHANGES

Since 1519 to date, the size, structure, and composition of homegarden flora have been modified. The domestic fauna in the homegarden from that time to date has been mainly of pigs, chickens, and turkeys. Depending on the region, there might have been ducks, pijijis, geese, and other small birds, mammals, and reptiles, like iguana (Mariąca-Méndez 2017: 475–508). Currently, the tropical homegarden fauna includes domesticated, semi-domesticated, and wild animals.

The homegarden in the Tuxtlas, Chiapas, a region integrated with the forest and the wild animals living there, was a place for hunting games for the family food or for its sale in local markets. The domestic fauna in other regions of the country has been reduced to chickens, pigs, turkeys, and dogs: in some cases adding cattle and draft and pack animals. Sometimes in zone near to the house, there is a special place for horses, donkeys, and sheep; there is also a place where family keeps the bales of straw or the corn stalk, which will serve to feed the cattle as well as draft and pack animals. Also, farmers keep wood and some agricultural implements in this place (Blanco-Rosas 2006; González-Jácome 2012: 522–535) (Figure 13.3).

The conversion of the homegarden into a monoculture fruit orchard or garden for vegetables or flowers directed toward the family enjoyment on weekends or into zones for the construction of rural neighborhoods for family members who married is a process that has extended to various places of the country. In both cases, often the plots replace trees and flora in general so reducing strata and diversity. The role of domestic fauna in the homegarden has changed. In Chan Kom, domestic fauna is reduced to turkey and an occasional pig. In villages and towns of Tlaxcala, chickens and turkeys continue to be part of the homegarden. They are consumed in the village festivals, family celebrations related to the completion of elementary school studies, the birthday of the family head, marriages, baptisms, confirmations, rituals, offerings, and ceremonies associated with religious beliefs

FIGURE 13.3 Chan Kom, Yucatán: Homegarden with Stone Wall (albarrada) to Keep Domestic Fauna Indoors.

and local customs (González-Jácome 2012: 522–535; LeCount 2001: 935–953; Moctezuma-Pérez 2010: 26–46; Oliveros-Espinosa 2015: 491–514; Sierra-Carrillo 2015: 413–428) (Figure 13.4).

In the 1970s and 1980s, the townspeople were able to pay their quotas for the construction of public works or the introduction of services to the community (electrical energy, road repair, construction, or repair of works for the introduction of drinking water) by working without payment in the civil works (mutual help), paying with money, or delivering domestic animals to the tax collectors for the respective quotas (González-Jácome 2003).

Currently, there are native trees found in Mexican homegardens that were absent in the pre-Hispanic agroecosystem. This is the case of the Mexican hawthorn, which after the conquest was used as a base for grafting pears and apples. Thus the hawthorn was incorporated as a component of the homegarden in addition to being planted on the raised field borders (agroecotones) between plots (González-Jácome 2003). This continued until the 1990s in southwest Tlaxcala where road construction broke the canals in the drained fields, so currently Mexican hawthorns and Mexican black cherries are disappearing due to urban expansion and the expansion of road construction (González-Jácome 2006, No 3, pp. 19–46, 2011; Ochoa 2002). See Table 13.1.

FIGURE 13.4 Tlaxcala, Matlalcueye Volcano Lowlands: Peach Orchard.

In temperate-cold and tropical areas of Mexico, domestic fauna was included in the agroecosystem, replacing wild animals and their meat. Hens and pigs were traded and generated monetary income for the families of small-scale farmers. The structure of the homegarden has been modified in present times. Now in the homegarden, the domestic fauna is mainly pigs, chickens, and turkeys, and fruit trees have waned to give space to shade trees and shrubs.

In the sixteenth and seventeenth centuries, the Mexican polyculture homegarden arose including various kinds of trees (fruit, wood, medicinal, and shadow), edible, medicinal, aromatic, and ornamental plants as well as tolerated green seasonal weeds and shrubs that were sold in markets. During the Viceroyalty, the Mexican homegarden included a great diversity of trees; and wild, tolerated, and cultivated flora; and domestic fauna. Vice royal homegarden included flora and domestic fauna from pre-Hispanic times and the Old World. It was a diverse agroecosystem that adjusted its ecologic, economic, social, and cultural components (food system, health, and related costumes) to different ecosystems in New Spain.

Homegardens were established with those of convents that were the most important for their market-directed production, like the S. Angel irrigated one in Mexico City that had 30,000 fruit trees, such as pears, peaches, olive, apple trees, poplars, and other trees that did not bear fruit that "[…] only serve for the recreation and amenity of said homegarden" (Abundis-Canales 2007, I: 459). Irrigation was derived from the Magdalena River, the Arroyo Chico stream, and three very large tanks, one that was fed from the S. Bartolomé Ameyalco spring. The lands were fertilized with the silt that was deposited in the streams that crossed the homegarden. The fruit was sold in Coyoacán, Pachuca, Toluca, Puebla, and Mexico City. In 1671–1672, the homegarden generated 9,000 pesos per year from the sale of its products.

The religious community's and the friars' food system was sufficed by a legume and vegetable garden. From the fruit, jellies, jams, and sweets were made for internal consumption to give away or to sell. A college was associated with the homegarden and had 14–15 young men living in the convent permanently. The number increased to 50 when it was fruit-picking time in the homegarden. The convent had 40 mules, 10 saddles, and 30 cargo mules to carry equipment and take the fruit to the markets. A lay manager looked after the homegarden, but he had to be an expert in the matter (Abundis-Canales 2007, I: 461).

13.5 DIVERSITY AND CHANGE

In towns of the municipality of Santiago Chazumba, in the Mixtec region of Oaxaca state, the current family homegarden is diverse, with the axis of the agroecosystem being the Blood's dragon fruit (*pitaya*) (*Stenocereus pruinosus*) and the sour prickly pear (*xoconoxtle*) (*Stenocereus stellatus*). Rojas Benitez (2019) studied 17 locations in the Oaxaca Mixtec, covering an area of 32,000 hectares. This agroecosystem is adapted to the conditions of the sub-humid semi-warm climate, with an annual average temperature of 18.4 °C at 19.5 °C, and an altitude of 1,700–1,990 masl. Homegardens do not require irrigation, and their production yields give significant economic benefits to families (Rojas Benitez 2019). These yields, monetary benefits, and the promotion of development programs to expand their monoculture are contributing to changes (Luna-Morales 2002: 25–42; Luna-Morales and Aguirre 2001: 18–24; Nolasco 2010: 4–5; Rojas-Pérez et al. 2017: 1–20; Rosas-Benitez 2019).

A stonewall (tecorral, albarrada) circumscribed the homegarden plots (*solar*) and the domestic fauna in the southeast of Mexico that had been enumerated and described by Mariaca-Méndez (2017: 492). Domestic fauna included: dog; chicken, pig, turkey, dove, sheep, cat, horse, cow, donkey, rabbit, mule, goose, guinea fowl, Peruvian duck, and stinger-less bees. Domestic fauna is used for food, agricultural work, pack animals, guard animals for protection of the houses and land (dog, cat, goose), animals to market (chicken, pig, turkey, duck, sheep, and rabbit), for recreation, and animals for the folk medicine (chicken and bees). There was a list of 69 wild birds that were in the backyards, 32 were for ornate (50.8%), 13 were for pets (15.8%), 13 were for market (15.8%), and 11 were used as food (17.5%), occasionally.

Dragon tree in homegardens of the studied towns was the most important for farmers. Homegardens have an area of 1000 sqm approximately and are for the family's self-sufficiency, but the dragon tree is the most important one for monetary earnings. Development programs in the region aim for farmers to eliminate other trees and plants from the system so that they become dragon fruit monocultivators. The price of dragon fruit on the market made its cultivation productive and lucrative. To date, those programs have not succeeded, despite developers who exert pressure to convince farmers.

In Yucatan, the homegarden surface had been reduced by the construction of houses for the emigrated married young men, who occasionally visit their parents. In Yaxcaba, the kitchen is in the outer courtyard, where it is a place to prepare food and also for all family members and friends to meet and socialize. Yaxcaba homegardens are inside or on the edges of the forest and have an area for bees with honeycombs in hollow tree trunks for honey production, which provides families with important monetary resources.

The money received for agricultural and nonagricultural activities constitutes the family income. In families with children, young men, women, and adults, seasonal work in tourist places is a source of income that allows them to improve their livelihoods; set up businesses, like shops, internet businesses, retail trade, etc.; and pay for the education of the youngest family members.

In Chan Kom, the homegarden belonging to aged owners and migrant young sons is a garbage dump, where some turkeys and pigs live. In the tropical regions, the homegarden maintains a structure similar to that of the forest, so it has subsisted over the centuries in regions where regional, state, or federal development plans did not boost livestock. Hernández X. et al. (2012: 80–81) wrote about the family homegardens in the Yucatecan towns of Tixcuytún and Corral. In these homegardens (*solares*), there are similar areas and structures among generations and among communities:

> (1) The grandparents of Tixcuytún previously built their Maya-type houses (oval, and made with natural materials) to the center of the site, while today they are in front, as their descendants prefer concrete homes. The reuse of the former as maize stores replaced the structures previously built ex profeso. (2) The old ka'anche', the raised table where vegetables are grown, was depleted by the cultivation in waste containers, the physical effort of its construction, and the sale of vegetables in the town shops. In both communities, fruit trees have two spatial arrangements: scattered or aligned. The alignment, especially of citrus fruits, is more evident, and widespread among the generations of Corral due to the direct participation of their inhabitants in commercial agriculture programs. (3) In both communities the use of the ya'ax jalal che '(*Pedilanthus tithymaloides*) plant is maintained as protection against bad winds, although the location within the site varies between generations and is conditioned by the presence of the Pentecostal Church in Corral. (4) The most notable change occurs in the ornamental area, due to the increase of ornamental species, and varieties, and the adoption of urban aesthetic models, which in Tixcuytún emulate the Tekax parks, and in Corral to the hotels of Playa del Carmen, Quintana Roo (process they call the homegarden plots (solares) conversion into gardens).

In the middle basin of the Mayo River, in Sonora, in northern Mexico, the homegarden occupies a small space next to the house, and its importance lies in the production of medicinal plants and some condiments. In the town of Chorijoa, the homegarden is about 70 sqm, with lemon trees, redwood (*Bursera simaruba*), penguins (*Arctostaphylos pungens*), orange trees, guavas, papaya, mangoes, silk cotton tree (*Ceiba aesculifolia*), tamarind, white stick (*Lysiloma candidum* Brandegee), tabachín (*Caesalpinia pulcherrima*), Mexican black cherries, *Opuntia*, and bougainvillea (Moctezuma-Pérez and Pérez-Sánchez 2018: 226).

In the middle basin of the Mayo River, the homegarden is irrigated with tap water, usually (Moctezuma-Pérez and Pérez-Sánchez 2018: 269). The agroecosystem has a few fruit trees and cacti (González-Jácome et al. 2014). When tomato is grown in the homegarden, it is fenced with barbed wire (González-Jácome et al. 2018). In places of the Mexican Central Highlands, like the towns of the upper basin of the Totolac River in Tlaxcala, the function of the homegarden have been reduced to the cultivation of medicinal plants, some ornamental ones, and those that are used for herbal teas, as occurred in southern Sonora (González-Jácome and Reyes-Montes 2015, 52–53: 21–42; Robles-Linares 2018: 269–296). In villages in the south of the State of Mexico, homegarden flora passed from the soil to pots, which are located in the corridors of the houses or in spaces in the homegarden, where herbal medicinal (Juan Nentvig 1764, in González-Jácome 2018: 39–48) and ornamental flora is planted (Juan-Pérez 2003; Moctezuma-Pérez 2013a, 2013b).

13.6 KITCHENS, THE HOMEGARDEN, AND OPEN SPACES

The food system and the space for the kitchen were articulated directly to the homegarden, which underwent major transformations. This is observed in the X-Mejía and Yaxcaba homegardens in the Yucatan peninsula, in towns of the middle basin of the Mayo River, in Sonora; in past decades, in Morelos, open kitchens were located in open places, outside the houses, and next to the homegarden. The old open kitchen is a place where tortillas and food were cooked on ground level, and smoke kitchens were widespread in rural areas of Mexico, mainly (Figures 13.5 and 13.6).

In the 1980s and early 1990s, in Chiconcuac, and the surrounding villages in the current state of Morelos, the kitchens were in spaces outside the houses, in oval or circular open constructions (*palapas*), without walls, supported by wooden columns, or thick tree branches, with palm roofs, or roof made of other materials of the region. In these spaces surrounding the stove or grill, there was a rustic table and sometimes one or two chairs; a place where family members consumed food and talked.

FIGURE 13.5 Yaxcaba, Yucatán: Open Smoke Kitchen in the Homegarden, 2008.

FIGURE 13.6 North Sonora, Mexican Border with USA: Open Smoke Kitchen.

In regions where temperatures were high, like in the Mayan area of the Yucatan peninsula, the open kitchens had a place dug in the ground used as an oven. These kitchens were located in the homegarden that had natural vegetation, ventilation, and freshness. Open kitchens allowed family and friends to talk, interact, and make important arrangements for the economy and social life of the participants. In the dry season, open kitchens are places where older generations met younger generations, and through informal talks, an important part of traditional knowledge was transmitted. The state of well-being of family members and close friends was known, the life of the neighbors was talked about, and also business was done among relatives and friends.

On one side of the homegarden was the ka'anche, a table with pots where Mayan farmers grew and experimented with plants brought from other places, to see if they adapted to the climate, soil, and environmental conditions, so they could be taken to the homegarden or the cornfield later (Figures 13.7 and 13.8).

In the 1980s and early 1990s, the homegarden with the open kitchen of Chiconcuac, Morelos, was modified to become a very crowded place for family members, leisure, and parties. In addition to eating food –usually cravings – business was done among landowners and neighboring home owners and their guests. The old homegarden had been turned into garden, and the old irrigation canals (*apaxtles*) were used to grow ornamental Japanese fishes. In outdoor kitchens, there were a gas stove, electricity, and a refrigerator for cold drinks, sometimes. In the rainy season, food was

FIGURE 13.7 Yaxcaba, Yucatán, Ka'anche in a Homegarden.

FIGURE 13.8 Ciudad del Este, Paraguay, Ka'anche in a Homegarden.

cooked in special spaces located inside the houses or next to them. Open kitchens coexisted with smoke kitchens, where functions were similar, although there were fewer appliances and a greater presence of family and friends of all ages.

In Chan Kom, a town southwest of Valladolid, in Yucatán, now there are abandoned homegardens and open kitchens whose owners are old and their children have emigrated to work in Cancun or Playa del Carmen, the tourist centers of Quintana Roo, or to other new touristic developments on the peninsula. The old owners no longer have the youth or strength to work, so homegardens are becoming garbage dumps. In the interior of the houses, there is a zone with a stove, a table, chairs, and a few kitchen utensils. In the homegarden, there are areas with trees where some turkeys and pigs live. The old people keep their cornfields, paying workers to plant, weed, and harvest the maize, but the cornfield has become a maize monoculture. From 2008 to 2012, there were some homegardens cared for the owners, but still the number was decreasing (González-Jácome 2012: 487–521).

In the 1940s, anthropologist Alfonso Villa Rojas (1945: 159, 202) described Chan Kom, a town located about 21 km southeast of Chichén Itzá, in the state of Yucatán. At that time, Mayan rebels populated the zone, because for almost a century they had fled the white domination of the peninsula (Goldking 1965; Villa-Rojas 1945). The Mexican anthropologist Villa-Rojas stated that practically the town is same as it was when the Spaniards found it upon arrival:

> [The town has] palm huts spread irregularly around sinkholes; next to them the small table with pots with the plant seedlings in the process of experimentation to see if they can be transplanted to the homegarden (*ka'anche*), the cages for their domestic birds, and as an important structure, the hollow log apiary; turkeys, and other chickens; fruit trees planted on communal land [...]". Villa Rojas also mentioned "within the town, the property is communal. Domestic plots are not privately owned, and anyone can make their home where it suits them best. Fruit trees belong to the individual who planted [...] fruit trees can be sold or rented [...].

Cahuich (2012) studied homegarden agroecosystem in X-Mejía, a small town in Campeche. The homegarden and the kitchen had functions articulated to the subsistence and socialization of families. The food system depended on what was grown in the homegarden that was explained by deficiencies in commercial services, difficulties in communications, and transport that went to larger populations, with shops and urban services. In the past as in the present in contemporary homegarden agroecosystem, cultivated plants are combined with some plants of the original vegetation of the area. Both cultivated and wild vegetation covered partially and in varying percentages the nutritional, medicinal, and ornamental needs of families (Figures 13.9 and 13.10).

In the middle basin of the Mayo River, among Guarijíos, the open smoke kitchen is located at one side of the house. The heat that characterizes these regions for several months of the years is

FIGURE 13.9 Cuaxonacayo, Tlaxcala: Warehouse for Firewood, and Bales of Grass, in the Homegarden.

FIGURE 13.10 Las Flores, Sonora: Carrying Firewood to Home.

the reason for this. Stove, oven, and water deposits made of clay pots, sometimes large, are in the smoke kitchen. In the towns of the municipality of Alamos, family members consume their food in a space, often under boughs or roofs supported by thick and tall branches, where they are protected from the sun. Table, chairs, and a spot to place food and for the clay plate to cook wheat tortillas with countless pots with plants around it are all in the eating space (González-Jácome et al. 2018; Moctezuma-Pérez and Pérez-Sánchez 2018: 225, 228).

13.7 CONSIDERATIONS AND CONCLUSIONS

The small and medium-scale homegardens include commercial plants, which generate monetary income to families (González-Jácome and Ramírez-Martínez 2010e; Ramírez-Martínez 2007). People's need for firewood and shade depends on where their agroecosystem is located, the climatic conditions, or the population's access to services, such as gas for domestic purposes. The homegarden is a sustainable agroecosystem, with the ability to reduce or expand its space. It is the environment to supply the needs of a family, number and destination of its production, to decide workforce that is required, the number and distribution of trees and associated plants, and the abandonment or revitalization of the system.

The elements that together make up the resilience of the system to the strong expansion of the market had not reckoned phenomena like the modernization of rural life. The migration that moves younger members away from their families and separates them from agriculture as a basic activity is one factor prompting change in the agroecosystems, as is the low income brought in by crops, in relation to the means to cover costs generated by modern life.

Social needs require money, which is an aspiration for young people associated with urban life and its culture, like fashion in clothes, shoes, ornaments, body paint, haircuts, and hairstyles, ideas about the world, and even eating habits, Cell phones are an urban must for them, also. All of the aforementioned have given rise to the separation of families from customs where they worked as a group and not individually. These processes are evident in towns like Chan Kom and Yaxcaba in Yucatán, or S. Francisco Tepeyanco in Tlaxcala, in Oaxaca, Sonora, and elsewhere in the country (Allyson 1983; González-Jácome 2003: 197–230, 2012: 487–521; Mariaca-Méndez et al. 2012; Moctezuma-Pérez 2010: 26–43, 2013a, b; Roldán-Botello 1979: 82–97).

TABLE 13.1
Homegarden Agroecosystem Characteristics

Author	Definition
FAO (1964)	Homegardens reproduce the multilayered structure and species diversity of forests; this allows to simultaneously maintain perennial and annual crops in small areas. Crops have different biological cycles; the family usually collects some product daily, even in small quantities.
Kimber (1966: 97–118)	The homegarden is a biological/cultural complex, which speaks volumes about how people express them in the plant world. Each homegarden results from thousands of decisions a person makes about plants and the use of nearby space.The homegarden is an index of the relationship' between man and a culture. In the space occupied by the homegarden, plants are combined, including fruit trees, shade and firewood trees, various plants for family food, and to sell at the market. The homegarden has a great diversity of plants, management, and production directions. It is flexible because the farmer can modify the plants in it without modifying the structure of the system.
Allyson (1983)	Both the tropical homegarden and the temperate /dry climate homegarden are very diverse. Their vegetal structure and strata have a wide range, although a homegarden in tropical vegetation is a little more diverse.
Níñez (1984)	Homegarden is a subsystem within small-scale production systems that aims at the production of household consumer items (plants, animals, and utilities), which cannot be readily obtained, are not available, or are accessible through permanent agriculture, markets, hunting; gathering, fishing, and raising cattle.
González-Jácome (1985)	The Mexican homegarden has pre-Hispanic origins and was widely distributed in Mesoamerica. Homegarden has a very wide range in the types of adjustments to the environment, which spans from the coastal regions to the highlands, from the semi-arid and dry regions to the humid and tropical ones. This is reflected in the number of terms used by the ancient settlers to describe them. The dynamics of a homegarden are associated with the production and reproduction needs of human group. The homegarden is a mechanism to alleviate the demand for food, money, and materials associated with the culture of the human group. It is based on traditional ecological knowledge.
Fresco and Westphal (1988)	Homegardens are made up of soil, crops, weeds, pathogens, and insects that convert resources, like solar energy, water, nutrients, and air into inputs (food, condiments, fuel, fiber, and pharmaceutical products).
Gupta (1989)	The researcher's training and gender can skew the perception of a homegarden.
Caballero (1992)	Homegarden fulfills ecological functions such as nutrient recycling, carbon sequestration, erosion control, increased uptake of water by infiltration, conservation, and management of genetic diversity of plants and animals. It is the refuge of different threatened species and protection against natural attacks, such as hurricanes.
Torquebiau (1992)	Homegarden has several sustainability attributes for its ability to satisfy a series of needs of farmers without negatively affecting the resource base, in many cases improving it, and for its potential to satisfy economic, social, and ecological conditions that contribute to its sustainability over time.
Gispert et al. (1993)	Homegarden is a vegetable reservoir adjacent to the houseroom, which reflects the cultural identity of a human group in relation to nature, where social, biological, and agronomic activities are practiced as well as constitutes a rich supply for self-consumption at the door of the home. Throughout history, the dynamics of homegarden is associated with the production and reproduction needs of human groups, which represents a mechanism to alleviate the demand for food and materials associated with eating habits based on traditional knowledge among human groups and nature.
Valstar (1999)	Homegarden has a low cost and does not depend on inputs or high-tech tools, so making it a suitable strategy to achieve food self-sufficiency.
González-Jácome (2003)	Homegarden is an agroforest system of ancient origin made up of trees, in addition to other crops and domestic fauna often occupying small spaces, which are located near houses. Both its horizontal structure and its vertical structure (strata) are biodiverse.

(*Continued*)

TABLE 13.1 (*Continued*)
Homegarden Agroecosystem Characteristics

Author	Definition
Del Ángel and Mendoza (2004)	Homegarden is a design that mixes different elements, such as cultivated plants, wild plants and livestock, based on the use of traditional knowledge for its management.
Vanderwarker (2006)	The incidental homegarden resulted from the taste of the forest fruit consumers, who threw the remains of plants and animals they consumed on the ground forming garbage dumps. The seeds generated areas in the forest with an abundance of these selected fruits. These places were spaces for the presence of fauna of different sizes, depending on the specific place, and the environmental, climatological, or feeding conditions for local wildlife.
	The incidental homegardens was an important resource in the diet of the first settlers and can be found as early as the Prehistoric period in Los Tuxtlas, in central Veracruz. Little by little they integrated into this type of homegarden; domesticated fauna as well as cultivated and tolerated flora intermingle with the natural one. When maize was domesticated, it was cultivated in this type of homegarden, which later gave rise to the cornfield, an agroecosystem where maize was the central axis.
Wiersum (2006)	Homegarden is a traditional, sustainable agroforest system with positive ecological and socioeconomic features. Its functions and composition depend greatly on socioeconomic conditions as well as household livelihood strategies.
	Homegarden has different types and different pathways of development that can be identified. Its changes in features are associated with social and professional changes of villages and reflected a search for a new balance in the relationships between cities and villages.
	There are two points of view on the trends in homegarden functions and composition; they represented different "norms" in assessing social sustainability of the system and differences in value judgments on ideal structure of homegarden. Homegarden types: (1) survival gardens (staple food crops, and complementary crops); (2) subsistence homegarden (part of multicomponent forming system in conjunction with permanent or shifting field production); daily supply of vegetables, herbs, species, and fruits for household needs and occasional sale; (3) market gardens. Specialized farming system or part of multicomponent farming system. Cash crops with possible complementary production of household products; and (4) budget gardens. Gardens of households with economic from the market bases in rural or urban employment; family needs on monthly purchased from the market cultivation of "hobby" products for household consumption, and ornamentals.
Mariaca et al. (2007)	Homegarden is a complex agroecosystem with traditional roots in which the family unit lives, where selection, domestication, diversification, and conservation processes are oriented to the production of flora, fauna, and mushrooms, eventually abutting it.
Cuanalo and Guerra (2008)	Small area of cultivated land that surrounds a home or that is within walking distance.
Aguilar-Støen et al. (2008)	Homegardens have high levels of biodiversity, maintained, and enriched with agricultural practices, particularly with the exchange of seeds and plants. The greatest diversity is found in young homegardens, and in those where the owners actively exchange material with other people.
Alayón and Gurri (2007)	They have ecological functions like nutrient recycling, carbon capture, and erosion control, plus increased water uptake by infiltration, and conservation and management of genetic diversity of plants and animals. They are the refuge of different threatened species and provide protection against natural attacks, such as hurricanes.
González-Jácome (2009)	The tropical homegarden (Tabasco) underwent changes in the second half of the twentieth century with respect to the flora, fauna, and family workforce, which is related to weekly, seasonal and/or temporary migration to workplaces such as Cancun where young and adult men work in activities that are better paid than what the sale of homegarden commercial products could provide.
Cano y Siqueiros (2009)	Family homegardens have characteristics of the environmental conditions in which they are found, reflecting the customs and culture of the inhabitants.

(*Continued*)

TABLE 13.1 (*Continued*)
Homegarden Agroecosystem Characteristics

Author	Definition
González-Jácome (2012)	In tropical regions, the same as in temperate-cold places where the family workforce has been reduced by the departure of some of its members to work outside the communities. Homegardens have been transformed into spaces for housing construction, thus reducing the area of trees and other crops, as well as turned into leisure homegardens and even as spaces where garbage has accumulated. Only a few domestic animals remain, generally turkeys and pigs, to feed the old people who continue to live there. In other cases, the homegarden is a traditional agricultural system close to the houses where the people who work and maintain them live. It provides some of their food, money from the sale of flowers, medicinal herbs, and fruits with commercial value (avocados, mangoes) to the market.
Mariaca et al. (2012)	Homegardens have ecological functions: nutrient recycling, carbon capture, erosion control, increased water uptake by infiltration, and conservation and management of genetic diversity of plants and animals. They are the refuge of different threatened species and protection against natural attacks, such as hurricanes.
Cano (2012)	Today, they evolve due to the continuous adaptation of the space they occupy, changes in their components, and the socioeconomic and cultural environment of the families who manage them.
Kantún et al. (2013)	Homegardens are a refuge for wild plant species that have disappeared from their natural habitat.
Nieto et al. (2013)	It shows a specific articulation of human beings with natural resources: water, soil, solar energy, plant species, and the rest of animal species.
Moctezuma (2013)	Among the main elements that have decreased the number and size of the homegardens in Tepeyanco, Tlaxcala, are: the reduction of space for cultivation; decrease in domestic animals and vegetable species due to change in irrigation technology; loss of consciousness to divide and organize the space of the homegarden instead turning it into a garden. Homegarden are immersed in processes that make them more vulnerable to anthropocentric factors, and their functions depend on the interest of their owners, with adults, and the elderly showing more interest than young people in their care and conservation.
Moctezuma (2014)	The homegarden in Tepeyanco has almost disappeared to become a space with houses and pots to grow some flowers and medicinal herbs, and there are some surviving trees. It has been converted into leisure gardens.
Cahuich et al. (2014)	Homegardens contribute to satisfying food needs preparing dishes, drinks, and sweets used in agricultural rituals.
González-Jácome (2018)	Homegarden has been described as a sustainable agroecosystem, with a multi-strata structure, of species, which maintains areas with perennial and annual crops. This agroecosystem had been described by geographers since the 1970s, and later by other specialists who have become interested in its ability to maintain as sustainable agroecosystem over time. However, despite its importance, several of its fundamental aspects have not been analyzed, especially with regard to its social and cultural components, which are linked to its economy, and its ability to generate food, contribute to the health of families, and get money and other important resources from it. In Mexico, the homegardens and the cornfield seem to have arisen in prehistoric time from the incidental domestication of important plants for ancient settlers, such as maize, squash, and fruit trees, like the Mexican hog plum (*Spondias* spp.) and gourd (*Leucaena* sp.). The homegarden is a system whit the combination of water management, fruit trees, and flowers that bring freshness and pleasure to their owners. In ancient times, like today, many homegardens were converted into leisure gardens. In fact, there are different types of homegarden, and the agroecosystem can be productive highly in monetary terms.

Sources: Alayón and Gurri (2008); Aguilar-Støen et al. (2008); Allyson (1983); Caballero (1992); Cahuich et al. (2012); Cano et al. (2014); Cano y Cisneros (2009); Cuanalo y Guerra (2008); Del Ángel and Mendoza (2004); FAO (1964); Fresco and Westphal (1988); Gispert et al. (1993); González-Jácome (1985, 2003, 2008, 2009, 2012, 2018); Gupta (1989); Kantún et al. (2013); Kimber (1966: 97–118); Mariaca-Méndez (2012); Mariaca et al. (2007); Moctezuma-Pérez (2012, 2014); Nieto et al. (2013); Níñez (1984); Torquebiau (1992); Valstar (1999); Vanderwarker (2006); and Wiersum (2006).

RECOMMENDED READING

Wiersum, K.F. 2006. Diversity and Change in Homegarden Cultivation in Indonesia. In Kumar, B.M. and P.K. Nair. 2006. *Tropical Homegardens. A Time-Tested Example of Sustainable Agroforestry*. Springer, Chapter 5.

This article is part of Section 1, Chapter 2 of Kumar, and Nair's book on the tropical homegarden. Throughout the text, the homegarden elements that had been presented in various writings of specialists in the agroecosystem, mainly located in Asia, are discussed. Among the points that are established, the most interesting one considers the homegarden as a traditional, sustainable agroforestry system, where its functions and composition depend to a large extent on the socioeconomic conditions surrounding it and on the lifestyle of the homegarden families and their strategies.

The author considered that there are different types of homegardens that he identifies according to the degree of intensity of that agroecosystem in relation to the production and consumption of food plants, their connection with other agricultural systems, and their approach to the market. The article provided readers with several tables of comparative information, and Table 2 in the article gave different types of homegardens in relation to the household. Nair considered that there are two proposals related to the homegarden, its changes, and direction: one that postulates the disappearance of the agroecosystem; and the other that considers changes in the system represent different ways to make them occur in relation to the sustainability of the agroecosystem and the differences in the value of the judgments on the structural ideal of the homegarden.

Ford, Anabel, and Ronald Night. 2016. *The Maya Forest Garden*. Routledge Taylor & Francis, London, and New York.

The authors defined the Mayan cornfield as a diverse agroforestry system, and as a polyculture. Today's Mayan homegardens reflect the sustainable strategies of the Mayan civilization, whose climax occurred in the Classic period (500 AD. 900 AD) and lasted long enough to be remembered by the Spanish in 1524.

The Mayan cornfield, like any other farming system, disturbed the natural environment but worked with the forest in an integral way for its creation and sustainability. The authors concluded that there is continuity in the sophisticated practices of the current Mayan cornfield, with the forest homegardens, which were connected to the settlement patterns of the Mayan ancestors.

The current Mayan cornfield is the result of the activities in the virgin forest, the human activities in pre-Historic times, the Viceroyalty, and recent activities. The role of the shaping of the forest environment had been ignored in the ecological history of the Mayan region and in the narrative of the people there.

The integral cornfield system was developed in the forest, and the forest was its homegarden. The authors argued that the conservation of the Mayan forest depended on traditional farmers, whose skills and knowledge maintained the forest, and continued to conserve it and its culture. The traditional Mayan system had the benefit of social sophistication and collective organization.

Today's traditional systems separate the cornfield from the homegarden; however, these have to be viewed as a single system. Together they show the complexity of their productive capacity and a remarkable diversity of family resources, which make their current management, soil improvement, and biodiversity possible.

14 The Current Mexican Homegarden

14.1 INTRODUCTION

The homegarden had undergone a series of changes over time, but its trees and the human management as well as adaptability to environmental, climatic, and economic conditions had given the agroecosystem its resilience and sustainability throughout time. That is why the homegarden agroecosystem has been, and continues to be, an important source of information for all those dedicated to the study of agriculture (Caballero 1992: 135–155; Del Ángel-Pérez and Mendoza 2004: 329–246). The homegarden differs according to time, local, and regional factors of the places where it is located (González-Jácome 2016c: 233–257; Kumar and Nair 2006; Mariaca-Méndez (cord.) 2012).

Currently, the small-scale and polyculture homegarden include several strata: herbaceous, shrubby, and arboreal, including the canopy. Their functions are multiple and varied that characterize and explain their diversity (Mariaca-Méndez 2012; Ordoñez 2018). The homegarden remains a traditional agroecosystem of great importance for the economy, food system, and daily life of families in various parts of the country, as occurs in Tabasco, Campeche, Oaxaca, Veracruz, Chiapas, and other regions of Mexico (Juan-Pérez 2005; Lerner-Martínez 2008; Mariaca-Méndez et al. 2012; Ordoñez-Díaz et al. 2018; Wilkerson 1983).

In 2006 and 2009, FAO organized international programs for the creation of homegardens intended to promote healthful eating and food to supply family pantries (Cano-Contreras 2015: 68–89; Lerner-Martínez 2008; Lerner-Martínez at al. 2009). From other instances, some research on homegardens was done in Yucatán (Flores-Guido 2012: 149–176), Campeche (Cahuich 2012), Veracruz (Blanco-Rosas 2006; Leyva 2017), the High Mixteca (Rosas-Benítez 2019); Tabasco, and Chiapas (Mariaca-Méndez et al. 2007, 2012). However, the results of these programs have yet to be evaluated to see the results and correct their shortcomings (Brownrigg 1985).

Homegarden resilience has deteriorated in several places, in part by the new worldview those small-scale farmers, and their offspring have of what is a culture of good living and life in general. The concept of a good life changes from one time to another, region, and local places. Also, the concept changes from rural to urban life, from one social group to another, and from one generation to another.

14.2 CHANGES IN THE HOMEGARDEN

In the Mexican homegarden, the size, structure, and composition of flora and fauna have been modified. The fruit trees grow and subsist next to shade trees, shrubs, herbaceous, and other plants destined for self-consumption, or to satisfy other needs such as wood and firewood, or as tree curtains that reduce the impact of strong sun rays and wind. The conversion of homegardens into gardens destined for family enjoyment on weekends, or into zones for the construction of rural neighborhoods for family members who marry is a process that has extended to various parts of the country. In both cases, often the owner of the plot replaces trees and flora in general, so reducing strata and diversity.

The role of domestic animals in the homegarden has also changed. In the current agroecosystem, there are pigs, chickens, and turkeys. Depending on the region there may be ducks, pijijis, geese, and other small birds, as well as mammals (Mariaca-Méndez 2017: 475–508). In the Isthmus of Tehuantepec, families can have iguanas and reptiles that are important both for family consumption

DOI: 10.1201/9781003198833-14

and for commercial food. They can include domesticated, semi-domesticated, and wild fauna species as well. The homegarden in the Los Tuxtlas region has been associated with the forest and the wild animals living there. It is a good place for hunting game for the family food system or for sale in local markets (Blanco-Rosas 2006).

In Chan Kom, domestic fauna is reduced to turkeys and an occasional pig. In villages and towns of Tlaxcala, poultry, particularly turkeys, continue to be part of the relics of ancient homegardens. These are consumed in the village festivals, in family gatherings related to the completion of primary school studies, the birthday of the head of the family, baptisms, confirmations, and parties associated with religious beliefs and local customs (González-Jácome 2012: 522–535; Moctezuma-Pérez 2010: 26–46).

In towns of Yucatan, such as Yaxcaba, the surface of the plot has been reduced due to the construction of houses, which are intended to create spaces available for children who reside in them only in certain seasons, since most of the year they live in places where they work and only return during holiday seasons, or when work is scarce in the hotels, bars, and restaurants (González-Jácome 2012: 522–535) (Figures 14.1 and 14.2).

In the past, the homegarden generated important ecological processes and economic success for families that as medium and large-scale merchants took their products to supply centers, convenience stores, tianguis, regional, or national markets to sell (Allyson 1983; González-Jácome 2017: 30–51, 2018; Moctezuma-Pérez 2013a, 2013b; Roldán-Botello 1979: 82–97).

With the money obtained from medium and large-scale trade in places like Tepeyanco, homegardens were converted into recreational gardens for families, their kin, friends, and guests. The owners prepared barbecue and drink beer, and have fun on weekends. These recreational gardens are systems with ornamental or medicinal herbs, trees, and some flowers. They are places to go to play and have fun.

14.3 CURRENT SITUATION OF SOME HOMEGARDENS IN MEXICO

The homegarden is in the process of extinction in various parts of Mexico, not for lack of economic success, its ecological functions, or because it is not profitable. Educational issues, social and cultural decisions and phenomena such as migration and emigration of a part of the family members influence their abandonment. Owners, and mainly young people, acquire urban ways of life, which are different from rural ways and which are stimulated by modernization and urban expansion over rural spaces, causing changes in land tenure and uses, stimulating children and young people to acquire a formal education, and employment in non-agricultural activities.

FIGURE 14.1 Tepeyanco, Tlaxcala: Cornfield and Homegarden, 2014.

FIGURE 14.2 Sonora: Homegarden in the Road to Mochibampo, 2013.

From the second half of the 1980s, Spanish and Mexican construction companies acquired land of the former endowed lands (*ejidos*) in Mexico to establish tourist complexes there, or to modernize those that already existed on such lands. In Central Veracruz, in towns like La Antigua, the properties tourist agencies want to buy are the old homegarden plots that are located in spaces next to rustic places with open constructions locally called *palapas* to be used for complexes related to the development of local, regional, or state tourism (González-Jácome 1988b: 1–60).

The homegardens of the ethnic Guarijío group in the upper basin of the Mayo River in Sonora are surfaces adjacent to the houses, with some medicinal plants, ornamental flowers, and occasional trees (Robles Linares 2018: 269–295). Likewise with the homegardens in the upper plains of the Totolac river basin, in Tlaxcala (S. Tadeo Huiloapan, S. Mateo Huexoyucan, and S. Francisco Ocotelulco)· The current homegarden has very small spaces, with one or two trees and some herbs for infusions. The morning drink is Melissa or chamomile that family members have before going to work. With leaves, flowers, or twigs boiled in water, medicinal herbal infusions are remedies for stomachaches, headaches, diarrhea, and minor conditions that, apparently for families, do not merit medical attention and medical care (Bilbao 1979; González-Jácome 2003).

Somehow, modernization and urbanization processes of the country have contributed to the extinction of the homegarden as a sustainable agroecosystem, which in reality is highly productive in several dimensions, including food sustainability (García and Bermúdez 2014). The resilience of the Mexican homegarden has been diminished by the new vision that small-scale farmers have of what is "modern" or "backward" agriculture, wellbeing, everyday life, and modern culture. It must added that these small-scale farmers have consumerist aspirations that are reinforced daily by the mass media, by their proximity to cities, and by urban ways of thinking, socializing, communicating, living, dressing, and eating. They all go hand in hand with the need to obtain greater profits and money to be able to follow such trends (Alvarado-Méndez and Delgadillo-Macías 2017: 21–31). Since the 1980s, the importance of the family homegarden in the food system and the nutrition of human populations have augmented.

The homegarden agroecosystem presence has been supported and partially subsidized by international organizations such as FAO, Mexican State programs, and those that have been carried out with financing and support of the federative entities. However, the projects have not been successful always. In several cases, causes of this are unknown partially (Del Amo-Rodríguez 1994, 2000a, 2001, 2008, 2009, 2011; Fernández-González 2017).

Fernández-González (2017) found in her research of the programs for family homegardens in the state of Tlaxcala:

1. The people who participated in the programs did not select the crops adopted by the programs, nor did they suit their tastes and needs.
2. The participants registered in the programs received monetary resources or some materials (e.g., water tanks, supplies of some kind).
3. Although the productivity was good, participants of the programs did not want to – or could not – consume crops that were harvested, or, had no place to sell the surpluses of the homegarden production. The producers argued that in the local markets they were not accepted.
4. The participants in the programs did not know how to cook or to eat the vegetables, which they disliked, so they did not incorporate them in their diet, which has been the main purpose of the programs. For example, local producers did not like broccoli, and they did not know how to cook it. The program participants do not eat salads either.

14.4 HOMEGARDEN DIVERSITY

The current homegardens are of ancient origin and show various types of management of crops planted in them, bringing to light some aspects of their diversity. In Ocotal Texizapan, a Nahua town south of Veracruz studied by Leyva (2017: 230–231), she found that the family homegardens are located in spaces adjacent to the houses. Most of the villagers cultivate both the cornfield and the homegarden agroecosystems nowadays, obtaining 80% of their food system from them. Also, they sell some of its output to earn income (15%) or exchange some of it (5%) for other things they need.

For the people of Ocotal Texizapan, the homegarden is: "[...] The place where plants are sown; structure and form reflect the family's needs, manifesting their traditional agricultural knowledge, worldview, ethnicity, climatic conditions, culinary tastes, land availability, local materials, and gender; besides the species' adaptation to the local context".

Its diversity explains the homegardens permanence as an agroecosystem and in large part its ability to be sustainable. The diversity of flora found in 20 homegardens studied by Leyva (2017: 231) consisted of 192 species, distributed in 74 families, whose abundance was of: Fabaceae (8%); Anacardiaceae (5%); Musaceae; Araceae; and Rubiaceae (4%). They are represented by 149 genera, of which the most cultivated are Musa, Citrus, and Quercus. This wealth of flora was similar to that of other neighboring communities in Veracruz.

M.A. Cruz (2003) reported a total of 193 species in homegardens of Soteapan in Los Tuxtlas, Veracruz. De la Cruz-Osorio (2009) identified 143 species in family homegardens of Francisco Villa; a municipality of Tihuatlan; and López-Acosta (2009) registered 123 species in Mozomboa, in the present municipality of Actopan. Homegarden richness does not reflect a high consumption, necessarily, but a diversity of flora, with parts of the crops destined for self-consumption, exchange, or sale.

Leyva (2017: 231) recorded an average of 368 plants per homegarden, with a maximum of 807 and a minimum of 134. Of 194 species, 88 plants were registered for food at different times of the year (45.36%). Live fences surrounding homegardens are cultivated with croto (*Codiaeum Rumph.* Ex A. Juss.), arnica (*Heterotheca inuloides* Cass.), tulip (*Hibiscus rosa-sinensis* L.), foxtail (*Sansevieria trifasciata* Prain), and izote (*Yucca filifera* Chabaud). These species are sown to avoid the prying eyes of neighbors and prevent domestic animals (chickens, ducks, or dogs) from destroying crops.

Homegardens in Ocotal Texizapan are conformed of an herbaceous stratum (45%), shrubs (29%), and trees (26%) of the total of the 194 identified species. The species grown in this agroecosystem are four-corner banana (*Musa* spp.), Roatan banana (*Musa paradisiaca* L.), chipile (*Crotalaria maypurensis* Kunth), orange (*Citrus sinensis* L.), malanga (*Colocasia esculenta* Schott), sweet potato (*Ipomoea sweet potatoes* L.), Annatto (*Bixa orellana* L.), chocho (*Astrocaryum mexicanum* Liebm. ex Mart.), mango (*Mangifera indica* L.), and cedar (*Cedrela odorata* L.) (Leyva 2017: 232).

The inhabitants of Ocotal Texizapan do not consume some of the edible plant species in the homegarden because people do not know how to cook them. Among these are crossheads, *Opuntia* cladodes, and flowers of the yucca plant (*Yucca filifera* Chabaud) (Leyva 2017: 232). In addition, there are 173 plants with other uses: 70 ornamental; 8 seasonings; 23 medicinal; 18 for shade; 13 are used like live fence; 5 are used for firewood; 7 are timber; 2 have artisanal purposes; 25 are used in construction; and 2 species are for fodder.

Leyva's (2017: 232) research concludes that 60% of the species present in the Ocotal Texizapan homegardens are native to the Americas. Of these, 35% are native to Mexico: yucca plant (*Yucca filifera* Chabaud); Mexican forest palm (*Astrocaryum mexicanum* Liebm. Ex Mart.); chili (*Capsicum frutescens* L.); macaw palm (*Acrocomia aculeata* (Jacq.) Lodd. Ex Mart.); Arnica (*Heterotheca inuloides* Cass); *Inga jinicuil* Schltdl. & Cham. Ex G. Don (*jinicuil*); wild tomato (*Lycopersicum esculentum* P. Mill.); vegetable pear (*Sechium edule* (Jacq.) Sw.); and others. Herbs and ornamental plants are cultivated in containers (buckets, cans, bottles, and pots) and nurseries placed in the homegarden or next to the house. Crops are sown according to what they require (soil, humidity, light), and fertilizer is obtained from the septic tank and is used for crop fertilization (Leyva 2017: 233). The same occurs in other places in Mexico, as can be seen in Figures 14.3 and 14.4.

FIGURE 14.3 Sonora, Las Flores: Plants Cultivated in Containers, 2011.

FIGURE 14.4 Yucatan, Yaxcaba: Homegarden and Nursery, 2008.

Domestic fauna is an important part of the homegarden. It generates monetary resources for small-scale farmers. A chicken was sold for $150.00–$200.00 MXN.[1] Poultry (chickens, turkeys, ducks) sleep in coops, which are simple pens made of *Bursera simaruba* or *Mangifera indica* branches. Poultry also sleep in trees. They are fed with maize as well as tortillas soaked in water or maize hominy water (*nejayote*). Families have an average of 25 chickens, but some own up to 50. Hen eggs are selling for $ 2.50 MXN a piece. Their highest consumption is from January to February, before the dry season. Lying hens "[...] ensure the regular provision of eggs, and meat for the family food system. Eggs require a lower investment for their production, and they provide vitamins, and minerals such as iron, calcium, and phosphorus" (Del Ángel-Pérez and Mendoza 2004; Lerner et al. 2009; Leyva 2017: 236).

Homegardens that present the greatest flora diversity belong to families that do not have cornfields. Those without a cornfield supplement their diet with game nutrients. The homegarden in Ocotal Texizapan has native species of high food potential, and its crops are essential to the families' food system. Regardless of their size, homegardens are natural reserves of diversity. The division of labor for agricultural activities is carried out taking into consideration sex.[2] Males do the tasks that require the most physical strength, such as slashing, digging holes, and pruning. Women are responsible for activities that require less strength, such as planting, watering, and weeding. Regularly, work in the field is done in the afternoons, when the sun and heat go down.

Folk or local knowledge about the homegarden includes the various ways of thinking, acting, and working in regard to cultivated plants. Among the folk idea is to punish the birds that eat citrus fruits (they are hunted and eaten). Every June 24, a person named Juan or Juana threatens to hit the *Psidium guajava*, coffee, mango, and *Inga* (*jinicuil*) trees, which do not produce fruit. The trees are told that if they continue to not bear fruit they will be cut down. To the trees that need protection, and to prevent them from drying up, a red cloth is tied to the branches (mainly this activity is carried out on Yam bean, guajava tree, and mango tree).

In other places, such as Tlaxcala until the 1970s, tying a red ribbon on a plant was to prevent "the evil eye" from affecting its growth or reducing its fruit production. This is a widespread practice in Mexico. Salt is put at the base of coconut trees so that the fruit will be juicy and sweet. This practice is found in several coastal regions of the state of Veracruz. The salt is placed in holes dug in the ground before placing the freshly germinated coconuts in them, and thus give rise to future palms. This practice was found in the 1970s in areas planted with coconut palms on the Pacific coast of Colima.

Banana plants and orange trees are planted according to the moon phases. The relationship of the moon phases with planting, pruning, care, and harvesting activities is found in several parts of rural Mexico today, as happens today in S. Felipe Ixtacuixtla, Tlaxcala (Pérez-Sánchez 2012, 2014), central and southern Veracruz (Leyva (2017: 233–234); and Chiapas (Mariaca-Méndez 2003).

14.5 HOMEGARDENS AND ORCHARDS FOR COMMERCIAL PURPOSES

There are homegardens and orchards of all scales, both according to their surface (from 700 m^2 onwards) and to the number of fruit trees in them, or if they are changing from polyculture to monoculture production. Monoculture orchards have been very productive, with crops destined for export, such as avocados. There are small and medium-scale homegardens where apples and pears are grown. Fruits that are exported from the United States and Chile are produced also in the Sierra de Puebla homegardens and orchards. Figures from SENASICA (2018: 163) showed that in 2016,

[1] Currency is variable; one dollar was equal to 18.78 Mexican pesos (10/01/2020). In May 2020 currency increases to almost 25 pesos per dollar, and in June 2020 currency was 23.00 Mexican pesos per dollar. In July 2021 currency was 18.60 Mexican pesos per dollar. In November 2021 currency was 20.5499 pesos per dollar.

[2] The term sex is used as in demography, to distinguish between men and women according to their biological birth, and the initial registration granted to the newborn. The concept of gender is cultural. It varies because it is a personal choice that can be made when an adult so requests and carries it out accordingly.

Mexico exported 23 thousand 145 million dollars worth of agri-food products to the United States. Of them, 278 million dollars were from organic products (1.2%). In January 2021 USDA considered that Mexican agricultural exports to USA will reach US 29.500 million dollars (USDA 06/01/2021).

Some agricultural products were brought to Mexico recently. For example, kiwis (*Actinidia deliciosa*), blueberries (*Vaccinium corymbosum*), and lychees (*Litchi chinensis*) are successful due to their commercial expansion in the urban public markets and supermarkets, reaching high prices in convenience stores and municipal markets.

An example of the prices that these fruits fetch happened in January 2019, when the apricots brought from Chile had a price of 190 Mexican pesos per kilo, and imported kiwis were sold at 120 Mexican pesos per kilo (32%–40% higher than the Mexican fruits). The price differences for imported fruits had more to do with economic and political factors, more than with the quality and taste of the fruit. This phenomenon has to be studied in depth and from several points of view, starting with its impacts on local flora, the structure of homegardens, and the economy of farm families.

Some orchards allocate their production for the elaboration of other products. For example, a well-known apple orchard in the Sierra de Puebla produces hard cider and distilled beverages made with different fruits from the region. The apple fair is several decades old and is an annual activity that contributes to the maintenance of local apple production and its industrialized derivatives. However, this agribusiness requires strong monetary investments. The specialized fairs in the diffusion of local productions boost the sale and consumption of the products derived from Mexican agroecosystems like the apple Fair. They also promote exchanges of seeds and plant varieties, among producers from different regions (Figures 14.5–14.7).

Some of agricultural products are exposed in fairs, and they have economic importance and diffusion, like the apple, the wine, maize, Opuntia, and amaranth fairs. Also, there are fairs for elaborated products, like *tamales*, *tortas*, or mole fairs. A better dissemination of them would support agricultural producers. This is being done in the case of maize. Agroecosystems, such as the homegarden and the orchard, are highly productive and can generate economic benefits for their owners easily, but the small-scale farmers should have in addition to market sales, agronomic, economic, and diffusion advice. By adding additional elements to their productions, the farmer's monetary earnings in the markets rose.

The studies of the homegarden in S. Francisco Tepeyanco show the capacity of the agroecosystem to generate monetary gains. This agroecosystem requires technical advice in several fields of knowledge so that it can continue to be an ecologically, economically, and socially sustainable, resilient, and welfare generating system. In Tepeyanco, the highly productive homegarden generated

FIGURE 14.5 Matlalcueye Lowlands, Peach Orchard.

FIGURE 14.6 Matlalcueye Lowlands: Cornfield, Peach Orchard, and Forest.

enrichment for the small-scale agriculturalists, which marketed their products themselves. At the same time, young generations abandoned agriculture and became urbanized; some continued studying and became professionals in different fields and others went to work in the tertiary sector. Homegardens became leisure gardens or their terrain used to building homes for married sons (González-Jácome 2003, 2012: 522–534, 2016: 344–356).

The homegarden had endured from early times to the present day due to its ecological characteristics as an agroecosystem, its diversity, its ability to feed human populations in times of crisis, and its possibilities to generate monetary resources. In short, the homegarden agroecosystem is diverse, sustainable, and resilient. The agroecosystem has been successful in ecological and economic terms, generating good living and wellness, but in several cases the agroecosystem disappeared due to social and cultural factors, mainly.

Currently, although it is subject to great pressures related to issues such as if it has the ability to generate wealth, the agroecosystem has survived the modernization processes, the creation of industrial areas, and the official support for powerful companies and government public policies, which subtract land and water from agriculture, use large-scale electricity, control markets, and support the development of transnational agribusiness companies. In addition to granting lower types of employment and salaries to Mexican nationals in contrast to the higher salaries paid to foreign personnel.

The flora found in the ancient and contemporary homegardens is characterized by its diversity, which included the original vegetation of the area, and its covering, even partial, plus the nutritional, medicinal, recreational, and ornamental needs of families. The need for firewood, shade, and diversity will depend on where the agroecosystem is located, and its physiographic, ecological,

FIGURE 14.7 Matlalcueye Lowlands: Mexican Black Cherry and Mexican Hawthorn Orchard.

and climatic conditions, as well as the availability of services such as gas for domestic purposes, grocery stores, tianguis, or markets. The need for firewood, shade, and diversity of products will depend on where the agroecosystem is located, and its physiographic, ecological, and climatic conditions, as well as the availability of services such as gas for domestic purposes, grocery stores, tianguis, or markets.

The homegarden has been combined with other agroecosystems and nearby ecosystems, and incorporated to family economic activities like commerce, work in touristic employments, and different works of the tertiary sector which give families monetary income (González-Jácome 2010: 141–158, 2012: 522–534).

In the twentieth century, in her study on Tepeyanco, Roldán-Botello (1979) shows how the small-scale farmers had the opportunity to market their products directly, and accumulated capital to acquire cargo trucks, and then a line of these, that took Tepeyanco products to the markets of Mexico City, Puebla, Tabasco, Campeche, lowlands of Chiapas, and various places on the Gulf of Mexico coast (García-Martínez 1987; González-Jácome 1976, 2002b, 2003b, 2007a). This line of trucks became so important that in the first decade of the XXI century, they took green chili from Calakmul, in Campeche, to the markets of Puebla, and Mexico City (González-Jácome, Campeche 2009; Molina-Rosales 2010).

The current variety in homegarden (Ordóñez-Díaz et al. 2018) allowed establishing a continuum that goes from polyculture to monoculture (homegarden to orchard), and to nonagricultural garden, as has happened in Tepeyanco. From 1970 to 1979, homegardens in the town occupied 52 hectares (4% of the total cultivated land). They were irrigated and generated ample economic gains to the owners who used nuclear family, kin, and close friends as free labor (mutual aid).

Agriculturalists applied seemingly simple agricultural techniques: using green manures (fallen tree leaves; cuttings of the herbs on homegarden land edges that had been removed from fields, and planted on the banks of the canals); planting seedlings and transplanting trees; applying muddy water (*agualodo*) obtained from the river to the seedlings (in pots of different sizes); grafting on the Mexican hawthorn; cutting the infected leaves of the avocado trees manually and burning them; clearing the homegarden of litter, then accumulating it to form the green manures; and cleaning the irrigation canals in December (González-Jácome 1976, 2003).

In 1979, homegarden trees in Tepeyanco were avocado, custard apple, big fig, black cherry, apricot, custard apple, plum; peach, fig, lemon, tangerine, apple tree, nance, and walnut (Roldán-Botello 1979). In addition, the homegarden had Devil's hand tree (*Chiranthodendron pentadactylon*), coral bean, blackberry, Tabasco banana trees, and blackberry bushes. By 1983, homegardens trees were avocado, apricot, custard apple, peach, plum, guava, fig tree, lime, lemon, apple, nance, orange, walnut, pecan, pear, Mexican hawthorn, white sapota, coral bean, palm, and blackberry (Allyson 1983).

By 2013, Tepeyanco homegarden agroecosystem was in a process of extinction. The eight homegardens that continued working had trees: avocado, big fig, coffee, black cherry, apricot, custard apple, plum, peach, guava, pomegranate, fig, Inga food (*Inga edulis*) (*jinicuil*), lime, lemon, mammee, tangerine, apple, passion fruit, nance (*Byrsonima crassifolia*), orange, walnut, pear, Mexican hawthorn, White sapota, Coral bean (*Erythrina* sp.), palm, butterfly bush (*Buddleia cordata*), and Tabasco banana (Moctezuma-Pérez 2013, 2018: 95–116).

González-Jácome (1996, 2003) described how the homegarden merchants obtained economic success by the superregional transportation of their products, those of their peers, and increased commercial routes to the south and southeast of the country. With the capitals earned, they sent their sons and daughters to study in Puebla City and Mexico City. The offspring studied and became professionals, but did not return to live in their hometown. When their parents died, they sold the homegarden or turned the productive agroecosystem into recreational gardens for weekends.

Moctezuma-Pérez (2013, 2018: 95–116) studied the process of conversion of these homegardens into ornamental gardens with potted medicinal and ornamental plants. In Tepeyanco, homegardens were turned into flower gardens with a greater diversity than that of the agroecosystem with trees and plants of all kinds, although they lost their productive role in the family economy and the town economic importance (Allyson 1983; González-Jácome 2012: 487–520, 2018: 43–85; Roldán-Botello 1979; Moctezuma-Pérez 2010, 2013, 2018: 95–116).

A similar situation was found in several homegardens in Mexico. Juan-Pérez (2005) described the process in towns in the State of Mexico, and Robles-Linares (2013, 2018: 269–296) described the Sonoran Guarijío homegardens. Rosas-Benitez (2019) studied a similar process in homegardens in Chazumba, Oaxaca, where the agroecosystem had Dragon fruit (*Hylocereus* sp.) as its main crop. Some homegardens had diversified crops, others mixed the cornfield with Dragon fruit trees, and the remaining ones used pots to cultivate certain herbs.

From 2006 to 2009, FAO programs, and the Tabasco, Puebla, and Tlaxcala state governments promoted the formation of family homegardens. The programs that were relatively successful selected vegetables to grow that were promoted and included among the everyday foods cooked and consumed by the inhabitants who participated in those programs. On the one hand, when the managers of the programs selected families that were not engaged in agriculture and did not consume the vegetables planted, the programs failed, because they did not like them and did not know how to cook or prepare them, as happened with broccoli, lettuce , and the Brussels sprouts (Fernández-González 2017). On the other hand, they found it difficult to take the small production of agricultural products to the market due to the cost of transportation, and the little profit they could obtain, besides not knowing the system to sell in the *tianguis*.

It is difficult to consider the cost of moving crops to regional markets because participants in the program did not produce vegetables before the program chose them. In large quantities, vegetables leave a profit after paying the freight for their transfer from town to market. Nor were they organized to bring together the productions of various families and collectively take them to sell at the tianguis (Castillo-Ramos et al. 2001: 229–245; Fernández-González 2017).

The evaluation of these programs, their real results with regard to family consumption, or the market price of what is grown would have to start with the selection of what will be grown in these family vegetable gardens. The selection of crops would have to take into account that these families are not farmers and do not consume vegetables nor legumes daily. Families agree to participate in these programs because they received free of charge supplies, water tanks, tools, and financial support. They do not have to repay. Importantly, they are not interested in cultivating what is offered to them, necessarily (Fernández-González 2017; Leyva 2017).

14.6 CONSIDERATIONS AND CONCLUSIONS

The current homegarden is in the process of extinction in several parts of Mexico, not because it is not an economic success, or because it is deficient in ecological functions, but because of the family, educational, economic, social, and cultural decisions made by its owners about obtaining better lifestyles than their rural ones. Somehow, the modernization and urbanization of the country have contributed to the extinction of the homegarden as a sustainable agroecosystem, although it is highly productive in several aspects.

Other factors are contributing to the extinction of the homegarden like the reduction of the surface destined to the agroecosystem; the lack of family labor to maintain it; problems with the system to market products, because agro-food business paid low prices to the farmers for their products; and the conversion of farmers' young sons and daughters to professionals or nonagricultural workers. The list is longer, especially if the results of new sociological and economic research reports are added. Agricultural work is hard, and profits are few, although in the case of homegardens and orchards the agroecosystem economy can leave important profits, especially this happens when orchards products are for exportation, as is in the case of avocado.

The homegarden has proven to be a sustainable agroecosystem, with the ability to reduce or expand its space; an ample scope to meet family needs; the quantity and destination of its production; the workforce that is applied; the number, places, and distribution of trees and flora associated with the system; the processes of abandonment and revitalization; and the power to resemble – or not – the structure of the forest, which had subsisted in this homegarden agroecosystem over centuries.

Together these elements form part of its resilience, although the homegarden has not been able to endure phenomena like the modernization of rural life, the migration that separates the youngest members from families, prompting them to abandon agriculture as a basic activity, or temptating new needs generated by urban and modern life, which require money to be fulfilled.

RECOMMENDED READING

Montagnini, F. 2006. Homegardens in Mesoamerica. Biodiversity, food security, and nutrient management. In: Kumar, B.M. and P.K. Nair. 2006. *Tropical Homegardens. A Time-Tested Example of Sustainable Agroforestry*. Springer, Chapter 5, pp. 61–84.

The chapter deals with Mayan culture, population, poverty, malnutrition, rural development, and homegarden. The agroecosystem provides the household with basic food and money-generating products of high value, whose returns are used in development projects aimed at achieving food security often, especially in the poorest zones of "Mesoamerica".

For the author, Mesoamerican homegardens are quite diverse, both in vertical and horizontal structure, and in the composition of species. They contain native and exotic plants emphasizing fruit trees. Domestic fauna is mainly chickens and pigs that add protein to the diet, which is generally deficient. Many communities are descendants of the ancient Mayans, who use techniques in homegardens that include handling ash residues to enhance nutrients as well as recycling and conserving them. They capture solar radiation. These multilayered agroecosystems are of minimal regional and global importance because they occupy relatively small areas. Managing strategies that promote nutrient recycling and maintain high species diversity should be encouraged to ensure sustainability of the agro-ecosystem in the region.

Atran, S; D. Medin; N. Ross; E. Lynch; J. Coley; E. Ucan Ek'; and V. Vapnarsky. 1999. Folkecology and commons management in the Maya Lowlands. *Proceedings of the National Academy of Sciences of the United States of America* 96: 7598–7603.

The article deals with three human groups living in the same Neotropical Peten rain forest ecosystem. The groups were: (1) native Itzaj Maya (Spanish speakers) who lived in the town of S. Jose; (2) immigrant Ladinos (Mestizos) living in the settlement of La Nueva S. Jose; and (3) the Q'eqchi-Maya immigrants who live in the hamlet of Corazal. The inhabitants have different behaviors, local knowledge, and social relationships with the forest.

Only the Mayans who are native to the area had a systematic awareness of ecological complexity. This involved fauna, flora, people, and management that favor forest regeneration and its resources. Spanish-speaking immigrants were closer to native Mayans in thought, action, and social media than Mayan immigrants. There is no primary "local Indian", or "immigrant" relationship with the environment.

The study showed a concern for self-interest. Rational and institutional constraints did not explain common behavior, and the cultural patterns of knowledge and access to relevant information were significant predictors. Unlike traditional accounts of the relationships between culture, local knowledge, and behavior, models offered are not synthetic interpretations of people's thoughts and behaviors, as they were emerging cultural patterns statistically derived from measurements of individual knowledge, and behavior.

15 The Cornfield (*milpa*) in Mexico and the Food System

15.1 INTRODUCTION

The cornfield has a great variety of types, a diversity that is explained by the ability of this agroecosystem small-scale farmers to adjust (ecological, social, and cultural adaptation) to different historical, physiogeographic, altitudinal, environmental, climatological, and social organization; agricultural family functions based on nuclear family, relatives, close friends, and mutual aid; family settings to the inheritance system of the land; cultural situations changing over time; and family adjusted to the particular and regional economic conditions that at any given time may affect cornfield development.

The variety and types of cornfield (*milpa*) in Mexico are as wide as the places where this agroecosystem is practiced, and ecologically, economic, socially, and culturally adapted, plus the physiogeographic, environmental, altitude, location, external processes, and events in which the region and the specific town are settled, like state or federal policies with respect to the areas surrounding the localities are pivotal. Ravines, hills, highlands, troughs, rivers, hilltops, canyons, plains, and places or zones with the available resources help to produce some adaptations. The interrelations of the small-scale farmers with the regional and national tianguis and markets affect decisions as well as social and cultural adaptations.

There are many elements of the cornfield agroecosystem that open new lines of research. The list is large, and expands as their components, and their managements in the agroecosystem are better known. An example is the cornfield research from specialized ecological perspectives such as that opened of chemical ecology perspective, that poses questions like the interactions of plants in the agroecosystem, about their consumers, on the role, and multiple functions played by secondary metabolites in the agroecosystem (Anaya Lang and Espinosa García 2006:13; Gliessman 2015: 161–172, 191–201).

15.2 WHAT IS A CORNFIELD?

The cornfield is a polyculture agroecosystem of millenarian origin where maize is the central axis of the system but is mixed with other crops (multifunctional management). The agroecosystem is integrated, and its crops intermixed with tolerated weeds, wild plants, mushrooms, epiphytes, domestic and wild fauna, microorganisms, trees, and bushes, which are located on the borders of the plot (Mariaca-Méndez 1988, 1997). Diversity in crops prevents the loss of the entire harvest in case natural disasters or climatic contingencies occur. The edges of the plots made up of hillsides where there are sown or are wild, some trees, shrubs, herbs and some grass that serve to protect the fields from the erosion of the wind, and the water that falls in the rainy season, or that runs through rivers, ditches, or nearby channels. In the central Mexican highlands, Agave and Opuntia are added to these earth edges. In this case, rows planted with Agave (*metepantles*) contribute to the control of erosion on slopes (González Jácome 2007a: 55–80) (Figures 15.1 and 15.2).

The cornfield, the homegarden, terrace, drained fields, and many other traditional agroecosystems provide organic matter for the soil that contributes to the greater diversity of pollinators, predators of certain pests, and microorganisms in the soil. They serve as plant defenses against predators in these agroecosystems. The properties of plants can be constitutive or induced.

DOI: 10.1201/9781003198833-15

FIGURE 15.1 Tlaxcalan Matlalcueye Lowlands: Cornfield, Crops in Squared Spaces, Homegarden, and Natural Vegetation.

FIGURE 15.2 Chiapas Highlands: Cornfield, Land in Fallow, and Forest.

Constitutive includes the development of new structures and chemical compounds through, among others, secondary metabolites, whose ecological role is well known and studied (Anaya-Lang and Espinosa-García 2006: 4–13).

Diversification has agro-food, plus ecological, climatological, economic, social, and cultural advantages, in addition to the ecological-chemical advantages in these traditional agroecosystems. Crop diversity adds a certain degree of resistance to agroecosystems against subsequent attacks by herbivores, pathogens, or parasites, and it improves the control of unwanted pests, and weeds. Diversified-agroecosystem production is intended for family consumption, although some of its products are directed to some degree toward the market (Kato et al. 2009).

The work applied to the cornfield agroecosystem is familiar, with the mutual aid of the family members, kin, and friends in planting, and harvesting times. The occupied surface is of different sizes ranging from a few meters located next to the house, to larger areas in plots located in different places, sometimes away from granted land (*ejido*), or on properties that farmers have in nearby places or in other towns.

The cornfield agroecosystem usually works interrelated with other agroecosystems and with natural ecosystems and the homegarden. The cornfield could be a conventional agroecosystem whose production is mainly destined for the market. Other types of cornfield are the mixture between the cornfield and the nearby natural ecosystems like the forest; the cornfield and the management of available water resources like lagoons, rivers, or canals; the cornfield with land edges or *metepantles* (Agave rows); the cornfield and very small surfaces destined for some medical herbs or vice versa and other combinations (González-Jácome 1992: 494–592; Juan-Pérez 2003; Márquez-Mireles 2006, 2007: 351–373; Robles-Cervantes 1998) (Figures 15.3–15.5).

The cornfield agroecosystem can be a commercial monoculture that is rotated every 2 or 3 years with alfalfa or is intermixed with orchard fruits like avocados, apples, or pears that take longer to produce fruit. The popular farmers' notion is "don't put all the eggs in one basket". In Progreso, Hidalgo Juan-Pérez (2003) studied how the cornfield is combined with a strawberry patch, parcel with commercial flowers, and a cornfield. The strawberries and the flowers are for the market, and the cornfield is destined for the family food system. Márquez-Mireles (2006, 2007: 351–373) studied the cornfield combined with commercial fruit agriculture, apple chili, and market crops

FIGURE 15.3 Cuaxonacayo, Tlaxcala Agave in Rows (*Metepantle*).

FIGURE 15.4 Tlaxcala Block Agave Rows (*Metepantle*), Land Edge, and Ditch.

FIGURE 15.5 Matlalcueye Volcano Lowlands Cornfield between Edges of Land with Natural Vegetation.

in Cruz de Piedra community. Both cases are in the state of Mexico (Juan-Pérez 2003; Márquez-Mireles 2006, 2007: 351–373).

In the 1970s, in S. Francisco Tepeyanco, Tlaxcala, the homegarden had an average of 300–400 m^2, and the cornfield plots from 400 to 700 m^2. The lands whose surfaces were larger than 1 ha were located in the flat areas next to the town, and their crops were watered with seasonal rains. Their surface area has been reduced owing to their conversion into industrial areas, residential units, or family recreation gardens.

In villages, such as Chorijoa, in the middle basin of the Yaqui River, in Sonora, Sánchez-Muciño, and Velasco-Orozco (2018: 235–268) described the homegardens located next to the farmhouses, and their average surface area is 70 m^2. These homegardens are associated with the *mawechis* or to the summer crops agroecosystem (*Verano*), next to which tributaries of the Mayo River pass (Robles-Linares 2013, 2018: 269–293). Maize crops on larger areas are monocultures directed toward the market (SAGARPA 2016).

According to SAGARPA in 2016, avocado orchards in Mexico occupied a total area of 174,398.03 ha. With a national average yield of 10.63 ton/ha, there is no information about the number of orchards established or accounted for in each of the 23 avocado producing regions in which this federal institution divided the country. SAGARPA figures indicated that in 2016, the total value of Mexican avocado exports reached 2 billion 227.25 million dollars. According to SIAP (Agro-food and the Fisheries Information Service) information in 2018, avocado production was 6.5% of the total earnings from national agricultural production and placed third after maize and sugar cane (Gallegos-Castillo 07/11/2019). From January to September 2019 (SIAP), Mexican avocado production was 1.64 millions tons (5.7% greater than in the same months in 2018). The production of avocado for export in Mexico has been a good business for the owners of the orchards. However, it has generated serious violence, illegalities, and problems between the towns, justice, and the federal judicial apparatus about the orchard properties and monetary earnings.

15.3 HOW THE CORNFIELD HAS BEEN DEFINED

Ortega-Paczka (2002: 97–108) defined the cornfield as "[...] maize plots alone or associated with other crops, mainly beans and squash". This definition showed the state of the current system. The cornfield can be watered with seasonal rains, and there are variations in its biodiversity, which depends largely on where the cornfield is found, and the environmental, physiographic, climatological, and altitudinal characteristics of its location. The cornfield has variations according to its distance from the urban centers, shops, and markets of different types where its produce will be sold, plus the roadway network, and transport to get the produce to these locales. The size of the family, the age of its members, and the people involved in cornfield management are factors that cause agroecosystem variations. According to Mariaca-Méndez et al. (2014: 323–359), the cornfield can be defined as:

> An agroforestry, and fauna system where the peasant family obtains maize, but the system is multiple cultivated, associated with interspersed species such as weeds, fauna, medicinal plants, fungi, firewood, and some other satisfiers that contribute to the family unit reproduction. The management, and conservation of the cornfield depends on the traditional knowledge of the group, its organization, its worldview, and the feeling and affection bestowed by the farmer to the system ("the heart").

Added to these definitions would be that the cornfield is an agroecosystem characterized by the use of an unsalaried family labor force and that in times of sowing and harvesting, the family farm received the help of kin and close friends (mutual aid), although some of its members may work in nonagricultural activities. These members of the family contributed to the economy with monetary resources. When the only members left on the family land are the elderly, who cannot afford to hire a salaried person to help them in the planting, weeding, and harvesting activities, the agricultural work by reducing the amount of weeding from two to one, or from one to cero, the number of crops are also reduced, and the cornfield becomes a maize monoculture, or to a non-cultivated land.

The cornfield managements had to be associated with other components of the environment and ecosystems such as forests, lagoon areas, riverbanks, ravines, watercourses, grasslands, earth edges with ruderal vegetation, or other agroecosystems like homegarden, orchard, gardens, or flatlands destined for commercial monoculture. However, the processes of modernization, industrialization, urban sprawl, and the construction, or expansion of roads have affected the size and diversity of the cornfield.

As a result of the above-cited purposes, the ancient agricultural lands and associated ecosystems have been destroyed, thereby reducing their diversity. In fact, the earth edges that surround the cultivation plots are relics of the old natural areas that before surrounded agricultural lands. In recent years, these small natural areas have disappeared to widen roads and because farmers decided o dry up the canals that before were crossed the agricultural lands, in order to increase the area for the crops (Ochoa 1989) (Figures 15.6–15.8).

FIGURE 15.6 Lacandon, Chiapas Slashes on Land for Future Cornfield.

FIGURE 15.7 Lacandon, Chiapas Freshly Slashed Ground Ready to Be Burned.

FIGURE 15.8 Lacandon, Chiapas Slashed Ground, and Burning Begin.

15.4 CHARACTERISTICS OF SOME CORNFIELDS IN MEXICO

Hernández X. (1998) stated that there are up to 100 different types of plants arranged in various associations with diverse components that could be grown in the cornfield. The most frequent crops in the system have been maize, beans, and squash with their varieties (Gliessman 2002; Aguilar et al. 2003). The changes in traditional cornfields, their crops, the varieties of grains, vegetables, fruits, roots, tubers, and bulbs are related to their surrounding ecosystems, the economy, culture, population, and social organization of the towns depending on each region. In addition, the cornfield includes domestic fauna that complements the food system, agricultural work (for plowing, carrying, or manuring), and economy of small-scale farmers' families. Added are wild fauna, fish, and crustaceans (hunted, fished, or collected in nearby places). In places like Soteapan, these foods can be partly consumed, and part of them sold to increase the family's monetary resources (Blanco-Rosas 2006; Foster 1940, 1967).

The agronomist Ortega-Paczka (2002: 98) considered that a "core" part of the phylogenetic resources of Mexico is made up of the production of the "traditional cornfields", considering that the cornfield is a plot planted with maize, grown alone, or with other crops, which are essentially beans and squashes. In Highland Chiapas, the cornfield was intermixed with crops of cabbage (AGJ 2012). In lowland of Matlalcueye volcano, in Tlaxcala, the maize was intermixed with lucerne crops (AGJ 2014). Ortega-Paczka (2002: 98) provided a list of the grounds of "traditional cornfields" for its economic, cultural, and scientific importance:

> (1) About a quarter of the cultivated area of Mexico is occupied by "more or less traditional" cornfields, which allowed the subsistence of more than 2 million small-scale farm families, and from which many more Mexican families subsisted; (2) the native plant genetic resources of the country's cornfields are a fundamental basis for the genetic improvement of maize, beans, and pumpkins; (3) the cultural interest of these resources are closely related to the history of Mexico, and Mexican rural life; (4) among the aspects of scientific interest is that maize had become a paradigm for many basic biological studies, due to some of its characteristics and the large pool of knowledge currently available. In addition, there is an international interest in knowing the local varieties of maize that are in the hands of Mexican farmers. Therefore, Ortega-Paczka (2002: 97–108) proposed to investigate advantages and disadvantages of in situ conservation of cultivars, linked to the seed selection by small-scale farmers in order to conserve and improve maize crops (Louette and Smale 2000: 25–41). These farmers selected the seed:

> The seeds are planted several in the same plot to maintain their identity, and to preserve the characteristics of each of these maize cultivars, because they crossed widely. This has been possible to corroborate in other parts of Mesoamerica, including in the regions of Chalco-Amecameca, and the North Sierra de Puebla [...].
>
> *(Ortega-Paczka 2002: 98).*

The continuation of studies in that direction is required, particularly because throughout the country small-scale farmers obtained their seeds from poorly called "Creole" maize, combinations of it, or improved maize in advanced generations. After there has been purchasing maize from the government company or from private companies for more than a generation, it is important to explain why the study of traditional agriculture is still relevant. One of its lines is to obtain knowledge about the types of cornfields found in the country and its regions (Figures 15.9 and 15.10).

FIGURE 15.9 Highland Chiapas Cacahuacintle Maize.

FIGURE 15.10 State of Hidalgo Maize Harvest Drying Inside the House.

Chemical ecologists Anaya-Lang and Espinosa-García (2006: 8–13) characterized a type of cornfield in the drained field agroecosystem of southwest Tlaxcala with the following features: (1) it is a polyculture (maize, beans, squash, fava, chili, alfalfa, and reed) with multifunctional handling; (2) the greater the variety of crops, the greater the wealth derived from harvested products, and the greater guarantee to avoid the total loss of the crop in case of environmental contingencies; (3) diversification has the advantages of improving the control of pests and weeds, increasing the contribution of organic matter to the soil and promoting greater diversity of pollinators and predators to curb some of the pests in crops and microorganisms in the soil (Anaya-Lang, and Espinosa-García 2006: 13; Gliessman 2015: 161–172, 191–201; Kato et al. 2009; Luna-Morales 2006).

From 1980s and part of the 1990s, cornfields located on the platforms of the drained fields were surrounded by canals. They had trees (*Alnus acuminata*, *Fraxinus uhdei*, *Populos alba*, *Populos nigra*), which functioned as barriers against the wind, modifying the climate of the plot. On the borders, which separated the plot from the canals, the vegetation included the bush known locally as a river jar (*Baccharis glutinosa*), which functioned as an insect trap, protecting the crops. The river jar attracted aphids and puffins (diurnal beetles of the genus *Macrodactylus*), collected on the leaves of the river jar (*Baccharis glutinosa* Pers.), rue (*Ruta graveolens*), common groundsel (*Senecio vulgaris*), and ruderal plants (González-Jácome 2008a).

The earth edges are home to other trees, mainly white and black poplars (*Populos alba* and *P. nigra*), ash trees (*Fraxinus* sp.), Mexican black cherries (*Prunus* sp.), and Mexican hawthorns (*Crataegus mexicana*). The ripe fruits of black cherry and hawthorn were for the daily diet of the settlers and birds. When they fell to the ground, together with the leaf litter, and combined with water mud, they fertilized the soil or rotted in the waters of the canals to later be applied to the soil of the plots (Figures 15.11 and 15.12).

The agroecosystem had to be seen as a collection of populations interacting in many kinds of organisms, including uncultivated species, animals, and microorganisms. Integrated pest management requires a simultaneous analysis of the population ecology of each member of the specific group of crop pests of the natural enemy complex and other populations of organisms with which the complex system interacts (Gliessman 2015: 161). There are some examples of the types of Mexican cornfield agroecosystems. It is shown in these examples what should be included in the management of cornfield agroecosystems and the local knowledge articulated to these complex and multifunctional agroecological systems.

FIGURE 15.11 Atoyac-Zahuapan Basin: Ditch Bordered by Alnus.

FIGURE 15.12 Tlaxcala, Atoyac Zahuapan Drained Fields (*Camellones*), and Ditch.

15.5 EMERGENCE OF THE FIRST CORNFIELDS (*MILPA*) IN MEXICO

The date establishing the age of the cornfield in Mexico is very difficult because there are so many variations of when its beginning was. These disparities due to different causes: the way in which domesticated and cultivated plants are included within the system; the environmental conditions of the places where domesticated maize has been found; the lack of archaeological studies and chronologies for the places in question; and the emergence of new studies with technologies for dating that are increasingly more accurate (Hurtado 2013). Data on the age of the process of maize domestication is modified as studies in this regard increase.

Currently, the oldest date on maize domestication corresponded to Xihuatoxtla, a rocky refuge in the middle basin of the Balsas River, in the present state of Guerrero. In that site, the oldest chronology for the domestication of teosinte *parviglumis* is from 8900 BP to 8610 BP (Piperno and Smith 2012: 151–164). The S. Andres, Tabasco study claims that *Zea mays* was cultivated in 7100 BP, although there is no certainty that the phytoliths studied were of maize or teosinte (McClung et al. 2014: 97–121; Pohl et al. 2007 (104): 6870–6875). In an article about the origins of food production in Mesoamerica, Piperno and Smith (2012: 151–164) wrote:

> Mexico, along with the remainder of Mesoamerica in smaller part, formed one of the world's great centers for the independent development of agriculture. Dozens of crop plants were brought under cultivation, and domesticated there in the prehistoric era. They include the most famous crop of the Americas, maize (*Zea mays*); two species of squash (*Cucurbita pepo*, and *C. argyrosperma*); the common bean, and small-seeded (sieva) lima bean (*Phaseolus vulgaris*, and *P. lunatus* [sieva type]); the pseudocereals *Amaranthus*, and *Chenopodium*; avocados (*Persea Americana*); at least one species of chile pepper (*Capsicum annuum*); and a number of important tree crops, including *Leucaena* spp. (White lead tree), and *Spondias purpurea* (the Mexican Hog Plum).

Piperno and Smith (2012: 154–155) gave a list of the plants domesticated and cultivated in Mesoamerica: (1) *Zea mays* (maize) 8700 BP, Xihuatoxtla shelter, Guerrero, in central Balsas; (2) pumpkin, zucchini (*Cucurbita pepo*), Guilá Naquitz cave, Oaxaca, wild ancestor unknown; (3) silver-seeded squash (*Cucurbita argyrosperma*), Romero cave, Tamaulipas 5085 BP (?), Xihuatoxtla 8700 BP; (4) common bean (*Phaseolus vulgaris*), Tehuacán Valley 2300 BP; Lerma-Santiago basin, Jalisco; (5) sieva lima bean (*Phaseolus lunatus*), humid lowlands of western Mexico; (6) runner bean (*Phaseolus coccineus*), Tehuacán Valley 5000 AC, cool and humid highlands of Mexico; (7) yearlong bean (?) (*Phaseolus dumosus*?) Guatemalan Highlands; (8) sword bean (*Canavalia ensiformis*) eastern Oaxaca 1600 BP; (9) foxtail millet (*Setaria macrostachya*) from 6000 to 4400 BP (?) in Ocampo caves, Tamaulipas; (10) *Amaranthus cruentus* (?) the chronologic sequence of Tehuacán needs clarification, Mexican Central Highlands; (11) *Amaranthus*, Highland Mexico (?); (12) *Amaranthus hypochondriacus* (?), Highland Mexico (?); (13) *Chenopodium berlandieri* (?) Southern Mexican Highlands; (14) ground cherry, green tomato (*Physalis philadelphica*) (?), it is not certain if the remains from Tehuacán belonged to this species; (15) vegetable pear, chayote (*Sechium edule*) (?), the humid forest eastern Mexico (?).

Piperno and Smith (2012: 154–155) included six fruit trees as domesticated and cultivated: (1) tree gourd (*Crescentia cujete*) by 3000 BP, in the Tehuacán Valley, and widely distributed in seasonal tropical forests of Mexico, but it is unclear if the Tehuacán remains represent wild or cultivated plants; (2) avocado (*Persea americana*) 3000 BP, in the Tehuacán Valley, and lowland seasonal forests of southern Mexico, probably Balsas region; (3) red white lead tree (*Leucaena esculenta*), sometimes from 6000 BP (?) to 3000 BP. In the Tehuacán Valley, lowland seasonal forests of southern Mexico, probably Balsas region; (4) green white lead tree (*Leucaena leucocephala*) 3000 BP in the Tehuacán Valley and lowland forests of central Veracruz; (5) Mexican hog plum (*Spondias purpurea*) (?), seasonal tropical forest of western central Mexico, and southern Mexico or Central America; (6) Mexican hog plum (*Spondias mombin*) 7000 BP (?) in the Tehuacan Valley, lowland seasonal forest.

Root Crops: (1) yam bean (*Pachyrhizus erosus*) (?) in the lowland seasonal forest. The rest of their list included: *Succulents*: (1) Maguey, Century plant (*Agave* spp.) (?). It is uncertain if *Agave* remains at Tehuacán were cultivated; (2) India fig (*Opuntia Ficus indica)* (?) in upland tropical habitat and Central Mexico; (3) *Opuntia* spp. (prickly pears) (?) it is uncertain if *Opuntia* remained at Tehuacán were cultivated. *Condiments and beverages*: (1) *Theobroma cacao* (chocolate) 3000 BP Puerto Escondido rock shelter in Honduras, its wild ancestor was native to tropical forest in Amazonia; (2) chile pepper (*Capsicum annuum*) presented throughout much of the Tehuacán valley sequence, chronology needs clarification and Central Mexican Highlands. *Industrial plants*: (1) cotton (*Gossypium hirsutum*) rarely was present throughout much of the Tehuacán sequence, and in the Yucatan peninsula, chronology needs clarification.

The list by Piperno and Smith list should be compared with information from other researchers, since it shows differences both in the chronology and in the list itself, which in other studies such as that of Earl Smith Jr. are much more extensive and detailed. It should be considered that there are a number of new studies in this regard. However, the article gave an idea of some of the plants that were domesticated and cultivated in the early times of Mesoamerican prehistory. It is worth noting the difference between domesticated plants with edible uses and fruit trees, among which the Mexican black cherry, Mexican hawthorn, and sapodilla were not included. According to Piperno and Smith (2012: 151), domestication and plant cultivation in the Central Mexican Highlands and the tropical lowlands emerged during the early Holocene period.

In the Zoque-Popoluca region of Los Tuxtlas, in Veracruz, the oldest date for domestication of maize is 4600 BC. Niederberger (1979: 132–142) in Zohapilco, in the Basin of Mexico, established that by 3090 BC±115, domesticated maize was consumed there already. From 2500 BC to 2200 BC, during the Formative of the Central Highlands, maize was the staple food. In 2000 BC, several agricultural villages were established (Uruñuela and Plunket 2012: 41–60). The polyculture cornfield articulated to these first villages corresponded to the Middle Preclassic (1600 BC to 600 BC).

On the S. Lorenzo site, in the coastal plain of Los Tuxtlas, Cyphers (2004: 23) stated that from 1200 BC to 800 BC (the apogee development of the site), the population reached 13 thousand inhabitants (about 2,000 families). Those natives lived on about 50 artificial islets, with an average area of approximately 10,000 ha^2. The total extent of the village was about 20 ha. The inhabitants lived in seasonal camps in the nearby mountains, where they sowed and reaped crops.

In the Formative (1500 BC–1200 BC), maize was cultivated on dry plains and riverbanks, where the slash and burn agroecosystem was employed. Also, wild grains were collected and ground in basalt mortars, along with tubers and macaw palm (*Acrocomia aculeata*). Aquatic comestibles from rivers, meanders, wetlands, and seasonal flood zones complemented their food system. This included fish, turtles, crustaceans, mollusks, and waterfowl (Cyphers 2004: 21–40). For many millennia, wild green edibles were part of the food system, along with cultivated plants and collected food.

In In La Joya, and Bezuapan sites, in the Los Tuxtlas, during the Formative, the cornfield arose associated with the hunting of different species and sizes of wild fauna, according to the season of the year and the cyclical periods of rain and drought (Vanderwarker 2006). This statement coincided with what happened in other places in the Central Highlands, more or less in the same period. The cornfield seems to have arisen associated with the incidental homegarden, and it was generated by the consumption of fruits and the formation of garbage dumps where waste was thrown. The homegarden and the cornfield appear with the domestication of maize, along with the consumption of fruits, mushrooms, birds, and terrestrial and aquatic fauna, all of which were fundamental in the first settlers' food system.

15.6 THE CORNFIELD IN THE CENTRAL HIGHLANDS IN 1519

From information in the Nahua and Spanish Vocabulary of Alonso de Molina (1571), some ideas were obtained on the Central Highlands cornfield agroecosystem in the sixteenth century:

> The cornfield was made up of plots planted with maize, or sowing plants (*centemilli*). The cornfield work included the following tasks: weeding so that the field was not covered with weeds (*aceceyoa; acicicyoa; zacacaua*); removing the earth (*actitlaza*); raking the grass (*zacapi*); rubbing or picking

> herbs by hand to pluck them from the roots (*momo*, *tzoa*); making ridges (*centlacuentectli*); making ridges for sowing something (storyteller; *cuenteca*); manuring the soil (*cuitlauia*; *zoquipachoa*); softening or molding the soil (*moleua*); piling soil on maize plants when they are tilled (*motlalhuia*); diging the land that is already tilled (*mopopoxoa*); performing the second tilling (*oppauia*); trampling the maize cobs after the harvest (*netitixiliztli*). There were lands that were allowed to rest for a year [...] so they did not get tired [...] (*moxiuhcaua*) (Molina 1571).

An article in *Mexican Archaeology* (2019 Special N° 84: 28–29) shows that the Mesoamerican cornfield of the Central Highlands followed an agricultural calendar that began in the fourth month (*Huey Tozoztli*), which ran from April 13 to May 2. That date marked the beginning of the rains, with a party dedicated to *Chicomecoatl*, where was a kind of "blessing" of the seeds that would be sown. The cobs were seven, organized in bunches (*ocholli*), which were collected in the previous harvest. They were adorned with red paper, and the maidens (*ichpopochtin*) took them to the *Cinteopan* temple of the goddess to be blessed, so the cobs became the hearts of the granary and the maize for next planting.

The peasants took the corncobs to home, placing them in their granaries until the next planting (March-April) of maize, squash, tomato, and sunflower. Chía was planted in July, and chili in November. In tropical areas, vanilla planting occurred in May, sweet potato in June, and cassava in August. In the sixth month (*etzacualiztli*), which ran from May 23 to June 11, the corncob and rainwater fest were celebrated. In the 11th month (*ochpaniztli*), the ripening of the maize, and its first fruits were celebrated. This festival was represented in the Bourbon Codex, where this ritual included the marriage of mother earth to the sun, from which the God of maize was born.

Two months later, in the 13th month (*tepelihuitl*), the harvest and the beginning of the dry season commenced. Rituals were done for Tlaloc, the god of rain, who was the god of the hills. The Bourbon Codex (p. 32) shows Tlaloc in his temple on the hill and beside him an offering of ears of maize decorated with ceremonial papers and dotted with drops of rubber sap. According to Broda (*Mexican Archaeology*, 2019 Special N° 84: 28–29), after 1591, this party would become the celebration of the day of the dead (Figure 15.13).

Human work applied to the cornfield agroecosystem, and changes in cycles have changed over the centuries. In the Central Highlands, a very important change arose with the arrival of the Spaniards, but it was followed by the terrible results of the war of the conquest in several places of what would be New Spain. The introduction of plants, animals, and technology during the sixteenth, and seventeenth centuries by the Spanish was paramount (Dunmire 2004; González-Jácome 2001: 315–318). The defeat of the Mexica modified land ownership, which changed from being functional to private. This included the towns and lands that were handed over to the Crown, and

FIGURE 15.13 Tlaloc Carrying a Basket with Maize Cobs on His Back.

then given to the conquerors as payment for their services in the enterprise of the conquest, and the royal grants bestowed for the duration of one or more lives, for the creation of private property, and the management of space and land in the Indian villages (Moreno 1987).

In antiquity, the slash and burn agroecosystem was preceded by the clearing of a strip of land all around the area (*guardarraya*) to be burned to prevent the dispersion of the fire toward other plots or toward the forest. The period of ground rest lasted up to 20 years, a term that had been shortened because the population has increased and agricultural land had been reduced (González-Jácome 2011; Hernández X. et al. 1987: 338–340; Hernández X. et al. 1995); Palerm 1955, 1972) (Figures 15.14 and 15.15).

The number of weeds decreased from three to two in places where the men of the family migrated. In 1900, three weeds were applied to the cornfields in Mexico, for which both the plow and the labors of two children were used (Kaerger 1986: 244–245). Currently, only one weeding is done, but female labor applied to the cornfield has increased, and though the number of plants grown in the cornfield has been reduced, the varieties of maize, beans, and squash remained to be staples in the food system of the villagers.

FIGURE 15.14 Lacandon Chiapas Cabbage Crops Next to Slashed Plot and Forest.

FIGURE 15.15 Lacandon Chiapas Burning Vegetation Before Planting Maize.

15.7 THE CORNFIELD AND CATASTROPHIC CLIMATOLOGICAL EVENTS, EPIDEMICS, DISEASES, AND FAMINES

The catastrophic weather events recorded before 1518 caused famines and mortalities in the Mesoamerican indigenous population. An extemporaneous frost and pollution of the waters caused the death of aquatic animals in the year ten-rabbit (1450) producing hunger and diseases. Another series of frosts caused the loss of crops and the shortage of seeds for the next planting. Hunger beset the small cities in the Tenochtitlan and Texcoco area; however, the latter suffered more epidemics, which caused the death of many inhabitants. The rulers Moctezuma Ilhuicamina, Netzahualcoyotl, and Totoquihuatzin took action as the heads of the Triple Alliance (Mexico-Tenochtitlan, Texcoco, and Tlacopan) and stopped raising taxes during the 6 years that the calamity lasted. They opened the storage rooms where the proceeds of previous years were kept and distributed the maize and beans among the people.[1]

The snowfall of 1450 cooled the temperature, and there was a "pestilential cold", which affected the respiratory tract and killed many people, especially the elderly. According to Mandujano et al. (2003: 9–21), "[...] it could be some type of influenza that periodically manifests itself with an epidemic character, and high mortality". Catastrophes generated food shortages. This situation lasted 5 years. In addition, "[...] various types of prolific infectious, and contagious diseases were added, furthermore the immune system of the affected population was depressed with the inability to obtain proteins" (Mandujano et al. 2003: 9–21). There are several cases of this type of event. In 1452 (*CeTochtli*: 1 Tochtli), a famine caused by weather events occurred, and war decimated the population.

The Chimalpopoca Codex recorded the calamity in the year 1 *Tochtli*, when everything "became rabbited" (*aconejado*), as Captain Guillermo Dupaix (1964 [1805, 1808]; López-Luján 2016: 432–435) wrote in his antiquarian expedition of Mexico: "[...] for an unfortunate sign (one-rabbit), or [...] he left with the Totonac. Then there was the war of the people from Chalco in all the fences of the mount (*quauhtenampan*); and because this ceased, war was no longer waged". There was a 3-year famine; there was no upkeep any longer. "The inhabitants emigrated to the hot land where the inclemency of the weather had not been felt". The inhabitants of the highlands sold their freedom in exchange for a handful of food, and others exchanged their children for maize, and the Totonacs, who took the highlands people to the center of Mexico, bought them. Torquemada (1723) described the event:

> Here was such great hunger, that these poor Mexicans went so far as to eat tulin roots, and other roots of wild herbs, for a lack of food. The hardship was such that they sold each other for the price of maize. Seeing the king, and his council that this was happening, and that it was a force to reckon with, because the Mexicans did not perish, they gave permission that they be sold as slaves, for the value, and price of a maiden, 400 ears of maize (whose shells make a hanega), or less, and for a young man 500 ears.

They allowed the inhabitants of Tenochtitlan to migrate to other places so that they could survive. As month 8 of the year in the Mexican calendar approached, food was distributed to the poor, but Moctezuma's reserves began to decline and finally ran out. The exodus was immense, and the crowd of those who emigrated on their own was added to that of the purchased crowd. The roads were strewn with the bones of those who could not reach their destination and died of starvation, and the "plague of heaven" that broke out (Mandujano et al. 2003: 9–21). This narrative described the effects of the years of drought, with insufficient agricultural production to meet basic food needs, the necessities of the population, and the urgencies of famine.

[1] Aguirre-Beltrán 1972 [1946]; Cook and Borah 1960, 1974; Gibson 1980 [1967]: 138–167; González-Jácome 1994: 1–12; Malvido 2003: 65–78; Mandujano et al. 2003: 9–21; Reff 1990: 237; Rosenblat 1945, 1967; Sanders 1976: 85–150; Santley 1990: 325–343; Sherbourn and Borah 1960; Sherbourn and Simpson 1948; Withmore 1992; Zambardino 1980: 1–27.

15.8 OLD WORLD PLANTS, FAMINES, AND THE CORNFIELD IN NEW SPAIN

In addition to the introduction of Old World plants, animals, agricultural managements, and technologies to New Spain, and the increase in biological and cultural diversity, a series of negative events befell in New Spain, like epidemics that hit the population. These were combined with recurring famines, which affected the indigenous population and the Spanish population, however to a lesser extent. Depopulation in some places was extreme, reaching 90%, like in Veracruz, in Tlaxcala, it reached 60%; in Yucatán, strong demographic impacts occurred. Bishop Diego de Landa (1983 [1566]: Annex 8) wrote that the Visitor Tomás López issued a set of Ordinances in the years 1552–1553, highlighting that:

> To remedy the daily famine, which in this land is usually due to the lack of providence of the natives [he ordered] that the chiefs not only take care of the common Indians planting the plots according to their family, so that there be surplus, but that they are forced to have where to keep the leftover, and that if the year were abundant, be renewed for the next. With this, a remedy to so many evils was prevented, as seen in this land with the sterility of a single year. We all feel them when it happens: but nobody remembers it if it is not when the need is being suffered. In order for this to have a better effect, he ordered the encomenderos to gave all favor, and help, [under] pain [that] they would be punished in their persons, and property. The impairment is not small, which when it happened, they have in taxes. He ordered the introduction of livestock, and cattle breeding among the Indians.

It is probable that among the greatest impacts on the population was how agricultural productivity was affected by natural disasters and famines, which were followed by plagues and diseases. Associated with them were the effects of the war and the conquest of central Mexico, such as the high mortality rate, the expansion of the Spaniards on the depopulated lands of the indigenous people killed in the war, epidemics, and famines. In the sixteenth and seventeenth centuries, the reduction of the Mesoamerican population was a demographic and economic phenomenon of great proportions. In some regions of the country, populations recovered only until the end of the nineteenth century.

With the Spaniards' arrival in the Central Highlands, agriculture and the food system underwent transformations, which are reflected in the information on what was bought and sold in the markets (Long 2003; Long and Vargas-Guadarrama 2005). In the *General History of Things in New Spain*, Bernardino de Sahagún (1830, Vol. III) described the trades, including that of the sellers. The friar wrote that Tlatelolco merchants sold wheat, flour, bread, chickens, and eggs from Castile.

The Spanish monopolized the best-irrigated agricultural lands. Charles Gibson (1967: 326) said that the Indians changed the food production system by replacing the lands with maize and bean crops for Agave crops. The indigenous cornfield in the basin of Mexico changed from irrigated land to seasonal areas.

Since 1590, Mexico City surrounded itself with: "[a] great belt of Agave plantations", formed by Acolman, Chiconautla, Ecatepec, Teotihuacan, Cuautitlan, and Otumba (Gibson 1967: 326). Hernández-Palomo (n/d: 326) considered that "The cultivation of Agave, and the production of Agave fermented mead (*pulque*), became a lucrative business, for both the Spaniards, and the vice regal authorities who entered it, and very soon took control of it". The expansion of Agave through large regions (Hidalgo, Tlaxcala, State of Mexico, and Puebla states) had its peak in the nineteenth century.

In 1697, the Italian traveler/merchant Francisco Gemelli Careri arrived in New Spain at the port of Acapulco, where the Fair was organized after the arrival of the Philippines Nao occurred (Dávila 2011). He remained in Acapulco to conduct business and then heads to Mexico City and from this city to the port of Veracruz to embark back to Europe. On the Mexico basin, he wrote in his personal diary that the weather was very mild, rains begin in July and continued on irrigating the cultivated land with maize:

> Precisely because of these rains, the lands give three harvests per year: an irrigated one; a seasonal one; and an adventurous one that is uncertain. Maize was sown from March to May. Mexico is, therefore, an excellent city in whose main square there are flowers, and fruits of all species throughout the year.

With the information of seven cases in New Spain, including the basin of Mexico, the Toluca Valley, Puebla, Rio Grande de Santiago, and Sonora, it was found that from 1535 to 1803, in addition to maize, beans, and squash were grown in the cornfield (García-Martínez 1987; Trautmann 1981). In plots, outside the cornfield other crops were wheat, beans, lentils, peas, legumes, cotton, peppers, herbs, Agave, and Opuntia. The recorded weeds were wild cactus and medicinal herbs. Maize production was in: (1) Querétaro, Morelia, and the Toluca Valley where irrigated maize in 3.56 ha produced 800 bushels (17,632 kg); (2) Sonora where the irrigated maize produced per bushel 7.56 liters of grain and 200–300 bushels (4,408 to 6,612 kg)/ha (Nentvig 1971 [1764]).

At the beginning of the nineteenth century, Alexander von Humboldt (1822, T. II, Book IV, Chap. IX: 276–277) wrote that the "prosperity" of New Spain depended on the laps from rainy to dry seasons, and "Sometimes maize, and cereals from Europe are exposed to partial flooding in the plains, many of which form circular shells closed by the mountains, wheat is planted on the slopes of the hills which weather is much more freshness [...]". In the north from the 26° to the 30° parallel, rains were rare and short-lived. Due to the "extreme dryness" from June to September, the inhabitants of New Spain required artificial irrigation, bleeding the rivers to conduct water from far away via ditches, which together with dams and water wheels were of "major importance" for Mexican agriculture.

Humboldt (1822, T. II, Book IV, Chap. IX; pp. 280–281, 313, 319, 326–327, 331) wrote that maize "suffers" much more than wheat from autumn frosts but is more easily replenished after droughts. In the valley of Mexico, maize produced 200 grains/each one sown. It was the food of Indians and Mestizos. The indigenous person of the tropical land preferred maize, cassava, and banana. The vegetable productions that were the basis of the food system of the Mexican people were banana, cassava, maize, and cereals.

Humboldt included a list of various crops cultivated in the cornfield in the Valley of Mexico: tomatoes (*tomatl*), which "[...] were formerly planted intermingled with maize [...]", *Trigidis pavonia* (*cacomite, oceloxochitl, cacomitl*), peanut ("pistachio of the land"), different species of peppers (*chilli*), sweet potatoes; Sunflowers with large flowers (*chimalatl*), native onion (*xonacatl*), Scarlet runner beans (*ayocotli*), pumpkins, and cress. Mexicans included in their crops as well the plants, vegetables, and fruit trees brought from Spain like turnips, cabbage, leafy greens, onions, leeks (long onions), chives, garlic, watercress, thistle (*cardillo*), and common Golden thistle (*Scolymus hispanicus*).

15.9 CHANGING THE NATIVE PRODUCTION SYSTEMS

Spaniards monopolized the best and irrigated agricultural land. Charles Gibson (1967: 326) explained how the natives changed the food production system in the Basin of Mexico by replacing maize and bean crops for *Agave* crops. The cornfield was switched from irrigated plain lands to seasonally rainwater-irrigated areas. Gibson (1967: 326) said that since 1590, Mexico City had surrounded itself with "[...] a large belt of Agave plantations, in Acolman; Chiconautla; Ecatepec; Teotihuacan; Cuautitlan; and Otumba". Hernández-Palomo (n/d: 326) as previously noted, the Spanish and viceregal authorities controlled the cultivation of Agave and the production of fermented Agave mead (*pulque*), which became a very lucrative business.

With the production of *pulque*, the Agave ceased to have many of its ancient and multiple uses to be destined then for the production of the fermented Agave mead beverage (pulque). This was sold in places called "*pulquerías*" or boxes, dominating by the aforementioned. The lands were bought, usurped, or leased from the indigenous owners, and Agave plantations (*magueyales*) began replacing the cornfields.

Intensive planting of Agave was mixed with the sowing of wheat and barley in the lowlands of the semi-terrace agroecosystem (*bancal*) (Donkin 1979; Patrick 1997; West 1970), which were located in the plains of Apam and Pie Grande, to the west of Tlaxcala state. Maize was grown to supply food for agricultural workers. The semi-terrace (*bancal*) agroecosystem was consolidated in the eighteenth and nineteenth centuries, with Agave estates (*haciendas pulqueras*). It had its background in terraces, with Agave rows (*metepantli*) as a subsystem, but its main crops (Agave, wheat, and barley) were destined to the market. Gradually, Agave estates were formed, and they would characterize the Central Highlands in the nineteenth century (Rendón-Garcini 1994; Tortolero-Villaseñor 1995).

The production of Agave fermented mead (*pulque*) was being carried out at the same time that public workers, officials, administrators, great merchants, miners, and Jesuits were hoarding land to devote it to the intensive Agave cultivation. On farms, the owners left spaces close to the workers' homes (*calpanerías*) so that they could plant their cornfields and stock up on maize (Hernández-Palomo 2004; Rendón-Garcini 1990).

15.10 CONSIDERATIONS AND CONCLUSIONS

In the twentieth century, among the small-scale Indian farmers of the Llanos de Apam (the Hidalgo and Tlaxcala plain), the basin of Zahuapan and Atoyac Rivers (Luna-Morales 2006), the slopes of the Tlaxcala Block, the Tlaxco mountain range, and the Matlalcueye volcano lowlands, Agave became part of the cornfield structure, occupying the terraces, semi-terraces, plains, and edges of farmlands.

Small-scale farmers had 50–100 Agave plants at different stages of growth (Tylor 1979). The Agave rows grew and plants in them matured at different times allowing the continuous supplies of mead to be fermented. The small-scale farmer producers sold *pulque*, along with other products obtained from Agave in their villages, directly to local consumers, but some output was taken to the tianguis where *pulque* was bartered a retailed (Corcuera de Mancera 1991; González-Jácome 2016: 1–51; Hernández-Palomo 1979, 2004: 239–257; Mendoza-Monroy 2017; Rendón-Garcini 1995).

Agave cultivation and pulque production grew after the sixteenth and seventeenth centuries and continued until the beginning of the twentieth century, causing numerous movements of popular discontent as a consequence of the imposition of taxes on pulque sales. In Tlaxcala, they were an antecedent of 1921 revolutionary movements, which occurred in the region from 1890 to 1910 (Rendón-Garcini 1995, 1996).

Agave rows were planted on the land edges of cultivated land (*metepantle*) in the towns, building terraces, and semi-terraces of the slopes where Agave, trees, and ruderal vegetation were combined by the farmers to protect the soil of the plots where the cornfield continued to be the crop that supplied maize and other plants to the indigenous and mestizo farmers' Agave rows (*metepantles*) were planted on the land edges of cultivated land of the towns, the small-scale farmers built terraces thus creating a biodiverse subsystem on the land edges of agricultural plots of the terraces, and semi-terraces of the slopes where Agave, trees, and ruderal vegetation were combined by the farmers to protect the soil of the plots where the cornfield continued to be the agroecosystem that supplied maize, and other plants to the indigenous, and mestizo farmers' food system. The land edges with the metepantles were used to grow medicinal plants, and Agave for pulque, for small-scale farmers families' food system.

RECOMMENDED READING

Rendón-Garcini, Ricardo. 1994. *Haciendas de México*. Fomento Cultural Banamex A.C.

The chapter of *Pulqueras y Mezcaleras* (pp. 209–244) in the book *Types of Haciendas and Hacendados* is recommended. The chapter deals with Agave plantations and the production of fermented Agave mead (*pulque*) on *haciendas* from the seventeenth to nineteenth centuries, which developed in the natural corridor that ran from the east of Puebla, crossing through Tlaxcala to reach the current state of Hidalgo and the north of the State of Mexico (Rendón-Garcini 1994: 209–244).

Pulque production was more intense in the Llanos de Apam. This region is semi-arid, the Agave required little water, and the workforce was paid per task. Pulque production ceased to be local with the arrival of the railway (1866), encompassing more distant markets, but never was it an export product. Mexico City was the main market for pulque.

The Agave rows (*metepantles*, *amelgas*) are planted on a dirt ridge, generally following contour lines and may have only Agaves, or include other plants. Between them are left free spaces (*entrecalles*) where barley, maize, or wheat was planted. Agave-free zones were cultivated with maize and crops for the self-sufficiency of the estate and of the workers settled there (*peones acasillados*). Estates had a building called *tinacal*, where the mead was processed to be turned into pulque. The *tinacal* was a gallery with tanks, barrels, and tools, and it had wall paintings, sometimes. Some estates had alembics to produce alcoholic beverages (*mezcal*) using the central part of the Agave (heart), which was cooked, and then ground to obtain mezcal.

Brading, David A. 2009, [1978]. *Haciendas and Ranchos in the Mexican Bajio: Leon 1700–1860.* Cambridge University Press, Latin American Studies.

The book is situated historically in the eighteenth century when the Bajio region became a great grain producer and important in the Mexican economy. There was a huge bonanza in silver mining and the population increased rapidly. This book examines the impact of changes in the structure of agricultural production and the rural society.

The Bajio plains had mesquites and grasses that were lusher than those of northern Mexico. In Celaya, the fertile zones attracted Spanish people who received land from the Crown (1571) to encourage farming, and maize required irrigation in the dry winter months.

The discovery of the silver mines in Zacatecas produced rich miners, the founding towns, and garrisons located along the road to protect commerce and people. Most of the lands were grasslands for the herds and cattle on the ranches. In 1630, most of the Bajio was occupied, and many sites were attached to the estates. Landowners rented much of their land to farmers, who often assisted the landowner's workers in the harvest, thus reducing the use of temporary hiring. The old miners invested their capital in the purchase and improvement of *haciendas* in the district of Leon where the countryside was cleared and transformed from wasteland to land for the cultivation of grains, and the agricultural land was irrigated then.

At the end of the eighteenth century, there were 62 indigenous villages in Guanajuato, although most of the Indians had lost their lands, scattered on estates, ranches, and towns. Indian homegardens and gardens remained near Querétaro city. Ranches were of small and medium scale, which frequently arose from a Royal Mercy, initially granted by the council, but which later had to be confirmed by the king.

16 The Cornfield Evolution in Mexico

16.1 INTRODUCTION

In time and space, there were and are numerous types of cornfields in Mexico. This is due to the characteristics of this old Mesoamerican agroecosystem. Among these are its ability to adapt to ecological, climatological, historical-geographic, and altitudinal conditions, and its survival in spite of excess or lack of water to irrigate crops or physiographic, demographic (total population, densities, and migration), social, cultural, and other situations. In this chapter, some cases will be treated in detail.

The contemporary types of cornfield are subject to specific external situations like belonging to some ethnic group, or the governmental, regional, or local support programs that small-scale maize farmers have received. Also, they are affected by their interrelations with demographic processes such as migration, emigration, return migration, and legal and illegal emigration; the forms of family and community organization that exist in each case; and their interrelations with the market. The description and study of this agroecosystem require more than one book; however, only some cases that exemplify its diversity and capacity for social and cultural adaptation to specific environments will be given here.

16.2 BACKGROUND

Over time, and according to the specific and historical conditions in which the cornfield has been invented and developed in Mexico, this agroecosystem shows agricultural, social, and cultural variations that correspond to historical, local, regional, or specific situations in relation to its physiogeographic, environmental, climatological, economic, social, political, and cultural conditions. The cases whose agroecological information is available are as many as the places where this agroecosystem is still practiced, with existing elements that are common to all of them. For obvious reasons, only some of the types will be included here as examples of their ability to adapt or adjust to the aforementioned conditions. This chapter will have the cornfield types in pre-Hispanic times, the Viceroyalty, the nineteenth and twentieth century, and up to now in the present twenty-first century, some of its aspects, only. The next chapter will deal with some contemporary types of cornfields.

16.3 EARLY MESOAMERICAN CORNFIELDS

The emergence of the cornfield in Mexico is connected with maize domestication. Information on the age of maize and its domestication processes is modified as the number of research in this regard increases. At the present, the oldest date on maize domestication corresponds to the rocky refuge of Xihuatoxtla, in the middle basin of the Balsas River, and in the current state of Guerrero. The date for the domestication of teosinte parviglumis is from 8900 BP to 8610 BP. In the S. Andrés site, Tabasco, it is claimed that Zea was cultivated by 7100 BP, although there is no certainty that the phytoliths studied were of maize or teosinte (McClung et al. 2014: 97–121; Pohl et al. 2007 (104): 6870–6875). Dolores Piperno and Bruce D. Smith (2012) in an article about the origins of food production in Mesoamerica wrote:

> Mexico, along with the rest of Mesoamerica in a smaller part, formed one of the world's largest centers for independent development of agriculture. Dozens of crop plants were grown, and domesticated there in the prehistoric era. They include the most famous crop in America, maize (*Zea mays*); two species

DOI: 10.1201/9781003198833-16

of pumpkin (*Cucurbita pepo* and *C. argyrosperma*); common bean, and small seed lima bean (sieva) (*Phaseolus vulgaris* and, *P. lunatus* [sieva type]); the pseudo cereals *Amaranthus*, and *Chenopodium*; avocados (*Persea Americana*); at least one species of chili (*Capsicum annuum*); and several important tree crops, including *Leucaena* spp. (White lead tree), and *Spondias purpurea* (the Mexican hog plum). Plant cultivation, and domestication in the highlands, and tropical lowlands emerged during the early Holocene period.

The data to establish the dates of the birth of the cornfield in Mexico have variations, which are huge. These are due to different causes: (1) the way in which domesticated and cultivated plants are included within the system; (2) the environmental conditions of the places where domesticated maize has been found; (3) the lack of archaeological studies and chronologies for the places in question; and (4) the emergence of new studies with dating technologies that are more and more accurate.

A summary of the chronology in the Veracruz studies indicates that in the Zoque-Popoluca region of Los Tuxtlas, the actual oldest date is 4600 BC. In the Central Highlands, the studies of Niederberger (1979: 132–142) in Zohapilco, in the Basin of Mexico, established that by 3090 BC ±115, domesticated maize was consumed, already. From 2500 BC to 2200 BC, during the Formative in the Central Highlands, maize was the axis of the food system. In 2000 BC, numerous agricultural villages were established. The polyculture cornfield articulated to these first localities corresponded to the Middle Preclassic (1600 BC to 600 BC) (Uruñuela and Plunket 2012: 41–60).

Returning to Veracruz, in the coastal area of S. Lorenzo, in Los Tuxtlas, Cyphers (2004: 23), stated that in the Formative period (1500 BC–1200 BC) the area of the village was about 20 ha; the inhabitants lived in seasonal camps and in the nearby mountains they sowed and reaped some plants. Maize was cultivated in the dry plains and riverbanks, with slash and burn. People collected wild grains, which were ground in basalt mortars along with tubers and coyol palm. Aquatic resources from rivers, meanders, wetlands, and seasonal flood zones complemented their food system. Inhabitants collected resources including fish, turtles, crustaceans, mollusks, and waterfowl. From 1200 BC to 800 BC, the population reached 13 thousand inhabitants (about 2,000 families), living in about 50 artificial islets, with an average area of some 10,000 m^2 (Cyphers 2004: 21–40).

In the Formative, at La Joya and Bezaupan sites, the cornfield arose associated with the hunting of wild fauna of different species and sizes according to the season of the year and the annual stages of rain and drought (Vanderwarker 2006). This statement coincides with what happened in other places in the Central Highlands, more or less for the same period. Summing up again, the cornfield arose associated with the forest and the incidental homegarden by the consumption of fruits and the formation of garbage dumps where the waste was thrown. The homegarden and the cornfield appeared with the processes of maize domestication, along with the consumption of other plants, fruits, mushrooms, terrestrial, and aquatic fauna, and birds, that were fundamental in the food system of the first settlers.

By 1200 BC, maize was the basis of subsistence in Cahal Pech (Belize), one of the earliest sites in the Mayan lowlands (White et al. 2006: 148). The ceramic representations gave evidence of the use of maize since 2900 BC, and studies show that the food system was complemented with food obtained from multiple estuaries, coastal, and land resources. Chisholm and Blake (2006: 161–172) found that in the Preclassic, the food system of these groups included: avocado (*Persea americana*), beans (*Phaseolus vulgaris*, *Phaseolus lunatus*), and maize (*Zea mays*), completed with resources obtained in estuaries, coast, and the surrounding land areas. White et al. (2006: 145) considered that maize-based food system intensified at the end of the Preclassic and that the grass provided food for both humans and their dogs.

On the Maya lowlands Classic (250–909 AC), Velasquez-García (2012: 102–112) described the cornfield as based on a maize crop that developed under arduous conditions due to the rapid growth of the "bad herbs", which competed with the maize making it difficult for the cobs to reach maturity.

Therefore, land was planted for 2 or 3 years and then was abandoned as others lands suitable for cultivation were sought. In the forest, the distance could be great to a source of water at the ground level of fertile ground. At the end of the Classic, with the population increase, it was difficult to find land suitable for cultivation, and the population density began to exceed the food system capacity of the land.

Once the land was found, it was measured and divided into quadrants 20 steps long, and a stone block was placed in each of its corners. In August, the forest was cleared using stone axes, starting with bushes and vines, leaving the trees at the last, after they were dampened by rain. In the rainy season, this process lasted several weeks. The men worked from sunset until noon and left the remains of weeds piled up. In March or April, they chose a "favorable" day to burn when the gods of the winds were implored, and the fire was applied at the extreme sides of the terrain, according to the direction of the wind, allowing the ashes to enrich the soil (Velasquez-García 2012: 102).

According to Harrison (2001: 73) after the first rains, usually in May, farmers selected and husked the cobs, taking 2 days. Then corn was sown. An average cornfield occupied about 12 days of labor; husking, sowing, making holes with the planter cane, and depositing in each one 4–5 grains of maize, along with bean or squash seeds. They could have placed sticks so that the plants could climb them. Ruz (1998: 103) thinks that the Maya used traps made with seashells and small eggs to scare away the birds.[1] The cornfield had from one to three weeds, which grew from May to September, to be harvested according to the maturation of each one.

From September to October, the maize plants bent over, leaving them with the top facing down to avoid moisture. From November to April or May, the maize was harvested, which was done little by little. The maize was stored in boxes lined with palm leaves[2] or bundles with full ears that were tied. The cobs were shucked by hand or rubbed together over nets or wooden grills (Morley 1992: 148; Thompson 1992: 277; Velásquez-García 2012: 103). Thompson (1992: 277) supposed that the temporary huts that had been raised in the cornfield clearings served to keep the farming instruments and be used as granaries.

Velásquez-García (2012: 103) considered that since the Mayans knew maize varieties that matured in 2½, 4, or 6 months, it was possible that in some places a second planting was carried out that would occur in September. They might have sown in different "types of place" [agroecosystems], which would allow them to obtain both crops every year. However, after 2 or 3 years of cultivation, the land had to be left fallow for about 5 years,[3] making it impossible to maintain the system in places like Tikal or Calakmul (Alayón and Gurri 2007: 243–260).

The meat obtained from game, of which there are scenes in codexes and ceramics, complemented the food system of Mayans. They knew several hunting methods: using the blowgun, making holes in the ground, using ropes, nets, lances, wooden spears, traps, baskets or fine nets to capture wild fauna, and in the case of birds captute them without damaging their feathers, and other method. Hunters asked imaginary owner and keepers of the forest for permission, and a they organized gratitude ceremonies for what they received (Velásquez-García 2012: 103–104).

In the Mayan Classic, the hunting fauna were: the jaguar that represented the underworld, deer and peccaries. Thei guardians were the twins with headbands (Ju'n Ajaw and Yax B'ahlam), and a being named Huk Si'p. The Maya kept deer and peccaries in captivity, fed them of *Brosimum alicastrum* (*ramón*) leaves, and sacrificed them at parties (Harrison 2001: 74). Fishing and capture of aquatic fauna were other complementary activities of agriculture that contributed animal protein to the food system.

1 In Tlaxcala, in the 1970s, small-scale farmers used tape recorders hung on tree branches to scare away birds; in the 1990s, they placed floppy disks hanging from the branches for the same purpose. There was one or another scarecrow, dressed in old clothes.

2 In El Rincón, Oaxaca, these boxes have been found in archeological sites. Also in the Viceroyalty, plants were transported in boxes or drawers so that they reached their destination in good condition.

3 In Mexico, the word fallow and the verb fallow are used to break the soil.

16.4 THE CORNFIELD IN THE SIXTEENTH CENTURY AND THE PLANTS THAT ARRIVED FROM SPAIN

The work applied to the cornfield of the sixteenth century in the Central Highlands can be reconstructed using information from the Spanish-Nahua Vocabulary written by Alonso de Molina (1571). This provides some ideas about the agroecosystem, the work to be done, and its management, which were described:

> The cornfield is made up of maize planted land, or plots with sowing maize plants (*centemilli*). They included the following tasks: weeding so that the weed does not covered the field (*aceceyoa; acicicyoa; zacacaua*), loosening up the soil (*actitlaza*), brushing the grass (*zacapi*), hand rubbing or picking herbs to uproot them (*momotzoa*), making ridges (*centlacuentectli*), making ridges to sow something (*cuentataca; cuenteca*), fertilize the soil (*cuitlauia; zoquipachoa*), soften or molding the soil (*moleua*), put soil around the root of the maize cane to support the impact of the wind when it is growing (*motlalhuia*), dig the land that is already tilled (*mopopoxoa*), perform the second till (*oppauia*), trample the cobs that were left in the cornfield after harvest (*netitixiliztli*). There were lands that were allowed to rest a year "[...] so they do not became exhaustrd [...]" or shear them (*moxiuhcaua*)
>
> *(Molina 1571).*

Human work applied to the cornfield agroecosystem and its transformations over the centuries had changed. In the Central Highlands, a very important alteration arose with the arrival of the Spaniards, followed by the results of the war of the conquest in several places of what would be New Spain, and then with the introduction of plants, domestic animals, and technology during the sixteenth and seventeenth centuries (Dunmire 2004; González-Jácome 2001: 315–318). The defeat of the Mexica modified land ownership, which converted from functional to private, including towns and lands that were handed over to the Crown, and some of them that were given to the conquerors as payment for their services in the conquest enterprise (Moreno 1987). Royal grants were received by conquerors for one or more lives, with the creation of the legal estate, the management of space, and land in the Indian towns (*pueblos de Indios*).

Pre-Hispanic agroecosystems underwent major changes after 1519, as seen in the slash and burn agroecosystem, where the burning was preceded by the opening of stripe guards, which prevented the dispersion of fire toward other plots or toward the forest. The rest given to the ground reached up to a bit over 20 years, a period that was shortened to the extent that the population had increased and agricultural land was reduced (González-Jácome 2011; Palerm 1955, 1972). The number of weeds decreased from three to two. Still in 1900, these plants were ripped of the cornfields, for which a plow was used and the labor of two children (Kaerger 1986: 244–245).

The cornfield in New Spain was an example of the diversity they could have. The information in written sources between the sixteenth and seventeenth centuries showed some of the variants that existed at the outset, as a result of the local environmental characteristics and the processes of adjustment. Idem with the plants sown in cornfields that changed from 1 year to the next.

The information of Bernardino de Sahagún (1830) was collected from 1540 to 1580, and that of Alonso Ponce in 1586. Both sources show that in the second half of the sixteenth century, fruit trees brought from the Old World soon were already widely distributed throughout New Spain and reached the Mayan area of Guatemala, forming an important part of agriculture, especially the homegarden and the cornfield.

After the conquest of the Central Highlands, the expansion, colonization, and conformation of New Spain Viceroyalty in the sixteenth, seventeenth, and eighteenth centuries, Mexican agroecosystems underwent several transformations. If other regions of New Spain, or of present-day Mexico were taken into account, the information would show local changes. In the nineteenth century, the cornfield suffered important changes derived in great part from diseases, epidemics, and the ensuing mortality. Family farms suffered the loss of some of their members and the lack of labor force for agricultural activities. Supply and demand, changes in land tenure, export markets, and the national market, all affected land possession and productive systems, including the cornfield and the homegarden (Linares, and Bye 2014: 151–178; Márquez-Morfin 1994).

The introduction of plants from Spain to New Spain and vice versa resulted from the process of interaction between the New and Old World, having impacts on economic activities, daily life, the food system, and most of all everything concerning medicinal herbs. The relevance of these herbs explained the interest of several doctors and botanists in collecting plants for Spanish herbalists. In Spain, herbariums and botanical gardens began in 1755. By royal order of Fernando VI, the Royal Botanic Garden of Madrid was created on the outskirts of the city – in Soto de Migas Calientes – on the banks of the Manzanares River. Morales and Blanco (2013: 55–68) explains that by 1881, the botanical garden moved to Paseo del Prado, in Madrid, where currently there are floral gardens, and the Spanish Botanical Garden descended from that of Joseph Quer (1762, T. 5: 516).

Information about some Mesoamerican plants is found in the writings of the botanist Joseph Quer, who was the first director of the Madrid Botanical Garden. In his book *Spanish Flora and the History of Plants that are Raised on the peninsula*, he wrote about plants like chia, and the uses the plant had in Mexico, saying that it was used to make oil. In addition, it had a high consumption as a very popular drink, especially in Holy Week: "[when] so much of this drink was sold, and thousands of reais were spent on it". The book tells how chia seeds were used to decorate clay vessels with animal figures, forming a green and leafy layer on the outside that lasts a long time, if the vessel contains water that feeds the seed.

From 1493 to 1623, the American plants that arrived in Spain had little impact on Spain's pharmacopoeia when the first news and descriptions about them appeared. The relevant works in this regard were that of Nicolás Monardes, considered to be the first treatise on American medical matters, which was published from 1565 to 1574, originally (López-Piñeiro and Pardo-Tomás 1994; López-Piñeiro and López Terrada 1997: 7).

Francisco Hernández's book: *The History of the Plants in New Spain* arose as a result of his expedition to Mexico (1571–1577), the book helped to the incorporation of the new American remedies into the Spanish pharmacopoeia, and therapy of the fifteenth and sixteenth centuries (López-Piñeiro and Pardo-Tomás 1994; López-Piñeiro and López Terrada 1997: 7). The influence of American plants and remedies on European botany and medical matters did not take effect until the early nineteenth century.

Joseph Quer himself (1762, T.5, pp. 75–76) described how the Franciscans took avocado (*Persea americana*) to Spain and sowed it in the homegarden of the convent of Bethlehem, where nuns of the same order resided. This convent was located on the outskirts of Valencia. Avocado trees were found from that same time in Catalonia, on the immediate shores of Mataró in a place called Calella. Several people interested in American plants planted avocados. Dr. Jaime Salvador took avocado to his homegarden in Seville, where Simon de Tobar cultivated the tree. It must be noted that the plants that arrived in Spain from New Spain were legumes, medicinal herbs, ornamental plants, condiments, and some fruit trees that adapted to the ecological and climatic conditions.

16.5 THE NINETEENTH-CENTURY CORNFIELD

In the first half of the nineteenth century, cornfield changes can be explained – at least partially – by the effects of the Independence war (1810–1821), including: mortality of the population; the destruction of bridges, roads, and natural resources; emphasis put on supporting products destined for export; and the entry of foreign capital from England, France, Germany, and the United States to the Independent Mexico during the mentioned century (González-Jácome 2004a, 2009b, 2010b, 2010e). With natural resources from Mexico, trade was reopened without Spanish intervention. The series of resources become important for these industrialized countries, which will modify and deforest forests, transforming Mexican landscapes and agricultural systems, including the cornfield, the homegarden, and the forest-cornfield, which becomes a coffee forest (González-Jácome 2009b: 1–15, 2010b: 521–554; Humboldt 1922, Book IV, Vol. 2: 256–264).

Among crops that become of great importance in the nineteenth century were coffee; vanilla; rubber; quinine tree bark (*Cinchona officinalis*) to cure malaria; chicle, which was extracted from sapota tree (*Manilkara zapota*); big grass (*zacatl*) (*Muhlenbergia macroura*), which was collected and sent to Hamburg for industrial use; Agave fibers to make nautical ropes; and fine woods like Lignumvitae (*Guaiacum Officinale*) and Elephant ear tree (*Enterolobium cyclocarpum*) to make ship parts. An example of a negative effect happened to deforestation of sapota tree (*Manilkara zapota*), which was used in the manufacture of railway sleepers and other industrial processes (González-Jácome 2010b: 256–264; Humboldt 1922, Book IV, Vol. 2: 256–264).

The emphasis on the export of the above products for the industries of the aforementioned countries changed the producer regions, having a destructive effect on the tropical forests of the south and southeast of the country. From the beginning, those articulated to the foreign trade with the aforesaid industrialized countries, mainly England, Germany, France, and the United States (González-Jácome 2004b, 2008b, 2009b, 2010b, 2010e).

The political independence of Mexico from Spain marked an end and a beginning. However, the beginning was of a new economic dependence of Mexico and then under the economic control of the above-industrialized countries (González-Jácome 2010b: 256–264; Humboldt 1922, Book IV, Vol. 2: 256–264). Those industrialized countries were expanding across all the continents looking for raw materials for their industries, cheap labor, and places to exploit minerals. The new agroecosystems continued to work together with the cornfield, whose products were destined for domestic consumption, whereas the maize was to feed the small-scale farmers and workers (Blanco-Rosas 2006).

A combination arose between an agroecosystem that was mainly destined for family food system (the cornfield) and the conventional monoculture agroecosystem (wheat, barley, or Agave), which existed since the sixteenth century but grew with its great expansion in the nineteenth century. Those conventional agroecosystems were related to commercial agriculture and directed to the industrialized European countries and the United States markets (sugar, tobacco, coffee, vanilla, rubber, quinine, and sisal) and industrial factories. This combination continues to date. However, the cornfield agroecosystem underwent changes to adapt or adjust to the unequal relationship with the capitalist economy that generated enormous inequality between small-scale farmers and those who participated in the export economy of agricultural products.[4]

With its agricultural products destined for export, conventional agriculture developed in Mexico, and articulated with industrialization, and mechanization in the English countryside and the United States agriculture. Its innovations come in the second half of the nineteenth century mainly through the estate owners and the government that supported the idea of modernizing the country and its economy (Nickel 1988).

The type of agriculture developed at that time was traditional, as the main way of life in the country remained traditional in its operation, having its own characteristics, some of which there is little known. An example would be the distribution of crops and plants in the cornfield and the impacts on the use and management of agricultural technology, especially with the dissemination of the plow in those years.

16.6 MARKETING AND CHANGE IN AGROECOSYSTEMS: AGAVE AND PULQUE

In the nineteenth century, with the expansion of the Agave as the main crop on the estate lands and its production directed to the domestic market, Agave ceased to have many of its ancient and multiple uses to be destined for the production of the fermented Agave mead (pulque). This was sold in stores (*pulquerías*, or boxes) and being a lucrative business, soon became controlled by Spaniards and viceregal authorities.

[4] Braudel 1984a, 1984b, 1984c; Chevalier 1963; Hobsbawn 1975, 1987; Humboldt 1922, Book IV, Vol. 2: 256–264; Palerm 1968; Wallerstein 1979 [1974], 2011.

The lands were bought, usurped, or leased from the indigenous owners, and Agave plantations replaced cornfields. Intensive Agave cultivation was associated with wheat and barley crops in the semi-terraces agroecosystem (*bancal*). In small zones of Agave rows, maize was cultivated near the small-scale farmers' houses (*calpanerías*) so that they could sow their cornfields and stock up on maize.

The semi-terrace agroecosystem was expanded in the Central Highlands in the eighteenth and nineteenth centuries, along with Agave estates (*Haciendas pulqueras*). It had its background in the pre-Hispanic terraces and the Agave rows (*metepantles*), and its main crops were: Agave for fermented mead (*pulque*), wheat, and barley were destined to the market. Little by little, the Agave estate owners who characterized the region organized the *pulque* trade, acquiring economic and political importance in the country. At the same time that the Mayors (*Corregidores*), great merchants, miners, and Jesuits were hoarding land to devote it to intensive Agave cultivation (Rendón-Garcini 1990, 1994: 209–243).

Among the indigenous people who cultivated Agave to produce *pulque* were those from Hidalgo, Puebla, the State of Mexico, and Tlaxcala. They were the small-scale farmers of the Llanos de Apam, the Zahuapan and Atoyac rivers basin, the slopes of the Tlaxcala Block, the Tlaxco mountain range, and the lowlands slopes of the Matlalcueye volcano. Those small-scale farmers cultivated the century plants in the Agave rows (*metepantles*) on a small scale (González-Jácome 2016).

Among the small-scale producers, Agave became part of the cornfield structure, occupying the earth borders on the terraces and farmland. Small-scale farmers had 50–100 Agaves at different stages of growth (Tylor 1979). The Agaves grew and matured at different times, which allowed the continuous supply of mead and *pulque*, plus other elements such as red and white maguey worms, the cuticle of leaves used in the wrapping of steamed foods, such as the lamb meat mixiote or the fish tamale. The small producers sold *pulque*, along with other artisan products that they made in their villages and sold directly to local consumers and to the local tianguis through barter or retail (González-Jácome 2016: 1–51).

This activity lasted until the twentieth century, ending as a result of numerous movements of popular discontent caused by the imposition of taxes on the sale of *pulque*. In consequence a revolutionary movement against taxes arise in the region from 1890 to 1910 (Rendón-Garcini 1995, 1996). The cornfield continued to be the main agroecosystem that supplied maize and other plants to the indigenous and mestizo farmers' kitchens. Agave became part of the edges of the plots (*metepantles*) both in the plains and on the terraces, where they were combined with other trees and ruderal vegetation (Figures 16.1 and 16.2).

From 1800 to 1899, the cornfield had other characteristics: (1) in the Basin of Mexico was a cultivation done in chinampas, with three crops a year, along with maize there were cultivated:

FIGURE 16.1 Cuaxonacayo, Tlaxcala Block: Agave Rows (*metepantle*) in Ravine La Carbonera.

FIGURE 16.2 Agave Rows (*metepantles*), Tlaxcala Bloch.

cabbage, cauliflower, carrot, zucchini, and tolerated edible wild green weeds that included: small green tomatoes (*tomatillo*), edible green seasonal herbs (*quelites*), aromatic herbs, and medicinal plants (Linares and Bye 2015, 2016: 9–12); (2) in the valley of Toluca (Alto Lerma), the raised fields where two crops a year were produced; (3) in the Tuxtlas, south of Veracruz, white, yellow, blue, red, and pink round grain maize and beans were grown, and in temporary lands, bean or fava bean, hard-shell pumpkin; (4) in lands with residual humidity, two crops a year were produced: chayote, sweet potato, banana, small green tomato, yucca, chives, and squash.

16.7 THE CORNFIELD: FROM THE END OF THE NINETEENTH TO THE FIRST QUARTER OF THE TWENTY-FIRST CENTURY

In 1900, the German agronomist Karl Kaerger (1986 [1901]: 242, 244, 248) described the state of Mexican agriculture in a report with a section on the cornfield in various parts of the country. He described: (1) diverse forms of maize were grown in the different regions of the country; (2) the plow was used on the slopes of the highlands, and planter cane (*coa*) was used in the tropical regions; (3) maize cultivation was carried out on a reduced scale almost always; (4) in the Highlands, maize without irrigation was grown in summer; (5) in the states of Puebla and Mexico, the preparation of the land was carried out in January, although it could be done in November or December; (6) if the soil was lumpy from a previous maize crop, it was broken up using a plow and beam harrow (it is plowed twice, and the soil is raked with the harrow), and moisture is conserved too; (7) the groove orientation depended on the amount of soil swept away by the summer downpours.

Kaerger (1986 [1901]: 242–247) described the stages of cultivation: (1) Sowing was done in March or April when small showers interrupted the dry season and made simple furrows. In several parts of the country, planting was done in June, at the beginning of the rainy season. (2) In each groove, holes were opened with the planter cane, at hand's distance. (3) Sometimes, the first row of furrows was crossed by a second. (4) The grooves had to cut into the moist soil, and 3–4 maize grains were placed in each hole. (5) The separation between furrows depended on whether the soil was good or bad: a good soil needed 1.25 of a distance between grooves, and 9 kg of seed/ha; a bad soil required 1.5 m of space between grooves and 11.5 kg of seed/ha.[5] (6) After sowing, the furrow was smoothed over with a beam to conserve moisture. (7) Three weeds were performed on the cornfield.

[5] A sowing bushel was equal to 75 liters of maize it comprises 56,180 m^2. This calculated in 1-ha is equivalent to 13.4 liters (100 liters of maize), with an average weight of 69–75 kilos.

The three weeds applied to the cornfield: (1) a month after planting, a plow passes between furrows piling soil on both sides of them; (2) one child removes the soil that fell on the maize plants with a trowel, and another child pluck herbs removing all of them with his hand; (3) in the second month of sowing, the second weeding was done, using a plow with two dumps, to which the elbow is added, it aims to pile up soil next to the plants. Meanwhile, another boy weeds the ground and raised the fallen maize plants. When a lot of soil was piled up on each maize plant, the continuity of the furrows was lost, and the field appears formed by small mounds of land. (4) From November to January, maize was harvested.

There were two methods of maize harvesting. In southern Mexico, the first one was carried out usually when the maize kernels were firm, and the maize plants were bent above the last ear of the upper end (stripping or sacked) to prevent moisture from entering the ear. Sometimes, the ears were cut from the plant and shelled to be taken to the farmer's house and ensiled. When the ears were ripe, they were shucked using a tramper and placed in baskets for storage. In the second method, the whole stems with maize ears and all were cut with a sickle and piled together forming cones (mogotes) located in the field, where they remain until the first frost fell.

In December, the maize ears are cut off and shelled. On February 2 (Virgin Candelaria Day), maize grains, pumpkin seeds, and beans were taken to church for blessing before the beginning of agricultural calendar. On February 2, the ceremonies carried out in the churches of the different towns, included the consumption of tamales. That ritual is done in the countryside, and in the cities whose cultural festivities are part of the culinary culture, although the urban population has already forgotten its origins, and today is a popular holiday (Figures 16.3 and 16.4).

The local differences of maize harvesting that Kaerger (1986 [1901]: 247–248) found were: (1) in Tabasco, to protect the cobs from weevils when cobs are almost mature, stalks are bent and cobs are

FIGURE 16.3 Maize Ear Picker (*Pixcador*) Made from a Turkey Bone. Valladolid Museum, Yucatan.

FIGURE 16.4 Calpulalpan, Tlaxcala: Cones (*Almeales*) Made with Dry Maize Canes.

left hanging from the stem for a month; (2) in Oaxaca, unshucked cobs are yanked off the plant and piled up for 1–3 months, before being husked and shelled. Husks (*totomoxtle*) are used as a fodder for cattle. The shelling can be manual or mechanical, with shelling machines. Rachis is ground and used as food for draft animals and dairy cattle. The grain is shaken in perforated cowhides, which are fastened to two bars, and the obtained pellets Kaerger (1986 [1901]: 248) are given as forage to cattle (Figures 16.5, 16.6, and 16.7).

Kaerger (1986 [1901]: 248) reported that maize yields were diverse, depending on several factors: type of soil (good or bad), type of year (bad or dry), the distance of planting between furrows, and if it was irrigated land. A very low yield might be obtained in bad or dry years with only 15–20 grains obtained per one that was sowed. In poor soils, a yield goes from 80 to 100 grains for one sowed. In very good and irrigated soils, the yield reached 800 grains for one grain sown. The production costs in the state of Puebla were from 1.40 to 2.50 pesos/metric quintal, without considering administration expenses, but in Celaya they were 50 cents/metric quintal, with an average harvest of 80 metric quintals/bushel. From 1898 to 1899, the country exported 3,560 tons of maize. In 1900, Mexico imported 7,042 tons of maize Kaerger (1986 [1901]: 248).

FIGURE 16.5 Mayo River Basin, Sonora: Straw Cones (*Almeales*).

FIGURE 16.6 Mayo River Basin, Sonora: Straw Cones (*Almeales*).

16.8 CONSIDERATIONS AND CONCLUSIONS

The historical and contemporary study of traditional agriculture is relevant for its contributions to the knowledge of resilient and sustainable agroecosystems, like the cornfield in Mexico and its regions. Some answers to the current problems were found in the results of studies on the cornfield agroecosystem in Mexico, its characteristics, development, and changes through the years. There is a great diversity of types of cornfields owing to their ability to adapt or adjust to ecological, social, cultural, historical, demographic, physiogeographic, altitudinal, environmental, and climatic situations. Family structure, social organization, the land inheritance system, and particular and regional economic conditions explain their resilience as well.

The long history and evolution of traditional agroecosystems in Mexico show their ability to be sustainable, resilient, and capable of providing food to the small-scale farmers who have invented, developed, and adapted them to various social and cultural situations and conditions. Similarly, these agroecosystems have demonstrated their ability to incorporate and interrelate with other natural systems, and even those that have been created by anthropomorphic action.

By the sixteenth and seventeenth centuries, the traditional Mexican cornfield agroecosystem had been formed. In the Mexican central Highlands, cornfield cultivation and its associations constituted a polyculture, where trees were mixed with cultivated and tolerated weeds and wild plants. Between the last century and the present one, the number of crops has reduced, with the consequential decrease in the biological and agrodiversity of the agroecosystem (Blanco-Rosas 2006; Cahuich 2012). From the remote past to date, the cornfield has been a complex agroecosystem, which has inside or on the banks and edges, herbs, mushrooms, shrubs, vines, and trees. The trees may or may not be fruit trees but are used to obtain firewood or fibers, to be windbreak curtains, to modify the light on the plot, or to have some other ecological utility in the agroecosystem.

RECOMMENDED READINGS

Yves Carrière, Y.; Z. Brown; S. Aglasan; P. Dutilleul; M. Carroll; G. Head; B.E. Tabashnik; and P. Sogaard Jorgensen. Crop rotation mitigates impacts of corn rootworm resistance to transgenic by corn. *Proceedings of the National Academy of Sciences of the United States of America*, July 20, 2020.

A group of 9 researchers conducted a maize monoculture study from 2011 to 2016 in 25 districts of Illinois, Iowa, and Minnesota. Western maize had an insect pest, and it developed resistance to genetically modified maize that produces insecticidal proteins derived from the *Bacillus thuringiensis* (Bt) bacteria. In maize monoculture with severe rootworm damage, it was reduced by rotating the maize with other crops, and by not planting the same type of Bt maize year after year in the same field.

The study results supported EPA recommendations to reduce the negative impacts of rootworm resistance to Bt maize by rotating it with other crops, and by diversifying the type of planted Bt maize. Transgenic crops that produce the insecticidal protein *Bacillus thuringiensis* (Bt) can suppress pests and reduce insecticide sprays; however, their effectiveness decreases when pests developed resistance. Farmers plant shelters with plants that are not Bt hosts to delay resistance to pests, but this practice has not been sufficient against the western maize rootworm.

The frequency of cornfields with problems decreased 92% from 2014 to 2016, compared from 2011 to 2013, when those fields were associated negatively with the rotation of maize with soybeans. The effectiveness of maize rotation in reducing Bt resistance problems did not change significantly between districts whose crops had, and have, still rotationally resistant rootworm populations. The results highlight the central role that crop rotation had in reducing the impacts of the resistance of *V. virgifera* to Bt maize.

Rotation is a very old agricultural practice that has been applied in various parts of the world, including Mesoamerica. In slash and burn agriculture, the land that was not under cultivation (resting), covering with natural vegetation during some years was another form of rotation, where crops and land were rotated with natural vegetation to regain soil fertility. Rotation became a common practice in England in the mid-seventeenth century, rotating cereal crops with turnips and clover, and making cattle selective breeding. This second agricultural revolution "[...] transformed English agriculture from a peasant subsistence economy to a thriving capitalist agricultural system, capable of feeding the teeming millions in the new industrial cities" (Overton 2011-02-17).

In the recent article by Yves Carrière, Y.; Z. Brown; S. Aglasan, et al. (2020), the information provided by these authors showed that crop rotation helps control the appearance of the worm that attacks maize roots. This research is important, because the maize in question has been genetically modified, and yet pests also affected maize when they gained resistance to rootworms population.

Merrill, W.L.; R.J. Hard; J.B. Mabry; G.J. Fritz, K.R. Adams; J.R. Roney; and A.C. McWilliams. 2009. The diffusion of maize to the southwestern United States and its impact. *Proceedings of the National Academy of Sciences of the United States of America* 106(50): 21019–21026.

Knowledge of the beginnings of agriculture in the southwestern United States had been transformed by recent discoveries establishing the presence of maize at 2100 cal. BC, documenting the processes by which maize was integrated into local food-seeking economies. The authors reviewed archaeological, palaeoecological, linguistic, and genetic data to evaluate the hypothesis that proto-uto-aztec (PUA) farmers migrating from Mesoamerica introduced maize agriculture to the region. Available data indicates the Great Basin as the homeland of the PUA, the breakdown of this linguistic community in the northern and southern divisions in 6900 cal. BC, and the dispersal of maize agriculture from Mesoamerica to the southwestern United States through the spread of one group to another in a southern Uto-Aztec linguistic continuum.

Multiple radiocarbon AMS dates established that maize (*Zea mays* spp. Mays) was introduced to the southwestern United States no later than 2100 cal. AC (calibrated calendar years AC). The arrival of maize in this region was preceded by a long period during which its ancestor, the wild teosinte (*Zea mays* spp. Var parviglumis), native to southwestern Mexico, was transformed through human intervention into a species productive in grain cultivation. The oldest known maize macrofossils, date from 4280 cal. BC, were recovered in the Guilá Naquitz cave, in Oaxaca. The authors proposed that the initial dispersal of maize from Mesoamerica to the southwestern United States began in 4300 cal. BC.

The earliest evidence of maize in the southwestern United States came from three open sites and two rock shelters in Arizona and western New Mexico, at an altitude from 700 to 2,200 masl. These sites have great diversity in their configurations. Clearwater and Las Capas are located in the middle of and along the St. Cruz River in the upper basin of the Sonora desert. The MacEuen Cave is in a deep canyon east of the Gila River Valley in the mountainous transition zone between the southern basin and the Colorado Plateau. The sites are (1) Old Corn, Colorado Highlands, Western New Mexico, from 2460 to 2060 BC; (2) MacEuen Cave, Gila Mountains, Eastern Arizona, from 2270 to 1940 BC; (3) Clearwater, Tucson Basin, Southern Arizona, from 2200 to 1960 BC; (4) The Layers, Colorado Plateau, Southern Arizona, from 2200 to 1940 BC; and (5) Three Fir Shelter, Colorado Plateau, Northeast Arizona, from 2470 to 1540 BC.

This article is based on the confirmation of the antiquity of maize arrival to the cultural area of Greater Chichimeca and the Southwest of the United States of America (Huckell 2006). This huge biogeographic region, which in ecological terms is in fact one, housed groups of horticulturists, growers, farmers, and numerous bands of semi-nomadic hunter-gatherers (Huckell 1996). The understanding of this frontier cultural area, which from the geohistorical point of view is part of the current border between both countries, is currently of geopolitical importance.

17 Some Types of Cornfield in Contemporary Mexico

17.1 INTRODUCTION

These cases are representative of other regions in the country, and each one described and explained the reasons for their adaptations and characteristics. Although they are insufficient to give an overview of the whole country, they have the advantage of having useful and detailed information on their structure and interrelationships with various ethnobotanical, social, economic, and cultural aspects of Mexican agriculture.

This section of the book is dedicated to describing eight cases of the Mexican cornfield (*milpa*) agroecosystem, with the main characteristics of each type and their combinations. The cornfield agroecosystem is very diverse, because it had and has the capacity to adapt to local and regional environmental, physiogeographic, historical, and climatological conditions. The cases in this section are: (1) the interrelations between the cornfield, the homegarden, and the forest in the Totonacapan region in Veracruz; (2) the March maize (*marceño*) agroecosystem in wetlands of Tabasco; (3) the combination of maize cultivation intermixed with a fruit orchard in Central Veracruz; (4) in Tlaxcala, the combination among the cornfield with irrigation and the cornfield depending on seasonal rains is described; (5) in the mountainous areas of Guerrero, the tlacolol or tlacolole agroecosystem and its relations with environment and religious ceremonies were studied (Serafino 2015: 207–227); (6) the tlacolol agroecosystem that was managed on the slopes of the hills in Guanajuato; (7) the Zoque-Popoluca cornfield in tropical southern Veracruz was described; and (8) the polyculture cornfield in the Sierra de Juárez, in Oaxaca that shows how the maize crop was adapted to altitudinal conditions

These cases are representative of cornfields in other regions in the country. Each one describes and explains the reasons for their adaptations and characteristics. They have the advantage of having useful and detailed information on the agroecosytem structure and its interrelationships with various ethnobotanical, social, economic, and cultural aspects of t the society in which the cornfield and the related food system es functioning.

These cases are representative of the cornfield in various regions of the country. Each one describes and explains the reasons for their historical, social and cultural adaptations, as well as their main characteristics. Although they are insufficient to give a general vision of the cornfield in Mexico, they provide useful and detailed information on the structure and functions of the agroecosystem, its interrelationships with some ecological, ethnobotanical, social, economic and cultural aspects of the society in which the cornfield is found, and the related food system.

17.2 BACKGROUND

The current Mexican cornfield presents changes and adjustments due to external conditions that are related to its local and regional history; its interrelations with urban society; and the economic, demographic, political, and cultural events that the agroecosystem and the small-scale farmers had suffered throughout time. The types of cornfield from the pre-Hispanic cornfield to the current agroecosystem have a common characteristic: their diversity. This results from environmental, climatic, social, and cultural adjustments based on traditional ecological knowledge (TEK) and traditional agricultural knowledge (TAK) accumulated during thousands of years of experience and practice. These can be explained initially through the different historical and contemporary types

DOI: 10.1201/9781003198833-17

of cornfield in Mexico. Doing a complete study of the topic is outside the limits of this book, as its proposal is to display important points, presenting some cases.

Taking into account the cornfield biogeographic characteristics broadly, it is possible to distinguish at least four types of contemporary cornfield in Mexico: (1) the one that arose in tropical and subtropical regions of Mexico, that is usually a polyculture; (2) the one developed in the semi-desert regions of the country, with differences in diversity according to the specific areas where it is located; (3) the one in the Mexican Central Highlands, with great environmental and climatic diversity, located in places where Mesoamerican civilization and irrigated agriculture emerged; and (4) the Mayan ones, with their different subtypes, which deserve specific studies to be better understood.

Originally, cornfield types shared the basic characteristics of the system, where they were located and their diversity crops adapted to the specific environmental conditions in which the agroecosystem arose and developed. These included the environmental characteristics; the availability of water and soils near the cultivated lands; the social, cultural, and demographic specificities of human populations sowing and working in cornfields; the available family labor force; and the relationship between cornfield, food system, the means of consumption, and the production for the market (Figures 17.1–17.6).

FIGURE 17.1 Chiapas Highlands: Cornfield and Secondary Forest.

FIGURE 17.2 Los Tuxtlas, Soteapan and Small-Scale Farmer.

FIGURE 17.3 Tlaxcala, Tepeyanco: Irrigated Farmland with Maize and *Chenopodium berlanderi (Huauzontle)*.

FIGURE 17.4 Tlaxcala, Matlalcueye Lowlands: House, Cornfield, and Edge of Land (*Metepantle)*.

FIGURE 17.5 Chiapas, Lacandon Region: Cornfield.

FIGURE 17.6 Tlaxcala, Cornfield Zone between Xiloxoxtla and Tepeyanco.

The human action of taking plants from one place to another increases diversity. The selection of plants transported by Men is based on taste, smell, and the size of the edible part. Also, of each plant, its relocation depended upon its specific characteristics such as the time for maturation, its possibilities of growth in a different site, the available space there for its sowing, and growth, its possibilities for exchange or commercialization, integration with the new agroecosystem, and the population's taste in food (Ayora 2010: 221–249).

17.3 THE ITINERANT CORNFIELD IN THE TOTONACAPAN

The physiogeographic, environmental, and climatological characteristics, defined the types of cornfields in Mexico. For example, their adjustments or adaptations to the altitude, gradient, and surface of the land; the annual seasonal rain; the availability of water; and the amount and quality of the land and soils available for agricultural activity, as occurs in the slash and burn cornfield agroecosystem. In addition to demographic issues, such as the total population, its population densities, and migration processes occurring in a region or ecological zone. The itinerant cornfield is the oldest type, and it developed in tropical areas. It has intermediate and long-term fallow periods that can take up to several years to recover soil fertility from the effects of the slash and burn process.

Slash and burn (r-t-q) agroecosystem management is applied in a rotating system of selected land for planting. In conditions of abundance of land available for cultivation, the fallow period worked in the medium and long term according to the recovery of soil fertility, its specific characteristics, and access to water (Palerm 1967: 26–52, 1972; Palerm and Wolf 1972), furthermore to the size and density of the human populations that inhabit the town and the region. To function without eroding the soil, this type of cornfield requires dispersed, and low-density human populations/ km^2. Generally, the itinerant cornfield is found in the tropical and subtropical areas of Mexico, of which there is good information on cases, corresponding to different times and spaces, thus allowing researchers to consider all of the relevant characteristics and processes of change (Hernández X. and Levy 1995).

According to INEGI information, in 2009, the Municipality of Papantla had five topoforms: hills with plains (48%), typical hill (25%), low mountains (19%), coastal plains (7%), and typical valleys (1%). The temperature ranged from 24 °C to 26 °C throughout the year, and rainfall ranged from 1,100 to 1,600 mm annually. The warm climate is sub-humid, with summer rains, 32% being more humid; and the warm humid climate with abundant rains in summer 28%. The dominant soils are regosol (67%), vertisol (15%), phaeozem (10%), and cambisol (5%).

The municipality belongs to the Tuxpan-Nautla hydrological region (100%), which includes the basins of the rivers of Tecolutla (58%), Cazones (36%), Nautla, and other tributaries (6%). Perennial water streams are Tecolutla River (53%), Estuary and Tumilco Lagoon (30%), Solteros Stream (6%), Joloapan River (5%), Cazones River (5%), and S. Marcos River. The intermittent streams are Troncones, La Antigua, Cuatro Arroyos, El Cepillo, Cañas, San Miguel, La Guásima, and Indio. In 2009, land use was agriculture (57%), grassland (31%), forest (9%), and urban area (2%). The geographical information referred to traditional agriculture as "continuous manual agriculture", which occupied 41% of the agricultural land in the municipality. This category on the record seems to correspond to the cornfield and the cultivated vanilla agroecosystem (INEGI, Prontuario, Papantla, 2009a: 1–2).

Ángel Palerm (1972a: 12–13) described the slash and burn agroecosystem in the Tajín since as it was before in the Porfirio Díaz long governmental period (Porfiriato), where the activities that were required to do it were the following: (1) cut down a section of the forest; (2) let the vegetation dry; (3) burn it; (4) sow with a planter cane (*macana*, *coa*); weed periodically, using the planter cane; (5) let the ground be when the soil is depleted and the yield decreases significantly; (6) abandon the land, giving time for the soil fallow and forest to regenerate; (7) cut down another section of the forest to restart the agricultural cycle. The time for the land to rest varies, depending on the recovery of soil fertility and the pressures that the populations exert on the land (e.g., population density, spatial distance between homes, and planting grounds, and the number and availability of family members to work in agriculture).

In the Porfiriato, the Totonacs lived in dispersed settlements. An important group settled in the Tajín, organized into extended, and patrilocal families. The forest was communally owned. Totonac farmers completed their food system by hunting in the forest ("high mountain") where there were deer and wild boar. Domestic animals meat for consumption (pork, turkey, and chickens) provided enough animal protein within their diet. Furthermore, hunting provided Totonac with some monetary resources. Thus, the meat of wild and domestic fauna complemented agricultural activity.

At the end of the nineteenth century (the *Porfiriato*), due to Colonization Laws, the Totonac territory was measured and divided into plots that were about 30 ha each. The legal estate was created of half a plot, and the other half was divided into lots, drawn up in a reticulated fashion, with "hypothetical" streets. The plots were put up for sale, with no rights granted to former occupants. In the future town center, several public buildings were installed, such as municipal agency, school, and chapel.

Cornfields were located next to the settlers' houses, and vanilla cultivation increased. Despite the struggle and defeat of the rebel Totonacs in this situation, the sale of plots continued (Kelly and Palerm 1952: 99–149; Palerm 1972a: 14–15). Political authority became dependent on the Papantla authorities.

In addition to population growth, with the new situation, the cultivation of larger land areas increased, the soil resting time was reduced, and as it was not enough, the "High Mountain" disappeared almost completely, as did the wildlife there. Consequently, "Low Mountain" (*acahual* and *zacatal*) replaced the high forest. In the absence of resources, the felling of the forest trees was increased, to sell the wood for constructions and for fuel. The cornfield saw its itinerant rotation reduced and was located on the land near the houses, that by then was privately owned. This reduced the years that were destined for soil fallow, previously.

In 1895, Totonacs from Papantla sold some of their homegarden products in towns in the same municipality. Among them were small avocado (*Persea gratissima*. Gaert var.) and large avocado (*Persea gratissima*. Gaert var.). Avocado was sold from May to September, depending on the location of the towns. It was produced and distributed in Chicualoque, Gutiérrez Zamora, Papantla, S. Lorenzo Cuapiaxtla, and Zozocolco de Hidalgo. Its price varied and avocado could be sold at 6 cents/dozen, or from 75 cents/100 pieces. In the towns in the Papantla municipality, Totonacs sold two varieties of soursop (*Anona glabra*. Lin.; *A. custara*) in the markets of Cuautitlan, Chicualoque,

Espinal, Gutiérrez Zamora, Papantla, and Zozocolco de Hidalgo. The annual full amount of the sales could reach 25 pesos (Carpología mexicana 1895: 17–18, 30–32).

In 1952, Ángel Palerm (1972a: 15–17) recorded information about the cornfield that indicated its yield in two annual maize crops; however, this occurred in the first 2 years, and then the depletion of the soil was excessive. Some small-scale farmers kept the cornfields in cultivation during 4 years, and others up to seven, but managed to maintain production only for 3 years, generally. With this, a new crop cycle was developed, which included maize replacement with vanilla, a creeper plant that requires certain types of trees to grow on (Lascurain et al. 2010). Planting trees to serve as a support for vanilla favored the partial regeneration of the forest. Vanilla produced from 10 to 12 years, and then the soil fallows for a similar period so that the soil regenerated completely, and the vanilla cycle could resume.

In 1952, each Totonac family required 1½ ha for the cornfield. A. Palerm (1972a: 16) calculated that due to roaming, for every 1½ ha of cornfield sown, 12 ha of land susceptible to cultivation were needed. Maize produced 100 units for each seed sown and gave two annual crops. In the cornfield, plants cultivated for family consumption were maize, plant bean and creeper bean, squash, chili, cassava, sweet potato or yam, potatoes, arrowroot (*Maranta*), sesame (planted on the banks of the cornfield), cacao, and some coffee. These plants were not grown to sell. Besides tobacco, pineapple, cotton, flowers (cartridge or gannet), assorted vegetables like green peas, green beans, cabbage, onion, chives, and garlic were grown to sell.

In addition, wild plants were collected, such as small green tomato (*tomatillo*), amaranth, and a vine called in Totonac *cahuayote* (*Gonolobus niger* [Cav.] R.Br. ex Schult) (UNIBIO; Biological Collections 2008: 12–03), *talayote* (*Inga spuria* Humb. & Bonpl. Ex Willd.), *chupipe*, or *uy*, whose edible fruit Totonac people liked. Sometimes cahuayote was sown in the cornfield or in the homegarden close to trees on which the vine climbed. Its crushed root was used as a foaming agent in the preparation of a beverage known as *popo* (smoke, smoky), which was a mixture of water, ground cacao beans, rice, cinnamon, and sugar. *Popo* was drunk in gourds (*jícaras*) made of *Crescentia cujete* and *C. alata*. Today, *popo* is still drunk, but it is served in disposable cups, accompanied with tamalehili atole and sour atole both prepared from maize when young corncobs are available.

From June to August in the high mountain (monte alto), the Totonacs collected Mexican wild hog plum (*Spondias mombin* L.) (*Jobo, ciruela de montaña*), which was also sown in the cornfield and in live fences, where it was reproduced by cuttings, which were cut when the tree was leafless and about to bloom. The hog plum fruits collected in summer were sold on a small scale in Papantla (Kelly and Palerm 1952: 141; Lascurain et al. 2010: 45). The cultivated Mexican hog plum (*Spondias purpurea* L.) was planted in the plots and commercial plantations. In addition, Mexican hog plum was used as forage and a living fence. The extract of its leaves and its bark were used as febrifuge. The infusion of its leaves was used in the washing of wounds, or for inflammations or burns. Its fruit syrup had astringent properties, and its ashes were used in soap making (Lascurain et al. 2010: 65).

The Tajín Totonacs had both, cultivated and wild fruit trees, which were grown in the cornfield and the homegarden next to the houses. In the cornfield, weeds provided food for family feeding. Among the fruit trees were papaya (cultivated in the cornfield and wild in the "high mountain"); several varieties of sapota, like black sapota tree (*zapote negro*), sapodilla tree (*chicozapote*), *mante*, warmth, *Domingo*, hair, and mammee apple; green or black skin avocados, long (*Kukakai*), tropical *pahua* avocado (*Persea schiedeana*), and fragrant avocado; wild and cultivated guavas; custard apple (seeded by birds); *Couepia dodecandra* [DC] Hemsl (*gurupillo de Castilla*); citrus fruits (orange, lemon, and lime); mango; different types of bananas[1]; *Jatropha curcas* L ("pine nut") that was grown for its oil (Kelly and Palerm 1952: 140–141). In addition, they have gourds (*pokgékiwi, pokdge, sacual*), which were cultivated in the courtyard, both by cutting and by seed. In the 1950s, wild gourds (*Akgpasas, Akgpili*) disappeared from the natural vegetation zones in the region.

[1] In Totonac: Banana (*Sekna*), Castilian banana (*Kanásekgna*), purple banana (*Tsutsokgosekna*), and macho banana (*Kawin*).

The trees had to be protected from the onslaught of leaf cutter ants (*Atta cephalotes*), a problem that was fought by small-scale farmers with little success. For this, maize hominy water (*nejayote*, *nixtamal* water) mixed with lime was applied to the entrances to the ants' nests. Fire was used to combat these ants, with torches made of thin branches, and placed on top of the nests. All family members participated in this activity to protect fruit trees and legumes, although according to Kelly and Palerm (1952), the results were not very satisfactory.

The Totonacs cultivated red grass (*Imperata contracta* (HBK), *paral* grass (*Panicum purpurascens* Raddi), and guinea grass (*Panicum maximum* Jacq.) for cattle fodder and fine grass (*Cynodon dactylon* [L.] Pers.) whose seeds were to feed their poultry (Kelly and Palerm 1952: 145). In his book on the biocultural indigenous regions of Mexico, Eckard Böege (2008: 167) gives the data of 97,810 ha that were of induced and cultivated grassland of the Totonacs, which meant that 31.15% of the total area of these indigenous territory (313,948 ha) was lands mainly dedicated to livestock.

In the 1950s, each complete cycle of vanilla required 24 years (12 cultivation, and 12 for fallow); In addition, cultivated vanilla (*Vanilla planifolia* Andrews) cultivated was sold in Papantla, providing small-scale Totonac farmers with a significant amount of monetary resources (Del Amo, Ortiz-Espejel and Andrade 2009: 5–28; González-Jácome 2009b). Harvest vanilla (*Vanilla pompona* Schiede) grew in the tropical deciduous forest from 500 to 1,000 masl but was of less commercial importance than the cultivated one (Lascurain et al. 2010: 125). The agricultural system worked successfully if the land was not divided between more than two children and allowed the regeneration of soil and forest. Therefore, the population increase had to be controlled, but in practice it did not happen.

The cornfield in tropical and subtropical regions of Mexico had specific local characteristics, which depended to a large extent on altitude, geographical gradient, relief, environmental and climatic conditions, the type of soils, flora, and fauna existing in the area. Its combination with local demographic, social, and cultural elements must be included. The itinerant cornfield worked successfully only with dispersed populations, low population densities, and a communal land ownership regime. At the beginning of the year 2000, the itinerant slash and burn cornfield was still found in the Mayan lowlands of Quintana Roo, Chiapas, and in some regions of Campeche.

In fact, the productivity of the soil in lands with tropical forest and slash and burn has been a hot topic for the researchers, whose discussion was based on the real possibility of the cornfield to obtain two annual crops, and a yield per unit of sustainable sown seed (Hernández X. 1982; Hernández X. et al. 1987: 338–340; Hernández X. and Levy 1995).

In general, in other parts of the tropics, the cornfield worked together with the homegarden and the pastures that are destined for livestock. Thus they form a fundamental part of the economy of the small and medium-scale farmers (Blanco-Rosas 2006; Cahuich 2012; Mariaca-Méndez 2015). However, livestock activity contributed to the deforestation of tropical and subtropical forests in an important way. Eckard Böege (2008: 111) gave the following figures on the existing hectares of deciduous forest in the Totonacapan: primary forest 1,695 ha; secondary forest with trees 2,112 ha; secondary shrub 1,508 ha, with no hectare recorded for the herbaceous secondary forest.

In 2010, SEMARNAT registered forest production within the Sierra Totonac Forest Region stating that in 2001, the wooded area of the region was 14 thousand ha. There were 34 registered forest plantations that had a granted land area (*ejido*) of 199 ha in the municipalities of Coahuiylán, Coatzintla, Coyutla, Papantla, Poza Rica, and Tihuatlan. The increase in forestry and livestock activity had affected the Totonac region agriculture, especially in the cornfield as a basic agroecosystem related to small-scale farmer subsistence. A study of the interrelation of these activities is needed, with demographic processes such as internal migration and emigration.

In his study of the Totonac population of Mecatlán, a town located in the mountains, Moctezuma-Pérez (2008) observed that the forest was managed in combination with coffee, but this management had problems when coffee prices fell, and IMECAFE disappeared. That institution had supported the natives by keeping the prices of the aromatic grain they produced and sold stable. With the disappearance of IMECAFE, as a prelude to the introduction into the country of the neoliberal economy, seasonal migration and emigration originated or increased. Seasonal migrations lead Totonac farmers to work in trades like masonry in cities and industrial centers where construction required this type of seasonal labor. The cornfield exists in the region, but the changes it has undergone in the last two decades must be studied in greater detail.

17.4 MARSH AGRICULTURE: THE CORNFIELD AND MARCH MAIZE (*MARCEÑO*) IN THE WETLANDS

The cornfield in the tropics continues to be of great importance in family food system and regional food production. In fact, it includes a wide variety of traditional agricultural systems in wetlands, which integrate cultigens, some medicinal plants, some other trees, and plants that cover the basic food system necessities and needs of farmers (Orozco-Segovia 1999: 111–122; Siemens 1980, 1989a; West et al. 1969). Among the various swamp agriculture systems, the cornfield, which is characterized by the cultivation of March Maize (*Marceño*), along with other plants, is found in waterlogged areas. *Thalia* leaves are a part of the *popal* ecosystem and are cut with the cane planter (*macana*). A vegetable bed is formed with them on which a maize variety with a short maturation period (*Mején*) is planted, or a 40-day maize variety (*cuarentano*) is sown (Orozco-Segovia 1999: 111–122; Gliessman 1980: 93–101; Mariaca-Méndez 1992: 69–76).

This type of cornfield adapted to the wetlands of the tropical and subtropical regions of Mexico, where small-scale farmers grow maize (*marceño*) planted in flooded areas during March. After slashing vegetation and burning it, cultivation begins when the swamp's water is at its lowest level. After burning, *Thalia* has dry upper leaves, but the part of the plant that is in the ground is green and retains moisture. A planter cane is used to open holes in the soil putting 4–6 maize kernels in each hole, and the seeds are not covered. When the rains last longer than what the farmers consider as "normal", or water is flooding the *Thalia* leaves bed, March maize harvest may be carried out from canoes or boats.

In Tabasco, the cultivation of March Maize agroecosystem, in waterlogged areas, began with the slash of the vegetation in the marsh ecosystem (*popal*). The vegetation is of hydrophilic broadleaf plants and herbaceous helophytes, jutting up one to three meters above the surface of the waters. After raking the leaves from these plants, a vegetable bed is formed, where maize and other associated species (pumpkin and cucumber) are sown. The March Maize (*marceño*) is adapted to these waterlogged conditions, where the systems maintain a high water level or seasonally for several months. The maize varieties are local and short-cycled known as *cuarentano*, or *mején*, that have a maturity period of 2½–3 months. Vegetation grows along with maize, serving as green manure (Guerrero-Herrera 2015: 51–52; Gliessman 1980: 93–101; Orozco-Segovia 1999: 118; Mariaca-Méndez 1992: 69–76).

From 1990 to 1991, 183 communities planting March Maize were studied in six municipalities of Tabasco, where this maize was cultivated. The planted areas ranged from 1 to 200 ha, and several of the farmers were senior citizens, who apparently did not guarantee the continuity of the system. The area of the crop is raked and swept to control the growth of the *Thalia*, forming a vegetable bed before burning the vegetation, where later March maize could be sown. Before harvesting, the plant is bent to protect the corncob, which is often harvested by boat. After that maize harvest is transported out of the swamp and stored.

The municipality of Cárdenas gave information about the productivity of the system in 1977. Maize production was 4.5 tons/ha measured in dry grain, including the two varieties Mején and Cuarentano. In the same year, the difference between the production of March Maize and mechanized maize was of 3.2 tons/ha in favor of the March Maize (Orozco-Segovia 1999: 120). In 1990–1991, the production of March Maize was on average 3.5 tons/ha and was achieved without the use of agrochemicals (Mariaca-Méndez 1992: 74).

Other agricultural systems developed in water bodies with vegetable associations of hydrophilic plants are, in addition to the cultivation of March Maize, drained fields and raised fields in the S. Juan river basin, where cacao was grown (Siemens et al. 1988, 242: 105–107). These systems of cultivation of maize and other food plants, which adjust or adapt to conditions of excess water during several months of the year, are found in other places outside of Mesoamerica, where waterlogged areas abound (Mariaca-Méndez 1992; Palerm 1968a, 1993; Siemens et al. 1988, 242: 105–107). In Central America, the cultivation of *chococito* maize is known, which, like *Marceño*, is produced in the wetlands of Choco, a swampy tropical region, which encompasses the Darién, from Panama to Colombia.

The March Maize agroecosystem was adapted to wetlands, where rainwater covered the banks of swamps and increases their surface for several months of the year. It is a "swamp agriculture" that before it was considered to be found only in tropical and waterlogged lowlands. A recent study by Albores-Zárate (2018: 117–144), in towns of the Toluca Valley, in the former lake region of Highlands of the Lerma river, showed the existence – with little evidence – of a *Marceño* agroecosystem in that region.[2] In Mexico, this March Maize agroe*co*system is part of traditional agriculture and have a pre-Hispanic origin. It is left the proposal that March Maize was found also in regions with wetlands and with climates that range from cold to temperate, to the tropical ones, where wetlands abound.

S. Mateo Atenco, in the Toluca Valley, is one of the old river towns that were affected by the desiccation of the Lerma lagoon system from 1942 to 1970. It is located in a sub-basin, which Albores-Zárate (2018: 117, 125,127) described as "lake area high with snowy volcanoes", near the Xinantecatl volcano (*Nevado de Toluca*), and the lagoon relic of the High Lerma River; a group of people grow March Maize (*marceño*). This is sown from March 19 to 31, at an altitude of 2,600 masl, being in danger of being strongly affected by frost. In other parts of the region, planting takes place in April or May, setting fires, and burning grass between crops fight frost. However, there is little information about this highland March Maize (*marceño*).

17.5 EL PALMAR IN CENTRAL VERACRUZ: THE CORNFIELD AND THE ORCHARD

In the 1980s, the town of El Palmar, in Central Veracruz, administrative forming part of the municipality of Puente Nacional had an agroecosystem that was combing the orchards with the cornfield. This agroecosystem and other similar combinations were aimed at improving food and obtaining monetary income. They were carried out in the second half of the twentieth century, among the best known was the Puebla Plan (1967). In the Huexotzinco-Temexmelucan region of the Poblano Valley (Cano and Winkelmann 1972), it created a rural development district with combined crops of maize, beans, peaches, apples, Mexican hawthorn, and apricot. The central idea was that small-scale farmers could have economically productive (fruit) crops, but that while they grew and matured for the fruits to be commercialized, families obtained their food system from the cornfield.

El Palmar is located at the northwest of the medium-size city of José Cardel and Cempoala that was located at the southeast of this city. On the south runs La Antigua River, which flows from the mountains to the Gulf of Mexico. The village is reached by a paved rural road, which leaves from highway 180 that goes to José Cardel city, and the port of Veracruz. In 2010, it had a total population of 1,074 inhabitants, 528 men (49.16%), and 546 women (50.83%), and 388 homes constituted the urban area of the place, of which 99.37% had electricity and 99.68% had piped water (INEGI, Sistema de Información Municipal 2015: 1–10).

El Palmar sits on a hill, at 32 masl, which is surrounded by lower and flat land, planted with sugar cane, maize monocultures, and chili (*chile Serrano*). In the upper zone, there are intermixed

2 It seems that small-scale farmers who are still in this region also talk about April maize, for the month in which the grain is sown, which depends on the rainfall regime.

cornfields with fruit trees, mangroves, and papaya (*Maradol*). In the low and plain lands, there are pastures for cattle, and in farmers' homes, there are some domestic animals, mainly pigs and birds.

In central Veracruz, in the 1990s, a type of cornfield was combined with elements of orchards. The cornfield was sown in strips with maize, sectioned by areas cultivated with rows of commercial fruit trees (mixed cultivation). The objective of this agroecosystem was to grow commercial fruit trees, but as long as they reached their size and age to achieve a production that was economically profitable, the spaces between the rows of trees owe maize to farmers for the family system food. That is, the cornfield agroecosystem became temporary intermixed with a commercial orchard but produced maize while the fruit trees grew.

The irrigation of this system was basic due to the high temperatures in the region[3] and the low annual precipitation.[4] The farmers watered both crops using donkeys, on which they loaded – in each of their loins – a can with 20 liters of capacity each, which were tied to each side of the donkey's back. A hose connected to a water outlet filled these cans. The cans had holes in the bottom, with which they were irrigating the crops manually (González-Jácome 1984: 1–31).

It was a system that – in a funny way – it could be call "donkey watering", paraphrasing the "hand watering", practiced in Oaxaca by pre-Hispanic farmers (Flannery 1976); or to the manual irrigation system that was done in the chinampas of the basin of Mexico, which was carried out through poles with a kind of spoon (*zoquimail*), to throw the water from the canals to the crops; or that it was also carried out by means of containers to fill with water manually and to throw it to the crops (González-Jácome 1984a).

The farmers of El Palmar argued that while the fruit trees grew and began to give them money for the sale of the fruits, this mixed agroecosystem would be important for their subsistence. Sale of fruits could be to the wholesale, by whole harvest before the ripening of the fruits, and by boxes of 20 kilos each one, at retail at the end of the mangoes harvesting. Maize provided them with food for their families, and with a "good harvest," they also obtained some monetary resources to subsist. Thus, the cornfield gave them food for the family, and the orchard production was mainly for the market, provided them with money for higher expenses. In addition, trees formed a curtain that protected maize from strong winds, hurricanes, and the sun.

The problem the small-scale farmers faced was the domestic market price of the mango because the price changes from year to year, and if the production of mango is high, the price in the market falls and it would be not profitable for the small-scale mango producers. A low price for mangoes is not acceptable for the producer, who in addition to the production costs has to pay the salaries of the fruit pickers and the transport of the mango with the traders who will take it to the urban markets. When the price of mangoes in the market is less than the total cost of production, the farmer sells the mango by the kilo, placing it in boxes made of wooden boards, in stalls that are on the edge of a main road.

Another problem was that in the region there were numerous villages dedicated to the cultivation of mango, and some of them – like Actopan – were large-scale mango producers. They produced *Manila mango* and another small one (about 5 cm.), locally called *Mango Niño*, very sweet, with a small pit that was consumed locally. Little by little, they began to introduce the *Mango Petacón*, because this variety had greater resistance in its packaging and commercialization, in addition to having better sales opportunities in the markets. These farmers sold most of their production wholesale to merchants who in turn took them to urban markets in Xalapa, Puebla, or Mexico City.

Other forms of the homegarden-cornfield or the cornfield-orchard have been described in previous chapters. In several regions of Mexico, the orchard–cornfield interrelation or the cornfield with other types of fruit trees exists. Another example of this is the parcel planted with maize and

[3] Average in January 21.7 °C, February from 18 °C to 26 °C, March from 20 °C to 29 °C, April 22 °C to 31 °C, May 24 °C to 33 °C, average in June 28.1 °C. Annual average: 24 °C–26 °C.

[4] Climate Data Organization, Mexico, 2014. Veracruz de Ignacio de la Llave, José Cardel 46160: Precipitation in millimeters, Annual average: 1 100–1 300 millimeters.

associated plants, with boundaries formed by trees that serve as a wind barrier, natural vegetation borders that are barriers against insects, hosts to pollinators, introduce nitrogen into the system, and are important in the conservation of species. The ecological role of trees integrated into the cornfield has been studied very little, particularly from on environmental history and ecological anthropology angle, where the subject has been ignored to date, practically (Farrell 1990: 169–183).

Since 2018, Veracruz producers have abandoned Manila mango fruit cultivation and replaced it with malanga cultivation, which is more lucrative for them. Actopan and Medellin were the two zones with the most important Manila mango producers in the entity. Currently, the Ataulfo mango, which is planted in Chiapas, has displaced the Manila mango in the market (Sierra Leor 2018). The small-scale farmers have stopped sowing soursop and sapota fruit (*chicozapote*).

17.6 ST. ISABEL XILOXOXTLA, TLAXCALA: COMBINING SEASONAL AND IRRIGATED CORNFIELDS[5]

On the low, west slopes of the Matlalcueye or Malinche volcano, there are several towns that most of them are of pre-Hispanic origin (Trautmann 1981). They arose from minor settlements that were congregated as towns of Indians (*pueblos de Indios*) in the sixteenth century, which when merging with larger populations, were divided into neighborhoods (*barrios*), and each of theses was composed of people from the ancient populations of the settlements. This New Spanish political-religious organization was based on a Council (*Cabildo*), with a town hall, prosecutors, and stewards. They created a rotating system of neighborhoods (*barrios*), each with a patron saint and a butler (*mayordomo*). Of this old system are five neighborhoods in St. Isabel and a system of stewardship, that is strictly religious now. Concurrent to this system is the granted land organization institution (*Comisariado Ejidal*), created in 1916, as a result of the agricultural land distribution land organized by the Mexican State (González-Jácome 1996).

In 1970, St. Isabel had a total of 1,614 inhabitants, and a population density of 413.84 inhabitants/km^2. There were 315 families of small-scale farmers, who combined agricultural activity with small-scale trade. Six of these farmers were large-scale merchants and owned both transport and trade networks, which allowed them to take their products to markets to the cities of Puebla and Mexico. There were some people who were workers in nearby factories or masons.

By 1990, the total population reached 2,764 inhabitants (708.71 inhabitants/km^2). According to information from INEGI, in 2010, the total population was 4,436 people (with an increase of 174.85% from 1970 to 2010). This demographic situation affects the land inheritance system (*xocoyotazgo*) and causes the abandonment of agricultural activity, directly.

In 1972, the town had 399.37 ha of land for cultivation, in addition to those occupied by public and religious buildings, roads, and homes (1.87 km) (Ochoa-Salas 1989). The information indicates that in 1983 the Xiloxoxtla Industrial Park was created, which occupies 30.44 ha of land and where 30% of the population of St. Isabel works. According to government information, the municipality was created in 1995 and had a total area of 5.87 km^2 and 10 water wells. In 1995, S. Francisco Tepeyanco and Xiloxoxtla were separated politically, both organized as independent municipalities.

In 1972, the seasonal cornfields were in the possession of 73.65% of the small-scale farmers. The other farmers applied their own monetary resources to pay for the excavation of wells destined for irrigation and for the acquisition of electric water pumps to be able to irrigate their lands and to

[5] This section will be a summary of information published about St. Isabel Xiloxoxtla, Tlaxcala. This information included S. Francisco Tepeyanco, Tlaxcala (González-Jácome 1976, 2003). The information was obtained from 1969 to 2016, at different stages of research. The botanical classification of the species was carried out by the biologist Miguel Ángel Martínez (+), of the Botanical Garden of the UNAM. From 1969 to 1971, Dr. Ángel Palerm supervised my fieldwork, and directed my BA, and MA thesis in Ethnology. Dr. Richard N. Adams, from the University of Texas at Austin, supervised my fieldwork in the summer seasons of 1977, and 1978; his guidance and suggestions were very useful. Dr. Stephen R. Gliessman, and his students, with a team from the University of California at St. Cruz, and a student from the Universidad Iberoamericana, from 1981 to 1983 performed fieldwork in the region. My gratitude to all-them.

increase the productivity of the crops. In 1975, of the 315 families of small-scale farmers, 83 owned wells with electric pumps for irrigation (23.35%). In the 1980s, a government program introduced irrigation by cement channels in the seasonal rain fields and the neighboring towns that are settled in the macro-region of the western slopes of the Matlalcueye volcano, located in the center and southwest of the state of Tlaxcala.

17.7 TRADITIONAL AGRICULTURE AND INDUSTRY

The distribution of the land in 1916 caused problems among the farmers of the villages in the region. Between Sta. Isabel Xiloxotla and San Francisco Tepeyanco the problems were serious, because the disputed lands were those that had access to river water. In 1983, a factory, and what would be the industrial park of Xiloxoxtla, with an area of 30.44 ha was built. After, factories were installed: six textiles, one of mineral products (Durametal), one of candles, and another of electro systems. With those sources of work, the authorities intended to silence the constant complaints of the residents of the aforementioned villages who continued dispute over the irrigated land located between the two towns.

Xiloxoxtla had agricultural land on the east banks of the Acuitlapilco lagoon and a nearby wetland, separated from the lagoon shore by a hill. This area was a swamp known as *La Ciénaga*, and its lands occupied the lower zone at 2,240 masl, near the Tepeyanco River. The town of the same name bordered S. Isabel at the southeast. Its farmers cultivated maize and vegetables in the agroecosystem of drained fields or ridges. From 1975 to 1983, the permanent area of the swamp had an approximate area of 39.70 ha (9.94% of the total area of the town), which increased in the rainy season and when the rains were very strong, mainly from August to September (Figures 17.7 and 17.8).

A canal that irrigated the lands of Tepeyanco started from Acuitlapilco lagoon, but it was abandoned in the 1970s. Another source of water for the southeastern farmlands of Xiloxoxtla was the foggara descending from the Matlalcueye volcano, through a square-shaped canal, about 2 km long, which drained into a reservoir belonging to the Molino de Flores de Santa Ana Tepeyanco estate. This underground canal led the water from the volcano's slope to the Tepeyanco River through a line of 13 underground in-line wells, a reservoir, and drains or ditches. The excess water went to a nearby ravine. The canal system was a qanat or foggara of pre-Hispanic origin.

The soil in this area is sandy-silt of recent sediments and local colluvium with silt to slightly crumbly sand, containing carbonate. It is deep, with high water permeability, and a reduced capacity to retain it. The vegetation in the wetland is ash trees (*Fraxinus uhdei*) and willows (*Salix babilonica*). In the homegardens next to the houses, there are White sapota (*Casimiroa edulis*),

FIGURE 17.7 Tlaxcala, La Loma: Acuitlapilco Lagoon View from La Loma, Xiloxoxtla, 2010.

FIGURE 17.8 Tlaxcala, Acuitlapilco Lagoon View from Tlacochcalco, 2010.

FIGURE 17.9 Xiloxoxtla, Tlaxcala: Housing for Domestic Fauna, 1972.

Mexican hawthorn (*Crataegus pubescens*), pear, peach, apricot, and European walnut (Tarango-Rivero; Macías-López; Alarcón; and Pérez-Moreno 2004). A tree called butterfly bush (*Buddleia cordata*) is abundant in the near ravine, and there is White cedar (*Cupressus benthamii*). Within the limits of land for cultivation is Mexican black cherry trees (*Prunus serotina*) and hawthorn trees. In addition, there are introduced species such as Australian pine (*Casuarina equisetifolia* L.), thunder, and eucalyptus. Farmers have dogs, poultry, and occasionally pigs, which are kept on the grounds surrounding the houses (Figure 17.9).

17.8 AGRICULTURAL CYCLE

The agricultural cycle begins in December, when the ground of the cornfield still is covered with some of the plant remains of the previous harvest, which was not used as food for livestock. At the beginning of January, the land is cleared, plowed, and leveled, using a wooden beam. Then the ground is allowed to rest; the leveling preserves humidity in the soil. In March, soil will be plowed again mixing the remains of weeds that had grown, and the soil will be ready to receive the rainwater that usually occurs at the end of the month.

FIGURE 17.10 Atoyac Zahuapan Basin: Soil Piled Up on Maize Plants to Protect Them from Wind (*Amogotar*).

At the beginning of the rainy season, maize, beans or broad beans, and squash were sown, the seeds of which were distributed in the terrain of the cornfield. At 20 days, the weeds grown larger than maize and therefore they have to be removed (first weeding), "so that the grass would not drown the corn" and it would grow. Weeding was carried out using the plow, and the cut weeds were used to feed the livestock and poultry.

A month later a second weeding was carried out, and the lower part of the maize plant was covered with a mound of earth (*amogotar*) so that the plant would not fall due to the winds that ran through the area at that time; it was also a protection for excess rainwater to run off. Then the herbs were allowed to grow because they could no longer "drown" the plant of maize. The herbs that were allowed to grow in the field were used for different purposes, including food. Still in the 1980s, this labor was done in the maize plants; however, it was stopped due to the time it took, the hard work that adding soil to each maize plant needed, and the lack of human labor that it was needed (Figure 17.10).

The weeds that grew in the third month after the beginning of the agricultural cycle, served, in some way, as a cover to protect the soil from the impacts caused by the beating of raindrops and the sun's rays. The abandonment of agricultural practices such as the second weeding modifies aspects in the cornfield such as the use of herbs to feed livestock and poultry in times when the farmer still does not receive money for the harvest. It is also unknown if the work of weeding or its lack of accomplishment modify and in what way, the type of herbs that will grow in the cornfield, and if they were tolerated after a selection.

17.9 SMALL-SCALE IRRIGATION IN TRADITIONAL AGRICULTURE

Before the 1980s, 74 wells supplied the irrigated land of St. Isabel with water where some crops intended for trade, mainly, were planted in 176 ha: tomato, green tomato, lettuce, radishes, round squash, fig leaf gourd (*Cucurbita ficifolia*), cabbage, cauliflower, turnip, parsley, purslane, and carrot. Farmers planted flowers for commercial sale: *Tagetes* sp. (*cempoalxochitl*), gannet, gladiola, and baby's breath. Sporadically some farmers grew chili, although it was affected by excess moisture. Among the varieties they sowed with irrigation was "crazy chili", named like that, because in the same plot there grew chilies with little itching and chilies with much itching. It was consumed by the same farmer families, mixed and cooked with eggs. This crazy chili reached the market rarely. Other varieties of chili were cultivated to sell in the market.

In those years, electricity was from 700 to 1,000 pesos/month,[6] a very costly price for small-scale farmers who has just begun to cultivate lucerne for sale. This fodder plant had a good price in the market, where it was sold wholesale (per zones en the parcel), per furrow, or retail (per kilo). Also in irrigated land, vegetables were grown for family consumption and the market. The electric pumps were removed from the terrain once the watering was over and taken to the house of the small-scale farmers to protect them from the weather and thieves.

In the 1970s, there were 113 parcels with irrigation in an area of 57.3323 ha (Average 0.507 ha/plot). Farmers planted two crops a year: one with maize, and the other with cash crops. At that time, they preferred to cultivate lucerne. This crop was cultivated for 3 or 4 years, and then farmers reverted to planting maize and associated plants. The average yield of irrigated maize was 2,096 tons/ha. The maximum yield reached was 5.1 tons/ha, but 8.25 tons/ha were produced in two plots. The top minimum production was 1.1 tons/ha, with an average minimum yield of 0.675 tons/ha. The rest of the maize was grown in seasonal cornfields, where the grass was planted with beans or broad beans and squash.

In the spring of 1972, in Xiloxoxtla, 182 seasonal plots were cultivated with maize (Average of 0.458 ha). They occupied a total of 83,272 ha that produced 101.79 tons in total, with an average of 1.22 tons/ha. These plots belonged to 56 small-scale farmers, each with an average of 3.25 plots, which gave them a total of about 3,965 tons of maize obtained in their seasonal plots. Maize was combined with squash, and if the quality and humidity of the soil allowed it, beans, also.

In 1972, the average size of the seasonal or irrigated plots was ½ ha–. The small-scale farmer had little more than three terrains, which gave him a total of approximately 1½ ha of agricultural land. With irrigation, two crops a year were obtained: one of maize, beans, or fava beans, and squash; the other with legumes and lucerne.

17.10 COMMERCIAL AGRICULTURE

By 1985, commercial crops on irrigated land in the neighboring town of S. Francisco Tepeyanco were: maize for ripe ears and maize tender were grown, lucerne, zucchini, round squash, onion, vegetable pear (*chayote*), fig leaf gourd (*Cucurbita ficifolia* Bouché) (*chilacayote*), wild chili, apple chili, green beans, beans, fava beans (sold green and dry), Goosefoot (*huauzontle*), tomato, wild and cultivated edible green weeds (*quintoniles*), purslane, Agave flowers (*gualumbos*), and Opuntia cladodes. The lands included a reticulate section for the homegarden, another section of ridges (located in the neighborhood of Atlamajac, that since 1995, became an independent village), an area of ejidal plots with and without irrigation, and another cultivation area located on the slopes of the surrounding hills.

The avocado cultivated in Tepeyanco homegardens had several routes for its commercialization.[7] In 1895, avocado production (*Persea gratissima*. Gaert.var.) was sold in regional markets (Apizaco, Chiautempan, Teolocholco, St. Cruz Tlaxcala, Zacatelco, Tlaxcala City, Tlaxco, Chignahuapan, Zacatlán, Cuapiaxtla, and Huamantla). The merchants took avocado to sell it in the markets of Tabasco, Atlixco, and the cities of Puebla and Mexico. In addition to avocado, they carried other agricultural products that they were selling and exchanging for others along the commercial routes of ancient origin by animal transport (mules). This changed after 1930 when the construction of the

[6] The exchange rate was 37.57 USD per 700 Mexican pesos, and 53.68 USD dollars per 1,000 Mexican pesos in February 2020. In May 2020 the exchange rate was between 24 and 25 pesos per USD. In July 2021 the exchange rate was between to 19 and 20 pesos per USD.

[7] The commercial relations between towns in Tlaxcala, such as Tepeyanco and Atlixco, are old. In 1895, avocado was produced in the municipalities of Chiautempan and Nativitas, and it was sold in the markets of Atlixco and Matamoros. This occurred from May to June, and September, and sales gave to producers from $125.00 to $260.00 Mexican pesos annually (Mexican Carpology 1895: 17–18, 30–31).

paved road was completed, gasoline-driven cargo transportation began to be used, and the quantity of agricultural products marketed changed.

With the paved roads, new rules were put in place to prevent the passage of wagons that did not have rubber wheels, because old wagons pulled by oxen or horses damaged the roads (Jacinto Zacapa-Rugerio interviews 1970–1972). In Tepeyanco, the owners of cargo trucks gathered the productions of several small-scale farmers and took everything to urban markets.

Sometimes the truck owners bought the productions of their neighbors, and to each farmer, they paid him or her the share of the products. The merchants charged an amount for the transportation and occasionally for the commercialization, which was arranged individually. The price established varied, depending on the personal relationship between the merchant and the small-scale farmer. Products, such as avocado, could be sold/dozen (retail) or percent (wholesale). The price varied according to whether it was large or small avocado, according to the months of production, and according to the prices paid in the markets.

17.11 TRADITIONAL AGRICULTURE AND MIGRATION

Farmers with greater resources obtained from trade rented the lands of those who migrated or emigrated from the village to work seasonally or permanently inside or outside the country. In general, migrants did not sell their land, unless they decided not to return to St. Isabel and live permanently in another place.

At that time, renting granted ejidal plots was prohibited in Mexico. However, it was a disguised and generalized trend in various parts of the country. Also, the plots that were not cultivated served for the small-scale farmer to obtain loans in private arrangements with the lender. This was done according to the amount borrowed, and for a certain number of years, the lender could cultivate the debtor's plots. The arrangement was made between the granted land farmer and the lender, and the conditions and interests varied from one deal to another, depending on the interpersonal relationship between them. Also, sometimes lenders had to pay the expenses to cover the loan arrangements negotiated by third parties. Those who made the arrangements went to work in the United States, and it was their relatives who were responsible for covering the amount arranged with the lender and its corresponding interest.

The Tlaxcalan population increased with the construction of several roads that crossed the region and its towns, bringing about a lack of agricultural land for the younger generations, the increase of schooling for children and young people, and a higher commercial income now desired in contrast to the hardness of agricultural work. Low prices paid for agricultural products, the high cost of private irrigation, the residential and industrial growth in the town and its surroundings lands reduced the number of inhabitants and the importance of the human population to be dedicated to agriculture. It should be added that, currently, the aspirations of young people are based on urban life ideas, customs, and new ways of relating to other young people and adults ways of thinking and having fun.

These new generations, mainly made up of young people, have gone out to study and work outside of their hometowns. In general, if they continue to live in their towns of origin, they work in the service sector. If they studied in higher education centers, they leave the towns where they were born and raised to go to live in the cities. Some children of wealthy farmers prefer to live in the cities of Puebla or Mexico, there are those who prefer to leave the country to live in other places, including the United States and Canada, although some of them emigrated to European countries or went to South America, depending on factors that may include meeting other countrymen who live in those places. At the beginning of the current century, there were Tlaxcalan living in Argentina, Chile, Spain, and Germany, and in places in the United States such as Hawaii and Alaska.

17.12 DAILY LIFE AND CORNFIELD SMALL-SCALE FARMERS

In Tlaxcala, the processes of modernization from the 1960s to 1980s included many aspects, like schooling and formal education, types of housing, clothing, ornaments, hairstyles, and footwear, and strongly including forms of interpersonal communication, diet, and new ideas about the world. In Tlaxcala, modernization included the construction of roads, superhighways, bridges, industrial zones, residential housing, and urban-type schools, with media rooms. These schools organized activities to inform about soil erosion control; support federal projects for this; support festivals, like carnivals, trade fairs, touristic centers; and disseminate information about all of them. Small-scale farmers have modified some of their customs and ideas about the world (Buendía-Rodríguez et al. 2008: 1–9; Castro and Montiel 2019; González-Jácome 2003; Nutini 1968, 1993).

In Xiloxoxtla, the number of couples who cohabit increased. They argued that the cost of weddings was very high. Young adult women sought information about how to prevent pregnancy or how to avoid cancer (Castro and Montiel 2019; González-Jácome 2003).

The cornfield that still manages numerous crops (polyculture cornfield) has lost economical and family system food importance along with the surface destined to it. In contrast, the monoculture cornfield has increased. In general, the cornfield is disappearing as an agroecosystem to open the way to barley, wheat, and maize for commercial forage monocultures. The landscape has changed dramatically, forest relics no longer exist, and adobe homes have been replaced by urban-type houses imitating various architectural styles – including those found in poor urban areas (simple rooms with small windows, and an access door) or American-style homes – and copying the style according to the places where the small-scale farmers had migrated to work, and with the money they could send to their relatives or bring back with them when they returned to the Tlaxcallan villages.

Industrial buildings and roads have grown at the expense of natural areas and agricultural land. In the region, there is a relic of the Acuitlapilco lagoon. It is 1 km long, and on its banks, there are crops with irrigation and moisture soils, even today. Former small-scale farmers have become workers, and young people look for jobs that take them away from the countryside and lead them to urban life in the cities of Apizaco, Huamantla, Tlaxcala, and Puebla.[8] Usually, cities and towns located on the roadsides of some road have numerous locales with small businesses, like the repair shops, mechanics, garage, shops selling aluminum booths, lumber, or places where cargo attachments are installed in the vans to transport agricultural products and merchandise.

17.13 STEEP PLACES AND THE SEASONAL CORNFIELD: THE *TLACOLOL (E)*, THE *HUAMIL*, THE *COAMIL*, AND *ECUARO*

There are several types of temporary cornfields agroecosystems adapted to zones with steep orography and dry weather. In different places in Mexico, anthropologists and agronomists have studied agroecosystems like Tlacolol (e), Huamil, Coamil, and Ecuaro. Among the investigators are Casas (1988), Chávez-Segura and Cruz-León (2016: 11), Franco-Gaona et al. (2017 (16): 3357–3370), J.V. Palerm (1997: 121–133), and Reynaldo Valencia (1981). These agroecosystems are characterized: (1) They are considering to be of pre-Hispanic origin; (2) They have been combined with elements from the Old World and Viceroyalty origins; (3) They use slash and burn as a generalized method to manage the soil and plants; (4) They are seasonal; (5) They are polyculture agroecosystems; and (6) They use the planter cane in the steep places, and the plow in areas where it can enter easily.

[8] González-Jácome 1976, 1986, 1988b, 1988c, 1989b, 1991, 2008b, 2008d, 2009b: 1–15, 2009c: 250–181, 2010a: 141–281, 2010b: 521–554, 2010e: 37–65; Mariaca-Méndez and Robins 2007b; Villalobos-Nájera 1998, 2008, 2009.

In his study on Tepoztlán, American Anthropologist Oscar Lewis (1944: 281–314) considered *tlacolol* as a form of horticulture "[...] more primitive than *milpa* [...] which, although it solved immediate, long-term problems produced population increase [...] incompatible with the need to clear more land each year [...]". Lewis considered *tlacolol* as an inefficient agricultural system because it wasted the soil causing erosion. In addition, its productivity was low, so the small-scale maize farmers preferred to migrate from Tepoztlán to Mexico City and live there due to better salaried jobs, which were very different from agricultural ones.[9]

González-Martínez (1992), in a study conducted in Loma Tendida, Valle de Santiago, in the Bajío, found tlacolol in areas with seasonal plots (90% of the total) located in steep or hilly places, in lands that were classified by the agricultural distribution as "non-cultivable", cornfields were planted with maize, beans, and squash using the agroecosystem known as *huamil*. The region had summer rains (June–September). In these lands, the use of agricultural machinery was not feasible, therefore the work of preparing the lands, loosening the soil, sowing, and harvesting are done manually. In addition to the *huamil*, in Loma Tendida, there are other agroecosystems in the plains but with irrigation and humidity that are seasonal crops.

The *huamil* emerged because of a need for farmers to extend the area for cultivation, which they did on the only land available: the hills, which used to be cattle-grazing land (*agostadero*). There were about 250 ha where more than 100 ejidal granted parcels were established. The *huamil* is found in other parts of Mexico, like in some places in Michoacán, where it is called *coamil* (*milpa* with *coa*), or *ecuaro*.

According to González-Martínez (1992), the *huamil* is not a traditional agricultural system because it was introduced in the area recently. This appreciation has to be reconsidered, because what is called "traditional" is built in modern times. Traditional agriculture results from local adaptations to physiographic, environmental, and climatological conditions, which are not always old. It is the folk or local agricultural knowledge accumulated and put into practice by the small-scale farmers, as it can be seen in the stages of the agroecosystem.

Huamil work starts at the beginning of March when the dry remnants of the previous harvest were cut, collected, and burned, along with the weeds that grew in the plot (equivalent to slash and burn). The ashes are used to fertilize the soil. These activities required 1½ month, then the farmer began to open holes in the ground (*jollar, ajollar*). The task starts making furrows with a hoe, deep and high so that they can store the rainwater. However, if it is feasible, a plow was used to furrow or loosen the soil. In this process, the stones are removed, and they are dumped at the lower limit of the terrain, sometimes to form stone fences. This work must be completed by the end of April (Gonzalez-Martinez 1992: 155–156).

As of May 10, the *huamil* is sown in dry soil before the rains begin (early June); maize (called Creole and early varieties), beans, squash, and occasionally sago (*Maranta arundinacea* L.) are grown. Sago reeds are consumed. This herb cutting reeds are a meter tall, and flowers from three to six weeks in August and September. Its rhizomes are washed, dried, and used to make flour. They are antidiarrheal. With the rains, the ground is "squeezed" (becomes compact) and farmers have to be loosened with a hoe. Ten days after maize is sown, it is necessary to do a weeding (1st half of June), using a hoe (*desquelite*). This work lasted about 5 weeks, with all-family members doing it (even wives and children). After 15–20 days, another weeding was done.

The weeds that grow and pull away from the furrows are purslane, red weed (*quelite Colorado* or *quelitilero*), *andán*, olives, and *lechoncilla*. Weeds were cut and stacked around the plant so that they help conserve moisture and rot thus serving as fertilizer. During the month of September,

[9] In *The Children of Sánchez*, Lewis describes the life of a family that emigrated from the countryside to Mexico City. He coined the concept of "anthropology of poverty", almost costing him expulsion from the country, because the readers considered that his vision of the Mexican was a slander that denigrated the nation. One of the characters in the novel was a waiter in a Mexican food restaurant, located in the Pink Zone of Mexico City.

maize was collected, and sago cane (*Maranta arundinacea* L.) is cut. In October, the family harvested beans, taking 4–5 days,

Then, farmers use a donkey to take the load to the house. The *huamil* produced three bushels of beans (210 kg of beans). In November, squashes are cut. A 2-ha *huamil* produced up to 1,000 pumpkins and moved by donkey to the house. Maize is harvested in the first days of the year taking about 15 days, with all of the family members collaborating in the work.

So maize, beans, and squash are not sold, they are for family consumption. Only in case of need, like illness or extraordinary expenses, they are marketed. From this harvest, seeds that will be used the following year in the planting season are selected and separated. The *huamil* in Loma Tendida did not require wage labor, fertilizers, or insecticides. In other words, it did not require agrochemicals for cultivation or a paid labor force, needing extra money to hire them to do the work.

Chávez-Segura and Cruz-León (2016) studied the traditional agroecosystem called tlacolole, in the Guerrero Central-Mountain region. They conducted a regional analysis, with an integrated perspective to study mountain farming. The authors included some aspects of history, ethnicity, physiography, and environment, which allowed them to characterize the strategies of rural life and the key elements of *tlacolole*. On the background check, they collected geographic and statistical information. They also gathered materials related to the physiographic characteristics, vegetation, demography, territory, and mountain agriculture of the municipalities of Chilapa de Álvarez, Ahuacuotzingo, and Zitlala, on which they gave statistical information that relates to the types of wild flora existing along with agricultural crops.

The aforementioned authors defined mountain agriculture, including *tlacolole*, "[…] as spaces where it is taught, learned, and experienced, where ones own culture resisted, and regenerated at family, town, and regional levels. Mountain farming existed, adapted and persisted". Also, they described it as a place of survival with historical functions and peoples' resistance as an axis that articulates the livelihoods of rural societies that are dynamic and constantly readjusting and adapting. That is, in addition to a food production system, it is an ideological system for the organization and maintenance of the societies that practice it (Chávez-Segura and Cruz-León 2016: 118).

In the region studied by Chávez-Segura and Cruz-León (2016: 118–119), *tlacolole* was characterized: (1) as an agroecosystem being worked by families; (2) the land breaks are organized by community consensus and in relation to livestock calendars, but in recent years it has abandoned transhumance and reduced its rest time, with cultivation becoming intensive; (3) it is sown on the mountain slopes and on stony terrain, with an inclination of up to 45°–50°, which requires a good balance of small-scale farmers with specialized techniques using animal traction.

(4) The agroecosystem is seasonal, and the rainwater is articulated to the work of breaking the soil, although there are springs that allow "micro-irrigation"; (5) the agroecosystem is articulated to seasonal migration and nonagricultural activities like the elaboration of handicrafts made with palm leaves and corn husks; (6) it produces varieties of maize with the colors of the maize kernels; (7) whole families carry out the harvest entirely before any of them migrate to Sinaloa and other parts of the country, seasonally; (8) in November, the maize harvested and sun dried on the roofs of the families' houses.

In the Sierra Madre del Sur, there are groups of small-scale farmers who belong to different ethnicities: Nahuas, Mixtecs, Tlapanecs, and Amuzgos. Also there are mestizos. They plant maize of various colors (white, yellow, blue, red, and pastel pink). White maize is broad and used to make pozole. The *Arrocillo* is maize with thin rachis and a high grain yield, and used to make tortillas. Yellows maize is used as fodder, although it can be for human consumption. Blue maize is used to prepare *tlaxcales*, *tortillas*, *tamales*, and *atoles*. Red maize has ritual uses, and pastel pink maize is handled the same as blue maize. There are tamales for special agricultural, social, and religious events, such as the *tzoalli*, made by the Nahuas of Guerrero that are related to the agricultural cycle (Dehouve 1995, 2007; Serafino 2012, 2014; Villela 2016: 46–53).

The average total surface area used by families is 1.7 ha/unit, and the area used for agriculture is 1.3 ha, a situation that explains seasonal migration. The average maize kernel production ranges from 1 to 1.4 tons/ha. Beans and squash are produced as well. Small-scale farmers also obtain maize husks (*totomoxtle*) and stubble. There are weeds that grow associated with maize and other plants in the cornfield. Maize and bean production is enough to feed small-scale farmer families throughout the year in the Sierra Madre del Sur (Chávez-Segura and Cruz-León 2016: 122–123; Serafino 2012, 2014).

There are several types of *huamil* agroecosystem, with variants that are adjusted to the specific characteristics of the places where they are found. A huamil characteristic is to be located in regions or areas with complex orography and steep landscapes as well as with small plains in the area, where it is feasible to introduce the plow or agricultural machinery. The farmers of Guerrero, Jalisco, Oaxaca, and Puebla use the term *huamil* to refer to the land that is resting (fallow) for a few years, where many of the weeds that are part of the early succession are observed (Altieri 1992: 262–2819). Juan V. Palerm-Viqueira (1997: 121–133) wrote about the *huamil* in the Bajío in Guanajuato:

> The Mexican indigenous agricultural systems emphasis is on the one hand, their adaptation to the environment, sustainability, and on the other, their persistence, and distribution. Second, based on ethnographic research carried out in Valle de Santiago, Guanajuato, we focus attention on the case of the "huamil" (*coa-milpa*) system of recent introduction to a region dominated by the so-called "green revolution". The coexistence of two agricultural systems (one traditional, and the other modern) recently introduced to the same rural space allowed the study of their complementarity, and interdependence without the persistence of the traditional being seen as simple, and as a curious historical residue condemned to disappeared it as soon as possible as the modern takes roots.

One of the *huamil* agroecosystem variants is *ecuaro*, a local adjustment that is practiced in Michoacán. Franco-Gaona et al. (2017 (16): 3357–3370) described the *ecuaro, equator, ekuaro, or ekuarhu (ekuaɹu*=yard) as a diverse and multifunctional traditional agricultural system. It can be classified in four types according to its location and the type of technology used in it: (1) *Ecuaro* type yard, solar, garden, or homegarden; (2) *Ecuaro* of agroforestry exploitation associated with livestock; (3) *Huamil* type *ecuaro*; and (4) the *ecuaro* of intensive land cultivation. In these types, the cornfield is cultivated, and the *ecuaro* or *uarhuti* maize is preserved, because it is at risk of disappearing. The authors consider that the *ecuaro* is:

> A traditional complex ethnoagroforestry system typical of the state of Michoacán, and the *Purhépecha* or *Tarascs*, which according to its growers can be part of the house, since it is interrelated with the home, the barn, the raising, and exploitation of major, and minor livestock, agriculture, and some forest species; or the ones located on the slopes of the hills with maize cultivation, and other associated species, without soil breakage, since its low depth, presence of stones, and high slopes prevent it.

Palerm-Viqueira (1997: 129)[10] considered that from 1930 to 1970 in Valle de Santiago, the land for livestock (*agostadero*) in the municipality was reduced by 46%, at the same time that the seasonal *ecuaro* agroecosystem land area grew by 50%. These changes in land use influenced the development of the *huamil*. Valencia (1981: 259) calculated that in 1981, about 8,000 ha of land for livestock in the Valle de Santiago municipality were availed with incipient *huamil*, which was occupying marginal lands difficult for agricultural work. Already established *huamil* followed the same phases and calendar as the seasonal maize crops in the region. The stages were: (1) in the spring, the soil is cleaned and prepared to receive rain water and with it begin to sow; (2) in June the rains arrived, and the cultivation process started; (3) the semi-terraces and seals were worked to avoid landslides during the rains; (4) weeding (*desquelite*) is done; and (5) in winter, improvements (extensions) were made in the structure of the huamil.

[10] Juan V. Palerm-Viqueira directed the project in Valle de Santiago, Guanajuato, and the research for theses in social anthropology at the Iberoamericana, and Metropolitana-Iztapalapa Universities. Among Palerm's graduate students were Laura González, and Reynaldo Valencia.

The *huamil*: (1) always includes associations of maize, with beans and squash at a minimum; (2) its varieties are called "Creoles"; (3) at least three different varieties of each crop were planted (e.g., one type of early and late maize: Northern conical, *Tabloncillo*, and *Celaya*; *higuerilla* beans, May flower, and Palace beans; Castilian pumpkin, zucchini pumpkin (*zapoma*), and *Indita* pumpkin (*Cucurbita moschata*); (4) this planting method is done to reduce weather risk, and the uncertainty it caused; (5) farmers seek and sow crops that are adapted to the specific environment of the huamil; (6) they select the seeds and keep them from 1 year to the next; and (7) farmers visit germplasm banks to stock up on some and experimented with other new varieties.

The yields/unit area vary from one *huamil* to another depending on its specific location and its age. One ha of young *huamil* had about 19 thousand maize plants sown and yields 800 kg of grain. One ha of mature maize accepted up to 35 thousand plants and yields about 1,800 kg/ha, also farmers planted squash, as well as medicinal herbs, and farmers allowed the growth of wild fruits and vegetation to obtain firewood. A temporary surface planted with hybrid maize has up to 45 thousand plants, and its yield can exceed 3,000 kg/ha. However, it should be considered that a family of small-scale farmers could not attend a surface area of more than 2 ha due to the volume of work required for its construction, maintenance, and cultivation (Valencia 1981: 62,147).

The families of small-scale farmers in the region had several economic activities that they combined: (1) the monoculture of cereals in granted plots that are from 4 to 8 ha. These lands are seasonal, and irrigated. The small-scale farmers have a conventional agricultural system that provides monetary income derived from the commercialization of its production and that gave them access to the subsidy systems of the Mexican State, like credit and assistance; (2) the production of the *huamil* agroecosystem occupied the family work force of children, women, and the elderly, generating assets for subsistence; and (3) the seasonal migratory work of adult males and more productive members who work in the agricultural fields of California produce monetary resources (Palerm-Viqueira 1997: 131).

The production of the *huamil* focused on family subsistence. It provides a margin of safety in the face of weather risks (e.g., a lacked of, or low or excessive rains), pests, and diseases that plague the crops, the loss of the commercial value of production (e.g. price fall, contraction of labor markets, lack of employment), having to use areas that are agriculturally not the best, but are productive, without deteriorating the environment. The *huamil* is not a "curious or exotic" historical survival or is it an isolated event disconnected from the regional or national economy (Palerm-Viqueira 1997: 131).

17.14 LOS TUXTLAS, SOUTHERN VERACRUZ: THE ZOQUE-POPOLUCA CORNFIELD

From pre-Hispanic times to date, there have been many types of cornfields. Each of them has different degrees of agrodiversity, structure, need for labor power, and relations with other agroecosystems, such as the homegarden or surrounding areas of natural vegetation; likewise with available water resources. Each type represents a local adaptation to the family organization, to the food system, with its needs and culinary tastes, and particular environments (e.g., climate, humidity, altitude, winds, cyclones, hurricanes, water resources, floods, and so on). In addition, soils must be adjusted to their type and characteristics, such as thickness and fertility, plus the availability of water or residual moisture and the inclination of the land. Micro local conditions include the distribution of plants in the space occupied by each cornfield. The tropics have a great variety of crops and weeds, many of them medicinal that cover the basic needs of farmers (Blanco-Rosas 2006).

The borders of the Soteapan municipality are to the north Catemaco and Tatahuicapan de Juárez municipalities; to the east, the Tatahuicapan de Juárez, Mecayapan, and Chinameca municipalities; to the south Soconusco and the municipalities of Acayucan, Hueyapan de Ocampo, and Catemaco; to the west the Soteapan municipality that occupies 0.67% of the surface of the state of Veracruz. Soteapan municipality has 87 locations that had a total population of 28,104 inhabitants in 2009 (INEGI, Prontuario Soteapan, Veracruz 30149, 2009a).

The physiography includes the Sierra de los Tuxtlas (79%) and a portion of the Veracruz coastal plain (21%). In the sub-province of the Tuxtlas, the topoforms were the volcanic mountain range with slopes (79%) and the typical hillside (21%). The temperature ranges from 20 °C to 26 °C, and rainfall ranges from 1,900 to 4,600 mm. The climate is warm and humid, with rains year-round (36%), but abundant summer rains (27%) and semi-warm humid weather the rest of the year (15%). The dominant soils are Andosol (52%), Acrisol (14%), Vertisol (12%), Luvisol (11%), Regosol (8%), and Cambisol (1%).

The Soteapan municipality has two hydrological regions: the Papaloapan (36%) and the Coatzacoalcos (64%). It has the sub-basins of the rivers S. Juan (14%), S. Andres (9%), Tecolapilla (8%), and Calzada (64%), plus the Catemaco Lagoon (5%). The perennial rivers are Ahuacapan, Cuitlaxoyo, El Bastonal, El Caracol, El Naranjo, Huatzinapan, Huazuntlán, La Margarita, Mezcalapa, Michapan, Soteapan, Osoluapan, Tapazulapa, Texizapan, Tilapa, Piedra Labrada, and Zoxoltepec as well as the intermittent stream El Naranjo.

In 2009, in the Soteapan municipality 36% of the land was used for agriculture, 36% for grassland, 18% was the tropical forest, and 8% mountain forest. The 2% occupy the urban zone that was in the volcanic area of slopes. The urban area has a warm humid climate, with rains throughout the year and abundant rains in the summer. The St. Marta area is located to the northeast of the municipality.

Blanco-Rosas (2006) provided information for several years in the region. This included the history of the cornfield since the nineteenth century of the Zoque-Popoluca who inhabited Soteapan, in Los Tuxtlas, southern Veracruz. There are several historical studies on the same region, making it possible to organize a table using data on its history and its characteristics, all described by researchers at different historical moments.

Blanco-Rosas (2006) indicates that agricultural activity in the cornfield has been associated with other tasks that are not articulated with the cultivation itself or with the management of other traditional or conventional agricultural systems, as is the case today (e.g., cornfield cultivation, and the sale of coffee). These activities have included: the production of ceramics, textiles, and clothing; small-scale trade; hunting to obtain meat, both for family consumption and sale; fishing and collecting fauna species for family consumption and sale. The small-scale farmer takes care of small livestock, barbecues, roasts pork, smokes of fish, and other edible fauna. In the 1960s, the Zoque-Popoluca cornfield was a polyculture, with up to 50 food species. For Blanco-Rosas (2006), the study of the cornfield opens with its definition:

> [The Zoque-popoluca cornfield is] a set of crops that covers different strata in the vegetation, but maize is the center of the agroecosystem, and the most widespread element towards other agriculture. The study of the cornfield in statistics, as well as an anthropological study, is difficult to follow, since there are no figures of cornfields, but only the monitoring of maize crops as a monoculture. The cornfield is a complex agroecosystem where, although maize is the center of the polyculture, it is associated with various plants, planted, sponsored, and tolerated; where each one fulfills an ecological, economic, or sociocultural function. This agroecosystem became the main provider of food for the diet communities.

There are several types of cornfield in Soteapan with cultivated plants that have changed over time. Cornfield polyculture farming is practiced in Los Tuxtlas, Veracruz, to date. Historical information showed its changes over time, demonstrating the flexibility of the agroecosystem to modify its crops according to the locals' tastes and the market, both of which have modified. Blanco-Rosas explanation for this draws attention to some aspects: the physiographic, environmental, and climatological characteristics of the places where maize is planted, the crops that accompany it; changes in the daily food system; and the supply-demand of agricultural products in regional open-air markets (*tianguis*) and markets.

The importance of Popoluca folk medicine, for which information is available, should be added. However, the tables with the data on the plants do not indicate the place of collection, so it is unknown if they were obtained in the cornfield, the forest, or somewhere else associated with the agroecosystems managed by the Popoluca ethnic group. A good example of this is the Leonti, Vibrans, Sticher, and Heinrich (2001: 1653–1669) study of ethno pharmacology, from a database with 4,488 reports on the use of 614 plant species, provided by 72 informants divided into five groups: healers, herbalists; midwives; bonesetters; soul loss experts; and herpetologists specialized in snake bites. The researchers found that the boundaries between these different groups are fluid. When deemed necessary, nearly 50 informants were interviewed a second time, to complete the relevant information the herbalist had gathered.

Throughout the Viceroyalty, and in the nineteenth and twentieth centuries, the cornfield was subject to local situations, as well as those related to external events: changes in land tenure and the expansion of the large estates; the presence and influences in the agricultural work force of these large estates; the relationship between the towns with private properties and the government on land laws; the social, economical, and cultural impacts of the Mexican revolution of 1910–1921; federal public policies, including taxes, and the plans and programs directed to small and medium-scale agriculture; and the results of the processes of industrialization and modernization in the estates and in the country.[11]

17.15 ST. CRUZ YAGAVILA, OAXACA, AND THE ALTITUDE CORNFIELD MANAGEMENT, AND SEED SELECTION

In the late 1980s and early 1990s, Tyrtania-Heidt (1992) studied the management of maize in a process of seed selection and mobility in different ecological floors of St. Cruz Yagavila, a Zapotec town in the Sierra de Juárez, or Sierra Norte de Oaxaca, which is 1,059 masl. The town belongs to the El Rincón region (Tyrtania 1992: 94), located west of the Juquila River in the Ixtlán de Juárez municipality. In 2005, St. Cruz Yagavila had a Zapotec population of 595 inhabitants (305 men and 290 women) distributed in 204 homes. The population of Yagavila decreased –9.02% from 1990 (654 inhabitants) to 2005 (595 inhabitants) as a result of emigration, which impacted the agricultural labor force (Tyrtania 1992: 94; INEGI 1990, 2005).

In 2005, the populations of the region (INEGI) had electricity (94.35%), piped water (95.48%), toilets and inside plumbing (97.74%), radios (60.45%), televisions (49.72%), refrigerators (22.03%), washing machines (5.65%), cars (2.26%), personal computers (0.56%), landline telephones (2.82%), and Internet (0.56%), but nobody had a cell phone. The town was reached by the dirt road that connected to S. Pedro Yaneri. Loggers began to build roads for automotive transport.[12] In 1979, the road, the CONASUPO governmental grocery store, the Health Center, and the National Indian Institute Boarding School (I.N.I.) appeared in S. Juan Yagila. However, the new roads could not be used in the rainy season (Tyrtania 1992: 95).

According to Tyrtania, in 1992, the towns west of Rincon were St. María la Luz, S. Miguel Tiltepec, St. María Josaá, S. Juan Yagila, Santiago Teotlaxco, St. Cruz Yagavila, St. María Zoogochi, S. Pedro Yaneri, and S. Juan Tepanzacoalco. They had a total of 3,623 inhabitants; the one with the highest population was St. Cruz Yagavila, with 654 inhabitants (18.05% of the total).

Yagavila is in the mountainous region of northern Oaxaca, where slopes fluctuate from 40% to 60% (Morales-Cruz 2009: 10). The main elevations are the Summit of the Pozuelos (3,200 masl) and the Naguetzi hill, both in the same physiographic province. To the east, the municipality joins the Sierra Madre del Sur to form the junction or intersection of the Zempoaltépetl mountain range.

[11] Blanco-Rosas 2011: 177–202; Barkin 1972, 1998; Barkin and King 1970; Barkin and Zavala 1978; Ewel 1980; González-Jácome 2011: 17–37; Palerm 1972, 1993a, 2006; Tortolero-Villaseñor 1995.

[12] The conditions of the towns and the accesses to them were very different in the second half of the 1990s. A group of graduate students and me entered the region to go to S. Juan Yaee in a Volkswagen that went behind a machine that broke the ground to make the gap in the hillsides of the sierra. The upper area was covered with *Pinus patula*. The machine belonged to a Canadian timber company that had cut down the trees that were piled up at the edge of the road to S. Juan Yaee, previously. This town and St. Cruz Yagavila were separated by a ravine in whose bottom runs the Cajonos River flows, a Papaloapan tributary (González-Jácome Field Journal 1990).

The altitude in the municipality goes from 200 to 3,200 masl. The weather varies from mild to cold. The highest rainfall is in the summer, and frosts are in winter. The old "royal road" initiated in St. Cruz Yagavila connects the towns of the west to those of the east of El Rincón and reaches the markets of S. Juan Yaee and S. Miguel Talea de Castro (Tyrtania 1992: 94).

Morales-Cruz (2009: 13–14) describes three ecological systems in the basin. In the high part is the pine-oak forest, whose vegetation is a mesophilic mountain forest, with a great diversity of species, including elements of humid temperate and tropical zones. In it, there are pastle, mosses, vines, flowers, and different species of fungi living on the tree trunks. Within the forest, epiphytic plants, mosses, lichens, bromeliads, orchids, and arborescent ferns *(Cyathea* spp.) are common. In the middle part of the basin, the flora includes *Ficus elastic*, *Alnus acuminata*, soft stick, *Ampelocera hottlei* (*Aguacatillo*), *Abies guatemalensis* (*pinabete*), *ayacahuite* (*Pinus ayacahuite*), and turpentine black pine (*Pinus uncinata*).

The sub-deciduous forest had vegetation in the area at lower elevations. From 50% to more than 75% of the trees shed their leaves in the driest time of the year. Frequently, this type of forest is distributed on hillsides with perennial species. The vegetation comprises: cedars (*Juniperus* spp.), mahogany (*Swietenia mahagoni* (L), silk cotton tree (*Ceiba pentandra*), salmwood (*Cordia alliodora*), white cedar (*corpo*), picho, candlestick tree (*Senna alata*), *macayo* (*Andira galeottiana* Standl.), tree of life (*Guaiacum sanctum* L.), bitter, oak, *huaya de monte* (*Ampelocera hottlei*), esquahuitl, Croton draco (*Croton draco* Schltdl.), Soursop or Sweetsop (*Annona reticulate* L.), Santiago, quickstick (*Gliricidia sepium* (Jacq.) Kunth ex Walp.), *pompo*, naked Indian tree (*Bursera simaruba*), break axe (*Aspidosperma*), nanchillo, trumpet tree (*Cecropia peltata*), woodpecker (*Cecropia obtusifolius* Bertol), willows (*Salix* spp.), sailor, chicken coop (*Azadirachta indica*), sea hibiscus (*Hibiscus tiliaceus*), white, royal jonote (*Heliocarpus appendiculatus*), primavera (*Tabebuia donell-smithii* Rose), and oje (*Ficus* spp.). In the lower part of the basin lies the induced grassland (Morales-Cruz 2009: 17–19).

The wildlife has numerous birds: quetzal (*Pharomachrus mocinno*); pavón (*Orephasis derbianus*); colibrí oaxaqueño (*Eupherusa cyanophrys*); colibrí cola blanca (*E. poliocerca*); colibrí cola rayada (*E. eximia*); and mammals like weasel, mole, coyote, badger, squirrels, armadillos, opossum, paca (*Cuniculus paca*), porcupine, skunk, monkeys, marten, miners' cat *Bassariscus astutus*), tapir, otter, deer (temazate), rabbits, raccoons, wild boar, anteater, chameleon, flying squirrels, jaguars, pumas, leopard, tiger, and jaguarondi (*Puma yagouaroundi*), which are found in the deeper areas of the forests (Morales-Cruz 2009: 18–19).

There are reptiles, such as vipers (Bothrops asper, deaf, rattlesnake, *Atropoides olmec*), lizards, iguanas, chameleons, and black iguana. There are insects such as black ant, yellow ant, and *Atta cephalotes*, *Trombiculidae* (*pinolillo*), flies, mosquitoes, cicada, ticks, wasps, bees, crickets, grasshoppers, the insect locally called "lazy", *Pyrophorus*, and fireflies, plus arachnids (spiders and poisonous tarantula) and nematodes (Morales-Cruz 2009: 18–19).

The aquatic fauna is caught or captured in the lagoon systems of the Juquila and Tepanzacoalco rivers, the streams of Yaviguile, Yego luya, Pisle, and one spring. This aquatic fauna includes prawns, shrimp (donkey, long hand, sweet), Hog nose mullet (*Joturus pichardi*), trout (*Lebiasina bimaculata*), peje (*Lepisosteus osseus*) pig, ahuja beak, snorers, crappie (*Kihnichthys ufermanni*), pepesca (*Bramocharax baileyi*), catfish (*Siluriformes letaluridae*), eels, juiles (*Rhamdia reddelli*), snail, crab, terrestrial, aquatic turtles, and the aquatic centipede (FAO, TOP/GUA/3506, 2018; Morales-Cruz 2009: 28–29).

El Rincón has physiographic variations according to the altitude. The region is an "ecological mosaic" as a result of the altitude and complicated physiography, which produce a great variety of climates, temperatures, soil qualities, humidities, flora, and fauna. In the lower zone called "hot ground" or *yudra'a*, located from 300 masl to 1,100 masl, the region is humid subtropical, with a long, cool summer, and an annual rainfall over 2,000 mm. The agriculture there is temporary and diverse, with species that are grown according to the altitude, like maize, soybean, squash, chayote, green bean (Christmas Eve, quarantine, *chirrión*), cat chili, *chinene* avocado, sweet potato, yucca, scallion, red onion, and yam bean (Tyrtania 1992: 33–34).

Cornfield weeds include Kwan-sleyi, deer and shrimp green weeds (*quelites*), purslane, *quintonil* (*Amaranthus hypochondriacus*), and wild pea. The medicinal plants are the naked coral tree (*Erythrina americana* Mill), marjoram (*Origanum vulgare*), and chamomile. Seasoning plants in the cornfield include Creole coriander, peppermint, and wormseed (*Dysphanias ambrosioide*). In 1992, cornfields had coriander, wormseed, *miltomate* (*Physalis* sp.), mint (*Menta spicata*), and other medicinal herbs and trees that were planted in the orchard or the homegarden, like zompancle or zompantle (*Erythrina americana*) (Tyrtania 1992: 99). In 2005, the plants collected in the forest were river and hill *tepejilotes* (*Chamaedorea tepejilote*), yellow mushrooms from white oak and royal *jonote* (*Heliocarpus appendiculatus*), and Mexican hog plum. However, the variety of species collected was being affected by deforestation (Morales-Cruz 2009: 28–29).

The fruits grown in the different altitudes of the basin are pineapple, sour sop, yellow and pink guava, papaya, banana, tangerine, lemon, grapefruit, mango, tamarind, cane, mammee, *cuajuinicuil* (*Inga edulis*), macho banana, cacao (*Theobroma*), passion fruit, and coconut. Those collected in the forest include wild pomegranate (*palao*), mountain melon, and wild grape (October). The flowers are lily, gannet, daisy, crouching, gladiola, baby's breath, roses, chrysanthemum, betulia (*Betula*), flucia (*Fuchsia*), doggy flower (*Antirrhinum majus*), and tulip (*Hibiscus* rosa-sinensis) (Morales-Cruz 2009: 14–15).

Wild vegetation consists of mustard, blackberry grass, wow, eagle stick (*Alnus acuminata*), soft stick, *aguacatillo* (*Persea schiedeana*), pinabete, ayacahuite; turpentine black pine (*Pinus uncinata*), large zarcina pine, red pine, Quercus thin leaf oak, white teaspoon oak (*Quercus Leucobalanus*), red teaspoon oak (*Q. Erytrobalanus*), thin leaf black oak, oak Senna leaf, large oak yellow, and oak broad-leaf, *huachichi*, ash, cross stick, wild orange, *Tepehuaje*, oak Senna leaf, *Zapotillo*, mate's hillock, *cuajiniquil* (*Inga* edulis), almond, blackberry, arbutus, pochote (*Ceiba pentandra*), and cedar (Morales-Cruz 2009: 14–15).

The temperate land, or *yunala*, locally known as "land above the town", goes from 1,100 to 1,600 masl. The climate is mild humid, with summer rains, starting in June, and lasting until September, with an average of 2,516.9 mm. May is the driest month in the year, and the annual average temperature is 18.6 °C; the hottest month occurs before June. This temperate land is – in fact – an ecotone, with winter rains and humidity, providing from 18% to 20% of its annual rainfall, which allows maize cultivation in the summer and a winter sowing (*tonamil*) (Tyrtania 1992: 36–37).

The cold land, known as *monte*, or *yundri'aga*, is located from 1,600 to 3,000 masl. The climate is mild-humid, with a reduced winter rain and little oscillation of the annual average temperature of 14.4 °C. The hottest time is before June. It can frost and occasionally at dawn in the winter, and the highest peaks can have snow, which melts in a few hours. The climatic changes in El Rincón are related to the summer winds, tornadoes, and hurricanes that occur in the Gulf of Mexico. These can affect the flowering of fruit trees, like mango and citrus, so to counteract this, farmers grow a variety of trees. The trees serve as a windbreak (Tyrtania 1992: 37–38).

In the late 1980s and early 1990s, Tyrtania in his research in St. Cruz Yagavila (1992) discovered how the maize grown by small-scale farmers there was selected and how they managed maize seeds, moving them from the coast (200–1,100 masl) to the town center at 1,059 masl (ecotone), and later from the town center to the upper part (1,600–3,000 masl), to return again to the middle area of the mountain (the ecotone). These movements aimed at getting the maize grown in them to tolerate those extreme zones, and maize acquires the necessary strength to defend itself against insects, which preyed upon it after it had been planted in the same ecotone zone for 1–2 years. This was a genetic management of the plant to reduce production losses caused by predators.

The soils in St. Cruz Yagavila are sandy-stony, and they are also laterized (*tepetates*). The main problem is the gradient of the land that causes the nutrient to be "washed" away by the copious rains. There are no managements for the recovery of eroded soils, but preventive practices are carried out like: doing a single selective weeding; accumulating biomass in the soil (green manure); including coffee husk, and biodegradable waste in green manure; leaving the trunks and plant branches on the ground, which quickly sprout in the cornfield; building stone and earth walls to handle the rugged topographic relief, and reducing erosion. Trees are planted on the edges that are horizontal strips on the plots, where ruderal vegetation is allowed to grow, prevent, and reduce soil drag (González-Jácome 1984b (146): 26–33).

Small-scale farmers in St. Cruz Yagavila carried out land management, which they call plow *arada*, to prevent landslides of sloping land. Following the contour lines horizontally, they build raised earth edges where humus accumulates. To do this, two horizontal strips are made with an iron hoe placed in an oak plow, which is pulled by two oxen. The removed humus accumulates between the two strips, and the plowing process distributes the humus evenly on the farmland. Then, the soils are left to rest to recover its fertility (Tyrtania 1992: 44).

The ecosystems and agroecosystems in El Rincón included: (1) pines or Quercus (30.4%); (2) mixed forest (28%); (3) low mount (6.6%); (4) chaparral (11%); (5) mesophilic forest (6.8%); and (6) scrub (3.2%). The total area of El Rincón was 1,821 ha in 1992. Natural vegetation covered 1,567 ha (86%), and agricultural land had some 259 ha (14%). The agricultural area was divided into: (1) cornfield 138 ha (7.6%); (2) reedbed 15 ha (0.8%); (3) banana plantation 10 ha (0.5%); and (4) coffee plantation 96 ha (5.3%). In 1992, the cornfield constituted 53.3% of the total agricultural area (259 ha) (Tyrtania 1992: 81).

St. Cruz Yagavila is a town with pre-Hispanic origin, with a stele 1.5 m high and 0.50 m wide that is dedicated to the "Rain Saint". The stele is located in the town square. In the dry season, when fields need rainwater, the stele is doused with a pitcher of water so that there will be rain, although it no longer responds as before" (Tyrtania 1992: 97). In the eighteenth century, Yagavila was dedicated to the cultivation of Opuntia, which hosted the Mealy bug insect that was the source of grana or cochineal, used as a paint on textiles, and works of art.In the nineteenth century, the oldest buildings in the town were built (parish, and municipal agency). In 1921, the first school and the teacher's house were built. From 1934 to 1948, the first stores emerged. By the year 1959, the potable water system was built. In 1973, more classrooms for the school, a basketball court, and housing for the teachers were built. In 1975, the Parish Dispensary was organized, and in 1976–1981, the governmental Cooperative a CONASUPO-COPLAMAR store was established (Tyrtania 1992: 97).

In 1992, the town had 143 homes, a dozen houses in ruins, and some houses under construction. In 1992, there were four stores and seven commercial establishments that were operating in private homes. The village had 161 domestic units. The houses had a 1 m^2 vegetable garden, with a stone-wall bordering where chives and vegetables were grown, fertilized with stove ashes. In addition, they had a chicken coop and a space for pigs, although the widespread custom was to let domestic animals roam free.

Surrounding the rooms was the homegarden, planted with banana, orange, lemon, lime, grapefruit, avocado, mammee tree, mango, papaya, sapodilla, pepper, custard apple, guava, Mexican hog plum, peach, cocoon, loquat, plum, fig, cacao, coffee, *cuanabiche* (*Hippocratea celastroi*), *kuando* (*Rivea corymbosa*), *Cestrum noctumum* L. (*huele de noche*), silk-cotton tree (*Ceiba pentandra*), and oil nut tree (*Ricinus communis*) (Tyrtania 1992: 206). In 1992, the homegarden occupied an area of 70.20 ha. Homegardens were part of the legal estate of the town, but in the 1940s, they began to be converted into coffee plantations (Tyrtania 1992: 96–97).

Cornfields were worked with slash and burn, organized in cycle with the fallow of the soil the time varying it was hot land (rotation every 2 years), or cold land (20% of the land is used annually) (Tyrtania 1992: 147). There are three types of fallow and one variant: (1) the short cycle, when secondary vegetation reaches the first phase of succession (chaparral) not allowing the restoration of soil fertility, so the fields become reed, banana fruit, or coffee plantations that replace natural vegetation of the second phase of succession (chaparral); (2) the medium cycle, when the secondary vegetation (chaparral) and is replaced by the low forest which does no allow the restoration of soil fertility that replace the natural vegetation in the second phase of succession; (2.1.) And a variant of the medium and productive fallow, when perennials are cultivated; and (3) the last cycle is long, when the vegetation grows and reaches its mature phase (forest or thicket), but is abandoned if the soil only produces weeds and rest is mandatory. Then, the soil is fertilized with ash (Tyrtania 1992: 149–151).

In 1982, a Mexican cornfield had an average yield of seasonal maize of 1,100 kilos/ha (Montañez (cord.) 1982). In the case of St. Cruz Yagavila, the yield varied owing to several factors. The maize managed there by the small-scale farmers could have been short cycle (*zxoa, kuidi*), which matures in 4 months, or long-cycle maize (locally called *guiá*, or maize from the mountain), which matures in 8 months. All the cornfield systems were polyculture; however, they included maize, beans, and squash, because the three were very important in the farmer family diet. Small-scale farmers distinguish four maize crops (Tyrtania 1992; 160).

1. *Yulaguia'h* is the warm weather storm cornfield where short-cycle maize (*zxoa kuidi*) is planted. The altitude of the land indicates the starting date for planting, which at 500 masl is from April to May. Seasonal sowing usually uses yellow (*zxoagachi*) and white (*zxoachikichi*) maize. From 1,100 to 1,300 masl, there is no temporary planting. The red maize variety (*zxoagachi*) is more productive, but many people do not like "red tortillas". This maize is obtained from the selection and mixing for 5 or 6 years of the *zxoamiú* maize grain with the *zxoapintu*. It is a complicated procedure practiced by few farmers. Some varieties, such as *zxashunazx*, are less frequent "[They] are probably in the process of disappearing [...]" (Tyrtania 1992: 157).

   In maize, sowing four seeds are placed/hole, with a distance between each hole of 0.60–0.80 m. The clearing of the land and the harvesting vary according to the altitude and the date the maize is planted. Each maize plant produces one cob. At 1,100 masl, the agricultural cycle begin in November, but at 1,300 masl, it began in mid-August (Tyrtania 1992: 156–157).
2. *Yulaguia'h*, or hot land *tonamil*, is grown after the rains and the same varieties of maize are planted. A "short maize" variety with a 5-month ripening cycle grew on these lands. This is a poor and unstable crop, which produces less after three or four successive cycles, and the fallow is lengthened in geometric proportion. The tonamil is planted after rain, and its cycle is the one that requires the least work (Tyrtania 1992: 157–158).
3. *Yulayechu*, this *tonamil*, called "misfortune" or "adventurer", is planted in the temperate land on the well-drained ecotone terrains. The small-scale maize farmer (*milperos*) cultivates a long cycle maize variety whose ripening cycle exceeds 6 months. At the end of the summer, when harvesting is done, the available food for farmers and their families is maize, green beans, sweet maize, and squash, that is cooked with brown sugar (*panela*) (Tyrtania 1992: 160).
4. *Yulaguia'h*, or mountain harvest, is called "cold land onset of rains". The higher the cornfield is, the longer the maize matures. Maize is sown together with mountain beans (big beans or Celaya beans) and squash. The deeper the farmer is inside the mountain, the earlier the maize is planted, because the forest has more moisture. In areas with more inclination, it is feasible to plow after slash and burn. The cornfield is planted for 2 years, and then the soil is left to fallow another 2 years or left unplanted for the establishment of chaparral ecosystem. In most fields, the resting lasts 4 years. Preferably, white maize (zxoachikichi) is sown placing four seeds in each hole, with a distance of 0.90 m between each hole (Tyrtania 1992: 160).

The local and regional classification of maize is as follows (Tyrtania 1992, Table 5; p. 163): (1) white *zxoachikichi*; (2) yellow *zxoagachi*; (3) purple *zxoamiú*; (4) black (dark purple) *zxoagassa*; (5)

white, purple, and red colored maize *pintu*; (6) red *zxoashná*; (7) pink buzachi; (8) purple and yellow maize *bayu* or double; (9) live red *shunazx*; (10) maize with a cob of 16 or more rows of kernels *lassi*; (11) and maize seeds from the La Luz ranch or other parts. Maize moved in two directions: horizontal, changing the varieties from one ecological floor, and vertical, changing the maize varieties from one ecological floor to another. The exchange of genetic material is carried out with the maize of cold land, and those of ecotone land (temperate) or vice versa. In 1986, Tyrtania (1992: 166–167) found 18 varieties of maize handled in the hot land and 12 in the cold land.

Maize productivity in St. Cruz Yagavila varied from one ecological floor to another. In seasonal cultivation of the hot land, a bad crop produced 676 kilos/ha, a regular harvest 944 kilos/ha, and a good harvest up to 1,579 kilos/ha. The yield average was 1,066 kilos/ha. In the *tonamil* in the hot-land, a bad harvest produced 490 kilos/ha, a regular one 672 kilos/ha, a good one 960 kilos/ha, averaging 707 kilos/ha. In the setback of the ecotone in the temperate land, a bad harvest produced 480 kilos/ha, a regular one 768 kilos/ha, but a good one 1,267 kilos/ha, averaging 838 kilos/ha. During the seasonal rains in cold weather, a bad harvest yielded 640 kilos/ha, a regular one 960 kilos/ha, and a good one 1,929 kilos/ha, averaging 1,120 kilos/ha (Tyrtania 1992: 170–172).

In cold land, the cultivation of 1 ha of maize required 161 days of labor/man, not applying agrochemicals and using metal tools to produce the 1,120 kilos of maize/ha. In St. Cruz Yagavila, the total annual maize consumption was 137,696.25 kilos (Tyrtania 1992: 172), insufficient to meet the needs of the population. Maize constituted 75% of the population's food. A person from 18 to 60 years old consumed a daily average of 0.5–1 kilo of maize, and the average annual consumption/person was 224 kilos (Tyrtania 1992: 172). In the 1950s, coffee cultivation began. This crop was destined for trade, to provide people with monetary resources.

The organization of agricultural work manages the following social elements and economic costs: mutual aid (*guzún, mano vuelta*); sharing work; profits and losses equally (*gasaquiá-gasaquiú*); and favors that do not require reciprocity (*laguetza, non-paid*). People had to collaborate in communal work (*tequio*), and there are *tequios* between communities, but salaries are paid only on workdays and a piece-by-day basis according to the amount of work invested. In 1983, agricultural workers employed in the cornfield or coffee picking, received 80 pesos/day/man, plus food. A woman was paid less for doing the same type of work (Tyrtania 1992: 182–183).

17.16 CONSIDERATIONS AND CONCLUSIONS

The types of cornfield in this chapter are not the only ones that exist in present-day Mexico. One of the characteristics of a cornfield is the capacity for biological and cultural adaptation to different environmental, climatic, and altitudinal circumstances, annual rains, and humidity, in addition to adaptation to physiographic conditions of the regions and their zones, adjusting to the tastes and needs of the small-scale cornfield farmers (*milperos*), their families, and their kin, plus the relations with the market and the greater society. Cornfield diversity can be understood as a continuum, which goes from polyculture to monoculture, and occasionally there is a return to polyculture, or there is an emergence of gardens, as happened in ancient times with changes from the forest to the forest-homegarden-cornfield, and later to the cornfield agroecosystem as occurs in small and medium-scale farms today.

The itinerant slash-and-burn cornfield has been reduced in extent more and more due to issues such as population growth, increasing population densities, a growing lack of land for cultivation, forest deforestation, and the conversion of forest into grazing areas for livestock or into land for monocultures that are aimed at trade. Migration and emigration had reduced the amount of family non-paid labor force for agricultural work. The slash and burn areas are shrinking, as they have caused uncontrolled fires, or because they produce high levels of pollution. Furthermore, the young people of several villages no longer wish to engage in agricultural work.

There is a population decrease that indicates processes of migration and emigration whose effects on agriculture would have to be studied. Likewise, commercial relations inside and outside the region need to be studied, especially regarding agricultural products like coffee, and all those that were introduced in the 1980s, which reduced the time for soil fallow, and are basically directed to commercial sale. In any case, St. Cruz Yagavila is very illustrative of what has happened in other places in Mexico with respect to the cornfield and its interrelations with other ecological and agroecological systems (Bacon 2001, 2005a, 2005b, 2008; Méndez et al. 2010; Moctezuma-Pérez 2008; Pérez-Castro 1988, 1996, 2007).

After 1980, in regions where coffee was interspersed with the forest (coffee forest), after 1980 local and regional coffee production entered into crisis, and seasonal migration and emigration increased. This was generated by the disappearance of institutions that contributed to a more or less stable price control, such as IMECAFÉ, to avoid the sudden changes in prices established for coffee in New York, which characterized the last two decades of the last century, and continues to the present. These crises in coffee prices and markets favored coffee from Vietnam and other countries in Asia and Africa over coffee produced in the Americas, including Brazil, Colombia, and Mexico (González-Jácome 2004: 1–31, 2009: 1–15, 2012; Moguel and Toledo 1999; Moguel 2011: 193–219; Pérez-Castro 1988, 1996, 2007). Initially, the impact on coffee-growing indigenous societies was economic, but now it has affected the organization of families, communities, and agriculture in these regions (Moctezuma-Pérez 2008; Pérez-Castro 1988, 1996, 2007).

Mexican traditional agroecosystems combined with other agroecosystems like the polyculture cornfield with conventional monoculture agroecosystems, whose production is destined for the market basically, have grown because they have been productive for the small-scale farmers' families who receive an important part of their food and monetary income from them. They need to be able to cover their daily needs and the extraordinary expenses related to the outlay for industrialized food, the payment of services (water, electricity, gas, land tenure payment, transportation, landlines, cell phones, and the education of children), the purchase of clothing, footwear, appliances, the payment of community fees, the costs of religious holidays, and family events.

Agricultural production requires funds to purchase seeds, fertilizers, and herbicides; to pay salaries; to cover services such as tractor rental when indispensable; to acquire tools; and to rent or acquire work animals. Expenses include the appropriation and management of agricultural technologies, like the annual application of chemical fertilizers, plus the expenditure of labor for the rotation of the 3-year or 4-year application of green manures, animal fertilizers, or for the application of muck or mud-water on the parcel.

Cultivated plants are grown together with weeds and combined with wild plants collected in nearby agroecosystems. The use of tractors alone or mixed with the use of draft animals (oxen, horses, and mules) depends on the type of soil and gradient, or monetary availability. Small and medium-scale farmers combine the use of irrigation systems of ancient origin along with irrigation from wells with electric pumps.

To do the plot work, there is a combination of the nuclear family, kin (an unpaid work force), and wage labor that may or may not be a part of the extended family. Agricultural work done by women and industrial in-bond factory work done at home by families had been increasing due to the exit of family members who go to work in nonagricultural paid jobs out of their towns and away from their families. The feeding of domestic and work animals with a mixture of industrialized foods induced or cultivated pastures, and ruderal vegetation (browsing) is the custom, if it is possible now.

These combinations do not discard the elements of traditional agriculture, rather they incorporate them and thereby increase diversity. This is a characteristic of traditional agroecosystems, but it will be lost if they disappear. Unfortunately, polyculture is replaced by monocultures, and the food system is modified by the massive introduction of industrialized foods into family consumption. The food system can be modified more quickly than it could ever have imagined (Holtzman 2006), and the collective memory associated with the traditional system fades to the extent that young generations leave, and communities to work in non-agricultural salaried activities.

The adaptation of traditional Mexican agroecosystems to environmental; climatic conditions; historical-geographical, and altitudinal situations; adaptations to flooded, tropical, semi-desert areas, with complex physiography; and others are characteristics of these traditional agricultural systems. There are several examples of these processes.

The adaptation of the cornfield agroecosystems to waterlogged areas, which generated agroecosystems like the March Maize (*marceño*). In this type of case, maize undergoes genetic changes so that it survives successfully in humid conditions and shortens its ripening period. It was thought that March Maize was an adaptation of maize to tropical conditions with excess humidity. However, it is still unanswered if March Maize has been cultivated in conditions of high humidity or in temperate-cold climates where maize has to have a short cycle to reach maturation before events such as frost and cold affect the zone.

The lack of adequate land for crop support is one consequence of the development of agroecosystems in mountainous or hard-to-reach lands. This is a historical fact because from 1519 Spaniards displaced Mesoamericans from the best agricultural plains-lands to the hills and zones with no water for irrigation. Among these agroecosystems is the *huami*l. This *huami*l agroecosystem has several advantages, including: (1) the system does not need capital investments, although it does require a lot of manual labor; (2) the work can be allotted throughout the year; and (3) the *huamil* agroecosystem is better adapted to the social and economic conditions of the region's population, which is rich in labor resources, but poor in capital goods. In contrast, monoculture hybrid maize requires capital to invest in the acquisition of seed, fertilizer, and machinery. Although *huamil r*equires little work focused on certain stages of the production process (sowing, weeding, and harvesting) (González-Martínez 1992: 142–159; Palerm-Viqueira 1997: 130–131; Valencia 1981: 62–117).

In St. Cruz Yagavila, changes in the cornfield agroecosystem that occurred in the twentieth century were: (1) an increase in cultivated area by 40%; (2) the increase in farmland varied according to the ecological floors; (3) the fallow cycle of soils was reduced by half; (4) a diversification of cultivated species as well as new crops were being introduced, and others became more important like coffee, sugarcane, banana, citrus, mango, and mammee trees; and (5) factors that impacted the region and the populations were the opening of roads and highways in the region, the presence of foreign companies logging in the forests, and the low coffee prices in the national and international market.

The processes of maize management at diverse altitudes and its variety have availed the maize coevolution with other plants in the cornfield (Benz and Long 2000: 459–465). The processes of adaptation versus counter adaptation, with respect to several issues, like resistance to pests and diseases, allowed the cornfield to exist in places like El Rincón and St. Cruz Yagavila at least 20 years ago without the need to apply agrochemicals to this agroecosystem. The mobility of seeds among the ecological floors and among the plots of the same floor is related to forms of social organization and work promoting it horizontal and vertical rotation entails specific types of soils, humidity, and maize kernels.

The cases described in this chapter showed that the role of diversity is extended to include the crops of the cornfield, the relationship of the cornfield with the forest, and the natural areas surrounding the town, plus the physiogeographic and environmental characteristics there.

The role of the combination of annual and permanent crops is important, a similar fact to that of the homegarden-cornfield. Also, diversity plays an important ecological and economic role in the functioning of the agroecosystem. It is the combination of agroecosystems and ecosystems, that attains their success, despite changes in the food system. To date, the cornfield agroecosystem is able to provide sufficient food for the family and to generate monetary income for small-scale farmers (Anjani et al. 2020; FAO 2001; Lachat et al. 2018; Rebato 2009; Wilk 2019?) (Table 17.1).

TABLE 17.1
History and characteristics of Zoque-popoluca Cornfield

Date	Crops	Agroecosystem	Other Economic Activities	Source
4600 BC	When vegetative agriculture is combined with grain agriculture, the Popoluca cornfield originates.	The cornfield arose and the weeds also.	No data	Blanco (2006: 49)
2500 BC	In Los Tuxtlas' coastal area, the maize varieties crossed, and the oldest ones disappeared. For Belize (Colha, Pulltrouser Swamp and Cob Swamp), the forest was cleared, initially cultivated with maize and cassava.	Cornfield included grasses, vegetative reproduction, and edible green weeds like purslane, and wild small green tomatoes.	No data	Blanco (2006:50) and Fedick (1996)
2500 BC– 1600 BC	Early Formative, or Early Postclassic.	Cornfield was a complex agroecosystem with plant diversity. Human population increases, and combined maize with beans and chili. Maize hominy (nixtamalization) (*mooxi*) appeared, and the grain was boiled with lime from wood ash or burned shells.	Agriculture is combined with hunting, fishing, and gathering. Women make clothing woven from Agave fibers and cotton.	Blanco (2006: 50–51)
2000 BC	Early Formative, or Early Preclassic.	Maize is cultivated in Los Tuxtlas. Tree pollen decreased, and maize pollen increases along with weed pollen associated with cornfields.	No data	Goman (1992)
1600 BC–600 BC	Middle Formative, or Middle Preclassic.	There are agricultural villages, with a polyculture cornfield and intensive agriculture. Agricultural tributary villages of urban centers were founded in the region (S. Lorenzo).	No data.	Blanco (2006: 50)
1400 BC	Middle Formative, or Middle Preclassic	There is archaeological evidence of human occupation in Catemaco, in Los Tuxtlas region. Tree pollen decreased, and maize pollen and weed pollen increased.	No data	Santley and Arnold II (1996: 225–249)
1000 BC–600 BC	Middle Formative, or Middle Preclassic, Classic, and Postclassic. In 1000 BC there is population growth (8.5 habitants/km^2	The greatest agricultural activity occurred in the Late Classic and Postclassic. There is an increase in tree pollen, which the authors related to the post-conquest population decline. They founded high percentages of herbaceous plants, especially Ambrosia.	No data	Byrne y Horn (1989: 181–193)

(Continued)

TABLE 17.1 (*Continued*)
History and characteristics of Zoque-popoluca Cornfield

Date	Crops	Agroecosystem	Other Economic Activities	Source
600 BC– 200 AD	Late Formative, or Late Preclassic.	Agriculture with water management in both low and high areas. Greater diversity.	No data	Blanco (2006: 51)
300 AD	Late Postclassic	There is population growth, greater deforestation, but patches of vegetation in the forest are left, which will later allow its regeneration. Tree pollen decreased, and maize pollen increases along with that of cornfield weeds.	No data	Santley y Arnold II (1996: 225–249)
200 AD–800 AD	Classic	Irrigated agriculture. There is slash and burn in the rain forest of the Mayan zone of the Yucatan peninsula and Belize. In the tropical lowlands, it is calculated that the minimum management unit in the traditional cornfield was 2.28 ha.	No data	Blanco (2006: 51) and Fedick (1996)
800 AD–1519	Postclassic	The Olmec centers collapse, and the chiefdoms arose. The one from Coatzacoalcos came into contact with the Spanish. Xoteapan paid to the Coatzacoalcos Chiefdom: (1) 4 tables of cotton tablecloths; (2) 10 cacao zones (4,000 cacao grains); (3) 5 turkeys; (4) a pitcher of honey; (5) annually 80 hanegas of maize; (6) every 2 years a half of hanega with beans.	No data	Blanco (2006: 52–53) and Gasco de Herrera AGN (1952: 570)

(*Continued*)

TABLE 17.1 (*Continued*)
History and characteristics of Zoque-popoluca Cornfield

Date	Crops	Agroecosystem	Other Economic Activities	Source
1519	Spanish conquest. They defeat the Zoque-Popoluca of Soteapan in Cuilomenequis (today Cuilona). The mandatory work not paid (*encomienda*) was created. The Spanish *encomenderos* in the region were Joaquín Martín de Valencia (1554–1567) and Luis Guillén (1597–1599). The settlers continue to pay the same taxes to the *encomenderos*.	No data.	No data.	Blanco (2006: 53)
1520–1521	The Coatzacoalcos chiefdom region had more than 50,000 inhabitants. However, they were depopulated because of smallpox and measles. Cotton, cacao, maize, birds, copper axes, and some gold jewelry were taxed to the chiefs.	They planted: maize, beans, pumpkins, yucca, and all January fruit trees.	They catch birds, oysters in the lagoons, fish, and shrimp. They made Ceiba wood paper. They made trusses (*bragueros*).	Relación Geográfica de Coatzacoalcos, 1580. En: Antequera, T.1°. Acuña (1984: 115–126).
1525–1528	Hernán Cortés established the first sugar mill in Santiago Tuxtla. Acayucan and Santiago become the economic centers of Los Tuxtlas. The native population decreases and the indigenous people who remained in the mountains are relocated to new towns. Some take refuge in inaccessible areas of the mountains.	No data.	A bovine cattle breeding base is established in the Sierra de Los Tuxtlas and the lower parts of the Papaloapan and Coatzacoalcos rivers.	Laborde (2008: 64)

(*Continued*)

TABLE 17.1 (*Continued*)
History and characteristics of Zoque-popoluca Cornfield

Date	Crops	Agroecosystem	Other Economic Activities	Source
1580	About 3,000 Indians remain. There were 66 "poorly populated and poorly ordered" towns including Xoteapa. The head of the Province of Coatzacoalcos was Villa St. Spirit that was changed to Acayucan due to epidemics and population decline. It is extremely hot and humid land. The Sierra de St. Marta or Soteapan, changes its name to Sierra de S. Martín. Located on the bank of the Cuitlaxcolapa river, it is 3 leagues from the sea; 10 leagues from the Minzapa and S. Martín mountain ranges; 90 leagues from Oaxaca (the road is all of mountain ranges, rivers, swamps, there is little plain). In Acalapa, there is a spring with a saline. There are several lagoons with a lot of fish (Minzapa). The main river is Coatzacualco; it has many good waters at its mouth, a lot of firewood. There are ranches near the sea.	They plant maize twice a year,; squash; chili; tomatoes; beans; sweet potatoes; yuccas; a lot of cacao; a lot of tobacco; a lot of cotton; and vegetables from Spain. Popolocas had medicinal plants: tobacco, mimizptle (dropsy), chamomiles (purge), zezequipatli (cecepatli), and suchimecatle. There have wild and cultivated trees of mamey, medlar, sapodilla trees (sapota, peruétanos), avocados, quaxiniquile (carobs), cuajilotes (axilotes), sapodillas, plantains, pineapple, Paqui (like hazelnuts), red Dragon fruit (*pitahaya*), wild grapes and hog plums of four genera all-different from those of Spain. They obtain wood from the cuacuahuite (*quocobite*). Only cacao and maize are taxed. The Popoluca people of western Coatzacoalcos prayed to Homshuk, the god of maize. Their rituals for the cornfield continue to be an axis of the life of the current human population.	They hunt the animals of the region: tigers, deer, temazatles (they are deaf), wild boars, rabbits, lizards, vipers and snakes. There are many deer to the coast and poultry. The Spanish had cattle ranches, mares and mules. They set up fisheries in the lagoons. Spaniards were engaged in the merchandise trade of Spain and cacao.	Relación Geográfica de Coatzacoalcos, 1580.En: Antequera, Tomo 1°.Acuña (1984: 115–126).
1619	The Spanish Crown recognizes Soteapan as a Republic of Indigenous people, and they receive titles from the Viceroy Marqués de Guadalcázar. They buy private land.	No data.	No data.	Blanco (2006: 53, 55).

(*Continued*)

TABLE 17.1 (*Continued*)
History and characteristics of Zoque-popoluca Cornfield

Date	Crops	Agroecosystem	Other Economic Activities	Source
1700–1793	They enter judgment with the Hispanic cattlemen of Chinameca by limits of earth that require for the cornfield, the fallow land, territories for hunting, fishing, and gathering.	Cornfield with fallow.	Hunting, fishing, and gathering.	Blanco (2006: 53–54)
1778	In the eighteenth century, the Chinameca Parish has 444 wild trees: cedars, balsam, quiapinole, Palo de María, guayacán, Mulato tree, zave tree, Red Dragon fruit, yellow Dragon fruit, totoloches, palm quaoque, nanche, black, and white and red Hicacos. Settlers continue with the cult and ancient rituals of the cornfield and the carnival party of the tiger. In addition to war ceremonies and defense of the territory.	Non-irrigated cornfield with two harvests a year (seasonally and tonamil). Maize, orange, banana, sugar cane, rice, beans, cotton, cassava, pumpkins, and sweet potatoes are planted. Medicinal plants: rue, star, immortelle, staph, sangria, Chinese grass, and Mulatto stick (to use the leaves, which are medicinal). Cultivated fruit trees: avocado, medlar sapodilla tree, Mamey tree, Sunday sapota, black sapota, yellow sapota, villain orange, files lemons, quaxiniquile, white and purple custard apples, coconut palm, guava, pineapple, papaya, watermelon, cantaloupe, banana, sugar cane,; Sisal Domestic animals were chickens, turkeys, pigeons, ducks, pigs, dogs, and cats.	Wild animals: deer, wild boar, rabbit, quagueguече, armadillo, squirrel, monkeys, badger, tepescuintle, tiger, fox, coyote, pheasant, partridge, chachalaca, doves, ducks, parrots, parakeets, macaws, cenzontle, calendars, cardinals, and Pepes.	Relación Geográfica del Obispado de Oaxaca. Relación de Juan Chinameca, Coatzacoalcos.

(*Continued*)

TABLE 17.1 (*Continued*)
History and characteristics of Zoque-popoluca Cornfield

Date	Crops	Agroecosystem	Other Economic Activities	Source
1906–1924	Revolution in the region. Manuel Romero Rubio, father-in-law of Porfirio Díaz appropriates the lands of the Old Republic of Indians of Soteapan. In 1906 the Soteapan popolucas enter the war to regain their lands. The towns of Buenavista, Tulin and Ocotal Grande were burned several times.	Zoque-popoluca farmers practice polyculture cornfield. They fight but return to cultivate their cornfields and do the rituals to Homshuk. They grow: maize, beans, squash, chayote, and 20 other plants.	No data.	Blanco (2006: 65)
1924	Zoque-popolucas' lands are converted into granted lands (ejidos) and receive endowments. The Popolucas fight to preserve their communal lands.	No data.	No data.	Blanco (2006: 65)
1940	No data.	Non-irrigated cornfield (tonamil), using slash and burn is planted with coa.The cobs are folded to prevent rain. Lands rest every 7 years. Cornfield had maize, beans, squash, chayote, sweet potato, cassava, tomato, Yam bean, Annatto, pineapple, papaya; and coffee that is planted at 1,800 masl. There is a secondary crop called *tapachol*. Food: in a brown pan they cook: beans, squash, and small portions of chicken meat, turkey, pork, and hunted animals. Posol had grown maize mixed with water; it is taken to the cornfields and on trips because it is easily preserved and can be transported.	They cut the honeycombs from the trees and take them home. They hunt deer, rabbit, and wild boar. Birds: chachalaca (with a bow and arrows without an iron tip, almost twice as long as its stubby bows). They catch crayfish. The evergreen bush (*Barbasco*) is used to stun fish.	Foster (1940: 459–462) and Foster (1942)

(*Continued*)

TABLE 17.1 (*Continued*)
History and characteristics of Zoque-popoluca Cornfield

Date	Crops	Agroecosystem	Other Economic Activities	Source
1950	Zoque-popoluca farmers had a system of beliefs and rituals related to the agricultural cycle of maize.	Farmers plant more than 30 plants including maize, beans, squash, chili peppers, green small tomatoes, yucca, taro, sugar cane, fruit trees (sapodilla, Soursop, papaya, banana, orange, and mango). They have domestic animals. Cornfield polyculture function tied to a system of beliefs and rituals related to the agricultural cycle of maize.	Farmers had self-supply economy and also aimed at commerce. They hunt wild animals.	Blanco (2006: 60)
1960	Perla del Golfo and Zapoapan were founded; both of them are mestizo towns. The municipality of Soteapan had 9,064 ha of farmland, of which 4,109 ha were resting, and 4,955 ha were in production.	Cornfield: maize 2,629 ha produced 2,020 tons (768 kg / ha); beans 881 ha (581 ton and 659 kg / ha); rice 123 ha (113 ton / ha); Sesame and pumpkins. Tubers in the cornfield: cassava, Yam bean, and sweet potatoes. Fruit trees: mango (11 ha (86 ton / ha); Roatan plantain (11 ha with 86 ton / ha), papaya, orange, lemon, plum, and cocoon. In little quantity sugar cane and pineapple. Coffee 1,158 ha (1,338 ton / ha); and Barbasco. Slash and burn cornfield with 7 years of resting. Includes *tapachole* (winter planting); and *kotumóc* (storm milpa); orchard with fruit trees; coffee forest; planting of pangola grass for livestock forage in Perla del Golfo and Zapoapan.	Farmers had commerce and cultivation of pastures for livestock.	IV Censo agrícola y ganadero.

(*Continued*)

TABLE 17.1 (*Continued*)
History and characteristics of Zoque-popoluca Cornfield

Date	Crops	Agroecosystem	Other Economic Activities	Source
1964	No date.	Milpa with slash and burn, storm cornfield with 7 years of rest. Includes tapachole and summer cornfield, orchard with fruit trees, and coffee forest. Cornfield 2,629 ha (2,020 tons and 500 kg / ha); beans 881 ha (400; ton / ha); pumpkin and rice 123 ha (113 ton / ha); sesame; and tubers in the cornfield (cassava, yam bean, and sweet potatoes). Fruit trees: mango (11 ha (86 ton / ha); roatan plantain 11 ha (86 ton / ha); papaya (20 trees, 15 in production, 5 kg / tree); orange (100 trees, 60 in production, 15 kg / tree); lemon (100 trees, 30 in production, 15 kg / tree), Mexican Jog plum (60 trees, 13 in production, 15 kg / tree), and cocoon. In little quantity sugar cane and pineapple. Coffee 1,158 ha (1,338 ton / ha), wild and cultivated cacao in the homegardens, and barbasco.	No date.	Báez, information of Census of agriculture and cattle raising.
1967	No date.	Itinerant slash and burn with 7 years of resting. Cornfield cultivated with maize, beans, and coffee forest.	No date.	Foster (1967)
2006	No date.	Polyculture cornfield.	No date.	Blanco (2006)

RECOMMENDED READING

Yang, C.J.; L.F. Samayoab; P.J. Bradburyd: B.A. Olukolue: A.M. York; M.R. Tuholskia; W. Wanga; L.L. Daskalskaa; M.A. Neumeyera; J.J. Sanchez-Gonzalez; M.C. Romay; J.C. Glaubitzg, Q. Sung; E.S. Bucklerd; J.B. Hollandb; and J.F. Doebley. 2019. The genetic architecture of teosinte catalyzed and constrained maize domestication. *Proceedings of the National Academy of Sciences of the United States of America* 116(12): 5643–5652.

The domestication of crops allowed researchers to understand evolution. The authors of the article studied the genetic architecture of maize domestication with a quantitative genetic perspective. They analyzed traits of domestication in a local breed of maize and a population of teosinte; finding a strong divergence in the underlying genetic architecture, including the change in genetic correlations between traits. Although selection intensities were low for all traits, indicating that selection under domestication may be weaker than natural selection. The analysis suggested that the total weight of the grain/plant did not improve and that the genetic correlations restricted the selection (Turner 2010: 169–174; Turner and Lawrence 2011: 445–463).

The author's study was done with an evolutionary quantitative genetic approach to understand the limitations gave by genetic architecture of trait variation in teosinte, the wild ancestor of maize, and the consequences of domestication in genetic architecture (Turner and Lawrence 2011: 445–463). Using modern populations of teosinte and maize varieties as representatives of the ancestor and the domesticated respectively, they estimated heritability, additive, and dominant genetic variations by environment, genetic correlations, and genetic covariance for 18 traits related to domestication.

The authors used estimated genomic ratios from genomic markers. They found a reduction in heritability in most of the traits, and the reduction was stronger in the reproductive traits (size, number of grains, and ears) than in the vegetative traits. The authors found greater exhaustion in the additive genetic variance than in the dominant genetic variance. Selection intensities during domestication were weak for all traits, and reproductive traits showed the highest values. Neutral divergence was rejected in 17 of 18 traits, suggesting that they were selected during domestication. Yield (total grain weight)/plant is the only trait in which selection did not appear to have improved in maize relative to teosinte. From a multivariate evolution perspective, a strong non-neutral divergence was identified between the genetic variance covariance matrices of the teosinte, and the local varieties (matrices G). The structure of the G matrix in the teosinte presented considerable genetic restriction in early domestication, and the G matrix of the local maize race indicated that the degree of restriction is more unfavorable for further evolution along the same path.

Based on the results, the authors proposed several inferences about the domestication of maize. First, the selection focused on cob architecture, as most of the features that define its architecture had an inheritable loss of variation and its morphology changed dramatically. Second, the difference in plant architecture is likely a consequence, or a correlated response to selection in cob architecture, to create a less branched plant with fewer and larger ears. Third, due to favorable genetic correlations between most selection targets, former farmers had multiple options for selecting what a corncob would become.

In this way, EL is positively correlated with GE in the teosinte, so the selection could have been applied to any of the traits, or even to both traits to speed up the process. Fortunately, the relative differences between plant traits are easily observable with the naked eye and do not require modern tools for selection. Finally, evolution can continue despite significant genetic restriction, because selection can be applied to traits with the least restriction.

Altieri, Miguel A. and J. Trujillo. 1987. The agroecology of corn production in Tlaxcala, Mexico. *Human Ecology* 15(2): 189–220.

The article is based on information obtained in the Llanos de Apan macro-region, Tlaxcala, in the area of the Huamantla plain, with some examples taken in the Atoyac-Zahuapan River Basin, and Tlaxco. This is important because in Tlaxcala the type of cornfield varies according to the specific characteristics of each region in particular. The Huamantla plain is the most important in the monoculture production of maize, barley, and wheat. In the nineteenth century, and a third of the twentieth, the production of pulque and wheat or barley gave it economic importance. The study by Altieri and Trujillo described the main components of maize agriculture in Huamantla, the cultivation patterns used, the resource management strategies, and the interactions of human factors with biological factors.

Altieri and Trujillo, based on the studies of Gliessman, García, and Amador (1981), and Farrell (1984) in the Atoyac-Zahuapan Basin, used measurements derived from some selected fields, which showed that the trees integrated in the cropping systems modified the aerial environment and the soil of maize associated with plants, which influenced their growth, and yield. As the distance from the trees decreased the surface concentrations of most soil nutrients increased. The authors found that certain tree species affected maize yields more than others.

In the ridges of plots, the abundance of arthropods also varied depending on their degree of association with one or more of the plant components of the system. In the plain of the slopes of the Matlalcueye volcano, located to the east of the Basin, the densities of predators, and the maize pest *Macridactylus* sp., they were highly dependent on the presence and phenology of adjacent alfalfa strips. Although the data were obtained from non-replicated fields, the authors pointed to some important trends, information that could be used in the design of new crop associations to achieve sustained soil fertility and low pest potential.

Although the agricultural prosperity of the region's population was never fully achieved, because the agricultural macro-region of the Atoyac-Zahuapan Basin was practically destroyed by the construction of roads that passed over canals filled with stone, earth, and cement, and the construction or expansion of industrial zones, the modernization of the rural towns and communities. The authors' study allowed to reconstruct an important part of what was once the best highly productive agricultural area in Tlaxcala (González-Jácome 2008). On the other hand, the plain region in Huamantla continues to be linked to commercial agriculture with monocultures, associated with the brewing companies located to the east, in Calpulalpan, and the Apam plains (González-Jácome 2016b).

18 The Cornfield and the Food System

18.1 INTRODUCTION

For governments in Mexico and international institutions like FAO (2014), the current state of the food system is articulated with agriculture and small and medium-scale farmer. Agriculture has played a very important role by increasing the family food system, monetary income, and health quality, in addition to benefitting the global economy (Appadurai 2001). This articulation is related to food security, environmental sustainability, and the ability of farmers' families to make innovations that promote agriculture with the intensification of production and therefore the wellness of rural life. In this process, innovation is very important to increase production, whose profits increase family economy, and to create more effective practices of crop management.

From this perspective, FAO (2014) considered that there are certain factors that constitute a challenge for human populations: (1) climate change; (2) the extensive degradation of soil, water, and other factors whose negative impacts limit intensive production in agricultural systems; and (3) the discoveries of agroecosystems that are truly sustainable and inclusive can withstand an increase in production to cover the future food system needs of the poor and to be accessible and affordable for them. These factors were known by ancient societies with traditional agroecosystems, which responded to them, as discussed in previous chapters. Therefore, the generated local knowledge (TAK) is important to take into account in current discussions and responses.

For this, FAO (2014a: vi) proposed that it is the small and medium-scale farming family that is closest to achieving sustainable food production. However, small and medium-scale farmers have been considered to be an obstacle to development, and they have not received government support (Maestre-Alfonso, Casas-Gragea, González-Jácome 2008; Maestre-Alfonso 2011). In 2014, there were 500 million small and medium-scale farmer families (9 out of 10 farmers), who produced 80% of the food on the planet.

In Mexico, the opposition of authorities to traditional agroecosystems is historical, and it arose in the Viceroyalty, accentuated after the industrial revolution in England, and impacted agriculture negatively because they have supported monocultures and the mechanization of agricultural work since then. The late nineteenth century and the twentieth century, particularly, ushered in a series of technological and ideological elements opposed to the operation of traditional agriculture. The expansion of monocultures and the use of agrochemicals and machinery on land with complex orography, as was the case in the Mexican Central Highlands, produce the elimination of raised land edges, irrigation ditches, canals, forest relicts, and edible green seasonal weeds, which resulted in the reduction of biodiversity. Monocultures required flat areas where agricultural machinery could enter easily.

As the aforementioned factors are shown to be true, this chapter proposes that others must be considered as well to clarify the processes that have occurred since the origins of agriculture. The current situation in Mexico offers information about adjustments and modifications which agroecosystems suffered hundred or thousands of years ago. Secondary consequences include: (1) changes in the food system over time; (2) relationships between food and good health or illness; (3) alterations of the composition of families and their participation in agricultural work caused by social and economical circumstances, bringing about commuting, migration, and emigration; (4) changes in the local and regional cultures, the aspirations of young people, and their relationships with the families; (5) impacts on rural societies of processes such as modernization and the construction of

DOI: 10.1201/9781003198833-18

roads and highways; (6) changes in rural and urban cultures of the different human groups in the country; (7) losses of local and regional traditional ecological knowledge (TEK) and the traditional agricultural knowledge (TAK).

In this chapter, there is information related to the food systems that have occurred over time in some regions of Mexico. It is hoped to conserve a part of the local and regional knowledge that articulates agroecosystems with food production, which fulfills the demand for consumption. Without this local traditional knowledge, it will become more difficult to reach the goal of agroecosystem sustainability and a fair food system.

18.2 BACKGROUND

The archaeological, paleoecological, and genomic studies that were dated indicate that the cornfield arose in prehistory, from the domestication of plants, especially maize, and the creation of the first agroecosystems. There is information about what happened in two types of ecosystems. In the first, older type, the region was semi-desert, with trees, herbs, nearby water sources, and rocky places where residents took refuge from the contingencies of the environment and climate. The cornfield was born, also, in forested regions, where hunting-gathering was pursued and in the incidental homegarden, where the collection and consumption of fruits, berries, mushrooms, and seasonal weeds produced the incidental emergence of the first agroecosystems and agri-food systems in Mexico.

The domestication of maize was developed linked to the collection of flora, and hunting in the forest. The hunters threw the waste of what they consumed onto the ground, gradually leaving areas in the forest where these remains accumulated ("garbage dumps"). Seeds, plant remains and consumed fauna accumulated in them; the seeds took root and the cuttings grew, emerging in the forest areas with the foods preferred by the villagers. Then, little by little, the incidental homegarden arose and the preferred plants in this incidental selection, caused by taste and their consumption, were domesticated.

This type corresponded more to the cornfields in tropical, and subtropical regions of Mexico. However, it shares many characteristics with the forest, the recently born incidental homegarden-cornfield agroecosystem in temperate, cold, or semi-desert regions of the country. Among its characteristics are the increase in biodiversity, and the combination of the homegarden-cornfield (huerto-milpa) agroecosystem with natural ecosystems; in addition to the mixture of domesticated and wild plants, which were included together in the food system of these first settlers.

18.3 THE CORNFIELD AND THE FOOD SYSTEM IN ITS EARLY STAGES

The verification of the conformation of the Mexican cornfield agroecosystem is based on chronologically organized information from archaeologists, ethnologists, palaebotanists, phytologists, ethno botanists, and geneticists.[1] The information was obtained from several food specialists from different parts of what was ancient Mesoamerica, and from the north of the country and the southwest of the United States. It began with a summary of the Prehistoric data, and then included information of the most important events affecting human food systems at different times through the years.

The first human groups that lived in the middle basin of the Balsas River were organized in bands that had 10–12 people, dedicated to hunting, fishing, and gathering. The bands lived seasonally in rock shelters and caves. In this basin, the studies were done in the rocky refuge of Xihuatoxtla, near the city of Iguala, Guerrero. From 8900 BP to 8610 BP, the domestication of teosinte (*Zea mays* spp. parviglumis) and the beginning of its dispersion from this river basin began (Figure 18.1).

This teosinte is endemic in the region, and it was collected near the shelters where the settlers lived. The mobility of hunter-gatherers hindered the emergence of agriculture, but not the gathering

[1] Benz and Austin 2000: 459–465; Benz et al. 2006: 73–95; Buckler et al. 1996: 612–622; Flannery et al. 1981: 48–93; Huckell 2006; MacNeish 198: 131–47; MacNeish 1997: 80–93; Pohl et al. 2007 (104): 6870–6875.

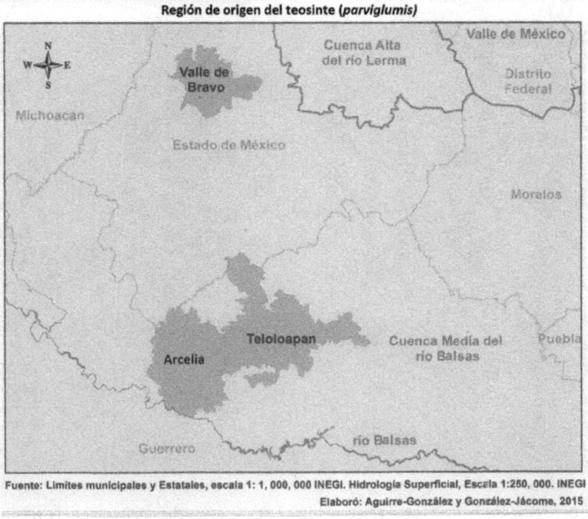

FIGURE 18.1 Map: The Original Region of Teosinte Parviglumis.

of flora. The selection of plants based on the tastes of the inhabitants began the process, which continued with the formation of "garbage dumps", where the hunter-gatherers strew the remains of plants and the bones of the fauna that they consumed (La Rochelle and Berkes 2003; Sutton 2010). These remains were located inside caves and shelters, or in the nearby areas.

Anthropogenic zones were formed where plants selected by the first settlers grew. Later, the first crops of teosinte arose coinciding with incidental domestication in the places where teosinte, squash, White lead tree pods (*Leucaena*), and fruits like the Mexican hog plum were collected. Selection was based on the tastes of the inhabitants, and consumption of these plants and their fruits was the first step in the invention of agriculture. Some decisions were related to the abundance or shortage of plants during the year.

The food system was based on plants collected for consumption, and the terrestrial and aquatic fauna that were hunted, fished, captured, or gathered. The fauna varied according to the places where the hunter bands were established seasonally. This situation changed as domestication, and plant cultivation began and advanced. In Tlacuachero, Chiapas, toward 2800 BC, maize, avocado, two varieties of beans, and Opuntia were cultivated, but the food system remained highly dependent on species obtained by hunting, fishing, and gathering (Smalley and Blake 2003; Blake 2006).

The first crops of which there is news were teosinte, Cushaw squash (*Cucurbita argyrosperma*), Mexican hog plum (*Spondias mombin*), and White lead tree (*Leucaena leucocephala*) which pods and seeds were consumed. By 7920 BP, there was cultivated teosinte in the central Balsas. Teosinte stems were sucked to obtain the sweet sap, and teosinte panicles and tufts were eaten fresh, toasted, or boiled in water. The teosinte is endemic in the region, and it was collected in places near the shelters where the settlers lived.

The hypothesis of Iltis (2006: 25) is pending confirmation. It is that a fermented beverage, similar to beer, was made with teosinte sap, and that it was consumed in celebrations.

Some flora of the landscapes at that time provided the locals with seasonal food. These included fresh and dried pods (toast) from the white lead tree (*Leucaena*) that were collected in two seasons of the year, and wild fruit and cultivated fruit of Mexican hog plum (*Spondias mombin*, and *S. purpurea*) that were consumed fresh (Chisholm and Blake 2006: 161–172). Centuries later, in the food of Tlatoani Moctezuma II, the Mexican hog plum was prepared with native onions and duck, which was one of the dishes that was recorded by the friar Bernardino de Sahagún (1956, Vol. 2, Book 13: 305–308).

FIGURE 18.2 Sonora, Middle Mayo River Basin Near S.Bernardo: Modern Barbecue Oven, 2013.

FIGURE 18.3 North Sonora, Near the Mexican-USA Border: Modern Oven, 2013.

The major contribution to the food system of the settler bands was the fauna of wild species obtained in the forest and in the first homegarden-cornfield. By 7200 BP, the human bands grew, and then the food system included maize, prickly pears, wild or cultivated Opuntia, and Agave. The pineapple, or heart of the Agave, along with the stem (*quiote*), and the parts of the leaves nearest to the pineapple were cut into pieces and cooked or roasted in kilns to obtain a sweet food (*mexcal, mescal*). This technique of roasting the Agave in kilns produced syrup, sweets, or a fermented beverage. This method has been used in other places in ancient times like Sonora (La Playa archaeological site) and the southwest of the United States of America. Roasted Agave was consumed for thousands of years (Huckell 2006) (Figures 18.2 and 18.3).

18.4 THE CENTRAL VALLEYS OF OAXACA: FOOD SYSTEM

By 8000 BP, in the central valleys of Oaxaca, near Mitla, the food system of the settlers included cultivated maize, squash, and pineapple. In a cave in Tlacolula, Oaxaca evidence was found that squash was domesticated before maize was domesticated in the Balsas River basin, as Smith (1997:

932–934; 2006: 12223–12228) had confirmed. Also, at that time wild and cultivated chilies were consumed and collected along with various types of wild seasonal green weeds. Piperno and Flannery (2001) found remains of wild and cultivated chili peppers in the Guilá Naquitz and Sylvia caves that date from 6250 BC, so it was there that the domestication of chili (*Capsicum annuum*) began. Also, they found maize rachis that were dated from 6229 BC in Guilá Naquitz (Piperno and Flannery 2001).

The studies of Kaplan and Kaplan (1992) and Perry and Flannery (2007, Vol. 104, No. 29: 11905–11909) conclude that a characteristic of the cornfield was the inclusion of tolerated plants. In Guilá Naquitz and Silvia from 6250 BC to 6000 BC, the food system included domesticated chilies, wild beans, avocados, Agave, *Opuntia* prickly pears, tropical sapodillas, wild onions, and berries. In caves there, investigators found evidence of the first tortillas, which were dated later in the Formative or Preclassic from 500 AC to 300 AC.

18.5 THE TEHUACÁN VALLEY FOOD SYSTEM

Richard MacNeish (1997: 81–82) considers that the evidence in the Tehuacan Valley shows an occupation sequence that covered about 12 thousand years. Its first phases indicated activities of regional flora and fauna removal. The human population was sparse and seasonal. By 7900 BP, remains were found confirming that maize was domesticated and that usually it was consumed; hunting, gathering, and fishing products complemented the food system (Mangelsdorf et al. 1967a).

In El Riego cave (6800 BC), in addition to maize, the roasted seeds of native cotton (*Gossypium hirsutum*, *G. barbadense*, *G. aridum*, *G. gossypiodes*) were eaten, along with cultivated avocado and cultivated sweet potato. It was selected for its taste, smell, and size from those found in natural vegetation. It was collected in the forest and taken to the first homegardens, which were located on riverbanks. Avocado domestication began around 6500 BC (Mhammed, Kaufman, et al. 1997: 279–286; Miller and Schaal 2005: 71–122).

From 6800 BC to 6500 BC, the bands had four to eight people living in caves, three, or four times a year depending on the season who moved their camps. The basis of their food system was hunting and gathering supplemented with wild and cultivated plants. The inhabitants were trappers of small animals: hares, moles, rats, snakes, turtles, birds, lizards, and small mammals.

In the El Riego cave (6500 BC to 5000 BC), Smith Jr. (1967: 133–142) identified the remains of plants consumed and used by villagers: mesquite (*Prosopis juliflora*), toasted gourd seeds (*Lagenaria siceraria*), prickly pear (*Opuntia* spp.), and *Hechtia*. Also inhabitants ate avocado, amaranth, acacia (pods), Mexican hog plum (*Spondias mombin*), Opuntia prickly pear (fruits, and cladodes), two types of cacti Agave, *Lemaireocereus hollianus* seeds, *Jatropha*, Cosahuico fruit (*Sideroxylon* ef. tempisque Pitt-Sapotaceae), wild squash (*Cucurbita pepo*), cultivated squash (*Cucurbita mixta*), and chili (*Capsicum annuum*). There is evidence of the consumption of foxtail grass seeds (*Setaria geniculata*).

Remnants of lead tree (*Leucaena esculenta (Moc. & Sessé ex DC.) Benth.*) and cotton from silk cotton tree (*Ceiba parvifolia*) were found. Recent studies on squashes show the spontaneous hybridization of legumes in the incidental homegarden through a process favored or precipitated by human action, which preceded domestication.

Lead tree (*Leucaena esculenta*) roasted seeds were eaten. That was a pre-domesticated plant consumed from the center to the south of what is the current Mexican territory. This plant underwent numerous independent transitions from its wild state to cultivation, and data on it dates it as an established crop at around 4000 BC (Hughes et al. 2007: 14389–14394). From palms (*Brahea dulcis*), people obtained fiber. Two unidentified grasses were found, whose uses are unknown.

For Earl Smith Jr. (1967: 232 Table 26), the earliest evidence of domesticated amaranths is from Tehuacán, about 5500 years ago, when there was a change in the food system of hunting and gathering. Along with squash and chili, amaranth made up 6% of the population's diet, but in the next millennium it reached 14% of the diet in their food system. Anderson (1967: 93) considers that the human remains of the Purrón and Coxcatlán caves indicated that in the dry season, the inhabitants of the valley subsisted on what they hunted, and in the dry season the inhabitants of the valley subsisted by hunting, and in the wet season on what they gathered. Then the first incidental homegardens were formed that combined domesticated and wild plants from several of which fibers were obtained.

Casas and Caballero's materials (1995, Table 2, p. 43) on the temporal variation of the elements of the human diet came from studies in the Tehuacán valley (MacNeish 1967), when populations in 3200 BC began to have more than 50% of their diet based on agricultural products, reaching 75% until the Venta Salada phase in 880 BC.

For Hughes et al. (2007: 14389–14395), the collected plants were important food or important fiber. Their branches and dried stems were used to make fire, and their gourds were used to make containers. These practices were parallel to those of other domesticated perennials, such as Agave and Opuntia. Hybridization in backyard homegardens played a central role in the domestication of plants and was a trigger for the cultivation of plants for those uses. The first incidental homegarden arose with the first cornfield. These systems and the domestication of the first cultivated plants formed a traditional agroecosystem, and the homegarden-cornfield produced a different food system from that of hunting-gathering system (Early 1992: 17–19).

Anderson (1967: 93) studied flaked stone instruments, like foliated projectile knives and points, scrapers, cutters, spokes shaves, chisels, and blades. They had mortars to crush their food and did not use polished stone objects. Because of their seasonal mobility, hardly any of their food was obtained from agriculture. Therefore, the process of plant selection subject to the taste of the inhabitants coupled with the accidental action of forming areas with the preferred plants is the most remote antecedent of the beginnings of agriculture and the food system in Tehuacan in that period.

From 3400 BC to 2300 BC, a sedentary lifestyle increased, and cultivated maize began to be consumed. In the dry season, the bands lived in hunting camps. In the rainy season, they lived in villages or in houses that were built on river terraces. Seventy percent of the food system was made up of plants and wild fauna. Twenty percent of it consisted of diverse varieties of cultivated maize plus Canavalia, squash (*Cucurbita pepo*), and Tepari beans (*Phaseolus acutifolius*). Also the bands grew cotton, whose roasted seeds were eaten. From 2500 BC to 1000 BC, clay plates (*comales*) appeared, with thick edges that were intended to roast seeds and meats. The thinnest clay plates were used to cook tortillas and appeared with the maize hominy (nixtamalization) process between 500 BC and 100 BC.

Agriculture was established and stabilized in the Tehuacan valley from 1600 BC to 1500 BC. Hunting, fishing, and gathering were the means to obtain nourishment for the food system. García Cook (1986) states that in 1590 BC, clay plates appeared to roast seeds and roast meats in Tlaxcala, but its use was abandoned in 1500 BC. From 1300 BC to 1250 BC, trunk-conical constructions arose, possibly for maize storage. From 1200 BC to 600 BC, artificial open cisterns (or rainwater deposits like *jagüey, xagüey*) appeared on hilltops. They were 10–12 m in diameter and were to accumulate water in the rainy season. The water was for domestic uses and to irrigate agricultural land. The residential terraces included houses, open cisterns (*jagüeyes*), seasonal irrigated homegardens, and zones for cultivation (Abascal and García Cook 1973: 117–124; and 1975: 199–210).

18.6 LOS TUXTLAS, VERACRUZ: FOOD SYSTEM

Cyphers (2004: 21–23) and Chisholm and Blake (2006: 161–172) wrote that in Los Tuxtlas, the food system included fauna resources caught in rivers, estuaries, meanders, and surrounding land areas of the flooded area and on the coast. Cyphers information (2004: 24, 27–28) indicates that the food system from 1200 BC to 800 BC, included maize, but that it was cultivated in few lands so people depended on commerce to supply them with food, which was brought to them in canoes from homegardens in Tenochtitlan (Veracruz) and Loma del Zapote.

The food system was made up of boiled fish, turtle, shrimp, collected plants, maize, tamales made with maize, tubers, and the fruits of the coyol palm (*Acrocomia aculeata*) that were ground in

FIGURE 18.4 Central Veracruz: Totonac Preclassic *Metate* without Legs.

mortars or on *metates*. The list added meat from game like wild white-tailed deer and domesticated dog meat that were roasted or cooked in kilns and smoked fish. In the Olmec sites of the Middle Preclassic (1150 BC to 900 BC), there is evidence in the maize representations that coincide with the increase in the importance of grain in the population's sustenance (Figures 18.4 and 18.5).

Maize iconography included elements like jade or quetzal feathers. The maize deity had anthropomorphic and feline features with almond-shaped eyes, ornaments of maize seeds, and maize ears on various parts of the head.

Studies at the La Joya and Bezuapan sites in Los Tuxtlas (Vanderwarker 2006) indicated that by 2000 BC, maize was grown in the region (Goman 1992). Archaeologists Santley and Arnold III (1996) obtained evidence of human occupation in the region since 1400 BC (ceramics and artifacts).

FIGURE 18.5 Museo Schuk Toak, El Pinacate, in North Sonora: O'odham Metate without Legs (1200 AD).

FIGURE 18.6 Woman Gridding Maize, Diego Rivera 1924. Mural, National Palace, Mexico City.

Byrne and Horn (1989: 181–193) found in their study on the clearing of forest and agriculture in the region near Catemaco, in the Sierra de los Tuxtlas, that it was occupied 3000 years ago, because there was maize pollen in that place throughout the period. Forest regeneration occurred after the conquest, owing to the depopulation of the region.

The cornfield agroecosystem in Los Tuxtlas arose in the Early Formative, and initially, it was associated with the forest. In Los Tuxtlas, the first incidental homegarden was related to the gathering of wild plant species and the hunting of wild fauna of different species and sizes, which occurred both in the forest and in the first homegarden-cornfield (huerto-milpa). The forest was cut down when the population increased and the land area needed for the cornfield (milpa) and its different crops increased, leaving some wooded areas that allowed the forest to recover, which occurred when the population decreased. These circumstances are explained by studies of pollen in the Sierra de Los Tuxtlas (Byrne and Horn 1989: 181–193; Laborde 2009: 65–78) (Figure 18.6).

18.7 TRAVELING SOUTH: THE FOOD SYSTEM IN SOUTHERN MESOAMERICA

The southern route of maize showed that in Chalcatzingo, Morelos, from 1100 BC to 1000 BC, the farmhouse had a kitchen-room and a warehouse for maize. By 1200 BC, maize was the basis of subsistence in Cahal Pech (Belize), one of the earliest sites in the Mayan lowlands. In ceramic representations, there is evidence of the use of maize since 2900 BC, and studies show that the diet was complemented strongly with food obtained from coastal and terrestrial resources (White et al. 2006: 148). The food system included maize (*Zea mays*) and cultivated plants, like avocado (*Persea americana*), and two varieties of beans (*Phaseolus vulgaris* and *Phaseolus lunatus*) (Brown 2010; Smalley and Blake 2003; Blake 2006).

The food system of the human population included a combination of domesticated and cultivated plants, with wild weeds and fruits; the food system was complemented strongly with resources obtained from forests, rivers, estuaries, marshes, lagoons, hillsides, ravines, and surrounding land areas, as well as resources taken from the coast (Chisholm and Blake 2006: 161–172).

In S. Andrés, a coastal region of Tabasco, Pohl et al. (2007), in addition to Rust and Sharer (1988), found and analyzed maize pollen dated in 7100 BP, suggesting that by that time grass was already grown in this region. These remains are earlier by about 800 years than those found in Oaxaca. These dates open new questions to be solved by researchers. White et al. (2006: 145), and Willey (1981: 385–415) stated that in other places in Guatemala there was archaeological evidence of the presence of maize pollen (*Zea Mays*) since 3500 BC.

For Clark (1994), cultivated maize was found in the S. Carlos site of Soconusco, Chiapas, from 3335 BC ±55, on the route of maize dispersion to the Maya lowlands, For White et al. (2006: 148), archaeological and archaeobotanical studies indicated that in 2900 BC, the food system included hunted and fished fauna, and collected coastal and terrestrial resources. Ceramic representations show that by 2900 BC, maize was used as food, and by 2450 BC ±70, there was cultivated maize in El Arenal Reserve, in Costa Rica. Cultivated maize was found from 2280 BC ±40, in El Gigante, a rock shelter in Honduras.

In Salinas La Blanca, Guatemala the cultivated maize was found from 1928 BC ±105. As of 1200 BC, maize was the basis of the subsistence system in Cahal Pech, Belize. It should be considered that the diversity of native maize arose from the adaptations of the grain to different environmental, altitudinal, and agronomic conditions. In Colha, Belize, toward 1000 BC, the food system included crops that were combined with the gathering of tubers, avocado, tomato, guava, cashew, and sapodilla, in addition to crustaceans in large amounts for food. As well maize was cultivated in the lowlands of S. Lorenzo, in Los Tuxtlas, Veracruz from 1200 BC to 800 BC (Cyphers 2004: 27–28).

Henderson et al. (2007, 104, No. 48: 18937–18940) found that cacao was grown in the Late Preclassic to make a beverage where mucilage was extracted from the cacao pod, mixed with water, and fermented. Subsequently, chocolate emerged extracted from cacao beans, which were dried, roasted, and ground to mix with water, sweeten with honey or Agave syrups, and aromatic flowers.

As maize traveled north and south on the continent, the inhabitants adopted and adapted the grain to the characteristics of the different places where it passed, thus emerging new races and varieties. The Guatemalan Mayan people hybridized the teosinte (*Euchlæna luxurians*) of Huehuetenango, with (*Tripsacum* spp.), from which *Zea luxurians*, formerly known as *Zea Guatemala*, was derived. This maize was planted in the cornfield along with various kinds of beans, tomatoes, chili peppers, squashes, and cassavas.

In the Maya Classic, the information related to the food system came from the glasses, polychrome dishes, and codex. By 500 AD in the Mayan zone, tamales were made with crushed maize ears and game. They were wrapped in corn shucks, cooked in kilns (*pib*), and then seasoned with sauces. In addition to tamales (*waaj*), *pinole* (*ch'aj*), and *atole* of various types (*sa'*, *sakja'*, and *ul*) were consumed. Those maize ear tamales were painted in the Dresden Codex, showing that they were associated with religious ceremonies, but it is unknown if they were part of the daily diet. However, if the type of tamale the Maya made was equal to that in the Codex, the Mayan settlers may have separate recipes, one for the nourishment of the nobility, for religious ceremonies, and a different one for the daily consumption of small-scale farmers (López-Arenas 2006: 3–8; Montero and Varela 2017: 183–191) (Figures 18.7 and 18.8).

In the same period, in other Mesoamerican regions, maize hominy process (nixtamalization) was invented, so the dough for tortillas, and possibly the tamales were made with maize dough, but the date when the tortillas appeared in the Mayan area is not known. In the Post-Classic period, a beverage made with dried ground maize dusted in water, cacao and a flower flavoring called *posol* was consumed (López-Arenas G. 2006: 3–8; Montero and Varela 2017: 183–191).

To the south, the food system followed by the villagers may have included food from exploited nearby resources or interchanged products of different human bands and tribes. This also happened in the marshes and lowlands of S. Lorenzo, in Los Tuxtlas, Veracruz from 1200 BC to 800 BC (Cyphers 2004: 27–28).

Other plants accompanied maize in the diffusion that started the cornfield agroecosystem. Studies of phytoliths, pollen, *Zea* starch grains, and microfossils put the dates for the expansion and diffusion of maize in Central and South America further back (Hernández 1977; Ortega 2003: 126; Wellhausen et. al 1952).

In 2007, the article by Pohl, Piperno, Pope, and Jones (2007: 6870–6875) on the Mayan collapse, the origins of agriculture, maize dispersal, and soils gave new dates for maize cultivation, without

FIGURE 18.7 Chiapas, Palenque Site Museum: Special Plate for Tamales.

FIGURE 18.8 Dresden Codex: Mayan tamales with Fish on Top.

discarding the idea that it was teosinte, possibly. This maize was found in S. Andrés, Tabasco in 7300 BP and on the Tabasco coast in 7100 BP. Maize arrived in Panama by 7800 BP. These dates generated discussions, to which the cited authors responded:

> Numerous accumulated data from tropical archeological sites show that macro botanical remains of maize (and many other crops) often fail to survive for a long time in the tropical forest, where moisture, and soil characteristics, create adverse conditions for their conservation. In contrast, when studying microfossils such as phytoliths, and starch grains that can withstand the forces of degradation over

time, from archaeological sediments, stone and ceramic tools, the evidence for maize farming is consistently earlier (7800–4500 BP or 7000–4000 BP) than that obtained from the macrophiles. [...] In summary, the last 20 years of research have made it clear that studying the origins, and dispersions of maize requires a multiple approach (for example, macro botanicals [ears, grains, stems], and data from microfossils, pollen, phytoliths, and grains of starch) of paleoecology, and archaeological research. Especially striking are the main differences that have been shown to exist with respect to the preservation, and visibility of these different types of maize fossils.

The differences in the chronologies modified the conceptions about the maize diffusion from the middle basin of the Balsas River to other regions in Mesoamerica and other cultural areas of the continent make us realize that the current chronology will be modified as new evidence emerges, giving a different series of characteristics in the structure of the first Mesoamerican cornfields than the present ones, which are: (1) the combination of trees (incidental homegarden) with flora including grasses, Agavaceae, Cucurbitaceae, Legumes, and others (homegarden-cornfield); (2) there was a combination of wild, and cultivated plants; and (3) the domestication dates varied from one place to another, which may indicate that there was a parallel process in different sites, combined with the action of diffusion from centers of origin, such as the middle basin of the Balsas River.[2]

18.8 THE BASIN OF MEXICO: FOOD SYSTEM[3]

In the basin of Mexico, one of the sites that have an older chronology is Zohapilco, in the Chalco lagoon. Niederberger (1976, 1979: 132–142) showed that by 3090 BC ±115, domesticated maize was consumed in Zohapilco. The studies on this lagoon basin indicated that by 2300 BC (equivalent to the Purrón Phase of Tehuacán), the plants grown and eaten were maize, squash, avocado, Opuntia cladodes and Opuntia prickly pears, cassava, coyol fruits, and roasted or battered Agave to obtain its sweet sap (*mexcal, mescal*). Fresh and dried gourd seeds, mesquite pods, and cottonseeds were roasted.

Plunket and Uruñuela (2016: 43) calculated that in the Formative (2000 BC), there were thousands of agricultural villages, which would be grouped in denser communities and later formed the villages, towns, and cities. Since that time, agriculture has been the way of life for human societies. Plunket and Uruñuela (2016: 41–46) calculated that in the Central Highlands, during the Formative or Preclassic (2000 BC to 100 AD), the food system was based mostly on agriculture, and houses had storerooms to reserve the maize. In Tetimpa, a site located in the eastern flank of the Popocatepetl (700 BC–50 AD), houses were surrounded by cornfields (Bird 1980: 30–41).

McClung, Ibarra, and Morán (2014: 106–107) estimated that the first chinampa agroecosystem arose from 1300 BC to 800 BC, in the Basin of Mexico. Chinampas were associated with the homegarden-cornfield. By the Post Classic, many plants were consumed in this region (Xaltocan, Iztapalapa, Tlatelolco, and Otumba). The food system included: maize, amaranth, squash, beans, some species of green small tomato (cultivated and wild), chili, purslane, the common sunflower, *Opuntia*, and the Mammillaria Haw cactus. The aromatic plants were wormseed and sage (which was medicinal and ornamental). Also, there were carobs (trees whose branches are attractors of insects); fruit trees, such as the Mexican hog plum (*Spondias mombin* L.); the Mexican black cherry, the Mexican hawthorn, avocado, and the Bilberry Cactus (*Myrtillocactus geometrizans*).

From 1400 BC to 300 BC, cultivated foods constituted 20% of the total diet: maize, squash, beans, and tepari beans. Wild plants and wild fauna formed 70% of the food system. From 1100 BC to 100 BC, in Chalcatzingo, Morelos, houses were built with kitchen-living rooms and rooms for maize storage. Since 900 AD, the size of maize cobs has increased. The food system was modified

[2] Buckler et al. 1998; Buckler and Holtsford 1996; McClung de Tapia et al. 2014; Miller and Schaal 2005; Pohl et al. 2007; Ranere et al. 2009; Piperno et al. 2009; Sanjur et al. 2002.

[3] This section is a summary that includes data from the Markets in Mexico, a book by Alba González-Jácome which is being edited by the Chapingo University in Texcoco, State of Mexico. See: Bedolla and Vanegas 1990.

with several inventions. In Oaxaca, from 500 BC to 300 BC, maize hominy tortillas appeared. The meat consumed was from domestic, wild, or captured animals, and fish from water reservoirs, rivers, or nearby lagoons (FAO 2018; González-Jácome 2011; Rojas-Rabiela 1985).

In the Formative or Preclassic, maize become the basic staple of Mesoamericans. The information on the antiquity of the chinampa agroecosystem in the Basin of Mexico from 1300 BC to 800 BC, indicated that the system arose associated with the cornfield. This gives the daily food system of the inhabitants a wide diversity of agricultural products, which were increasing together with the protein derived from the species of fauna that were obtained by hunting, fishing, and capture in ecosystems nearby the human settlements (Benz 1997).

From 1200 BC to 100 BC, in the Coapexco site, to the southeast of the basin, wells for maize storage appeared. The consumption of protein from wildlife hunted or caught in the forests, lakes, or natural zones was, and has remained, very high. In the archaeological sites of Terremote Tlatenco, Cuanalan, and Temamatla, located in the Basin of Mexico, the group of researchers led by McClung of Tapia (2014: 106–107) found evidence of cultivated plants and edible wild weeds that the residents ate in the Formative period.

The food system then included: maize (*Zea mays* L.); Amaranth (*Amaranthus* L.); Sunflower (*Helianthus annuus*); chili (*Capsicum annuum* L.); *Opuntia*; squash (*Cucurbita* L.); *Chenopodium* L. (*quinilla, cenicilla*); Wormseed (*Dysphania ambrosioides*); Creeping vinagrillo, sorrel (*Oxalis corniculata* L.); Silky grass, *motilla, triguillo* (Setaria Beauv.); purslane (*Portulaca oleracea* L.); bramble; raspberry; blackberry (*Rubus* (Tourn.) L.; small green tomato (*tomatillo*); *jaltomate*; *acahualera* (*Jaltomata procumbens* (Cav.) J.L. Gentry); tomato (*Physalis* L.); eggplant (*verenjena, berenjena*); donkey tooth (*Solanum* L.); and bean (*Phaseolus* L.). Also, it had: sage, a weed typical of the pine/oak forest (*Salvia* L.); suaeda (*Mexican Suaeda* (Standl) Standl); and *Crotalaria* L., which is a nitrogen fixative.

The fruit trees whose people consumed were Garambulla cactus (*Myrtillocactus geometrizans*); avocado (*Persea americana*); Mexican Hawthorn (*Crataegus mexicana* Moc. and Sessé ex DC.); and Mexican Black cherry (*Prunus serotina* ssp. capuli (Cav.) Mc Vaugh).

For Manzanilla (2013), in the Teotihuacan Formative period (2500 BC–1000 BC), agriculture was practiced in the slash and burn agroecosystem. Studies and excavations of Manzanilla et al. (1993) carried out in the Teotihuacan residential complexes showed that from 200 AC to 400 AC (Tlamimilolpa Phase), the residential units were organized around squares, with a temple and domestic units. The human population of each group had specializations like textile weaving, the production of bifacial; ceramics; and baskets, as well as unique regional specializations, like making grindstones. There were high-level specializations, like painting codexes or creating music, also. In the Middle Formative (1000 BC–400 BC), obtaining resources formed an "economic symbiosis", and food of many types that arrived from other parts of the basin. The grain storage units were concentrated in Loma Terremotes.

During the Classic, around 600 BC, the food system of the villagers was composed of maize, beans, squash, chili, amaranth, Opuntia, purslane, tomatoes, avocado, toasted gourds seeds, the common sunflower, bilberry cactus, and the small green tomato formed the food system of the villagers in Teotihuacan and neighborhoods. The people used wormseed and oregano as aromatic herbs. The fruits they consumed were from Mexican hawthorn, Mexican black cherry, and Mexican hog plum, and white sapodilla trees. The edible fauna consisted of: dogs, turkey, deer, hare, rabbit, several rodents, skunk, wild boar, armadillo, duck, heron, geese, pigeons, quail, frogs, turtles, lizards, and various species of fish (McClung et al. 2014: 106–107).[4] Some problems with the climate and their impacts on agriculture appear to have occurred during the rise and fall of Teotihuacan (García-Espinosa 1974: 35–69).

[4] Escalante-Gonzalvo 2004: 63; González, Javier et al. 1993, Vol. II: 661–728; Manzanilla 1993, Vol. I: 548–566; McClung and Rattray 1987: 57–74; Valadez 1993, Vol. II: 729–831.

For the food system, fishing and hunting were very important activities in the Basin of Mexico, and their quarries complemented the food system of the human populations. The diet included fish and seasonal water birds, and those that lived in the lagoon system permanently: Mexican duck (*Anas platyrhynchos diazi*); Cinnamon Teal (*Anas cyanoptera*) (*cerceta café, cerceta canela, Chilcanauhtli*); Pied-billed Grebe (*Podilymbus podiceps*) (*Acintle, zambullidor pico pinto, zambullidor pico grueso*); Eared Grebe (*Podiceps nigricollis*) (*pato luna, Nacaztone*); Ruddy duck (*Oxyura jamaicensis) (pato tepalcate, Yacatextli*); Great Egret (*Ardea alba*) (garza o garzón blanco, Aztatl); Snowy Egret (*Egretta thula*) (*garza nívea, garceta pie dorado, Aztatl*); Green Heron (*Butorides virescens*); Black-crowned Night-Heron (*Nycticorax nycticorax*) (*Acacahoactli, perro de agua, Pedrete, corona negra*); White-faced Ibis (*Plegadis chihi*) (*Atotola, Ibis cara blanca*); Common Gallinule (*Gallinula galeata*) (*Quachilton, polla de agua, gallineta frente roja*); American Coot (*Fulica americana*) (*Yacacintli, gallareta, gallareta americana*); Killdeer (*Charadrius vociferus*) (*Amacozcatl, tildío, chorlo tildío*); Black-necked Stilt (*Himantopus mexicanus*) (*Comaltecatl, monjita, candelero americano*), and American Avocet (*Recurvirostra americana*) (*Icxixoxouhqui, picocurvo, avoceta americana*).

Manzanilla et al. (1993) propose that during the Teotihuacan Classic (AD 200–AD 600), a "regional redistribution network" of products was formed that included foodstuffs. The inhabitants of the basin participated in it, who "offered surpluses" of their productions to the Teotihuacan priests and goods. There were no markets yet, but possibly there was "direct barter". As a result, there were warehouses for centralized storage, located in the rooms at the south of the complexes of the three temples in Teotihuacan.

In Oztoyahualco, Manzanilla et al. (1993) found ceramic vessels containing large amounts of pollen and macrophiles of plants that were very important in the region's economy, plus medicinal plants. The redistribution of raw materials from the southern and southeastern regions may have been part of a "restricted circulation" route. These products were cacao, honey, copal, seashells, feathers, hides, jadeite, serpentine, hematite, cinnabar; malachite, and others. Possibly the flow was under the control of the priests and emissaries of the temple, who lived neighboring the local groups in Matacapan, Kaminaljuyú, and perhaps Alta Vista and Timbangato (vertical control of ecological floors).

Bernardo Rodríguez-Galicia's book (2017), based on Archaeo-zoology studies, highlights the importance that fishery seafood and freshwater lagoon systems had in the everyday food system, and in the symbolism, rituals, and communal banquets of the inhabitants of Teotihuacan in the central Highlands. In Teopancazco, a neighborhood located in the southeast of the city, remains of some 14 varieties of fish, from the coastal lagoons of Nautla, Veracruz was found. Down the rivers of these regions, the goofy fish (*Joturus pichardi*) went from the mountains to the coast to spawn around November 2. It was an easy catch. The red snapper (*Lutjanus campechanus*) was netted in the sea (Rodríguez-Galicia 2017: 27).

The daily and ceremonial consumption of fish is depicted in several images in codexes. In the Plates XXIII, XXVII of the Dresden Codex (Postclassic, 1100 AC to 1200 AC), a woman appears carrying a plate with tamales and fish on top of them. In the same codex, there is a plate of crushed tamales with a fish on top. The plate with tamales and fish has been interpreted as an offering being carried by the woman (López 2006: 3–8; Mariaca et al. 2010: 165–167).

In the sixteenth century in the Basin of Mexico, many species of wild fowl were added to the list of the food system: the howling duck; the kingfisher duck; *opipixcan*; the bright duck; *pipixcan*; *cuacoztle*; *quapachcanauhtli*; *quauhcilin*; goose; *mociño*; *toltecolotli*; Northern Pintail (*Anas acuta*) (*pato golondrino, Tsitzihoa*); duck swallow, *tzonyayauhqui*, *xacacintli*; American Wigeon (*Anas americana*) (*xalquani, chalcuán*); blue-winged teal (*Anas discors*) (*cerceta aliazul, Metzcanauhtli*); northern shoveler (*Anas clypeata*) (*pato cucharón, cucharón norteño, Yacapatlahoac*); green-winged Teal (*Anas grecca*) (*cerceta aliverde, Quetzalcocolton*); American white pelican (*Pelecanus erythrorhynchos*); great blue heron (*Ardea herodias*) (*garza morena*); tricolored heron (*Egretta tricolor*) (*garza flaca, garceta tricolor*); sora (*Porzana carolina*) (*gallineta de ciénaga*); spotted sandpiper (*Actitis macularius*) (*Alzacolita, playero*); and migratory geese, like the *Branta canadensis*.

Plants brought from the Old World to New Spain modified some dietary likings or added others to already existing ones. Based on information from the *Geographical Relations*, Machuca (2013) stated that by 1580, in the Jalisco Balsas River, tamales were wrapped in banana leaves instead of other plant leaves, like fresh or dried cornhusks. After 1519, the banana became an important part of the food system in tropical regions. In Apatzingán, people ate dry, raw, cooked, or roasted banana. In the town of Tancítaro, people ate banana cooked in stews, which were consumed in lent. In the Huaxteca region, tamales were wrapped in a leaf called *platanillo* (*Heliconia*) locally.

18.9 THE TLATELOLCO MARKET (1536–1577) AND THE FOOD SYSTEM

Bernardino de Sahagún (1830, III: 241–281) provided an idea of what the people in the towns in the basin of Mexico consumed in their daily diet. The Mexican food system was based on cornfield products and other cultivated plants, 20 edible seasonal green weeds, and 7 wild plants, plus homegarden products. Those seven wild edible seasonal green weeds were insufficient for all family's needs. In contrast, the wild-collected plants included various trees and flowers destined for herbalism numbering a total of 120 collected plants. The figure would be different for the Mayan zone or for other New Hispanic regions.

Bernardino de Sahagún (1830) in his description of the Tlatelolco market wrote where food was obtained, the regions nearby, and the products that came from long-distance trade. In addition, there were products brought by Spaniards from the Old World, as will be seen below. Without a doubt, the Tlatelolco market was one of the largest and most important ones in the Central Highlands of Mesoamerica. From 1536 to 1577, Sahagún did his research, but market already sold products that originally with the Spaniards in the Basin of Mexico in 1519.

Sahagún arrived in Mexico in 1529, and wrote his *General History of Things in New Spain*, with the collaboration of four students from the College of Tlatelolco.[5] The research and writing of the book took place from 1540 to 1577 and was an example of a rigorous methodology to obtain, compare, and organize information. The Florentine Codex and its illustrations complemented it. Sahagún (1956, T. III, XVIII: 139–140; T.III, XX: 138, 140–141; XXII: 144; XXVI: 156) produced a detailed list of the raw and cooked foods and beverages that were sold in the Tlatelolco market:

Tortillas that were made with cornfield maize and had various sizes, thicknesses, colors, and flavors. Some were filled with ground meat or chili or eggs, and others were made of hominy-maize-dough mixed with honey and cooked under the embers, and "[...] many other forms of tortillas"; beans, chili stews; tamales (fish, frog, or chicken); *tlacoyos*[6]; hot or cold atole (*atolli*); a prepared cacao beverage mixed with water, corn kernels, aromatic species, and honey; stews; pumpkin seeds boiled with honey; cooked hard shells of the pumpkin (*cascos de calabaza*); syrup; sweet fermented Agave mead (*pulque*); venison; and fresh fish, which arrived from the Gulf coast by way of relay runners (Sahagún 1956, T. III, XXII: 142–144).

Sahagún (1956, T. III, XXII: 142,144) described the products grown in the cornfields that were sold in Tlatelolco: sweet maize canes; goldfinches (xilotes); green corncobs (elotes) and roasted corncobs; shelled maize to make tamales and tortillas; tortillas made of dough mixed with honey; honey-covered corn kernels. Since those times, esquites cooked or roasted corn kernels mixed with epazote and chili are still consumed; and honey-wrapped corn kernels and rosettes or popcorn, which is consumed and has spread to the world, where it has become a very important industrialized product.

Other food products from the cornfield were squash, chayote, yam bean, wild potatoes, sweet potatoes, edible roots (*tocimatl*), small green wild tomato (*tomatillo*), edible green wild and

[5] Sahagún informants were: Antonio Valeriano, a native of Azcapotzalco; Antonio Vejarano, from Cuautitlan; Martín Jacobita, from Tlatelolco; and Pedro de San Buenaventura, and Andrés Leonardo, who were born in Tlatelolco, also.

[6] Rather rectangular corn dough patties, stuffed with cottage cheese, and ground beans, or fava beans. They are cooked on a clay plate or fried, and then spread with a green chili sauce, and more cheese.

cultivated weeds (*quelites*) that were sold in the market. Medicinal herbs were obtained from the wild areas near the villages, and those herbs were taken to the Tlatelolco market to be sold. There were sellers of cultivated edible herbs that had been grown and harvested [...] in the field at the time of the rains [...] and merchants of chili leaves, crushed fruit drinks, porridge, sorrels, hot lunches, pulley, and herbs [...] good to eat" (Table 18.1).

Friar Bernardino de Sahagún described in Volume III (Book X, Chapter 14; pp. 35–36), the activities of the cooks that included "using yeast", work that was not of pre-Hispanic origin and that was related to wheat, and the making of bread. In native cuisine, Agave fermented mead (pulque) was used to ferment the corn dough. About the cooks, he wrote:

> Female cook. The woman who knows how to cook well, has the job of understanding the following things: doing good eating, making porridge, kneading, knows how to throw yeast, for all of which is diligent and hardworking, knows how to make plain, round, and well-made tortillas; and on the contrary, make them prolonged and thin, or make them with folds or rolled with chili and know how to dough the beans cooked in the tortilla dough, and make these of meat as dumplings and other stews they use. The one that is good in this trade, knows how to try them if they are good or not, and is skilled and experienced in all kinds of stews, understood and clean in their trade, and makes them cute, and tasty. The one that is not such, and the trade is not well understood, it is painful, and annoying, because it stews badly, it is dirty, and it hides, eaten, sweet, it bakes the tortillas badly, and the stews of its hand are smoked, salted or stews, and she is rude, and crude in everything.

Sahagún (1830, III: 241–281) consigned the list of sixteenth-century cornfield weeds in the *General History*. The list sums more than 60 herbs destined to the food system. In contrast, the number of collected wild plants is greater, totaling close to 100, which were used for medicinal purposes. Most of the herbs for herbalism in the basin of Mexico were obtained on the banks of the lagoons system, moors, mountains, and on nearby ravines and hills. There were some cultivated plants and weeds that were for the same purpose, but they were destined to feed the cornfield farmers and their families for the most part. The farmers sold the surpluses, the same as rural populations of Mexico do today, but the application of agrochemicals and the loss of natural areas have reduced the number of weeds and natural ecosystems available to farm families.

The fruits of the homegardens in several regions inside and outside the lake system in the Basin of Mexico reached the Tlatelolco market and were cacao, pineapple, various types of sapodilla, "[...] pears (*peruétanos*), mammies, plums of all types [...]"; guavas, Mexican hawthorns, Mexican Black cherries of any kind; yellow, red, white, pink, or green Opuntia fruits, and "[...] small, sweet tomatoes, which are sold as fruit [...]". Aromatic herbs, roots, and medicinal animals or parts of them were marketed, too (Sahagún 1956, T. III, XXII: 142,144). Fruits remain important in the current Mexican food system, but some fruits, like the Mexican black cherry, Mexican hawthorn, or the Mexican hog plum, have lost importance, and their consumption is almost restricted to rural areas.

Vendors traded big beans (*ayecotlis*) (*Phaseolus coccineus*), and yellow, red, white, small, and variegated beans along with wide, large, small, green, and dry chilies, which were produced in summer. Food merchants sold chia, wide, large, and small green, and dry chilies, which were produced in summer, "[...] those of various feet, and those harvested [...] after the first frost" (Sahagún 1956, T. III, XVIII: 136). There were vendors of different types of squash seeds, including those that "[...] are roasted, and wrapped in a dough mixed with salt, that are appetizing to eat"; and sellers of yellow, red, and well-ripened tomatoes (Sahagún 1956, T. III, XVIII: 137). Roasted seeds, mixed with salt, and chili, have lost popularity and have been replaced by walnuts, pistachios, blueberries, sunflower seeds, and others; however, peanuts are still an important part of national consumption.

Spaniards arrived with various plants, animals, and tools, which they introduced into the life, economy, culture, and agriculture of native farmers in the Central Highlands and the Mayan zone. In the Tlatelolco market, fish, game, cattle meat, geese, ducks, birds, quail, eagles, opossums, and game were sold. The animals of Castile (birds, cows, pigs, sheep, goats, hens, and chickens) were added to the list. The meat was sold roasted or cooked on coals, dried jerky, or cooked in under kilns (barbecue type) (González-Jácome 2011: 437–438, 450).

The inhabitants on the banks of the basin lake system hunted and captured a series of birds that were permanent residents of the lagoons, such as heron (*xoxouhquihoactli*), brown teal (*chochopitli*), water chicken, coot, egret, and others that were named on previous pages. In areas of Lake Texcoco, Spirulina seaweed (*Spirulina maxima*) was gathered to make local "green cheeses", which were sold in the Tlatelolco market, and which were palatable to the Spaniards (Francisco Hernández 1574). Duck, quail, and chicken farmers sold eggs for omelette or stewed in casserole. They raised poultry like the "chickens of the land" (turkeys) and Castilian chickens, or acquired them from others, also, to market (Sahagún 1956, T. III, XXIV: 147; Francisco Hernández 1574).

Creole cuisine emerged in the Viceroyalty, with the fusing of cornfield elements, and the vegetable gardens with Spanish dishes. The convent nuns were an important source of invention in this mixing in the New Spain food system: like *Poblano* chilies with a creamy walnut sauce (*nogada*); red, yellow, green, or black moles cooked or stewed with beef, pork, chicken, mutton, or turkey cooked with bacon; chickpea, potato, maize, squash, and carrots; and cheese-filled tortillas (quesadillas). Added to the list of foods and beverages are mixtures like milk with chocolate, fruit sorbets, candied fruit candies made with dry bread, fried and drizzled with brown sugar syrup, and fresh cheese (*torrejas*).

Fusion in the food system reached urban and rural areas with the widespread use of pork fat, bacon, butter, olive oil, meats, cane sugar, grains like wheat and barley, vegetables, and fruits of the Old World that were incorporated quickly into the system. Mesoamerican beverages underwent several changes as well, with the introduction of distillation, which was added already existing fermentation process from ancient times. Also, the way of cooking was mixed, for example, adding fried foods, cooking in metal pots, jerking (there was dehydration of plants such as squash), or candying in the preparation of sweets. Likewise, utensils were introduced in kitchens, like metal pots, metal spoons, and knives, and kitchen fittings (cabinets, hutches, tables, raised burners), and new ways of preserving food were introduced like special rooms for hanging sausages with ventilation to avoid decomposition (*fresqueras*).

The mixed of food systems was not exclusive to America or Mesoamerica, but it was a universal happening. The Hispanic food system underwent fusion, too. Several components of the cornfield (maize, tomatoes, peppers, and squashes) were taken to Spain, leading to the creation of new foods, which were incorporated into the Hispanic diet, like talos and small red chilies stuffed with some type of fish-*piquillos*. They were peppers derived from poblano peppers, adapted to the soils of the Basque country. Padrón peppers were sautéed in olive oil and seasoned with sea salt and tomatoes. Fresh or cooked tomatoes were used in the preparation of various dishes and salads, as they use to date. Only the popularity of the tomatoes was far reaching, but the green ones no (López-Piñero and López-Terrada 1997).

In the sixteenth century, the food system in the Mexican central highlands included: 11 species of chili; *Amaranthus* sp. (*huauzontle*); epazote; purslane; chayote; nine species of red tomato; green tomato; 18 varieties of Agave; *Opuntia* cladodes, and prickly pear (there are seven species of prickly pear in Mexico); Giant barrel cactus (viznaga: *Echinocactus platyacanthus*); the Mexican hawthorn; at least three varieties of Mexican black cherry (*capulla*, *elocapulin*, *xitomacapulin*); yellow and red Mexican hog plum (*Spondias* spp.), also called "plums of the earth"; White sapodilla and Mexican blackberries (*amacapulin*). Some of the Agaves were sources of fiber. The list included vegetables (some of which are no longer consumed in Mexico) and flowers like various types of orchids, which were for decoration and flavoring chocolate (González-Jácome 2011: 476–480; Hernández 1984 [1574], Vol. 3).

The inclusion of new elements in the native food system began with fruits, legumes, and herbs, which before 1519 were unknown in Mesoamerican cultures. Hispanics were afraid of using some Mexican plants in their food systems; therefore, their diffusion was hampered. Ethnic groups consume prickly pears and Spaniards, although the latter has certain fears as to their properties and effects, because there were those who thought that "[…] it hurts and causes sorrows […] and it tends to constrict the belly […]". The Hispanic physician Juan de Cárdenas (1571: 127, 127v) wrote:

If any fruit from the Indies, with a very just title, deserved the reputation of sweet, tasty, healthy, and gifted it was the one called the natural *nochtle*, which matches our Spanish prickly pear (*tuna*). It is a cold fruit, and humid in the second degree, the pulp or interior part of which it is composed is sweet, tender, delicate, and watery, within which some grains, like those of grapes, are sown, and dispersed, it is not now time to try in particular the properties of the prickly pear plant as well as the fruit that is harvested from it, we only understand that if there is fruit in the world that can be more safely eaten, that is prickly pear [...].

From what I say, the experience is seen every day not so much among city people who use various maintenance, as among the rural people, who, because many times they lack maintenances, and tended to give themselves over to the tunas, and that the satiety They find it is, that they happen to urinate every while, and on the other hand they are constipated so that we have seen them reach the point of death, and from what I say there is no soldier from Nueva Galicia, who in himself does not have enough confusion of his life had the experience, because of this disappointment, the cause is now requested.

For the doctor, the cause of the constipation was in the seeds of the prickly pear, "[...] that going through the guts they stop the feces, closing, and attacking those routes" (Cárdenas 1571: 127v, 128). The doctor lists that other fruits that constipate (quince if eaten upon eating other foods, like olives) or relax the belly (quince if eaten at the end of eating other foods) do not have these effects, despite having prickly-pear seeds (*granillos*), like guava eaten without the skin.

Taking the Chichimeca indigenous groups as an example to explain the reason why the Opuntia prickly pears that they consumed in abundance did not restrict them, Juan de Cárdenas (1571: 131) thought that as the Chichimecas ate the Opuntia prickly pears (tunas) with their skin and seeds; as all that was thick and sticky " went through the guts " without problem. From here he recommended that the prickly pears not be eaten in large quantities, and be accompanied by other "delicacies" so that with this humidity the seeds pass through the belly.

That Spanish doctor in his study of the characteristics of food consumption plants that were unknown to Europeans explained how they should be managed before consuming them to avoid health problems. Some were quickly included in the diet of the Spanish, like the Caribbean cassava, used in the manufacture of bread (Juan de Cárdenas 1571: 131v). On the qualities of maize, Juan de Cárdenas (1571: 131, 139v, 140,140v) wrote:

Maize is one of the seeds with the best title to be estimated in the world, and this for many reasons, and causes. The first is because of its generality, I mean that because it is a seed that grows, and can be cultivated in hot, or cold soil, in the dry, in the wet, in the mountains, on the plains, in winter, for irrigation, and in the storm season, and it is grows, and benefits. The second is because of its abundance, which is like saying that from a sown hundred hanega are taken two hundred, and they not with too much work, but easily and restfully, not waiting almost for one year to the next, like wheat in Spain, which is planted in October, and is harvested in June and July, but not maize, as it is harvested in three or four months, and even in some parts within fifty days it is reaped and stored. The third, because of the ease and speed with which it is kneaded and seasoned, because we see and know of the wheat that needs to be added, ground, sifted, and cooked, kneaded, and then smoked and cooked, and even left overnight, for better and without less harm to eat it, pouring salt, yeast, hot water, and looking for a very warm oven, and appropriate according to the amount of bread that is kneaded, and maize needs none of this [...].

18.10 THE CORNFIELD AND THE NINETEENTH-CENTURY FOOD SYSTEM

In 1900, Karl Kaerger (1986 [1901]: 249–251) a German agronomist listed and described maize fare that was part of the daily consumption in Mexico. They included great varieties and tortilla-based dishes made with hominy maize dough. This maize is boiled in water with lime (calcium oxide) and left in it overnight before the maize kernels are washed and ground on a metate. Kaerger included in the diet: beans cooked without fat; boiled maize; various types of tamale; bean-filled dough patties; small, thick tortillas (*chalupas*); quesadillas; tortillas with dough sweetened with sugar and then toasted; atolli mixed with cacao, sugar, and water (*champurrado*), chocolate from Tabasco, candied fruits; and cornmeal bread, which was consumed in Tampico, Ciudad Victoria, and Monterrey. Tamales were given to hacienda guests or sold to passengers at railroad stations.

In Mexico there is no single food system; there are regional cuisines and variants, in addition to the food systems that are organized for religious and civil festivities in the towns and regions of the country. However, there are some foods found in various parts of the country. Among them are pozoles; beef broths, pork, poultry, fish, or seafood; barbecued lamb; grilled or roasted pork and beef, tamales, moles; white or red rice; pasta soups, atole, hot chocolate, and other local drinks derived from cacao, herbal teas, and citrus, or seasonal fruits, depending on the region and others factors.

The list would be incomplete if it does not include tacos of all types, sizes, fillings, and sauces. In addition, there are tacos made with corn dough, and tacos made with wheat flour. In Sonora, for example, flour tortillas come in different sizes, including the so-called *sobaqueras*, which measure up to 60 cm in diameter and are cooked on special metal plates (*comales*). These comales are made with parts of cargo trucks that no longer serve for these transports, and they are adapted and used in the cooking of these large tortillas, which are sold folded and packed in plastic bags and are typical of the northwest of the country (Figures 18.9–18.13).

FIGURE 18.9 Sonora: Wheat Tortillas (*Sobaqueras*) and Process to Make Them.

FIGURE 18.10 Sonora: Wheat Tortillas (*Sobaqueras*).

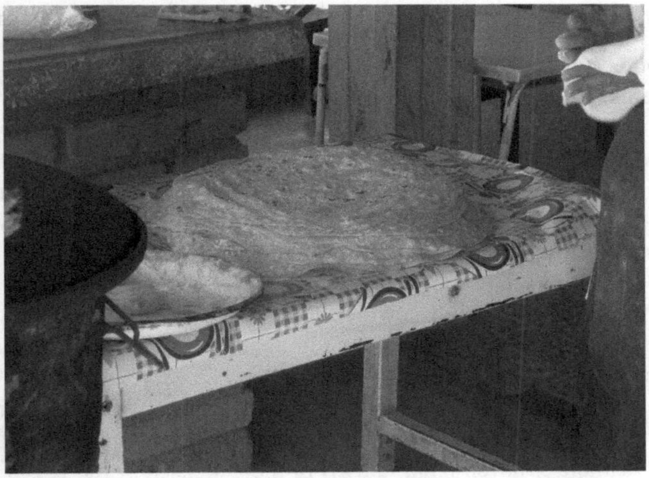

FIGURE 18.11 Sonora: Wheat Tortillas (*Sobaqueras*).

FIGURE 18.12 Sonora: Iron Plate (*Comal*) Where Wheat Tortillas are Cooked.

FIGURE 18.13 Sonora: Iron Plate (*Comales*) for Cooking Wheat Tortillas (*Sobaqueras*).

18.11 THE CONTEMPORARY MEXICAN FOOD SYSTEM AND THE CORNFIELD

Researchers on food systems, chefs, tourism specialists, and politicians have been studying Mexican food systems. In Mexico and all of the American continent, maize is the prime ingredient for many foods. There is a long list of Mexican dishes with maize: tamales; tortillas; tacos; *enchiladas*; *tlacoyos*; *lolos*; *gordas*; *picadas*; *sopes*; *huaraches*; *pupusas*; *arepas*; *quesadillas*; fresh corn on the cob maize ears that are toasted, boiled, or roasted corn kernels (*esquites*) with wormseed (*epazote*); popcorn (rosettes); and *pinole*. In addition, many other ways exist of roasting, boiling, cooking, grilling, and consuming it, for which people have local and regional terms (Goody 1982).

These food products are generally known as food cravings (*antojitos*). However, many maize products are part of the daily diet for large populations in the country, like tacos, *quesadillas*, *gorditas*, *huaraches*, and *tostadas*.

The ceremonial food system varies according to the specific region, culture, and state in the country. In Tlaxcala from 1970 to 1980, the adult women in the families supported their relatives and the godparents of their children (*compadres*) in the expenses of the feasts for the stewards of saints (*mayordomías*) and collaborated by preparing the food and the tortillas that were to be consumed on that holiday. To do this, women arrive very early at the home in question, began to pluck, and highlight the turkeys, hens, and chickens that would be consumed on that fest and then boiled them in clay pots or large aluminum pots and pans.

Women took hominy maize to the mill to be ground and began to make tortillas on a huge clay plate (comal), especially for that job. The women of the house grilled and ground the chilies on a metate or they took them to the mill to be ground and cooked the mole in a large clay casserole (*cazuela*) (Katz, Hediger and Valleroy 1974: 765–773; Nelken-Terner 1968). Hours later, they added the chicken broth. Other women made an enormous clay casserole of red rice, which is fried, added ground tomato to it, then added chicken broth, and garnished with sliced carrots and peas.

The women of the steward's house were in charge of serving the mole with a piece of chicken or turkey, rice, tortillas, or tamales to the guests. At stewardship festivals, it was customary to give attendees and musicians lunch, prior to the start of the religious activities. At noon, the whole town was invited to eat mole at the steward's house. Although not all the inhabitants attended, a significant number of guests arrived, so the casseroles with food were enormous, usually, as were the tortillas or "silly" tamales, to consume with the mole, which is expensive to make. The steward's friends and relatives collaborated with a quota to pay for everything that is needed, and they bring soft drinks, beer, or liquor to accompany the food which is brought. This luxury spending can be extremely costly, so the collaboration of family, relatives, and close friends is essential.

As in the case of stewardships, food is ritual, and sometimes has pre-Hispanic origins. Foods are: tamales; mole; pozole; broths with meat and vegetables; soups; cornbread; *atoles*; and products made with cornmeal, which include various types of polenta; the piki or cornbread of the Hopi; and breads and corn tarts that are sweet and can be served hot or cold; the corn tart (*pastel de elote*) can be accompanied with eggnog (*rompope*) or ice cream. In fact, the corn tart has become a gourmet dessert in Mexican restaurants.

There is a long list of salty and sweet foods, which are consumed both in the daily and ceremonial diets of regional and national populations. For example, popcorn toasted, with salt or caramel; maize kernels boiled with in salt and seasoned with some fragrant herb, and chili; corn-on-the cob boiled with salt and some aromatic herb, which is consumed seasoned slathered with cream or mayonnaise; and drizzled with chili podwer; roasted or boiled sweet maize as a side dish with turkey or duck. Before the drying of the lagoon system in the Valley of Mexico, duck was a common food among the settlers. However, currently it is a gourmet food and is eaten in mole and *pipianes*, roasted and called Peking or Mandarin duck.

FIGURE 18.14 Sonora: Northern Bayo Beans with California Chili, 2013.

FIGURE 18.15 Sonora: Maize Tamales, 2013.

The list of foods is long, including craving foods (*antojitos*) (Iturriaga 2013), like toasted maize tortillas, triangular pieces of fried, or toasted maize tortillas, salted or not, served alone as a snack, or with melted cheese, and sliced chilies in vinegar (*nachos*). Nachos arose at the Victoria Club in Piedras Negras, Coahuila, in 1943, where the maître, Nacho Ayala, prepared the first nachos using golden and baked maize tortilla triangles with Wisconsin cheese and slices of chili peppers. Ms. Sinan, an American woman who was accompanied by a group of U.S. military wives, tried them, loved them, and soon nachos spread to northern Mexico and the southern United States. Since 1995, the Nacho Festival or Nacho Fest had been held in Piedras Negras (Jácome 2010: 36) (Figures 18.14 and 18.15).

Lead-tree pods (*Leucaena esculenta*) are important in the elaboration of regional dishes. Some are pods in mole (*huaxmoli* or *guatzmole*), which is eaten in the states of Morelos and Oaxaca; pods in green tomatoes; pods with flower sauce; toasted seeds; omelete with pod seed in chili sauce; and other dishes. Lead tree pods continue to be important in the diet in various regions of the country. Nowadays, there is an annual fair which is held in Yautepec, state of Morelos. Cooks

FIGURE 18.16 Sonora: Boiled Maize (*Elotada*) in Las Flores Town.

FIGURE 18.17 Sonora: Food Stand on the Mochibampo – S.Bernardo Bus Station.

take pod recipes there and prepare them. There are cash prizes for the best-prepared tastiest dishes (Figures 18.16 and 18.17).

Regional fairs have been organized to promote maize, other plants, the cornfield, and the homegarden, whose herbs are important in the daily diet of the average Mexican. Some of the fairs are maize, Opuntia prickly pears, amaranth, apple, grape, cantaloupe, watermelon, cotton, grape, cheese, apple, sweets, walnut, tamale, national craving-foods, and Mexican food. In the case of fruits, like apples, distilled beverages are included; at the apple fair, wines are promoted. There are fairs of other products, such as honey, and prepared foods that are not part of these agroecosystems but are part of national daily diets like cakes, tacos, tortas, lamb barbecue mole, tortas, Agave mead (aguamiel), Agave fermented mead (*pulque*); traditional foods, and many others that are promoted at 32 state fairs that feed the nationals (Jácome 2010).

Currently, different products are made from fruits that have been consumed since ancient times. One of those fruits is the Mexican hog plum (*jobo*, *hobo*), which although it is bought very little now, continues to be consumed in different ways: as a fresh fruit; in sweets; jellies; pickles; dried;

dehydrated such as prunes; with or without salt; candied; as jam to use in the preparation of tamales, atoles, and sauces (Cruz-León and Rodríguez-Haros 2012: 77–101; Velez-Coria and Cruz-León 2012: 111–123). One use in fashion is the fresh Mexican hog plum ground with water, sugar, and ice to make a drink in gourmet restaurants. For another drink, the ripe fruit is ground and mixed with sugar cane, alcohol, and mineral water to make an alcoholic beverage kwon as *toros* or *toritos* (Cruz-León and Rodríguez-Haros 2012: 77–101; Velez-Coria and Cruz-León 2012: 123).

Local, non-alcoholic beverages are consumed, made with maize flour, roasted, and ground maize (*pinole*), maize dough, or cornstarch, that may be mixed with water, sugar, cacao, cinnamon, or flavorings. One example of these traditional drinks is *cacahuatole* (also called cacao, or "ravine drink"), which is made with water, cacao, roasted, and ground fava beans, maize dough, cinnamon, anise, and sugar, like the one sold at St. Inés regional market of Zacatelco, in Tlaxcala, and Cholula, in Puebla. Women organized as producers make this drink and sell it in the markets. They began to sell this drink in the fixed markets and street markets (tianguis) of Cholula and Zacatelco. Due to the acceptance it has had among the population, the cacahuatole spread rapidly to neighboring towns and outside markets (*tianguis*).

Regarding products for the Mexican diet obtained from agroecosystems, like homegardens, and cornfield, Willias and Harvey (1997) said "[...] the kitchen of the homegardens shows variations of 0.3 to 0.9 ha and, therefore, contributed greatly to subsistence". In the Mayan lowlands, an ethno historical analogy suggested less production. Caballero (1992) stated that instead of just grains, a polyculture homegarden produced nutritious food, dietary supplements, medicinal herbs, ornamental plants, and ritual objects. He concluded that this agroecosystem produced around 11% of the necessary caloric intake of the national average per capita.

This could be true for certain types of homegarden, but it would be necessary to specify the place and the time in when that occurred, because in the studies by Blanco-Rosas (2006) and Cahuich (2012), it is proved that the homegardens of Soteapan (Los Tuxtlas, Veracruz) and X-Mejía (Campeche) had higher productivity, better distribution of their crops and a direction to self-sufficiency, and the marketing of some of their products if there was an excess. The earnings depend on the harvests and the market prices.

In general, the food systems in contemporary Mexico are organized regionally and are based on local native ingredients, which include plants, mushrooms, terrestrial and aquatic animals, birds, insects, aromatic leaves, flowers, and fruits. These ample lists have been marked as cultural heritages.

Supported by CONABIO, and advised by Robert Bye Boettler, Edelmira Linares, and Clarissa Jiménez, renowned specialists in botany, Diana Kennedy paid many visits to public markets, collecting and analyzing the ingredients intended for food in 31 states of Mexico, obtaining 1,179 recipes, preparing 688 field notes, analyzing the ingredients that were collected by her, and publishing four books through the project "Documentation of the biodiversity of Mexican gastronomy: rescue of Diana Kennedy's culinary archives". The indigenous ingredients that are used in food in Mexico still are:

> Black, white, bay beans, May flower, big blue beans (*ayocotes*); squash, chilacayote, squash blossoms, pips, zucchini; chilies (*pasilla, ancho, poblano, chilaca, serrano, guajillo, jalapeño, ancho, cora, tree, rattlesnake, chipotle, yellow, güero, colorado, miahuateco, mora, morita, chilpaya, habanero, amashito, and pasilla mixe*). Also *xoco, chaya*, Annatto, White leaf, Holy leaf, *cuajilote, ahuahtli*, wow, wild chayote, and coral bean flowers (*Erythrina coralloides*)[7], *Opuntia* cladodes, sour *Opuntia* prickly pear (*xoconostles*); mushrooms (*trompillos*, blue mushrooms, *clavito, cuitlacoche,* and bird's foot). Some fauna ingredients are *chicatanas*, deer, *jumiles* (*Euschistus taxcoensis*), dogfish, eel, *acamayas,* and catfish.

[7] This tree exists in thickets and dry forests, as well as in oak forests from 1,180 to 1,900 masl. It is used as a living fence, shade, and to make crafts; its flowers are edible and have many medicinal uses. The flowers are prized in stews, whether boiled or as seasoning in sauce; the petals were used as toys for children.

18.12 THE CORNFIELD, EDIBLE GREEN WEEDS, AND THE FOOD SYSTEM

Botanical vestiges of the pre-Hispanic presence of tolerated seasonal edible green weeds in the cornfield in the basin of the Mexican Highlands had been studied very little. McClung de Tapia et al. (2014: 113–114) provided a list of those weeds: (1) herbs eaten cooked were: *Amaranthus*; Amaranth pigweed (*quiltonili*); Goosefoot (*uauhzontli*) (*Chenopodium berlandieri* ssp. Nuttalliae); Cucurbita; green bean (*exotl*) (*Phaseolus*); purslane (*itzmiquitl, itzmiquilitl*) (*Portulaca oleracea*); and chipilin (*chipilli, chepiles, chipilín*) (*Crotalaria longirostrata*) that was found in Xaltocan, only, a Postclassic site; (2) herbs that were eaten raw, cooked, grilled, or roasted: cornfield tomato (*mil-tomate*), which is a plant that was sown together with maize; shrubby seepweed, seablite (*romerito, iztaquilic, iztaquilitl*) (*Suaeda torreyana* S. Watson); Creeping woodsorrel (*xoxocoiolli, agritos, vinagrillo*) (*Oxalis corniculata*), sunflower (*Helianthus annuus*) that was ground in the countryside and plots, out if sunflower was intentionally cultivated is unknown, its seeds were ground for bread or roasted. Foxtail (*Setaria*), whose seeds were eaten; (3) herbs that were eaten raw: goats (*Calandrinia*); creeping false holly (*xaltomatl, jaltomate, tomatillo, acahualera, equelite, quelites cojudo*); (*Jaltomata procumbens*) whose fruits were eaten; and the raw root, cooked, or roasted (*tomato tomatli, xitomatl*) (*Solanum*); and cornfield tomato with small green fruits (*miltomatl*) (*Physalis*) (Figures 18.18 and 18.19).

Among the condiments used were: wormseed (*epaçotl, epazote*) (*Dysphania ambrosioides*), which is a fragrant herb that was used to make sauces and was considered medicinal; and chilies (*Capsicum*) to season sauces, but since they were eaten together with their seeds, they are difficult to find in archaeological samples. The researchers recovered seeds in six of the ten sites studied. The peppers found in the sites were cultivated. They found a total of 27 selected genus in the Post Classic, 13 were registered in the Epi Classic, 17 in the Classic, and 22 in the Formative. The taxa that appeared in the sites were ten, of all the period: *Amaranthus, Opuntia, Myrtillocactus, Chenopodium, Phaseolus, Salvia, Zea mays, Portulaca, Crataegus,* and *Physalis* (McClung de Tapia et al. 2014: 115).

Currently, the weeds (tolerated and encouraged) that grow in the cornfields are not the product of the carelessness of the small-scale maize growers (*milperos*), or they are harmful to crops. They are tolerated because they have ecological importance in the system, and they have uses that maintain the system, or because they are integrated into the food system and the culture of the small-scale cornfield cultivation families (*milperos*).

Rendón-Aguilar et al. (2014) proposed that the roots of the weeds form a mesh, which prevents the soil from sliding off, reduces erosion, holds moisture, and participates in the nutrient cycle in the soil. Some of the weeds give shade, others serve as "plants traps" being feeding herbivores that could become a pest, hosting beneficial insects, or repelling parasites, thus promoting beneficial interactions for the agroecosystem.

FIGURE 18.18 Florentine Codex, Book XI, f.123v. Women selling *Leucaena esculenta* (*Guajes, Huajes*).

FIGURE 18.19 Francisco Hernández (1574), Tomato (*Miltomatl, Miltomate*).

Ecologically, weeds play a very important role in traditional agroecosystems (homegardens, orchards, gardens, cornfield, terraces, and semi-terraces). They are considered to be pioneers of succession in those systems and arose when one of those systems was abandoned. In the sixteenth century, they were consumed raw or cooked. Currently, they are collected in the rainy season like it occurs in places like the north of the Toluca valley where the farmers call them "seasonal green weeds (*quelites*)". These weeds are found in the first furrow that was opened in the ground when plowing of the cornfield began (*besana*) and on the banks of streams and canals.

As traditional agroecosystems developed under diverse conditions, ecological (altitude, soil, humidity, and temperature), biological, and cultural (agricultural practices), frequently, cultivated plants evolved and adapted to the local, particular conditions of each place. That adaptation was due to the complex forms of management to which they were subjected, so it is common to find great intraspecific variations in many weed species. In cultural terms, weeds meet various needs: as food (*quelites*); as herbalism, for many weeds are used as medicine; as fodder; as ornaments for ceremonial uses; in construction; as insecticides; among other needs. Weeds are seasonal, so they are sold in the markets during the seasons when they are found in the cornfield.

Biologists Rendón-Aguilar, Bernal-Ramírez, and Sánchez-Reyes (2017) collected and organized a list of useful weed species in the cornfields of S. Bartolo del Llano, State of Mexico, giving an idea of the richness of these species and of the uses that small-scale farmers grant them. The town is located at 2,540 masl, to the northeast of Ixtlahuaca, and in the north of the Toluca valley. The predominant climate is temperate sub-humid, with an annual average temperature of 14.8 °C.

S. Bartolo del Llano had a total population of 12,227 inhabitants in 2015, of which 30.5% were indigenous population; the majority of inhabitants were dedicated to agriculture, livestock, and commerce. The list of seasonal green weeds by Rendón-Aguilar et al. (2017) included: Amaranth pigweed (*Amaranthus hybridus* L.), which is food and forage, and the pod is bird feed; mustard (*Raphanus raphanistrum* L.) is human food and forage; mouse ear (*Drymaria glandulosa* Bartl.) is food; chicken grass (*Commelina coelestis* Wild.) is medicinal and used for constipation; yellow nutsedge (*tolpate, panochita, coquille*) (*Cyperus esculentus* L.) tubers are food and forage; sulphur cosmos (*Cosmos bipinnatus* Cav.) that is ornamental; turnip or heart (*Brassica rapa* L.) is food and forage, and its pod is food for birds; (*Dalea leporina* Ait. Bullock) is medicinal and was used for stomachaches; clover (*Medicago polymorpha* L.) is food; lion's paw (*Geranium seemanii* Peyr) is medicinal and used to relieve stomachache, colic, baby rash, and for wounds; and purslane (*Portulaca oleracea* L.) which is food (Bye 1981 1 (1): 109–123; Mapes-Sánchez 2010: 217–222; and Vibrans 1997 (38): 21–67).

In Xocotitlán, a town in the municipality of Ixtlahuaca, north of the Toluca valley, edible green weeds are known locally as "seasonal green weeds (quelites)" and generally are consumed by the cornfield farmer families (*milperos*), generally. González-Jácome and Reyes-Montes (2016: 21–42) provided a list of weeds found in the Xocotitlán cornfields in 2013. These plants became rare because of the increasing application of herbicides in crop fields. The edible seasonal green weeds were turnip (*Brassica campestris*), mallow (*Malva sylvestris*), amaranth pigweed (*Amaranthus hybridus*), purslane (*Portulaca oleracea*), cruet (*Rumex acetosa*), and ashy milkweed (*nexhautli*) (*Chenopodium berlandieri* subsp. *nuttalliae* spp.). Other weeds grow near the cornfield *besanas* and on the banks of the streams, like sneaks *Calandrinia micrantha* and watercress (*Nasturtium officinale*). Occasionally, the pore leaf (*Porophyllum ruderale*) and the small green tomato were found.

In Mexico, around 500 species of tolerated green weeds are consumed (Mera et al. 2011, Linares and Bye 2015). They have been selected according to the food traditions of the different people and regions. Among the most common weeds are the pápalo (*Porophyllum ruderale* subsp. *Macrocephalum*), purslane (*Portulaca oleracea*), amaranth pigweed (*Amaranthus* spp.), shrubby seepweed (*Suaeda nigra*), ashy amaranth (*Chenopodium berlandieri* subsp. Berlandieri), Amaranth (*C. berlandieri* subsp. *nuttalliae*), Alaches (*Anoda cristata*), wormseed (*Dysphania ambrosioides*), *chaya* (*Cnidoscolus aconitifolius*), the holy leaf, or *acuyo* (*Piper auritum*), and the *chepiles* (*Crotalaria* spp.) (Linares and Bye 2015).

Tolerate green cornfield weeds are eaten raw in salads usually, although purslane is cooked with pork and Amaranth pigweed is cooked in stews and mixed with meat or chicken. Also, seasonal green weeds can be eaten in soups, in raw snacks, and in tacos with avocado and salt. The tolerated green weeds harvested in the cornfield are sold in local markets, like the Friday market in the city of Toluca. At the moment, it is divided into several sections and distributed in different areas of the city. Some of the wild cornfield weeds are found and sold in local markets and tianguis. In recent years, some of them began to be cultivated and sold in urban supermarkets: purslane, amaranth pigweed, watercress, Mojave seablite (*romeritos*), and the small green tomato (*tomatillo*) (Table 18.1).

18.13 THE MEXICAN CORNFIELD AND THE FOOD SYSTEM TODAY

Presently, there is an annual selection of what plants will be planted. The choice is made according to local soil, water, climatic conditions, and the target clientele for the products that will go to the market. In Tlaxcala, broad bean is planted instead of beans, at this time. This substitution does not mean that the bean is not a fundamental element in the cornfield or in the diet. It is related to conditions that allow broad bean growth and development that are dependent upon the annual rainfall, soil erosion, or the loss of soil fertility. The bean is a profitable crop on the market, and the consumption of beans – both green (*ejotes*) and dry –continues to be important for local diets.

Beans maintain their commercial importance in regional markets, like in Tlaxcala, and the Highland Chiapas, where the bean is the basic staple crop intended for both self-sufficiency and commercial sale. However, beans are susceptible to rainy conditions, market demand, and price. Broad beans can compete with beans in the market, in terms of prices, but it is less demanding as a crop and supports torrential rains, floods, and early frosts better.

Adaptations or adjustments allow the success of crop productiveness throughout the annual agricultural calendar. In places far from urban, industrial, or commercial centers the intervention of cornfield protective deities is still requested and appreciated throughout the ritual calendar year, articulated to the different management of the cornfield and maize cultivation (Alcorn et al. 1984, 2006; Blanco 2006; Dehouve 2007; Márquez-Mireles 2000; Serafino 2012). Gradually the pre-Hispanic cornfield included the plants that were being domesticated, and of which there are archaeobotanical, and genetic studies that confirm their ages, which are not necessarily the same in various places, like the Basin of Mexico, or the markets in northern, and southern Mexico.

Studies, such as those of Staller (2006), proved that from the earliest times the food system generated in the cornfield was complemented with meat from domestic and wild fauna obtained by hunting in the forest and homegardens, plus fishing and gathering. Wooded areas or the incidental homegarden served to attract mammals and birds (Blanco-Rosas 2006; Vanderwarker 2006). The same thing in the cornfield was an attractor for edible fauna, contributing significantly to the food system of human populations. Where there were coasts, mangroves, lagoons, marshes, and rivers, fishing provided an important number of species for food consumption, the same as the gathering of shellfish, like oyster, prawn, and crab (Gómez-Nava 2005; Méndez-Guevara 2004).

There were, and are, many types of cornfield, each with degrees of diversity, representing local adaptations to food needs, culinary tastes, particular environments (climate, humidity, altitude, strong winds, cyclones, flooding, early frosts, and so on), in addition to adjustments to soil type, availability of water, residual humidity, or land inclination. These micro conditions influenced the type of agroecosystem that people developed in a region and elements, like the distribution of plants in the cornfield plot, were included. It is necessary to know the biological and genetic capacity of plants to distribute them at different altitudes and in distinct climatic conditions. In tropical regions, the cultivation of tubers is essential because they are basic in the daily diet of people in those regions.

The cornfield in the tropics included a wide variety of cultigens, medicinal herbs, and trees associated with the cultivated and tolerated flora, which had multiple functions and covered the basic needs of small-scale farmers. In the Central Highlands, the vegetation on the edges of the cornfield land has a vital role in family herbology and in the biological control of insects. These cornfield systems tied to the agricultural lands require specific studies lest they disappear completely.

Cultivated plants are multifunctional (Blanco-Rosas 2006; Cahuich 2012; Mariaca-Méndez 1988). Some of the maize parts are considered medicinal, such as the inflorescence and panicle, which is boiled in water and drunk to prevent kidney disease. Other parts of the plant are dry, and they are used in fire to cook food (reeds and dried plant roots, and rachis or corncobs). One way to preserve the maize kernel is by boiling the ears, then drying, shelling, and storing them until the following Easter. This conservation system is called *guachales*, and the boiling water is given to domestic animals, or it is thrown on the houseplants, like in Chihuahua.

In Mexico, a way to use the water with lime in which maize was boiled (*nixcomel*) is to give to pigs and poultry. For pigs, the water might be mixed with old tortillas cut into pieces to fatten them quickly. Corn shucks are green or dry and are used to wrap tamales. Maize kernels, tender or hard, and dry are used to make tamales, dough, *esquites*, or corn-on-the-cob, which is slathered with mayonnaise or cream, and cheese, and dusted with chili powder. Boiled maize kernels with chili and wormseed (*esquites*) are sold in plastic cups at street stands. Dried maize canes are used to make fences to protect vegetables and other plants from domestic animals.

Corncobs are used as fuel for the kitchen fire. Mariaca-Méndez found that the practice of burning corncobs does not exist in Chiapas, because this part of maize is considered sacred. The maize fungus (*huitlacoche*) is not consumed there either. Mariaca-Méndez's studies show that the inhabitants in the south and southwest of Mexico manage domestic animals have revealed that they feed their hunting dogs the boiled hominy water mixed with sun-dried tortillas or pieces of leftover tortillas from that day.

The management that the inhabitants in the south and southeast of Mexico feed their hunting dogs with the boiled hominy-water mixed with pieces of sun-dried tortillas or leftover tortillas from the same day. The idea is that this food is very nutritious, and hunting dogs deserve it. Apparently, this practice was very widespread in Mexico, when all women cooked maize with water and lime to hominy it at that time (González-Jácome 2004, 2008c: 1–17, 2018 (60): 9–69; Long 1997, 2009: 127–136; Long and Vargas 2005).

The importance of maize in the food system because of its characteristics as a plant, its versatility to be consumed at any stage of its development (when young and tender until mature and dried) and its storage possibilities led it to spread throughout Mesoamerica, in the north and the south of the continent, and later to other countries in the world. On its journey, other plants, like the squash, beans, tomatoes, and chilies, joined maize in this long process of diffusion; maize was popularized, but no information was given on the hominy process (nixtamalización), where maize grains gave new nutritional characteristics and new possibilities of various ways to prepare it.

The aforementioned plants continue to be of great importance in the diets, kitchens, and daily diets in many countries in the world today. The eating habits of people and domestic animals were modified, as well as the cultures of those people. Cultural uses refer to maize as non-industrialized human and animal food. A series of food products that are important in the popular food system, and daily diet in Mexico, and on the rest of the American Continent, are industrially processed. Commercial packaged cereals that combine maize, oats, and rice with nuts, raisins, chocolate, almonds, dried fruits, sugar, honey, and a long list of flavorings are popular for daily consumption. Also, there is popcorn seasoned with salt, caramelized and ready to pop in the microwave, in a pot, or in a pan with a lid. Popcorn consumption has become important in many nations.

Industrialized foods in Mexico include: precooked maize kernels for *pozole*; corn flour for making dough; cornstarch and envelopes with cornstarch prepared with flavorings to make *atole*; *tamales*, *tortillas*, *sopes*, and precooked *tlacoyos* to complement with refried beans, cheese, sauce, and cream, which are all packed in plastic bags; roasted and ground maize to which they add cocoa, cinnamon or vanilla, and sugar (*pinole*); toasted *tortillas* that can be cut into strips to put in soup or into triangular pieces to make *chilaquiles* or *nachos*; maize products sold packaged, and in packets containing hot sauce; tamales ready to heat in the microwave; dried corn husks to wrap tamales; and salty or sweet rosettes or popcorn ready to pop in the microwave or special appliances for that.

There are other industrial maize products sold, such as the "coffee cream" made with maize, and all kinds of packaged tortillas (white, blue, or green maize with *Opuntia* cladodes added; tortillas to make flutes, *sopes*, tacos, *tlacoyos*, and *huaraches*). The list gets longer as new prepared foods emerge. The list of products includes canned processed foods, like canned beans in foil-lined plastic bags, ready-to-eat sauces in cans, or grass bottles. There are: cheese-filled chilies packed airtight plastic bags; empanadas filled with meat or corn kernels and cheese; fish fillets packed and ready to cook in the microwave or electric ovens. TV dinners ready to heat up in microwave or electric ovens are not as popular now as the foods mentioned earlier. Industrialized foods have been adapted to national tastes but offer a few variants to buyers, besides they are very high priced for the general population.

18.14 CONSIDERATIONS AND CONCLUSIONS

In this book, the cornfield is defined as a polyculture agroecosystem of ancient origin where maize is its central axis, and it is associated with other crops like cultivated herbs, tolerated seasonal green weeds, and trees, in addition to domestic fauna. Currently, polyculture has decreased in relation to the number of crops that are planted or left to grow wild in the field. In extreme cases, where there is a lack of family labor, the cornfield is used for maize cultivation, exclusively. This grass has ancient and current forms of consumption. The cornfield arose articulated to areas near where there are relics of forest and herbaceous plants, or where the original homegardens were gathering and hunting places.

Residents' tastes in consuming certain types of plants and their fruits, or obtaining sugars by sucking the stems of grasses such as teosinte, were decisive in the selection of future domesticated crops (Iltis 1983: 886–894). In numerous populations of ancient Mexico, maize was the central axis in daily life and ritual diet. Maize cultivation and food systems lasted into the Viceroyalty, the nineteenth and twentieth centuries, and to date.

Maize grown in any food agroecosystem is part of national consumption; any polyculture agroecosystem includes maize always. In various parts of the country, small-scale farmers plant different varieties of maize, which are recognizable by their color, size of the ear, quality of the kernels, and the destination for each one of them. Small-scale farmers grow different species of beans, squash, chilies, cultivated herbs, tolerated weeds, and other plants. The diversity of the cornfield depends on the region of the country in which this agroecosystem is located, the environmental characteristics of the region, climatic conditions, access to water for irrigation and the customs, and culinary tastes of the small-scale farmer families.

The cornfield provides food for part of the year including the associated domestic and wild fauna in it that is added to the small-scale maize growers' food system. Until the 1950s, the Zoque-popolucas in Ocotal Chico, Soteapan Municipality, obtained an annual total of 341.5 kilos of red meat from a cornfield. The total included another 368.5 kilos of foul flesh obtained from the forest fowl. In the bush, meat from large mammals was obtained, which reach about 100 kilos a year. About 460 kilos of shrimp and snails were obtained from the river (Blanco-Rosas 2006).

Old and recent evidence and studies prove that the diet based on cornfield products is complemented with domestic animal meat and wild fauna flesh from other systems, such as the forest or the ecotones formed by ravines, lagoon areas, such as lake shores, rivers, and canals. Activities, like hunting, fishing, and gathering, depend on the specific place, the regional context, the characteristics of the human population living there, and the cornfield areas that small-scale farmers possess. Some small-scale farmers growing maize sell part of the bushmeat obtained, and thereby supplement their family income.

The hunting and capturing of fauna in the cornfield continue today and are important for farmer's families since the cornfield agroecosystem provides them with animal protein for their diet. The zoque-popoluca cornfield included wood pigeons, chattiest, partridges, toucans, *chuco* birds, quails, woodpeckers, and parakeets, providing peasant families with a total of 345.5 kilos of game-fowl flesh per year. In his study on the biodiversity of the zoque-popoluca cornfield, Blanco-Rosas (2006) calculated that in Soteapan, the capture of partridges, pigeons, and other wild birds in the cornfield plots gave families enough animal protein to satisfy their basic annual food needs. Industrialized foods have been replacing these sources of protein and do not contribute to an improvement in the nutrition of small and medium-scale farmers, necessarily.

The cornfield agroecosystem is always linked to other cultural systems and ecosystems. A partial explanation of these interrelationships and combinations is that since the nineteenth and twentieth centuries, the cornfield has lost its interrelationships with other agricultural and natural systems. Also, it has been lost spaces like land edges, hedges, rows of Agave (*metepantles*), wooded strips that limit the farmland, zones with wild vegetation mixed with cultivated, that together with the crops were an important part of the annual needs of the small scale farmer families. Deforestation includes forested areas, and also these apparently small agroforested zones, which extent is unknown, and many of which has now been lost, but which were part of the agroecosystems, and of the biological and cultural diversity in the traditional Mexican agriculture.

The cornfield agroecosystem is isolated by the processes of industrialization, modernization, and urbanization. Traditional agriculture has received negative impacts as a consequence of public policies to support extensive agriculture with export monocultures; the lack or insufficiency of monetary capital; the impact on soils and plants from the application of agrochemicals and genetically modified seeds (e.g., the loss of seasonal green weeds); the decrease of local and regional varieties of flora and fauna adapted to specific environmental, climatological, altitudinal, physiographic, and cultural situations, and to family and local needs; the low prices that local agricultural products have in the market; and other issues that have left traditional Mexican agriculture aside.

The traditional agroecosystems need to be re-studied to highlight their importance in issues such as sustainability and resilience. It is necessary to observe and understand the components, management, processes, and operations focused on the maintenance that small-scale farmers grant to agroecosystems, as they contribute to the sustainability of the food system. The cornfield is not just a family pantry that feeds families a few months a year, nor is it just the land where crops are. It is an articulating axis of several ecosystems and agroecosystems, which evolve with them: relicts with natural vegetation, land edges, ecotones, the dry land or irrigation systems, plots for maize, and other flora.

The cornfield and its traditional food systems remain key components in the wellness of small and medium-scale farmers. The cornfield is associated with health, daily, civil, social organization of families and communities, workforce, formal education, religious and ceremonial life of mestizo, and ethnic-groups populations that are agriculturalists to this day. This agroecosystem is in the process of disappearing in various areas of the country. Its loss has numerous consequences, and for this reason, its study, care, and maintenance continue to be fundamental for thousands of small-scale farmers in Mexico (Lazos 1994: 79–109; López and Vizcarra 2016: 113–136) (Table 18.1).

TABLE 18.1
Ways of Consumption of Wild and Tolerated Edible Weeds. Friar Bernardino de Sahagún, 1830, Vol. III

Nº	Name	Uses	Raw	Cooked	Roasted	Total
1	Nanacatl	These edible mushrooms grown in the fields. However, they cause vomiting or gases if they are undercooked or eaten raw, and they can be deadly. This effect is counteracted with enema.		X		1
2	Tzontecomanatl	Large round mushrooms that grow together on one stalk, some tall and others low.		X		1
3	n/d	Wide, round mushrooms used in dishes.		X		1
4	n/d	Round white mushrooms easily cooked or roasted on clay plates. They are very tasty.		X	X	2
5	Cimatl	A wild weed with a root called *quavecoc*, *cimatl*, or *eimatl*). It produces large beans similar to broad beans. They are boiled for 2 days.		X		1
6	Huauhquilitl	A very green annual plant similar to amaranth with thick and thin branches, and broad leaves. The stems and seeds are called den (*Huautli*). It is cooked with salt, has an ashy taste, and is like Spanish ones. When it is cooked, it is squeezed and used to make tamales (*quiltamalli*) and tortillas.		X		1
7	Quilonilli	It has broad leaves and is edible when small. When it is large, it is cooked with the *tequixquitli* and then squeezed before using. Its black seed is called *pitzitl*.		X		1
8	Itzmiquilitl	Brown weed with large branches, round and flat leaves.		X		1

(*Continued*)

TABLE 18.1 (*Continued*)
Ways of Consumptionof Wild and Tolerated Edible Weeds. Friar Bernardino de Sahagún, 1830, Vol. III

Nº	Name	Uses	Raw	Cooked	Roasted	Total
9	aioxochquilitl	Flowers, lumps, or limbs of pumpkin branches. The skin is removed from above them before cooking. Note of Sahagún (1830 Vol. III: 245). "In Oaxaca they are eaten together with the guides and green pumpkin stems, mix some balls of maize dough with salt, and some cooked green chilies. It is the food of poor peasants and very delightful because it is dominated by openness, simplicity, and the good humor of the countryside".		X		1
10	Axoxoco	Weed with long, wide leaves. It is tasty and sour.		X		1
11	Mizquilitl	It is attic, very green, with bladed leaves. It is tasty to eat.		X		1
12	Acuitlapali	Long, brown weed grows at the water's edge. It is good to eat.		X		1
13	Tziunquilitl	It grows by the water's edge. It has blue and harp leaves. It is good to eat.		X		1
14	conalquilitl	It grows in the mountains; it is ash colored; its root (*tacanalli*) is eaten.		X	X	2
15	mamaxtl	It grows at the water's edge; it is tasty.		X		1
16	n/name	It is like a nettle, it is edible weed.		X		1
17	ueiquauhquilitl	Wild smoothies before cooking are bitter. They are tasty.		X		1
18	Etenquilitl	Herb of beans, which spill when they are picked. It is eaten.		X		1
19	tlalaioquilitl	They are wild pumpkins.		X		1
20	xaltomaquilitl	It is eaten cooked.		X		1
21	Tzitziquilitl	It is dark green, with flowers and seeds, and very good to eat.	X			1
22	Eloquilitl	It is very green and tender; it has flowers; it is very tasty.	X			1
23	quauheloquilitl	It is wild; it is born in the Opuntia *tunales*. It is very tender and good to eat.	X			1
24	mozoquilitl	It is green, fluffy, and very tender. It is tasty.	X			1
25	tzaianalquilitl	It is made in water; its branches are hollow and harp; it is good to eat.	X			1
26	achochoquilitl	It is light green; it is born near the water; it is good to eat. "They say of this grass that if the boys or girls eat it they become powerless to beget; but after they grow up everyone eats it ". Sahagún, Volume III, p. 247.	X			1
27	Tzonquilitl	It is light green; it is born near the water; it is good to eat.	X			1
28	Iztacquilitl	It is low and flat, it has a salt flavor.	X	X		2

(*Continued*)

TABLE 18.1 (*Continued*)
Ways of Consumptionof Wild and Tolerated Edible Weeds. Friar Bernardino de Sahagún, 1830, Vol. III

Nº	Name	Uses	Raw	Cooked	Roasted	Total
29	Tepicquilitl	With long and pointed leaves. If you eat a lot of this herb, it gives cameras.	X			1
30	Ecoquilitl	They are the leaves and branches of beans. They are a little rough and fluffy. They provoke burp (regoldar).	X			1
31	Vitzquilitl	They are thistles of the earth and have thorns; the leaves below are ashy, and the ones above are green. Inside they have lint, like the thistles of Castile, make them at the water's edge. It is also a garden weed. They are good to eat.		X		1
32	qaavitquilitl	Wild thistles that are born in the mountains. They are prickly and itch those who eat them.		X		1
33	chichicaquilitl	Edible grass that grows near water and in sweet, tilled land. It has white roots; it is very tender and somewhat bitter.		X		1
34	Coiocuexi	The stems of this herb do not have thorns, and flourishes, is bitter, it makes the voice hoarse in children. Girls do not eat it.		X		1
35	Popoiauh	It is a black and green bush; it is kneaded with maize; and tortillas are made from it.	X	X		2
36	Exixi	It has tiny leaves; its seed is yellow; it is chewy; and if you eat a lot, it produces blisters and it is too hot. It is used to make tortillas and tamales and atole or porridge (for those with chambers of matter and blood, since it purifies the intestines).	X	X		2
37	xoxocoiolli	Perennial plant with an acidic flavor is used as a condiment. It is sour.	X	X		2
38	xuxocoiopapa-tla	Herb with tall, slender feet, broad, rounds leave. The leaves at the tip of the foot are tasty to eat cooked.		X		1
39	xoxocoiolcue-cuepoe	Weed with fat round feet, rare leaves. It is born when the rains begin. They are tasty like tomatoes.	X			1
40	xolxococolol-vivila	It is a stopped herb with small and round leaves it is tasty.		X		1
41	miccaxoxocoili	Herb with a thick, hairy foot; the leaves are wide, very acidic, and cause an unpleasant sensation that is experienced in the teeth and gums when eating *(dentera)*.		X		1
42	quauhxoxocoili	They are soft to eat.		X		1
43	quananaca-quilitl	Herb that the chickens of Spain eat. They are the *serrajas* (*Sonchus oleraceus*) de Castilla. Indians say there were none in these lands before the Spanish came. Now there are so many that the whole earth is filled with them, because the seed flies and they have multiplied everywhere.	X			1

(*Continued*)

TABLE 18.1 (*Continued*)
Ways of Consumptionof Wild and Tolerated Edible Weeds. Friar Bernardino de Sahagún, 1830, Vol. III

Nº	Name	Uses	Raw	Cooked	Roasted	Total
44	cultivated xonacatl	Little onions from this land. They are planted, Indians are horticulturalists, and they eat the onions of Spain.		X		1
45	wild xonacatl	Wild onions are found in these fields and burn a lot.		X		2
46	Maxten	Wild onions are onions that grow a lot, together, faint and flourish are bland. Indians eat the root or the head.		X		1
47	papaloquilitl	Round-leaved, fragrant, and tasty herb. It is found in hot lands.	X			1
48	Aiauhtonv	Bush wild and low. It is found in the fields and mountains. Its flowers are fragrant.		X		1
49	no/name	It is edible; Sahagún named it as "the sweet potato".		X		1
50	Xícama	Indians also eat its leaves.	X			1
51	tolcimoquilitl	Herb with very beautiful and very delicate flowers. The root is edible.		X		1
52	xaltomatl, xaltotomatl	Fruit from the plant called *xaltomaxilhuitl*. There are black and white, very juicy, sweet, and round. The root is edible.	X	X	X	3
53	coiotomatl	Fruit like that of the small tomatoes called *miltomatl*. It has yellow coverage, they are sweet, and they slightly lock the throat. If you drink a little of the root of this herb, it is medicinal and it cleanses the intestines. Women who breastfeed drink it because it purifies their milk.	X	X	X	3 It is medicinal also.
54	Atlitiliatl	It has a sweet, blackish strawberry that declines to bittersweet. The leaves are good for the sick steam bath (*temascal*) to add to the water.	X			1
55	Tlalilotl	Edible weed.	X			1
56	Tlalaiotl	Edible weed.	X			1
57	iacanalxivitl	Medicinal herb that takes away the fever. It is ground with maize and drunk with water.	X			1
58	coanenepille	Herb to purge. After drinking it, you eat bird broth or an atole (*yolatolli*).		X		1
59	teonanacatl	Heal gout and fever.		X		1
60	tlatlalayotli	Its leaves are edible; the root is medicinal.	X	X		2
Total	60					

Source: Friar Bernardino de Sahagún, *Historia de las Cosas de la Nueva España*, Tomo III, pp.244-249, 257.

RECOMMENDED READING

Lentz, David l.; M. DeLand Pohl; J.L. Alvarado; S. Tarighat, and R.A. Bye. 2008. Sunflower (*Helianthus annuus* L.) as a pre-Columbian domesticate in Mexico. *Proceedings of the National Academy of Sciences of the United States of America* 105(17): 6233–6237.

Mexico is recognized as a plant domestication center in the world, with evidence such as the domestication of squash (*Cucurbita pepo*) in 8000 BC, maize (*Zea mays*), chili (*Capsicum annuum*), common beans (*Phaseolus vulgaris*), and cotton (*Gossypium hirsutum*). Sunflower (*Helianthus annuum*) is added to the list after several discussions about being domesticated first by people in eastern North America.

Recent research shows that its initial domestication occurred in Mexico. Thus there is information from the Ocampo cave, in Tamaulipas (Flacco Phase 2900 to 2200 cal BC); then at the site of S. Andrés, Tabasco (2875 BC to 2575 BC). Data about sunflower domestication extends to El Salvador in the first millennium BC. It was also found in the Del Gallo Cave, in the state of Morelos (290 ±40 cal BC) affiliated with the Ticumán culture, south of the Basin of Mexico.

Sunflower was a very important element in the Mesoamerican diet of the third millennium BC, when maize was a fledgling crop or was grown in areas where maize was less reliable. In the post-classical Aztec era, the sunflower seems to have been less used as food and was used more as an ornamental plant and as a symbolic component in ritual activities or medicine – all those uses that were common to current Mexican indigenous cultures. Table 3 deals with the uses of sunflower achenes in 11 indigenous groups in Mexico: Nahua, Raramuri, Tepehuan, Seri, Totonac, Otomi, Huastec, Tzeltal, Mixe, Zoque, and Popoluca. The information archaeological and linguistic was compared, showing the changes through time in the uses of sunflower.

19 Diversity, History, and Geography of Maize in Mexico

19.1 INTRODUCTION

In addition to the study of ancient and current agroecosystems, another important group of researchers has dedicated themselves to investigating ancient agriculture, domesticated and cultivated plants, in addition to maize, and races that are currently distributed throughout the country. Hernández X. and Alanis (1970: 3–10) found five races of maize in the Sierra Madre Occidental, and Aragon-Cuevas et al. (2006) studied 35 races of maize in the state of Oaxaca, which represented 70% of the known races in the country.[1]

19.2 BACKGROUND

Barrera Bassols and Astier Orozco (2009: 1-6) studied 15 varieties of maize adapted to mountain conditions, through the genetic recombination of six races in S. Francisco Pichátaro, Michoacan. These maize varieties adapted to the mountains included: (1) white young maize (cold, wet slope); (2) small pink Toluca maize (cold, and wet hillside); (3) glassy pinto maize (cold, wet slope); (4) chocho maize (*malpaís*); (5) Blue maize (*malpaís*); (6) mealy red maize (homegarden); (7) black uaroti maize (homegarden); (8) yellow cuchitan maize (cold and wet hillside); (9) mealy *uacash* maize (homegarden); (10) mealy sky blue maize (homegarden); (11) uaroti maize (homegarden); (12) eight furrow maize, small, yellow (temperate and sub-humid plans); (13) mealy purple maize (homegarden); (14) cuchitan pinto maize (cold and wet hillside); and (15) large, white Pichátaro maize (sub-humid and dry temperate plains). Six of these local varieties are cultivated in homegardens (40%). Barrera-Bassols and Astier (2009: 2) presented a table with photographs that explain the combinations.

These variations are local and regional adjustments that human populations carried out and that depend largely on the environment in which they live, family needs, the market, supply/demand, and – to a certain extent – the arbitrariness of prices, and their oscillations throughout the year (Mapes-Sánchez 1987: 345–379). In addition, it was necessary to consider the family needs and the amount of money that they had available to cover their family and social expenses, and those related to agricultural festivals and celebrations.

19.3 SOME CHANGES IN AGROECOSYSTEMS AND WEEDS

Since the arrival of Spaniards in Mesoamerica, Old World plants entered and were incorporated into pre-Hispanic agriculture, enriching the diversity of the cornfield, the homegarden, and the food system. Along with them, weeds came in, of which there is no information. In the sixteenth century, the native flora incorporate: lime (1518); orange (1518); banana (in the 1520s); wheat and sugar cane (1523); turnip; cauliflower; lettuce; carrots, and radishes (1526); grape (1529); eggplant (1530); pea; cucumber; chickpea; olives; plum; pomegranate (in the 1530s); yellow lemon (1531); apple; pear;

[1] Aragon-Cuevas et al. 2006; Benz 1999a, 1999b, 2011: 2104–2106; Bruce 1997: 16–23; Flannery 1986; Hernández X. 1978: 3–12; Martínez-López et al. 2018: 111–121; Wellhausen et al. 1952.

DOI: 10.1201/9781003198833-19

and quince (1536); date and almond (from 1530 to 1540); nectarine; barley; millet; broad bean; Spanish walnut (in the 1500s); alfalfa (mid-1500s); cabbage (1538); fig and garlic (in 1530s); lemon (1531); oats (1575); spinach; artichoke; asparagus; onion; lentil; melon; watermelon; peach; cherry; watercress; oregano; mint; and parsley (introduced before 1580) (Dunmire 2004: 126–130; García-Paris 1991: 33–193).

In the sixteenth century, the native cornfield (*milpa*) had at least nine varieties of squash: Indian (*ayotli*); *pipian* (*cuicuilticayotli*); yellow (*coxticayotli*); yellow (*chilacayotli*); *tamallayotli* (*tzonayotli*); red pulp (*quauhayotli*); hairy (*tzonayotli*) whose seeds were eaten; white (*iztaczilacayotli*); and *tlilticayotli*. Also, there were native onion (*xonacatl*) and numerous cultivated and seasonal edible green weeds (*quelites*), such as *papaloquelite* (*Porophyllum ruderale*), of which Francisco Hernández cites 31, whose Nahua term have the word ending *quilitl* (González-Jácome 2011: 476–480) (Table 19.1).

In the seventeenth and eighteenth centuries, plants that arrived in New Spain (mid 1600s) were rice (mid-1600s); broad beans (*Vicia faba*) (in the 1690s); tamarind (after 1700); sorghum (in the 1700s); coffee (1730); celery and marjoram (from 1750 to 1760); and mango (early 1800s). The list provided by Dunmire (2004: 126–130) included plants that reached northwest New Spain through Sinaloa. The plants that Spaniards brought to the New World were those that they required in their food system but had nothing to do with the tastes of Indian ethnic groups.

According to Patiño (2002: 5–14), the Spanish prejudices about native fruits were very strong, to the point of being considered causes of dysentery, since they were laxative and cold mostly, with the exception of avocado, which was hot. Juan de Cárdenas (1945 [1571]: 32v-33) considered that "[…] they lack the flavor, and virtue that European fruits have". The European fruits planted in the New World lost their sweetness after being cultivated in these humid lands. It seems that Europeans did not like the taste of native plants, including fruits, and that is why they brought plants from the Old World (Patiño 2002).

19.4 MAIZE TRAVELLED THE WORLD

In Spain, maize arrived in the Guadalquivir valley, Granada, and Malaga in the early sixteenth century. In 1549, corn bread was sold for two maravedies and wheat bread for three (Anonymous 2017). In the homegardens in Valencia, maize was introduced at the end of the sixteenth century, and by the eighteenth century, it had displaced the other summer cereals (López-Piñero, and López-Terrada 1997; García-Paris 1991: 40). Beans, squash, chili, and tomato followed maize. That is a simple but basic association of what the Mexican cornfield was so in Spain, maize displaced millet and mixed with native crops, such as broad beans.

Currently, it is called: "pre-Columbian association", and "homegarden-garden/or homegarden and garden with maize" in Castilian. This cultivation system was known by local names then. The old, local names that in Spain were designated to name maize, and its products corresponded to the linguistic variants in Castilian, Basque, Galician, or Asturian; they are maize, mahíz, máis, maínzo, meiz, millo, milloca, paínzo, panizo, adeza, arto, artúa, and dacxa (García Mouton 1986: 121–146).

In the sixteenth century, there was a strong emigration to Guipúzcoa (Gipuzcoa) to the American continent. Owing to seasonal agriculture and several years of drought, there was not sufficient production of wheat or barley for food. The Basques took maize to the Cantabrian coast in 1576 (García-Paris 1991: 40), beginning their maize production in the form of *talos*. *Talos* were eaten as bread in the Basque farmhouses in Guipúzcoa, northern Navarra, and the French Basque Country; the leftover of *talos* was eaten as a kind of soup, dipped in milk. *Talos* is a typical food and is a kind of large maize tortilla, with the difference that maize was not nixtamalize previously (hominy processed), so it has a brittle, sandy texture, and is less nutritious. The dough is made mixing cornmeal with water. Wheat flour and salt can be added. Then it is grilled. It can be eaten with different filling, like a taco but made with a hard tortilla.

The hominy maize process (*nixtamalization*) did not reach Spain, so the grain was handled dry and crushed like flour, with which breads and corn bread cakes (*boroña*). *Talos* are from the Basque country, Navarra, and the southwest of France. They are a sort of hard tortilla, to which a component was added to give it flavor, such as roast sausages. Other forms of maize consumption were *gachas,* tortes, fariñas, farrapas, stuffed empanadas, and polenta. This is because those who took maize to the Iberian Peninsula explained how to make dough with it but did not include the maize hominy process (nixtamalization) in their explanations. Physician Juan de Cárdenas (1571: 140v) wrote about the process of cooking maize:

> Maize is ground on a stone (*metate*), kneaded, and made into bread without using salt, yeast, or any other thing than a little cold water. At the moment it is cooked on a casserole or clay dish (*comal*), it is so hot you eat it with great pleasure (*tortilla*), and it is a gift to the world, and especially because it take brevity to make it [...], that I do not know what can be said of bread (*tortilla*), that along with being of his so good, and of so much sustenance, it is so easy, and cheap to season.

Maize arrived at the herbariums of Europe in the late sixteenth century and was considered an ornamental plant (García-Paris 1991: 38–39). From 1515 to 1525, maize was cultivated in Coimbra, Portugal, as an edible plant for humans and animals. In the mid-sixteenth century, maize was the basic food in the Miño and Beira provinces. By the end of the century already, it had entered the southwest of Portugal. In Germany, the naturalist Jerome Bock mentioned maize in 1539. In 1542, Leonhard Fuchs described the grain in various German herbariums where it was known as "Turkish grain".

From 1535 to 1550, the Portuguese introduced maize in the Cape Verde islands, and by 1550, on the western coasts of Africa, where it was used to feed slaves who would be brought to America (García-Paris 1991: 40–42). In the Pierandrea Mattiolli herbarium, maize was mentioned by 1570. In the herbarium of the English naturalist Gerarde, maize was mentioned in 1597 (García-Paris 1991: 38–39).

Maize continued its journey through various countries of the Old World. In the early 1500s, the Portuguese introduced maize to the western coasts of Africa (García-Paris 1991: 40–42). Around 1550, maize reached the Veneto in Italy. In the middle of the seventeenth century, it passed to Austria and the Balkans. Miracle (1966) considered that maize was introduced in Turkey at the same time than the Turks spread its cultivation to Russia, probably (García-Paris 1991: 40).

In 1561, the grain was cultivated in Mozambique, where it was for Portuguese residents. In 1634, Pieter van den Broecke reported on his trip to Cape Verde, Guinea, and Angola that it was a "[...] fertile land with all kinds of crops such as the so-called mile masse or Turkish wheat".

In various parts of tropical Africa, linguistic evidence suggested that many areas of that region-received maize through the Sahara, perhaps with the Arab merchants. In 1526, maize reached the Coptic Christian regions of Ethiopia and Eritrea (Anonymous 2017). By 1573 maize had arrived in Iraq, traveling from Tripoli to Aleppo (Syria) and Jerusalem. Then, using the routes of the lands of Fertile Crescent, maize arrived in Baghdad, Iraq. In 1605, there was maize grown in Angola, where Dutch merchant Pieter van den Broecke reported its presence. From the 1500s, the Portuguese and Arab merchants from Zanzibar took maize to South Asia (Brandolini 1970).

Through the Silk Road merchants introduced maize to the northwest of the Himalayas, from where it spread to neighboring regions (Dowswell, Paliwal and Cantrell 1966). From 1615 to 1627, maize arrived in the Kathmandu valley in Nepal, and apparently its introduction occurred during the reign of Jagatjyotir Malla. By 1687, the grain was grown in Thailand, where an envoy from the King of France reported that from 1687 to 1688, the Siamese cultivated maize (*bled de Turquie*) as a garden vegetable for domestic uses (Anonymous 2017 [1498]; Humboldt 1978 [1822]).

An often discussed theme is the idea that there were contacts in pre-Columbian times between the New and Old World, including Asia (Marszewski 1978), and so that was when the primitive forms of maize arrived in Asia. Anderson (1945) and Stonor and Anderson (1949) maintained that the Himalayan region was a secondary center of maize origin in that region, because in Sikkim and Bhutan there were some types of maize that did not exist in any other part of the world, like primitive Sikkim maize (Paliwal 2001). Mangelsdorf (1974, 1986) considered that there was not insufficient evidence to support the theory of the Asian origin of maize because there was no presence of maize in India or Asia during pre-Columbian times (Galinat 1988).

In 1778, maize and wheat were introduced to the British colonies in Australia, where those grains were intended to supply ships (Anonymous 2017). Ho Ping-ti (1956) considered that maize was introduced into China via two routes, one by land through Burma-Yunnan, and the other by sea. In 1555, maize was in the Chinese district of Kung-hisien. By the 1750s, its cultivation was widespread in the Fukien, Hunan, and Szechuan provinces in southern China. In the eighteenth century, the greatest expansion of maize occurred (García-Paris 1991: 43). Suto and Yoshida (1956) reported that by 1580, Portuguese mariners introduced the grain to Japan. Maize spread as a food crop throughout South Asia from 1550 to 1650, where it was an important crop in Indonesia, the Philippines, and Thailand. The Spanish took maize to the Philippines in the sixteenth century and by the seventeenth century, it had spread to all the islands of the archipelago, where it was used as a reserve to counteract famines caused by poor rice harvests (García-Paris 1991: 42–43). The Portuguese took maize to Zanzibar (1643), Reunion (1690), and Mozambique (1750) (García-Paris 1991: 42).

In less than 300 years, maize had traveled around the planet, becoming an important food for human populations in countries involved in trade, but maize was used as fodder in others countries (Dowswell, Paliwal and Cantrell 1996; Paliwal 2001). Trade routes were very important in spreading spices, cultivated plants, and seeds, in addition to a large number of luxury products.

After 1492, the European expansion over the seas generated the exchange of crops, domestic animals, technologies, and cultural elements between the Old and New World. Thist included grains such as maize and wheat, and a large number of plants that modified the landscapes and agricultural and livestock activities, plus generating changes in the food systems in the regions where maize arrived.

19.5 SOME CURRENT USES OF MAIZE

Current maize uses are summarized in scientific, industrial, and cultural usages that include the importance of this grass in the food system of various human populations on the planet.[2] Cultural uses refer to maize as non-industrialized food for human and animals. Industrialized food is a phenomenon that modifies the food system and eating habits of people and animals, in addition to the cultures of peoples. There is a series of industrial food products that are an important part of the popular food system and daily diet. Industrialized food is spreading all around the world. Cereals are a good example. Maize and rice are combined with nuts, raisins, chocolate, dried fruits, sugar, honey, and a long list of popular flavorings and all for daily consumption. The type of popcorn marketed is adapted to national tastes. There is salted or caramelized to toast in the microwave, in a popcorn maker, in a pot, or a pan with a lid.

19.6 CULTIVATED MAIZE FOR FORAGE

Maize is an important food for animals in Asia, Africa, Latin America, Europe, and the former Soviet Union. In 2009, in the south and east of France, some 50 thousand hectares were planted with "creole" maize, which since 1826 has been used as food for ducks there (Balmaceda 2014). In the south and east of France, the average of maize productivity is 8 tons per hectare. In 2015, in Garona and the irrigated areas of Landes, maize production reached 17.5 tons per hectare. On March 15, 2014, by unanimous vote in the Senate, a decree by the French Ministry of Agriculture

[2] Scientific: (1) gene banks; (2) botanical and ethnobotanical studies; (3) ecological and agroecological studies; (4) genetic research; (5) historical, social, and cultural studies; and (6) new technological developments. The industrial uses of maize are contemporary: (1) the production of biodegradable plastics (PLA); (2) the industrial production of syrups (like corn syrup); (3) the industrial production of feed for poultry, and pigs and livestock; (4) the industrial manufacture of tortillas; and (5) the production of biofuel (ethanol) made from maize.

that prohibited the marketing, use, and cultivation of genetically modified maize MON 810, produced by Monsanto (La Jornada 30/05/17) was approved.

One of the most important uses of maize in European countries is for forage, and a large part of its production is destined for that. This has facilitated the expansion of transgenic maize on most of Europe, although the effects on humans are not known, or how they affect the care of native seeds or alter agroecological processes, but none of this is discussed (Álvarez-Buylla et al. 2011).

In European countries, hybrid maize was introduced in the 1950s, such as in Spain. However, currently, Spain has the largest number of hectares cultivated with transgenic maize on the Continent where it is used for forage. Now places exist there where this grain is planted for human consumption (in Galicia and the Basque country). The commercial hybrid maize of the 1950s in Spain replaced local and traditional varieties (Rolmarta 2007).

19.7 THE CURRENT CONTEXT OF MAIZE IN MEXICO: SOME FIGURES

In the 1950s, Wellhausen et al. (1952: 22–39) recognized the existence of 25 races, four sub-races, and more than 2000 maize varieties, which were widely distributed in the states and their regions. Rafael Ortega-Paczka (2003: 133) considered that there are now from 41 to 60 races. CONABIO established the existence of 60 races. In 1970, Efraím Hernández X. and Alanís (1970: 3–10) found five races of maize in the Sierra Madre Occidental. Some breeds were concentrated in Oaxaca, where Aragón-Cuevas et al. (2006) found 35 races of maize in 2005 that constituted 70% of the diversity on the national scale (Figures 19.1 and 19.2).

Figures on maize production in Mexico vary according to the source used. For SAGARPA-SIAP in 2016, about 18.2 million tons were produced, and maize cultivation covered 7.76 million hectares. In that same year, the annual per capita consumption was 196.5 kg of white maize in food, beverages, and tobacco. 75.59% of the surface of agricultural land was not mechanized, and 65.06% did not have technology to apply to better plant health. Only 30.16% of the land planted with maize received technical assistance. For SAGARPA-SIAP, in 2017, 3.55% of maize production used gravity irrigation; 0.19% had water pump irrigation; 45.25% had some other type of irrigation; and the rest was seasonal rainwater.

In small-scale production, maize is planted in the cornfield agroecosystem, considering that this concept ranges from one plant to polyculture. Small-scale farmers (92% of all producers) and own farms of less than 5 hectares contribute 56.4% of total production. 52% of this production is for self-consumption, and its yields fluctuate between 1.3 and 1.8 tons per hectare, but keep in mind that maize is grown together with other plants (at least squash, beans, or broad beans). It is this figure that has been used as one of the justifications to end traditional Mexican agriculture (Sagarpa-SIAP 2016).

In 2016, the medium and large-scale maize producers (7.9%) had farms larger than 5 hectares. They contributed 43.6% of the national production, and their yields ranged from 1.8 to 3.2 tons per hectare, but in monoculture. They only allocated 13.55% of their production to self-consumption; the rest went to grain industrialization processes. SAGARPA-SIAP figures for 2016 showed that in 2013 Sinaloa had a yield of over 9 tons per hectare, followed by Aguascalientes (over 8 ton/ha) and Chihuahua (over 7 ton/ha), the national average being 3 ton/ha. In 2010, SAGARPA-SIAP data indicate that 7.8 ha were planted with irrigated maize, of which 7.1 ha were harvested, having an average yield of 3.1 tons per hectare. In the rainfed lands (7.3 ha) a yield of 1.9 ton / ha was obtained.

According to the Ministry of the Economy (2012: 9–15), the total maize production in 2010 was 24.7 million tons, decreasing 3.7% compared with 2009. National maize production was different. From 1985 to 1991, it averaged 12.8 million tons; from 1991 to 2000, it increased to 17.5 million tons; from 2000 to 2010 period, the increase reached an average of 21.7 million tons. In Mexico even before NAFTA went into force, national production was insufficient to cover domestic consumption. In 2012, there was a deficiency of 7.9 million tons to meet the demand, so yellow maize has been purchased from the United States to cover the difference.

FIGURE 19.1 Maize (*Tlaolli*): Francisco Hernández, 1574.

The Ministry of Economy (2012) indicated that since Mexico's entry into NAFTA (1992), maize imports from the United States have been increasing. In 2004, maize imports reached 6 million tons (1/3 of what is national production), and almost entirely of yellow maize, which is industrialized. The Mexican government gradually stopped supporting small-scale farmers. In 1995, the federal budget for the countryside was 6.4% of the total budget, but in 2000 it was down to 2.9% (–43.31%). In 1993, agricultural credit was applied to 6.5 million hectares, but in 2000 it only supported 750 thousand hectares (–11.54%). Since 1992, the Mexican small-scale farmers had used resources such as those of the "*Oportunidades*" Program (*Progresa* until 2018, and none to date) to support their agricultural economy (Márquez-Mireles 2006).

In 1994, Zedillo, the Ministry of Economy (2012: 12–15) supported maize marketing with 2,700,000 pesos, which in 1999 was reduced to 528 million pesos (–19.6%). In 1998, CONASUPO disappeared, and its functions as an intermediary were transferred to the private companies Cargill (Monsanto), to Archer Daniells, Maseca, Mimsa, Portimex, and El Porvenir. Transnational enterprises: (1) set harvest prices; (2) concentrate grain distribution; and (3) import US maize into Mexico (contaminated with transgenic). The very low prices supermarkets paid small-scale farmers for their agricultural products affected them negatively.[3] From 1994 to 2000, basic grains (maize, beans, wheat, and sorghum) decreased in value (price) between 45% and 55%.

[3] Ernesto Zedillo was the President between 1994 and 2000; Espinosa et al. 2016: 383–409.

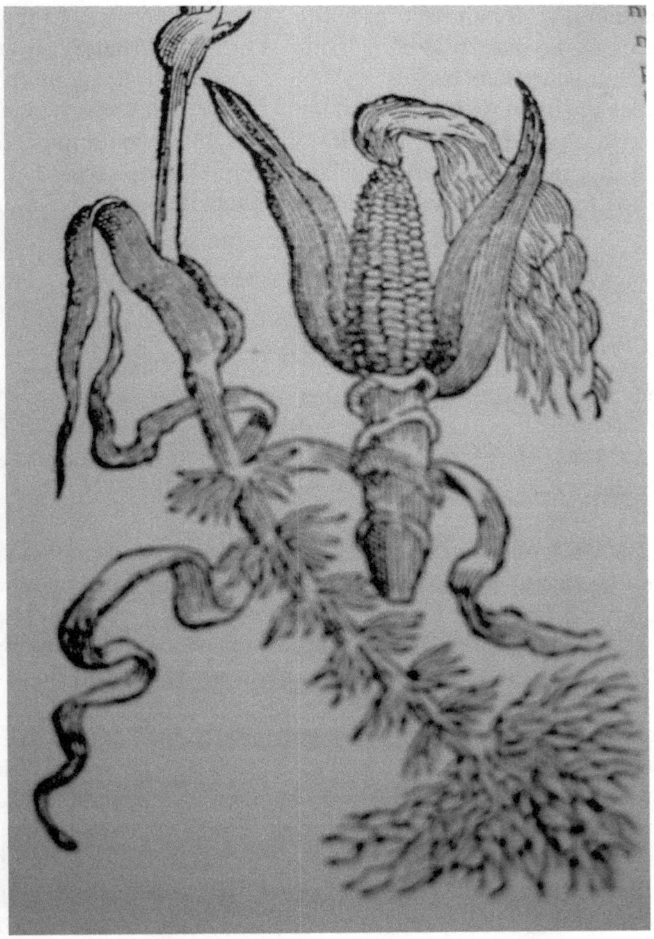

FIGURE 19.2 Maize (*Tlaolli*): Francisco Hernández, 1574.

From 1994 to 1999, total grain purchases from the United States and Canada amounted to 58.8 million tons, the equivalent of two years of national production, and $757 million was paid. This was equal to PROCAMPO's budget for 10 years. In 2014, theoretically, the Mexican government would allocate more resources to farmers through projects to support the countryside, but they were intended for medium and large-scale farmers. Only after several decades of green revolution, a large part of Mexican small-scale farmers are consumers of agrochemicals now, although the government stopped subsidizing their use directly. Currently, farmers pay 63% more for seeds and 57% more for fertilizers than before. When politicians require a vote for their campaigns and popularly elected positions, they provide fertilizers to the peasants; thus its use may be related to local, regional, federal, or state politics.

With NAFTA (1992–2020), the Mexican government opened the borders to US maize, and in two years, it stopped collecting 120% of the grain tax burden, whose import quotas are allowed, but only 3% are charged. Small-scale farmers found that maize costs less in the market than what they invested in their cultivation, so they stopped planting it and dedicated their production to self-sufficiency solely. The question of maize, in any of its historical moments, is fundamental in countries like Mexico, where it continues to be a basic product in the food system, daily diet, and culture of the inhabitants.

The present situation due to industrial uses (PLA and ethanol) turned maize into a strategic product, whose prices in the international market has increased sharply. However, this increase does not benefit small-scale farmers because they have to purchase the grain for their consumption at higher prices. Hence the importance of maintaining the cornfield, which in addition to providing grain to the small farmer's food system, includes other plants sown in the agroecosystem, which complement his diet throughout the year, in addition to providing medicinal plants, ornamental plants, firewood, shade for their homes, and various products to sell in the tianguis or in local and regional markets.

19.8 TOLERATED EDIBLE GREEN WEEDS IN THE CORNFIELD AGROECOSYSTEM

The control of weeds growing together with maize has been considered a problem, but its effects have been offset by the cultivation of plants such as squash (Gliessman 1980, 5: 93–101). In general, in modern conventional agroecosystems weeds are considered pests and are eradicated with the application of herbicides.

Weeding applied to the cornfield has different purposes. The first one weeding is done a month after maize had been planted, and it is to remove the herbs that choke the newly sown grain; they are withdrawn and used to be fed to domestic animals. In general, this weeding is still done in order to have a good maize harvest. The second weeding was done about 20 days after the first, and it could be done with plow and by pouring soil on the plants to support them; the cut weeds are selected to be part of human food or serve as fodder for domestic animals, depending on which herbs came out in the cornfield.

Formerly a third weeding was done, which is no longer carried out; after harvest the growing weeds are cut together with the maize plant and the farmer has the option of leaving them in the field, as a mulch, to protect the soil, or moving them home to chop them little by little and give them as fodder to livestock. In December, the land is plowed and flattened to retain moisture; both herbs and crop debris are plowed together with the soil to compost it, preparing it for the next harvest.

The ecological role of weeds in the cornfield agroecosystem includes serving as soil fertilizers. Some weeds are useful to feed poultry and other domestic fauna, or are used in herbalism. Other tolerated weeds help to control the weeds that suffocate newly sown seeds. Agricultural weeds are allopolyploids, "[...] resulted from the hybridization between two related species, and are facilitated by the migration promoted by us [...]" (Espinosa-García and Sarukhán 1997; Mack 1991; Vibrans 1999, 2002, 2009: 558–561; Pimentel 1997). Espinosa-García and Vibrans (2009: 23–32) proposed the general principles for the construction of a model for weed management in Mexico.

There is little information on agriculture and weed management in ancient times. In the sixteenth century, the descriptions provided by Bernardino de Sahagún (1839, III [1567]) showed how the cornfield provided food mainly, while the forest, ecotones, ravines, land edges, lagoon areas, riverbanks, and other natural areas close to population settlements, provided plants, which were collected for herbalism were wild weeds. A few medicinal herbs were planted in the cornfield, and wild weeds in the cornfield were destined for fodder (Espinosa-García and Díaz-Pérez 1996a: 83–94).

There are several questions to discuss about the problems faced by current small-scale maize growers, who – in general – have lost both control of natural areas (due to urban expansion, development of industrial areas, opening of roads and highways, and other matters) and knowledge of the components that were in these systems in detail. There are a few towns that continue to conserve the cornfield but have lost their natural areas or their relicts like in St. Isabel Xiloxoxtla, Tlaxcala (González-Jácome 1976),[4] or Progreso Hidalgo, in the State of Mexico (Juan-Pérez 2003).

[4] The classification of collected wild plants was done by Miguel Ángel Martínez Alfaro (+), who was one of the most important ethno botanists in Mexico.

There are a few ruderal plants left on the edges of roadsides and highways, the earth edges, and the earth edges of the canals, where farmers can collect them, but currently their are little known by the younger generations of small-scale farmers families. The Atoyac-Zahuapan Basin was characterized by the cornfield agroecosystem as a polyculture. From the 1970s to 1980s, plots were drained fields, with a combination of platforms, canals, and earth edges with trees, and herbaceous flora.

The Teenek of S. Luis Potosí cultivated maize, beans, squash, chili, and tomato in their cornfields. Herbaceous plants like purslane and edible wild green weeds were used as food. According to Quevedo, Cervantes, Noriero, and Zepeda (2017: 5–19), other crops were sown at intervals with the maize, a process that prevented the presence of weeds. Gliessman (1980, 5: 93–101) studied the role of squash as a cover crop and weed controller (Amador 1980; Gliessman et al. 1981).

These small-scale farmers cultivated maize with a wide variety of polyculture and agroforestry designs, which resulted in ecological processes that included the management of pollinators and insect-attracting plants on edges of land (*metepantles*) with vegetation that combined Agave, wild weeds and fruit trees, like the Mexican black cherry (*capulín*), and Mexican hawthorm (*tejocote*), medicinal flora, and aromatic herbs.

On the platforms of the agroecosystem, maize was intermixed with beans or broad beans, squash, and other crops selected by the farmer according to his needs and market prices. Those land edges are ecological corridors forming a system of useful boundaries, interspersed in the crop fields, that contributed to diversity and in addition to what was previously written are refuges for wildlife and plant life.

In Southwest Tlaxcala, small-scale farmers took care of the fertility of the soil by applying green manures mixed with domestic animal manure and mudddy water (*agualodo*) from canals and trenches. Farmers also used muddy water from canals to fertilize soils. Canals were cleaned every year in the winter, and on the heap of muddy water, over one to three or four years, plant remains, urine and detritus of domestic fauna accumulate; they can be covered with plastic to heat the heaap, and repel flies and mosquitoes. Later, it was ground to a fine powder that was applied to each maize plant manually.

In an article, published in the *Ethnoecological magazine*, Espinosa-García and Díaz Pérez (1996a) reported their research on weeds in which they conducted 130 surveys in five regions of the Valley of Mexico.[5] They found that 100% of the interviewees used weeds from the cultivated plots as fodder for livestock. 66.7% used them daily, when they were available. 64.4% used them because they could not afford commercial forage.

Plants were fed to livestock whether they were simply cut; chopped or ensiled along with maize stubble; scrambled with grass, grain, or bran; dry and cold. They were not considered "harmful" to domestic animals. Their use was due to the high cost of commercial forage and the shortage of pastures and land to cultivate forage in the region. Weeds can be nutrients for the agroecosystem, as well.

The percentage of mention in all the paces and the names of the edible green seasonal weeds used as fodder were: (1) 85.4% annual bush Sunflower (*Simsia amplexicaulis* (Cav.) Pers.); (2) 66.9% (*Tithonia tubiformis* (Jacq.) Cass); (3) 52.3% Bur clover (*Medicago polymorpha* var. vulgaris (Benth.) Shinners); (4) 41.5% Spanish needle (*Bidens odorata* Cav.); (5) 40.0% Fat hen (*Chenopodium album* L.); (6) 36.9% Rough pigweed (*Amaranthus hybridus* L.); (7) 33.1% Bitter dock (*Rumex obtusifolius* L.); (8) 30.8% Sow thistle (*Sonchus oleraceus* L.); (9) 23.1% White clover (*Trifolium repens* L.); (10) 21.5% Field mustard, turnip: *Brassica campestris* L.); (11) 21.5% *perilla* (*Lopezia racemosa* Cav. (21.5%); (12) 19.2% little mallow (*Malva parviflora* L.); (13) 16.9% Mullein, bitter (*Gnaphalium arizonicum* A. Gray); (14) 15.4% purslane (*Portulaca oleracea* L.); (15) 12.3% Bell, little bell (*Ipomoea purpurea*) (L.) Roth); (16) 12.3% Rocket salad, Garden-rocket (*Eruca sativa* L.); (17) 12.3% Linden leaf sage (*Salvia tiliifolia* Vahl.); (18) 14.0% Virginia pepper

[5] Tizayuca, Hidalgo; Tepotzotlán-Cuautitlán; Xochimilco-Milpa Alta, Mexico City; Texcoco, Amecameca-Tlalmanalco-Ayapango, State of Mexico.

weed, Poor man's pepper (*Lepidium virginicum* L.); (19) 7.6% Tansy Mustard, *Nabillo, cilantrillo, culantrillo, pipiyahuitl* (*Descurainia impatiens* (Cham. & Schl.) O.E. Shulz); (20) 6.1% Creeping woodsorrel, Creeping oxalis (*Oxalis corniculata* L.); (21) 6.1% Nettle leaf, *Hediondilla, hierba del perro, quinoa negra, pie de ganso* (*Chenopodium murale* L.); (22) 5.4% *Teosintle, assese* (*Zea mays* sp. mexicana (Schrod) Iltis); (23) 4.6% Arizona beggar ticks (*Bidens aurea* (Ait.) Sheff); (24) 4.6% Coat buttons (*Tridax coronopifolia* (Kunth) Heml.); (25) 3.8% Lemongrass, *almaraduz*, ant grass, *mesquitillo*, speck, rattle (*Dalea foliolosa* (Ait.) Barnevy); (26) 3.8% common Four o'clock, Garden four o'clock, Marvel of Peru, False jalap (*Mirabilis jalapa* L.); (27) 3.1% Pale Mexican poppy (*Argemone ochroleuca* Sweet); (28) 3.1% Postrate knotweed, Door weed, Knotgrass, Mat grass (*Polygonum aviculare* L).

With percentages of less than 1.5%, the list continued until reaching number 38, thus giving indications of the importance of those weeds in crops. From this list, numbers 22–26 were mentioned in Tepoztlán, only, and number 27 in Texcoco. In Amecameca, 8% of jaltomate (*Jaltomata procumbens* (Cav.) J.L. Gentry) was used in human food. Those weeds have various ways of being used by humans in other parts of the country. In Tlaxcala, they are used as herbal infusions, or the flower of the thistle (number 27) was collected and sold to laboratories to manufacture medicines for heart control. Numbers 5, 6, 13, 19, and 21 were used as food for humans. *Mirabilis jalapa* L. was considered medicinal, and in the nineteenth century, it was exported in large quantities to Europe, where it was used in the manufacture of medicines (Aguirre-Beltrán 1943, 1947, 1955, 1963, 1965; González-Jácome 2004b: 1–39; Vibrans 1997: 21–67).

In villages and towns of Tlaxcala, poultry, particularly turkeys, continue to be part of the relics of ancient homegardens. These are consumed in the village festivals, in family gatherings related to the completion of primary school studies, the birthday of the head of the family, baptisms, confirmations, and parties associated with religious beliefs and local customs. In Chan Kom, domestic fauna is reduced to turkeys and an occasional pig (González-Jácome 2012: 522–535; Moctezuma-Pérez 2010: 26–46).

19.9 CONSIDERATIONS AND CONCLUSIONS

Local adaptations or adjustments and diversity allowed the success of crops throughout the annual agricultural calendar. The intervention of the protective cornfield deities is requested and thanked throughout the ritual calendar, regarding different management of the cornfield and cultivation of maize (Serafino 2012). Evidence gives information on the cornfield food system, which was supplemented with meat from wild animals obtained by hunting, fishing, and gathering, and from domestic animals (Staller et al. 2006). Currently, especially in remote regions, domestic fauna play an important role to obtain monetary resources needed for: the payment of family, and communal parties and celebrations; urgent unforeseen needs; payment of doctors, fees, and medicines in cases of health problems or accidents; and other sudden and unexpected expenses small-scale farm families have. In San Luis Potosí there are groups of "chiveros" who take the herds to graze in areas more or less far from their towns. They do not consume the meat of the goats, they sell them to obtain money (Trenti-Very 2022).

Another important point is the combination of domesticated components of traditional agriculture (flora and fauna) and elements of the natural environment adjacent to population settlements. In their study of the Guilá Naquitz cave and Silvia rock shelter, northwest of Mitla, in Oaxaca et al. (2007: Vol. 104, No. 29: 11905–11909) showed that by 6000 BC, the food system included: wild chilies, domesticated chili peppers; wild beans; avocados; Agave; Opuntia prickly pears; tropical sapodilla; wild onions; and berries. In those same sites, the authors found evidence of the combination of cultivated plants with edible weeds and wild plants. The collection of wild plants destined for herbalism is an important pursuit from several points of view, including those related to the very evolution of plants, their uses, and biodiversity.

Human labor and technology applied to the cornfield has changed through the centuries. In ancient times, slash and burn were preceded by the opening of cleared guard strips, which prevented the spread of fire to other plots or to the forest. The rest given to the soils it could regenerate lasted just over 20 years; a fallow soil period was shortened as the population increased. In the twentieth century, the number of weeds decreased from three to two, and in some cases to one. Possibly that was the result – among other factors – of the use of the plow to clean the furrows in the cultivated lands a month after sowing, and a second time manually when the herbs around the maize grew to be as big as the maize itself. The female labor force applied to the cornfield had increased by that time (González-Jácome 2011; Márquez-Mireles 2000; Palerm 1972).

The number of plants cultivated in the cornfield has changed, but the varieties of maize, beans, and squash continue to be essential in the agroecosystem and in the diet of human populations. Beans are a fundamental element in the cornfield and have commercial importance in regional markets, but it is replaced by broad beans, sometimes. To this day, maize is a staple crop intended for family self-sufficiency and the market. However, its varieties are related to the specific uses that small-scale maize growers give to each one.

The concept of the cornfield is very broad. There are places where the inhabitants call a maize plant a cornfield. Currently, this concept ranges from a single maize plant, to the maize monoculture, or to the polyculture system. The cornfield agroecosystem includes its interrelationships with other systems on which the functioning of ecological aspects depends (like control of soil erosion, irrigation water, wind, insects; relationships between cultivated and not cultivated plants, and edible green seasonal weeds); economic relationships (crops for self-sufficiency and for the market); social (family workforce, nuclear family, kin, and friends that help in cultivation at different times of the cycle); and cultural (forms of mutual aid, food system, and conservation of crops) (Figure 19.3).

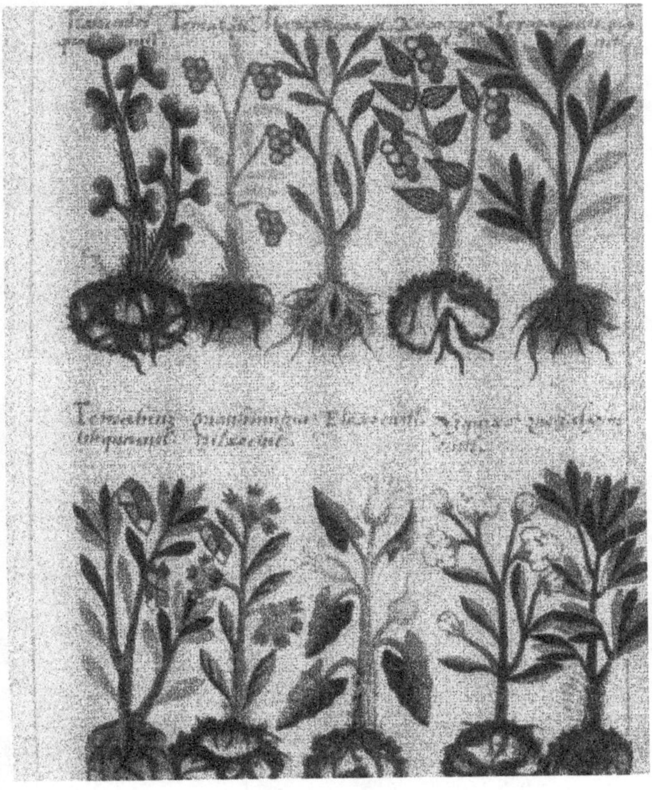

FIGURE 19.3 Badiano Codex, Medicinal Plants, 1552.

TABLE 19.1
Ways of Consumption of Wild, Cultivated Herbs, and Wild Edible Green Weeds

No	Name	Uses	Raw	Cooked	Roasted	Total
1	Nanacatl	These mushrooms are born in the fields; they are good to eat. But they cause vomiting or chambers if they are raw or undercooked. They can be deadly. Its effect is counteracted with enema.		X		1
2	Tzontecomanatl	Large round mushrooms that grow together on one foot, some tall and others low.		X		1
3	Mushrooms	Wide and round mushrooms as dishes.		X		1
4	Mushrooms	Round white mushrooms that are not hard to cook and are also roasted on clay plates, and they are very tasty.		X	X	2
5	Cimatl	Wild weed with a root (*quavecoc*, *cimatl*, or *eimatl*) produces some large beans, like broad beans. They are boiled for 2 days.		X		1
6	Huauhquilitl	Very green smoothies with thin and tall branches, broad leaves. The stems and seeds are called Huautli. It is cooked with salt and tastes of ashes and is like those of Spain. It is squeezed, and they make tamales with it (*quiltamalli*) and tortillas.		X		1
7	Quilonilli	It is of broad leaves, when it is small it is edible. When it is large, it is cooked with *tequixquitli* and the water is squeezed. Its black seed is called *pitzitl*.		X		1
8	Itzmiquilitl	Brown weed with large branches as well as round and flat leaves.		X		1
9	Aioxochquilitl	Flowers, lumps, or limbs of pumpkin branches. The skin is removed from above them before cooking. Note of Sahagún (1830 Vol. III: 245). "In Oaxaca they are eaten together with the guides and green pumpkin stems, mix some balls of maize dough with salt, and some cooked green chilies. It is the food of poor peasants and very delightful, because it is dominated by openness, simplicity, and the good humor of the countryside".		X		1

(Continued)

TABLE 19.1 (*Continued*)
Ways of Consumption of Wild, Cultivated Herbs, and Wild Edible Green Weeds

No	Name	Uses	Raw	Cooked	Roasted	Total
10	Axoxoco	Weed with long, wide leaves. It is tasty and sour.		X		1
11	Mizquilitl	It is attic, very green, with bladed leaves. It is tasty to eat.		X		1
12	Acuitlapali	Long, brown weed, grows at the water's edge. It is good to eat.		X		1
13	Tziunquilitl	It grows by the water's edge. It has blue and harp leaves. It is good to eat.		X		1
14	Conalquilitl	It grows in the mountains; it is ash colored; its root (*tacanalli*) is eaten.		X	X	2
15	Mamaxtl	It grows at the water's edge; it is tasty.		X		1
16	No/name	It is like a nettle, it is edible weed.		X		1
17	Ueiquauhquilitl	Wild plant drinks that are bitter before cooking, but they are tasty.		X		1
18	Etenquilitl	Herb of beans, which spill when they are picked. It is eaten.		X		1
19	Tlalaioquilitl	They are wild pumpkins.		X		1
20	Xaltomaquilitl	It is eaten cooked.		X		1
21	Tzitziquilitl	It is dark green, with flowers and seeds, as well as very good to eat.	X			1
22	Eloquilitl	It is very green and tender; it has flowers; it is very tasty.	X			1
23	Quauheloquilitl	It is wild; it is born in the Opuntia *tunales*. It is very tender and good to eat.	X			1
24	Mozoquilitl	It is green, fluffy, and very tender. It is tasty.	X			1
25	Tzaianalquilitl	It is made in water; its branches are hollow and harp; it is good to eat.	X			1
26	Achochoquilitl	It is light green; it is born near the water; it is good to eat "They say of this grass that if the boys or girls eat it they become powerless to beget; but after they grow up everyone eats it ". Sahagún, Volume III, p.247.	X			1
27	Tzonquilitl	It is light green; it is born near the water; it is good to eat.	X			1
28	Iztacquilitl	It is low and flat; it has a salt flavor.	X	X		2
29	Tepicquilitl	With long and pointed leaves. If you eat a lot of this herb, it gives cameras.	X			1
30	Ecoquilitl	They are the leaves and branches of beans. They are a little rough and fluffy. They provoke regoldar.	X			1

(*Continued*)

TABLE 19.1 (*Continued*)
Ways of Consumption of Wild, Cultivated Herbs, and Wild Edible Green Weeds

No	Name	Uses	Raw	Cooked	Roasted	Total
31	Vitzquilitl	Thistles of the earth, they have thorns; the leaves below are ashy and the ones above are green. Inside they have fluff, like the thistles of Castile, which are found at the water's edge. It is also a garden weed. They are good to eat.		X		1
32	Qaavitquilitl	Wild thistles that are born in the mountains. They are prickly and itch those who eat them.		X		1
33	Chichicaquilitl	Edible grass that grows near water and in sweet, tilled land. It has white roots; it is very tender and somewhat bitter.		X		1
34	Coiocuexi	The stems of this grass are not thorny, and flourishes, is bitter, it makes the voice hoarse in children. Girls do not eat it.		X		1
35	Popoiauh	It is a black and green bush; it is kneaded with maize and tortillas are made from it.	X	X		2
36	Exixi	It has tiny leaves; its seed is yellow; it is chewy; and if you eat a lot, it produces blisters; and it is too hot. It is used to make tortillas and tamales and atole or porridge (for those with chambers of matter and blood, since it purifies the intestines).	X	X		2
37	Xoxocoiolli	Perennial plant with an acidic flavor that is used as a condiment. It is sour.	X	X		2
38	Xuxocoiopapa-tla	Herb with tall, slender feet, broad, rounds leave. The leaves at the tip of the foot are tasty to eat cooked.		X		1
39	Xoxocoiolcue-cuepoe	Weed with fat round feet and rare leaves. It is born when the rains begin. They are tasty like tomatoes.	X			1
40	Xolxococolol-vivila	It is a stopped herb with small and round leaves; it is tasty.		X		1
41	Miccaxoxocoili	Herb with a thick, hairy foot; the leaves are wide, very acidic and cause an unpleasant sensation that is experienced in the teeth and gums when eating (*dentera*).		X		1

(*Continued*)

TABLE 19.1 (*Continued*)
Ways of Consumption of Wild, Cultivated Herbs, and Wild Edible Green Weeds

No	Name	Uses	Raw	Cooked	Roasted	Total
42	Quauhxoxocoili	They are soft to eat.		X		1
43	Quananaca-quilitl	Herb that the chickens of Spain eat. They are the *serrajas* (*Sonchus oleraceus*) de Castilla. Indians say there were none in these lands before the Spanish came. Now there are so many that the whole earth is filled with them; because of seed flies, they have multiplied everywhere.	X			1
44	Cultivated xonacatl	Little onions from this land. They are planted, Indians are horticulturalists, and they eat the onions of Spain.		X		1
45	Wild xonacatl	Many wild native onions grow, which are less hot than those from the Old World.		X	X	2
46	Maxten	Many wild native onions grow, which are less hot than those from the Old World, together, faint and flourish are bland. Indians eat the root or the head.		X		1
47	Papaloquilitl	Round-leaved, fragrant, and tasty herb. It is found in hot lands.	X			1
48	Aiauhtonv	Bush wild and low. It is found in the fields and mountains. Its flowers are fragrant.		X		1
49	No/name	It is edible; Sahagún named it as "the sweet potato".		X		1
50	Xícama	Indians also eat its leaves.	X			1
51	Tolcimoquilitl	Herb with very beautiful and very delicate flowers. The root is edible.		X		1
52	Xaltomatl, xaltotomatl	Fruit from the plant called *xaltomaxilhuitl*. There are black and white, very juicy, sweet, and round. The root is edible.	X	X	X	3
53	Coiototomatl	Fruit like that of the small tomatoes called *miltomatl*. It has yellow coverage; they are sweet; and they slightly lock the throat. If you drink a little of the root of this herb, it is medicinal and cleanses the intestines. Women who breastfeed drink it because it purifies their milk.	X	X	X	3 Medicinal, also.

(*Continued*)

TABLE 19.1 (*Continued*)
Ways of Consumption of Wild, Cultivated Herbs, and Wild Edible Green Weeds

No	Name	Uses	Raw	Cooked	Roasted	Total
54	Atlitiliatl	The plant produces a sweet, blackish and bittersweet strawberry. The leaves are good for the sick, and are added to the water used in the steam bath (temazcal).	X			1
55	Tlalilotl	Edible green weed.	X			1
56	Tlalaiotl	Edible green weed.	X			1
57	Iacanalxivitl	Medicinal herb that takes away the fever. It is ground with maize and drunk with water.	X			1
58	Coanenepille	Herb to purge. After drinking it, you eat bird broth or an atole (*yolatolli*).		X		1
59	Teonanacatl	Cures gout and fever		X		1
60	Tlatlalayotli	Its leaves are edible; the root is medicinal.	X	X		2
Total	60		24	43	5	72

Source: Friar Bernardino de Sahagún, *Historia de las Cosas de la Nueva España*, 1830, Vol. III, pp. 244–249, 257.

RECOMMENDED READING

Ranere, A.J.; D.R. Piperno; I. Holst; R. Dickau; and J. Iriarte. 2009. The cultural and chronological context of early Holocene maize and squash domestication in the Central Balsas River Valley, Mexico. *Proceedings of the National Academy of Sciences of the United States of America* 106(3): 5014–5018.

The article gave the results of the archaeological research based on molecular evidence, indicating that the wild ancestor of maize is native to the dry tropical forest in the Balsas Central basin in the current state of Guerrero of southwestern Mexico. The study was developed near the Iguala Valley, showing a long sequence of human occupation and plant exploitation dating back to the early Holocene for the first time.

One of the excavated sites was the rock shelter of Xihuatoxtla, where well-stratified deposits were found, and a set of Clovis stone tools with flake-cut bifacial points, simple flake tools, numerous hand stones, and grinding stone bases that were dated with radiocarbon to at least 8,700 calendar years BP.

Residues of starch grains and phytoliths were found in the rock shelter soil, in addition to flaked stone tools that were directly associated with sediment phytoliths, which gave evidence of domesticated maize (*Zea mays* L.), and squash (*Cucurbita* spp.) corresponding to contemporary contexts, being stratigraphic below the 8,700 calendar years BP.

Radiocarbon determinations, the stratigraphic integrity of the Xihuatoxtla deposits, and the characteristics of the stone toolsets associated with the maize and squash remains indicated that these plants were domesticated in the early Holocene. Early agriculture in this region of Mexico appears to have been practiced by small bands of farmers who moved from their settlements seasonally and engaged in various subsistence activities. Research in the Xihuatoxtla rock shelter indicated that plant exploitation and cultivation were important activities from the earliest stages of the site's occupation. The palaeoecological data from nearby lakes indicated that the settlers used the lacustrine resources early, probably, and that they were an important component of subsistence in seasonal settlement cycles (Alexandrovskiy et al. 2004: 9–17).

Perry, L.; and K.V. Flannery. 2007. Precolumbian use of chili peppers in the Valley of Oaxaca, Mexico. *Proceedings of the National Academy of Sciences of the United States of America* 104(29): 11905–11909.

Excavations at Guilá Naquitz Cave and Silvia Rock Refuge near Mitla, Oaxaca, Mexico, found remains of 122 chili peppers dating from 600–1521 AD. The chili peppers appeared to belong to at least 10 cultivars of the species *Capsicum annuum* or *Capsicum frutescens*. The specimens were well preserved and made it possible to evaluate the criteria be used to separate wild and domestic chili peppers, and to distinguish among cultivated races. They also allowed evaluating the reliability of starch grains to document the presence of chili peppers in archaeological sites where macrobotanical remains are not preserved.

In the archaeological record, one of the criteria to distinguish domesticated chili peppers from wild ones is the presence of a non-deciduous pericarp or fruit. Wild peppers are often loosely attached to the peduncle, and this makes their dispersal by birds easier. Perry and Flannery (2007) found fruits that were still attached to the peduncle, even when the color of the pericarp was yellow or orange, indicating that the fruit was not green when it was harvested. Apparently, the shelf life in wild Capsicum species may depend on the degree of maturity and, therefore, the time of harvest.

Another characteristic is fruit size, which is often used to determine if there was domestication. Wild Capsicum fruits are typically 1 cm in diameter and length, and the largest fruits can be attributed to cultivation and the effects of human selection. As large fruits have not been documented in wild populations of chili peppers, size is probably a reliable indicator of their relationship to cultivation, and domestication.

Most of the occupants of Monte Alban IIIb, IV, and V in the cave area lived in the urban area of Mitla, or near it, the evidence from Guilá Naquitz, and Silvia suggests dispersal in the field. The set of plants seemed to included those grew in foothill fields near the caves, including fresh chilies, as well as plants brought from elsewhere, possibly dried chilies and tropical fruits. The remains seem to confirm that although many Zapotecs from Mitla were full-time farmers, they made frequent forays into the "desert", to harvest useful wild plants. Guilá Naquitz evidence suggested that plants collected near the cave were temporarily stored in grass-lined wells, and then brought back to Mitla, in baskets or net bags, fragments of which were found in the caves.

The 10 different cultivars of chili peppers suggest a complex and spicy cuisine similar to that of modern Mexico, the remains of the cave included multiple varieties of beans, not all were grown in the same fields. Along with maize, squash, cotton, and other cultivars, the impression is that they had an economy with multiple crops in contrasting environments, including tropical fruits in the warmer parts of the valley, obtained through trade or the market.

20 Soil, Water, and Erosion in Traditional Agricultural Systems

20.1 INTRODUCTION

One of the most debated and studied aspects is the role of soils, deforestation, and droughts in agriculture in regions like the tropical lowlands of the Mayan area. For a long time, these lands were considered not very sustainable for slash and burn agriculture. Then these would be the reasons, partially, for the abandonment of the Classic cities and the relocation of the settlers in other regions. In several studies, it was proposed that environmental and climatological conditions that lasted for several years and were exacerbated by the lack of rain, drought, increased forest deforestation, and soil erosion in the Maya lowlands cause the so-called Mayan Collapse, mainly in the Late Classic (Santley et al. 1986: 123–159).

20.2 BACKGROUND

The idea that soils in Mesoamerica had low productivity was the stimulus to seek questions and answers about subsistence in antiquity. This could provide ideas for the improvement of contemporary agriculture, through sustainable agricultural systems (Altieri 1983; Gliessman 1998). The importance of ancestral knowledge about soil began with the ethnographies of the farmers of yore, combined with information in Codexes and documents where data in this regard were recorded. The ethnopedology of Native American peoples can – in some way – remedy the lack of written records for Mesoamerica, since there is some information on the knowledge of soil held by the Mexica, and there are documents and reports from the Viceroyalty.

20.3 THE MAYA LOWLANDS

On this subject, there are numerous studies. Beach et al. (2002: 372–397, 2006: 166–178, 2010; Lohse 2010: 372–397, 2011: 109–118) in their study on the anthropomorphic impacts on the soils of the central Maya lowland established three stages where erosion and soil degradation had accelerated impacts: (1) in the Preclassic period (1000 BC to 250 AD); (2) in the Late Classic (550 AD to 900 AD); and (3) in the last decades of the twentieth century. The main paleosol (*Eklu'um*=in Maya black soil) at these sites is a well-developed Mollisol or Vertisol that began to form in the early Holocene and was buried under a weaker paleosol in the Preclassic or Classic period (250 AD–900 AD). This second paleosol in turn is below the one corresponding to the Classic and subsequent sediments.

In Cancue'n, an archaeological site in Guatemala, the greatest erosion occurred in the Late Classic (550 AD–830 AD), when the depressions showed from 1 to 3 m of accelerated soil degradation. This corresponded in time to the Late Classic and to the process called "Maya Collapse", which modified the political and social situations of the population in cities, resulting in their emigration to the Yucatan peninsula.[1] Contemporary land degradation concurs with the occupation of the "post-revolutionary Mexican pioneers", who came to occupy these lands in relatively recent times (Beach et al. 2006: 166–178; Molina 2011).

[1] Abrams et al. 1996; Douglas et al. 2015: 5607–5612; Lohse 2006 (65): 166–178; Luzzader and Beach 2009 (99): 1–26; Santley 1990; Shaw 2003; Turner II and Sabloff 2012.

DOI: 10.1201/9781003198833-20

In Mexico, this corresponds to the so-called "march to the sea", started in the 1950s and continued in the 1970s, where one of the objectives was to redistribute the country's population to sparsely populated regions. Another objective was to deliver land to small-scale farmers who lacked it and claimed some (Revel-Mouroz 1980). Other objectives were associated with occurrences like the exploitation of forests (deforestation?), the expansion of grasslands, and the raising of cattle in the tropical regions of the country.

In critical history, according to Wallerstein (1979, 2011), the borders of the past and the present do not exist and are artificial. It must be considered that for the Mayans, their agriculture, their reservoirs built in the Classic (near 550 AC–850 AC), their cities, and their lives depended on rain. Seasonal droughts lasted from 5 to 7 months, but the population adapted to it by building large reservoirs in urban centers, taking care that the water was clean in the dry season. From 800 AC to 900 AC, water was scarce, and the leaders were unable to maintain the support of farmers, who migrated from the interior cities in the north of the Yucatan peninsula in search of more reliable sources of water.[2]

The Mayan Collapse affected cities like Yaxchilán, Palenque, Tikal, and others. The migrant populations from the Maya lowlands sought better soils for slash and burn agriculture, and clean water for domestic and agricultural purposes but located near prosperous commercial settlements. In this direction, the prevailing idea among researchers is that slash and burn agriculture depleted soils, and bad weather produced recurrent droughts, thus both reduced agricultural production. This emigration to coastal places close to water sources, where the mercantile towns and commerce were, has been called "urban diaspora". Turner II (2010: 169–174); Turner II, and Sabloff (2012, 109 (35): 13908–13914) considered that in the Terminal Classic, complex interrelations between Man and his environment took place:

> The large-scale alterations of the Mayan landscape and the demands on ecosystem resources and services generated high-stress environmental conditions that were amplified by increased climatic aridity. Coinciding with this stress, the flow of trade shifted from land traffic through the peninsula to maritime traffic around it. These changing socio-economic and environmental conditions generated increasing social conflicts, decreased control by the Mayan elite, and led to decisions to move to other places in the peninsular region instead of incurring the high costs of maintaining human and environmental systems in their place. After abandonment, the environment of the central Maya lowlands recovered largely, although it was altered from its pre-Mayan occupation state; the population never recovered [...].

Aridity was an exogenous element of the Late Classic, affecting the conditions of the human environment in the Mayan lowlands. The scale of the changes in the landscape and the tensions for the resources generated by inhabitants amplified the problem of the aridity and the climate, and this event affected its environmental impacts (Del Amo and Cayetano 2011). This amplification was apparently much larger than that experienced in previous drought episodes, after which cities had recovered. The water storage tanks were in the central places, but there was no rain to fill them to cover the needs of the population. In addition, since it was not being replenished, it became contaminated, causing hemorrhagic fevers and mortality in the population.[3]

Ford (2002: 814–820) and Marcus (1993: 111–172) considered that the Mayans had extensive knowledge about flora and several of them were water purifiers. Scheffran et al. (2012) wrote, "Climate instability interferes with the supply of food and water, which in turn affects public health and worsens the conflict [...]". Fletcher (2009) and Lucero et al. (2015, 2017) argued that agrarian societies in the north of the Mayan zone, including the Yucatan peninsula, had "low urban intensity", which helped to maintain forest diversity. In addition, agriculture and water systems of small

[2] Andrews 1998: 16–23; Lucero 2017; Santley et al. 1986: 123–150; Scarborough 1993; Scarborough et al. 1991: 658–662; Shaw 2003; Turner II and Sabloff 2012, 109 (35): 13908–13914.

[3] Acuña-Soto et al. 2005: 405–409; Gill 2000; Lucero 1999: 34–49, 2017; Scarborough 1993, 2012; Scarborough et al. 1991; Scheffran et al. 2012.

scale allowed the formation of settlements and towns that used their secondary forests to cover their needs, from which they obtained resources for more than a millennium.

Until the arrival of the Spanish in Yucatan, the survival of the Post-Classic Mayan society was achieved by semi-dispersed settlement systems, the production of food through various types of small-scale agroecosystems, water supplies (sinkholes (*cenotes*), chuntuls, and other natural deposits), and forest resources obtained on a small scale. It should be remembered that until the seat of the Classic Maya on the peninsula, the settlements were scattered and located around sinkholes to have a supply of water for domestic use. For the Spanish, this was a problem to be able to control and evangelize the native population, which they tried to solve with the congregation of the peoples in places that were close to the sinkholes (Mariaca-Méndez et al. 2010: 6–7).

Recent studies on soils quality in traditional agroecosystems, such as the slash and burn cornfield and homegarden, indicated that anthropomorphic action on them was not the cause of their deterioration – as had been considered in studies on the Mayan collapse – but, on the contrary, had enriched them with human managements. The role of phytopathogens in soils is a very important issue to include in the study of Mayan agroecosystems (García-Espinosa 1978, 1994: 30–44). The slash and burn agroecosystem for the cultivation of maize and for the cornfield food system was not the only agroecosystem. There were other traditional agroecosystems (e.g. the swamp agriculture) that were developed in these regions.

Swamp agriculture practiced by tropical agricultural societies and hypotheses about its role in human societies are logical and important. Studies by specialists in various areas of knowledge generated well-founded hypotheses about the key role that swamp agriculture in wetlands, shoals and petenes had in sustaining Mayan agroecosystems and human societies.

Gurri and Morán (2002: 41–47) summarized what they considered limiting factors for the Yucatan peninsula: slash and burn agriculture; patchy procedures for the distribution of water deposits; in past times the pressures generated by high human population densities; the concentration of sinkholes in the hands of few people; the rupture of communities as a result of adult male emigration (patrilineal); and insufficient land for crops.

Through time, humans have understood, adapted, or adjusted the management of soil, water, flora, and fauna, and agroecological practices to survive in the face of drastic external changes. That is, the people who managed these agroecosystems were able to develop resilience and sustainability in them through a series of agroecological practices and management based on local and regional knowledge of environment, climate, water, soils, flora, fauna, and domesticated plants that were cultivated. So these traditional agroecosystems have resilience and capacity to be sustainable, which are characteristics of traditional agricultural systems (Cooper and Duncan 2016; Fedick 2010: 47–74; Ford and Nigh 2016; Nations and Nigh 1980: 1–30).

20.4 SOILS IN THE SLASH AND BURN CORNFIELD AND THE HOMEGARDEN

The UN (2015) calculated that the current Mesoamerica has 26% of its soils degraded by the processes of chemical pollution, erosion, salinization, and extreme weather events. In the Mayan area, there are several matters related to human uses and management of soil and their impacts on agricultural activity. One of them is the Classic Maya collapse, and another is the soil management in the Mesoamerican homegarden intended for subsistence, and how this agroecosystem is an effective form of soil management. In general, agroecosystems require deep soils with human contributions that Anthropomorphic soils can develop according to the management carried out in them (Harrison 1982: 119–130).

The slash and burn cornfield has generated extensive discussions in this regard because for a long time it was considered to be carried out in tropical soils poor in nutrients, and that slash and burn generated soil degradation (Beach et al. 2002: 372–397, 2006: 166–178; Fedick 2010: 47; Lohse 2006: 166–178).

Recent studies indicate that human contributions to ancient agricultural landscapes can be identified as generators of soils with better qualities for agriculture than had been assumed, previously.[4] In other words, archaeological studies on the slash and burn cornfield in tropical lands showed that the ancient levels of the Mayan population had gone beyond the subsistence capacity of the simple slash-and-burn cultivation in soils where maize was grown with beans and squash. They suggested that there were alternative or additional crops of tubers and breadnut or Maya nut (*Brosimum alicastrum*) (Flannery 1982; Meiners et al. 2009: 7–16; Puleston 1968, 1971 (36): 322–335; Siemens and Puleston 1972: 228–239).

In addition, they carried out new (or re-examined) tests on intensive cultivation practices. These included hillside terrace cultivations and the conversion of wetlands into raised and drained fields for cultivation. These agroecosystems – such as wetlands for swamp agriculture – were created to compensate for the soil erosion in the itinerant slash-and-burn cornfield (Beach et al. 2002: 372–397; Siemens 1983b, 1989a, 1989b, 1998; Siemens et al. 1988: 105–107).

According to new evaluations and proposals, human management applied to the cornfield and the homegarden was probably partially the cause of the formation of anthropogenic Mollisol soils rich in organic matter with great agricultural potential. These soils have been found in sites in the lowlands of the Mayan area and archaeological cities like Chunchucmil and Mayapán (Luzzadder–Beach 2000: 493–510; Luzzadder-Beach and Beach 2009: 1–26). With this, the concept that the soils of the Mayan lowlands were poor, with a low state of nutrients; were little depth; and were located on limestone rocks was modified, because it considered those soils highly limited and poorly suited for plow farming, unsuitable for agriculture, and unsuited for commercial production.[5]

Archaeologist Scott Fedick (1996b, 2010: 47–48) and several researchers[6] showed with new soil studies that in the Maya lowlands, the soils were neither highly degraded nor poor in nutrients, as tropical lands in other parts of the world were often characterized. In fact, Maya lowlands stand out because in terms of rainfed agricultural potential, they have some of the best soils in the world. It was also evidenced that there is great local and regional variability in land resources, forming a complex mosaic of lands with different agricultural potentials and management results (Gómez-Pompa et al. 2003).

Numerous studies have been done on human action on soils of agroecosystems, like the homegarden and the Mesoamerican fenced homegarden (*solar*). This labor has the purpose to optimize the available agricultural area and discover the possibilities to achieve intensive production, with high inputs and appropriate polyculture techniques. To do this, traditional small-scale farmers have applied various techniques and management of soils and flora, such as the vertical arrangement of crops distributed in various strata (Atran 2003; Atran et al. 1996; Barrera et al. 1977; Killion 1992; Nations and Nigh 1980) (Figure 20.1).

20.5 ARCHAEOLOGICAL EVIDENCE

This brief review of Mesoamerican soil history attempts to show human impacts on patterns and trends in different settings and various studies. One issue is that of human impacts in the form of clays, paleosol sediments, or organic lake sediments. Brenner et al. (2003: 77) said that the climate and events such as hurricanes, floods, or droughts, in the Maya lowlands had impacts on the vegetation that were greater than those on the Peten savannas. The climatic conditions caused by drought-incited responses from the residents in prehistory, around 5610 BC. In 3400 BC, the sedimentation of inorganic clays in Mayan soils increased. This has been established by pollen from economic

[4] Del Amo and Cayetano 2011; Beach et al. 2000 90 (4): 493–510; Cook and Lohse 2006 (65): 166–178; Beachet al. 2010 (92): 372–397.

[5] Beach 1998a; Fedick 1992: 47, 2003: 144–147, 2010: 47–74; Hammond 1978; Nations and Nigh 1980: 1–30; Turner 1978.

[6] Beach et al. 2002; Cook et al. 2006 (65): 166–178; Beach et al. 2020 (92): 372–397.

FIGURE 20.1 Yaxcaba, Yucatán: Mayan *Solar* with a Stone Wall (*Albarrada*) Surrounding the Homegarden and the House.

taxa and disturbance in depressions. There is compelling evidence that deforestation and soil erosion began around this time.

Hansen et al. (2002) and Beach et al. (2002), in the chronology of their studies on the Peten in Belize and Guatemala, affirmed that the conservation and management of soil through the construction of terraces began in the Preclassic. However, the development of the process was slow and did not generalize until the Classic and Late Classic. For Beach et al. (2003a, 2003b; Smith 1995a, 1995b), impacts on pioneer agriculture of the Preclassic increased the surface of many depressions and valleys (Luzzadder-Beach and Beach 2009: 1–26). But in the Late Classic, soil conservation was expressed in the emergence of a trend toward sustainable agriculture.

There is little evidence at various Late Classic Maya sites about conservation, but a large part of it is related to soil erosion. Soil loss in the Maya lowlands ceased in the collapse of the nineth and sixteenth centuries. It was then when the forest returned, soil was formed, and sedimentation slowed. Pre-Hispanic homegardens in the Maya lowlands appeared to have contributed to this, becoming increasingly urbanized landscapes. According to some scholars of the process, the distinction between the forest and the homegarden became blurred probably (Folan et al. 1979; Smyth et al. 1995). This process is what is called in earlier chapters of this book the forest/incidental homegarden.

In the Central Highlands of Mexico, human activity that altered soils began in the Preclassic, with the construction of selected sites for housing and the clearing of forests for agriculture. In Central and northern Mexico, soil conservation began in the Early Preclassic. The domestication of maize began from 9000 BP to 8000 BP, but evidence of the major changes in land use that were human-induced, and may have accompanied maize cultivation, is later.

The pre-Hispanic collapse, which occurred in the Maya lowlands, did not affect the populations of central Mexico. In fact, the depopulation in central Mexico occurred after the arrival of the Spaniards and possibly had a greater environmental impact due to the mortality caused by the war, epidemics, and subsequent famines (Santley 1990: 325–343). These devastated the native populations in the sixteenth, and seventeenth centuries, so much so that they could not maintain the soil conservation systems and other procedures owing to the entry and spread of livestock that favored erosion. The General Archive of the Nation (AGN) abounds with information on the lawsuits between indigenous peasants and the owners of large livestock expanses that show the dimensions that this problem reached (Chevalier 1963; Melville 1999; Gibson 1952).

Erosion on the upper foothills, low hills, and slopes of the Mexican Central Highlands shows widespread and eroded soil profiles popularly known as *tepetate*, which still erode at nearly five times the rate of areas with soil relict (Etchevers et al. 1998; Werner 2012). In both the Central Highlands and the Maya lowlands, soil erosion had causes related to the "new pioneers" who had provoked erosion in modern soil due to inappropriate modern mechanized techniques, ignorance, or indifference toward the newly deforested fields, which were converted into pastures for livestock, like in Tabasco and Campeche since the 1970s (Molina-Rosales 2010; Rentería-Garita 2006).

There are studies that show very high levels of erosion that have been traced along the Mexican Pacific coast. These show a low productivity of maize that is associated with soil loss. According to Maas et al. (1988), this is only 49–70 tons/ha/year. Millward and Mersey (1999), in their studies on soil erosion, and GIS modeling to optimize the time of human actions in land use, in addition, a study of 137Cs under a deciduous forest showed a soil loss of 13.2 ton/ha/year (García-Oliva et al. 1995).

Lal (1990) considered erosion a problem in tropical soils because their upper layers have a high percentage of the main soil nutrients. Currently, in central, southern, and southeastern Mexico, there are severe rates of soil erosion, but studies are indispensable to make complete evaluations. The history of soils in Mesoamerica is related to be able to make complete evaluations. Likewise, researchers seek to learn about what ancient societies, with their very high urban and rural populations both in central Mexico and in the Maya lowlands, did with respect to the use of agricultural land, because its current productivity is low and its soils are depleted (Williams 1989; Turner 1978).

20.6 VICEROYALTY AND EXPANSION TOWARD THE HILLS: THE SOIL EROSION

In the Viceroyalty, the creation of human settlements and their expansion toward the hills are associated with unfavorable factors, like the presence of eroded areas often. The clearing of land should be seen, as an articulated project where the conversion of agricultural land into pastures for sheep did not help in the recovery of the soil (Melville 1999). According to Trautmann (1981), urbanization on agricultural soils and eroded areas is associated with the creation and expansion of human settlements. The building of cities such as Tlaxcala and Puebla accentuated deforestation by the construction of houses, textile works, religious buildings, and structures destined for the sociopolitical administration, as shown in the following paragraph written by Mariano Fernández Echeberría y Veytia in 1780 (1931), that described the foundation of Puebla on April 16, 1530:

> Already assigned the place where the new town was to be built, it was up to the Franciscans to add, and attract Spaniards, who would populate in it, and request workers, and materials for the building of the houses [...] that they obtained from the indigenous people that were in charge of cutting the woods, and taking them to the place where they talk about making the town, where due to the proximity to the river (which today is called S. Francisco) there was an abundance of stone, mud, and fagin[7], which were the materials they talk about to make the houses, and they worked on them without any stipend.

The introduction of cattle and sheep herds in Tlaxcala was since 1521 (Gibson 1952; González-Jácome 1991). Generally small cattle ranches with 500–2,000 sheep operated in conjunction with an agricultural area or with textile works. The indigenous sheep farmers moved the herds during the dry season. In St. Ana Nopalucan, a town of the Atoyac-Zahuapan river basin, the sheep farmers and their flocks returned to the community every year, for the festivity of S. Peter and S. Pablo (Chevalier 1963; Gibson 1952; Castañeda-Castillo 2014; Juárez-Choque 2013).

Every year in June, herds invaded farmlands in Tlaxcala, particularly to the south and east of the Province, leading to litigation between Indians and Spaniards. This situation stabilized with

[7] Fagin: do military engineers especially for linings use beam of very tight slender branches. There were also to crown, burn, etc. (RAE 2019).

FIGURE 20.2 Tlaxcala: S. Manuel Morcom Textile Factory, XIX–XX Century.

population decline, the result of epidemics, and migration to the north of New Spain (Chevalier 1963; González-Jácome 1994; Martínez-Saldaña 1997). In the semi-arid areas of New Spain, cattle, sheep, goats, draft and pack animals, and plow were introduced.

The effect that grazing had on soil, water, and vegetation in Tlaxcala at that times has been little studied, but it is clear that the vegetation cover had to change because of that activity. With the introduction of cattle and sheep, manure as a component of vegetable waste modified agricultural productivity, hence the vegetation cover had to have deteriorated in those soils which were characterized by the shortage of perennial herbaceous plants to protect them, especially during the torrential rains from August to September (González-Jácome 1993 (27): 141–150).

Textile works brought a decrease in the availability of water from rivers and springs, and contaminated the rivers by washing the wool and applying dyes to their products. This textile production began in the sixteenth century, in the works that were located by the Atoyac and Zahuapan rivers, and their tributaries like the Tequixquíatl. With the appearance of factories in the nineteenth century, terraces were built near the rivers, and the cotton fabrics produced were put to dry there. Now the Tlaxcala textile industries are public offices, stores, resorts, residential areas, or dilapidated (González-Jácome 1989: 86–107, 2008b: 9–24) (Figures 20.2 and 20.3).

FIGURE 20.3 Tlaxcala: La Trinidad, Textile Factory, Nineteenth Century.

On the changes of land use brought about by the erosion in Tlaxcala, there is information on the ridges system in the Atoyac-Zahuapan basin, the Acuitlapilco Lake, and Apizaco, and on the management of fertile plains along the Zahuapan, like that of the river basin plain at S. Pablo Apetatitlan. It is necessary to consider the earth banks of intermittent ravines, plains, and streams, whose use changed due to abandonment or the expansion of housing, and government facilities on them, like in Apetatitl´n. The canals between platforms became areas for tule extraction, or duck hunting, or redfish, and waterfowl. Their groves gave rise to the increase of wild fauna, which complemented the diet of the people of the region (Gemelli-Carrieri 1983 [1697]; González-Jácome 2008).

20.7 MEXICAN AGAVE FERMENTED MEAD ESTATE *(HACIENDA PULQUERA)*, AND SOIL EROSION CONTROL

In the nineteenth century, Agave and *pulque* estates were relevant in controlling soil erosion and good for the regional and state economies. Due to the lateral extension of its roots Agave supported thin soils, and it was an excellent inhibitor of soil layer erosion, keeping soil particles within a dense network formed by its roots. The *pulque* estates managed a combination of cash crops (wheat-Agave, or barley-Agave) through the semi-terraced (*bancal*) agroecosystem, which in the nineteenth century reached its maximum and extension (González-Jácome 2013: 1–51; Parsons and Parson 1990; Patrick 1977; Rendón-Garcini 1990, 1995).

The terraces or semi-terraces (*bancales)* are land edges *(metepantles*) combined with a plain zone for cultivation (planting area, footprint, or platform), with a slight inclination following the contour lines on the lower slopes of the hills, protected with a row of Agaves (*metepantles*), planted to avoid erosion (West 1970: 361–369). Occasionally, canals were built next to the land edges to retain humidity. The planting area had a surface 15–25 m wide that allowed the introduction of seeders and harvesters, depending on the slope, but the steeper the slope, the smaller the planting area. The land edges were 3–4 m wide and 1.0–1.5 m high. Agaves were planted on the earth edges every 3.0 m (Figure 20.4).

The semi-terraces allowed two harvest periods, and the Agaves were used for fermented mead (*pulque*) production. The second half of the nineteenth century was the scene of the development and boom of Agave large-scale production. The northwestern plain of Tlaxcala, which geographically forms part of the macro-region of the Llanos de Apam, underwent great transformations with

FIGURE 20.4 Tlaxcala: Apam Plains with Relics of Edges of Land (*Metepantles*).

the introduction of the railway, which allowed the transportation for commercialization of Agave fermented mead (*pulque*) to the markets of Mexico City and Puebla City (Rees 1976; Rendón-Garcini 1996; Villalobos-Nájera 2012).

From 1893 to 1894, Agave fermented mead constituted 11.7% of the total cargo on the Interoceanic Railroad, which traveled in a west-east direction, but from 1894 to 1895 it amounted to 13.5% of the total cargo (Rees 1976; Villalobos-Nájera 2012). In non-*pulque* production areas, Agave played a role more related to agricultural work than to the commercialization of Agave fermented mead on a large scale (González-Jácome 2013, 2016; Rendón-Garcini 1990, 1995, 1996).

Alfonso Luis Velasco (1892) described the uses of Agave to obtain fiber (*ixtle*), alcohol, fuel, and worms for food. In addition to being a plant for erosion control, Agave continues to be important in the production of honey, mead, sap, and *pulque*. The cuticles of the leaves (mixiotes) are used to wrap lamb in when making barbecue in earth kilns. With the cuticle, they make a litter bag to fill with pieces of lamb or chicken, adding to the meat a chili sauce and then steam it. Agave flower buds (*hualumbos* or *gualumbos*) are consumed and are a gourmet food today.

In the 1970s, in an interview with the textile engineer Carlo Clérico, an Italian who worked as the director of a textile company in St. Ana Chiautempan, currently closed, he told me that with *pulque* the personnel in charge of the maintenance of the machinery washed the gears, parts of looms, gummers, and other machines, which gave better results than the use of chemical products that in addition to being very expensive they were export products, did not always leave them in good condition, and wore them down more quickly. Instead, a liter of pulque was much cheaper and more accessible in the region. In addition, he told me that there were the Tlaxcalan workers who had taught him how to use pulque for this activity that was fundamental for this textile factory.

Until the first half of the past century, products derived from Agave were specialized and sold on a small scale in towns. One was St. Isabel Xiloxoxtla, a town to the northeast of the Acuitlapilco ex-lagoon. It sits 9 km south of the capital city of the entity. There, the settlers were engaged in agriculture, commerce, and the manufacture of articles made of Agave fiber, such as sacks, belts, bags, and ropes of all long and thick. This activity disappeared in the 1950s with the introduction of synthetic fibers on a large scale, which was cheaper. This town was famous for the Agave fiber (*ixtle*) articles that it produced and sold. There were three families specialized in this activity. By the 1970s, there was only one old man who still had the utensils to shred the Agave stalks, wash and dry the fibers, and make the aforementioned products (González-Jácome 1976, 2003, 2016).

At the end of the nineteenth century, and mainly in the twentieth, several areas of the Tlaxcala Block macro-region were cleared for mines to obtain granite, sand, and non-metallic minerals. This contributed to the deforestation of slopes in places like Cerros Blancos (White Hills). From the 1970s to 1980s for the extraction of tincal, those areas became zones for indiscriminate extraction of non-metallic minerals for industry in other places in the Tlaxcala Block besides Cerros Blancos. Those areas with laterized soils were used to obtain granite, gravel, and sand.

In Tlaxcala, the inadequate construction of roads and highways increased the formation of canyons, mainly in the Hueyotlipan–Apizaco and Tlaxcala City–Puebla City areas, and with the introduction of heavy agricultural machinery, deforestation, monoculture, and cattle and sheep grazing increased the processes of soil erosion (Figures 20.5 and 20.6).

Other fact was the emigration of peasants with their families and the abandonment of agricultural lands, as occurred in the Viceroyalty with the departure of 400 families from Tlaxcala to populate the north of New Spain (Martínez-Saldaña 1997). This farmer families with abandoned their lands, in places like the Cerros Blancos, and their abandoned terraces, that left the soil without the layer of vegetation that protected it, and abandoned soils degraded rapidly, worsening the problem (González-Jácome 1989, 1992, 2016; Hilger 1973; Werner 1998, 2012).

FIGURE 20.5 Tlaxcala: Soils Cut with Heavy Machinery by the Construction of the Calpulalpan-Tlaxcala Highway, 2017–2018.

FIGURE 20.6 Tlaxco, Tlaxcala: Eroded Soils, 2016.

20.8 MANAGEMENT OF EROSION IN THE PUEBLA–TLAXCALA VALLEY

Mexican geographers and the National Institute of Statistics and Geography (INEGI) consider the Matlalcueye, Malitzin, or Malinche volcano as the dominant element of the Tlaxcala landscape, so that explains the integration of the entity with the geological unit known as the Neovolcanic Axis, and to a larger subdivision called Lakes and Volcanoes of Anahuac (Hilger 1973; INEGI 2007; Montero 2012; SPP 1981).

In this region of central Mexico, volcano and the Zahuapan River make up the entity's rural and urban landscapes. La Malinche occupies 20% of the total state surface. The foot of its southern slope has an altitude of 2,170 masl, and the crater, to the west, is about 3,100 masl (Hilger 1973; Montero 2012; SPP 1981; Werner 1986, 2012). The volcano and its slopes form the eastern limit with the macro region of the Atoyac and Zahuapan rivers basin (González-Jácome 2008, 2016).

Gerd Werner (2012) said that there was a lot to learn, and understand about the processes of management, conservation, and use of soil, particularly the hardened Duripan. In Tlaxcala Duripan (*tepetate*), zones are the product of a combination of natural and human factors, such as climatic components, general environmental conditions, inappropriate agricultural management, loss of natural vegetation, rain falling on bare soils, and a lack of protective elements to replace said loss (Etchevers et al. 1998; Wilken 1969, 1987).

Buendía-Rodríguez et al. (2008: 1–9) and Werner (1992) affirmed that most of the eroded soils in Tlaxcala corresponded to the hardened areas in the municipalities of Tlaxco and Terrenate, constituting 80% of the soil surface that was the most severely eroded.

Terraces and semi-terraces (*bancal*) that include land edges, rows of stirrups (*cajetes*, or *estribos*), and *metepantles* are anti-erosion systems, and although they have disappeared from various conventional agricultural systems, in the new terraced or semi-terraced areas are being created in the north and southwest of the Tlaxcala Block. After the 1970s, two relevant proceedings have occurred: (1) the elimination of elements considered inadequate with respect to the introduction of modern technology, as, for example, the indiscriminate eradication of land edges and their canals in low hillside areas to increase the area under cultivation; and (2) the use of agricultural machinery unsuitable for the type of thin soils found in that regions of Tlaxcala (e.g., combines and very heavy tractors), the opening of roads and cargo trucks passing them have great erosive potential. The construction of these roads does not have a soil protection zone, so there is no control on the formation of canyons, the lost of soil, and soil erosion.

In the second half of the last century there was an increase in population densities and the need to open new lands for cultivation of barley monoculture. In addition the modernization of rural populations, the construction of residential areas, and the growth of non-agricultural employment in the tertiary sector acquired great importance. In other words, since the 1970s, the modernization and urbanization of the rural areas was an important governmental program developed in Tlaxcala. Also, it must be considered that young people no longer want to engage in agricultural activity, and as they have a higher educational level, they have become urbanized, their aspirations are different, and therefore they are looking for jobs in urban centers or else they emigrate.

The rains in Tlaxcala occurs in the summer, their collection, and storage is important. Since pre-Hispanic times, and as part of the components of the agricultural system, farmers have used *jagüeyes*, which are artificial open reservoirs for rainwater. In the 1960s and 1970s, farmers drew rainwater from these open cisterns (*jagüeyes*), using the water for domestic work, for washing clothes, as well as for filling water troughs for livestock. These reservoirs provided them with water in the dry season, too. But, in Cuaxonacayo, when a deep well was installed to supply the population with drinking water, the locals covered the open cistern (*jagüey*, *xagüey*) that was in the center of town putting a kiosk and ornamental plants on it (Pérez-Sánchez 2012).

In the hills surrounding the Huamantla valley, open cisterns (*jagüeyes*) are used, still, especially as watering holes for animals. Some have been lined with membranes, which prevent the liquid from seeping into the subsoil. Canals and ditches have disappeared in the last 20 years to make way for areas destined for roads or to increase the surface of the land. Open cisterns (*jagüeyes*) were common in the semi-desert areas of Tlaxcala, Hidalgo, and Puebla. Their origins are pre-Hispanic (García Cook 1975 (1): 199–210), but they were maintained and used in the Viceroyalty, where they enabled agricultural and livestock farming on the estates (Ajofrín 1763). As noted above, some of these deposits continue to this day (Cruz et al. 2018; González-Jácome 2013) (Figures 20.7–20.9).

Many of the open cisterns (*jagüeyes*) that are still in operation had an irregular shape; they are artificial constructions and made in hard ground (Duripan, *tepetate*), which prevented water from leaking into the subsoil. Membranes are currently used to cover the floor of the open cistern and prevent water seepage. Among the *metepantles* is the water that runs through the ditches and that is diverted to the deposits by means of canals built manually. If the water tank reaches its maximum capacity, the liquid flows through a spillway into the trenches; if it has no outlet, the board "breaks" or fractures. Open cisterns have variations in their structure. Some have vegetation on the borders, others lack it; the predominant flora is of Montezuma cypress (*Taxodium mucronatum*), Agave, Opuntia, grass, and numerous weeds.

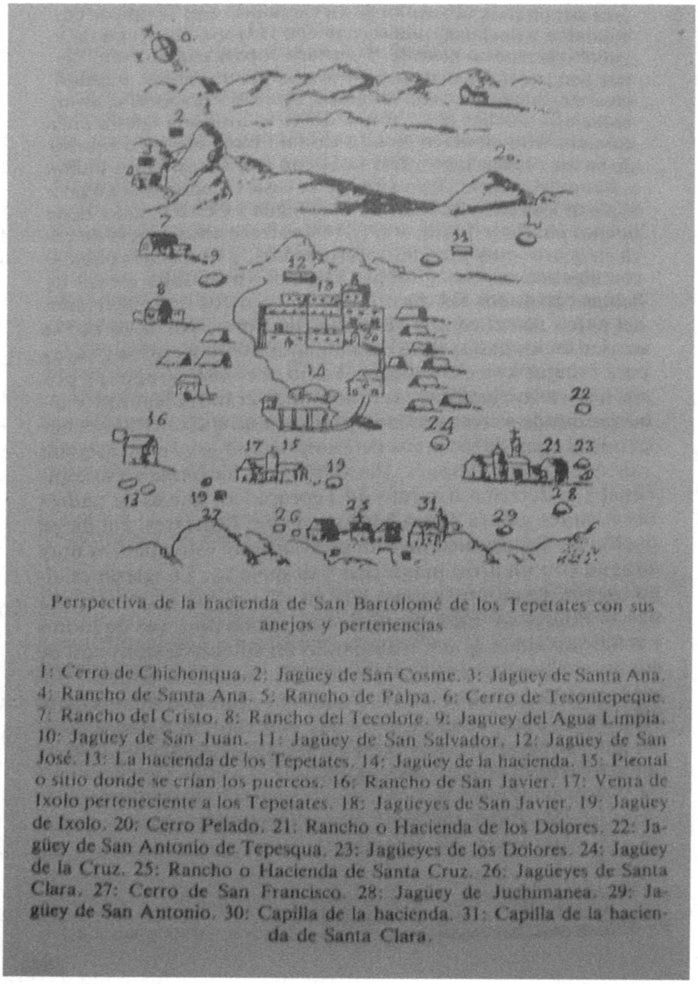

FIGURE 20.7 Map: Apam Plains: A System of Open cisterns (*Jagüeyes*) in 1763.

FIGURE 20.8 Tlaxco, Tlaxcala: Open Cistern (Jagüey) and Metepantle.

FIGURE 20.9 Stirrup (*Cajete, Estribo*) After been cleaned during Winter.

The antiquity of Agave lines (*metepantles*), earth edges, terraces, and open cisterns (*jagüeyes*) is articulated with the use and management of soil and water, illustrating how human populations invented and reinvented structures, shapes, and elements to live with wellness. The solution, and farmers activities articulated with it, is a first step in evaluating the problem, to focus on the scientific study of the basic techno-environmental aspects of the twenty-first century, for the elaboration of concrete and functional proposals that lead to the proper management and conservation of vegetation, soils, and water, for their different uses, particularly agricultural.

20.9 DISCUSSION ON THE PRODUCTIVITY AND KNOWLEDGE OF SOIL IN MESOAMERICA

Knowledge of ancient soil is obtained from the study of the different types and contexts of the soil and their elements, together with geo archaeological, chemical, and biophysical evidence, which shows the existence of a typology of soils in traditional agricultural systems. Samples came from the Formative and the Mesoamerican Classic, with many types of agricultural systems developed for tropical areas, semi-desert, with excess or lack of rainwater, with constructions such as open cisterns (*jagüeyes*), chultuns, and tanks to retain and store rainwater, canals of different types and dimensions, dams and dykes in rivers or alluvial fans, walls and earth edges on the terraces – with or without canals. There were many other traditional systems where the soils and water were managed, indicating extensive control of the landscapes, both rural and urban.

An example of this knowledge is the use of contour lines, ravines, and troughs, which allowed a combined management of soils, with rainwater, which is also a primary objective of study in traditional agriculture. These systems are the oldest in various places in the Mesoamerican cultural area, such as the Mixteca or the Cerro Gordo in Teotihuacan. Although there is extensive knowledge related to Mesoamerican soil management and use, it is accepted that this knowledge is in an initial phase. There is a lack of specific and comparative studies about the soils on terraces, platforms, earth edges (*bordos*), canals, stirrups (*cajetes*), open cisterns (*jagüeyes*), jollas, hedges (*setos*), and stone lines (*ringleras*), which could give some ideas about how ancient people managed the soils before the arrival of the Spanish to the New World (Galindo-Escamilla 2008a; Galindo-Escamilla et al. 2008b).

20.10 SOIL IMPROVEMENT IN THE PRE-HISPANIC ERA

Mesoamericans applied green manure, human debris, and animal manure to the soil. The soils of wetlands, chinampas, and terraces were fertilized with excrement (like bat dung) or with organic matter (like periphyton, formed in the Yucatecan *petenes* ecosystem). Periphyton has been studied in the wetlands of the Yucatan peninsula.[8] Periphyton is a community composed of algae, bacteria, fungi, and invertebrates, mainly, and is found in freshwater ecosystems throughout the world. It is rich in phosphorus and nitrogen (Tavera and Novelo 2011: figure 2, Research, Ecological Reserve El Edén, Quintana Roo 1999; Weitzel 2005). The "Silt" is a biological indicator of water quality; therefore, it has generated interest in the management of wetlands, because is used to monitor and regulate the concentration of nutrients. Many wetlands – natural and artificial – are used for economic and ecosystem purposes as well as for the improvement of water quality in human settlements (Fedick et al. 2000 (27): 131–152; Ibarra et al. 2003; Tavera and Novelo 2011: 40–45).

The agricultural management to improve soils found in the Late Classic homegarden agroecosystem in the northwest of the Yucatan peninsula is shown through the study in Makabil, where the application of peripheral from the wetlands to the homegardens improved their agricultural productivity. To these homegardens, the abundant shells of amphibian snails (inhabitants of the wetlands associated with freshwater periphyton) were applied to the soils of the homegardens of the archaeological studies. The periphyton and the shells were found together and distributed in a chosen area, showing that they were not the result of being dragged by floods (Morrison and Cózatl-Manzano 2003).

In a study on Teotihuacan, results on erosion in the valley's soils were based on "[…] two kinds of records: profiles located in the edafo-alluvial sequences at the bottom of the valley (S. Pablo River, and El Tular); and soil sediment profiles associated with pre-Hispanic occupations proof of Mexica occupation surfaced at Cerro S. Lucas, and in fillings of the Moon pyramid" (Rivera-Uria et al. 2007: 203–217). The researchers produced maps with the distribution of settlements in different periods of occupation. They carried out "[…] select analyses of the properties that constitute the memory of the soil; Radiocarbon, and AMS dates were obtained from certain horizons to determine the time frame. Comparisons were made with stable isotopes of carbon, and available phosphorus" (Rivera-Uría et al. 2007: 203–217).

The results showed that excessive population putting pressure on the space and the resources of the ancient inhabitants of the valley, also, originated erosion. They showed that the oldest paleosol, which corresponded to the initial population of the region, indicated a more stable environment, with "[…] good development and state of conservation". The overlaying soils were poorly developed and showed signs of higher erosion rates. The materials within the Pyramid of the Moon showed that part of the soil used by the builders was from the surrounding areas. It was observed that in the Mexica house the areas with different activities showed different degrees of disturbance, using soil brought from other areas to level the Duripan (*tepetate*) that was present on the floor of the house.

The study on the Teotihuacan valley concluded, "[...] the application of pale pedological techniques associated with geological studies contributes to the understanding of environmental degradation in natural geosystems as a consequence of human activities". In other words, the greatest

[8] Fedick et al. 2000 (27): 131–152; Novelo and Tavera 2003; Gómez-Pompa et al. 2003; Ibarra et al. 2009 (80): 763–769; Tavera and Novelo 2011: 39–45.

FIGURE 20.10 Tlaxcala Block Macro Region: Stirrups (*Cajete* or *Estribo*).

erosion coincided with the highest population densities. More information on the agricultural practices used by the residents lacked.

In later historical times, the chinampas of the basin of Mexico were fertilized with human and bat excrement mixed with muddy water (*agualodo*) obtained in the canals that was deposited with seedlings in cub-shape quadrangular boxes made with muddy water (*chapines*), of different sizes according to the plant that will be planted in them, where the seeds were planted, before being taken to the farmland. Human feces were obtained from places like the Tlatelolco market, deposited in flat-bottomed canoes, and taken to the south of the basin of Mexico for accumulation and management.[9]

Recent research findings have begun to clarify the history of soil for agricultural uses, its interrelationships with water, and the human management applied to improve them, and increase agricultural productivity (García 1994: 30–44, 2007: 328–341). In the Tlaxcallan terraces, a natural fertilizer consisting of plant remains, dead animals, insects, micro-organisms, and soil was applied to the land, which remained in the water accumulated in the parallelepiped-shaped tanks locally known as stirrups (*cajetes* or *estribos*), during the rainy season (González-Jácome 1989: 91–117, 2014: 1–30).

It remained there for a year, and before the following rains, the contents of the tanks were emptied into the cultivation fields. The antiquity of this practice is not known, but stirrups (*cajetes*) have been studied by archaeologists, who determined them to be of pre-Hispanic origin (Donkin 1979; González-Jácome 1989: 91–117, 2014: 1–30; Moreno-Calles et al. 2016; Pérez-Sánchez 2012, 2014) (Figures 20.10–20.13).

[9] Cox and Atkins 1972; Palerm 1972, 1977; Rojas-Rabiela 1983; Rojas et al. 1974; West and Armillas 1950.

FIGURE 20.11 Totolac Highland River Basin: Stirrups (*Cajetes or Estribos*).

FIGURE 20.12 Tlaxco, Tlaxcala: Soil Erosion, 2016.

With the expansion of estates, the management of the silt-board system (*lama–bordo*; also called trench or silt-board type terraces) increased. That system was formed by the accumulation of earth, stone, and vegetation along the canals that carried water to the land for cultivation. It is found in ravines, where soil can accumulate. Those earth edges and stone may or may not have trees planted on the upper part. The content accumulated in the canal was deposited in the cultivation fields for the following agricultural cycle (González-Jácome 1989: 91–117, 2014: 1–30, 2016: 111–144).

FIGURE 20.13 Tlaxcala Block: Metepantle with Opuntia and Mexican Hawthom.

From a social and cultural perspective, changes in agricultural soils were initiated by two factors: (1) those related to population dynamics, which included migratory processes, seasonal or permanent migration, return migration, population densities, and their distribution by sex and occupation (INEGI 2005; Rivera-Uria et al. 2007: 203–217); (2) the in legislation on the use of land, and issues such as government support raising livestock, or the expansion of exportation crops, having as its results modifications in the use, and management, or abandonment of soils intended for agricultural activity, originally.

Among the texts that have been written about traditional agricultural systems and soil management, many aspects are discussed, like their relationship with water: (1) the modifications and management that excess water areas undergo during several months of the year (e.g., wetlands – swamps, *petenes*, shallows, lagoons, river banks, waterlogged areas, raised and drained fields, ridges, and inland chinampas[10]). The management of rainwater, its ways of accumulation, retention, management, and distribution throughout the year is basic in areas that lack water.

(2) The modifications made to the soils in mountainous, hilly, and high elevation areas through the construction of terraces associated with canals and land edges systems. (3) Traditional agricultural systems that use small areas in forests, where the soil is better suited for agriculture, like the *banquetera* agroecosystem of central Veracruz, that is located on the steep slopes of a ravine (Servín-Segovia 2000, 2002). (4) Texts on agriculture in alluvial soils, such as that developed on the banks of rivers (e.g., tecallis, studied by Silvia del Amo et al. 1999, in the Balsas River basin[11]; or lines with fruit crops on the banks of the river in Tehuacán).

There are several references in the studies of ecologists, geographers, and anthropologists[12] on contemporary homegardens and the use of their soils. Those homegardens had lacked a series of specialized analyses in the study of soils and associated pathogens, since they focused mainly on the management of vegetation. However, those assessments do disclose some basic aspects of soil management by small-scale farmers, especially those related to the reconstruction of eroded soils.

[10] González-Jácome 2008; González-Jácome and Velasco-Orozco 2015; Harrison and Turner 1978; Palerm 1955, 1972; Siemens 1989a, 1989b; Siemens and Puleston 1972; Wilken 1969.

[11] Siemens and Puleston 1972; Siemens 1989a, 1989b; Palerm 1955, 1972; Palerm and Wolf 1961; West et al. 1969.

[12] Caballero 1992; Farrell 1989: 169–183; Fenegan 2004; Ford and Nigh 2016; Gliessman 1989: 160–168; González-Jácome 1985, 2004a, 2011a, 2016: 111–144; Kimber 1966; Mariaca-Méndez 2012; Moctezuma-Pérez 2013a, 2013b.

20.11 ELEMENTS IN THE HISTORY OF SOIL IN MEXICO

There are examples of native conservation technologies, as well as of modern mechanized tillage, which is often done on top of traditional farming systems. This had happened in several cases, such as on ridges and terraces with stirrups (*cajetes*) in their land edges. In these, the agricultural machinery destroyed what had been a successful adaptation of traditional agriculture to the slopes – as occurred in the Calpulalpan plain, or in La Recova, west and center of the state of Tlaxcala – where they filled in with earth and stones in canals that carried the water, and stirrups that stored both rainwater and runoff. By destroying the old erosion control systems of the soils on these slopes, erosion was accelerated. Due to the abandonment by the population (emigration to northern Mexico, or increase in population mortality caused by epidemics), some slopes have been eroded since the sixteenth century, like the Cerros Blancos in the entrance of the Totolac River, and the confluence with the Zahuapan River (INEGI 2005).

The information on the change of soils began with the first agroecosystem to build terraces and build canal systems in their platforms to modify natural hollows and create deposits for open rainwater cisterns (*jagüeyes*), which allowed agriculture to be carried out, and that from very early on were altering the natural landscapes. After the conquest, the high mortality resulting from epidemics and migrations both caused human populations to drop precipitously in New Spain. In addition, new forms of environmental change (soils, water, flora, and fauna, mainly) arose, which came from the raising of livestock, the burning of vegetation to maintain grazing lands, the introduction of the plow, and the developing of new agroecosystems.

The uses of land by diverse human populations and ethnic groups over thousands of years have generated scholars to have divergent opinions on the causes of severe soil erosion in the central highlands. In some cases, the erosion was so strong that farmers eroded what was left of the original soils to take advantage of the softer sediments below the surfaces dominated by the cemented C-horizons. There are studies that put the blame for erosion in the basin of Mexico and the Central Highland on the pre-Hispanic populations with intensive agriculture which caused sedimentation, the depopulation of the sixteenth, and seventeenth centuries, the abandonment of agricultural soils, and the disastrous effects of grazing cattle, sheep, and goats in large regions of New Spain (Rincón-Maurer 1999; Melville 1994).

The great problems of the soil destined for agriculture directed to self-consumption or to the market, and its erosion in regions such as those of Tlaxcala are centered on the fact, currently, that they are in a worse state than most of the soils were in ancient Mesoamerica. On the other hand, it turns out that the terrace systems and land edges planted with rows of Agave (*metepantles*), canals with different configuration, inclination, and management, and natural or artificial receptacles to concentrate and retain rainwater (*jagüeyes*, *chultunes*, tanks, and modern reservoirs) are systems that control erosion to a certain extent (Figure 20.14).

The landscapes in the Central Highlands are diverse and complicated owing to the history of the soil. There are rich studies on the region that analyses its soil erosion, popular taxonomies, and its soils management through traditional ecological knowledge (TEK). The studies of the soils and their erosion were developed from those focused on knowing the geological processes and the role of human populations in the formation of ravines to the syntheses that try to give some broader meaning to the complex interactions of human beings with the soils (Bocco 1993; Heine 2003; Werner 2012).

In the Puebla-Tlaxcala region, Heine (2003) used several lines of evidence to correlate severe soil erosion with intensive land use throughout history. This had been high during intensive land use in periods like now, but it decreased during the abandonments by the population, which occurred in Tlaxcala in the sixteenth and seventeenth centuries. Córdoba and Parsons (1997) maintained that the decreases in population and the abandonment of the countryside produced two episodes of accelerated soil erosion. A strong flood in Toltec times (750 BC to 1250 BC) hit the region, leaving relics in the form of soil pedestals. The Mexica restored the region with terraces, containment dikes,

FIGURE 20.14 *Agave mapisaga* (*Maguey Pulquero*): in Earth Edge, Matlalcueye, Tlaxcala, 2011.

and soils throughout the 1500s. The Viceroyalty brought about depopulation and lack of maintenance of the terraces and dikes that accelerated the loss of soil.

Remains of ancient terraces included stone and support walls, floor sequences, and artifacts. All remains can provide dates for construction. Donkin (1979: 18) wrote about the terraces in the Americas saying that they had begun to be built around 2500 BC, both in the Andes and in the Central Highlands of Mexico. Accelerated soil erosion started before 3000 BC, but the need to build terraces was earlier. Donkin's work on Mesoamerica and the Caribbean Islands, and his research in northern Mexico, the Maya lowlands, and central Mexico put the origins of the terrace in northern Mexico around 1000 BC. Dura and Roney (1998) and Dura et al. (1999) gave strong evidence for a group of terraces dating from 1050 BC, according to radiocarbon dates obtained from remains of maize and squash, in addition to Late Preclassic artifacts.

On the Maya lowlands, Hansen et al. (2002) reported finding in the late Middle Preclassic (near 600–400 BC) terraces in Nakbé, Guatemala. In the Central Highlands, Mountjoy and Gliessman (1988) noted that terraces with stirrups started at around 1000 BC. Despite the 3000-year-old evidence of terraces, most began in the Mayan world in the Classic Period (Beach et al. 2003a), while terraces which are widely spread geographically, culturally, and chronologically throughout central Mexico were later (Donkin 1979). García-Cook (1986) and Merino-Carreón (1997) considered that the first terraces in Tlaxcala were located from 1200 BC to 800 BC.

Terraces have a surface for planting that maintains soil humidity, favoring the growth of crops by the infiltration of water. This reduces soil erosion through slowing land flow, by capturing eroded soils on thin-mantle slopes, or by redirecting water to fields in lower areas or watercourses. Many studies provided evidence of excavation for each of these functions. Others have developed terrace taxonomies, using factors like the position of the landscape, its size, connection to drainage, and irrigation, plus the materials to form the basis for the classifications.[13]

Beach and Dunning (1995) stated that terraces that were poorly constructed or lacked maintenance were prone to mass loss and erosion in gullies. This happened because their the rain sweeps the soils toward the low zones, or by cascade can weaken them, or for reasons related to altitude and land inclination, the lack of a vegetation cover to protect the soil from the impacts of heavy rainfall, especially during the strongest part of the rainy season (Wilken 1969). This situation was evident in places such as the Cerros Blancos, which had been left unused since the late sixteenth and early seventeenth centuries, owing to the Tlaxcala emigration to northern New Spain (Martínez-Saldaña 1997).

[13] Treacy and Denevan 1994; Beach and Dunning 1995; Beach et al. 2002, 2006 (65): 166–178, 2010 (92): 372–397.

The construction materials used in the pre-Hispanic terraces reflect the limitations of the methods: (1) the weight and size of the stones used to make dams on the tropical slopes of Mesoamerica, where these could be the most persistent remaining anchors of a landscape, that included soil and vegetative berms, land edges, hedges, with vegetation (*metepantles*) also. This was determined by excavation and mapping of those tropical slopes. Examples of this are the works of J. Neely (2017) on the Lencho Diego ravine and on the Purrón dam in the Tehuacán valley, and the discussions that arose in this regard (Neely and Castellón Huerta 2014).

In the basin of Mexico and in the macro regions of the Tlaxcala Block, or on the western slopes of the Matlalcueye volcano, in the state of Tlaxcala, there were earth edges made of vegetal walls, such as Agave, Opuntia, trees, weeds, and ruderal vegetation used in terraces, and *metepantles*, which also served to collect the sediments around them. In these, the lands, aqueducts, and canals irrigated the platforms of the contour terraces, configured with earth edges, vegetation walls, or of earth or rock walls that helped to maintain the floors of the platform (Palerm and Wolf 1957; Donkin 1979; Evans 1992; González-Jácome 1989, 2014, 2016). In this sense, there is a lack of studies on the formation of silt, sediments like periphyton, and their annual application to contiguous soils on platforms (Córdoba and Parsons 1997).

The history of soils in Mesoamerica has to do with their manipulation for crops. In this sense, there is a great contribution from ethno edaphology studies, chemical tests, and archaeological excavations, which have been aimed at discovering what people in ancient times could have been doing on their lands to achieve sufficient agricultural productions to cover their nutritional needs (Williams 1989). Ancient populations were numerous, especially in central Mexico, as were the ancient Maya in the lowlands (Turner 1976), where agricultural productivity today is low and soils are depleted.

There are a number of soil management studies done on small-scale farmers, such as those carried out in Tlaxcala by Cruz (1949: 44–96); González-Jácome (1989, 2014: 1–30, 2016); Mountjoy (1985); the agroecological studies of Mountjoy and Gliessman (1988); the anthropological studies of Pérez-Sánchez (2012, 2014); and those specialized studies in soils, carried out by Werner (2012) and other researchers. Those studies showed how in agroecosystems with terraces, without walls or covered cant, with tracks that have vegetative land edges and/or canals that divert runoff to stirrups or deposits, farmers cleaned them annually, and applied the muddy water to the cultivation fields (terrace platform), to improve their productivity, or to recover eroded soils.

Smith and Price (1994) suggested that the Post Classic agricultural terraces in Mesoamerica were the result of decisions made in households, in conditions in which populations were growing. Doolittle (2000: 254–308) considered that terracing in dry land systems had to be adapted to land planning and water management. In the archaeological landscape farmers have problems, because over time gradual construction can appear to result of central planning, especially after many centuries. Regrettably, there is no detailed historical information for all sites regarding water management to control erosion in soils intended for crops. Terraces in these land "tend to be concentrated in areas with andesite surface rock which weathers rapidly and produces soils high in fertility" (Doolittle 2000: 301).

20.12 CONSIDERATIONS AND CONCLUSIONS

There are rich archaeological studies on the Mexican Central Highlands and the history of their soil erosion, its causes, popular taxonomies, and soil management (McClung et al. 2003: 270–282; Rivera-Uría et al. 2007: 203–217). The region has a diversity of natural and cultural landscapes, as well as an ecological diversity (ecological mosaic). The area included semi-arid zones, rugged relief areas, basins, and cold plateaus in the Central Highlands. Human populations reached high densities of different ethnic groups, and there was a diversity of languages and cultures. The region has been important for scholars who study soil and soil erosion. Their studies included the processes in the formation of canyons and hardened soils (*tepetates*), such as those by Bocco (1993), Buendía-Rodríguez et al. (2008), O'Hara et al. (1993: 48–51), and even those that deal with the complex interactions that take place among soils, agriculture, and their human management (González-Jácome 2014: 1–30).

In the 1990s, studies showed that soils in pre-Columbian societies were severely impacted by the environment because of intensive land use. Studies by historians focused on European colonization, the arrival of livestock in the central and northern regions, and about soil degradation in New Spain, and up to the present time. Rincón-Maurer (1999) and E. Melville (1994) considered that the greatest impacts on soil in the last 500 years were caused by the use of land for sheep and goats herding. Human management induced these changes.

Rincón et al. (1997) described the fields as having a few remaining pedestals of eroded soil that preserve small portions of the old 70 cm of topsoil, above the current surface of most of the fields. These Duripan soil pedestals testify to the surface, age, and profile of the soil (Rincón 1999: 664). Córdoba and Parsons (1997) show that even without the presence of livestock, these soils are prone to erosion. Also, that the cattle ranch has varied over time and from place to place.

The studies by Butzer (1992a, 1992b), Butzer and Butzer (1993), Denevan (1992), Cook (1949: 86), McAuliffe et al. (2001), and Williams (1972) focused on the Mexican central regions and the understanding the causes of soil erosion, the formation of hardened soils (*tepetate*), and the exposed C-pre-Columbian petro calcic horizon, which covered a wide propagation surface and the Bajío soils where they found no evidence to confirm the early impact of livestock, but it is greater now than it was in pre-Hispanic times (Werner 1992).

For the Mayan zone, Ford, Clarke, and Morales (2015: 481–495) as well as Ford and Night (2016: 18–19) considered that during the Maya Classic (500 DC–900 DC) the combination of the forest with the homegarden and the cornfield contributed to the diversity of resources for families, at the same time that it improved soil conditions. The cycle of this combined agroecosystem had the potential to produce sufficient food for important populations. For these researchers, agriculture was based on the quality and sustainability of the land while maintaining the forest environment in order to provide families with a wide variety of products to pay taxes and tributes to the ruling elites.

One conclusion about soil erosion comes from the information summarized in this book. This considers that the greatest cause from 1519 to the current era is by anthropomorphic uses. The soil management of farmers, the cattle rising of ranchers, and the cattle handling of itinerant groups are examples. This must include the total size of population at different times, population densities, and the public political focuses and practices about the population and land tenure distribution in the country.

In any landscape, over time there are a number of changes, sometimes slow, sometimes drastic, and there are various causes of soil erosion. In the Puebla-Tlaxcala region, Heine (2003: 67) used multiple lines of evidence to correlate severe soil erosion with intensive land use, off and on throughout history. He considered that erosion was high due to intensive land use in periods like the present and was low during the abandonments of land due to population decline, such as in the sixteenth and seventeenth centuries. In contrast, Córdoba and Parsons (1997) argued that declines in population and population abandonment of land produced episodes of accelerated erosion.

RECOMMENDED READING

Montgomery, David R. 2007. Soil erosion and agricultural sustainability. In: *Proc. Natl. Ac. Sci. U.S.A.* (PNAS), August 14, Vol. 104, No. 22: 13268–13272.

This article deals with information obtained from numerous studies confirming that erosion rates in conventionally plowed agricultural fields average 1 or 2 orders of magnitude higher than rates of soil production, erosion under native vegetation, and long-term geological erosion. The general equivalency of erosion under native vegetation indicated that, considered globally, soil production on slopes and erosion evolve to balance geological and climatic forces, while conventional plow-based agriculture increases erosion rates accordingly, which is unsustainable. In contrast to what occurs with net rates of soil erosion in conventionally plowed fields (1 mm / year), which can erode a typical hillside soil profile on time scales comparable to the longevity of major civilizations, untilled agricultural land produces erosion rates much closer to soil production rates and could therefore provide a basis for achieving sustainable agriculture.

Beach, T.; N. Dunning; S. Luzzadder-Beach; D.E. Cook; and J. Lohse. 2006. Impacts of the ancient Maya on soils and soil erosion in the central Maya Lowlands. In: Elsevier, Catena 65, 2006: 166–178, Science @ Direct.

The text deals with the soil degradation in the Maya lowlands of central and southern Belize, Guatemala, Honduras, and Mexico, mainly on its sedimentation, and erosion from before 1000 BC to the Mayan collapse around 900 AD. The report on the Blue Creek and Cancuén sites synthesizes other findings. More than 100 excavations were carried out in the upland and depression sites, core lakes, and wetland sediments. The sediments were studied in the field and in the laboratory using radiocarbon dating, a battery of soil chemistry tests, stratigraphic analysis, magnetic susceptibility, elemental analysis, and tests of artifacts.

The objective was to determine when sedimentation and soil erosion occurred to identify stable surfaces and correlate them with the state of knowledge about past land use. The findings gave three epochs of accelerated soil erosion and identified two main paleosols. These waves of soil erosion occurred in the Preclassic (c.1000 BC–250 AD), the Late Classic (550–900 AD), and in the last decades of it. The main paleosol (*Eklu'um*) at these sites is well-developed Mollisol or Vertisol that began to form in the early Holocene and was buried in the Preclassic or Classic (250–900 AD).

At some sites, the *Eklu'um* paleosol is below sediments with a weaker paleosol, which in turn is buried below Classic, and later sediments. There is higher than expected soil erosion in the Preclassic and lower than expected soil erosion in the Late Classic when population peaked, and land use was the most intensive. In Cancuén, Guatemala, most of the soil erosion occurred in the Late Classic Maya (550–830 AD), when it was intense but short-lived. The depressions registered from 1 to 3 m of degradation in two centuries. A third era of accelerated soil loss and degradation emerged with the rapid changes in land use brought by new pioneers in recent decades.

González-Jácome, Alba. 1993. Management of land, water and vegetation in traditional agro-ecosystems in Central Mexico. In: *Landscape and Urban Planning. An International Journal of Landscape Ecology, Landscape Planning and Landscape Design*, 27 (1993), 141–150.

The article dealt with some of the characteristics of traditional Mexican agroecosystems in relation to the management of soil, water, and vegetation in Central Mexico. It discussed some of the approaches and managements practiced by small-scale farmers that were faced by them as a result of a declining agricultural familial work force, processes like migration, and the pressures of a society that was supporting industrialization as a way to obtain better ways of life. There are two parts dedicated to erosion and the agricultural practices to maintain soil fertility in some lands in the Puebla–Tlaxcala Valley.

Lal, R. 1990. "Erosion measurement and evaluation". In: *Soil Erosion in the Tropics: Principles and Measurement.* McGraw Hill, New York; pp. 183–223.

21 Nature, Ideology, and Culture in Small-Scale Farmer Societies

21.1 INTRODUCTION

The interrelationships between nature and culture have been studied for more than a century, and there are at least three perspectives. The first considered that there is a dependence on the culture of nature. It seems to result largely from anthropological studies of the 1930s and 1940s carried out in the so-called "primitive" societies. They were tribal, were unaware of urban planning, and had a total population smaller than that of industrialized societies; their economy was not capitalist; and they did not have a political organization based on the State. Anthropologists studied rituals, ceremonies, paraphernalia, and events articulated to magic and religion of these tribal societies where the complex kinship system was the social system that replaced a complex political system (García-Mora (cord.) 1987–1988, 2013; Morehart 2002 [1998]: 1–27).

The second perspective (1960–1970) considered that humanity with its science, technology, and culture dominated nature. Hence scientific and technological advances were essential to achieve change. It was thought, and still is thought, that these were able to reduce humanity's dependence on nature more and more successfully. This idea remains open or is underlying contemporary societies, making it difficult to accept the existence of natural phenomena and events that cannot be controlled as easily or as quickly as people would like, although they are generated or accelerated by humanity, at least partially. An example is the lack of understanding and deficiencies in the control of the social, economic, and political impacts of events such as volcanism, droughts, floods, hurricanes, earthquakes, tsunamis, climate change, desertification, epidemics, and others.

The third perspective arose in the 1980s and argued that nature and cultures are interrelated systems. Therefore, man must handle nature with greater knowledge and care because his permanence on the planet and the subsequent evolution of the human species depend on it. Hence, the importance of conducting diachronic studies that describe, analyze, and understand the cultural mechanisms that have developed in rural societies related to the use, management, and care of their natural resources, their own environment, and the climate affects all human populations on the planet.

This perspective does not seem to be sufficient to achieve the central objective, which is to contain or reduce the negative impacts of human action on different ecosystems, regions, agroecosystems, and places over the entire planet.

This perspective included the studies on fantastic beings invented by these societies. Those beings are articulated to the care of the different natural environments and contribute to the social mechanisms that control and preserve the natural resources of local societies. These studies compare some changes that had occurred in the relationship between culture and nature, human beings and their ideas about the world, and the care of the environment in various regions of Mexico. These relationships go beyond being a merely ideological association.

21.2 BACKGROUND

For some of the researchers that argued about the relationship between nature and culture (Ellen 1996: 101), the role of the environment is fundamental, especially in simple societies where subsistence is linked deeply to the natural resources available in specific habitats (Abasolo-Palacios 2003; Ramírez-García y Cruz-León 2018: 141–173; Velasco-Orozco 2002). Other students of these relationships and events considered that through culture humanity imprints its stamp on nature and

DOI: 10.1201/9781003198833-21

controls it, using instruments derived from technological, scientific, and traditional knowledge. In other words, it is Man who orders and organizes the relationship between environment and culture through what had been defined in sociology as a "social construction of reality" (Berger and Luckmann 1966). This construction began at the local scale reached the global level and did not coincide necessarily with advances in scientific knowledge (Appadurai 2001).

21.3 IDEOLOGY, COSMOLOGY, AND WORLDVIEW IN TRADITIONAL AGRICULTURE

Within the studies and considerations on traditional Mexican agriculture, it is necessary to include the importance that the ideology, cosmology, worldviews, ceremonies, rites and celebrations have had, and continue to have human groups, which have been and are still elaborated by the peoples about nature and in particular about plants like maize. This grass is a staple food, with multiple uses and articulations with myths, rituals, legends, and festivities. It is a key element in the daily and ceremonial life of rural societies, including indigenous and mestizo small-scale farmers in present-day Mexico.[1]

Archaeologists, anthropologists, and historians have treated agricultural and its ceremonial aspects.[2] In an archaeological investigation in Ceibal, in the Passion region, a site was found that was built around 950 BC, in the Middle Preclassic, that showed there was a close relationship between sedentary lifestyle and ceremony. By 700 BC, a substantial residential complex was built. In the Late Preclassic (500 BC–300 BC), the inhabitants had long-lasting residences and staged burials; therefore, researchers referred to that as an "advanced sedentary lifestyle" (Inomata et al. 2015: 4268–4263). They did so by considering several questions:

> Substantial formal brain complexes seem to have been built only in a small number of important communities in the Maya lowlands and groups with different levels of sedentary lifestyle, probably gathered for their constructions and for public rituals performed in them. These collaborative activities probably played a central role in the social integration of diverse groups with different lifestyles and, eventually, in the development of fully established sedentary communities.
>
> Groups with different levels of sedentary lifestyle probably came together for their constructions and for public rituals held in them. These collaborative activities probably played a central role in the social integration of diverse groups with different lifestyles and, eventually, in the development of fully established sedentary communities.

The worldview and its relationship with rituals and ceremonies are part of the social, political, and religious organization of the peoples' cultures and their relationships with spaces used for rituals. These ideological/cultural, and social elements included legends, rituals, ceremonies, festivals, pilgrimages, and ideas about the cosmos and the world. Those ideas and customs were reflected in and applied to everyday life and festive social life. There are some events that are reinforced by the execution of dances and songs, ingesting something, like alcoholic beverages or drugs (Ashmore 2010: 297–322; Domenici 2010: 387–424; Morehart 2002 [1998]: 1–27; Ortíz-Díaz 2010: 425–446; Tourtellot and Hammond 1998: 185–206).

Some of these elements have Mesoamerican origin, others have a vice-regal origin, and others are a mix of both worldviews social, and cultural stories. There are several created recently. Current groups and human societies have adjusted them to their tastes and needs reinventing the vast majority. They are included as traditional because they are learned generationally, transmitted orally, and learned through practice. It is important to understand that they change over time (Ashmore 2010: 297–322; Domenici 2010: 387–424; Ortíz-Díaz 2010: 425–446; Tourtellot and Hammond 1998: 185–206).

---

[1] Broda 2004a, 2004b; García-Mora 1987–1988, 2013; González-Torres 2007: 45–80, 2008: 179–219; Inomata et al. 2015: 4268–4263; López-Austin 1992, Vol. 29: 261–283, 1994, 1996, 1997: 471–507, 2003: 17–35, 2012; Morehart 2002 [1998]: 1–27; Salazar 2006, No. 228; Sierra 1988; Tourtellot and Hammond 1998: 185–206; Tuz 2013.

[2] Blanco-Rosas 2006; Broda 1995: 16–21, 1997, 2004a, 2004b; Broda and Good 2004; Broda et al. 2001; Espinosa 1996; Inomata et al. 2015: 4268–4263; Márquez-Mireles 2000.

A case related to water, its engineering, and political operations articulated with ceremonies and rites is found in studies on the vital role of water in times of drought. Particularly, the case between the Classic and Late Classic Maya had been studied in relation to the continuous droughts that impacted agriculture and the supply of water to the interior of the cities in the area. It studied the associated ideological aspects, such as the fifth (*quincuncial*) relationship with the four cardinal points and the center from where communication to the underworld occurred through caves, tunnels, and sinkholes that abound in the lowlands of the Mayan area.

The ceremonials and rituals related to requests for rain, the reduction of negative impacts of hail and snow on crops, or festivities to ask for or to give thanks for a good harvest are innumerable and continue to be a vital part of various cultures (Albores and Broda 1997). These festivities are articulated to the communal work related to the organization of these events, like obtaining the resources and elements that will be part of the celebration (including prepared food, drinks, flower decorations, plants brought from nearby places, candles, altars, spaces for dances, songs, and rituals, special clothing, and decorations) (Albores-Zárate (cord.) 2015; Barros 2008; Quevedo-Pérez et al. 2017, No. 56: 5–19).

There are other related issues that were dedicated to the use and erosion of soils, water management, and droughts. For example, those that occur every year to request water in times of drought or to stop the torrential summer rains (Konrad 1985: 312–330, 1996: 98–130; Márquez-Mireles 2000). In the towns of the Central Highlands, Oaxaca, and Chiapas, the fall of lightning on a man could kill him, or failing that, turn him into a "time controller" (*granicero, tiempero*), a person who acquired the necessary qualities from the Lightning to manage the clouds and control the rains, hail, hurricane-force winds, and torrential rains that can destroy the cornfield plants and could destroy the town's crops (Albores and Broda 2003; González Jácome 2003: 467–501).

The future time controller (*granicero, tiempero*) had to learn a series of prayers, and how to make requests to the Rayo's helpers (*cheneques, chaneques*). The Lightning (Rayo) has helpers who fall to the ground in the form of serpents along with the rain; They live in the clouds, which are organized in the form of flocks of sheep, which help the time controller to carry the rains out of town. The *tiempero* learns the prayers and rites of another local "granicero", or time controller of a neighboring town. The "weathermen" charged for his work, which consisted of walking the area he had to protect and stopping every so often for praying to the rain deities (which could be a Catholic saint, such as S. Isidro) to carry the storm to another place. The last of the *graniceros* in Xiloxoxtla, Tlaxcala, died in the early 1980s.

21.4 CEREMONIES, MYTHS, RITES, AND BELIEFS CONCERNING TRADITIONAL AGRICULTURE

The worldview of human populations was attuned to agricultural systems and the daily life of the peoples who employed those systems over time. There are social and cultural traditions that have to do with what human groups live and perceive on a daily basis or in cycles (e.g., seasonally and annually). Also there is a series of issues that tend to change over time. For example, what is considered a resource, and the ways to relate it, like group norms, ideologies, rites, myths, legends, sacred landscapes, ritual spaces, beliefs, patterns of ritual uses of agricultural spaces, ceremonials to propitiate or stop climate impacts, beings, and characters related to the environment, climate, and disastrous events (floods, volcanisms, landslides, earthquakes, hurricanes, and others).[3]

Worldviews and their associated ceremonies had been very important since pre-Hispanic times, but that did not mean that they were invariable, since they transformed according to circumstances and elements available to face them, likewise. An example is the Chaak-Chaak ceremony,

[3] Albores-Zárate and Broda 1997; Domenici 2010: 387–424; Florescano 2003, 2012; González-Jácome 2004b: 153–163; González-Torres 2008; López-Austin 2012; Márquez-Mireles 1998; Navarrete 1997: 56–61; Ortíz-Díaz 2010: 425–446; Rituales 2006; Servín 2000, 2002.

previously dedicated to the request for rains, except now modified in places such as El Naranjal in Quintana Roo, to request protection from hurricanes (Márquez Mireles 1998). This town was strongly affected by hurricanes, especially Gilberto (1988) which devastated the region (González-Jácome 2004b: 153–163; Konrad 1985: 321–330, 1996: 98–130; Márquez-Mireles 1998). Ideas about the world – and about the local and regional environment – are fundamental, and together with climatic conditions that occur seasonally or throughout the year (droughts, rains, hurricanes, storms, frost, hail, and others) propagate a series of social and cultural behaviors that are aimed at reducing environmental and climatic risk (Alcorn et al. 2006: 600–609; Blanco-Rosas 2006; Hodell et al. 2000: 13–38; Márquez-Mireles 2000).

Family and community are in charge of maintaining solidarity (family and kins mutual aid, and community mutual help), ceremonial, ritualistic, and belief systems along with complementary activities (hunting, fishing, gathering, non-agricultural wage labor, processing, and selling crafts) that provided food and resources to small-scale farmer families (including monetary and those derived from) contributing to risk reduction.

There are some calculations about this. Casas and Caballero (1995: 37), based on information from the studies of Richard MacNeish (1967) and Kent V. Flannery (1986), in the Tehuacán Valley and Oaxaca, calculated that for 1995, 15% of the food system and the diet of small-scale farmers was made up of hunted animals and gathered animals; in addition to the wild plants, they had gathered (Smith Jr. 1969).

An important part of the money earned by family members, domestic animals, and hunted animals is directed to rituals consumption, including agricultural ones. The same occurs with maize and with cultivated or collected plants, which are part of agricultural rituals in peasant communities. Some cultivated plants are used directly in rituals and ceremonies that are important to the rural and urban population like the *cempoasúchil* (*Tagetes*), which give significant profits for those small-scale farmers who cultivate it for commercial purposes.

21.5 MAIZE AND THE WORLDVIEW OF SMALL-SCALE FARMERS

The study of farmer societies in Mexico that relate nature with culture in a diachronic and synchronous perspective allows considering that this relationship changes over time in the same society. That is, human society constantly adjusts to a series of conditions that includes numerous factors. As with the ideology, worldview, rituals and ceremonies that are related to environmental and climatic events, such as negative and surprising impacts on natural resources, agroecosystems and human populations.

These resources are articulated with the ideological forms of use and management that had been created, recreated, known, and developed by the group in question and – in some cases – can be articulated with the market and technology.

The degree of influence that the environment (nature) and culture have in each case is variable, as is the influence of ideological factors in said relationship. It must be taken into account that farmer societies need a natural area where they can obtain resources, the use value of which is essential for their survival, and that if they were to have exchange value, their impact on the family economy would be reinforcing. This is the case of herbalism in sixteenth-century Novo Hispanic society where the forest, moors, lagoon banks, rivers, mountains, and stony ground were places where medicinal plants were obtained that the indigenous society utilized to cure different diseases.

Currently, herbalism is in vogue and provides huge monetary gains for pharmacies, but little is known about its economic importance in local peasant economies (Robles-Cervantes 1998). Some elements from case studies can be compared to place them in the local environmental history, to obtain information about what had happened in rural societies, and to analyze the results. In general, it can be said that supra-local factors are the ones that impact the relationship between nature and culture most drastically. However, there is a wide range of sociocultural mechanisms that cushion or modify these impacts.

Among the local mechanisms, the series of legends, myths, stories, tales, rules of conduct, and ideas shared by a group can be mentioned, in addition to other ideological elements that constitute the group's worldview and that are transmitted from one generation to another through family and local groups socializations. Some of them are specific to certain communities, others are included at the regional level, and others with different components are found and practiced in larger areas as well as have very old origins in several cases, although they have made adjustments at different historical moments.

Maize had been associated with the cosmogony and worldview of the Mesoamerican peoples. From the Classic and Post Classic, there are several representations of maize and its associated deities; anthropomorphized ears or corn were made. Some examples are the god of maize represented in the Madrid Codex (BNAH ff. 28, 68a), the Borgia Codex (f.53), and the murals of the Red temple in Cacaxtla and in the Tancah temple in Quintana Roo. Among the representations where the plant is humanized are the murals of Cacaxtla and the sculptural panel of the Foliated Cross in Palenque, Chiapas, from the Classic Maya era (Navarrete 1997) (Figures 21.1 and 21.2).

The best-known representations are the early Classic polychrome ceramic vessels in Tikal, the stucco figures from Mayapán and Chichén Itzá, which confirm that maize deities were part of daily life and ritual in the Central Highlands as well as in the lowlands and highlands of the Mayan zone. The *Popol Vuh*, a sacred book that narrated maize mythical origins, its close relationship with the creation, and survival of Men, lay down a worldview that was generated around this plant, whose myths and rituals are news today. In the Central Highlands, narrations such as the Legend of the Suns and the maize delivery to humanity show the central position of maize in the cosmos of pre-Hispanic societies.

Added to the list are sculptural representations, codexes, deities associated with maize, stories and legends about the origin of maize, its delivery to humans by cultural heroes. This plant is the axis in the food systems of human populations in Mexico, several countries of the American Continent, and many places in the world where maize arrived. The cosmogony of small-scale farmers had been modified by the Christian religion and their practices, or by the mass media; the reason why, in some way, they had been mixed with pre-Hispanic ideas and practices to form another idea of the universe, which is different from the previous ones (Figure 21.3).

It is here that ideas about the cultural heroes who brought the grain to humans are reorganized and found thereby allowing their existence (Annals of Cuauhtitlán, Codex Chimalpopoca 1945: 1–118, 145–164, and facs). There are several pre-Hispanic legends, about maize origin, agriculture, and its association with gods, such as Quetzalcoatl (Figures 21.4 and 21.5). The legends appeared both in the basin of Mexico and in the Mayan zone, and they included several deities associated with maize; among them are the best known Quetzalcoatl, Chicomecoatl (goddess of what is eaten,

FIGURE 21.1 Palenque, Chiapas: The Foliated Cross.

FIGURE 21.2 Cacaxtla, Tlaxcala: Humanized Maize Plant.

FIGURE 21.3 Dance of the *Tecuanes*, Guerrero: Tiger Mask, for Requesting Good Rains and Harvests.

FIGURE 21.4 Teotihuacan, Tlaloc Carrying on His Back a Basket with Ars of Maize.

FIGURE 21.5 Cacaxtla, Tlaxcala: Mural of the Red Temple, Anthromorphized Maize.

and drunk), Centéotl, Chicomecoatl, Tlaloc, or Xilonen (goddess of the ear of maize who was represented as being male).

In the Mayan area, the best-known representations are the early Classic polychrome ceramic vessels in Tikal and the stucco figures from Mayapan and Chichén, Itzá, that confirm maize deities were part of daily life and rituals in the lowlands and highlands of the Mayan zone. .

21.6 NATURAL RESOURCES, SPECIAL FOODS, CHARACTERS, AGRICULTURAL FESTIVALS, AND CELEBRATIONS 1980–2020

Despite the influence exerted by the mass media, the expansion of modern electronic media, social networks, and the articulation of rural societies with urban societies, small-scale farmers, and village people have their beliefs. Yet, in the fabulous, mysterious, or supernatural linked to agriculture, partaking in social behaviors associated with the conservation of agricultural practices, or with the preservation of natural resources that belong to their communities. In several towns, there are norms or regulations that govern the use of communal resources of forests, lagoons, riverbanks, cultivation areas, or places far from the home plots, the crossroads, and others. The cases presented occurred from the 1970s to date.

The small-scale farmer's worldview is associated with agricultural and social activities, which unite participants through festivals, ceremonies, rituals, dances, requests to deities to receive rain on time or to reduce risks from storms, hurricanes, floods, and other catastrophic events, which can end a good maize harvest. There are celebrations to ask for a good harvest and to give thanks for it, which continue to be important in the daily life of the communities and their inhabitants. Maize as a staple is part of the daily diet in the non-materialistic culture of agricultural and rural societies in Mexico.

The cornfield has been associated with a series of material and ideological elements in many cases, like the food system and the characters that emerged in specific regions and places (Homshuk among the Zoque-Popoluca of the Tuxtlas, and Dhipak among the Teenek of the Huaxtec). The underlying idea of these characters is that Men were created from maize, that the gods gave them its seeds, and together with the signals that nature sent them, they received the knowledge on how and when to sow.

Requests and thanks concerning maize expressed through the holidays and celebrations are added to the agricultural worldview and beliefs that surround the system, contributing to its social and cultural management in agricultural communities (Báez 1983: 383–412, 1992: 303–332; López-Austin 2003: 17–35; Wonderly 1946, 1947).

The belief in supernatural beings associated with agricultural success and the successful survival of small-scale farmers and their families continue in various regions and towns of the country. There are examples in the highlands of Chiapas, in the Huaxtec, Veracruz, Yucatán, Campeche, Quintana Roo, or in the Sierra de Guerrero, Tlaxcala, Sonora, among other places.

21.7 XOPILAPA, VERACRUZ CENTRAL, AND CONSERVATION OF THE ENVIRONMENT

Xopilapa is a small town located in the municipality of Tenampa that is about 114 km south of Huatusco in central Veracruz. The town is part of the hydrological basin of the La Antigua River and is located at 431 masl. Basalt hills and steep hills, ravines, and canyons located on sedimentary rocks surround the small settlement (ranchería). Xopilapa is seat in a ravine that crosses the Xopilapa River, a tributary of the Los Pescados River that joins La Antigua River at the sea level. Xopilapa is near to the Encanto lagoon and a waterfall. The climate is warm and, sub-humid with summer rains; the soils are Vertisols and Leptosols; and there is a forest and a grassland zone in the upper entrance to the ravine in whose lower basin are the river and the town (Prontuario 2009).

Close to Xopilapa is the Cotlamanis, an orographic prominence in whose caves there are vestiges of the Totonac culture. Due to these archaeological remains, the population considers Cotlamanis to be a *mictlán* of the Totonacs, reason why the inhabitants did not allow the looting of archaeological materials in them. Mainly foreigners or outsiders who rappelled and slept in them had looted the caves. In his study on the mentioned town, Servín-Segovia (2000, 2002b) detected the inhabitants, who planned to make a small museum with the sherds that have been found, in the house of one of the settlers.

Servín-Segovia (2000, 2002a, 2002b) detected that the inhabitants believed in the existence of *cheneques* and conceived them as male dwarfs or small elves who live in the water or on the riverbank; they own the fish and take care of the forest and the lagoon. The community, according to its authorities, had established penalties for offenders who hunted or fished outside the authorized enclosures and used unpermitted technology to hunt or fish there. For example, fishing using dynamite or putting poisonous plants in the water to kill fish is prohibited. Penalties included jail, verbal reprimand, or monetary fine (Servín-Segovia 2000, 2002a, 2002b).

In Xopilapa, the forest is cared for by a character known locally as *Juan del Monte*, conceived as a male, who in order to communicate with hunters acquired the voice of their relatives or of known people. He took care that certain wild animals were not killed. They continue to believe in this mythical character and in his punishments. The inhabitants use their belief in him to control hunting in the forest since only what is eaten can be killed.

There is a lagoon called El Encanto where a spirit appears, who is thought to be a fish-shaped light, that takes care that there is no excessive fishing and punishes those who do so. Inhabitants have controlled the authorized fishing methods (Servín-Segovia 2000). A mermaid is the being in charge of caring for the lagoon; she is a woman with a fishtail. Today, inhabitants continue to believe in her – and her punishments. Villagers are allowed to fish only what they are going to eat. The idea of the existence of mermaids in lagoons seems to have come from the Old World. Now it is widespread in Mexico. Mermaids remain in the memory of fishermen and small farmers who fish in the lagoon to complete their food system. When the lagoons of the Lerma River existed in the Toluca Valley, artisans made clay mermaids, which are still made by artisans from Metepec, a town in the State of Mexico, for sale to tourists and those interested in local crafts.

21.8 LOS TUXTLAS IN SOUTHERN VERACRUZ: HOMSHUK, MAIZE, AND SEED SELECTION

In his research on Los Tuxtlas, a southern region of the state of Veracruz Blanco-Rosas (2006), studied the Zoque-Popoluca municipality of Soteapan. This is where Homshuk lives. He is a legendary hero, short (less than 90cm), with maize hair, who aged as maize matured in the cornfield (Elson 1947: 193–214). His person was linked to grain life and received the prayers of the faithful. After the 1960s, he was materially represented as an egg-shaped being. Formerly, he was the maize god. Currently, his main function is to maintain diversity in maize.

Apparently, the myth was related to the conservation of maize seeds and varieties, but it was becoming extinct, so the Jesuits of the Catholic Church revived it with the aim of promoting processes of social and cultural adaptation (development, democracies, and ecological protection). Homshuk was the god of maize in a tale of the Zoque-Popoluca, but the belief in him has extended through the Nahua and Popoluca zones of the Isthmus of Tehuantepec and other regions, although with different names.[4]

The central idea is that he made the domestication of maize and its production possible. In his work, Homshuk was aided by ants and maize guardians, who were in charge of uploading the remains of the cobweb where the proto-men found Homshuk in the form of an egg. This egg resulted from the love affair of a young woman with a musician; she became pregnant, angering the owners of the rain who, as punishment, made the musician disappear. From the egg a child was born, and the angry mother ground it together with the dough and threw it into the water. The remains of it emerge as an egg and the grandparents take it home to eat it, but when they wanted to eat it, the egg hatched and asked the mouse and the bats for help.

The story tells of Homshuk exploits and adventures in surviving and using magical means. The maize was born from the sacrifice, and Homeshuk was immolated the so that the men lived. The Popolucas were "the children of maize". The ants announced the start of the rainy season ("water time"); those that are winged blacken the sky, announcing the end of the drought and the beginning of the sowing of the grass. The land of the anthill is the best place to germinate the seeds of chilies and various plants of the cornfield (Hoppe et al. 1975: 489–505; López-Austin 1992: 263).

21.9 EL NARANJAL, QUINTANA ROO, AND AID AGAINST HURRICANES

El Naranjal, Quintana Roo, is a small community located within El Eden Ecological Reserve. It is located northwest of the Yucatan peninsula, about 90km from the tourist center of Cancun. There, Chaak, is still considered the god of water. The villagers organize a ceremony dedicated to the

4 According to López-Austin (1992: 263), the Tsotsiles call it Kox; the popolucas Homshuk, the Nahuas Sintiopiltzin, in some places is called Tamakustsiin. When it is named in Spanish, it is usually called Maize Saint or Maize Spirit (See: Elson 1947: 193–214; Köhler 2007).

divinity that decided the amount of rainwater that will fall that year, and also when it would rain (Chaak-Chaak). A Mayan priest led the ceremony. This involved searching for a place in the jungle, where a table with food is set up, and a drink made from the balche tree bark is prepared. The priest invokes Chaak to send the rain on time and in sufficient quantities for the cornfields. Then the inhabitants of the Naranjal consumed the food and drank the balche.

The Yucatan peninsula has been ravaged by hurricanes since ancient times (Konrad 1985: 321–330, 1996 (1): 98–130). Since 1988, as a consequence of the devastation by Hurricane Gilbert (category 5) that struck Cozumel on September 14, 1988, the inhabitants of El Naranjal decided that Chaak was a protective deity against hurricanes to take care of the inhabitants and their crops. In the hurricane season, when cyclones threatened to be more dangerous, the region required help; therefore, the date of the deity-related ceremony was changed. In El Naranjal, the people's belief in dwarves has disappeared from local mythology, but they were short men who lived at the entrance of the savanna and were in charge of making sure that an external event or the people did not spread destroying the forest and its fauna.

21.10 ST. MARTA CHENALÓ, AND THE GODS OF THE HILLS IN THE HIGHLANDS OF CHIAPAS

In St. Marta Chenaló, a town in the Highlands of Chiapas, people believe in the existence of 'Anjel, who is a male, and the owner of the mountains. Currently, he is thought to protect the soul of maize (*x'ob*). There is a mass for him in the church and a ritual meal for him in the cornfield, which includes four components: *vok'ich* (a cooked maize without lime, with chili, coriander, and onion), *'ul* (*atole*), beans, and hard-boiled eggs. The gods of the hills are currently seen as "Ladinos". They are the deities who take care of agriculture and hunting among the Tsotsil people. Among their functions are: keeping the soul of crops; sending the rain; and owning game for the hunt. In the communities near S. Cristóbal Las Casas, these deities represent an alternative to beings of indigenous beliefs, which are fused with Christian saints (Köhler 2007).

In Zinacantan, the gods of the hills are conceived as elderly men (*totilme'il*). Köler (2007) interpreted these as a result of local Juarist veterans who, according to popular belief, hid in the hills with their families when their French enemies arrived. For that, they asked the lord of the land for permission. As the Zinacantecos were traveling merchants, it is probable that they were recruited to be part of the Chiapaneca contingent that fought on May 5, 1862, in Puebla. This battalion was formed in late 1861 and had a battalion of 1 thousand soldiers; the Chamula also participated in these events (Trens 1957: 630–633).

21.11 THE MATLALCUEYE, OR MALINCHE VOLCANO WESTERN SLOPES

In Tlaxcala, it is thought that the Malinche, Malitzin, or Matlalcueye volcano was a beautiful woman, with long hair, wearing an ankle-length green skirt. She wore large earrings and *huaraches*. She made sure that people did not deforest the pine and fir forest. The tales and stories related to this mythical character have disappeared, although there are some memories about it. In the 1970s, state government institutions built a mountain camp where high-level athletes were trained. Now a tourist hotel is in use in La Malinche and its wooded area. Forest wardens control access to the forest.

Before the 1970s and 1980s, in St. Isabel Xiloxoxtla, green or blue-green snakes were thought to live in the clouds and carry water from them to the cultivation fields. They were the ones who took care of the rainwater. In the 1970s and 1980s, they were replaced by flocks of sheep (clouds) that carry water to crops (Linares and Bye 1996: 55–56). Weather specialists (*tiemperos*) lead them to lands that require rainwater (González-Jácome 1988: 467–501). In Xiloxoxtla, the people believed in the nahuales also; the male character who became predatory animals such as wolves or coyotes who lived in and cared for the ravines. In the same year, a male character dressed in black, called the evil one, replaced *nahuales*. This character is associated with acts considered bad by the villagers, and not necessarily associated with the ravines. Ideas about *nahuales* existed in Mesoamerican societies, and they are also found in current societies, as in Chiapas (González-Jácome 1988: 467–501; Wonderly 1946).

In St. María Acuitlapilco, a town located on the edge of the federal highway that goes from the city of Tlaxcala to that of Puebla and the ex-lagoon of Acuitlapilco, it is believed that there is a female character there called the mermaid. She is thought to be a woman with a fishtail, who took care of the Acuitlapilco lagoon. With the disappearance of the lagoon, the mermaid became a woman who scares those who approach the area that used to be the lagoon at night, especially nocturnal drunks and thieves (González-Jácome 1988: 467–501).

The local people used to believe in the existence of the *cheneques* (González-Jácome 1988: 467–501), male dwarfs, similar to goblins, who lived in the lagoon and took care of the fish, and the fauna that inhabited it. These characters disappeared from that popular mythology after lagoon dried up.

The town of S. Isidro Buen Suceso is located on the western slopes of the volcano. La Malinche is believed to be a woman who cares for the forest of the volcano. She wears a tattered dress that symbolizes the destruction of the forest. She impedes the excessive felling of the pine-oak-fir tree forest. She arose with the law that turned the volcano into a protected National Park (1980). In the same town, people believe in a male character that appears in the ravines around the volcano. He is the protector of traditional charcoal burners and prevents people from outside the communities from making charcoal. He arose in 1980, when the Malinche forest was turned into a national park (Montero-García 2012; Romero-Contreras 1988).

21.12 THE XINANTECATL VOLCANO IN THE TOLUCA VALLEY IN THE STATE OF MEXICO

In the foothills of the Nevado de Toluca (*Xinantecatl*), the town of Raíces is located. This region is part of the Valley of Toluca. Villagers think that there are elves, which are described as little men who take care of the trees (pines). The myth of their existence continues to this day. Their job is to prevent deforestation. In Raíces, people believed in the existence of "the girl of the volcano", a female character that arose when a disaster occurred, that woman killed a girl. This girl watches over the water in the river where clothes are washed. The myth of her existence continues to this day (Abasolo-Palacios 2003).

There is a strong tradition asking for favors (*peticiones a los Santos*) to mitigate natural events and their impacts on rural communities. The tradition is associated with Catholic saints and deal with agricultural activities, mainly. Sometimes they are mixed with other ideological beings and elements, the origin of which is not discussed in this work, but they continue for today.

21.13 CONSIDERATIONS AND CONCLUSIONS

With the arrival of the Spaniards in Mesoamerica and the catechization processes of the friars and missionaries, some pre-Hispanic rituals, festivals, and ceremonies were modified, created, or recreated, being accepted as part of the new religious culture. Several Catholic festivals were mixed with the indigenous ceremonial and were related to the agricultural cycle: February 2 (Virgin of Candelaria); April 25 (S. Marcos); May 3 (St. Cruz); May 15 (S. Isidro); June 24 (S. Juan); August 15 (The Assumption of the Virgin); September 28 (S. Miguel); October 4 (S. Francisco); October 18 (S.Lucas), and from October 31 to November 2 (All Saints Days, and Day of the Dead). Broda (2018) found that from November 2– 4 another religious ceremony is held at the top of the hills, to thank the storm and the harvest. This ceremony coincides with the celebrations of the dead, where an abundance of vegetable offerings and meals are displayed.

There are ethnic groups that continue to cultivate a great diversity of plants in their fields and that carry out a series of celebrations and rites related to the request for rain and a good harvest. Also there are others to give thanks for the fruits received at the end of the agricultural cycle and to bless the seeds for the next cycle. In general, these small-scale maize farmers (*milperos*) are located in places that are remote from the middle-size cities and regional centers, like in rural towns: Hopelchén or Calakmul, in Campeche; in rural communities in Quintana Roo; in the Totolac river basin and the Tlaxcala Block in the state of Tlaxcala; in rural areas of Highlands of Chiapas, in the Sierra de Juárez, or Sierra Norte of Oaxaca; in the Guerrero Sierra; in the Sierra Negra of Puebla; in the Huaxtec area of Veracruz, Tamaulipas, and Hidalgo estates; in the Sierra Madre Occidental in Nayarit; or in the Mayo River upper communities.

The list is long, and several rural communities in the country have suffered the effects of emigration, a phenomenon, and process that has not been fully studied, especially as to relationship and impacts on agriculture (González-Jácome 1988b, 1988c, 2009c, 2010d, 2011b, 2012a, 2018d; INEGI 2005).

In the worldview of these rural populations, non-material, ideological culture, worldview, and cosmogony had created, recreated, reinforced, and maintained legends, rites, and beliefs related to maize, its origins, and the way of seeing and understanding how the grain is an essential part of the small-scale farmer's world. The colors of tortillas made with maize have different functions, and some of which are used in rituals and special ceremonies. Among the best known are the colored ceremonial tortillas of the state of Hidalgo and the colored tortillas of the Mazahua Ethnic group (Marcelo et al. 2019, No. 19, March–June).

Contemporary studies on rural societies in Mexico show the existence of: (1) a series of beliefs in imaginary characters and beings that live in natural areas (forests, hills, lagoons, ravines, and others), who observe and monitor their use and punish those who abuse their resources by extracting them for purposes that exceed those of the group's subsistence; (2) supernatural beings associated with physiographic elements of the local landscape, such as hills, volcanoes, lagoons, ravines, watercourses, waterfalls, ravines, and the fauna that exist in them; (3) beings that directly intervene in climatic events, such as heavy rain, storms, tornadoes, and hurricanes, and whose presence or lack of it causes the absence of food or the destruction of crops.

To a great extent, communities have been losing their natural resources and the existence of these beings, and the ideology that characterizes them has become mythical and is transformed or disappears. Many of these characters remain in the cultures of the communities but have become elements whose work is to frighten the inhabitants. That is, they have lost their original function as regulators of human activities that affect the environment. In several cases, the characters have acquired new functions. It has been found that people outside the community promote and use some of these myths by adapting them to favor national economic policies or for political purposes.

In several cases, beliefs in the existence of these beings that watch over and regulate the use of natural resources in the communities are reinforced by norms, rules, and regulations accepted by all members of the community. The penalties applied to transgressors are socially regulated by internal norms and, in extreme cases, may mean the departure of the offender from his community of origin or the loss of his rights over the land. By substituting beliefs for socially accepted norms, the ideology associated with these supernatural beings loses strength and may eventually disappear.

This chapter is an incipient effort to collect, describe, and understand the interrelationships between nature and culture, particularly the aspects of ideological culture about the worldview that small-scale farmers have about these interrelationships. Much remains to be studied on these questions. There is no doubt that for a long time they have regulated the relations between Man and his environment. The processes that followed modernization in Mexico in the 1980s have produced three different effects: (1) the disappearance of belief in these characters and other fantastic and mythical beings; (2) the substitution of festivals and ceremonies that are mixed with the Catholic saints and that continue to this day; and (3) the change of attributions and functions of these characters that were created by the peasant cultures of Mexico.

Indigenous or mestizo small-scale farmers are aware of the differences between Catholic saints and their own ceremonies, which have been learned and maintained through many generations. Festivals, ceremonies, and rituals are distinguished clearly, between those that are Catholic and those that come from "custom". Dances, music, and offerings are ways to relate and their inhabitants with those supernatural elements that can help – or not – to reduce the negative impacts caused by environmental, and climatic-risk events, natural disasters, poor harvests, diseases, and other catastrophic occurrences that impact small-scale farmers and their families (Table 21.1).

TABLE 21.1
Daily Life in Mesoamerican and New Spain Times: Worldview, Agricultural Ritual, Ceremonies, and Offerings Associated with the Environment

Period	Place	Data
Late Preclassic to Late Postclassic (After 500 DC) References Aramoni (1992); Domenici (1992); Wonderly (1946, 1947).	El Ocote forest. Caves Media Luna, Cajetes, del Sapo, Apastes, José Juan, Sorpresa, and Laberinto del Diablo.	Utensils and offerings in caves, such as El Ocote, with almost total lack of surface water and arable land limited to small areas near the sinkholes. They are divided into two: one associated with sites; and the other with outdoor locations. The first relevant colonization of the region began in the early Late Classic. Half Moon Cave: 519 black bowls with a white border, stacked in groups, along the front of a stepped platform. The bowls contained ropes, copal, shells, palm thorns, and vessels. Roofs with smoked columns, calcite castings, and niches, or collapsed blocks of the cave ceiling.
Early Classic to Late and Terminal Classic. Reference Domenici (1992).	Outdoors site El Carpintero, right bank of La Venta River.	A rocky shelter remodeled partially, and artificially, at the top of one of the hills on the right bank of the La Venta River. Dozens of smoked Venta bowls were deposited, and a tripod decorated with the face of a supernatural animal from whose mouth a shark-like fin protrudes. Fine orange bowls appear, placed on top of older offerings that included metates and their cylindrical pestles.
Late Postclassic Reference Ortiz-Díaz (2010: 430).	Villa Alta, Oaxaca, site St. Lucía Xaca. Sierra de Juárez, (Sierra Norte de Oaxaca).	Squares and residential terraces at the top of hills. In a cavity in the earth (cista) that it is circular generally and covered with flagstones. In the Xaca square, there was a newborn in a fetal position accompanied by 25 small black stone beads, a green bead, and two eccentrics flints. Among the offerings of a grave was a metate.
1550 DC ±40 Reference Ortiz-Díaz (1992: 431).	Caxonos river basin, Sierra de Juárez, Oaxaca mountain range, S. Francisco Caxonos plain, near the source of the river.	Site formed by a small architectural complex with two mounds that come out of an embankment at the top of the hill. In the center of the plaza, a burial with a looted offering consisting of a bowl with skeletal remains was found. On the fourth terrace to the south, a set of cists and stone boxes was found. In one of the cists was the sculpture of a 45–50-year-old male, in a sitting position, with objects from his wardrobe, turquoise mosaics, ceramics, and a gold pendant representing an eagle knight. In addition, two stone boxes one with ceramic remains and the second with the remains of an infant.

(Continued)

TABLE 21.1 (*Continued*)
Daily Life in Mesoamerican and New Spain Times: Worldview, Agricultural Ritual, Ceremonies, and Offerings Associated with the Environment

Period	Place	Data
1666 Reference Ortiz-Díaz (2010).	Caxonos river basin, Villa Alta, Sierra Juárez, Oaxaca.	Yaguinissi Mountain, in certain festivals part of the townspeople climbed to the top of the mountain and celebrated in front of it ancient worship ritual activities.
Reference Domenici (1992 [1685 File, 1801 File]).	Two caves in Jijipila, Chiapas.	They are sacred sites and were the *nahuales*, and the sorcerers lived.
1700 Reference Ortiz-Díaz (2010)	Caxonos river basin, Villa Alta, Sierra de Juárez, Oaxaca.	In the hills, fields, and houses, the natives carried out pagan acts on a daily basis, not only on the hilltops, but also in the fields, rivers, and even inside their houses.

RECOMMENDED READING

Alcorn, Janis B., Barbara Edmonson, and Cándido Hernández-Vidales. 2006. Thipaak and the Origins of Maize in Northern Mesoamerica. In Staller, J.R., R. Tykot, B. Benz; *Histories of Maize-Multidisciplinary Approaches to the Prehistory, Linguistics, Biogeography, Domestication, and Evolution of Maize*. Elsevier. Academic Press, Amsterdam, Boston, Heidelberg, London, NewYork, Oxford, Paris, San Diego, San Francisco, Singapore, Syney, Tokio; pp. 600–609.

Mesoamerica's northern border passes through southern Tamaulipas and crosses through San Luis Potosí to the west, and to the east, it reaches the present Tampico. The region is in the middle between the Atlantic and Pacific coasts. This area is ecologically diverse, and maize has adapted to it. The grain is a symbol of cultural identity; it is the basis of the food system and plays a very important role in the ritual of the annual agricultural cycles of Mesoamerica.

The earliest evidence of agriculture among the Teenek ethnic group in the Huaxtec is 7000 BC, but permanent towns are dated 500 BC, in the south of the area, and along the Veracruz coast. Pre-agricultural societies depended on the resources obtained from the sea, the river, and the forest for their subsistence (3000 BC–1700 BC). Then agriculture arose and began to be of importance. The conquest of the Teenek began in 1554 and was so brutal that the area was depopulated as the population fled into the forest to avoid slavery.

This text includes stories about maize, conversations about a small-scale farmer's daily life, relationships with other Mesoamerican traditions, and rituals linked with growing maize. It has a discussion about, and conclusion of, the observations made, and it includes an extensive list of references.

Morehart, Christopher T. 1998. *Plants of the Underworld: Ritual Plant Use in Ancient Maya Cave Ceremonies.* FAMSI, Los Angeles County Museum, 2002; pp. 2–27.

Among the ancient Maya, caves were considered sacred areas of the natural landscape, perceived as access points to the underground world, and they were used for ceremonial activities. There are several interpretations about the types of rituals in the caves: agricultural fertility, relationships between the underworld, and the deities associated with rain and agriculture, or political rituals related to social, economic, and political power (López-Arenas 2006: 45–52; López-Austin 2003: 17–35).

The study of the uses in the Mayan caves include: (1) the observations of durable artifacts; (2) otanical remains; and (3) and the investigation of ritual practices in archaeological sites using archaeobotanical databases.

This article discusses the plant remains recovered from ritual contexts based on preliminary data from pale-ethnobotanical research in caves located in western Belize: Actun Nak Beh, Barton Creek Cave, Actun Chapat, and Actun Chechem Ha. The author considers that the degree of proximity of the caves to the sites resulted in differential patterns of the archaeobotanical record, reflecting different ritual activities. The article emphasizes the symbolic nature of the use of plants through ethnographic data of contemporary Mayan plants, and foods used in rituals, because many of them were selected for their specific and associated symbolic elements. Therefore, each has its place due to its cosmological and mythological importance.

22 Some Final Considerations and Conclusions

22.1 INTRODUCTION

It is an impossible task to cover the extensive bibliography on traditional agriculture, and there are many studies about it, its histories, and the discussions and problems that arose around it. This happens: with the site where plant domestication had its origins and its diffusion (maize in particular); the emergence of the first traditional agricultural systems; whether these systems change or not, and how; if the homegarden, and the cornfield arose in Mexico, and how; if the slash and burn is responsible for soil erosion; if the slash and burn cornfield is capable of producing sufficient food for urban populations with high densities.

It is relevant to see if the first direct use of the soil and the alteration of it arose with the advent of agriculture. In the early phase (8700m BP–6000 BP), hunters and gatherers used the products of the land; plants and seeds were collected; and the early domestication of teosinte, squash, and other species was carried out incidentally (Piperno and Pearsall 1998: 312; Smith 1997: 932–934).

Although some millennia later, ways were developed to use natural elements, such as muddy water, remains of flora, insects, and sediments trapped in stirrups (*agualodo*), to fertilize the soils, or to regenerate them. These methods had different names and methods of extraction and application to soils, which still are basic elements in the soils sustainability, and agricultural productivity.

Although they ways of obtaining and uses of muddy water are currently being abandoned, they have been successful through time in their original objectives, such as capturing rainwater in catchment basins, or extracting it from the bottom of canals, and *petenes* in Yucatán.

Among these agricultural practices are the bowls or stirrups (*cajones, estribos*) of Tlaxcala (Mountjoy and Gliessman 1988, Bilbao 1987, González Jácome 2008), the chapines of the chinampas of Xochimilco (Espinosa-García 1994, 1997), the "jollas" (pots, holes) in the edges of artificial terraces in the Mixtec (Guevara et al. 2008: 6–16, Nichols 2012: 1–24), or the extraction and application of muddy water and periphyton obtained in the Mayan wetlands (Fedick et al. 2000: 131–152, Morrison and Cózatl-Manzano 2003, Novelo and Tavera 2003, Tavera and Novelo 2009: 763–769, 2011: 39–45).

Studies of traditional agriculture along with ethnobiological (Ellen 1986, 1999), ethnobotanical (Casas, Torres-Guevara and Parra 2017), and plant economics, or those related to the genome of plants such as maize, enriched knowledge about agriculture in general (Hernández X. 1978, 1981, 1986, 1987a, 1987b, 1988, 1993, 2013b). In particular, traditional agricultural systems that are related to the change produced by humans in different Mesoamerican regions are better known (Palerm 1955, 1972, 1976, 2007; Palerm and Wolf 1961, 1968, 1972). Local systems show the particular adjustments that farmers make to manage particular physiographic, environmental, and weather conditions (Figures 22.1 and 22.2).

DOI: 10.1201/9781003198833-22

FIGURE 22.1 Yucatán: Sowing with cane planter (*coa*) in stony soil, 2008.

FIGURE 22.2 Yucatán: Cornfield (*milpa*) in Stony Soils, Next to Secondary Forest, July 2008.

22.2 BACKGROUND

Since its birth as a formal discipline in the nineteenth century, anthropology had been in charge of the study of human societies and their cultures. Anthropology had emphasized the role of the economic activities that these societies carry out to survive, and the changes they had undergone through time. In capitalist economies, the first studies of human societies were considered as tribal or primitive, because they did not have a state or human populations with high densities living in cities, with a division of labor that kept some human groups away from agricultural work. They had no political or administrative institutions that could be, or be considered, similar to those developed by the urban and industrialized societies of these times.

Early anthropological studies considered the study of the economy, markets, and agricultural systems to be fundamental, particularly those that anthropologists and agroecologists found operating in simple societies (Arellanes and Casas 2011: 93–123; Argueta 2016: 38–46; Arvizu Barrón 2013; González-Jácome 2019; Molina-Luna and Campos 2016). Thus, there are studies, such as that of homegardens, carried out by Malinowski on the Trobriand Islands around 1926. Archaeologists and historians (Hesiod (VIII BC); Xenophon (IV BC) also put their ideas, investigations, and writings on the accumulation of scientific knowledge about the invention of agriculture, the first steps in the formation of agricultural systems, and their physical characteristics, social organization, and development (Childe 1995 [1936]). In the twentieth century, these research interests and those of cultural geographers, botanists, and agronomists also will join.

The invention of agriculture radically transformed Men, human societies, and their culture. Agriculture allowed sedentary life and population increase. In addition, it modified the relationship between humanity and nature. The new technological inventions and new forms of organization of work and of obtaining food, with an emphasis on its consumption, that initiated and shaped other cultural elements, including both material and ideological as well as mechanisms of obtaining it.

For anthropology, the study of the origins of agriculture is important; of cultivated plants, weeds, or those others, which being wild are tolerated and collected, because they are associated with subsistence, herbalism, or the ecological functioning of agricultural systems. The origins of agriculture and its evolution through time are linked to the emergence of a sedentary lifestyle, the formation of the first urban societies, population growth and its densities, and the creation of the state. These facts are part of the questions and concerns of archaeological, historical, and anthropological studies, and the answers that are given as research and knowledge evolve. The emergence of interdisciplinary studies, including data from computing networks, is important in this direction.

Some ancient societies converted subsistence agriculture to intensive agriculture. The role of irrigation in that transformation had been widely discussed, being fundamental in the creation of civilization and State.[1] Intensive agriculture allowed the generation of surplus and the formation of new social groups that were far from direct food production. With this, a new conformation of human societies and their cultures began. This intensive agriculture was based on the production of cereals, which could be preserved and stored for longer periods; besides, they could be transported over long distances.

In those several thousand transformation years, exchange and trade, which were very early activities in human history, had a fundamental role. Among other issues, they allowed the movement of plants from one site to another and experimentation with new crops, in addition to the fact that they encouraged the movement of agricultural products through extensive regions (e.g., cacao from Xoconusco carried to the Mexican Central Highlands). Some productions became luxury items; others were more widely consumed among the populations (e.g., wheat flour, olive oil, and wine in the Mediterranean region).

Peasants cultivated to be self-sufficient in their food system, to pay tribute, to obtain monetary resources, and to exchange and trade. Cultivated plants that are intended for self-sufficiency vary less over time than commercial crops, which adjust cyclically to market conditions. In this sense, the change that occurred cannot be understood, as each emerged agroecosystem has its own history of adjustments to different environments. However, without the existing interrelation between culture and agriculture, and without the study of the process, it is not possible to understand human history or the agroecological history of human societies.

22.3 SMALL-SCALE TRADITIONAL AGRICULTURAL SOCIETIES: FOREIGN IMPACTS

In the 1980s, with the beginning of the neoliberal reform in the Mexican economy, several local impacts occurred in the societies of small-scale farmers in Mexico. In 1982, the government of Miguel De La Madrid instituted the neoliberal model in the national economy. Several support programs to the field and those in charge of supporting peasants' economies were eliminated, although a few of the former were modified. In 1982, the General Rural Credit Law was changed with the

[1] Flannery 1976, 1982; MacNeish 1967, 1981; Palerm 1953, 1955b, 1961, 1972a, 1972b, 1976a, 2007; Sanders 1956, 1968, 1979; Siemens 1978, 1983a, 1983b; Steward 1949, 1955; Turner II 1983; West 1950, 1969; Wilken 1969; Wittfogel 1956.

issuance of the Regulatory Law of the Public Service of Banking and Credit, which was revised on January 14, 1985, modifying the BANRURAL system, whose decree of January 13, 1986, governed its operations and services. Among the institutions that disappeared were IMECAFE and FINASA (ICP) (Mestries e Igartúa 1988: 87–165).

The Rural Bank (Banrural 2003: 6) governors and several economists expressed the opinion that these laws were intended to "[…] initiate a process of change to […] transform the production plant, raise the general level of efficiency of the economy, develop the export sector, and clean up public finances." Neoliberal reform had been discussed by specialists who assure that the years from 1934 to 1982 were characterized by strong state interventionism that led to a growth rate of 6.1%. Calva (2003: 3) considered that after 1983, an economic opening to the exterior and a thinning of the functions of the State in economic development were carried out, generating a decrease in the Gross Domestic Product (GDP), the deterioration of wages, and problems that led the country to the most serious financial crisis it had ever had.

The Solidarity Pact of 1987 caused an "agricultural disaster", which resulted in an excessive increase in food imports from 1982 to 1994. Trade openness and the exchange rate policy lowered the real prices of a large number of products and the profitability of the agricultural sector. At the same time, there was an abrupt suppression and reduction in sector development programs. For Calva (2003: 4), the national program supporting small-scale farmers (*Procampo*) was unable to compensate for the fall in grain prices that occurred from 1993 to 1994.

Along these lines of the national economic framework, it is useful to see case studies where small-scale farmer societies have had to face various problems, among them achieving self-sufficiency and obtaining greater monetary resources. At the same time, they market their products on larger scales than local and regional scales, because competition of their agricultural products with large hoarders and distributors is impossible in practice. These hoarders and distributors have spread throughout rural areas and pay very little and only on established dates by them. Marketers, convenience stores, and supermarkets do not buy from small agricultural producers.

Thus, various economic, social, and cultural phenomena have been generated – or reinforced – and the most shocking are the loss of food sufficiency in the country; drastic changes in the food system promoted the increasing number of industrialized foods consumed; the abandonment of the countryside by young people mainly, who abandon their towns and the agricultural activity; added to the increase in demographic processes such as migration and emigration, originating the departure of some members of the small-scale farmer family members, searching for better-paid waged work in non-agricultural activities or seasonal agricultural work through internal or international migration (legal or illegal) to the United States or Canada (Caloca-Rivas 1999; INEGI 2005).

Traditional agriculture in peasant communities has had to adapt to new circumstances. The range of adjustments includes: (1) the combination of traditional agriculture for self-consumption, with conventional agriculture for the market; (2) traditional agriculture obtaining resources in natural areas (forests, ravines, rivers, lagoons, estuaries, flooded areas, and others) belonging to the towns; (3) traditional agriculture combined with conventional agriculture and natural resource management; (4) traditional agriculture subsidized with migration money; (5) traditional agriculture mixed with small-scale livestock raising; (6) traditional agriculture with small-scale trade; (7) traditional agriculture mixed with salaried activities; (8) the abandonment of traditional agriculture and the urbanization of land used for cultivation previously; and (9) the increase in the female labor force in agricultural tasks that were carried out by men traditionally.

Regardless of the adjustments undertaken by the small-scale farmer society, in the case of those that articulate traditional agriculture with the use of natural resources, it was found that the ideological manifestations of the populations have not disappeared, but like the economy in rural areas and small-scale agriculture, they have been modified. The ideological characters involved have acquired new characteristics or new functions and save some of them that have disappeared. These beings not only contribute to the maintenance of social solidarity, but also avail the conservation of the natural resources of towns and regions where changes have generated poverty and emigration (INEGI 2005).

22.4 SMALL-SCALE FARMER SOCIETIES, CULTURE, AND AGRICULTURE

If one starts from the simple and obvious idea that the study of agriculture, culture, and its forms of interrelation with human societies is a central objective in ecological anthropology, one must consider the existence of various anthropological currents that study agriculture. Each one has a particular perspective, emphasizing different forms of relationship between culture and agriculture, and all of them contribute with information concerning agrarian societies and agroecological studies. Among these trends, ecological anthropology, human ecology, cultural ecology, political ecology, and many other branches and disciplines provide numerous studies, questions, and answers to the relationship, among environment, culture, agriculture, and food system (Mintz 1996).

In each current, the focus on agriculture puts human beings as the central axis, part of a social group, having some type of interrelationship with the environment. However, social groups do not necessarily connect with agricultural studies, ecological theory, or economics. Furthermore, some currents are synchronous, and others are diachronic in their approach, which make the relationship difficult. The combination of culture, agriculture, and the food system can explore and provide important knowledge about interrelationships between humans; their cultures; agriculture with its biological, social, and cultural components; the food system and the health, including human rituals associated with the environment; and the management of environmental, climatic, and economic risks.

The combination of agricultural studies and traditional agricultural studies in ecological anthropology can provide important knowledge about the interrelationships of humans with nature, agriculture, peasant culture, environmental and climate management, management of natural, human, and environmental resources, in addition to their ideology associated with these elements. It would be necessary to consider the management that each peasant culture creates, transforms, and considers regarding different types of risks to which agricultural societies are subject, and the way in which they conceive and confront such risks.

From these studies arises the need to generate interdisciplinary research, which can address the complexity of these traditional agricultural societies and their importance, as a way to obtain basic traditional agricultural knowledge (TAK) for several disciplines and practices such as agroecology, which are focused on the obtaining of sustainability and resilience by human agricultural societies (Agroecology 1973; Altieri 1983, 1991, 1992; Altieri and Nicholls 2012; Gliessman 2015). Traditional Mexican agriculture is a good example of how small-scale farmer societies have been working with different agroecosystems, and that their success through time is the result of diverse practices, management, and interrelationships with the environment and local conditions to which they adapted.

Anthropologists have studied the times when human societies were becoming sedentary, creating new subsistence systems and new forms of sociopolitical organization, of control, with a population increase and the creation of settlements, which gradually became towns, villages, and sometimes cities (Borah and Cook 1974). Along with these origins, cultivated plants appear too, which will feed increasing numbers of people. In this history, agriculture and the role of irrigation in it were closely related to the steps that man was taking to bring about the creation of permanent settlements to form the first States, and what it is now called civilization.

Agriculture, its evolution, and history explain the emergence and development of different forms of social, economic, and political organization, and their relationships with the material and ideological culture of human societies. It can add other questions to the ecological and cultural study of what people cultivate or collect that can be better for the understanding of the mechanisms of adjustment among humanity and society, culture, and nature. These issues include various aspects: food system daily diet, population structure; horizontal population movements; family, local, and regional economy; interrelationships with the market, family health, and disease; local and regional economy, local technology, impacts of the larger society in rural societies and of public policies on rural societies, ideas about the environment and ideological mechanisms for conserving resources, forms of social organization, and land inheritance systems.

Agriculture, its evolution, and history explain the emergence and development of different forms of social, economic, and political organization, and their relationships with the demographic and cultural dynamics of human populations (Borah and Cook 1974). These studies on traditional agriculture, whether or not associated with human, social, and cultural evolution, or carried out synchronously, have been enriched by research that comes from genetics, biology, ethnoecology, ethnobotany, ethno entomology, geography, and new research focused in ethnoagronomy, ethnobiology, and many others in the last decade (Bulmer 1965: 1564–1566).

Stephen R. Gliessman (2015: 289) proposed a list of the elements that traditional agroecosystems have in common, that they: (1) do not depend on purchased external income; (2) make extensive use of available and renewable local resources; (3) emphasize the recycling of nutrients; (4) have beneficial or minimally negative impacts on land and the environment; (5) are adapted or tolerant to local conditions rather than to massive environmental control; (6) take advantage of the full range of micro environmental variations within the land, region, and farming system; (7) optimize productivity without sacrificing the system's capacity and human ability to use resources optimally; (8) maintain spatial and temporal diversity and continuity; (9) preserve the biological and cultural diversity; (9) rely on local cultivars and incorporate wild plants and animals; (10) use production to meet local needs first; (11) are independent of external economic factors, relatively; and (12) are built on the local knowledge and culture of the inhabitants.

In traditional Mexican agroecosystems, there are variations with respect to Gliessman's (2015: 289) proposals, mainly due to several factors: (1) public policies do not favor traditional agriculture, or the small-scale farmers involved in it; (2) small-scale farmers are submerged in the market both for the sale and for the acquisition of goods and services managed by entrepreneurs, agribusinesses, and industries dedicated to the sale of agricultural inputs and food; (3) they lack monetary resources and affordable and low-interest bank loans to develop an agriculture different from that supported by the Mexican State; (4) a significant part of the peasantry and the families compose it, have abandoned agriculture, and migrate or emigrate to non-agricultural remunerated activities or work in the agriculture of other countries; (5) the expansion of agribusiness by arrogating the best agricultural and irrigated lands places traditional agroecosystems in disadvantageous physiographic, environmental, and economical situations.

Hence, Gliessman's proposal on the general characteristics of traditional agroecosystems provides an adequate framework for comparisons among regions, states, and countries of the components that allow achieving a better position with respect to sustainability and resilience.

These comparative studies reinforce agroecology and its foundational characteristic of having inter, multi, or transdisciplinary research that uses factual information from studies carried out in various fields of knowledge that include environmental, climatic, physiographic, social, and cultural components. In this sense, the new concept of agroecology expressed in the words of Stephen R. Gliessman (2013 (8): 19–26) is representative of what traditional agriculture studies have affirmed:

> Agroecology is more than a way to practice agriculture, such as organic or ecological production. Agroecology is also a social movement with a strong ecological base that fosters justice, relationships, access, adaptation, resistance and sustainability. Agroecology aims to unite the social and ecological cultures that helped human society create agriculture for the first time.

If the approaches used in traditional agriculture research are evolutionary, they allow understanding what has happened through time with agriculture and culture (Gliessman 2013 (8): 19–26), and to the adjustment mechanisms created for this, the sustainability of the agroecosystems over time, and the components that make them sustainable. Some authors have handled the idea of co-evolution between culture and agriculture that they are interdependent. It must be considered that the approaches to traditional agriculture are varied and highlight different aspects of rural and national problems.[2] In these different points of view, there is usually a lack of historical understanding of the processes related to traditional agriculture.

To this can be added that peasant families have modified their structure, composition, and organization, to adjust to economic and political changes, and to environmental, climatic, and food system risks. The urbanization of the countryside in Mexico, the direction of its crops toward commercial monocultures, the deforestation of large areas throughout the country, the lack of resources for irrigation, and the eventualities of weatherand rain have caused great agricultural losses to farmers, discouraging young generations from engaging in this activity.

The intergenerational connection that was an important part of the diffusion, recreation, and configuration of local, regional, and national cultures has been broken. A series of motivations, expectations, real, or socially acquired needs regarding consumption have captured the mentality of young people in rural families, especially those in small-scale farmer families. With more studies than their parents and grandparents, they prefer to go out to work in salaried activities that produce more resources than those obtained from hard agricultural work.

22.5 RELATIONS WITH OTHER DISCIPLINES

Relations with researchers of disciplines other than that of one researcher are very useful to carry out the study of traditional agriculture and its interrelations in an interdisciplinary or multidisciplinary way with socioenvironmental agroecosystems, climate, crops, agricultural technology, management, agricultural culture, forms of social organization, technology; work; and relationships with horizontal and vertical mobility.

In traditional agriculture, all of these studies are more useful when they can be done in a model whose unit of analysis includes disciplines that can move from the study of the plot and the family unit to broader systems: locality, region, and trade in agricultural products on national and international scales. This analysis requires approaches whose logical structure is compatible, using common terminologies and collaborating with specialists from other disciplines. It is of utmost importance to strengthen inter-institutional relations in research and studies of mutual problems.

In the last two decades, simple or combined models have emerged to use the information of researchers from different fields of knowledge whose studies are related to the problems being handled. This does not require more than a limited body of researchers who are going to do fieldwork, and who will combine their information with the studies provided by other researchers. That is, using interdisciplinary, inter-institutional, and international networks, or the published materials of studies, and projects that can be obtained through academic networks. The handling of large amounts of information has its own parameters and ways of being used, saving cost in time, money, organization, and administration of large interdisciplinary projects.

A series of new research methodologies have been added to the existing ones. There are cases where the forms to solve problems would be considered "traditional" in scientific research if their solutions were found using the hard and social sciences, which have been neglected for the sake of novelty. It is a fact that fieldwork in situ faces real problems of insecurity and crime in Mexico today. Therefore, the study of traditional agriculture and its indisputable and necessary interrelations with the people who practice it, who live from it, and who face numerous dilemmas, and problems – economic, social, and cultural – require being discussed and readjusted as to its feasible methodologies considering the current conditions in the Mexican countryside.

For small-scale farmers and their families, the undoubted importance of traditional agricultural systems, such as the homegarden, the vegetable garden, or the cornfield, is basic to achieve adequate

[2] In a recent interview that Mexican television broadcast (Channel 14 09/24/18), Antonio Turrent said that "Mexico has an obligation to recover its agriculture, stop importing transgenic", and Alejandro Espinosa said that "we must recover the food sovereignty that begins with the purchase of maize, rice, beans, and others basic products of Mexican food". See also: Toledo (1985).

and healthy food. In countries like Mexico, people who buy non-industrialized agricultural food produced by small-scale farmers have been on the rise in recent decades and have to be considered. It is a fact that organic food has become important, and the number of organic product markets has been growing, although slowly, and mainly in cities. Universities sponsor some organic open-air markets (tianguis).

Small-scale farmers, who are the core of traditional agriculture, have not received in their favor political, legal, or financial support to contribute to their existence. The technical support received by small-scale farmers has been aimed at replacing these traditional farming systems with modern ones, for commercial crops, using agrochemicals, and energy derived from oil.

With the increase of convenience stores, supermarkets, grocery stores, and local markets, the sale of industrialized food is increasing. Thus, directly or indirectly, agrodiversity has been reduced, soil erosion and contamination have increased, water and air quality have deteriorated, and forested areas have been reduced. Álvarez-Buylla and Piñeiro (2013) argued that these phenomena produced disaster situations, like the desertification of large territories, climate change due to their negative effects, and the contamination of soil, water, and air. To be added are the still largely unknown effects of transgenics and their relationships to health and diseases (Esteva and Marielle 2003).

Although there is great – or apparently great – concern about the increase in poverty and the poor diet in high percentages of human populations, the dilemma is that there is no concern about the disappearance of traditional agricultural systems, which for thousands of years have been feeding millions of people on the planet. Several questions arise: Why think that modernizing and being modern is to destroy the past? Why do people not acquire the basic knowledge to think about what is, or is not convenient for human beings, their economy, their health, and their education? Why think that poverty is fought with government handouts or support policies of any other kind that do not protect what humanity has already and that has proved to have historical legitimacy and success?

22.6 SOME REASONS WHY TRADITIONAL AGRICULTURE IS IMPORTANT

1. Traditional agriculture is called that because this term describes the way in which local and regional environmental and agricultural knowledge is transmitted and diffused. Traditional agriculture the reduced quantity and quality of energy used in the agroecosystem, and predominates in the world's agricultural lands with favorable or marginal climates for production. It includes the different worldviews of the human groups that practice it (physical and metaphysical aspects) (Hernández X. 1988: 674). For all this, traditional agriculture has an economic, social, and cultural importance and was practiced in larger agricultural geographical areas of the country (Hernández X. 1988: 674).
2. Governmental policies and urban people's judgments that have been making comparisons between traditional agriculture and conventional or modern agriculture have discouraged studies of traditional technology in Mexico (Hernández X. 1988: 674). For this reason, it is urgent to know the traditional agricultural systems that are being modified, and the causes of their persistence, the social environment in which they operate, and their social function.
3. Most traditional farming systems are based on a diversity of crops that associate and interact in time and space, which makes it possible for small-scale farmers to optimize and secure their crops, even if they have technologies considered to be of backward. Polyculture and agroforestry systems represent the level of plant diversity that has been achieved by small-scale farmers over time and in space (Altieri 1991; Chang 1977; Clawson 1985; Toledo et al. 1985).
4. This situation is the result of their ancient origins and their long permanence through time (biological and cultural coevolution) (Altieri and Anderson 1986; Gliessman et al. 1981; González Jácome 1985: 521–538, 1989: 91–117). Therefore, they represent experiences (trial/error/vicarious learning) (Bandura 1977; Berkes 1993; Moctezuma-Pérez

2016: 79–9) that were accumulated in the interaction between small-scale farmers and their local environments (local knowledge), transmitted generationally, and with none – or scare – access to external inputs (capital and technology driven by oil derivatives instead of human and animal energy). In the same way, traditional ecological knowledge (TEK) and traditional agricultural knowledge (TAK) have been passes on (González-Jácome 2016a: 95–116, 2016b, 2018a, b).

5. Comprehension of cultural and ecological bases of agroecosystems (accumulated traditional ecological and agricultural knowledge) would prompt planners, development agents, and scientists from various disciplines to stop taking actions that harm agricultural land resources, reduce diversity, and negatively affect cultures of the places where they work (Altieri 1991; Klee 1980; Gliessman 2015; Rambo and Sajise 1984; Roades 1984).
6. TAK is being lost as the outcome of the development of modern mechanized agriculture and its technological recommendations that ignore the environmental, cultural, social, and economic heterogeneity of traditional agriculture: traditional agricultural production systems; family non-paid labor; management practices, many of them with ancient origins; the ecological logic behind such practices; technology adapted to the physiographic characteristics of the cultivated lands; diversity in crops; flora and fauna; soils and water resources; the combination of agroecosystems with natural systems; and their vestiges of flora, fauna, soils, and water sources.[3]
7. The characteristics and features of traditional agriculture provided important information to develop more appropriate and sensitive agricultural strategies like agroecology in the face of complex agroecological and socioeconomic processes. This is the basis for the design of technologies that meet the needs of peasant groups and their local agroecosystems. These traits include farmers ability to avoid and manage different risks, folk biological taxonomies adjusted to local and specific conditions, and symbiotic mixtures of crops that make agricultural production more efficient (Altieri 1991).
8. For agroecologists, the following questions are important: (1) the knowledge that small-scale farmers possess about the physical environment (physiographic local and regional characteristics soil, climate, flora, fauna, local, and regional ecosystems); (2) local (folk) taxonomies to classify, including biological and native classification systems; (3) the knowledge that small-scale farmers possess about production practices and management; (4) the experimental nature (folk knowledge) that is the basis of traditional ecological and agricultural knowledge (Chambers 1983; Williams 1980; Williams and Ortiz-Solorio 1981).
9. Local ethnobotanical knowledge of small-scale farmers is related to the use, maintenance, and preservation of areas of "naturalized" ecosystems within, or in areas attached to their properties (e.g., forests, meadows, lakes, slopes, streams, swamps, ravines, land edges, hedges, boundaries, riverside groves with local and wild flora and fauna, and others). From them, collecting of food supplements, construction materials, medicines, organic fertilizers, fuels, religious objects, and others is basic (Bedolla and Vanegas 1990; Gliessman 1983 (9): 991–999; Ibero 1998–1999; Mintz 1965, 1996; Toledo 1980; van der Ploeg 2008).
10. In Mexico, the maintenance of areas surrounding towns that are used for the collection of firewood, plants, animals, and minerals is a phenomenon of pre-Hispanic origin. Currently, as a result of high population densities and land inheritance systems, these areas have been reduced to relics of vegetation and water deposit, canyons, desert areas, and others. Harvesting medicinal and edible weeds and edible mushrooms continues to be an important activity in various parts of the country. Information on the relationships of small-scale

[3] Aeppli and Schoenhals 1973: 15–18; Altieri 1991; Alverson 1984; Conway 1985; Gliessman 1978a, 1978b, 1980, 1981, 1982, 1984, 1987, 1989, 1990, 1998, 2002a, 2002b, 2004, 2010, 2015; González-Jácome 1976, 1984b, 1985, 1986, 1988a, 1988c, 1993, 1995, 1999, 2001a, 2002a, 2002c, 2003b, 2006, 2008a, 2010d, 2011a, 2011b, 2012, 2013a, 2016a, 2016c, 2018a, 2018b.

farmers with their local and regional society, with the larger society, and with the labor market, the quantity, characteristics, and destination of production should be included (Bedolla and Vanegas 1990; Gliessman 2013: 19–26; González-Jácome 2018a–2018c).

11. It is essential to consider the local knowledge (folk knowledge) managed by small-scale farmers, particularly that related to the food self-sufficiency of rural families in a region (Gliessman 1984: 160–171, 2016: 23–35; González-Jácome 1976, 2003, 2008, 2016; Toledo et al. 1985).
12. There are traditional agricultural systems managed by small-scale farmers that are adapted to local environmental conditions, which allow generating sustained yields, satisfying their food system, and fulfilling the rest of their subsistence needs, despite the harsh conditions and the low use of external inputs. This performance is explained, at least partially, by the complex management of apparently simple systems for soil and water conservation, and to the high levels of agrobiodiversity that influence the functions of traditional agroecosystems positively (Altieri and Toledo 2005: 365–385; Gliessman and Rosemeyer 2010; Gliessman 2015: 161–172).
13. It should be added that small-scale farmers maintain a collective memory bank folk knowledge about the environment, climate, physiography, family, kin, their communal social group, and food (Moctezuma-Pérez and Thome Ortiz 2015a: 69–84). All of this is articulated to the economic, social, and labor organization to achieve the successful survival of agricultural activity. This store of local knowledge explains biological and sociocultural adaptation (adjustments), integrating itself with scientific knowledge and the construction of traditional environmental and agricultural knowledge (González-Jácome 1988: 1–73, 1989: 91–117, 1999: 69–91, 2007: 59–95, 2009a: 71–111, 2009b: 179–204, 2010: 141–158, 2016: 233–257, 2018a: 19–22, 2018b: 23–50).
14. For all this, traditional agricultural systems are the optimal scenarios for evaluating the properties of stability and sustainability, thereby obtaining the basic criteria that allow generating, designing, and managing alternative agroecosystems that, according to Gliessman (2013), start from "[…] a deep understanding of what a holistic and ecological vision of the food system can become […]", with the goal of reestablishing the sustainability of food systems (Altieri 1991; Gliessman et al. 1981, 2013: 19–26).

Finally, it can be concluded that maintaining knowledge of traditional agricultural systems is not a superfluous or insignificant matter. It is articulated with the idea that the construction of a successful future for humanity derives, among other issues, from the local, regional, and scientific knowledge acquired over thousands of years by our predecessors, from the successes and failures in the environmental, economic, social, and cultural adjustments that they made at different times, and in many parts of the world, and of the gains and losses that history can teach us about it.

Also, to encourage human contribution to the conservation of biodiversity and cultural diversity, to avoid, or at least reduce, the deterioration of terrestrial and marine ecosystems of anthropogenic impacts that affects the environment and climate negatively. Hence it is essential to seek food systems for self-sufficiency and that provide optimal living standards.

This would have to be in accordance with what the different human societies consider as adequate to meet their basic needs. In this sense, there is the necessity to listening to small-scale farmers and taking into consideration their local knowledge about the use and management of the environment, the impacts of climate, agriculture, and their culture. All of this is important to sustain an agroecology that ensures life in the planet. In this sense, agroecological alternatives will be sustainability mechanisms articulated to the health of people, access to fair markets, better income from the sale of their products, food and health systems, to obtain justice in various aspects of family life (economic, legal, social, educational, and others) and higher quality standards of life.

Traditional agriculture has been a way of seeing the world, knowing it, to live and survive in it successfully. For this, it is necessary to maintain the local/regional ecological (TEK) and agricultural knowledge (TAK) generated over several millennia that are the results of small-scale farmers' agricultural practices and management of small and medium-scale agroecosystems which have been successfully productive. This is a task to be done, as we are still in an incipient stage of stimulating this important practice.

The contributions of researchers in different areas of knowledge, especially those related to the environment, traditional agriculture and interactions between Men and Nature, or between environment, society, and culture, have increased in the last decade. This has been combined with the concepts of traditional, folk, or local knowledge, together with the goal of having important agroecological systems to achieve agri-food, and culturally and ecologically sustainable and resilient systems.

RECOMMENDED READING

Gliessman, R. Stephen. 2015. *Agroecology. The Ecology of Sustainable Food Systems.* CRC Press, Taylor & Francis Group. Section V. The Transition to Sustainability, Chapters: 22, 23.

These chapters deal with the new field known as *sustainability science* in relation to the food system in a complex perspective initiated with the individual farm and directed to impulse their interrelated components toward the obtaining of sustainability. From this point of view, Chapter 22 (Converting to Ecologically Based Management) is directed toward the understanding of small-scale farmers' agricultural practices, which converted farms into being more sustainable.

In Chapter 23 (Indicators of Sustainability), the author proposed a comparative table with data from the properties of natural ecosystems, sustainable agroecosystems, and industrial agroecosystems, including yield production, productivity, diversity, output stability, flexibility, and human displacement of ecological processes. Data is given, also, on resilience to external human inputs, autonomy, interdependence, and sustainability as properties given to these agroecosystems that are applicable to the farm scale in a time frame from short to medium term.

For Gliessman (2015: 288–289), traditional agroecosystems are examples of sustainable function. In the search for sustainability in agroecosystems, indicators and the current meaning of measured progress are discussed; and what are the characteristics that traditional agroecosystems have in common. The author proposed a list of them.

Van der Ploeg, Jan Douwe. 2008. *The New Peasantries: Struggles for Autonomy and Sustainability in an Era of Empire and Globalization.* London, Sterling: Earthscan.

The book written by Jan Douwe Van der Ploeg analyzed the situation, role, and importance of the peasantry in the era of globalization, of agricultural markets, and multinational agri-food industries ("Empire"). Ploeg showed how agricultural families are in a dependency on globalized markets and have adopted forms of resistance, or distancing from the logic of capitalist production through various practices that in the European peasant societies of the 1960s and 1970s were described by anthropology, sociology, and disappeared.

Ploeg considered this phenomenon to be one of the options to the economic, social, food, and ecological crises to which capitalist globalization of markets, and agricultural production systems leads. He considered that there are limits to the transition from the mode of peasant agriculture to the model of agricultural entrepreneur, due to the essential differences between economic theory, and the practices required by the biological, but also social and human nature of agricultural activity.

Ploeg's concept of the "peasant principle" has endogenous technological innovations such as agroecology, and innovations, such as territorial cooperatives in Friesland, peasant seed networks, or citizen markets initiatives that are interdependent", and give visibility to the peasants, contrary to the structures of the "Empire" that keeps the peasants anonymous, invisible, and exploited. Ploeg proposed the concept of the "peasant principle" that has endogenous technological advances that are derived from experience through years, and of numerous human generations involved in their adjustments and advances. The peasant principle is associated with higher income of the peasant traditional agriculture, and from corporate agriculture in terms of resource use efficiency, product quality, quality of life, and social inclusion.

Ploeg considers that peasant agriculture depends on the sustained use of ecological capital and seeks to defend and improve peasant livelihoods. Furthermore, it is multifunctional, and the workforce is familiar or is mobilized within the rural community through reciprocal relationships with the land, and other means of family agricultural production is the basis of the reproduction of the family and the property of the land, and it is marketable, although the small-scale farmers try to be apart from the world and its consumerism.

Glossary

Acahual: Secondary plant community, which arises after the original low forest vegetation was slashed, and burned, and then left to resting for several years, to reach a natural process of regeneration of natural vegetation. Its floristic composition varies according to its state of development in the years of rest and according to its type of soil, humidity, temperature and other ecological and anthropomorphic factors.

Adaptation (sociocultural adaptation): In anthropology, the concept of adaptation refers to the social and cultural aspects that adjust to specific conditions of the environment, climate, changes in landscapes, and others. It is not, as in natural sciences, a phylogenetic process, so it does not necessarily remain fixed in human societies. It is always subject to processes of change that have their own causes, rhythm, and effects.

Sheep adaptation to weather conditions: The sheep in Yucatán had several problems of adaptation to the sweetly climate, which is why they did not flourish. However, their introduction was successful in temperate-cold places, such as Tlaxcala, Texcoco, and Querétaro (González-Jácome 1989b: 86–107, 1991, 2008b: 9–24).

A few years after merino sheep were introduced in those places, the structure and size of their wool changed from long to short, reducing the quality of garments woven with it. In Tlaxcala, the textile industry in the Viceroyalty and, later that of the nineteenth century, had problems when English looms replaced pedal looms because they were adapted to weaving the long wool of European sheep, and not short wool. Novo Hispanic wool textiles with short pile did not reach the quality of the European wool textiles, so only coarse fabrics were weaved for immediate use, but their price in the market was lower. This favored the market for fine European fabrics that entered the country through New Spain ports and via piracy (González-Jácome 1989b: 86–107, 1991, 2008b: 9–24).

Amaranthus: *Amaranthus cruentus* was grown in Mesoamerica, while *Amaranthus hypochondriacus* was cultivated in the Andes. The Spanish took the plant to Europe and used it for human food and fodder. In the New World, it was used as a dye to make religious objects intended for rituals and ceremonies. The Mendocino Codex shows how the Tenochcas (Mexicas, Aztecs) collected it as a tribute, as it was one of the four most important crops at that time. It is grown, currently, from the United States to Argentina (Early 1992: 17–19).

Agave: The center of origin and diversity of Agave is Mexico, where 75% of the species of this genus grow and 57% are endemic (García-Mendoza 2011; Gentry 1982). Also, the center of diversification of this product of human selection and management is Mexico, where the first farmers generated their genetic diversity (Colunga-García-Marín et al. 2017).

According to Rzedowski (1978), some areas were located in microenvironments where humidity conditions could be lower, like the Zapotitlán valley or more Southern Caltepec. Both zones were covered with xerophyte scrub. The region is the center of evolution of this type of flora, the Agave species, like the columnar cactus, which reaches 15 m in height. Earl Smith Jr. (1967: 133–142) found that 29.1% of the 253 species collected in the valley are endemic, with at least three genera of phanerogams, which are exclusive to this arid zone.

Agricultural evolution: The concept of the evolution of agriculture refers to the changes that occurred during thousands of years, from the Late Pleistocene and Holocene until 2000 BC. That is, since the arrival of the first settlers in Mexico, the beginning of agriculture went through the processes of domestication and dissemination of plants (maize, beans, squash, Mexican hog plum, gourds, and others). In this period, the first agricultural systems arose incidentally, the secondary result of the consumption of certain plants that were of the taste of human palates.

Agroecology: This is defined by Stephen R. Gliessman (2013 (8): 19–26) as "[...] more than a way to practice agriculture, such as organic or ecological production. Agroecology is also a social movement with a strong ecological base that fosters justice, relationships, access, adaptation, resistance and sustainability. Agroecology aims to unite the social and ecological cultures that helped human society create agriculture for the first time".

Agroecosystem: In this book, they are mentioned:

Cornfield (*milpa*). Agroecosystem where maize is the central axis crop. It is mainly a polyculture.

Chinampa. Artificially built agroecosystem on the banks of lagoon systems. It combines artificial islands, which are rectangular in shape and intermixed with a system of canals through which water circulates. The roots of trees (*Salix*) planted in the banks of the platforms protected the soil from water and wind erosion. Cornfields were established on the platforms, and with the extration of muddy water from the canals, peasants formed seedlings (*chapines*) which produced three crops a year.

Homegarden. Polyculture agroecosystem with fruit, shade, timber, cultivated, and wild trees and other plants. In Mexico, it can produce maize, vegetables, aromatic herbs, medicinal herbs, flowers, and Opuntia. The homegarden may include cultivated and wild flora.

March Maize (*Marceño*). Agroecosystem adapted from waterlogged lands during several months of the year, in which short-cycle maize is planted.

Orchard (Fruit tree agroecosystem). Flower garden (cultivation of flowers); Garden (cultivation of vegetables).

Irrigated land. Land with irrigation from natural means or by artificial systems.

Metepantle, metepantli. Row of Agaves that divide the strips for cultivation. Agaves are sown on land edges. They may or may not have an annex canal to retain rainwater, and they are intermixed with other plants. *Metepantles* are found mainly in hilly areas.

Semi-terrace (*bancal*) developed on the low slopes of hills that descend to the plains. The land is divided into strips for the cultivation of wheat and/or barley, and the dividing lines are rows of Agave that are dedicated to the production of fermented Agave mead. In the Apam area, in the states of Hidalgo, and Tlaxcala, the rows of Agave are for an Agave a monoculture, and there is no land for cultivation between them); and terraces, which is the agroecosystem developed on hill slopes using a staggered combination of cultivation platforms. On the banks of the platforms, there may be a system of *metepantles* that are the land edges, which on the top have Agave, fruit trees, and wild weeds that prevent the ground from sliding down toward the lower platforms of the agroecosystem.

Aldeas: In northwest Spain, they are sets of houses of the family members plus their annexed lands.

Angostura project: The results of this hydroelectric project in Chiapas were published in 1993, in the book by Ángel Palerm. *Regional Planning and Agrarian Reform*, edited by Gernika, and the Iberoamericana University. Participating in the project. Carlota Díez-Loredo, Shoko Doode, Raúl Gómez-Montero, Bolivar Hernández, Virginia Molina-Ludy, Hugo Trejo, and Vicente Villanueva, who was do the fieldwork coordinator during the development of the Angostura Project. The fieldwork lasted seven months, and then the *Report of the Sociocultural Aspects of the Population Affected by the Angostura Project. Study and Recommendations* was delivered to the Federal Electricity Commission (CFE). The materials obtained in the field were analyzed in a seminar of "Applied Anthropology", organized by Palerm at the Iberoamericana University.

Annual flood: In central Veracruz state, from the town of Cardel to the Laguna Verde region, during the 1950s–1960s, the annual flood was managed by villagers through the construction of houses with high two-gabled roofs, where they hung from the rafters, baskets, and recipients of different sizes made with tule (*jabas*), where fowl, small pigs, and domestic animals were kept, in addition to their clothes and household goods. The cattle owners of the region took herds to higher areas on their ranches, where the feeders were. Maize from the previous crop was deposited in special rooms and inside large hermetically sealed metal cylinders, which were built on cement piles.

Atole: Drink made with maize flour diluted in water or milk, seasoned with some flavor. It can be sweet, salty, or without flavor. There are types of medicinal atole (González-Jácome 2018, No. 60: 9–69).

Avocado and its domestication process: The avocado ancestors arrived from the Sierra Nevada, in California, to the Mexican Highlands, where the tropical climate prevailed until the Tertiary. In those regions 15 thousand years ago the cold weather contributed to the extinction of avocado trees there. The plant migrated south, through the Sierra Madre Occidental, Durango, and Sonora.

Toward 10 thousand BC, the climate in central Mexico was colder than it is today, and the Tehuacán Valley was adequate for adaptation of the avocado. The inhabitants of the valley had consumed it since they were semi-nomadic and began the process of selection and care of the avocado in the forests, thus beginning its domestication. In the Mayan area, the fruits of trees such as avocado, Mexican hog plum, and sapodilla had been consumed for 3400 years (Colunga and Zizumbo 2004; Galindo-Tovar et al. 2013: 65–70; Millar 1996; Schroeder 1968).

The domestication process of the avocado in the Valley of Tehuacán was long, initiated in 6500 BC with its collection in the forest and the selection of its best fruits. The avocado began to be sown near the houses when the weather changed, a variability that began unintentional modifications of the biophysical environment when the selected avocado trees (e.g., larger seeds, a smell of anise) were taken to more favorable habitats (Galindo-Tovar et al. 2013: 67; MacNeish 1964). Avocado domestication was a process that culminated until 900 BC (Smith 1966, 1969).

Bands: In archaeology and anthropology, a band is a human group formed from 10 to 30 related and unrelated people. These people were organized to live together and carry out survival activities, including feeding group members. They moved through areas where they found food, they were hunter-gatherers, and the bands were itinerant and moved through a territory seasonally. They could grow and have more than 30 members over time they increased in number as they became sedentary and learned to cultivate the land. The bands settled in camps near rivers and water sources. Benz and Long (2000: 459–465) Consider that together with the sedentary lifestyle and the first permanent settlements from 10000 BC–9000 BC to 4000 BC, bands took about 6000–7000 years to establish themselves, and to disperse.

Banana: In Mexico, this includes Tabasco and Dominico varieties that are for consumption as table fruit. The Macho variety is cooked, fried, and used to make sweets. Tabasco, Macho, and Dominico bananas are sold in tianguis, municipal markets, and supermarkets, which occasionally have the Purple or Roatan varieties. The banana varieties Valery, Pear, Cavendish Giant (giant Dwarf), and Manzano are found in markets sporadically, and they are unknown outside their cultivation sites (SAGARPA, SIAT n/d).

Banquetera: An agroecosystem where the cultivation of coffee trees and mango trees is done on hillsides of ravines with rugged orography using the small flat patches in the forested areas, where these microzones have deep, fertile soils suitable for agriculture. These patches are between the forest slopes of La Antigua River in Central Veracruz. In this agroecosystem, the production of the "banquetera" is commercial, and the small-scale farmers sell the coffee in the city of Huatusco and the mangos on the roadside (Jorge Servín-Segovia 2000, 2002).

Besana: The first furrow opened in the ground when the small-scale farmer starts plowing. Besana means the land to be ploughed.

Bimbalete, cigoñal: It is a simple machine, a lever that serves to raise water from a river, canal, or a well that captures rainwater. This water is used for irrigation and domestic and animal uses (Palerm-Viqueira el al. 2002).

***Brosimum alicastrum* (Ramón):** The fruit of this tree was used to make flour for tortillas. This food was eaten in times of famine. This flour is still made, and tortillas and cupcakes are made with it currently in places like Ocosingo, a town in the state of Chiapas.

Cacahuatole: It is a popular drink of cacao mixed with sugar and cinnamon; the mixture is made by hand and then whipped until a thick foam is obtained, which is what is drunk. Cacahuatole ("agua de barranca") is served in containers made with halves of the fruit (gourds) of the Crescentia alata tree; the gourds are painted on the outside with bright colors, such as red. It is sold in the markets of Tlaxcala and Cholula.

Cajetes, estribos, trenches, bowls, or stirrups, are canals with ground divisions: They were systems located in high areas of the hills that served to capture rainwater and control its downward flow to irrigate the fields. Muddy water (*agualodo*) is formed throughout the year in these canals which until the 1980's served to fertilize or restore eroded soils, being applied annually.

Carnival masks: From the Erythrina trunk, masks were made for dancers to wear at carnival celebrations. In the 1970s, Erythrina trees were preserved within the boundaries of the farmers' homes, and on borders of the drained fields in the Atlamajac neighborhood, which at that time was administrative and depended on the Tepeyanco municipality, in Tlaxcala.

Carrying capacity: The carrying capacity of a biological species in an environment is the maximum population size that the environment can support indefinitely in a given period, taking into account the food, water, habitat, and other necessary elements available in that environment.

Centers of origin of agriculture: The centers of origin of most of the current cultivated plants were in the botanical areas or regions where powerful type formation processes were active. The concept was developed according to Darwin's, and it is defined as a space where an origin or the differentiation of a particular species or population occurred. These species or populations when dispersed – or differentiated – by different phenomena (migrations, geographic or reproductive isolation, or mutations) gave rise to other species or forms, and so on. For example, the center of origin of pine is China, but its diversification centers were Mexico, Guatemala, and Honduras. Vavilov affirms this, because of 111 total species of pines in the world, 49 (44%) of them are found in Mexico.

Chilaquiles: They are small toasted or fries tortillas triangles garnished with a chili sauce, cream, and onion slices.

Chinampa: The chinampa agroecosystem was an intensive traditional agricultural system that emerged in the basin of Mexico before the arrival of the Spanish in this region. Its probable antiquity is the ninth century (Parson 1982). The chinampa is an artificial island formed in a lake, originally with a rectangular surface (platform) surrounded by canals. The base of the chinampa was stakes placed in a rectangular shape and filled in with branches, stones, earth, foliage, dry grass, and muddy water (*agualodo*) from the bottom of the canals. Platforms were located in a network of canals that surrounded the rectangular plots, which were about 100 m long and 20 m wide on average (Armillas 1971: 653–661; Gibson 1964: 300–334).

To achieve all three crops, small-scale farmers (*chinamperos*) used seedlings planted in soil in the form of buckets made of muddy water (*agualodo*) obtained from the canals, which besides being an important nutrient, prevented salinization of the soils and controlled the appearance of harmful microfauna in the soils and plants (García-Espinosa 1978, 1994, and 2007): The chinampa gave three crops a year.

Originally the ethnic groups the Chalcas, Xochimilcan, and Culhuas built the chinampas, but when the Mexica arrived in the basin, the chinampas reached their maximum expansion and splendor. Irrigation was by infiltration of the water from canals that surrounded platforms (Palerm 1973). The chinampa was used for the cornfield agroecosystem, which was surrounded by Salix that supported the soil with their roots and allowed the passage of light to the plot.

Chocolate: Henderson et al. (2007, 104, No. 48: 18937–18940) founded that cacao was grown in the Late Preclassic to made a beverage where the mucilage was extracted from the cacao pod, mixed with water, and fermented. Subsequently, chocolate emerged, extracted from the cacao pods, cacao beans were dried, roasted, and ground to mix with water, sweeten with honey, or maguey honey, and add aromatic flowers. Currently, chocolate is made and later mixed with water, milk, atole, or ground maize : In this last case, local drinks are produced, such as posol or tescalate, among others.

Chorote: In the state of Hidalgo, Mexico, *chorote* is a fermented beverage that is consumed still. It is a drink made with prickly-pear juice fermented in water.

Chultun: A "chultun" (*chultunob*' or *chultuns*) is a bottle-shaped cistern used as an underground storage chamber to stock rainwater or grains. It is built where a natural sinkhole had existed (e.g., the *Puuc* hills, which sit hundreds of feet above the Yucatán Peninsula aquifer). It has a small circular mouth where rainwater penetrates after the first rains washed the *chultun*. In some archeological sites, like Tenam Puente, it is located in the lower part of a pyramid or in the center of the stone-covered square.

Collecting crops by boats: Collecting crops by boats is an activity in wetland areas that are carried out from the United States to South America, at least. In 1905, the Minnesota Historical Society photographed the harvests by boats on the Cuchillo, Mouse rivers, and Lake Devils of the Hidatsa ("the people of the willows") of North Dakota.

Commercial routes: (1) This route runs from the center of Tlaxcala to the Gulf Coast; (2) from the center of Tlaxcala to Atlixco; and (3) from Tlaxcala to Puebla, and Mexico Cities. In the nineteenth century, there were two main commercial routes, one to the north crossing through the Sierra de Puebla to reach Papantla and the Gulf coast. The other went south to Puebla and Atlixco. In 1930, with the opening of roads and the dawn motor vehicle system, the route from Puebla to Mexico City became of great importance.
Data from interviews with Jacinto Zacapa Rugerio (1970–1972), who in his youth and in his senior years, had been a muleteer trading with ironworks: Zacapa traveled two routes, one toward the Sierra Norte de Puebla and another toward Puebla and Atlixco.

Congregations (*congregaciones*): The Spanish Crown in the middle of the sixteenth century established a resettlement policy for indigenous populations. In theory, Bishop Landa in Yucatan considered that Congregations would prevent Indians to live apart from each other by mountains, because the places where they live presently are no the old ones and friars need to have them together for doctrine and Mass.

Bishop Landa ordered: "Therefore, I command that all the natives of this province come together in their villages and make houses together, drawn in the form of villages all of a partiality, and header in a comfortable, and convenient place, and make their houses of stones, and of lasting work, each neighbor house in itself, within the trace that is given to him, and that some crops are not planted within the town, but that they cut everything, but it would be some fruit trees, [...]." (Ordinances of Tomás López (1552–1553), Appendix Document No. 8; Landa 1983 [1566]: 220).

The place where the Indians used to live were abandoned and the friars rearranged them and brought together several settlements in a single town, for doctrine. This was done with great zeal because the land is mountainous and the friars cannot go to so many places, that is why they have the most concerted Indians, and when they go to indoctrinate them, or to say mass they find the majority gathered (Relations of Cacalchén, Yaxa, and Sihunchen: TI-339).

Convent homegarden: Some plants, technologies, and elements of agroecosystems arrived through the convents, from where they spread to the New Spain villages to mix with the native homegarden and generate new crops that modified the food system. Although the structure of the homegarden remained based on trees and their combinations with other plant strata, the introduction of trees from the Old World increased homegarden diversity. In the north of the country, the role of the Jesuit missions was key in relation to agriculture and the introduction of plants for the food system. The knowledge that the northern groups had of the forest and herbalism was extremely broad. By 1764, the Jesuit Juan Nentvig (1764:

83–97) had compiled a list of 43 medicinal plants used by the Sonoran Ópatas (González-Jácome 2018b: 39–48).

Coral bean tree (*Colorín*), (*Erythrina coralloides*): This tree lives in thickets and dry forests, and oak forest from 1,180 to 1,900 masl. It is used as a living fence for shade and for craft making. Coral bean flowers are edible and have many medicinal uses. Their flowers are highly prized in stews, whether boiled or weathered in sauce, and their petals are used as toys for children:

Cosmogony: A mythical story that seeks to explain the origin and the evolution of the Universe, especially the solar system and the world. It is also a scientific theory dealing with the origin and evolution of the universe.

Change processes: I use the concept of change processes for those that occurred in agriculture and in the first agroecosystems for periods that span hundreds of years. These change processes included the presence of a wide variety of domesticated plants; the inhabitants who abandoned the seminomadic life and became sedentary; the ability to subsist on foods derived more and more from agriculture. This period began in the Formative or Preclassic period, around 2000 BC ±100 AD, continuing with the Classic (±100 AD to 650 AD), Epiclassic (±650 AD to 900 AD), Post Classic, and Historical (±900 AD to 1519), the arrival of the Spaniards in Mesoamerica, the Viceroyalty, the nineteenth and twentieth centuries, and the present century.

Chronology: Mesoamerica and Mexican Historical Periods: (1) Early Preclassic or Early Formative 2500 BC–1000 BC; (2) Middle Preclassic or Middle Formative 1000 BC–400 BC; (3) Late Preclassic or Late Formative 400 BC–AD 200; (4) Early Classic AD 200–600; (5) Late Classic AD 600–900; (6) Early Postclassic AD 900–1200; (7) Late Postclassic AD 1200–1492; (8) The New Spain Viceroyalty (Colonial Period), 1521–1821; (9) Nineteenth Century 1800–1899 (The Independence from Spain; The 1rst Empire; The 1rst Mexican Republic; The Reforma; The Second Empire; The Restored Republic, and The Porfiriato); (10) Twentieth Century, 1900–2000 (The Mexican Revolution; The Agrarian Distribution, and the Creation of Granted Lands for the peasantry (*ejido*); The Formation of the Post-revolutionary State; The First Post-Revolutionary Governments, and the Creation of Political Parties in Mexico; The Formation of Post-revolutionary State, and the 6-years Governmental Periods (Staller, and Carrasco 2010: 4).

Cross pollination: Cross-pollination occurs when pollen is transported from one plant to another through an external vector (biotic or abiotic). There is cross-pollination when insects (bees, butterflies, wasps), birds (hummingbirds), or bats, water, or wind intervene.

Cross pollination within the Mayan area: Almost all of the taxonomic categories of plants used as disturbance and deforestation indicators include species used by the Maya, which were resources for their subsistence probably. Pollen registration has serious limitations when trying to evaluate the contribution forest management and polyculture in homegardens made to the livelihoods of the ancient Maya. Many native food plants used by the Maya would be absent in the pollen register of sediment cores. Virtually all pollen has representation in wind-borne sediment cores, while animals pollinate most species of food plants, used by the Maya. Ford and Nigh (2014), using data from 18 homegardens in the Maya-cultivated forest of central-western Belize, confirm this. Another study by Fedick and Ford (1990) estimates that of the 20 dominant forest and crop species in homegardens, only the *Brosimum alicastrum* was pollinated by the wind, while the others were pollinated by bats, beetles, bees, moths, and other insects. Of the 37 species of plants grown in the homegardens, only one of seven species has been pollinated by the wind (Fedick 2010).

Culture and agriculture: Culture has been observed and recorded as an important element by agronomist Efraím Hernández X. in his book on *Agricultural Biology* (1985). He considered that the cultivation of a wild species requires the modification of the genetic scheme that resulted from processes of natural selection to a scheme adapted by human management and its purposes. Therefore, the manipulation of genotypes is a cultural activity, where inhabitants adjusted biological diversity to their needs (artificial selection).

Culture and cultural ecology: Julian H. Steward defined the concept of culture in Cultural Ecology (Steward 2014 [1955]: 58–60) as a construct that is defined in each study or investigation based on its questions, objectives, and explanatory needs. Culture is not static; it is adaptable and modifiable in relation to physical conditions, and adaptation results from discoveries and inventions that come from abroad.

Culture is a heuristic concept that can be applied to any science in order to elaborate means, principles, rules, or strategies to help find the most effective and efficient solution to the problem analyzed by the individual. Therefore, culture is a flexible concept, which can be understood as a mechanism of adjustment to a specific environment. However, different cultures can be found in similar environments are related to factors such as historical development, their relationships, and interactions with other human groups, or their relationships with federal, national, and even international political-administrative institutions.

Between the physical environment and culture, there are cultural patterns: These cultural patterns result from the interrelationship between the physical environment and human activity (like the collection of goals, values, sets of beliefs, and knowledge).

Cultural disease: In medicine and medical anthropology, cultural disease is a culture-bound syndrome, culture-specific syndrome, or folk illness. It is a combination of psychiatric and somatic symptoms that are considered to be a recognizable disease only within a specific society and its culture. Among these cultural diseases, there are the folk concepts of cold/hot, evil eye, pinwheel, fontanel pulled down, and others.

Cultural ecology: In Mexico, this anthropological theory and method applies to the study of simple and complex human societies, adding to them diachronic and synchronic dimensions, and the study of certain national problems that have called the attention of specialists of numerous disciplines. It has been interdisciplinary availing academic networks to obtain information related to different research problems. It is historical, ecological, anthropological, and ethnological. It includes the knowledge of various branches of natural sciences and social sciences, articulating it with local knowledge through fieldwork *in situ* (González-Jácome 1997: 167–188).

Cultural ecology is a methodological theoretical approach to ecological anthropology that is diachronic, synchronic, and interdisciplinary. It was initially proposed by the American anthropologist Julian H. Steward (1955, 2006: 5–9), who, in 1937, was appointed advisor in anthropology of the Office of Indian Affairs in the USA, and who, in April 1940, participated in the meeting in Patzcuaro, Mexico, where the Inter-American Indian Institute (III) was created.

Development program (Oaxaca): In the Oaxaca development project directed by Silvia del Amo Rodriguez, a person who led the cooperative, and he aspired to a position in regional politics and used the project's funds to acquire a vehicle intended for personal mobilization. This unleashed problems and lawsuits within the members of the cooperative, and the project that was successful, initially, in the end was finally unfinished.

Diversity: Polyculture and agroforestry systems represent the level of plant diversity achieved by small-scale farmers over time and in space (Altieri 1991; Chang 1977; Clawson 1985; Toledo et al. 1985). Altieri 1991, Chang 1977, Edens and Haynes 1982, and Toledo et al. (1985) called the obtaining of diverse products that resulted from diversity as "multiple use strategy". These researches attained with "traditional management" (Altieri and Merrick 1987) were based on local knowledge.

In a recent interview that Mexican television broadcast (Channel 14 09/24/18), Armando Bartra said, "The cornfield strategy is equal to promoting biological, and cultural diversity".

Diachronic methodology: It is applied to the study of evolution and the processes of change through which agriculture has passed. Everything diachronic is studied and analyzed in situ, currently. Traditional agriculture systems are found in many parts of the world, still, including the different types and states of reduction, disintegration, or change to other types of agriculture, including those that combine with capitalist and neoliberal economic developments. Two examples are France and Spain, where traditional, small-scale agriculture, and livestock raising, or permaculture, have become excellent businesses. There are also changes to become organic, ecological, biological, or agroecological agriculture.

Domestication: Casas et al (2017: 86, 88) define domestication as. "[…] process [by which], human beings modify the forms, and functions of organisms to meet their needs, especially food, for example by medium of larger and sweeter fruits, but also aesthetic, utilitarian, and medicinal needs, mainly by artificial selection of the most attractive individuals. In addition to direct selection, the management of phenotypic, and genetic variability present in populations leads to the action (deliberate or not) of other evolutionary forces such as gene flow, gene drift, and inbreeding, since domestication is commonly accompanied by changes in the effective size of the populations, their reproductive biology, and their dispersal capacities. The mechanisms, and criteria with which domestication operates are deeply linked to culture, which is why domestication is an eminently bicultural research problem […]. It is an evolutionary process that involves diversification".

For the domestication of plants, and animals, in the Americas, see the book compiled by Casas, Torres-Guevara and Parra-Rondinel (2017).

Domestication index: Gentry (1982. 276) said that the index of domestication is equal to the set of traits or characteristics that differentiate domesticated plants from their wild ancestors. These traits, when subjected to similar selective forces, are an expression of evolutionary convergence.

Domestication syndrome: Bautista-Lozada et al (2012: 253–267) used the "Domestication Syndrome" as a methodological tool for the study of evolutionary changes resulting from natural selection, and artificial selection. The latter changed the frequency of the genetic and phenotypic characters of the populations, and their divergences from their wild ancestors. Artificial selection and agronomic management modified the balance in susceptibility owing to attack by herbivores vs. productivity and palatability. There are studies showing that plant domestication status is associated with the degree of chemical resistance to herbivorous insects. The level of divergence between cultivated plants and their wild relatives depends on agronomic, traditional, and industrial management that meets different types of needs and artificial selection criteria that differ. Traditional agronomic management in the domestication of species depends on the cultural context of the local environment. Industrial agronomic management solves specific problems of the relations between the plant, and its biotic or abiotic environment. Studies on the variation of traits of resistance to herbivores or pathogens should focus on how and what components of the domestication process (including the anthropic environment in which the species develop) influence the secondary metabolites diversity and their ecological consequences.

Drained fields: They are an intensive subsistence farming system developed where the swampy environment is deliberately modified for agricultural use. They are built on poorly drained soils with a high water table. In the pre-Hispanic era, drained fields in central Veracruz were used for the cultivation of cacao and other crops (Palerm 1954: 2–15).

In the southwest of Tlaxcala, the streambeds are full of sand, so there is a constant threat of flooding and deposition. For this, the Tlaxcalan small-scale farmers had sophisticated practices to conserve and recover the soils. In the south and southeast of Mexico, drained fields are permanently flooded or for several months of the year. In central Veracruz, they received torrential rains and annual river floods, but in the summer the flooded area is reduced (Wilken 1969, Vol. LIX, No. 2: 215–241).

They are made up of a combination of platforms and canals. The maintenance of the canals, ditches, platforms, banks, and earth borders that surround the fields is carried out once a year in

December. These systems are highly productive; crops are rotated in the plots; maize along with beans and squash are sown from February to April; and maize is harvested from August to October; beans are harvested from June to August and squash from April to June. Onions are grown from September to November and will be harvested from November to December. Also from November to December lettuce, radishes, and coriander are sown, and they would be harvested from January to February next year (Wilken 1969, Vol. LIX, No. 2: 215–241).

Ecological floor (*Piso ecológico*): Ecological floor or thermal floor is that which has a level of relief above sea level, which allows the modification of characteristics and climatic conditions of a geographical area. It is known as thermal floor or biotic floor, according to the criteria used in its application. The modifications result according to altitude, relief, weather, and temperature.

Ecological floor is a term coined by John Murra (2009), after his research in Peru, and is a concept used in anthropology where the altitude above sea level is an explanatory factor of the location of the settlements, their historical, ecological, economic, social organization, and politics. In 1969–1972, Murra considered and applied the concept of ecological floor to the number of vertically ordered ecological floors and the resources they offered in the Andean cultural area. The population exchanges potatoes, maize, and vegetables for products of the intermediate zone.

Ecological mosaic: The ecological mosaic is a set of habitats formed and maintained as a result of altitude, climate, ecological characteristics of the region, variations in the frequency, magnitude, and extent of disturbances to which it is subjected. This produces a diversity of complex patterns in the composition, structure, and surface distribution of the human population living in this ecological floor. For example, the forests of the Yucatan Peninsula redistributed every year after the passage of hurricanes and floods, which affect the human population, its social and economic, and daily activities.

The disturbances that result from human activities (agriculture, livestock, dam and road construction, and others) modify the ecological processes in the landscapes, fragmenting the habitats and affecting the natural process of moving of individuals (the dispersion). The new habitats are barriers for some species and corridors for others, affecting landscape processes such as complementarity, supplementation, neighborhood effects and the relationships between donor and recipient habitats.

The concept developed in cultural ecology by Ángel Palerm (1980 [1967]: 120–126), William T. Sanders (1979: 1–10), and Scott L. Fedick (1996), among others is called "ecological mosaic". It is based on the concept of the cultural core of Julian H. Steward (1955). Ecological mosaic means the existence of an area that has a territory with ecological and cultural diversity. It is symbiotic region, with a continuous and interrelated development, and an economic activity based on a temporary and irrigated agriculture, which is feasible in large part for its extension. The criteria to define the mosaic are physiographic, ecological, and cultural types; symbiotic areas; and key areas.

The application of the concept to Mesoamerican studies has been highly productive; see, for example, the book edited by Scott L: Fedick (1996) on the Mesoamerican mosaic. In the case of Yagavila, Oaxaca, the altitude and climate make up the ecological mosaic, where the different human population groups and towns are located.

Economic symbiosis: It is a concept to explain the regional macro adaptations among the environment, climate, physiography, altitude, precipitation, and others with Mesoamerican societies (Palerm 1997, [1957]; Sanders 1956, 1968).

Edible insects (*Jumil, chumil, chinche de monte, xotlinilli*): There are several ways to eat them: roasted; ground; mixed with salt and pepper; or used as a powder to season food. Also, they can be used to prepare tomato and green chili sauces. The most usual way to eat them is rolled up in tacos. Before 1519, there was a tradition of collecting *jumiles* for the festival of the dead. A pilgrimage was organized, and it went up the Huixteco hill in Taxco, where a temple dedicated to this insect has been found. The current tradition is that every October the *Jumil* festival is held, beginning on the first Monday after the Day of the Dead.

Ejido: From 1916 to the 1970, the land for agricultural purposes was granted to small-scale farmers (him or her) by governmental authorities. This land could not be sold, rented, or bequeathed, since it was owned by the State. But in practice, farmers who were gifted with them do these activities despite being illegal. The state endowed this land for agriculture and livestock activities to small-scale farmers (*ejidatarios*); which on average, they received five hectares per person. The State created administrators who in the villages were obligated to work in accordance with the instructions received from the authorities. These administrators were called Ejidal Commissariat (*Comisariado ejidal*).

Ejidatario: This is a small-scale farmer that is defined as an Indian, a mestizo peasant, or a small-scale farmer endowed with land whose surface area was of 5 hectares average. The land was not privately owned, so it could not serve as collateral to get bank loans. The distribution of land for agriculture began in 1916 and ended in 1992. In the 1990s, with President Carlos Salinas De Gortari, land endowed holders (*ejidatarios*) were allowed to sell their land, and many homegardens were sold to build tourist hotels and businesses, as happened in La Antigua, Veracruz, whose homegardens were famous for the diversity of fruit trees planted there. National and foreign companies wanted to buy the land to established touristic complexes in them.

Emigration: It is the departure of people from a town, city, or country for life or temporarily to reside in another place different from the one where they were born and live alone, with acquaintances, friends, or with their nuclear family.

Epidemics: In Mexico, they have great importance, and their study has been widely treated by social anthropology, history, historical demography, and the medical history.

Among the researches are Aguirre-Beltrán 1972 [1946]; Carrieri 1983 [1697]; Cook and Borah 1960; Cook and Simpson 1948; Borah and Cook 1974; Gibson 1980 [1967]: 138–167; González-Jácome 1994: 1–12; Malvido 1993: 63–111, 2003: 65–78; Mandujano et al: 2003. 9–21; Reff 1990; Rosenblat 1945, 1967; Sanders 1976: 85–150; Santley 1990: 325–343; Siemens 1990: 65, 72, 77, 178, 180, 197–198; Viesca 1982: 157–165; and Withmore 1992.

Esfoyaza: This was a Spanish party, which lasted until one or two in the morning, where guests sang, laughed, and told stories. The party was organized to shuck partially the ears of maize and tie them in groups called *panoyas*. Then they were placed in the barn so that the air protected them from insects. The party took place at night in the house of one of the farmers. There were snacks (sardines, ham, cheese, pears, apples cooked with burgundy, and *garullas* full of pears and apples.

Ethnobiology: The term ethnobiology appeared in academic texts in 1935. A wide variety of definitions have been given to characterize what it encompasses. In Mexico, Maldonado-Koerdell (1940) defined ethnobiology as "[…] that branch of knowledge that is responsible for the study of the use of plants, and animals of any region, by a defined human group, that inhabits or comes to it for get them". For him "[…] ethnobiology must identify, described and classified organisms that have or have had a cultural value for a human group; to know its distribution, as well as to specify its cultural value and modes of use".

One of the ethno biologists who had the greatest influence on Mexican and Latin American ethnobiology was Efraím Hernández X., who in 1976 pointed out that "[…] this relationship between Homo, and organic entities should be studied through time, and in ecological, environmental, and cultural terms".

For Posey (1986, 1997), a renowned American ethno biologist who worked in Brazil, ethnobiology "[...] refers to the description of the knowledge and beliefs of a certain social group in relation to the environment that surrounds it and to which it is inserted". He stated that "Ethnobiology". It is essentially the study of knowledge and the concepts developed by any society regarding biology. "[…] It is the study of the role of nature in the belief system, and adaptation of men to certain environments".

For ethnobiological studies, see the British anthropologist R.F: Ellen (1993, 1986: 191–192, 1996b: 91–117, 1999: 91–117) on the Nuaulu animal categories among the natives of Central Seram of Maluku Island in Indonesia; on a review of Swidden agriculture in Indonesia (1994); on modes of subsistence, huntering, and gathering to agriculture, and pastoralism (1996a: 101–146); on "Human worlds are culturally constructed"; on Nuaulu plant names, and categories (1986a, 1986b).

Ethnobotany: This branch of botany studies the relationships between human groups and their flora environment, that is, the use and exploitation of plants in different cultural spaces over time.

Hernández X: (1976: 13–18) defined ethnobotany as the scientific field that studies the interrelationships established between Men and plants over time and in different environments. In Mexico, these interrelationships begin with the arrival of the Asian populations and their entrance into the territory. These Asians brought previous knowledge of hunting and gathering acquired in other areas over a long period of time. The first settlers initiated the processes leading to the use of resources through agriculture, the domestication of flora species, and some fauna species.

The use culminated in an agricultural and urban planning stage at the time of the Spanish conquest. This caused strong acculturation during the Vice royal epoch and led to the current situation in which the traditional uses of natural resources changed to be used as raw materials and monetary earnings.

Many centuries later in the twentieth century, with the "green revolution", the expansion of the necessary infrastructure to configure a national socioeconomic unit, the promotion of industrial activities, and the formation of urban centers with a high concentration of human populations occurred, causing strong impacts on local and regional flora and fauna. Government institutions supported the change of the agricultural Mexican model, from the traditional one, of Mesoamerican origin, to the American one, driven by energy derived from oil, intensive use of agrochemicals, conversion of the polyculture agroecosystems in monoculture ones, and intensive use of machinery instead of family workforce.

The elements of human-plant interrelations, the reason for the study of ethnobotany, are determined by two factors: (1) the environment (ecological conditions); and (2) culture. By studying these factors through the dimension of time, it can be seen that they change quantitatively and qualitatively. The environment due to modifications of the components of said environment; and the actions of Men and culture due to capital accumulation, and sometimes due to the loss of human knowledge.

Ethnoecology: Is the cross-cultural study of how people perceive uses the environment, its use, its management, or its exploitation. Traditionally, it has focused on linguistic analyses of terms for flora, fauna, habitats, and other ecological phenomena to reveal underlying structures of the human mind that influence human behavior. According to Nazarea (2006: 38), ethnoecology is a research of the systems of perception, knowledge, and use of the natural environment, and the underlying history and politics that are represented in the managerial aspects of culture, without moving away from distribution problems, access, and power that shape the knowledge systems and resulting practices.

Ethno-pharmacology: The study of ethno pharmacology by Leonti, Vibrans, Sticher, and Heinrich (2001, 53: 1653–1669), according to the authors in their analysis of phytochemical and pharmacological data on the use of plants by the Popoluca, showed that there are gaps in knowledge of the bioactive components of the species used. The authors (2001, 53: 1667) considered that "[…] the lack of pharmacological or phytochemical data on a given species should not be interpreted as evidence of lack of activity [...] with respect to indigenous medicinal plants". Information from the study on the plants collected and their medicinal uses among the Popoluca form a basis for further studies, including the criteria of this ethnic group for selecting species as medicinal.

Evolutionism: This has been one of the great paradigms of contemporary science. At the beginning of twentieth century, cultural ecology discussed evolutionism taking into consideration whether if it was unilineal or multilinear. That is, if all human societies had evolved in the same way, or if there was more than one evolutionary line that explained their basic differences as noted before. From this point of view, there had been at least three evolutionary models in human societies: (1) hunter/gatherer societies; (2) agricultural societies of the agro-hydraulic type, previously called Orientals; and (3) capitalist societies. For more information, see Julian H. Steward (1955).

Holistic: The concept of holistic arose in philosophy by Jan Smuts, in 1926. In cultural ecology, the holistic is understood as the system being more than the sum of the parts that compose it. It is useful in the study of complex systems, such as social and cultural ones. The concept had been used in ecology (Golley 1993).

Fairs: The fairs to promote agricultural products are becoming an important way of diffusion, like the apple, wine, and maize, Opuntia and amaranth fairs, and others with more elaborated products, like tamales or mole. A better dissemination of them would support agricultural producers. This is being done already in the case of maize. By adding additional elements to the productions, prices in the markets are raised.

Fariña: This is a bread made with the crushed or ground maize kernel.

Farrapas: They are made with corn flour cooked with water and dipped in milk, butter, and salt, and egg yolks, optionally. When the farrapas are served, sugar and butter may be added. They are cornmeal porridge.

Fermented Agave mead estates (*Haciendas pulqueras*): The region where the fermented Agave mead (*pulque*) estates were formed was the natural corridor that went from the east of Puebla, the north of Tlaxcala, the south of the state of Hidalgo, and the north of the State of Mexico, highlighting in it the Apam plains region. These were highlands, with sloping hillsides, scarce and irregular rains, extreme temperatures, frosts, frequent droughts, strong winds, and periodic hailstorms. Among the best known of the Agave estates were Zuapayuca; Venta de Cruz; S. Antonio Ometusco, and S. Juan Tlacatecpan in the State of Mexico; S. Bartolomé del Monte; S. Lorenzo Soltepec; Ixtafiayuca in the state of Tlaxcala: S. Bartolomé de los Tepetates; Los Arcos; Tetlapaya, S. José Titicuentla, Chavarría, S. María Tecajete, S. Antonio Tochatlaco, and S. Lorenzo Espejel in the state of Hidalgo (Monterrubio 2007; Rendón-Garcini 1990, 1994: 209–243).

Fermented beverages: They are still made from Mexican fruits. There are distilled ones. In the eighteenth century, the Jesuits of the missions established in the northwest of New Spain found fermented beverages made with battered fruit parts or baked and boiled in water Agave parts, such as the head of the *Agave bovicornuta*. The missionaries waged a battle against these drinks because of their poor quality, and this harms health. The list is long, especially if we include maize fermented beverages (González-Jacome 2018: 37).

In Veracruz, native Mexican hog plum (jobos, hobos, ciruela Mexicana, ciruela de hueso grande) is used to make alcoholic beverages such as the "toritos", and also these Mexican hog plums are sold in Mexico City markets at the late summer (González-Jácome 2018: 21–74).

Fifth (*quincunxes*): It is a five-piece geometric arrangement, which is made up of four elements that form a quadrilateral, and a fifth element that is added at the intersection of its diagonals. It has been used to describe and explain the five Mesoamerican cardinal points (north, south, east, west, and center).

Food imports: In Mexico, the phenomenon of European imports – as is the case of Spanish ones – is historical; its impact on food is important and not exclusive to this country, although it originated in the sixteenth century with European expansion through Asia, Africa, and the New World. It has been reinforced in recent times as part of the globalization process and the growth of trade controlled by transnational agro food enterprises. An example of what happens is the case of the olive oils from Baja California, Mexico, whose production is low and their prices are much higher than Spanish olive oils and some Italian ones, so they cannot compete in the national market.

Food (Popular):

Flautas: They are thin, oval tortillas filled with shredded beef or lamb, grated pork or chicken, fried, and garnished with cream, and crumbled cheese.

Huaraches: They are thick oval tortillas spread with re-fried beans and covered with shredded meat or chicken, hot sauce, sliced lettuce, sliced onion, and crumbled cheese.

Sopes: They are small, round, thin tortillas that can be fried or cooked on clay plates or metal griddles. Sopes are covered with green or red chili sauce, lettuce, shredded cheese, and slices of onion.

Tacos: They are tortillas filled with some stew, beefsteak, meat, cheese, chicken, beans, or other previously cooked foods and rolled up. They can be fried or cooked on a metal griddle. Tortillas can have different sizes and thicknesses for tacos. There are small and thin ones or medium-sized ones filled with cheese, stuffed chili, liver and onions, potatoes with chorizo, shredded chicken, squash blossoms, or a variety of other fillings. There are tacos left to sweat in large baskets, and the client adds chilies or chili sauce before eating them.

Tamales (with Mexican black cherries): They were made of corn mead dough filled with the fruits of the Mexican black cherry tree crushed and placed inside a cornhusk. They are steamed, and they are sweet. They are still sold in the tianguis of Tlaxcala and Puebla.

Tortas: They are a popular food made with a roll similar to a French roll, cut in half and filled with cold cuts, fish, omelette, sausages slices, cheese, lettuce, ham, and others, like a sandwich.

Gourmet food: In the twentieth century, the Agave flowers (*gualumbos, hualumbos*) and coral bean (*Erythrina americana*) flowers. The flowers are edible, dripped in egg batter, and cooked, boiled, or fried.

Ethnic origins: The families of Spaniards based in Mexico consumed collard greens (*Brassica oleracea* var. sabellica) cabbage until a few years ago. Nowadays with the alimentary fashion of vegetarians and vegans, it has gained adherents and is consumed as a salad, but known as kale.

Zapotec families in the Isthmus of Tehuantepec may have iguanas and reptiles that are important both for family consumption and for commerce. Iguanas are sold alive or already cooked and are found in the markets of Tehuantepec and, in general, of the Oaxaca Isthmus, along with dried and raw shrimp, fish, vegetables, and fruits.

In Mexico, there are cooked foods (*cocidos*), in common, but with regional differences, including some components of native and Spanish origin (olive oil, olives, capers, raisins, garlic, and saffron). In Veracruz beef, pork or chicken, onion, garlic, cabbage, sweet potato, banana Macho, carrot, maize, chayote, or squash, and parsley are added to a broth. The "norteño cooked" broth has beef tail, shank, celery, onion, carrots, cabbage, squash, and coriander.

Food system (aquatic species that were edible): Fishing and capturing aquatic species was a basic activity for Mesoamerican groups, including farmers, to complete their food systems. Places such as the Mexico basin with its system of five lakes, one of brackish water, and others of fresh water allowed not only fishing and capturing of aquatic fauna, but also the hunting of birds that lived permanently or seasonally in the lagoon system (Sahagún 1956, T. III, XXIV: 147).

González-Jácome (2011: 428–436) Figure 10 is about the technology used to hunt, fish, and capture edible aquatic fauna; Figure 11 (2011: 439–444) refers to edible birds hunted and captured in terrestrial biomes. Complementary information. del Olmo and Roldán (2007),

Espinosa-Pineda (1994), Rodríguez-Galicia (2017), and Rojas-Rabiela (1985). Data from these authors included the hunted animals and fish that were obtained in the lagoon system of the Mexico basin or that arrived from the coast of the Gulf of Mexico. On the consumption of Spirulina algae, see Francisco Hernández (1574).

Food system: Craved food (*antojitos*). These foods are known as craved or yen foods, generally. Maize is the raw material for these foods, such as tamales, tortillas, tacos, enchiladas, tlacoyos, lolos, gordas, picadas, sopes, huaraches, pupusas, arepas, quesadillas, fresh maize ears that are toasted, boiled, or roasted; esquites (boiled corn kernels), popcorn (rosettes), and pinole. There are many other ways of roasting, boiling, cooking, and consuming maize, for which people are use local and regional names. Many of them are tortillas in different sizes and shapes like the small ones called *sopes*, with a border around to avoid the chili sauce dripping out and topped with chopped onion and ground cheese.

Food system: Cattle Ranchers. From the 1950s to the 1970s, the cattle ranchers of central Veracruz did not leave any wild animal alive. As soon as they knew of the presence of one, like a deer, they went out armed to kill it and eat it. On the walls of their houses, they had the remains of hunted animals (antlers, tanned skins, and mounted heads). In these cases, the meat of wild animals did not cover any food needs, since these ranchers were rich, and they had a good and varied diet with an abundance of beef, pork, chicken, fish and shellfish, milk, cheeses, vegetables, eggs, and maize hominy tortillas. The mentality and the food system of these ranchers with respect to wild animals were very different from those of small-scale farmers.

Food system: Wild and cultivated green edible seasonal weeds also called edible foliage herbs (Spanish. *bledos; Nahua. quilitl or quelites*). These edible green plants could be wild, cultivated, or tolerated. They grew in the cornfield and were very important in the diet of the cornfield small-scale farmers (*milperos*), and their families.

Food system: Weeding food. It is cooked for important weddings and family parties. The traditional weeding fare includes red rice, mole with pieces of chicken and turkey, tortillas, and tamales accompanied by soft drinks, beer, and other spirit or aromatized drink (beverages produced by distillation with the presence or absence of aromas, of fermented or macerated natural products of vegetable substances, with the addition of sugars or other sweeteners or other agricultural products, "alcoholic beverages"). Families who consider themselves modern and urbanized have modified these foods, replacing them with pasta, mole, baguettes, soft drinks, and other spirit or aromatized drink.

Those who participated in the event are given a clay pot with mole, and chicken or turkey pieces to take home when they leave the festivity: The turkey or chicken pieces given to the guests are selected according to the importance given to each gets. Sometimes, guests are served half a turkey or chicken on their plates. The guest ate what they can, and the rest of turkey along with the mole is put in a small clay pot and delivered to them when they leave the event. They received tortillas, also upon leaving. Tamales are very important in the weeding fare in many places in Mexico, and there is a special types of tamales for this occasion (González-Jácome 2003).

Folk classifications related to fauna: Altieri (1991) attaches particular importance to folk classifications related to animals, especially insects and birds. Terms common among farmers and indigenous ethnic groups. For Altieri (1991), human groups tolerate several agricultural pests because they consume them (e.g., in the central Mexican highlands, the consumption of ant eggs "escamoles", and Agave worms).

Folk taxonomies related to flora: Folk taxonomies on ethnobotany are the best known and documented, as can be seen in studies of Mexican peasants, and the number of plant species they know and recognize. (See Alcorn 1984; Toledo et al. 1985.)

Fossilized in-line wells, and canals: These pre-Hispanic wells and canals were found by archaeologist James Neely and his team in in the Tehuacán Valley, where they were located in lines.

The systems studied were the S. Marcos well, the Purrón dam, and the "fossilized" canal systems in the valley, some of which are still working. In a recent archaeological investigation, Neely (1995 (12): 38–40, 2017: 136, 142–143), and Neely et al. (2017: 134–147) summarized the control of irrigation water studies of the *Tehuacán Archaeological and Botanical Project*, which in the 1960s was directed by MacNeish.

A similar foggara system was explored in December 1976, by Ángel Palerm on the lower slopes of the Malitzin volcano, and the St. Isabel Xiloxoxtla municipal borders with S. Francisco Tepeyanco, where in addition to touring them, he collected pre-Hispanic pottery in their wells and canals. In those years, researchers did not want to admit that the qanat or foggaras in Mexico had a pre-Hispanic origin.

This fossilized canal system served to confirm the proposal of Ángel Palerm about the pre-Hispanic origin of the Xiloxoxtla foggara, in Tlaxcala, but which required further study. This system is of native origin and was not brought by the Spaniards, as has been affirmed by numerous researchers (González-Jácome 2003). This system to conduct water has a wide presence in numerous arid and semi-arid areas of the planet, still (see qanats in this glossary).

Gacha: It is a food composed of flour cooked with water and salt that can be seasoned with milk, honey, or other seasoning. It is served as a filled patty or porridge.

Garbage regulations and its shipment to the seas in Mexico: The regulations that control the shipment of garbage to the seas in Mexico are not followed. However, research institutes in different universities are generating products that pollute less or that disintegrate in a short time, instead of forming gigantic islands that float in the sea and are mortal for marine fauna. For example, beverage straws made of rice or those made of *Opuntia* cladodes (*nopales*) derivatives. However, the low cost and the propensity to use plastic containers or non-recyclable materials sustain their popularity with merchants and the general public. PLA packaging made with maize has been used in the USA for several years, already, but its promotion in Mexico has been zero, practically.

Garulla: There are Corbatees, filled with a shelled grape; the grape grains are detached from the cluster.

Good life (*Buena Vida*): It is a concept used by several groups of small-scale farmers in Mexico, including mestizos and indigenous. It means achieving a way of life without financial deprivation. Each peasant family, each community, each region of the country, or each ethnic group has its own and different concept of what is a "good" or a "better" way of life. In general, peasant societies that are located near cities consider that urban life is better than the one they have. However, this is not shared by all small-scale farmers, be they mestizos or indigenous. Therefore, it is necessary to define this concept, according to each human group and agricultural society under study.

Governmental programs: *Sowing Life* (*Sembrando Vida*). The current government has put efforts and resources of the people of Mexico in the *Sembrando Vida* project that aims to introduce fruit trees in the forested regions of the country. The projects where fruit trees are introduced in forested areas have antecedents in Mexico. The current government proposes what is called Sembrando Vida, which focuses on planting fruit trees mixed with forest trees. However, care must be taken not to burn or cut down the forest to plant the fruit trees. In smaller cases that occurred before in Puebla and Tlaxcala states, the result was not long lasting. President Andrés Manuel López Obrador indicates that peasants working for the "Sembrando Vida" program will receive wages for their work. This payment is the aid that the government is giving to the peasantry.

The minimum wage salaries (*salario mínimo*) paid by governmental federal institutions include those for the construction of highways and roads, cleaning of canals, or clearing forests of leaves and fallen branches to avoid fires. In areas such as the Monarch Butterfly Reserve in Michoacán and the State of Mexico, the residents of several towns work clearing the forests and taking care of butterflies for which they are remunerated. However, deforestation in the region has not stopped, reducing the area occupied by the Monarchs.

Green revolution: It is the name of the program aimed at increasing agricultural productivity and food, which was developed from 1960 to 1980. It arose in the United States and spread to many countries, including Mexico. It was based on the accelerated increase in food production through the selective crossing of species, the use of fertilizers, pesticides, and new irrigation techniques. It was to increase food production of the countryside without expanding the cultivated land, but maximizing the yield of the already exploited areas. The key foods of this revolution were rice, corn, and wheat.

Holiday bread (*Pan de fiesta*): It is a wheat bread decorated with fruit jelly strips or with nuts dried fruits like figs and nuts. This bread comes from Viceroyalty time and is still made in Tlaxcala. It is bought and eaten by the attendees of the festivities in various rural and urban localities of the Mexican Central Highlands but reaches as far away as the northern borderlands also (Villalobos-Nájera 2012, 2013).

Homegardens: The most important homegardens, or at least the oldest known and studied, are found in tropical regions. These systems have been developed and nurtured over generations, lasting because farmers constantly experiment and innovate in them, to adjust them to their family needs, and to the market needs (Velasquez Morlet et al. 1988: 73).

Homegardens in current Tlaxcala (vegetable gardens): Claudia Fernandez-González (2017) shows and analyzes the results of some programs focused on the creation of the called family homegardens in several communities of Tlaxcala. The townspeople were involved in the project and agreed to participate because the state government provided them with water tanks and tools for the creation of vegetable gardens. However, after they received those things, they abandoned the program and sold what they had received.

The government employees who were to start up the gardens claimed that the production was to be of vegetable unknown to the participant's families, and not part of their diets. Furthermore, they would not consume them or could they sell them. Also, the production was not profitable enough to take to the markets to sell. In fact, they were not homegardens with trees but vegetable gardens only.

The gardens did not improve the feeding of the families there. A homegarden with trees takes much longer to mature than chosen in these state programs. For a homegarden to begin to bear fruit, a number of years are required, and these programs do not contemplate times longer than those of a state or federal government term.

Between 2014 and 2015, the Ministry of Public Education (SEP) in Tlaxcala organized a course for the training of secondary school teachers on how to create, organize, and maintain a homegarden on school land and how to take the idea to the students' home. Several experts in agroecology were invited as speakers in the course but only 14 professors attended and just 12 finished it. Unfortunately, they were not homegardens with trees or simply gardens for vegetable cultivation.

Homegarden names: After the conquest, we can infer that the Mayan homegarden continued, but having different terms established in the old dictionaries for it. One of them is Kuchil, used in the sixteenth century for the homegarden, that Barrera et al. (1980: 343) define based on the Motul Dictionary as "[...] place or seat or cover where it is placed, or store anything that is not natural there [...]". For the Spanish-Maya Dictionary of Vienna (seventeenth century), the homegarden is "[the] seat or abode where one lives [...]". The S. Francisco Dictionary (1866) describes the homegarden "[as a] site or home plot".

For the Yucatan Peninsula, in the twentieth century, and the first quarter of the twenty-first, it is registered with several names: family garden (Barrera Marín 1980); homegarden (Caballero 1992); and dooryard garden (Stuart 1993). The term solar is the most common and popular used by current Mayan populations. In the Yucatan Mayan language, the terms used are: *Inn luumel* (my land); in *Pisté*, and *Pach nah'* (everything around the house); and *Inn wotoch* (my house, my land). In Yaxcaba, the plot is described as *ac tan'nah* (the front of the plot); ix nah (inside the cultivation land house); and *ta'an cab* (outside the house) (Mariaca-Méndez et al. 2007: 119–138; Vara Morán 1980; Terán and Rasmussen 1994; Acosta 1995; Osorio Hernández 1997; Terán et al. 1998). In fact, the term solar as a synonym for plot or residential place was also found.

Homegarden, sixteenth century: Mariaca-Méndez et al. (2010) found that in places cited in the sixteenth-century *Geographical Relations*, the pre-Hispanic homegarden lacked stonewalls (*albarradas*, *tecorrales*) and connected directly with the forest. Stonewalls arise in the Viceroyalty along with the new land tenure, congregational policy, and the need to establish property limits.

Hórreo: Is a granary, an elevated barn whose structure was based on raised posts. It has a square or rectangular floor, covered with two or four pitched roofs covered with clay or slate tiles, and walls with slits for air to pass. The posts have a disc-shaped circular section, a bit before the floor of the barn, sometimes inclined to prevent animals from getting into the barn. The hórreo can be made of wood, masonry, or stone.

Human generation: In anthropology, it is considered to cover a 25-year period.

Hunting garden model: Emslie (1981: 306), Linares (1976: 331), and Neusius (1996: 276) proposed a model where crops attracted many varied animals, and the settlers hunted them in their homegardens. Then, homegardens provided the settlers with animal protein, and by hunting they were protecting their crops at the same time. Neusius (1996: 276) argued that, when agriculture became a predominant activity for subsistence, hunting became opportunistic, not selective, with occurrences related to other labors. Smaller fauna lived in the fields, ecotones, homegardens, raised edges of land, and areas with secondary vegetation. Linares (1976: 347) argued that the most selective strategies of settlers in the forest/homegardens were used for hunting larger species, like white-tailed deer, and peccary, and that hunting in the homegarden helped reduce the exploitation of aquatic fauna (Vanderwarker 2006: 23, 148–181). This proposal is valid for regions such as Los Tuxtlas, in central Veracruz, but is not applicable to places such as the basin of Mexico, where aquatic fauna, migratory birds, and permanent fauna inhabitants of the lagoon system remained a primary source of protein until the twentieth century.

Hydroelectric projects: The 20 most important hydroelectric plants in Mexico are distributed: five in the northwest; two in the north; five in the west; two in the center; and six in the southeast. Among the most important of them are Aguamilpa (Nayarit); Aguaprieta (Jalisco); Angostura (Chiapas); Expansion Temascal (Chiapas); Bacurato (Sinaloa); Chicoasen (Chiapas); Feeder (Sinaloa); El Cajón (Nayarit); El Caracol (Guerrero); Huites (Sinaloa); Infiernillo (limits between Michoacán and Guerrero); Ixtapantongo (Mexico City); Malpaso (Chiapas); S. Rafael (Nayarit); Zimapán (Hidalgo); and others, which produce little more 22% of the electricity in the country (Ramos-Gutiérrez and Montenegro-Fragoso 2012).

There are cases, such as that of the La Parota hydroelectric plant, in the state of Guerrero, where some 25,000 farmers were affected and opposed its construction for 15 years, but lost the fight despite the fact that some of the farmers had been murderer, and that the Latin American Water Court issued a ruling in their favor: This is one of the cases where institutions support entrepreneurs and developers over farmers (Barkin 1972, 1998; Barkin and King 1970; Barkin and Zavala 1978; Blanco-Rosas 2011: 177–202; Ewell and Poleman 1980; González-Jácome, Velasco-Orozco, Moctezuma-Pérez, and Cruz León 2018; Molina-Ludy 1976).

Jagüeyes. These open cisterns were natural containers formed in the hollows where water accumulated, and stagnated during the rainy season (Cruz-León et al 2020). In pre-Hispanic times, villagers enlarged them and occasionally coated them with mortar or stones (García Cook 1973: 9–15, 1986; Montero-García 2012), like in Tlaxcala, where they are dated at

1200 BC (García Cook 1973: 9–15, 1986; Montero-García 2012). In the eighteenth century, jagüeyes were the main source of water supply for estates and ranches like the Apam Plains in the state of Hidalgo and places in the state of Tlaxcala (Ajofrín 1958 [1763]).

However, open cisterns continue to be useful, now as a water site for cattle: Some have been closed, and the spaces have another function today (Pérez-Sánchez 2012, 2014).

However, the current use of open cisterns for domestic purposes has declined with the introduction of water wells for drinking water that is very deep sometimes to supply well water for human populations (Galindo-Escamilla, Palerm-Viqueira, Tovar-Salinas, Rodarte-García 2008: 21–31). Various state programs in places such as the state of Tlaxcala, included jagüeyes and Ferro cement cisterns, have supported the creation of rainwater collecting systems (Quispe-Limaylla 2008: 82–91).

Jollas: The "jollas", pots or holes, are agroecosystems where the small-scale farmer makes holes in the raised boarders of natural, or artificial terraces, or canyon banks in eroded soils where rainwater accumulates. In these holes, muddy water (*agualodo*) is formed. Then different plants that are used to feed the inhabitants are planted in them, like legumes, maize, and sometimes fruit trees. Jollas are found mainly in the Mixtec of Oaxaca (Galindo-Escamilla 2008: 17–20; Rivas-Guevara, Rodríguez-Haros, Palerm-Viqueira 2008: 6–16).

Juan Andrade Torres: Dr. Andrade Torres was born in Tabasco and studied a doctorate in Social Anthropology at the Universidad Iberoamericana. He was a professor at the Universidad Juárez in Villahermosa, Tabasco, but his family lived in Forth Worth. When he retired from the university, he went to live in Texas with his relatives. On a trip to Mexico City, he gave me an interview about the lives of migrants in Forth Worth; Juan Andrade recently passed away.

Juan de Navarro: He was a Franciscan born in Spain, who lived in the convent of Querétaro, in Mexico, and he died in Mexico. He is the author of Natural History in five volumes, of which only the fifth volume is preserved, that he titled American Garden (1801). This book contains the descriptions and color drawing of 517 Mexican plants. The list of names and drawings of the Illustrations by Juan de Navarro is in *Mexican Archeology* (Vol. X, No. 57, 2002: 16–17). It is difficult to identify all the plants cited by Juan de Navarro (1801). Francisco Hernández (original 1574) can complete the list with information in 1959, Volume II, Books 2, 6, 8–10, 50–51, 130–131, 119–120, 134, 166–167, 172, 294, 297, 301, 319–320, 364, 365, 369, 376–377, 389, 406. Data by Friar Agustín Farfán (1579, 1592) was compared with Juan de Navarro's data (Galeote 1997: 119–161; González-Jácome 2011. figure 1: 409–413).

Juan Netvig: In 1764, the Jesuit Juan Nentvig drew up a list of medicinal plants from Sonora. This information indicates the importance of herbalism in the simple, ancient societies of Mexico. Data include the common name, the name in Ópata, the medicinal use, and the application of it.

Ka'anche: The ka'anche is a table with pots to experiment with plants. It is found in homegardens of numerous tropical regions of the continent, including Paraguay.

Kin: Extended relatives of a nuclear family, kin, in-laws, close friends, and compadres.

Ladino: Ladino is a person with no Indian origins, a creole, and a foreman.

***Leucaena esculenta* (gourds):** Hughes et al. (2007: 14389–14394) propose that their management resulted in 2–6 local species and a total of 13 species. Their hybridization was extensive, complex, involving nuclei encoded in the DNA sequences and diversity in the hybrids resulting from the juxtaposition of species under cultivation.

Recent studies on bottle gourds *Leucaena esculenta* (*guaje*), a pre-domesticated plant, show spontaneous hybridization of legumes in incidental homegardens through a process-favored – or precipitated – by human action that preceded domestication: The roasted seeds of *Leucaena esculenta* were eaten and consumed from the center to the south of what is the current

Mexican territory. The plant underwent numerous independent transitions from its wild state to cultivation. As an established crop, it is pinpointed around 4000 BC.

Maize hominy procedure (nixtamalization): This is a process where maize kernels are softened, and the glume is removed. The process initiates with maize kernels being boiled in water with lime. After boiled grains in this water, the grains and the water rest for several hours and then they are drained, so the glume is removed easily, and the kernels are ground to form dough.

Vargas (2007a: 98–104) who does research on the history of human nutrition considers that evidence of the nixtamalization process indicates that lime sediments at the bottom of archaeological vessels – the discovery of stone utensils used to grind maize (*metates*), clay griddles (*comales*) for making tortillas (1250 AC), and other implements necessary to process maize and convert it into dough – are not sufficient evidence to set a safe date for maize hominy procedure until the Classic period (200 AC–900 AC).

Maize hominy (nixtamalization). Vargas (2007), a physical and medical anthropologist, says that the evidence of the maize hominy (nixtamalization) process requires further studies, as they incomplete still. Although they found lime sediments at the bottom of archaeological vessels, the finding of metates, clay griddles for making tortillas (1250 AD), and other implements necessary to process maize and convert it into dough do not attest a safe date for nixtamalization, until the Classic period (200 BC–900 BC).

March maize (*marceño*): It is short ripening maize that is sown in swampy areas during the month of March. The races are Cuarentano and Mején.

Cuarentano maize races (2008/2009) are Zapalote Grande/Tepecintle found in Huimanguillo municipality at 25 masl (Guerrero-Herrera 2015: 51–52).

Mején maize races (2008/2009) are Tuxpeño/Tepecintle and Tuxpeño/Olotillo, Tepecintle, Olotillo, Tuxpeño, Tuxpeño/Olotillo, Tuxpeño and Tuxpeño/Zapalote: These races were found in Huimanguillo municipality with an altitude of 20–151 masl, also in Centla at 11 masl (Guerrero-Herrera 2015: 51–52).

March to the sea (*Marcha al mar*): Mexican President Adolfo Ruiz Cortines said "[which] places were closer to a decline in population of the Altiplano, and took the inhabitants from the mountains to the hot land" (Revel-Mouroz 1980: 19). In fact, it was to proliferate the populations in the country's tropical regions, whether they had access to the sea or not. It was one of the two main axes of the colonization in Mexico proposed to alleviate the effects of the agricultural distribution. The other axis was the northern Mexico dry lands colonization.

Macana: Name given in Tabasco, Mexico, to the cane planter (*coa, macana*) with a metal tip.

Mass media: The mass media disseminate fashions, customs, and ideas, which are accepted by rural people and modify their codes of conduct. An example occurred in S. Francisco Teolocholco, a town on the western slopes of Malinche volcano, where young people challenged the political and social control exercised by the town's elders, and with the posture that they had studied more, and white formal studies plus life experience more important than those of the elderly, they took public places in the elections. They modified the ideas and customs of the population.

Alberto Almeida, a communicologist, and a doctoral student in social anthropology at Iberoamericana, in the 1980s, did several studies on television, its introduction, and impacts on the families in the town of Españita, Tlaxcala, selecting cases of families that had televisions in the 1980s or had access to them. His thesis was unfinished because of a serious health problem that affected him. However, the result of his studies was that television modified the culture of young people.

Maíz chococito: It is a maize variety that can be cultivated under seasonal flood conditions. It is planted in the Choco Eco geographic region, which includes southern Honduras, the

Darien bordering Panama, the Pichincha province in Ecuador, and the Department of Chocó in Colombia.

Maize (races): At a conference organized by CONABIO, Dr. Sarukhan said that in Mexico now, there are 59 races of "cultivated maize". To date that is, the most accepted number, but it will vary from one finding to another, because new races, varieties of maize are discovered, arise, or disappear.

Maize, races in Mexico: Currently, there are between 59 and 60 races of maize in Mexico; some of them are:

Cuarentano maize races (2008/2009) are Zapalote Grande/Tepecintle founded in Huimanguillo municipality at 25 masl (Guerrero-Herrera 2015: 51–52);

Mején maize races (2008/2009) are Tuxpeño/Tepecintle, Tuxpeño/Olotillo, Tepecintle, Olotillo, Tuxpeño, Tuxpeño/Olotillo, Tuxpeño, and Tuxpeño/Zapalote: These races were found in Huimanguillo municipality with an altitude of 20–151 masl, and in Centla at 11 masl (Guerrero-Herrera 2015. 51–52).

Oaxaca, Chiapas, and the coast of the Pacific Ocean are Nal Tel types Zapalote Chico/Large, Reventador, Planoncillo, and Chapalote.

And maize that Migrated from the Central Highlands through places with higher altitudes of the types belonging to the Conical group, mainly, Palomero Toluqueño, Arrocillo Amarillo, Conic, and Chalqueño.

In Peru, there is the endemic pattern of maize that occurs marginally (30 of the 48 races): To date, unless other information emerges, Mexico is considered to be the only center of domestication and diversity of maize and its wild relatives.

Mescal, mezcal, mexcal: There is a difference between mexcal (or mescal) and mezcal, as mexcal is produced white the Agave leaves, and the stem (*quiote*) cut into pieces and cooked in earthen ovens of various types to obtain a cooked section of the leaves that has a sweet substance from which honey, Agave sugar, or a fermented beverage are produced. The juice is drawn directly from the baked piece of Agave. Later, maguey sugar is made. The word mezcal (written with Z) refers to the distilled alcoholic beverage, which is obtained from the catch up (*cajete*), heart, pineapple, or center of the plant (*jima*) that is broken into pieces and baked in ovens also. Then it is ground, squeezed, and fermented to obtain a liquid that is distilled.

Mesoamerica: The term Mesoamerica refers to the pre-Hispanic territory of the central western, central, southern, and southeastern parts of what is present-day Mexico, and the adjacent regions in Belize, Guatemala, and part of Central America, where ancient societies and their cultures developed. Its geographical limits were in the north a border that moved according to the advance the nomadic, semi-nomadic, and sedentary groups that settled there. This border included the Sierra Madre Occidental, the Sinaloa, Pánuco, Moctezuma, and Lerma-Santiago rivers. To the south, the border reached the Nicoya peninsula and the Nicoya Gulf in Costa Rica.

Based on historical and archaeological materials, Mesoamerica was a cultural area defined by Paul Kirchhoff as a period of about 100–120 years, which occurred some 60 years before, and some more or less 60 years after the Spaniards entered the territory. Archaeological studies extend the application of the concept to human societies that were organized and lived in this cultural area before 1518–1519. This generated, developed, expanded, and modified the characteristics originally described by Kirchhoff (González-Jácome 2000: 121–151). In studies by American academics, the concept of Middle America is used, that is based on geological and geographical elements, that existed until the entry of the Spanish and the formation of New Spain.

Mesoamerican chronology: The chronology of those times is modified if it is based on the remains of human, animals consumed, ceramics, basketry, materials used in the technology of those times, or objects of them (spearheads, scrapers, lacquers, knives, cores, utensils made of fauna bones, and other items). The following periods are denoted. Prehistoric (from the arrival of Man to the American continent until 2000 BC; Preclassic, from 2000 BC to ±100 BC; Classic from ±100 BC to 650 BC; Epiclassic, from 650 AC to 900 AC; Historic or Post classic from 900 AC to 1519 AC (in the Central Highlands).

Metate: Square stone with three legs used in rural areas of Mexico and Guatemala to grind maize, other grains, tomatoes, and chilies. A special type of metate with four legs was used to grind cacao and make chocolate, but in this case fire was placed under the metate. The oldest metates found have no legs. They correspond to the initial Formative.

Metepantle: The word *metepantli* (*metepantle*) is of Nahua origin. *Metl* means maguey, and *pantli* means file, line, or wall. The metepantle is the name of the rows of Agave in a crop. It is an agroecosystem of pre-Hispanic origin, unlike the Agave files in semi-terraces (*Bancal*) that are a nineteenth-century agricultural system.

The *metepantle* can be a subsystem located in various agroecosystems, like the hillsides of hillocks, or mountains where terraces and semi-terraces are constructed: It consists of a combination of flat areas for cultivation, limited by raised ground borders where lines have Agaves planted in them, or planted with Agaves intermixed with trees (Mexican hawthorn or Mexican black cherries) and weeds:

In some cases, earth borders have canals next to them where rainwater accumulates. Sometimes they had associated canals next to the raised borders. The composition, structure, and importance of the borders in the agricultural systems require a lot of ecological, agronomic, and cultural studies because they have been considered so unimportant within agricultural activity. Soil erosion control using Agaves was combined with other plants, whose leaves and fruits fell to the crop plot, adding organic matter to it.

Mexican basin, waterfowl: Espinosa-Pineda (1994: 17–18) wrote that the Mexican basin waterfowl that lived in it throughout the year included species related to the crane, water hens, coots, day and night herons, and three or four types of ducks. The migratory species were geese, lesser scaup, cinnamon teal, comb duck, surface ducks, scuba divers, arboreal, mergers, pelicans, cormorants, anhinga, crane, other coots, rails, more than 30 plovers, avocets, squatters, varied and abundant least sandpipers, diverse gulls, and even some sea swallows. A wide collection of herons of different sizes specialized in fishing at different depths of the lagoon existed, like storks, ibis, 15 species of eagles, kingfishers, and birds related to the parrots. There were more than 100 species of waterfowl, which were migratory, mostly.

For the upper Lerma River basin, see Beatriz Albores-Zárate (1992, 1995) where she describes the pre-Hispanic methods for waterfowl hunting in the high lerma lagoon system.

Mexican, middle-class food system: In Mexico, the phenomenon of European imports is historical. Its impact on food is important. It is not exclusive to this country as it originated in the sixteenth century with European expansion through Asia, Africa, and the New World. It has been reinforced in recent times as part of the globalization process and the growth of trade controlled by transnational food companies. An example of what happens is the case of olive oils from Baja California, Mexico. They are high quality but of small productions, or their prices are much higher than Spanish olive oils; some Italian brands hence cannot compete in the national market.

Mictlan. It is the place where the dead go.

Migration types: In Mexico, migration can be daily, weekly, biweekly, monthly, seasonal, internal, international, return migration, or emigration, and they are all related to salaried economic activities. Commuters are people who live in one place and work in another where they go daily to their work activities.

Mixiote: It is a kind of tamale made with lamb, and hot sauce, wrapped in Agave leaves, and steamed in kilns ovens.

Mole: It is a thick sauce prepared with different types of chili peppers, tomatoes, spices, and other ingredients, including chocolate, peanuts, and sesame seeds that are roasted before they are ground in a metate. It varies according to the region. Nowadays, the chilies, onion, and garlic are boiled in water and then blended in a blender. The other ingredients are roasted in oil before being ground in a blender as well. Then both mixtures are combined and chicken or turkey broth is added.

Mozomboa: It was a cattle ranch in central Veracruz whose land was given on dowry to small-scale farmers when the agricultural land distribution program went into effect in the 1930s. The portion that was left to the original owners continued to be used for livestock raising, horse breeding, and an agriculture until the 1970s. A town was created with the name of Mozomboa that was formed as part of the Municipality of Actopan. An important part of the owner's land, and rented land was planted with fodder maize. The homegarden was close to the farmhouse, and there were shade trees, water tanks, and secondary forest zones. All near the ranch constructions. These were located on the land destined for cattle (AGJ 1980).

Actopan, Veracruz, and mango orchards: For many decades, Actopan was famous for its production of different types of mangoes. The orchards (*huertas*) were planted with *Manila* mango, *Niño* mango, *Ataulfo* mango, and *Petacón* mango trees. They were introduced to the region at different times. They had different production dates. The produce was directed to the markets of the port of Veracruz, Xalapa, Puebla, Tlaxcala, and Mexico City. The problem private orchard owners faced was the strong winds (*nortes*) and hurricanes, which affected the trees during the flowering period every year; therefore the production was reduced considerably by the impact caused by these strong winds and hurricanes. This production was sold wholesale in the orchard itself and in boxes for retail at stands on the roadside from Veracruz to Xalapa. The Manila mango is no longer the best-selling mango in Mexico City and has been replaced by the Ataulfo mango produced in Chiapas (AGJ 1980).

Multisite ethnography: The postmodern criticism of anthropology and its research methods based on fieldwork and local or regional studies changed, giving rise to "multisite ethnography", which is related to processes such as globalization. See Appadurai on global cultural flows (2001); Bauman (2012) on the understanding of individuals and practices rooted in distinct time and space, which must be understood in a more fluid way; Marcus (1995) on his proposal to analyze realities much more linked and holistic, starting from a particular place, from which his context and the circulation of objects, identities, and cultural meanings in distinct spaces and times are constructed; Massó and Santos (2017) on cultural constructions directly related to the world system, the architecture of the local framework in relation to the global framework –which has served to contextualize ethnographies in their more traditional form, usually – seeming less appropriate than the study of the incidents between different spaces, and the cultural construction of a group, its possible practices, speeches, and discourses.

Mutual aid/mutual help: They are social systems of community, volunteer, or voluntary cooperation. Mutual aid functions among family groups, and it is related to agricultural non-salaried work during planting and harvesting. Mutual aid includes other work for the families like helping to build a new house for some family member, helping with expenses for baptisms, weddings, stewardship, and others; it all includes unpaid work from family members, close friends, and volunteers. Those who receive mutual aid have a moral obligation to reciprocate likewise when necessary to the one who helped them:

Mutual help functions by working without payment to benefit the community. It includes the repairing of roads, cleaning of irrigation canals, repairing in public buildings, helping to set up infrastructures for the introduction of electrical energy, or drinking water and others. The federal government and some state governments use this form of free work for actions such as repairing roads, schools, and other public buildings, sometimes paying minimum wages for this work, since those who participate in them stop working on their own land to take part.

Nahual, nagual: In some indigenous and mestizo rural zones of the country, the nahual is conceived as a kind of wizard or supernatural being that has the ability to take on an animal form. The term refers both to the person who has that capacity and to the animal itself that serves as a guardian animal for a person.

Nuclear family: It includes father, mother, and children. Extended family includes: nuclear family plus grandparents, aunts and uncles, and spouses of children. It may include other family members, even compadres and close friends, as well.

Ovens: The native Mesoamerican population had two types of ovens. The more antique one was a pit, and the cooking method used in it is called barbecue. The other one was built on the ground (kilns).

Palapa: In Mexico, a *palapa* is a large gazebo-type structure made of sticks or logs with a palm-frond roof. It is a common construction in tropical and very hot places.

Paleo-Indian period: According to Götz and Emery (2014), the Paleo-Indian period was between 14 thousand BC and 8 thousand BC. So it is the oldest part of Prehistory in Mexico, with settlers arriving who traveled from Beringia to the south of the continent. These migrants organized in hunter-gatherer bands specialized in hunting mammals from the Terminal Pleistocene. They arrived with the lithic complex of Clovis-type spearheads, domesticated dogs, and knowledge of fire and its usage.

Palerm V., Ángel (1918–1980): He was a historian from the National Autonomous University (UNAM) of Mexico and a student at the National School of Anthropology (ENAH), where he was trained in fieldwork and did his first anthropological research directed by Isabel Kelly, who had been a student of the famous American anthropologist Alfred Kroeber. Palerm launched Cultural Ecology in the ENAH (1965–1968) and Iberoamericana University (1965–1968), schools where he was a tenured professor. He was a professor at the University of Texas in Austin (1970s), also. He studied a PhD in Regional Planning and did research in the agricultural agrarian reforms of various countries. He focused on the study of the ancient agricultures of complex societies (González-Jácome 1997, 2001a, 2006, 2007b, 2018: 410–414).

During the 1950s, he had done several studies on Mexican agriculture, among them the cornfield system, and vanilla production of the Totonac of Papantla, Veracruz (Kelly and Palerm 1952). He also did studies of Cuicuilco in the basin of Mexico and of the Acolhuacan region in Texcoco, where he was accompanied by the archaeologist Pedro Armillas (Palerm, Armillas and Wolf 1956: 396–369) and the anthropologist Eric Wolf (Palerm and Wolf 1955, Vol. 8, No. 3; pp. 265–281, 1968: 343–353). In the 1950s and 1960s, findings related to the irrigation systems in those places began to be discussed, as canals that they found in Cuicuilco and described were the oldest known to date in the Mexican Central Highlands (Palerm 1955b).

Palerm (1955, 1956, 1961, 1968, 1972a [1957]: 5) proposed to understand agriculture as a historical discipline and as a social science that started with the very presence of humanity in the New World, as well as the origin and development of his cultures and high ancient civilizations, which emerge from the discovery and evolution of agriculture.

In his discussion with Karl Wittfogel about irrigation in the basin of Mexico, Palerm (1972b, 1976a, 2007) concluded that the magnitude of irrigation water control systems is related to specific conditions (physiography, altitude, soils, water quantity, and other components of the environment) where the flow rates and seasonal circumstances of the rivers are taken into account by the builders of the systems to generate works such as canals, dams, and others: Although in environments such as those in the central Mexico basin, the systems did not have the same difficulties as those in Mesopotamia, Egypt, or China, they were in situations of great complexity, and the irrigation systems involved enormous constructive work, in addition to coordination with the sociopolitical organization there to be able to carry out the systems.

In Mexican archaeology, researchers from the National Autonomous University of Mexico (UNAM) with the academic advice of the American archaeologist William T: Sanders created an academic group whose theoretical- practical framework was that of cultural ecology. Other researchers studying agriculture and irrigation systems were West and Armillas (1950: 165–1820) and Pedro Armillas (1971).

Patrilocal families: Newly married sons and their wives will live with their parents. When couples have no sons, arrangements can be made, such as that the daughter's husband will live with the in-laws.

Peasant family (Family farm): Alexander V. Chayanov published his work *Organization of the Peasant Farm* in 1925. The Russian economist discusses in it the peasant family and their economy, based on numerous works carried out by him and his colleagues between 1880 and 1925. He had published an earlier and shorter version in 1923, with the title *The Theory of the Peasant Economy*, which was published in Berlin. Chayanov's 1925 work reached the American academy four decades later when published in English in 1966 and reaching interested Latin American readers, not until the late 1970s–1980s.

In the English version, the concept of family farm appeared (Chayanov 1966: 271–272) but referred to as an economic unit or peasant domestic unit regarded as a unit of production and consumption. The unpaid work that is performed by all the members of the family is counteracted by a balance to be established between their basic finances and the satisfaction that the family demands by compensating their un-remunerating hand labor.

The peasant farm is an agricultural unit that may be linked to other non-agricultural activities (handicrafts, commerce, industry), but the family continues to function around agricultural activity and the unpaid work of its members: This concept does not fall within the parameters of the formal economy because it does not provide economic benefit, but it establishes the balance between work and consumption.

Chayanov (1966). The Theory of Peasant Economy; Daniel Thorner, Basile Kerblay, and R.E.F. Smith (eds.), The Economic Association Translation Series, Homewood, Illinois).

Peony (*peonía*)/Cavalry (*caballería*): The peony (*peonía*) and the cavalry (caballería) were measurement units related to the land that the kings granted to the Spanish soldiers for their contributions in the war of conquest, including their participation in combat, the money, and plus other things that soldiers contributed to the organization of the expeditionary group. The peony was for the foot soldiers and the cavalry for the soldiers who brought their horses to the war. Peony was the land that in the name of the Crown was given in usufruct to the soldiers and the indigenous principals that were allied to them and that were living in the communities. In return, they promised to pay – in war or in peace – the maintenance of a man of arms and his horse, which amounted to 60 bushels or 3,863 areas (in Cuba there were only 1,343 areas) (*Diccionario de la Lengua Española*; Madrid, 1939: 206).

After conquering a country, a peony was the portion of land that was assigned to each foot soldier to settle on it. In the Indies, it corresponded to the land that could be plowed in a day (Diccionario de la Lengua Española; Madrid 1939: 972).

Periphyton: It is organic matter that was obtained from the wetlands. Periphyton is a community composed mainly of algae, bacteria, fungi, and invertebrates found in freshwater ecosystems throughout the world. It is rich in phosphorus and nitrogen. The agricultural management to improve soils found in the Late Classic homegardens in the northwest of the Yucatan peninsula is revealed in the study of Makabil, where the application of periphyton from the wetlands to the homegardens did indeed improve their agricultural productivity.

"Silt" is a biological indicator of water quality; therefore, it has generated interest in wetlands since it is used to monitor and regulate the concentration of nutrients. Many wetlands, natural and artificial, are used for economic and ecosystem purposes, or to improve the quality of water in human settlements (Tavera and Novelo 2011: 42–43).

On periphyton, see Tavera and Novelo (2011: 39–45). For the archaeological study of diatoms in the El Eden ecological reserve see Fedick; Morrison; Andersen; Boucher; Eyebrow Acosta, and Mathews (2000 (27): 131–152).

Petenes (*Akalche*): They are circular formation ecosystems located near the coast of the Yucatan peninsula. Also, they are called islands. Normally, these sites are fed by a freshwater spring that is located in the center of the ecosystem. Petenes have tall flora composed of sapodilla trees, mangroves, palms, and ferns among others. They have a large variety of fauna, such as alligators, as well as a variety of bird species. In these ecosystems, maize can be cultivated under seasonal flood conditions like the maize grown in March (*marceño*) in waterlogged areas of Tabasco (Orozco 1999: 111–122).

Piles of stalk cones (*almeales*): In Tlaxcala, corncobs are collected, ensiled, and left in the field in piles of stalks (cones) called *almeales* until the beginning of preparations for a new planting.

Piñata: Originally, it was a clay pot adorned with colored paper and filled with fruit. These pots were struck with a long stick and broken in games where children and adults participated separately, trying to hit the pot while blindfolded. Today, they are used in Christmas celebrations and children's birthday parties, still, but they are made of paper Mache and they are filled with industrialized candies and sweets, as well as small toys made with plastic and confetti.

Plan Puebla: The Puebla Plan (1967–1969) had as its central objective "[…] simultaneously to attack two development problems -the food shortage, and the low income to which those who live from agriculture are subject- by a massive increase in the yields of a basic crop among small producers […] (Plan Puebla 1967–1969: 4–5). For this program credits were provided, higher fertilizer inputs were applied, more plants were planted per hectare, and small-scale farmers were taught steps, and techniques to apply in this production strategy. The plan obtained higher yields in maize production, but what the results (yields) of the other crops were is unknown. Nor are results known regarding weeds, soils, and water in the region".

Popal: This ecosystem is found in areas flooded in various months of the year or permanently. It includes plants such as *popal* (*Thalia geniculata* L.), *platanillo* (*Heliconia* sp.), *cerillo* or *sibal* (*Cladium jamaicensis*), *Canna* sp., *Typha domingensis* Pers. (Nea), various genera of Poaceas, Ciperaceae, and others (Gliessman 1980: 93–101; Orozco-Segovia 1999: 118; and Mariaca-Méndez 1992: 69–76).

Practicing subjects: This is an old idea developed in applied anthropology, where research subjects (otherness) are included as a fundamental part of the studies on equal terms with researchers. Canadian anthropologists have developed a similar idea through their concept of traditional ecological knowledge (TEK) (Clark and Dickson 2003: 8059–8061; Turner II et al. 2003: 8074–8079; and Turner et al. 2003: 8080–8085).

Pre-Hispanic orchards: Before 1519, there were only Mexican black cherry, Mexican hawthorn, and avocado orchards in the Highlands of Mexico. In contemporary times, the most important orchards are the avocado mono cultivation agricultures (*huertas*) in Michoacán, which have become very productive and profitable.

Present studied trough the past: Immanuel Wallerstein's proposal (1979) considers that "The past serves to understand and diagnose the present with historical tools [...] the debate is to understand the present through the past". This author states, also, that history can be seen and understood as a social science, as a discipline open to all social sciences. If researchers apply his proposal to traditional agricultural studies, and not to history only, but to geography and ecology also, then other scientific disciplines and sciences can be involved. Seeing and understanding agriculture in Mexico this way to be able to walk down the paths of agricultural knowledge of the human populations involved to make the construction of new elements and their applications to the rural world by the practitioners of traditional agriculture possible (Golley 1993). This attitude accepts using the interdisciplinary, multi-disciplinary, transdisciplinary, or sustainability sciences.

Polenta: A thick mush made of maize flour, water, and salt, with sauce, meat, or other ingredients added to it.

Polymers: They arrive in Mexican agricultural lands between 1930 and 1942. They were used in the manufacture of bags for grains, mainly replacing those made of natural fibers such as Agave. Polymers are used to make bags and plastic containers for chemical fertilizers, herbicides, and other agricultural products. In 1938, DuPont began producing nylon, and P. Shlack made nylon from caprolactam. After World War II, there was a great boom in the manufacture and commercialization of polymers with the appearance of epoxy resins in 1947 and ABS in 1948. In 1953, Karl Ziegler and Giulio Natta discovered stereo regular polymerization and received the Nobel Prize for Chemistry in 1963. In the 1950s, linear polyethylene, polypropylene, polyacetal, polycarbonate, polyoxide, and other copolymers emerged (*History of plastics*; www.abc-pack.com/enccyclopedia/historia-de-los-plasticos/).

Porfiriato: The Porfiriato was the period in Mexico when Porfirio Díaz was president of the country from November 1876 to May 1911. Due to the enormous economic, political, and social inequalities in Mexico and the re-election of the president several times, the Mexican revolution of 1910–1921 originated.

Pulque bread: In the 1930s and 1940s, the aromatic thin branches and leaves of Mexican black cherries were used to wrap the "pulque" bread, but after the 1970s, they were replaced by paper or plastic bags them.

Quanat, foggara, in-line-well-system: The Qanāt (Arabic), or foggara, is an ancient type of water-supply system. It was developed and is used still in arid regions of the world. A qanāt taps underground mountain water sources trapped in and beneath the upper reaches of alluvial fans and channels the water downhill through a series of gently sloping tunnels, often several kilometers long, to the places where it is needed for irrigation and domestic use. In Egypt and Asia, the development of qanats began about 2500 or 3000 years ago, probably. New qanats are seldom built today, but many old qanats are used in Iran and Afghanistan still, chiefly for irrigation (Encyclopedia Britannica. www.britanicca.com/technology/qanat).

The system is known in many parts of the world: In Asia, it has the name of Karez, and these wells and channels are found up to Turfan in China. The channels are cleaned often to remove earth from them. Wells and canals are in the inhabitants' homes. The system is known by the names of foggara, campfire, qanāt, filter galleries, skylights, and other local names. Before the recent discovery by Neely and his colleagues (2017), it was thought that Spaniards had brought them at different times to New Spain (Palerm-Viqueira 2004, Vol. 1, No. 2). In this water collection system, the structure is underground and is built to reach an aquifer, capture its water, and lead it to reservoirs from where it flows through canals to the fields.

In Tehuacán, the system has been called in-line-wells or square wells. James Neely and colleagues who supported his projects in the Tehuacán Valley and the Purrón Dam have studied it. James Neely states "[...] in-line wells is a foggara, which has very ancient origins, and is dated to 1050 BC ±50, but which could be older in the section where its construction began":

A foggara system was found by Ángel Palerm in the municipal border between St: Isabel Xiloxoxtla and S. Francisco Tepeyanco, in 1969–1970, Palerm lowered himself down in that foggara in Tlaxcala. At the bottom of the wells, he collected surface fragments of pre-Hispanic ceramics Aztec III (1403–1455 AD), and Aztec IV (1507–1519 AD), which were later classified by an archaeologist. The existence of these pre-Hispanic "fossilized" channels in-line-wells founded in the Tehuacán valley, serves to confirm the proposal of Ángel Palerm about the pre-Hispanic origin of the St. Isabel Xiloxoxtla foggara (González-Jácome 2003). These systems still have a wide presence in numerous arid and semi-arid areas of the planet.

Rejolladas. They are cavities similar to sinkholes and possibly originated from them by the loss of the water mirror or by the dissolution of carbonates. Gómez-Pompa et al. (1987) describe them as places with great plant diversity, where the trees have greater height, coverage, and diameter at chest height (DAP) than the species that surround the place and are more important. Due to the high percentage of relative humidity they have, these ecosystems have species that exist only in latitudes with the highest rainfall. In Yucatan, there are wild *Brosimum alicastrum*, *Pouteria sapota*, Elastic Castile, ferns, mosses, and sellaginellas. It means that Yucatán has species that would not exist in soils without these characteristics (Flores Guido 2010).

Resilience: This concept can be considered as the internal resistance of a community, including its social, economic, political and / or cultural system to receive and accept the drastic changes that come from outside the system, and transform to adjust them to their needs. This includes local knowledge about the society, its environment and its culture. It is the capacity that a system possesses and develops to overcome external circumstances that are adverse to it, and to be able to achieve a successful development in accordance with society, its culture and its people.

Usually, catastrophic natural phenomena and external economic, social, and political changes (especially drastic ones) occur and affect human societies that live in rural areas and their environment at the local and regional scales mainly. In this sense, national policies that affect the economy, agriculture, and in general, the daily life of small-scale farmers must be taken into account.

Resilience has its central axis in the core of the system and has the capacity to continue maintaining human society with its changes over time (sociocultural adjustments or adaptations). This is what anthropologists call the sociocultural change, which includes different types of components (environmental, historical, economic, political, religious, social, cultural, and ideological). Thus in this way resilience is composed of local and regional knowledge, both in matters related to the environment (TEK) and to traditional agriculture (TAK), among others factors.

Resilience is the internal resistance of human populations and their communities in the face of drastic changes generated from outside that arise from the culture of the human groups; therefore, the rupture of the transmission of local and regional knowledge, both environmental and agricultural, if it modifies human communities, is what happens between one generation and another. That is why the conservation of traditional knowledge through traditional agriculture and the results of studies on it are essential. When there is more cultural diversity, there are more mechanisms of resistance.

Risk: It is the danger caused by eventualities that can occur in environmental, climatological, economic, health, or social circumstances at time. It is a disruption of the inhabitants of a place and their daily routine activities throughout the year. It is the possibility of harm to a person, a human group, or a human society. Environmental, climatological, and economic risks cannot be avoided because they do not depend on the human societies where they occur, but it is possible to know the strategies used by the human populations that were affected by them to regain their stability.

Royal grant (*Merced real*): The king granted land to the conquerors as a reward for their services. The Spanish Crown granted these lands that could be delivered to conquerers and their descendants through the system of grants them for one or more lives.

Sarteneja (*jaltun*): These are natural holes in stones or rocks where rainwater accumulates. There are several types of *sartenejas* and places to find them, such as in Tekax in the South of the Yucatan peninsula (Méndez 1999).

Self-sufficiency and self-consumption: There are differences between the concepts of self-supply and self-consumption. Self-supply is associated with agriculture whose production is destined for the family food and family consumption, almost exclusively. Self-consumption implies both family supply and the sale of products to the market. That is to say, self-supply is functioning within two different economic systems: one non-capitalist (informal, or substantive economy), and the other the formal capitalist economy. For more depth on this problem, see Ángel Palerm's (1978) book on the M-D-M formula.

Semi-terrace (*bancal*): It is an agroecosystem where the Agave is the main crop, and when it is planted on hillsides, ground is piled up to form land edges that follow the contour lines and prevent the soil of the upper parts from sliding down to less graded areas. Agave was cultivated with wheat or barley.

Sex/gender: Although the term sex is replaced by the word gender, currently, here, the term sex means that of people when born and not individual sexual preferences, which are modified socially and culturally later.

Sheep adaptation to weather conditions: The sheep in Yucatán had several problems of adaptation to the fiercely sultry climate, which is why they did not flourish. However, their introduction was successful in temperate-cold places like Tlaxcala, Texcoco, and Querétaro (González-Jácome 1989b: 86–107, 1991, 2008b: 9–24):

Silt edges system (Sistema lama/Bordo): Among the Mixtec, it is called *Coo yuu* and it consists of cultivating in terraces located at the foot of the mountain, in ravines, on high and low hills, ridges, and mountain peaks. For this, methods are used to conserve the soil, improve productivity, and cultivate native plants used traditionally, which contribute to improving family nutrition and recovering traditional agricultural knowledge (TAK).

The agroecosystem is of pre-Hispanic origin, and 284 silt edges systems associated with pre-Hispanic sites in Nochixtlán, Teposcolula, and Tlaxiaco were recorded: In the sixteenth century, agriculture began to decline, coinciding with the abandonment of the silt/boards system, forest exploitation, construction of agricultural soils in the valleys, relocation of villages, increase in goats, and the interruption of the rest cycle of farmland.

That system was developed in the terraces built in the natural drainage canals or ravines and river drains, which extended from the mountain to the bottom of the valley: These agricultural terraces were designed to trap water, avoid soil erosion (Spores 1969), reduce the degradation of the environment, and create productive lands on steep slopes (Pérez-Rodríguez 2003: 8). In the Mixtec, the GEF-Mixtec Northwest Project of the state of Oaxaca was carried out. Its area of incidence covered 1'709, 482.29 hectares, including the Mixtec and a part of the South Sierra. The districts of Coixtlahuaca, Tlaxiaco, Nochixtlán, Teposcolula, Huajuapan, Silacayoapam, Juxtlahuaca in the Mixtec, and Putla in the South Sierra were included. This region borders the states of Puebla and Guerrero.

Pre-Hispanic systems can be reused if they are maintained: They produce excellent crops of maize, cereals, and vegetables (Spores 1969). For Kirkby (1972), the silt/edges terraces had a yield of up to 2 ton/ha in well-irrigated land. For 2010, SAGARPA reported a yield of 0.88 ton/ha of grain maize in rainfed crops in the District of Huajuapan de León (SAGARPA 2011) and of 1.19 ton/ha for the entire state of Oaxaca (Galindo-Escamilla 2008. 17–20; Kirkby 1972; Santiago-Mejía, Martínez-Méndez, Rubio-Granados, Caquera-Huerta, and Sánchez-Escudero 2018 (15), No. 2, April/June; Spores 1972).

***Spondias mombin* and *Spondias purpurea*:** *Spondias mombin* wild, currently, and *Spondias purpurea*, commonly known as Mexican hog plum (jobo, xocote, red plum, abal, coconut, jondura, poon, bighorn plum, or bony plum) are cultivated. Domesticated Mexican hog plum is classified as *S. mombin* in Tehuacán (Smith Jr. 1967: 133–142), and according to Miller and Schaal (2005), several species of it were domesticated there. According to the *Species Plantarum* 1762 (CONABIO), *Spondias* is a very useful tree whose fruits are edible and consumed ripe or raw, dehydrated, or pickled in alcohol or brine. With those fruits, refreshing drinks, vinegars, jellies, firm jellies, atoles, cakes, and sauces are made. *Spondias* leaves are for fodder or for fattening pigs; its wood is used to make paper; its leaf and bark extract is used as a febrifuge; its leaves are used to make and infusion used to wash wounds, inflammations, and burns; its bark extract is used to cure rust, dysentery, and infant flatulence; its wood ashes are used to make soap; its fruit extract or syrup cures chronic diarrhea; its crushed fruit with alcohol is a diuretic and antispasmodic; and the tree is an orchid guide.

Recent studies on domesticated plants found in edges and homegardens in ancient Mesoamerica, like the native plum (Spondias sp.), show in their domestication the presence of haplotypes, unique in agricultural habitats. The Mexican plum supports that traditional agriculture was an important reserve of genetic variability of cultivated species, at the time that native populations of the wild ancestor were declining (Miller and Schaal 2005: 12801–12806)..

I thank Dr: Artemio Cruz León for the ethnobotanical information he provided. *Spondias purpurea* is one of the three species of Spondias native to Mexico and Central America. *Spondias mombin* L. var. mombin (jobo) is a generalized taxon that occurs from Mexico to Paraguay, originating from humid tropical forests. The tree grows to about 30 m, and this species differs from *S. purpurea* in the flower, the color of the fruit, the structure of the inflorescence, the leaf size, and the characteristics of the bark. In Mesoamerica, *S. mombin* var mombin occasionally is cultivated for its fruit and for the construction of the live fences, although it is not as common as the cultivated one.

Spondias radlkoferi Donn: Sm. is the third species of the Spondias native of Mesoamerica, that is morphologically similar to *S. mombin* var. mombin, although it is different in the shape of its fruit, its blooming time, and its cultivation is rare (Miller and Schaal 2005 Vol. 102 No. 36: 12801–12806). In central Veracruz, *Spondias mombin* is sown on road banks or on the extreme land edges of the crops. Generally, the fruit of wild Spondias is bright red, smaller, and more acidic than the cultivated one, and has less pulp surrounding the seed (Miller and Schaal 2005: 12801–12806).

Wild populations of hog plum in the dry tropical forests of Mexico and Central America have a restricted natural habitat. It is estimated that less than 2% of these Mesoamerican forests remain. Unlike those grown, wild Spondias reproduce from seeds and native population plants with a structured age and present a variety of young and ripe fruits. Morphology and the method of reproduction of wild and cultivated hog plum indicate that *Spondias purpurea* has been a genetically altered species since its domestication. In this process, trees with large, fleshy, and sweet fruits were selected, and cuttings easily reproduce them. *Spondias purpurea* is one of the three species native to Mexico and Central America, a tree native to humid tropical forests (Miller and Schaal 2005 Vol. 102 No. 36: 12801–12806).

Talo: It is a typical food of the Spanish and French Basque Countries and northern Navarra eaten on holidays, like in December. It is a crusty maize flour tortilla and eaten with hard cider apple sausage.

Tamales. Tamales in Mesoamerica have been common and ceremonial food since at least 500 BC. Originally, they were made with tender maize kernels crushed, supplemented with wild animal meat, and seasoned with a spicy sauce. Deities were offered tamales on special occasions. The Mayan lords had special clay dishes to eat them on. Today it is not known when or how the common (*macehuales*) and principal people began to consume tamales made of maize dough when the grain began to be nixtamalized, but the times when this occurred varied from region to region.

Currently, there are special tamales, which in Tlaxcala are called "fools, or silly" tamales because they do not contain meat, sauce, or mole, but sometimes they have anise mixed with the dough. They have different names according to each locality, and they are special to eat with mole, instead of tortillas.

The basin of Mexico: Until 1519, it was made up of a lagoon system that included four freshwater lakes (Xaltocan, Zumpango, Xochimilco, and Chalco) and a lake with brackish water (Texcoco). The Tenochcas had formed a freshwater lagoon built using a system of road-dikes, which allowed the existence of this lagoon where the Tenochcas built the city of Tenochtitlan. In the basin of Mexico, the abundance and permanence of the vital liquid in the lagoon system made possible fishing, hunting, and capturing of fish, birds, batrachians, algae, aquatic flora, and others to be very important activities from ancient times until the extinction of the lagoons. This last process ended until the first half of the twentieth century (Aréchiga-Córdoba 2011: 227–270; Correa-Ortiz 2011: 203–226; Espinosa-Pineda 1994: 17–22, 1996).

Tianguis: Weekly outdoors markets. They are of ancient origin and exist in many countries around the world. In Mexico, there are rural and urban street markets. In rural areas, vendors often include producers of fruits, vegetables, tortillas, already prepared fare, beverages, handicrafts, and more. In urban street markets, vendors at huge supply centers purchase the merchandise.

Tlacolole: The tlacolole, or tlacolol, is an agroecosystem developed in hilly areas, on mountain slopes, and the land edges of ravines. It is a rain-irrigated cornfield whose production is sold allowing families to obtain additional income besides what they obtain from the products of the cornfields on flat land:

Tlacoyo: It is a flat piece of rhombus-shaped maize dough filled with beans or ground beans. It can be filled with cottage cheese. It is fried and served with thick cream, hot sauce, and cheese on it.

Tortillas: In Oaxaca, the first tortillas made with maize dough appeared in 500 BC. In Tlaxcala, they appeared toward 400 BC. The thinnest clay plates (comales or griddles) to cook tortillas on arose with maize grounded after being cooked in water with lime (nixtamalization) and to quit the pericarp from the corn kernels. This occurred in Mesoamerica between 500 BC and 100 BC (Nelken-Terner 1968. "Pre-Hispanic grinding implements. Methodological essay", Ma. Thesis (Archaeology). National School of Anthropology, and History (ENAH), Mexico).

Torrejas: They are a dessert of Hispanic origin that arrived in New Spain with wheat bread. There were very popular in the middle-class food system until the 1970s–1980s. Currently, they are found as a dessert in Mexican food and gourmet food restaurants. They are made of hard bread slices, arranged on a plate along with raisins, walnuts, and almonds, bathed in water or sweetened milk, and sprinkled with crumbly fresh cheese.

Traditional agricultural knowledge (TAK): Originally, the traditional ecological knowledge concept (TEK) came from engineering and its use by Canadian anthropology. The author of this book proposes the acronym TAK for traditional agricultural knowledge to provide elements for the studies in traditional agricultural systems. For this purpose, the use of several parameters is proposed: (1) that people in local society and study centers want to share their folk knowledge about the environment, culture, and agriculture, and their management; (2) that they belong to the local group maintaining the practices related to the management of the environment, culture, and agriculture; (3) that they received the knowledge from their parents, grandparents, other relatives, or by observation; (4) that they allow researchers to accompany them in these activities and answer their questions voluntarily; (5) that they add to the study that is carried out the information obtained from observation in situ, historical, bibliographic, hemerography, and cartographic sources, plus font from fields of knowledge, and scientific advances according to the main study objectives (interdisciplinary). [The use of technological advances facilitates access to the information of agriculture, regions, landscapes, and so on from other researchers of various fields of knowledge. Many of them are using proxy and maxi proxy nets to search for information that was achieved through disciplinary, interdisciplinary, or multidisciplinary studies formerly. Researchers can use framework models such as the one proposed by the sustainability sciences also.]; (6) that it is necessary to compare the mentioned components to have new questions, get (on) some answer(s) to the problems under study, and organize some conclusions that will be realized through different types of writings, records, or direct contributions of the populations included in the research and studies (applied science).

There is a difference between traditional knowledge (scientific, structured, formally organized) and local knowledge (folk, generationally transmitted), concepts, and models which we will not discuss in this book but will explain in this note briefly. It would be necessary to ask how these two models or constructs can be combined and what for. That is why we propose the concept of TAK, but the answer to what we want it for is at the discretion of researchers and institutions where they work or those who support them in their research.

Traditional behavior patterns and traditional agriculture: Behavior patterns are called traditional and included ideas about the world, elements of material, and non-material culture that are transmitted generationally, but that are modified (changed) when adjusted (sociocultural adaptation) according to time, the needs of a human group, and its generations and their contacts with other societies and their cultures. So humans apply the innovations observed to their daily manners; behavior; ideology; forms of social, religious, and political organization and their economics; and generational interests.

On the contrary, the concepts of evolution and change are immersed in the description and analysis of agricultural systems or agroecosystems, which have been and are yet part of the studies on traditional agriculture, in this case those on Mexico: The processes of change occur over time, although changes may or may not be accepted and applied by all members of human society (community). It is only over time that we know if they were incorporated – or not – into the social life of a population, and how they fit its needs.

Traditional food and TV: Mexican cultural TV touts food recipes in cooking programs where experienced chefs or traditional cooks show the viewer how to select foods, cook them, and consume them. They show the importance of maize, as part of customary diets, and present special food for parties and ceremonies. The efforts of Channel 11-of the IPN- channel 14 and channel 22, as well people specialized in traditional Mexican food, such as Yuri de Gortari (+) and Edmundo Escamilla (+), are remarkable about the history of oral traditions and the production of "traditional foods". There are other channels on open television that feature programs related to food, but without the idea of preserving traditional knowledge about it, and apparently, are only a "fashion", to entertain the public.

Tuba: It is a palm beverage refreshment, which arrived in the seventeenth century from the Philippines to New Spain during the Viceroyalty. Currently, tuba is sold on the streets of Colima.

Vavilov: One of the forerunners in the study of agriculture is Nicolai Ivanovich Vavilov (1992 [1926a]; 1926b, 1927, 1931). He was a Russian botanist, geneticist, and agronomist who between 1924 and 1940 conducted numerous studies on the origins of agriculture and cultivated plants. To do this, he proposed several centers of origin of world agriculture. With the support of several collaborators, he carried out 98 investigations, including fieldwork in situ and the collection of materials for laboratory analysis:

These studies looked to obtain information on agricultural techniques and were done in the Mediterranean countries of Morocco, Algeria, Tunisia, Egypt, Portugal, Spain, Italy, Greece, Asia Minor, Syria, Palestine, Sicily, Sardinia, Crete, Cyprus, and the Rhodes islands. Vavilov and his collaborators investigated Abyssinia, Eritrea, Persia, Afghanistan, western China, agricultural Mongolia, Japan, Korea, Formosa, India, and the former agricultural regions of Transcaucasia and Turkestan. In the New World, they studied Mexico (Yucatán), Guatemala, Colombia, Peru, Bolivia, and Chile.

Vicarious observation (observational learning—of others observation): Albert Bandura, a Canadian psychologist, called vicarious observation to the learning process that a human being exercises by observing what others do and learning from it without the need for the trial of success or failure from doing it him/herself. Vicarious observation, or the observation of others, was the result of first-hand intergroup contact (Bandura 1977, 2002: 209–238). In agriculture, it is very important because the small-scale farmer do no need to lose the crops, and the money that it cost to him by experiment with them or new methods without the experience of what happened to other farmers (Moctezuma-Pérez, in Rebolledo 2016: 79–94).

Vitzo: It is a fermented beverage made with pounded parts like the head of the Gentry *Agave bovicornuta*, or that are backed and boiled in water. This beverage is called *vitzo* in the Alta Pimería, or *chahuí* in southern Sonora among the Guarijíos.

Vulnerability model: With three cases, Turner et al. (2003: 8080–8085) used a vulnerability model (framework) that included the coupled human-environment system, explaining the interactions that affected the system's responses to hazards and their vulnerability. The three case studies included the tropical south of Yucatan, the Yaqui Valley in northwestern Mexico, and the Pan-Arctic. The cases illustrated the role of external forces in the remodeling of systems and their vulnerability to environmental hazards. The claim holders had different capacities to respond to changes and dangers. The vulnerability model directed attention to the interactive parts of the coupled system, allowing identifying the gaps in information and understanding that were relevant to reduce vulnerability in systems in general:

The lack of a historical vision in the applied framework limited the better understanding of the processes of population adjustment in the cases studied, and the responses given over time to environmental and climatic hazards.

Wild holotype (also called type specimen): It is the sample of an organism used for the description and designation of a new species. Ideally, it should be typical of its taxon, although in the case of fossil taxa, frequently the holotype is a partial sample (Cela-Conde, and Ayala (2013), *Human evolution. the path of our species*. Madrid. Editorial Alliance).

Xocoyotazgo: From 1970 to 1980, the land inheritance system in Tlaxcala left the house and most of the land to the youngest son, with the obligation to care for the parents in their old age (*xocoyotazgo*). Older children sought economic activities, such as commerce or employment in nearby industries, or they migrated temporarily outside the town to raise capital with which to acquire land and build their houses. Daughters inherited some implements, such as the mother's sewing machine. However, the ejido land of the family became insufficient, gradually (from smallholding to group holding), so the men left agriculture to devote themselves to other activities to make a livelihood.

Zoochory: In Botanic, it is a form of dispersion in which the propagator is an animal the agent carrying the part of a plant that is capable of the vegetative reproduction, originating another similar individual. Zoochory dispersion and anemochory dispersion (wind dispersion) allow knowing in greater detail the processes of the invention of agriculture and the creation of diversity (López-Acosta et al. 2014: 44–58).

ABBREVIATIONS

AC:	After Christ
AD, AP:	After Present
BP:	Before Present
BC:	Before Christ
BANRURAL:	National Bank of Rural Credit
CINVESTAV:	The Center for Research and Advanced Studies of the National Polytechnic Institute
CONACYT:	The National Council of Science and Technology
CONCYTEY:	Council of Science, Innovation and Technology of the State of Yucatan
CONASUPO:	National Company of Popular Subsistence
COPLAMAR:	Commission for the Plan for Depressed Zones and Marginalized Groups (1965–2001)
DICONSA:	Rural supply program
ECOSUR:	The South Frontier College
FINASA (ICP):	National Sugar Finance (Promoter of the Papaloapan Basin)
FOMIX:	Mixed Fund for the Promotion of Science and Technology Research CONACYT
GDP:	Gross Domestic Product
IMECAFÉ:	Mexican Coffee Institute
INEGI:	National Institute of Geography and Statistics
km:	kilometers
r-t-q:	r, brush the shrub vegetation, grass, and wild plants which grew in the field; t, cut the trees that were not selected to remain on the plot; q, burn the vegetation that was in the land, after guarding the trees to be left in the cornfield and the boundaries of the land to be burned with a canal excavated in the land around them

LINEAR AND AREA MEASURE

Hectare:	2.471 acres
Acre:	0.405 hectare
1 meter:	0.5468 fathoms
1 square meters (sqm):	1.09 yards
1 foot:	0.3333 yards
1 mile:	1760 yards
1 kilometer:	0–6214 miles
1 square mile:	2.59 square kilometers
Meters above the Sea Level:	masl
Millimeters:	mm
OPORTUNIDADES (2000–2006):	Program that replaces PROGRESA in the Vicente Fox government
PIDER (1970–1982):	Investment Program for Rural Development
PROCAMPO:	National Program supporting small-scale farmers (*Programa de apoyo al campo*)
PROC. NATL. ACAD. SCI. U.S.A. (PNAS):	Proceedings National Academy of Sciences U.S.A.
PROGRESA (1997–2001):	Education, Health and Food Program
PRONASOL (1988–1995):	National Solidarity Program
S:	San
SAG:	Secretariat of Agriculture and Livestock
SRH:	Secretariat of Hydraulic Resources

St:	Saint
SEMARNAT:	Secretariat of Environment and Natural Resources
TGA:	Teosinte Glume Architecture
UNAM:	National Autonomous University of Mexico
UNIBIO:	Biodiversity Computing Unit, Institute of Biology (Biological Collections)
UIMQROO:	Maya Intercultural University of Quintana Roo

Historical Sources

Alva Ixtlilxóchitl, don Fernando de. 1965 [1610 to 1640]. *Obras Históricas*. Publicadas y anotadas por Alfredo Chavero, 2 Vols. México, Editora Nacional. Vol. II *Historia Chichimeca*.

Anales de Cuauhtitlán, en el Códice Chimalpopoca. 1945. Primo Feliciano Velázquez translator. México, Universidad Nacional Autónoma de México (UNAM), Instituto de Historia; pp. 1–118, 145–164, facsimile.

Anónimo. Siglo XVI. Diccionario de Motul. John Carter Brown Library, UAE: Rhode Island, Brown University.

Ajofrín, Francisco, fray. 1958 [1763]. Diario del viaje que por orden de la Sagrada Congregación de Propaganda Fide hizo a la América Septentrional en el siglo XVIII el p[adre] fray Francisco de Ajofrín. Imprenta y Editorial Maestre, Madrid. In: Archivo Documental Español; 2 Vols.

Burgoa, Francisco de. 1672. Geográfica Descripción de la Parte Septentrional del Polo Ártico de La America, y Nueva Iglesia de las Indias Occidentales, y Sitio Astronómico de esta Provincia de Predicadores de Antequera Valle de Oaxaca: En Diez y Siete Grados del Trópico de Cáncer: Debajo de Los Aspectos, Radiaciones de Planetas Morales, la Fundaron Con Virtud de Celestes, Influyéndola en Santidad, y Doctrina. Vol., I. Facsimile: Tecnológico de Monterrey. En: http:hdl.handle.net/11285/573957

Broecke, Pieter Van Den. 2000 [1634]. *Journal of Voyages to Cape Verde, Guinea and Angola*, 1605–1612. La Fleur J.D. Ed. Hakluyt Society.

Cárdenas, Juan de. 1791 [1571]. Primera Parte de los Problemas y Secretos Maravillosos de las Indias Compuesta por don Juan de Cárdenas Médico. Dirigida al Ilustrísimo Señor don Luis de Velasco, Virrey desta Nueva España. Con Licencia. En México, casa de Pedro Ocharte. Año de 1591 (Facsimile).

Cárdenas, Juan de. 1988 [1571]. *Problemas y secretos maravillosos de las Indias*. Madrid, Alianza.

Cervantes de Salazar, Francisco. 1971 [1557–1564]. *Crónica de la Nueva España*. Madrid, Biblioteca de Autores Españoles. Ed. Atlas, 2 Vols.

Clavijero, Francisco Javier. 1971 [1780]. *Historia antigua de México*. México, Ed. Porrúa.

Códice Borgia. 1963. Three Vols. Facsimile; Comments from Eduard Seler, Germany, 1rst Edition 1906. México, Fondo de Cultura Económica (F.C.E.).

Códice Borgia. F.53. In: Codice_Borgia_Digital_Facsimile.pdf

Códice Dresden. Facsimile de "Original de una Pintura Mexicana", Real Library, Dresden (74 pp.). Jay Kisla Foundation. Miami Lakes, FL. Arthur Dunkelman, Director. FAMSI photos from Justin Kerr. Consulted 21/08/2013, in Biblioteca Digital mundial 1200–1250. There is an edition in Arqueología Mexicana.

Códice Florentino. 1979. Facsimile from Manuscript 218–220, Palatina Collection, Library Medicea Laurenziana. México, Gobierno de la República. Edición facsimilar en Archivo General de la Nación (AGN), 3 Vols.

Códice Madrid (Trocortesiano, Matritense Maya), Museum de América, Madrid. 1985 [original Posclásico Tardío]. Facsimile by Thomas A. Lee, Universidad Autónoma de Chiapas, Tuxtla Gutiérrez.

Códice Madrid. BNAH ff. 28, 68a, Museo de América, Madrid, España. It has 112 Pages, divided in two sections: Códice Troano, and Códice Cortesiano. The two sections were put together in 1888.

Códice Magliabecchiano. XIII.3. 1540. Manuscript, Mexican Post-Colombian. National Central Library, Florencia. Reproduction from FAMSI – Akademiche Druck–Verlagnstalt– Graz– Codex Magliabechiano. Lam. 73r.

Cortés, Hernán. 1960 [1524]. Cartas de Relación. México, Ed. Porrúa.

Cortés, Hernán. 1993 [1524]. Cartas de Relación. Madrid, Clásicos Castalia.

Díaz del Castillo, Bernal. 1960. Historia Verdadera de la Conquista de la Nueva España. México, Ed. Porrúa.

Dupaix, M. 1964. Antigüedades de México. Basadas en la recopilación de Lord Kingsborough (Edic. facsimilar). México, Sria. de Hacienda y Crédito Público.

Emmeret, Emily W. 1940. Edición facsimilar del Manuscrito Badiano.

Farfán, Fray Agustín. 1579. Tratado Breve de Anathomia y Chirvgia. México.

Farfán, Fray Agustín 1592. Tratado Brebe de Medicina. México.

Fernández de Oviedo, Gonzalo. 2011 [1526]. Sumario de la Natural Historia de las Indias. Estudio, edición y notas de Álvaro Baraibar. Universidad de Navarra, Iberoamericana-Verbuert, 2010. Biblioteca Indiana, 26, Centro de Estudios Indianos (CEI). ISBN 978-84-8489-554-1 (Iberoamericana).

Fernández Echeberría y Veytia, Mariano. 1931 [1780]. Historia de la Fundación de Ciudad de la Puebla de los Ángeles en la Nueva España. Su Descripción y Presente. Estado de Puebla, Colección 5° Centenario, 2 Vols.

Gemelli Carreri, Giovanni Francesco. 1983 [1697]. Viaje a La Nueva España. México, Universidad Nacional Autónoma de México (UNAM).
Gómez de Cervantes, Gonzalo. 1944 [1599]. Memorial de Gonzalo Gómez de Cervantes para el Oidor Eugenio Salazar, Oidor del Real Consejo de las Indias, S. l., 1599. Editado con el título: La vida económica y social de Nueva España al finalizar el siglo XVI. Prólogo y notas de A.M. Carreño, México, Antigua Librería Robredo de José Porrúa e Hijos, 1944. Ed. crítica y traducción de W.L. Gram, Alburquerque, The University of New Mexico 1984.
Gómez Ortega, Casimiro. 1779. Instrucción. Sobre el modo más seguro y económico de transportar plantas vivas. Archivo General de Indias (AGI), España.
Hernández, Francisco. 1984 [1574]. Historia Natural de Nueva España. Obras Completas, 2 Vols. México, Universidad Nacional Autónoma de México (UNAM).
Humboldt, Alejandro de. 1978 [1822]. Ensayo Político sobre el Reino de la Nueva España. México, Ed. Porrúa S.A.
Humboldt, Alejandro de. 1822. Ensayo Político Sobre el Reino de la Nueva España. París, Edición Príncipe, Vol. II, tomos III y IV.
Kaerger, Karl. 1986 [1900]. Agricultura y colonización en México en 1900. México, Universidad Autónoma Chapingo y Centro de Investigaciones y Estudios Superiores en Antropología Social (CIESAS).
Landa, fray Diego de. 1983 [1566]. Relación de las Cosas de Yucatán. México, Mérida, Yucatán, Ed. Dante.
Lienzo de Tlaxcala. Códice Histórico Colonial del siglo XVI. Copia 1773 de Juan Manuel Yllanes del Huerto. Su Historia y su Contexto. 2016. Tlaxcala, Gobierno del estado, SEP, USET, Cultura, INAH, BNAH.
Linneai, Caroli. 1762. Species Plantarum, Exhibentes Plantas Rite Cognitas, and Genera Relatas, CUM Differentiis Specificis, No minibus Trivialibus, Synonymies Selectis, Locis Natalibus, Secundum Sistema Sexuale Digestas. Tomus 1. Edition Secunda, Aucta. HOLMIE. Impensis Direct, LAURENTII SALVII. Vol: T5. Consulted in: Biodiversity Heritage Library (BHL). www.biodiversitylibrary.org
López, Tomás. 1552–1553. Ordenanzas. In: Apéndice Documento N° 8. Landa 1983 [1566]: 220). Michelon, Oscar (ed.).1976 [1866]. Diccionario de San Francisco. Biblioteca Lingüística Americana, Vol. II, Graz, Austria.
Molina, fray Alonso de. 1970 [1571]. Vocabulario en Lengua Castellana y Mexicana Compuesto Por El Muy Reverendo Padre Fray Alonso de Molina, de la Orden del bienaventurado Nuestro Padre San Francisco. México, Ed. Porrúa.
Mota y Escobar, Fray Alonso de la. 1987. Memoriales del Obispo de Tlaxcala. Un Recorrido Por el Centro de México a Principios del Siglo XVII. A. González-Jácome (Introduction and Notes), México, Secretaría de Educación Pública (SEP). Hay edición de 2012.
Motolinía, fray Toribio de Benavente. 1996 [1568]. Memoriales. El Colegio de México (COLMEX), Centro de Estudios Lingüísticos y Literarios, Biblioteca Novohispana.
Motolinía, fray Toribio de Benavente. 2001. Historia de los Indios de la Nueva España. México, Ed. Porrúa.
Navarro, Fray Juan de. 1992 [XVIII Century]. Historia Natural o Jardín Americano. México, Universidad Nacional Autónoma de México (UNAM), Instituto Mexicano del Seguro Social (IMSS), Instituto de Salud y Seguridad Social de los Trabajadores de la Educación (ISSSTE)
Navarro, Fray Juan de. 2002 [siglo XVIII]. Historia Natural o Jardín Americano. In*: Arqueología mexicana*, Vol. X, N° 57; pp.16–17.
Nentvig, Juan J.S. 1971 [1764]. Descripción geográfica, natural y curiosa de la provincia de Sonora por un amigo del servicio de Dios y del Rey Nuestro Señor, Año de 1764. Edición preparada con una introducción histórica, notas, apéndice e índice analítico por Germán Viveros. México, Archivo General de la Nación (AGN), Vol. XXXI.
Nuttall, Zelia. 1920. "Los Jardines del Antiguo México". In: Memorias y Revista de la Sociedad Científica Antonio Alzate; Tomo 37, N° 4, 5, 6. México; pp. 193–213.
Nuttall, Zelia. 1923. "The Gardens of Ancient Mexico". Annual Report of the Smithsonian Institution, Washington DC; pp. 453–464.
Nuttall, Zelia. 1983. Códice Magliabechiano. In: The book of the Life of the Ancient Mexicans. University of California Press, Berkeley and Los Ángeles.
Observatorio Meteorológico Central (Compilador). 1895. Carpología Mexicana: directorio general sobre la producción de frutas en las municipalidades del país. Obra compilada por disposición de la Secretaría de Fomento. México, Oficina Tipográfica de la Secretaría de Fomento.
Pomar, Juan Bautista. 1891 [1582]. Relación de las Antigüedades Políticas y Religiosas de los Indios dirigida al Rey Nuestro Señor en 9 de marzo de 1582. In: Relaciones de la Nueva España. Madrid, Historia 16. Conocida como Relación de Texcoco, publicada por García Icazbalceta, Nueva Colección de Documentos para la Historia de México, Vol. III, México 1891. (Publicaciones históricas digitales, UNAM, mx.).
Ponce, Alonso fray. 1873 [1584–1587]. Relación breve y verdadera de algunas cosas de las muchas que sucedieron al padre fray Alonso Ponce en las provincias de la Nueva España. Siendo Comisario General de

Aquellas Partes. Escrita por dos religiosos, sus compañeros. El uno de los cuales le acompañó desde España a México, y el otro en todos los demás caminos que hizo y trabajos que pasó. Madrid, 2 Vols.

Quer y Martínez, Joseph. 1762. Flora Española. Historia de las Plantas que se Crían en España. Madrid, Joachin Ibarra. 6 Vols. Ed. Maxtor, Valladolid, España, 2012.

Relaciones Histórico-geográficas de la Gobernación de Yucatán. In: De la Garza, M.; A.L. Izquierdo; M. del C. León y T. Figueroa. 1983 [XVI-XVIII Centuries]. Relaciones Histórico Geográficas de la Gobernación de Yucatán (Mérida, Valladolid y Tabasco). México, Universidad Nacional Autónoma de México (UNAM), 2 Vols.

Sahagún, fray Bernardino de. 1956 [1567]. Historia General de las Cosas de la España. México, Ed. Porrúa. 4 Vols (12 books).

Sahagún, fray Bernardino de. 1830 [1567]. Historia General de las Cosas de la Nueva España, que en doce libros y dos volúmenes escribió El R.P. FR. Bernardino de Sahagún, de la Observancia de San Francisco, y uno de los Primeros Predicadores del Santo Evangelio en Aquellas Regiones. Edic. Carlos María de Bustamante, Imprenta del Ciudadano Alejandro Valdés, calle de Santo Domingo, y esquina de Tacuba.

Sahagún, fray Bernardino de. 1990. Historia General de las Cosas de Nueva España. Edición de J.C. Temprano. Madrid, Historia 16.

Solís, Antonio de. 1970 [1683]. Historia de la Conquista de México. Madrid, Espasa Calpe, 4ª edición.

Shull, Geoge H. 2012 [Published in the late ninetenth century]. "The Composition of a field of Maize"; American Genetic Association (Marian Breeder's Association), pp. 296–301. Consulted in: http://jhered.pxford journals. Org/at Duke University, on September 23, 2021.

Torquemada, F. Juan de. 1723. Primera parte de los veintiún libros rituales y monarquía indígena, con el origen y guerras, de los indios occidentales, de sus poblaciones, descubrimiento, conquista, conversión y otras cosas maravillosas de la misma tierra. Madrid, 2ª edición.

Torre, Tomás de la, Fray. 1544–1545. "Diario de Viaje". In: Viajeros en Tabasco, Tomo 1. Ciprián Aurelio Cabrera Bernat (comp.). Gob. Del estado de Tabasco, Instituto Estatal de Cultura de Tabasco, Colección Independencia Nacional, 2ª edición 2011.

Velasco, Alfonso Luis. 1890. Geografia y Estadística de la República Mexicana Vol. XI. México, Oficina Tipográfica de la Sec. de Fomento.

References

Abascal, R. y Á. García Cook. 1973. Sistemas de cultivo, riego y control de agua en el área de Tlaxcala. Sociedad Mexicana de Antropología, *XIII Mesa Redonda*, Sept. 9–15; pp. 117–124.

Abascal, R. y Á. García Cook. 1975. Sistemas de cultivo, riego y control de agua en la región de Tlaxcala. In: México, Mesa redonda, Sociedad Mexicana de Antropología (SMA) *Arqueologia*, I; pp. 199–210.

Abasolo-Palacios, V.E. 2003. La altitud y la agricultura en Raíces, Estado de México. PhD. Thesis, Social Anthropology, Universidad Iberoamericana.

Abasolo-Palacios, V.E. 2004. Personal communication.

Aboites-Aguilar, L. 1998. *El agua de la nación. Una historia política de México* (1888–1946). México, Centro de Investigaciones y Estudios Superiores en Antropología Social (CIESAS).

Abrams E.M.; A. Freter; D.J. Rue; J.D. Wingard. 1996. The role of deforestation in the collapse of the Late Classic Copan Maya. In: L.E. Sponsel; T.H. Headland; and R.C. Bailey (eds.); *Tropical Deforestation: The Human Dimension.* Columbia University Press, New York; pp. 555–575.

Abundis-Canales, J. 2007. *La Huella Carmelita en San Ángel.* México, Instituto Nacional de Antropología e Historia (INAH), 2 Vols.

Acevedo, M.L. 1980. Movimientos campesinos indígenas. El caso de la presa Cerro de Oro. *Anales de Antropología*, Vol. 2: etnología y lingüística. México, Universidad Nacional Autónoma de México (UNAM), Instituto de Investigaciones Antropológicas (IIA), pp. 223–232.

Acle-Tomasini, G. 2000. Gente de razón: Educación y Cultura en Temoaya, Estado de México. Thesis, Doct. Social Anthropology, México, Universidad Iberoamericana.

Acosta-Naranjo, R. 2002. *Los Entramados de la Diversidad. Antropología Social de la Dehesa.* España, Diputación de Badajoz, Departamento de Publicaciones, Colección raíces.

Acosta-Ochoa, G. 2008. La cueva de Santa Marta y los cazadores recolectores del Pleistoceno final-Holoceno temprano en las regiones tropicales de México. Thesis, Universidad Autónoma de México (UNAM), Instituto de Investigaciones Antropológicas.

Acosta-Ochoa, G. 2010. Late-Pleistocene/early-Holocene tropical foragers of Chiapas, Mexico, recent studies. *Current Research in the Pleistocene* 27: 1–4.

Acosta-Ochoa, G. 2011. El Poblamiento de las Regiones Tropicales de México Hace 12,500 Años. *Anales Antropología* 45: 227–235.

Acuña-Soto, R.; D.W. Stahle; M.D. Therrell; S. Gómez-Chávez; M.K. Cleaveland. 2005. Drought, epidemic disease, and the fall of classic period cultures in Mesoamerica (AD 750–950). Hemorrhagic fevers as a cause of massive population loss. *Med Hypotheses* 65: 405–409.

Adapon, J. 2008. *Culinary Art and Anthropology*, New York, Berg.

Aeppli, H. y E. Schoenhals. 1973. Los suelos de la cuenca Puebla–Tlaxcala. In: *Comunicaciones 7.* Fundación Alemana para la Investigación Científica. México, Puebla; pp. 15–18.

Aide T.M.; J.K. Zimmerman; R. Maydee and Marcano H. 1996. Forest recovery in abandoned cattle pastures along an elevational gradient in Northeastern Puerto Rico. *Biotropica*, 28: 537–548.

Aiuvalasit, M.J., Neely, J.A. and Bateman, M.D. 2010. New radiometric dating of water management features at the prehistoric purron dam complex, Tehuacán Valley, Puebla, Mexico. *Journal of Archaeological Science* 57(6): 1207–1213.

Agroecología. 1973. R-3. México, Secretaría de Recursos Hidráulicos.

Agüero, R.J.C. 1998. Desarrollo y Transformación étnica y social entre los mazatecos de la presa Miguel Alemán: Caso los mazatecos de Playa Vicente, Veracruz. BA Thesis, Sociology. Xalapa, Universidad Veracruzana.

Aguilar, J., C. Illsley, and y C. Marielle. 2003. El sistema agrícola de maíz y sus procesos técnicos. In: G. Esteva and C. Marielle (coords.); *Sin Maíz no hay País.* Consejo Nacional para la Cultura y las Artes (CONACULTA), Dirección General de Culturas Populares e Indígenas, México, D.F. pp. 83–122.

Aguilera R.S. 2004. Población, Desarrollo y Uso de los Recursos Naturales en el Valle de Uxpanapa, Veracruz. BA Thesis, Sociology. Xalapa, Universidad Veracruzana (UV).

Aguirre Beltrán, G. 1943. La medicina ilusoria de los negros. *Anuario de la Sociedad Folklórica de México*, 1: 192.

Aguirre Beltrán, G. 1947. La medicina indígena. *América Indígena*, 7: 107–127.

Aguirre Beltrán, G. 1995 [1952]. *Problemas de la población indígena de la Cuenca del Tepalcatepec*. México, Memorias, Instituto Nacional Indigenista (INI), 2 Vols. (Reedición 1995), Fondo de Cultura Económica (F.C.E.).

Aguirre Beltrán, G. 1955 [2ª ed. 1980]. *Programas de salud en la situación intercultural*. México, Instituto Indigenista Interamericano (III), Instituto Mexicano del Seguro Social (IMSS).

Aguirre Beltrán, G. 1958. *Cuijla. Esbozo etnográfico de un pueblo negro*. México, Fondo de Cultura Económica (F.C.E).

Aguirre Beltrán, G. 1963. *Medicina y Magia: el proceso de aculturación en la estructura colonial*. México, Instituto Nacional Indigenista (INI).

Aguirre Beltrán, G. 1965. El rol de la medicina en las regiones de refugio. *Homenaje a Juan Comas*. Libros de México, 1: 23–77.

Aguirre Beltrán, G. 1967. "Un postulado de la política indigenista". *América Indígena*, Vol. XXVII. In: *Aguirre Beltrán: obra polémica*. México, SepInah 1976, ediciones La Casa Chata; pp. 21–28.

Aguirre Beltrán, G. 1972. [1946]. *La población negra de México*. México, Fondo de Cultura Económica (F.C.E.).

Aguirre Beltrán, G. 1975. Etnocidio en México: una denuncia irresponsable. *América Indígena*, XXXV (2): México; pp. 405–418.

Aguirre Beltrán, G. 1986. *Antropología Médica*. México, Centro de Investigaciones en Antropología Social (CIESAS).

Aimers J.J. and D.A. Hodell. 2011. Societal collapse: drought and the maya. *Nature* 479: 44–45.

Alarcón, Ch. P., M. Olivo y L. Solís. 2001. Diversidad gastronómica de los pueblos indios de México. *Etnoecológica* 6(8): 100–102.

Alavez-Chávez, R.G. 1992. La Fauna en la Mixteca y sus Pronósticos. In: M.A. Vásquez-Dávila (coord.) *Como predecimos el clima. Augurios Animales*. Instituto Tecnológico Agropecuario de Oaxaca. Casa de la Cultura Oaxaqueña. Colección Como hacemos; pp. 9–19.

Alayón, A.J.; and Gurri, F.D. 2007. Flujo y balance energético en los agroecosistemas de campesinos tradicionales y convencionales del sur de Calakmul, Campeche, México. In: González, A.; S. del Amo y F.D. Gurri (eds.); *Los nuevos caminos de la agricultura: procesos de conversión y perspectivas*. México, Universidad Iberoamericana, Plaza y Valdés; pp. 243–260.

Albores-Zárate B.A. 1992a. El modo de vida lacustre en el Alto Lerma. Thesis, Doct., Social Anthropology. México, Universidad Nacional Autónoma de México (UNAM), Facultad de Filosofía y Letras.

Albores-Zárate B.A. 1992b [1995]. *Tules y sirenas: el impacto ecológico cultural de la industrialización en el Alto Lerma*. México, El Colegio Mexiquense, Gobierno del Estado de México, Secretaría de Ecología.

Albores-Zárate B.A. 1992c [reedición 2003]. *Graniceros, Cosmovisión y Meteorología Indígenas de Mesoamérica*. México, El Colegio Mexiquense AC, Universidad Nacional Autónoma de México (UNAM).

Albores-Zárate B.A. 2016. (coord.). *Flor-flora. Su uso ritual en Mesoamérica*. El Colegio Mexiquense, Fondo Editorial del Gobierno del Estado de México.

Albores-Zárate B.A. 2018. Aspectos adaptativos del maíz de temporal en una zona lacustre de altura. El caso de San Mateo Atenco, Estado de México. In: L. Reyes-Montes; J.M. Pérez-Sánchez; S. Moctezuma-Pérez (coords.). *Sistemas agrícolas tradicionales. Biodiversidad y cultura*. Estado de México, El Colegio Mexiquense AC.; pp. 117–144.

Alcorn, J.B. 1984. *Huastec Mayan Ethnobotany*. Austin, University of Texas Press.

Alcorn, J.B.; B. Edmonson and C. Hernández V. 2006. Thipaak and the Origins of Maize in Northern Mesoamerica. *Histories of Maize. Multidisciplinary Approaches to the Prehistory, Linguistics, Biogeography, Domestication, and Evolution of Maize*. Amsterdam, London, New York, Oxford; Elsevier, Academic Press; pp. 600–609.

Alexandrovskiy, A.L.; M.P. Glasko; N.A. Krenke; O.A. Chichagova. 2004. Buried paleosols of floodplains and paleoenvironmental changes in the Holocene. *Revista Mexicana de Ciencias Geológicas*, 21/1: 9–17.

Allen, M.F.; S.L. Fedick; A. Gómez-Pompa; J.J. Jiménez O. 2003. The Maya Lowlands; a case study for the future? Conclusions. In: Gómez-Pompa, Allen, Fedick and Jiménez (eds.). *The Lowland Maya Area. Three Millennia and the Human Wildland Interface*. Food Products Press, Haworth Press Inc, New York; pp. 623–634.

Allyson, J. 1983. An ecological analysis of home gardens (*huertos familiares*) in two Mexican villages. Thesis, Ma. University of California, Santa Cruz.

Altieri, M.A. 1983 [reedition1987]. *Agroecology. The Scientific Basis of Alternative Agriculture*. Westview Press, Boulder, Colorado.

Altieri, M.A. 1991. ¿Por qué estudiar la agricultura tradicional?. *Clades*, N° especial 1°/03. http://www.clades.org/rl-art2.hym

Altieri, M.A. 1992. Ecología y manejo de malezas. Chapter 14. In: Altieri M.A.; *Agroecología. Bases para una agricultura sustentable*. Nordan Comunidad; Montevideo; pp. 262–281.

Altieri, M.A.; C. Nicholls. 2012. Agroecología. Única esperanza para la soberanía alimentaria y la resiliencia ecológica. Una contribución a las discusiones de Rio+20 sobre temas en la interface del hambre, la agricultura, y la justicia ambiental y social. Sociedad Científica Latinoamericana de Agroecología (SOCLA).

Altieri, M.A.; M.K. Anderson. 1986. An ecological basis for the development of alternative agricultural systems for small farmers in the third world. *American Journal of Alternative Agriculture* 1:30–38.

Altieri, M.A.; L.C. Merrick. 1987. In Situ Conservation of Crop Genetic Resources through Maintenance of Traditional Farming Systems. In: *Economic Botany*, 4 (1):86–96.

Altieri, M.A.; D.L. Glaser; and L.L. Schmidt. 1989a. Diversification of agroecosystems for insect pest regulation: experiments with collards. In: *Agroecology. Researching the Ecological Basis for Sustainable Agriculture*. Springer-Verlag: 70–82.

Altieri, M.A.; J. Trujillo.; L. Campos; K. Klein; C.S. Gold; and J.R. Quezada. 1989b. El control biológico clásico en América Latina en su contexto histórico. *Manejo Integrado de Plagas* 12: 82–107.

Altieri, M.A.; V. Toledo 2005. Natural resource management among small- scale farmers in semi-arid lands: building on traditional knowledge and agroecology. In: *Annals of Arid Zone* 44(3 y 4): 365–385.

Alvarado-Méndez, M.C.; y J. Delgadillo-Macías. 2017. Procesos de cambio en la actividad agrícola e incidencia en la fragmentación territorial en el sur del estado de Tlaxcala. *Rev. de Geografía Agrícola, estudios regionales de la agricultura mexicana*. Estado de México, Texcoco, Universidad Autónoma Chapingo, Direc. de Centros Regionales, N° 58, enero-junio; pp. 21–31.

Álvarez-Buylla E.R. y A. Piñeyro N. 2013. *El maíz en peligro ante los Transgénicos. Un análisis integral sobre el caso de México*. México, Universidad Nacional Autónoma de México (UNAM), Universidad Veracruzana (UV), Unión de Científicos Comprometidos con la Sociedad.

Álvarez-Ríos, G.D.; C.J. Figueredo-Urbina; A. Casas. 2020. Physical, chemical, and microbiological characteristics of pulque: management of a fermented beverage in Michoacán, Mexico. Multidisciplinary Digital Publishing Institute (MDPI), *Foods* 9 (361): 1–17.

Alverson, H. 1984. The wisdom of tradition in the development of dry-land farming: Botswana. *Human Organization* 43: 1–8.

Anaya-Lang, A.L. y F.J. Espinosa-García. 2006. La química que entreteje a los seres vivos. *Ciencias*, julio-sept. N° 083. México, Universidad Nacional Autónoma de México (UNAM); pp. 4–13.

Anderson E.J. 1967. The Human Skeletons. *The Prehistory of the Tehuacan Valley. Environment and Subsistence*, Vol. 1. The University of Austin Press: 91–113.

Andrade-Torres, J. 2004. Construcción de una identidad: los trabajadores migrantes de Santa Bárbara, Guanajuato y su Barrio en la Ciudad de Forth Worth, Texas. Tesis, Doct. Social Anthropology; México, Universidad Iberoamericana.

Andrade-Torres, J. 2008. Entrevista sobre la vida de los migrantes en Forth Worth, Otoño.

Andrade, F.B.S; Del Amo R.S.; B. Ortíz-Espejel. 2011. Memoria, territorio y significación ambiental: el caso del Totonacapan. In: Argueta, A.; E. Corona y P. Hersch (Coords). *Saberes colectivos y Diálogos de Saberes en México*. Universidad Nacional Autónoma de México (UNAM), Centro Regional de Investigaciones Multidisciplinarias (CRIM), Instituto Nacional de Antropología e Historia (INAH); pp. 309–328.

Andrews, A.P. 1998. El comercio marítimo de los mayas del posclásico. *Arqueología mexicana*; Sept./octubre, pp. 16–23.

Andrews, A.P.; G.W. Andrews; and V. Robles. 2003. The northern Maya collapse and its aftermath. *Ancient Mesoamerica* 14: 151–156.

Angulo, F.F. y L. González-Álvarez. 2008. El fenómeno de El Niño en México, un estudio de caso: la cuenca del Papaloapan, Veracruz. In: Centro de Estudios mexicanos y Centroamericanos (CEMCA), Open Editions Books; pp. 567–590.

Anónimo. 1985. *Antología de la Planeación en México*. México, Gobierno Mexicano, Vol. 1.

Anónimo. 2017 [1498]. Christopher Columbus wrote about maize that was grown in Castile.

Appadurai, A. 2001. *La modernidad desbordada. Dimensiones culturales de la globalización*. Buenos Aires, Fondo de Cultura Económica (FCE).

Aragón-Cuevas, F.; S. Taba; J.M. Hernández-Casillas; J.D. Figueroa-Cárdenas; y V. Serrano-Altamirano. 2006. Actualización de la información sobre los maíces criollos de Oaxaca. Instituto Nacional de Investigaciones Forestales Agrícolas y Pecuarias (INIFAP). Hoja de cálculo SNIB-CONABIO Proyecto N° CS002 México D.F.

Aramoni, D. 1992. *Los refugios de lo sagrado. Religiosidad, conflicto y Resistencia entre los zoques de Chiapas*. México, Consejo Nacional para la Cultura y las Artes (CONACULTA).

Aréchiga- Córdoba, E. 2011. El agua expropiada: la subordinación hidrológica de Xochimilco durante el siglo XX. In: M. Rutsch y A. González (comps.); *Culturas y Políticas del Agua en México y un caso del Mediterráneo.* México, Instituto Nacional de Antropología e Historia (INAH) y Universidad Iberoamericana; pp. 227–269.

Arellanes, C.; Y. Casas F. A. 2011. Los mercados tradicionales del Valle de Tehuacán-Cuicatlán: Antecedentes y Situación Actual. *Nueva Antropología* 4(74): 93–123.

Archaeology Laboratory. n/d. Table of data on the age of maize in the Americas. Canada. University of British Columbia.

Armillas, P. 1949. Notas sobre sistemas de cultivo en Mesoamérica. Cultivos de riego y humedad en la cuenca del río Balsas. Anales del Instituto Nacionalde Antropología e Historia (*AINAH*), México, Instituto Nacional de Antropología e historia (INAH), Vol. III: 85–113.

Armillas, P. 1971. Gardens on Swamps. *Science* 174: 653–661.

Argivo, R. 1994. *Sastun: My Apprenticeship with a Maya Healer.* San Francisco, CA, Harper; pp. 187–188.

Argueta, V.A. 2016. El Estudio Etnobioecológico de los Tianguis y Mercados. In: México. *Revista Etnobiología* 14(2): 38–46.

Argueta, V.A.; E. Corona; y P. Hersch (coords.). 2011. *Saberes colectivos y Diálogo de Saberes en México.* UNAM- CRIM, INAH.

Arvizu Barrón, E. 2013. Mercados regionales de Huixcolotla y Zacapoaxtla Puebla: un enfoque de desarrollo rural. Tesis Doct. Colegio de Postgraduados (COLPOS), Montecillo, Estado de México.

Ashmore, W. 2010. Sacred Landscapes, Political Spaces. In: E. Ortiz-Díaz (ed.); VI *Coloquio Pedro Bosch Gimpera. Lugar, espacio y paisaje en arqueología: Mesoamérica y otras áreas culturales.* México, UNAM-Instituto de Investigaciones Antropológicas (III); pp. 297–322.

Atran, S.; A.F. Chase; S.L. Fedick; G. Knapp; H. McKillop; J. Marcus; N.B. Schwartz; and M.C. Webb. 1993. Itza maya tropical agroforestry. *Current Anthropology* 34(5): 633–700.

Ayala, M. de la Luz. 1999. La pugna por el uso y la propiedad de los montes y bosques novohispanos. In: *Estudios sobre historia y ambiente en América. Vol. 1 Argentina, Bolivia, México, Paraguay.* B. García-Martínez. y A. González-Jácome (comps.). El Colegio de México (COLMEX), Instituto Panamericano de Geografía e Historia (IPGH), pp. 75–92.

Ayora, S. 2010. Conexiones cosmopolitas y saberes locales: la diseminación de la cocina yucateca. In: Bueno, C. y S. Ayora (coords.); *Consumos globales: de México para el mundo.* México, Universidad Iberoamericana, Universidad Autónoma de Yucatán; pp. 221–249.

Báez-Jorge, F. 1972. Estructura Social de los Zoque-Popoluca de Soteapan, Veracruz. Thesis Ma., Social Anthropology, Universidad Veracruzana.

Báez-Jorge, F. 1983. La cosmovisión de los zoques de Chiapas (reflexiones su pasado y su presente). In: L. Ochoa y T.A. Lee Jr. (eds.); *Antropología e historia de los mixe-zoques y mayas. Homenaje a Frans Blom.* México, Universidad Nacional Autónoma de México-Brigham Young University; pp. 383–412.

Báez-Jorge, F. 1992. Homshuk y el simbolismo de la ovogénesis en Mesoamérica. In: Antropología mesoamericana, Homenaje a Alfonso Villa Rojas. V.M. Esponda-Jimeno; S. Pincemin Deliberos y M. Rosas-Kifuri (comps). Gobierno del estado de Chiapas, Consejo Estatal de Fomento a la Investigación y Difusión de la Cultura, Sistema Integral para el desarrollo integral de la familia, DIF-Chiapas, Instituto Chiapaneco de Cultura; pp. 303–332

Báez, J.F.; J.L. Ramos; E. Gonzalez; M. Bravo; y J.A. Ruiz. 2013. *Chapulines de los pastizales de México.* Instituto Nacional de Investigaciones Forestales, Agrícolas y Pecuarias (INIFAP)-Centro de investigación Regional Norte Centro (CIRNOC). Campo Experimental Pabellón. Folleto Técnico N° 47. Pabellón de Arteaga, Aguascalientes, México.

Balmaceda, D. 2014. *Historia de Letras, Palabras y Frases.* Ed. Sudamericana.

Barabas, A. y Bartolomé, M. 1974. Hydraulic development and ethnocide: The Mazatec and Chinantec peoples of Oaxaca, Mexico. In: *Critique of Anthropology*, London, Spring, N° 1.

Barabas, A. y Bartolomé, M. 1992. Antropología y relocalizaciones. *Alteridades* 2(4): 5–15.

Bandura, A. 1977. *Social Learning Theory.* New York, Prentice Hall.

Bandura, A. 2002. Environmental sustainability by sociocognitive deceleration of population growth. In: P. Schmuch and W. Shultz (eds.); *The Psychology of Sustainable Development.* New York, Kluwer: pp. 209–238.

Barba-Pingarrón P.L. 1987. Estudio de áreas de actividad. *Coba, Quintana Roo. Un análisis de dos unidades habitacionales mayas del horizonte clásico*: Manzanilla, S.L. (comp.). México, Universidad Nacional Autónoma de México (UNAM), Inst. de Investig. Antropológicas (IIA), Serie Antropológica 82: 69–115.

Barkin, D. 1970. *Desarrollo económico regional (Enfoque por cuencas hidrológicas de México).* México, Siglo XXI Editores.

Barkin, D. y T. King. 1972. *Los beneficiarios del desarrollo regional*. México, SepSetentas.
Barkin, D. y A. Zavala. 1978. *Desarrollo regional y reorganización campesina: La Chontalpa como reflejo.*
Barkin, D. 1998. *Riqueza, pobreza y desarrollo sustentable*. México, Ed, JUS.
Barrera-Bassols, N.; M. Astier; y Q. Orozco. 2009. El concepto "tierra" y el maíz en S. Francisco Pichátaro, Michoacán. *BioDiversitas*; Boletín Bimestral de la Comisión Nacional para el conocimiento y uso de la biodiversidad (CONABIO). México, N° 87:1–6.
Barrera-Bassols, N. y N. Floriani. 2018. *Saberes Locales, Paisajes y Territorios Rurales en América Latina*. Colombia, Universidad del Cauca.
Barrera M.A. 1980. Sobre la unidad habitacional tradicional campesina y el manejo de recursos bióticos en el área maya yucatanense: árboles y arbustos de las huertas familiares. *Biótica* 5(3): 115–129.
Barrera M.A. 1982. Los petenes del Noroeste de Yucatán: su exploración ecológica en perspectiva. *Biótica* 7(2): 163–170.
Barrera V.A.; J.R. Bastarrachea M.; W. Brito S.; R. Vermont S.; D. Dzul G. y D. Poot. 1980 [1773]. *Diccionario Maya Cordemex maya-español y español-maya*. México: Ed. Cordemex, Mérida, Yucatán.
Barrera, J.F. 2002. El taladrador de las ramas del café robusta: la otra broca del café. In: *Tres plagas de café en Chiapas. México*. El Colegio de la Frontera Sur (ECOSUR), Tapachula, Chiapas; pp. 81–84.
Barrett, J.W. and A.K. Scherer. 2005. Stones, bones, and crowded plazas: Evidence for Terminal Classic May warfare at Colhá, Belize. *Ancient Mesoamerica* 16: 101–118.
Bartolomé M. y A. Barabas. 1990. *La presa Cerro de Oro y el Ingeniero Gran Dios*. México, INI/CA, (2 Tomos).
Bartra-Vergés, A.; R. Cobo; y L. Paz-Paredes. 2013. *La Hora del Café. Dos siglos a muchas voces*. México, Consejo Nacional para el conocimiento y uso de la Biodiversidad (CONABIO).
Basauri, C. 1940. *La población indígena de México*. México, 3 Vols.
Basurto-Peña, F.; D. Castro-Lara; L.M. Ovando; y T. Juárez-Castro. 2015. Etnobotánica de las calabazas cultivadas (*Cucurbita* spp.) en los valles centrales de Oaxaca, México. *Agriproductividad*; pp. 46–53.
Bautista Lozada, A.; F. Parra Rondinel; y F.J. Espinosa-García. 2012. Efectos de la Domesticación de Plantas en la Diversidad Fitoquímica. In: J.C. Rojas y E. Malo (eds.). *Temas Selectos en Ecología Química de Insectos*. México, El Colegio de la Frontera Sur (ECOSUR); pp. 253–267.
Beach, T. and N. Dunning. 1994. Soil erosion, slope management, and ancient terracing in the Maya Lowlands. *Latin American Antiquity* 3(1): 51–69.
Beach, T.; N. Dunning; S. Luzzadder-Beach; J. Hagemen; and J. Lohse. 2002. Upland agriculture in the Maya Lowlands: ancient Maya soil conservation in Northwestern Belize. *Geographical Review* 92(3): 372–397.
Beach, T.; N. Dunning; S. Luzzadder-Beach; D.E. Cook; and J. Lohse. 2006. Impacts of the ancient Maya on soils and soil erosion in the central Maya Lowlands. *Catena* 65(2): 166–178.
Beach, T.; D. Cook; N. Dunning; S. Luzzader; and J. Lohse. 2011. Carbon isotopic ratios of wetland and terrace soil sequences in the Maya Lowlands of Belize and Guatemala. *Catena* 85(2): 109–118.
Beals, R. 1942. The Western Mixe Indians of Oaxaca, Mexico. *América Indígena*, II(1): 45–50.
Beals, R. and E. Hatcher. 1943. The Diet of a Tarascan Village. *América Indígena* México, III(4): 295–304.
Beals, R. 1967. The structure of the Oaxaca Market System. *Revista Mexicana de Estudios Antropológicos* 21: 333–342.
Beals, R. 1968. Un sistema tradicional de mercado. *América Indígena* 27: 556–580.
Bedoya, A.R. 2009. Plantaciones forestales como mecanismo para la recuperación de zonas degradadas en la Estación Biológica La Selva, Costa Rica. Thesis, Ma. Universidad Estatal a Distancia, Vicerrectoría Académica Escuela de Ciencias Exactas y Naturales, San José, Costa Rica.
Bedolla, A.G. y J.E. Vanegas. 1990. *La comida en el medio lacustre. Chimalhuacán*. México, Centro Comunitario Culhuacán, Fideicomiso del Fondo de Participación Ciudadana para el desarrollo Social de Iztapalapa, Universidad Autónoma Metropolitana-Iztapalapa (UAM-Iztapalapa), INAH (3a edición 2007).
Bellón, M.R.; A.F. Barrientos-Priego; P. Colunga-GarcíaMarín; H. Perales; J.A. Reyes-Agüero; R. Rosales-Serna; y D. Zizumbo-Villareal. 2009. Diversidad y conservación de recursos genéticos en plantas cultivadas. In: *Capital natural de México. Estado de conservación y tendencias de cambio*. México, Consejo Nacional para el conocimiento y uso de la Biodiversidad (CONABIO), Vol. II; pp.: 355–382;
Benavides-Castillo, A.; and L. Manzanilla. 1987. Introducción: estudio de centros urbanos. In: Manzanilla, S. (comp.), *Coba, Quintana Roo. Un análisis de dos unidades habitacionales mayas del horizonte clásico*. México: Instituto Nacional de Antropología e Historia (INAH), Serie Antropológica 82, Universidad Nacional Autónoma de México (UNAM), Inst. de Investigaciones Antropológicas: 11–24.

Beadle, G. W. 1932. Studies of Euchlaena and its hybrids with Zea I chromosome Behavior in E. mexican and its hybrids with Zea mays. *Z. Abstam. Vererbungsl* 62: 291–304.
Beadle, G. W. 1939. Teosinte and the origin of maize. *J. Hered* 30: 245–247.
Beadle, G. W. 1972. The mystery of maize. *Field Mus. Natural History Bulletin* 43: 1–11.
Beadle, G. W. 1978. Teosinte and the origin of maize. In: D.B. Walden (Ed.). *Maize Breeding and Genetics*. Wiley, New York; pp. 113–128.
Beadle, G. W. 1980. The ancestry of corn. *Scientific American* 242: 112–119.
Benz, B.F. 1997. Diversidad y distribución prehispánica del maíz Mexicano. *Arqueología Mexicana* 5(25): 16–23.
Benz, B.F. 1999a. On the origin, evolution, and dispersal of maize. In: M. Blake (ed.); *Pacific Latin America in Prehistory: The evolution of Arcaic and Formative Cultures*. Pullman, WA, Washington State University Press; pp. 25–38.
Benz, B.F. 1999b. *Pacific Latin America in Prehistory*. E.M. Blake (ed.), Washington State University Press, Pullman; pp. 25–38.
Benz, B.F.; and A. Long. 2000. Prehistoric Maize Evolution in the Tehuacan Valley. *Current Anthropology* 41(3): 459–465.
Benz, B.F. 2001. Archaeological evidence of teosinte from Guilá Naquitz, Oaxaca. *Proceedings of the National Academy of Sciences of the United States of America*. 98(4): 2104–2106.
Benz, B.F.; L. Cheng, S.W. Leavitt; and C. Eastoe. 2006a. El Riego and Early Maize Agricultural Evolution. In: *Histories of Maize. Multidisciplinary Approaches to the Prehistory, Linguistics, Biogeography, Domestication, and Evolution of Maize*. Amsterdam, London, New York, Oxford, Elsevier, Academic Press: 73–95.
Benz, B.F.; and J.E. Staller. 2006b. The Antiquity, Biogeography, and Culture History of Maize in the Americas. In: *Histories of Maize. Multidisciplinary Approaches to the Prehistory, Linguistics, Biogeography, Domestication, and Evolution of Maize*. Amsterdam, London, New York, Oxford, Elsevier, Academic Press: 665–673.
Benz, B.F. 2011. Archaeological evidence of teosinte domestication from Guilá Naquitz, Oaxaca. *Proceedings of the National Academy of Sciences of the United States of America*. 98(4): 2104–2106.
Berlin, B., D.E. Breedlove; and P.H. Raven. 1973. General principles of classification and nomenclature in folf biology. *American Anthropologist* 75: 214–242.
Berlin, B.; D.E. Breedlove; and P.H. Raven. 1974. *Principles of Tzetzal Plant Classification*. New York and London, Academic Press.
Berger, P. and T. Luckmann. 1966. *The Social Construction of Reality*. Harmondsworth, Penguin.
Bilbao J.A. 1979. Sistemas y prácticas agrícolas en una comunidad de Tlaxcala. Presentation, 43° International Congres of Americanists. Canadá: Vancouver.
Binford L.; G. Carrasco; S. Arana; y S. Antillana de Rojas. 2004. *Rumbo a Canadá. La migración canadiense de trabajadores agrícolas tlaxcaltecas*. México: Ediciones Taller Abierto.
Bird, R. McK. 1980. Maize Evolution from 500 B.C. to the Present. *Biotropica* 12: 30–41.
Bird, R. McK. 1982. South American Maize in Central America? *Presentation*, 44° International Congress of Americanists. Manchester, England.
Blake, M. 2006. Dating the Initial Spread of Zea mays. In: *Histories of Maize. Multidisciplinary Approaches to the Prehistory, Linguistics, Biogeography, Domestication, and Evolution of Maize*. Amsterdam, Boston, Heidelberg, London, New York, Oxford, Elsevier, Academic Press: 55–72.
Blanco-Martínez, M. y J.O. Moncada-Maya. 2011. El Ministerio de Fomento, impulsor del estudio y el reconocimiento del territorio mexicano (1877–1898). In: *Investigaciones geográficas*, N° 74, abril; versión On-line.
Blanco-Rosas, J.L. 1986. Uxpanapa, corazón del Istmo: Proyecto de una ciega modernización, la realidad de la selva, nuevos días, nuevos retos. In: *Fundamentos*, Revista de Iniciativa N° 14. Xalapa; pp. 2–16.
Blanco-Rosas, J.L. 2003. Mito, rito y tecnología en la producción de semillas y grano de la milpa Popoluca de Soteapan, Veracruz; mss.
Blanco-Rosas, J.L. 2006. La erosión de la agrodiversidad en la milpa de los zoque- popoluca de Soteapan: Xutuchincon y Aktevet. Thesis, Social Anthropology, México: Universidad Iberoamericana AC.
Blanco-Rosas, J.L. 2011. Los afectados chinantecos de la presa Cerro de Oro: persistencia de los conflictos y riesgos socioambientales. In: *Culturas y políticas del agua en México y un caso del Mediterráneo. Una mirada desde la antropología*. M. Rutsch y A. González-Jácome (comps.). México, Instituto Nacional de Antropología e Historia (INAH) y Universidad Iberoamericana (UIA); pp. 177–202.

Böege, E.S. 1988. *Los mazatecos ante la nación. Contradicciones de la identidad étnica en el México actual.* México, Ed. Siglo XXI.

Böege, E.S. 2008. *El patrimonio biocultural de los pueblos indígenas de México. Hacia la conservación in situ de la biodiversidad y agrodiversidad en los territorios indígenas.* México, Instituto Nacional de Antropología e Historia (INAH) y Comisión Nacional para el Desarrollo de los pueblos indígenas de México.

Böehm B. y A. Pereyra. 1974. *Terminología Agrohidráulica Prespánica Nahua.* México, Instituto Nacional de Antropología e Historia (INAH).

Borah, W. and S.F. Cook. 1974. *Ensayos sobre historia de la población: México y el Caribe.* México, S. XXI Eds.

Borejsza, A.; C.D. Frederick; and R.G. Lesure. 2011. Swidden agriculture in the "Tierra Fría?" Evidence from sedimentary records in Tlaxcala. *Ancient Mesoamerica* 22(1): 91–106.

Bose, E.R.; D.R. Foster; A.B. Plotkin; B. Hall. 2003. Geographical and historical variation in hurricanes across the Yucatán Peninsula. In: *The Lowland Maya Area: Three Millennia at the Human Wildland Interface.* A. Gómez-Pompa; M. Allen; S. Fedick; J.J. Jimenez-Osornio. Haworth Press, N. York; pp. 495–516.

Brandolini, A. 1970. *Maize.* O.H. Frankel y E. Bennett (eds). *Genetic resources in plants: their exploration and conservation.* USA, Philadelphia, IBP Handbook N° 11, F.A. Davis Co. pp. 273–309.

Braniff, B. 2009. Comercio e interrelaciones entre Mesoamérica y la Gran Chichimeca. *Caminos y mercados de México.* In: J. Long T. and A. Attolini L. (comps.). México, Universidad Nacional Autónoma de México (UNAM), Instituto de Investigaciones Históricas (IIH), Instituto Nacional de Antropología e Historia (INAH), Serie Historia General, 23; pp. 27–50.

Brewbaker J.L. and C.T. Sorensson. 1990. New tree Crops from Interspecific *Laucaena hybrids* spp. *Advances in New Crops*; Janick J. and J.E. Simon (eds.). Portland, OR, Timber Press: 283–289.

Briones, P.C. y Á. Moreno F. (coords.) 2019. Hongos de México. *Arqueología mexicana*, N° 87.

Broda, J. 1983. Ciclos agrícolas en el culto: un problema de la correlación del calendario mexica. In: A.F. Aveni y G. Brotherston (eds.); *Calendars in Mesoamerica and Peru: Native American Computations of time.* Oxford, Bar, International Series 174: pp. 145–165.

Broda, J. 1995. Observación y cosmovisión en el mundo prehispánico. In: *México antiguo: antología de arqueología mexicana.* México, Secretaría de Educación Pública (SEP), Biblioteca Para la Actualización del Maestro, Raíces. Instituto Nacional de Antropología e Historia (INAH); pp. 16–21.

Broda, J. 2004a. ¿Culto al maíz o a los santos? La ritualidad agrícola mesoamericana en una perspectiva histórica. In: J. Broda y C. Good (coords.); *Historia y vida ceremonial en las comunidades mesoamericanas, los ritos agrícolas.* México, Universidad Nacional Autónoma de México (UNAM)-Instituto de Investigaciones Históricas (IIH); pp. 61–81.

Broda, J. 2004b. Los muertos y el ciclo agrícola en la cosmovisión mesoamericana: una perspectiva histórica y comparativa. In: N. Leonardini, et al. (comps.); *Imagen de la muerte.* Lima, Universidad Nacional Mayor de S. Marcos; pp. 245–261.

Broda, J. 2004c. *Historia y vida ceremonial en las comunidades mesoamericanas, los ritos agrícolas.* México, Universidad Nacional Autónoma de México (UNAM)-Instituto de Investigaciones Históricas (IIH), Consejo Nacional para la Cultura y las Artes (CONACULTA), Instituto Nacional de Antropología e Historia (INAH).

Broda, J.; S. Iwaniszewski y A. Montero (coords.). 2001. *La montaña en el paisaje ritual (Estudios arqueológicos, etnohistóricos y etnográficos).* México, Univ. Nacional Autónoma de México (UNAM), Inst. de Invest. Históricas (IIH), Esc. Nacional de Antropología e Historia (ENAH), Instituto Nacional de Antropología e Historia (INAH), Universidad Autónoma de Puebla (UAP).

Broda, J. 2018. Cosmovisión y percepción del espacio en Mesoamérica y los Andes: Arqueología e Interdisciplina. *Revista Cuadernos del Instituto Nacional De Antropología y Pensamiento Latinoamericano;* Buenos Aires, Argentina, Series Especiales, INAPL; pp. 1–20.

Broda, J. and C. Good E. (coords.). 2004. *Historia y Vida Ceremonial en las Comunidades Mesoamericanas. Los ritos agrícolas.* México, Instituto Nacional de Antropología e historia (INAH), Universidad Nacional Autónoma de México (UNAM).

Braudel, F. 1984a [1979]. *Civilización material, economía y capitalismo Siglos XV-XVIII.* Vol. 1. Las estructuras de lo cotidiano. Madrid, Alianza Editorial.

Braudel, F. 1984b. [1979]. *Civilización material, economía y capitalismo Siglos XV-XVIII.* Vol. 2. Los juegos del intercambio. Madrid, Alianza Editorial.

Braudel, F. 1984c. [1979]. *Civilización material, economía y capitalismo Siglos XV-XVIII.* Vol. 3. El tiempo del mundo. Brownrigg, L.A.

Bruman, H.J. 1944. Some observations on the early history of the coconut in the New World. *Acta Americana* 2(3): 220–243.

Bruman, H.J. 1945. Early coconut culture in western Mexico. *The Hispanic American Historical Review* 25(2): 212–223.

Bruman, H.J. 1947. A further note on coconuts in Colima. *The Hispanic American Historical Review* 27(3): 572–573.

Buckler E.S. and T.P. Holtsford. 1996. Zea systematic: ribosomal its evidence. *Molecular Biology and Evolution* 13: 612–622.

Buckler, E.S.; E.M. IV; D.M. Pearshall, and T.P. Holtsford. 1998. Climate, plant ecology and Central Mexico archaic subsistence. *Current Anthropology* 39(1): 151–164.

Buendía-Rodríguez, E.; F. Islas-Gutiérrez y V. Guerra-de la Cruz. 2008. Identificación de áreas erosionadas por medio de imágenes Landsat ETM, en Tlaxco y Terrenate, Tlaxcala, México. *Terra Latinoamericana* 26: 1–9.

Bulmer, R. 1965. Review of Navajo Indian ethnoentomology. *American Anthropologist* 67: 1564–1566.

Bulmer, R. 1997. The "Natural" vegetation of the Mexican Bajío: In: Archival Documentation of a 16th-Century Savanna Environment. Elsevier Science; pp. 161–172.

Butzer, K.W. 2012a. Collapse, environment, and society. *Proceedings of the National Academy of Sciences of the United States of America* 109: 3632–3639.

Butzer, K.W. and G.H. Endfield. 2012c. Critical perspectives on historical collapse. *Proceedings of the National Academy of Sciences of the United States of America* 109: 3628–3631.

Bye, R. 1981. Quelites-ethnoecology of edible greens-past, present and future. *Etnobiology* 1(1): 109–123.

Bye, R. y E.M. Linares. 1987. Usos pasados y presentes de algunas plantas medicinales encontradas en los mercados mexicanos. *América Indígena* 47: 199–230.

Bye, R. y E.M. Linares. 2000. Los quelites, plantas comestibles de México: una reflexión sobre intercambio cultural. *Biodiversidad* 33: 11–14.

Bye, R. y E.M. Linares. 2010. La dinámica de un mercado periférico de plantas medicinales de México. El tianguis de Ozumba, Estado de México, como centro acopiador para el mercado de Sonora (mercado central). In: J. Long y A. Attolini (eds.). *Caminos y mercados de México.* Universidad Nacional Autónoma de México (UNAM)-Instituto Nacional de Antropología e Historia (INAH).

Bye, R. and E.M. Linares 2015. Quelites (quilitl). *Biodiversidad Mexicana*; México, CONABIO.

Bye, R. and E.M. Linares. 2016. Traditional markets in Mesoamerica A Mosaic of history and traditions. In: R. Lira; A. Casas y J. Blancas (Eds.). *Ethnobotany of Mexico: Interactions of People and Plants in Mesoamerica.* New York, Springer.

Byrne, R. and S.P. Horn. 1989. Prehistoric agriculture and forest clearance in the Sierra de Los Tuxtlas, Veracuz, Mexico. *Palynology* 13: 181–193.

Caamal, A. y Del Amo, R.S. 1987. La milpa múltiple como punto de partida del manejo de la sucesión secundaria. *Turrialba* 37(1): 195–210.

Caballero, J. 1992. Maya homegardens: past, present and future. *Etnoecológica* 1: 135–155.

Caballero, J.; L. Cortés y A. Martínez-Barac. 2010. El manejo de la biodiversidad en los huertos familiares. In: V.M. Toledo; *La biodiversidad de México. Inventarios, manejos, usos, informática, conservación e importancia cultural.* México, México, Consejo Nacional para la Cultura y las Artes (CONACULTA), Fondo de Cultura Económica (FCE); pp. 220–234.

Cabrera, R. 2005. Nuevas evidencias arqueológicas del manejo del agua en Teotihuacan. El campo y la ciudad. In: M.E. Ruiz-Gallut y J. Torres-Peralta (eds.), *Arquitectura y urbanismo: pasado y presente de los espacios en Teotihuacan*; Memoria, 3ª Mesa Redonda. México, INAH; pp. 121–162.

Cahuich, C.D.R. 2012. "La calidad de vida y el huerto familiar, desde la percepción ambiental de las familias de X-Mejía, Hopelchén, Campeche". Tesis Doct. Ecología y Desarrollo Sustentable. El Colegio de la Frontera Sur (ECOSUR), Campeche, Campeche.

Calnek, E. 1972. Settlement Pattern and Chinampa Agriculture at Tenochtitlan. *American Antiquity* 37(1): 104–115.

Caloca-Rivas. R. 1999. "Migración y Desarrollo Autogestivo en San Lucas Tecopilco, Tlaxcala". Tesis Doct. Ciencias Sociales (Antropología Social), Universidad Iberoamericana.

Cámara-Barbachano, F. 1966. Tianguis y mercados en Oaxaca. *Summa antropológica* en homenaje a Roberto Weitlaner. México, INAH; pp. 273–280.

Campos-Goenaga, M.I. 2012. "Sobre tempestades con remolino y plagas de langosta. Siglos XVI al XVIII en la península de Yucatán". *Relaciones* 129: 125–160.

Canabal-Cristiani, B. 1997. *Xochimilco. Una identidad recreada.* México, UAM- Xochimilco, CIESAS, UACh.

Cancian, F. 1972. *Change and Uncertainty in a Peasant economy.* The Maya Corn Farmers of Zinacantan. Stanford Univ. Press.

Cano-Contreras, E.J. 2015. Huertos familiares: un camino hacia la soberanía alimentaria. Revista digital *Pueblos y Fronteras* 10(20): 68–89.

Cano-Gallego, J.A. y D. Wilkenmann. 1972. Plan Puebla. Análisis de beneficios y costos. Centro Internacional de Mejoramiento de Maíz y Trigo. *El Trimestre Económico, F.C.E*39(156(4)): 783–796.

Caran, S.C.; Neely, J.A.; Winsborough, B.M.; Ramírez-Sorensen, F., Valastro Jr. S. 1996. A late Paleoindian –early archaic water well in Mexico: possible oldest water- management feature in the new world. *Geoarchaeology* 11(1): 1–36.

Carot, P. y M.A. Hers. 2011. Imágenes de la Serpiente a lo Largo del Antiguo Camino de Tierra Adentro. In: Bonfiglioli et al. (eds.). *Las vías del noroeste III: Genealogías, Transversalidades y Convergencias.* México, UNAM, I.I.A.

Casas, A. y J. Caballero. 1995. Domesticación de plantas y origen de la agricultura en Mesoamérica. *Revista de Cultura Científica.* México, UNAM, Facultad de Ciencias N° 40, oct-dic; pp. 36–44.

Casas, A.; J. Torres-Guevara y F. Parra (coords.). 2017. *Domesticación en el Continente Americano, Vol. 2, Investigación para el manejo sustentable de recursos genéticos en el Nuevo Mundo*; México, UNAM, Universidad Nacional Agraria La Molina (UNALM) del Perú.

Casas, A.; F. Parra-Rondinel; S. Rangel-Landa; J. Blancas; M. Vallejo; A.I. Moreno-Calles, S. Guillén; I. Torres-García; A. Delgado-Lemus; E. Pérez- Negrón; C.J. Figueredo; J.M. Cruse-Sanders; B. Farfán-Heredia; L. Solís; X. Aguirre-Dugua; A. Otero-Arnaiz; H. Alvarado-Sizzo; and A. Camou-Guerrero. 2017. Manejo y domesticación de plantas en Mesoamérica. Una estrategia de investigación y estado del conocimiento sobre los recursos genéticos en México. In: Casas, A. J. Torres-Guevara y F. Parra-Rondinel (comps.); *Domesticación en el Continente Americano*, Vol. 2, Investigación para el manejo sustentable de recursos genéticos en el Nuevo Mundo; pp. 69–102.

Casas F.A. 1988. La montaña de Guerrero. Intervoew. *Revista de Cultura Científica.* México, UNAM, Facultad de Ciencias.

Casillas, L. y L. A. Vargas. 1984. La alimentación entre los mexicas. In: López- Austin, A. and C. Viesca (eds.), *Historia general de la medicina en México.* T.1, México antiguo. México, UNAM.

Castañeda-Castillo, X. 2014. La elaboración de pan como estrategia adaptativa de las famiias de San Juan Huactzinco, Tlaxcala. Thesis, Ma. Social Anthropology. México, Universidad Iberoamericana (UIA).

Castello-Yturbide, T. 1986. *Presencia de la comida prehispánica.* México, Fomento Cultural Banamex.

Castillo-Ramos, M.I.; A. Espejel Rodríguez y R. Márquez Romero. 2001. Promoción y participación en el programa huertos familiares de traspatio en el Edo. de Tlaxcala, México y su impacto en el desarrollo rural. In: *Contraste Regional*; (Journal, Electronic), UNAM; pp. 229–245.

Centro de Investigación y de Estudios Avanzados del Instituto Politécnico Nacional (CINVESTAV). 2016.

Chacón, J.C. 1978. El concepto del mal y buen monte: su relación con el potencial alelopático en agroecosistemas tradicionales en la Chontalpa. Thesis, Colegio Superior de Agricultura Tropical (CSAT); Cárdenas, Tabasco.

Chamoux, M.N. 1997. La cuisine de la Toussaint chez les Aztèques de la Sierra de Puebla (Mexique). *Babel.* Internationales de l'imaginaire, N° Spécial, nourriture, 7: 85–99.

Chang, J.H. 1977. Tropical agriculture: crop diversity and crop yields. *Economic Geography* 53: 241–254.

Chase, D.Z. and A.F. Chase. 2014. Path dependency in the rise and denouement of a classic Maya city: the case of Caracol, Belize. In: A.F. Chase and V.L. Scarborough (eds.), *The Resilience and Vulnerability of Ancient Landscapes: Transforming Maya Archaeology through IHOPE.* Archeological Papers of the American Anthropol. Association, N° 24, Hoboken, N.J., Wiley-Blackwell, pp. 142–154.

Chávez S.J. and R.G. Thompson. 2006. Early Maize on the Copacabana Peninsula: Implications for the Archaeology of the Lake Titicaca Basin. In: *Histories of Maize. Multidisciplinary Approaches to the Prehistory, Linguistics, Biogeography, Domestication, and Evolution of Maize.* Amsterdam, London, New York, Oxford, Paris, S. Francisco, Sydney and Toronto: Elsevier, Academic Press: 415–428.

Chayanov, V.A. 1966. *The Theory of Peasant Economy.* D, Thorner, B. Kerblay and R.E.F. Smith (eds.); The American Economic Association Translation Series, Homewood, Illinois.

Chevalier F. 1963. *La formación de los latifundios en México.* México, Fondo de Cultura Económica (FCE).

Chevalier, J.M. and A. Sánchez-Bain. 2003. *The Hot and the Cold. Illes of Humans and Maize in Native Mexico.* Canadá, University of Toronto Press.

Childe, G.V. 1995 [1936]. *Man Makes Himself.* (C.A. Walt & Co) Pitman Publishing, London. (Spanish 1995, Fondo de Cultura Económica).

Chisholm, B. and M. Blake. 2006. Diet in Prehistoric Soconusco. In: *Histories of Maize. Multidisciplinary Approaches to the Prehistory, Linguistics, Biogeography, Domestication, and Evolution of Maize.* Amsterdam, London, New York, Oxford, Paris, S. Francisco, Sydney, and Toronto: Elsevier, Academic Press: 161–172.

CIMMYT. 1988. *Maize production regions in developing countries*. Mexico, Maize Program.

CIMMYT. 1993–1994. World maize facts and trends. Mexico, CIMMYT.

Clark, J.E. 1994. The development of Early Formative Rank Societies in the Soconusco, Chiapas, Mexico. Ph.D. University of Michigan-Ann Arbor.

Clark, W.C. and N.M. Dickson. 2003. Sustainability science: the emerging research program. *Proceedings of the National Academy of Sciences* 100(14): 8059–8061.

Clawson, L. 1985. Harvest security and intraspecific diversity in traditional tropical agriculture. *Economic Botany* 39: 56–67.

Clement C.R. 2017. Panorama de los recursos genéticos en Brasil con énfasis en la Amazonia. In: A. Casas; J. Torres-Guevara, and F. Parra-Rondinel (eds.) *Domesticación en el continente americano*; Vol. 2, Investigación para el manejo sustentable de recursos genéticos en el Nuevo Mundo. México, UNAM, UNALM; pp. 27–42.

Cline, H.F. 1966a. *Colonial Culhuacan, 1580–1600. A Social History of an Aztec Town.* Albuquerque: University of New Mexico Press.

Cline, H.F. 1966b. Introductory notes on territorial divisions of Middle America. *Guide to Ethohistorical Sources. Handbook of Middle American Indians*, Vol. XII, University of Texas Press.

Comisión Ejecutiva del Papaloapan (CODELPA). 1977. *Presentación.* In: *Recursos Naturales de la Cuenca del Papaloapan*, Vol. 1: IX.

Coe, M.D. 1964a. The chinampas of Mexico. *Scientific American* 211: 90–98.

Coe, M.D. 1994. *Mexico. From the Olmecs to the Aztecs.* New York, Thames and Hudson Inc.

Coe, Sophie D. 1994. *America's First Cuisines.* Austin, University of Texas Press.

Coe, Sophie D. 2004. *Las primeras cocinas de América.* México, FCE.

Colunga-García Marín, P.; D. Zizumbo-Villarreal. 2004. Domestication of plants in Maya lowlands. *Economic Botany* 58(Supplement): S101–S110.

Colunga-García M.; I. Torres-García; A. Casas; C.J. Figueredo-Urbina; S. Rangel- Landa; A. Delgado-Lemus; O. Vargas; D. Cabrera-Toledo; D. Zizumbo-Villarreal; X. Aguirre-Dugua; L.E. Eguiarte; and G. Carrillo-Galván. 2017. Los agaves y las prácticas mesoamericanas de aprovechamiento, manejo y domesticación. In: *Domesticación en el Continente Americano*; A. Casas (ed.); México UNAM-UNALM, Vol. II; pp. 273–309.

Conklin, H.C. 1954. The Relation of Hanunuo Culture to the Plant World. PhD. Thesis, Yale University.

Conklin, H.C. 1979. An ethnoecological approach to shifting agriculture. In: A.P. Nayda (ed.), *Environmental and Cultural Behavior: Ecological Studies in Cultural Anthropology.* The Natural History Press, New York.

Conway, G.R. 1985. Agroecosystems analysis. In: *Agricultural Administration* 20: 31–55.

Conrad, H.S. 1905. *The waterlilies: a monograph of the genus Nymphaea.* Publication No. 4. Washington, D.C., The Carnegie Institute of Washington.

Cook, O.F. 1921 [1919]. Milpa agriculture, a primitive tropical system. *Annual Report.* Smithsonian Institution, Washington DC.

Cook, S.F. and W. Borah. 1960. The population of central México in 1548: an analysis of The Suma de Visitas de pueblos. *Revista Iberoamericana*, N° 43, Berkeley.

Cook, S.F. and L.B. Simpson. 1948. *The population of central México in the sixteenth century.* Iberoamericana, Berkeley and Los Angeles.

Cooper, J. and L. Duncan. 2016. Applied archaeology in the Americas: evaluating archaeological solutions to the impacts of global environmental change. In: C. Isendahl y D. Stump (eds.); *The Oxford Handbook of Historical Ecology and Applied Archaeology* (Online Publication 02/2016).

Córdoba, E.C. 1997. Landscape transformation in Aztec and Spanish Colonial Texcoco, Mexico. PhD Thesis, Geography, University of Texas, Austin.

Córdoba, E.C. and R.J. Parsons. 1997. Geoarcheology of an Aztec dispersed Village on the Texcoco Piedmont of central Mexico. *Geoarcheology: An International Journal* 12(3): 177–210.

Correa-Ortiz, H. 2011. El canal nacional en 1929: un quiebre del sistema lacustre de la cuenca de México. In: M. Rutsch y A. González (comps.); *Culturas y Políticas del Agua en México y un caso del Mediterráneo.* México, INAH y Universidad Iberoamericana; pp. 203–225.

Cortés-Villavicencio, S. 2012. Migración Internacional y Cultura en Tonatico, Estado de México. Un Análisis de la Participación de los Migrantes en la Fiesta. Thesis, MA., Social Anthropology. Mexico, Univ. Iberoamericana.

Cox W.G. and M.D. Atkins. 1979. *Agricultural Ecology. An Analysis of World Food Production Systems.* San Francisco, W.H. Freeman and Co.

Crew, T.E. 1985. Raised Field Agriculture in Tlaxcala, Mexico: With A Focus On Nutrient Cycling. B.A.Thesis, Environmental Studies, University of California, St. Cruz.

Crew, T.E. and S.R. Gliessman. 1991. Raised field agriculture in Tlaxcala, Mexico; an ecosystem perspective on maintenance of soil fertility. *American Journal of Alternative Agriculture* 6(1): 9–16.

Cruz-García, S. 1992. Anuncios de los animales para los zapotecos de Xanica, Sierra Sur de Oaxaca. In: M.A. Vásquez-Dávila (coord.); *Como predecimos el clima. Augurios Animales.* Instituto Tecnológico Agropecuario de Oaxaca. Casa de la Cultura Oaxaqueña. Colección Como hacemos; pp. 4–8.

Cruz-León, A. 1989. *Los instrumentos agrícolas de la Zona Central de Veracruz.* México, Texcoco, Universidad Autónoma Chapingo (UACH).

Cruz-León, A. 1997. *Y sigue la yunta andando.* México, Texcoco, UACH.

Cruz-León, A. y T. Martínez S. 2001. *La tradición tecnológica de la tracción animal.* México, UACH.

Cruz-León, A.; Á. Pita D. y B. Rodríguez H. 2012. *Jocotes, Jobos, Abales o Ciruelas Mexicanas.* Estado de México, Texcoco, Chapingo, UACH.

Cruz-león, A. y A.G. Gutiérrez-Jiménez. 2012. Distribución Geográfica del Género Spondias en México. En: A. Cruz-León et al. *Jocotes, Jobos, Abales o Ciruelas Mexicanas.* Estado de México, Texcoco, Chapingo, UACH; pp.: 31–38.

Cruz-León, A. y Rodríguez-Haro, B. 2012. Cultivo. In: A. Cruz-León et al. *Jocotes, Jobos, Abales o Ciruelas Mexicanas.* Estado de México, Texcoco, Chapingo, UACH; pp.: 77–101.

Cruz-León, A. y O. Vélez-Coria. 2012. Evidencias del Proceso de transformación de los frutos de ciruela: las recetas. En: A. Cruz-León et al. *Jocotes, Jobos, Abales o Ciruelas Mexicanas.* Estado de México, Texcoco, Chapingo, UACH; pp. 111–123.

Cruz-López, C. 1949. Estudio agrológico regional del Estado de Tlaxcala. *Ingeniería hidráulica en México* 3(1): 57–97; (3): 78–136; (4): 44–96.

Cuevas F.H. y M. Navarrete. 2005. Los huracanes en la época prehispánica y en el siglo XVI. *Inundaciones*; Xalapa, Universidad Veracruzana, Instituto de Antropología; pp. 39–49.

Cyphers, A. 2004. La vida en los orígenes de la civilización mesoamericana. Los olmecas de San Lorenzo. In: P. Gonzalbo-Aizpuru y P. Escalante G. (coords.); *Historia de la vida cotidiana en México*, Vol. 1, México, El Colegio de México (COLMEX) y Fondo de Cultura Económica (FCE); pp. 21–40.

Dahlin, B.H. 2002. Climate change and the end of the classic period in Yucatan: resolving a paradox. *Ancient Mesoamerica* 13: 327–340.

Daltabuit, M. y Ríos, A. 1992. Cambio de la dieta familiar en Yalcobá, Yucatán. *Anales de Antropología*, 29(1): 23–33.

Darch, J.P. 1983. *Drained Field Agriculture in Central And South America.* Oxford, England, BAR International Series 189.

Dávila, J. 2011. Gemelli Careri en Nueva España. *Destiempos* N° 28, enero- febrero; pp. 14–44.

Del Amo, R.S. 1986. Las plantas tropicales y su uso en la medicina tradicional. *México Indigenista*, marzo-abril; pp. 64–67.

Del Amo, R.S. 2000a. *Las lecciones del Programa de Acción Forestal Tropical. Programa de Acción Forestal Tropical, AC.* México, Sría. de Ecología, Recursos Naturales y Pesca y Consejo Nacional para la Enseñanza de la Biología. Ed. Plaza y Valdés.

Del Amo, R.S. 2000b. El conocimiento tradicional en el florecimiento comunitario y en la conservación en el PROAFT. *Revista Geografía Agrícola.* Estudios Regionales de la Agricultura Mexicana N° 1: 17–26.

Del Amo, R.S. y L.M. Cayetano. 2011. Paisaje, memoria y cultura. Una trilogía para la conservación y el bienestar de las comunidades totonacas. In: Reyes E. y S. Barras G. (eds). *Saberes ambientales campesinos. Cultura y naturaleza en comunidades indígenas y mestizas.* Univ. de Ciencias y Artes de Chiapas (UNICACH), Universidad Autónoma de Madrid (UAM) y Agencia Española de Cooperación para el Desarrollo (AECID); pp:97–136.

Del Amo, R.S.; Aguilar L. y M. Delgado R. 1999. Los tecallis. Un cultivo hidropónico tradicional. In: A. González-Jácome y S. del Amo-Rodríguez (comps.); *Agricultura y Sociedad en México: Diversidad, Enfoques, Estudios de Caso.* México, Universidad Iberoamericana, Plaza y Valdés, Programa de Acción Forestal Tropical (PROAFT); pp. 77–101.

Del Amo, R.S. y A. Gómez-Pompa (coords). 1994. *Plan de Acción Forestal Tropical de México.* 4 Vols. (Vol. I. Sintético y Líneas de Acción; Vol. II. Documento Básico; Vol. III. Perfiles; Vol. IV. Anexos). México, Secretaría de Agricultura y Ganadería. Subsecretaría Forestal y de Fauna Silvestre.

Del Amo, S.R. y E. Malvido. 1990. Las tesis hechas por los médicos mexicanos en el S. XIX acerca de las plantas medicinales. *Quipu* 7(3): 345–381.

Del Amo, S.R. y S. Moctezuma-Pérez. 2008. Bienestar o calidad de vida en una comunidad de la Sierra totonaca, Veracruz. *Iberoforum* II(5): 1–15.

Del Amo, S.R.; B. Ortiz-Espejel y B. Andrade. 2009. Saberes tradicionales, transdisciplinariedad y participación ciudadana en el Totonacapan. En: Valverde, B y H. Bernal M. (coods.). *Investigaciones multidisciplinarias en la Sierra norte de Puebla.* Colegio de Posgraduados- Puebla, Universidad Intercultural del estado de Puebla, ALTRES COSTA-AMIC; pp. 5–28.

Del Ángel-Pérez A.L. and M.A. Mendoza. 2004. Totonac homegardens and natural resources in Veracruz, Mexico. *Agriculture and Human Values* 31: 329–246.

Del Río, J. 1990. La transformación ecológica indiana. In: J. Fernández-Pérez, I. González-Tascón; La agricultura viajera. *Cultivos y manufacturas de plantas industriales y alimentarias en España en la América Virreinal.* Madrid, España, Consejo Superior de Investigaciones Científicas.

Del Olmo, G. y E. Roldán-Velasco 2013 [2007]. *Aves Comunes de la Ciudad de México.* México, CONABIO.

De la Cruz-Osorio, J.C. 2009. El huerto familiar como sistema agroforestal en Francisco Villa, Tihuatlán, Veracruz. Thesis, BA, Forest Sciences; México, Universidad Autónoma Chapingo.

Demarest, A.A. 1992. Ideology in Ancient Maya Cultural Evolution. In: *Ideology and Pre-Columbian Civilization.* Arthur A. Demarest and Geoffrey W. Conrad (eds). Santa Fe, School of American Research Press, pp. 15–36.

Demarest, A.A.; P.M. Rice, and D.S. Rice. 2004. After the maelstrom: Collapse of the Classic maya kingdoms and the Terminal Classic in wetern Peten". In: A.A. Demarest; M. Rice y D.S. Rice (eds.); *The Terminal Classic in the Maya Lowlands: Collapse, Transition, and Transformation.* University Press, Boulder, Colorado; pp. 102–124.

Denham, T.; S.G. Haberle; C. Lentfer; R. Fullagar; J. Field; N. Porch; M. Therin; B. Winsborough, and J. Golson. 2003. Multi-disciplinary Evidence for the Origins of Agriculture from 6950-6440 Cal BP at Kuk Swamp in the Highlands of New Guinea. *Science*, June.

Dehouve, D. 1995. *Hacia una historia del espacio en La Montaña de Guerrero.* México, Centro de Investigaciones y Estudios Superiores en Antropología Social (CIESAS), y Centro de Estudios Mexicanos y Centroamericanos (CEMCA).

Dehouve, D. 2007. *La ofrenda sacrifical entre los tlapanecos de Guerrero.* Universidad Autónoma de Guerrero, Chilpancingo; Centro de Estudios Mexicanos y Centroamericanos; Plaza y Valdés.

Descola, Ph. y G. Pálsson (eds). 1996. *Nature and Society. Anthropological Perspective.* Londres, New York, Routledge.

Dewey, K. G. 1981. Nutritional consequences of the transformation from subsistence to commercial agriculture in Tabasco, Mexico. *Human Ecology* 9(2), 151–187.

Diccionario de la Lengua Española. 1939. Madrid.

Diskin, M. 1969. Estudio estructural del sistema de plazas en el valle de Oaxaca. *América Indígena* 29: 1077–1099.

Dikov, N.N. 1987. Beringia. El paleolítico de Kamchatka y de Chukotka en relación Con el problema de los pobladores de América. In: A. González-Jácome (comp.); *Orígenes del hombre americano (seminario).* México, SEP; pp. 47–57.

Doebley, J.F. 1990. Molecular evidence and the evolution of maize. *Economic Botany* 44(Suppl): 6–27.

Doebley, J.F. 2004. The genetics of maize evolution. *Annual Review Genetics* 38: 37–59.

Domenici, D. 2010. Patrones de Uso Ritual Del Espacio Hipogeo en la Selva El Ocote (Chiapas). In: E. Ortiz-Díaz (Ed.); VI *Coloquio Pedro Bosch Gimpera. Lugar, espacio y paisaje en arqueología: Mesoamérica y otras áreas culturales.* México, UNAM-Instituto Investigaciones Antropológicas; pp. 387–424.

Domínguez-Virgen, J.C. (coord.). 2017. *Megaproyectos, Desplazamiento Forzado y Reasentamiento Involuntario en México: Testimonios y Reflexiones.* Instituto de Investigaciones Dr. José María Luis Mora, CONACYT, Sociología Contemporánea.

Donkin R.A. 1979. *Agricultural Terracing in the Aboriginal New World.* The Wenner- Gren Foundation for Anthropological Research: Arizona: The Univ. of Arizona.

Doolittle W.E. 1990. *Canal Irrigation in Prehistoric Mexico. The Sequence of Technological Change.* Texas, The University of Texas Press.

Doolittle W.E. 2000. *Cultural Landscapes of Native North America.* Oxford University Press.

Dowswell, C.R., Paliwal, R.L. and Cantrell, R.P. 1996. *Maize in the Third World.* Boulder, Colorado: Westview Press.

Drenan, R.D. 1995. Chiefdoms in Northern South America. *Journal of World Prehistory* 9: 301–340.

Duch, G.J. 1992. Condicionamiento ambiental y modernización de la milpa en el Estado de Yucatán. In: *La modernización de la milpa: Utopía o realidad*. D.V.; C.H. Rasmussen, L. Arias, and S.C. Terán (eds), México, Mérida, CICY; pp. 81–96.

Dunmire, W.W. 2004. *Gardens of New Spain. How Mediterranean Plants and Foods hanged America*. Texas: The University Press.

Dunning, N.P. 1997. The paleoecology and ancient settlement of the Petexbatún region, Guatemala. *Ancient Mesoamerica* 8: 255–266.

Dunning, N.P. and T. Beach. 2011. Farms and forests: spatial and temporal perspectives on ancient Maya landscapes. In: I.P. Martini and W. Chesworth (eds.); *Landscapes and Societies*. Springer, Dordrecht, The Netherlands; pp. 369–389.

Dunning, N.P.; Beach, T. and Luzzadder-Beach, S. 2006. Prehispanic agrosystems and adaptive regions in the Maya lowlands. In: L.J. Lucero and B.W. Fash (eds.), *Precolumbian Water Management: Ideology, Ritual and Politics*. Tucson, Arizona, University of Arizona Press, pp. 81–91.

Durán, R. y M. Méndez. 2010. *Biodiversidad y Desarrollo Humano en Yucatán*. CICY, PPD, MAM, Conabio, Seduma.

Ellen, R.F. 1986. Review of Swidden agriculture in Indonesia, by M.R. Dove. *Journal of Tropical Ecology* 2: 191–192.

Ellen, R.F. 1993. The Cultural Relation of Classification: An Analyses of Nuaulu Animal Cathegories from Central Seram. Cambridge Studies in Social and Cultural Anthropology 91. Cambridge, Cambridge University Press.

Ellen, R.F. 1996a. Human worlds are culturally constructed. In: *Key Debates in Anthropology*. T. Ingold (ed), Routledge, London; pp. 101–146.

Ellen, R.F. 1996b. *Nuaulu Plant Names and Cathegories: A Systematic Ethobotanical Inventory* (Cited in D.L.), Medin and S. Antrand eds.

Ely L.L. Enzel Y.; Baker V.R.; Cayan D.R. 1993. A 5000-year record of extreme floods and climate change in the southwestern United States. *Science* 262: 410–412.

Elson, B. 1947. The Homshuk: a Sierra Popoluca text. *Tlalocan* 2(3): 193–214.

Emslie, S.D. 1981. Birds and prehistoric agriculture: the new Mexican pueblos. *Human Ecology* 9(3): 305–329.

Entrena-Durán, F. y Jiménez-Díaz, J.F. 2013. La producción social de los hábitos alimenticios. Una aproximación desde la sociología del consumo. *Revista Ciencias Sociales* XIX(4): 683–693.

Escalante-Gonzalbo, P. 2012a [1a ed. 2004]. La vida urbana en el periodo Clásico mesoamericano. Teotihuacán hacia el año 600 DC. In: *Historia de la vida cotidiana en México*. Vol. 1, *Mesoamérica y los ámbitos indígenas de la Nueva España*. México, FCE y COLMEX; pp. 41–136. 2012b. (Coord.). *Historia de la vida cotidiana en México*. Vol. 1, *Mesoamérica y los ámbitos indígenas de la Nueva España*. México, FCE y COLMEX.

Espinosa C.A.; A. Turrent F.; M. Tadeo R.; K.Y. Mora G.; B. Zamudio G.; N. Gómez M.; M. Sierra M.; J. Zaragoza E. y R. Valdivia B. 2016. Anexo. El abastecimiento de semillas de variedades mejoradas y nativas de maíz ante el marco jurídico, transgénicos y soberanía alimentaria de México. Una reflexión crítica sobre las condiciones sociales de generación del conocimiento científico en torno al maíz. In: López y Vizcarra (comps.). *El maíz nativo en México. Una aproximación crítica desde los estudios rurales*. México, Universidad Autónoma Metropolitana-Lerma, Ciencias Sociales y Humanidades y Juan Pablos Ed; pp. 383–409.

Espinosa-García, F. J. 1981a. Adiciones a la flora arvense del Valle de México. *Boletín de la Sociedad Botánica de México* 41: 27–32.

Espinosa-García, F. J. 1981b. Las malezas ¿una maldición? *Naturaleza* 5:297–307.

Espinosa-García, F.J.; y R. Díaz-Pérez. 1996a. El uso campesino de plantas arvenses en el Valle de México. *Etnoecológica* III(4–5): 83–94.

Espinosa-García, F.J.; y J. Sarukhan. 1996b. *Manual de Malezas del Valle de México*. México, UNAM y Fondo de Cultura Económica.

Espinosa-García, F.J.; H. Vibrans. 1999. The need of a national weed management strategy in Mexico. In: Van Devender T.; F.J. Espinosa-García; B.L. Harper-Lore and T. Hubbard (eds.) *Invasive Plants on the Move. Controlling them in North America*. USA, Arizona-Sonora Desert Museum Press, Tucson, Arizona; pp. 23–32.

Espinosa-Pineda, G. 1994. Las aves acuáticas, un medio prehispánico de interpretación del cosmos. *Ciencias*, N° 34, abril-junio. México, UNAM; pp. 17–22. 1996. *El embrujo del lago. El sistema lacustre de la cuenca de México en la cosmovisión mexica*. México, UNAM-Inst. Investig. Antropológicas, e Inst. de Investig. Históricas.

Esteva, G. y C. Marielle. 2003. *Sin Maíz no hay País.* México, CONACULTA y Museo Nacional de las Culturas Populares.

Estrada-Lugo, E. 1989. Perspectiva Interdisciplinaria en la investigación etnobotánica a partir del servicio: plantas medicinales. In: *Memorias*, 3er Coloquio de Medicina Tradicional Mexicana "un saber en discusión". Mayo-agosto. UNAM, México; pp. 211–228.

Etchevers B.; D.A. Pérez O.; H. Brito V.; M. Vargas I. y U. López N. 1998. La fertilidad de los tepetates del eje neovolcánico en los estados de México y Tlaxcala. In: H. Navarro G., H. Poupon and M. A. Pérez O. (eds.). *Aptitud productiva en suelos volcánicos endurecidos (tepGetates).* ORSTOM–CP. Montecillo, Estado de México; pp. 17–36.

Evans, T.S. 1990. The productivity of maguey terrace agriculture in central Mexico during the Aztec period. *Latin American Antiquity* 1(2): 117–132.

Ewell, P.T. y T.T. Poleman. 1980. *Uxpanapa, reacomodo y desarrollo agrícola en el trópico mexicano.* Xalapa, Veracruz, INIREB.

Fabila, G.; A. Fabila; M. Mesa-Apodaca; O. Soberón, M. 1955. *Tlaxcala: Tenencia y Aprovechamiento de la Tierra.* México, Centro de Investigaciones Agrarias (CDIA).

Farrell J. 1990. The influence of trees in selected agroecosystems in Mexico. *Agroecology. Researching the Ecological Basis for Sustainable Agriculture*; S.R. Gliessman (ed.). Springer-Verlag, pp. 169–183.

Fedick, L.S. and A. Ford. 1990. The prehistoric agricultural landscape of the Central Maya Lowlands; an examination of local variability in a regional context. *World Archaeology* 22(1): 18–33.

Fedick, L.S. 1995. Indigenous agriculture in the Americas. *Journal of Archeological Research* 3: 257–303.

Fedick, L.S. 1996. *The Managed Mosaic: Ancient Maya Agriculture and Resource Use.* University of Utah Press, Salt Lake City.

Fedick, L.S.; B.A. Morrison; B.J. Andersen; S. Boucher; J. Ceja-Acosta and J.P. Mathews. 2000. Wetland manipulation in the Yalahau region of the northern Maya lowlands. *Journal of Field Archaeology* 27: 131–152.

Fedick, L.S. 2003. In search of the Maya forest. In: C. Slater (ed.); *In Search of the Rain Forest.* Duke University, Durham; pp. 133–164.

Fedick, L.S. 2010. [1992]. Theory and method in the analysis of ancient Maya agricultural landscapes: the household model of agricultural production. In: E. Ortiz-Díaz (ed.); VI Coloquio Pedro Bosch Gimpera. *Lugar, espacio y paisaje en arqueología: Mesoamérica y otras áreas culturales.* México, UNAM-Instituto de Investigaciones Antropológicas (IIA); pp. 47–74.

Fedick, L.S. 2012. The Maya Forest: destroyed or cultivated by the ancient Maya? *Proceedings of the National Academy of Sciences of the United States of America* 107(3): 953–954.

Feinman, G.M. and L.M. Nicholas. 2004. *Hilltop Terrace Sites of Oaxaca, Mexico: Intensive Surface Survey at Guirúm, El Palmillo, and the Mitla Fortress.* Field *Fieldiana.* Museum of Natural History; Anthropology New Series N° 37, Pub. 1530, Sept 30.

Fernández-González, C. 2017. Huertos familiares establecidos por programas gubernamentales en Tlaxcala: lineamientos para promover su persistencia. BA Thesis, Agroecología. Universidad Autónoma Chapingo. June, 20.

Fernández-Nava, R.; C. Rodríguez J.; M.L. Arreguín S.; and A.D. Rodríguez J. 1998. Listado Florístico de la Cuenca del Río Balsas, México. *Polibotánica* 9: 1–151.

Ferraz da Rocha Santilli. 2009. "Agrodiversidad y los derechos de los agricultores". Tesis, Sociol. Rural. Argentina, Universidad de Buenos Aires. In: http://www.biblioteca.pucpr.br/tede/tde_busca/arquivo.php?codArquivo=1457

Fish, S.K. 2000. Hohokan impacts on Sonoran desert environment. In: D.L. Lentz (ed.); *Imperfect Balance. Landscape Transformations in the Precolumbian Americas.* Columbia University Press; pp. 251–280.

Flannery, K.V. (ed.). 1976. *The Early Mesoamerican Village.* New York, Academic Press.

Flannery, K.V.; J. Marcus and S.A. Kowalewski. 1981. The preceramic and formative of the valley of Oaxaca. *Supplement, Handbook of Middle American Indians*, Vol. 1, J. Sabloff (ed.). Austin, The University of Texas Press; pp. 48–93.

Flannery, K.V. 1982. *Maya Subsistence. Studies in Memory of Dennis E. Puleston* New York, London: Academic Press, Studies in Archaeology.

Flannery, K.V. 1986. *Guilá Naquitz. Archaic Foraging and Early Agriculture in Oaxaca, Mexico.* Academic Press, Inc.

Fletcher, R. 2009. Low-density, agrarian based urbanism: a comparative view. *Insights* 2(4): 1–19.

Flores-Guido, J.F.G. 2012. Diversidad florística, usos y origen de material genético de las especies de los huertos familiares de la Península de Yucatán. In: R. Mariaca-Méndez (comp.), 2012. *El huerto familiar del sureste de México*. México, Secretaría de Recursos Naturales y Protección Ambiental del Estado de Tabasco, El Colegio de la Frontera Sur; (ECOSUR); pp. 149–176.

Flores-Guido, J.F.G. 2013. Los Mayas de Yucatán: La pérdida, intercambio y legado. Revista Digital de la UNAM, 01/06/2013, Vol. 14, N° 6.

Foster, G.M. 1940. Notes on the Popoluca of Veracruz. Instituto Panamericano de Geografía e Historia (IPGH), Pub. 51; pp. 359–369. 1942. *A Primitive Mexican Economy*. Monographs of the American Ethnological Society N° 5. Seattle, University of Washington Press.

Foster, G.M. 1945. Sierra Popoloca Folklore and Beliefs. *American Archaeology and Ethnology* 42(2): 177–250.

Foster, G.M. 1967. The Mixe, Zoque and Popoluca. *Handbook of Middle American Indians*, Vol. 7, Ethnology, Part 1. University of Texas at Austin; pp. 448–477.

Foster, G.M. 1975. Medical anthropology: some contrasts with medical sociology. *Social Science and Medicine* 9: 427–432. [1974].

Foster, W.M. 1948. The folk economy of rural Mexico with special reference to marketing. *Journal of Marketing* 13: 153–162.

Foster, W.M. 1975. The Mixe, Zoque and Popoluca. In: R. Wauchope (ed.), Vol. 7–8, Section II; *Handbook of Middle American Indians*. Univesity of Texas Press; pp. 448–477.

Florescano, E. 2003a. Metáfora del grano y la mazorca. In: *La Palabra y el hombre*; revista de la Universidad Veracruzana, julio/septiembre; pp. 7–26.

Florescano, E. 2003b. Imágenes y significados del dios del maíz. In: G. Esteva y C. Marielle (coords.); *Sin maíz no hay país*. México, CONACULTA: 36–55.

Florescano, E. 2012. *Quetzalcóatl y los mitos fundadores de Mesoamérica*. México, Ed. Taurus (Pasado y Presente).

Florescano, E., and E. Malvido. 1982. *Ensayos sobre la historia de las epidemias en México*. México, Instituto Mexicano del Seguro Social (IMSS), 2 Vols.

Food and Agriculture Organization (FAO). 1974. Soil Map of the World. Paris: FAO/UNESCO, Vol. 1.

Food and Agriculture Organization (FAO). 1992.

Food and Agriculture Organization (FAO). 1997.

Food and Agriculture Organization (FAO). 1999.

Food and Agriculture Organization (FAO). 2001.

Food and Agriculture Organization (FAO). 2006a. *Crear y manejar un huerto escolar*. Programa para la alimentación y desarrollo. UNO (ONU).

Food and Agriculture Organization (FAO). 2006b. Seguridad Alimentaria. Dirección de Economía Agrícola y Desarrollo. Roma.

Food and Agriculture Organization (FAO). 2009.

Food and Agriculture Organization (FAO). 2010.

Food and Agriculture Organization (FAO). 2014. *Perspectivas de la agricultura y del esarrollo rural en las Américas: una mirada hacia América Latina y el Caribe*. Costa Rica, CEPAL, FAO, IICA.

Food and Agriculture Organization (FAO). 2015.

Food and Agriculture Organization (FAO). 2018. Top/GUA/3506.

Ford, A. 1990. Maya settlement in the Belize River Area: variations in residence patterns of the Central Maya Lowlands. In: T.P. Culbert and D.S. Rice (eds.), *Prehistoric Population History in the Maya lowlands*. Albuquerque, University of New Mexico Press; pp. 167–181.

Ford, A. 2002. Dominant plants of the Maya forest and gardens of El Pilar: implications for paleoenvironmental reconstructions. *Journal of Ethnobiology* 104: 814–820.

Ford, A. and S.L. Fedick. 2009. Prehistoric Maya settlement patterns in the Upper Belize River area: initial results of the Belize River archaeological settlement survey. *Journal of Field Archaeology* 19: 35–49.

Ford, A., and R. Night. 2009. Origins of the Maya forest garden: Maya resource management after the Holocene thermal maximum. *Journal of Ethnobiology* 29: 213–236.

Ford, A., and R. Night. 2010. The Milpa cycle and the making of the Maya Forest garden. In: Archaeological Investigations in the Eastern Maya Lowlands: Papers the 2009 Belize Archaeology Symposium. J. Morris; S. Jones; J. Awe; G. Thompson; and M. Badillo (eds.); *Research Reports in Belizean Archaeology*, Vol. 7, Institute of Archaeology, Belmopan; pp. 183–190.

Ford, A.; K. Clarke; and P. Morales. 2015. La Conexión del Pasado y Presente de los Antiguos Mayas: Uso de las Tierras Bajas, Distribución de la Población y la Densidad en el Periodo Clásico Tardío. 2014. XXVIII *Simposio de Investigaciones Arqueológicas en Guatemala*. Museo de Arqueología y Etnología, July 14 to 18. B. Arroyo, L. Méndez-Salinas, and L. Paiz (eds.); pp. 481–495.

Ford, A. and R. Night. 2016. *The Maya Forest Garden. Eight Milennia of Sustainable Cultivation of the Tropical Woodlands*. Routledge, Taylor and Francis Group, London and New York

Ford, R.I. 1978. Ethnobotany: historical diversity and synthesis. In: R.I. Ford (ed.); *The Nature and Status of Ethnobotany*; Anthropological Papers, Museum of Anthropology, University of Michigan, N° 67; pp. 33–49.

Fowler, C. 1977. "Ethnoecology". In: D. Hardesty (Ed.); *Ecological Anthropology*. New York, Wiley; pp. 53–59.

Franco-Gaona, A.; B. Ramírez-Valverde; A. Cruz-León; D.M. Sangerman- Jarquín; J.P. Juárez-Sánchez y G. Ramírez-Valverde. 2016. El ekuaro: un sistema agroforestal tradicional michoacano. *Revista Mexicana de Ciencias Agrícolas*, N° especial 16; mayo-junio; pp. 3357–3370.

Frake, C.O. 1962. Cultural ecology and ethnography. *American Anthropologist* 64: 53–59.

Galeote, M. 1997. Nombres indígenas de plantas americanas en los tratados científicos de Fray Agustín Farfán. In: BFUCh XXXVI (1997): 119–161.

Galinat, W.C. 1988. The origin of corn. *Agronomy* 18: 1–31.

Galindo-Escamilla, E. 2008a. Captación de agua de lluvia y retención del suelo en jollas en el parte de aguas de las cuencas de los ríos Actopan y Amajac. *Boletín del Archivo Histórico del Agua (AHA)*, Año 13, N° especial; pp. 17–20.

Galindo-Escamilla, E.; J. Palerm-Viqueira; J. Tovar y R. Rodarte. 2008b. Organización social de una fuente de agua: los jagüeyes. *Agrociencia* 42(2): 233–244.

Galindo-Tovar, M.E.; H.E. Lee-Espinosa; J. Murguía-González; O.R. Leyva Ovalle; I. Landero-Torres. 2013. Domesticación y distribución geográfica de *Persea americana* Mill. en la época precolombina. *Revista de Geografía Agrícola*, enero-diciembre, N° 50–51: 65–70.

Gamio, M. 1916. *Forjando Patria*. México, Ed. Porrúa.

Gamio, M. 1945. "La producción agrícola y la industrialización de los ejidatarios. *América Indígena*. México, Instituto Nacional Indigenista (INI), Vol. V, N° 4: 303–308.

Gamio, M. 1948. *Consideraciones sobre el problema indígena*. México, Ediciones Especiales, III.

García-Acosta, V. (comp.). 2000. *La Diversidad Cultural de Ángel Palerm*. México, CIESAS.

García-Cook, Á. 1972. Investigaciones arqueológicas en el Estado de Tlaxcala. *Comunicaciones*. Proyecto Puebla-Tlaxcala, N° 6. México, Puebla, pp. 21–26.

García-Cook, Á. 1974. Una secuencia cultural para Tlaxcala. *Comunicaciones*, Proyecto Puebla-Tlaxcala, N° 10. México, Puebla, pp. 5–22.

García-Cook, Á. 1978. Tlaxcala: doblamiento prehispánico. *Comunicaciones*, Proyecto Puebla-Tlaxcala, N° 15. México, Puebla; pp. 173–187.

García-Cook, Á. 1986. El control de la erosión en Tlaxcala: Un Problema Secular. *Erdkunde*, 40: 251–261.

García-Cook, Á. 2014. *Arqueología e Historia de Tlaxcala*; mss.

García-Cook, A.; y R. Abascal. 1973. Sistemas de cultivo, riego y control del agua en el área de Tlaxcala. In: México, *Arqueología I*, septiembre; pp. 9–15.

García-Cook, A.; y L. Merino-Carreón. 1997a. El cultivo Intensivo: Condiciones Sociales y Ambientales que lo Originaron. In: *Antología de Tlaxcala*, Vol. 4. México, Instituto Nacional de Antropología e Historia (INAH), Gobierno del Estado de Tlaxcala, pp. 283–303.

García-Cook, Á. 1997b. *Antología de Tlaxcala*, Vol. 2. México, Instituto Nacional de Antropologçia e Historia (INAH), Serie Arqueología.

García-Espinosa, R. 1974. Situaciones climáticas durante el auge y la caída de la cultura teotihuacana. *Boletín del Instituto de Geografía*. México, UNAM, 5: 35–69.

García-Espinosa, R. 1978. Reflexiones sobre el papel de los sistemas de cultivo en la incidencia de los fitopatógenos del suelo en el trópico húmedo. In: S.R. Gliessman (ed.); *Seminarios regionales sobre agroecosistemas con énfasis en el estudio de tecnología agrícola tradicional*. H. Cárdenas, Tabasco, México, Colegio Superior de Agricultura Tropical (CSAT); pp. 127–131.

García-Espinosa, R. 1994. Root pathogens in the agroecosystems of México. Etcheverts, B. (Ed.). *Transactions* Vol. 4a, 15th World Congress of Soil Science. International Soc. of Soil Science. México; 30–44.

García-Espinosa, R. 2007. Control biológico y supresividad. *Microbiología agrícola para el siglo* XXI. Ferrera, C.A.; Alarcón, A. (Comps.). Ed. Trillas, Colegio de Postgraduados (COLPOS), Montecillo, Texcoco, Estado de México: 328–341.

García-Espinosa, R. 2010. *Agroecología y Enfermedades de la Raíz en Cultivos Agrícolas*. Montecillos, México, Colegio de Postgraduados (COLPOS).

García M. and G. Bermúdez. 2014. *Alimentos sustentables a la carta. De la tierra a la mesa*. CALMIL–CONABIO, México.

García-Martínez, B. 1987. Los pueblos de la sierra: *El poder y el espacio entre los indios del norte de Puebla*. México, El Colegio de México (COLMEX).

García-Martínez, B.; y A. González-Jácome (comps.). 1999a. *Estudios Sobre Historia y Ambiente en América I. Argentina, Bolivia, México, Paraguay*. El Colegio de México (COLMEX), e Instituto Panamericano de Geografía e Historia (IPGH).

García-Martínez, B. 1999b. El Monte de Mixtlán: Una reflexión sobre el contrapunto entre poblamiento y naturaleza en el México colonial. In: *Estudios sobre historia y ambiente en América. Vol. 1 Argentina, Bolivia, México, Paraguay*; pp. 55–74.

García Mouton, P. 1986. Los nombres españoles del maíz. *Anuario de Letras, Lingüística y Filología*; México, UNAM, Vol. 2; pp. 121–146.

García-Mora, C. (coord.)1987–1988 [e-book 2013]. La antropología en México. Colección Biblioteca del INAH. México, Instituto Nacional de Antropología e Historia. 15 Vols.

García Paris, J. 1991. *Intercambio y Difusión de Plantas de Consumo entre el Nuevo y el Viejo Mundo*. Madrid, Ministerio de Agricultura, Pesca y Difusión.

García-Payón, J. 1979. La zona arqueológica de Tecaxic Calixtlahuaca y los matlatzinca: etnología y arqueología. W. Tommasi de Magrella y L. Manrique-Castañeda (eds.). 2a. parte. Toluca, Biblioteca Enciclopédica del Estado de México, tomo 30.

García-Quintanilla, A. 2005. Saak y el retorno del fin del mundo. La plaga de langosta en las profesías del katún 13 Ahau. *Ancient Mesoamerica*, 16. Cambridge Universty Press; pp. 327–344.

García-Rodríguez, Y.; Á. Bravo-Monzón; Y. Martínez-Díaz; G. Torres-Gurrola y F.J. Espinosa-García. 2012. "Variación Fitoquímica Defensiva en Ecosistemas Terrestres". En: J.C. Rojas; y E.A. Mateos (Eds.); *Temas Selectos en Ecología Química de Insectos*. México, El Colegio de la Frontera Sur; pp. 217–252.

García-Zamacona, G. 1999. "Algunas ténicas agrícolas tradicionales en agricultura moderna: el sistema de cajas en el Bajío". En: A. González-Jácome y S. del Amo-Rodríguez; *Agricultura y Sociedad en México: Diversidad, Enfoques, Estudios de caso*. México, Univ. Iberoamericana, PROAFT y Plaza y Valdés; pp. 249–266.

Gentry, S. 1982. *Agaves of Continental North America*. EUA, The Univ. of Arizona Press.

Gerhard, P. 1991. *La Frontera Sureste de la Nueva España*. México, UNAM.

Gibson, C. 1952. *Tlaxcala in the Sixteenth Century*. California, Stanford University Press.

Gibson, C. 1980 [1967]. *Los aztecas bajo el dominio español* (1519–1810). Madrid, Siglo XXI editores.

Gibson, C. 1991 [1952]. *Tlaxcala en el Siglo XVI*. México, Fondo de Cultura Económica (F.C.E.)

Gill, R.B. 2000. *The Great Maya Droughts: Water, Life, and Death*. Univ. of New Mexico Press, Albuquerque, NM.

Giordano-Sánchez-Verín, C.A. 2000. Obrajes y economía en Tlaxcala a principios del siglo XVII. Thesis MA., History, Universidad Iberoamericana. 2002. "Tlaxcala y sus Obrajes a Principios del Siglo XVII". Mss, 21 pp.

Gliessman S.R. 1978a. Sustained yield agriculture in the humid lowland tropics. INTECOL, Newsletter 7: 1.

Gliessman S.R. 1978b. Memorias del *Seminario regional sobre agroecosistemas con énfasis en el estudio de tecnología agrícola tradicional*. H. Cárdenas, Tabasco, México, Colegio Superior de Agricultura Tropical (CSAT).

Gliessman S.R. 1980. Aspectos ecológicos de las prácticas agrícolas tradicionales en Tabasco, México: aplicaciones para la producción. *Biotica* 5: 93–101.

Gliessman, S.R.; R. García-Espinosa; and M. Amador. 1981. The ecological basis for the application of traditional agricultural technology in the management of tropical agro-ecosystems. *Agro-Ecosystems* 7: 173–185.

Gliessman S.R. 1982. Nitrogen distribution in several traditional agroecosystems in the humid tropical lowlands of southeastern Mexico. *Plant and Soil* 67: 105–117.

Gliessman S.R. 1984. An agroecological approach to sustainable agriculture. In: *Meeting the expectations of the land: Essays in sustainable agriculture and stewardship*, J.W. Berry and W. Coleman B. (eds.). San Francisco, CA: North Point Press; pp. 160–177.

Gliessman S.R. 1987. Species interactions and community ecology in low external-input agriculture. *American Journal of Alternative Agriculture* 11: 160–165.

Gliessman S.R. 1989. Allelopathy and agricultural sustainability". In: C.H. Chow and G.B. Waller (eds.); *Pythochemical Ecology: Allelochemicals, Mycotoxins and Insect Pheromones and Allomones*. Institute of Botany, Taipei, Taiwan; pp. 69–80.

Gliessman S.R. 1990. *Agroecology. Researching the Ecological Basis for Sustainable Agriculture*. New York, Berlin, London, Springer-Verlag, Ecological Studies 78.

Gliessman S.R. 1998. *Agroecology. Ecological Processes in Sustainable Agriculture*. Sleeping Bear Press, Ann Arbor Press.

Gliessman S.R. 1999. Un enfoque agroecológico en el estudio de la agricultura tradicional. In: *Agricultura y Sociedad en México. Diversidad, Enfoques, Estudios de Caso*. Plaza y Valdés, PROAFT, Universidad Iberoamericana; 25–31.

Gliessman S.R. 2002a. Agroecología. *Procesos Ecológicos en Agricultura Sostenible*. Costa Rica, CATIE, UADY, AGRUCO, PROTRÓPICO, GTZ, UCSC.

Gliessman S.R. 2002b. Making the conversión to Sustainable agroecosystems. Getting from here to there with agroecology. *California Certified Organic Farmers Newsletter* 19 (3): 6–8.

Gliessman S.R. 2004. Integrating agroecological processes into cropping systems research. *Journal of Crop Improvement* 11: 61–80.

Gliessman, S.R.; F.J. Rosado-May; C. Guadarrama-Zugasti; J. Jedlicka; A. Cohn; V.E. Mendez; R. Cohen; L. Trujillo; C. Bacon y R. Jaffe. 2007. Agroecología: promoviendo una transición hacia la sostenibilidad. *Ecosistemas* 16 (1): 13–23.

Gliessmans, R.S.; and M. Rosemeyer (eds.). 2010. *The Convertion to Sustainable Agriculture. Principles, Processes and Practices*. CRC Press, Taylor & Francis Group, Boca Raton, London, New York.

Gliessman S.R. 2013. Agroecología: Plantando las Raíces de la Resistencia. *Agroecología* 8(2): 19–26.

Gliessman S.R. 2015. *Agroecology. The Ecology of Sustainable Food Systems*. CRC Press Taylor and Francis Group, 3rd edition.

Golley, F.B. 1993. *A History of the Ecosystem Concept in Ecology. More than the Sum of the Parts*. New Haven and London, Yale Univ. Press.

Götz, C.M. and K.J. Emery 2013. Introduction. In: *The Archaeology of Mesoamerican Animals*; Vol. 1, Atlanta, Lockwood Press; pp. 1–23.

Goldkind, V. 1965. Social stratification in the Peasan community: Redfield's Chan Kom reinterpreted. *American Anthropologist* 67: 863–84.

Goldsmith, E. y Hildyard N. 1992. La política de la construcción de presas. *Alteridades* 2 (4): 119–133.

Gómez-Nava, Y.A. 2005. Lanchas, alijos y redes. Economía y cultura de los pescadores en Isla Arena, Campeche. Thesis MA., Social Anthropology. México, Universidad Iberoamericana.

Gómez-Pompa A., H.L. Morales, E. Jiménez A., and J. Jiménez A. 1982. Experiences in traditional hydraulic agriculture. *Maya Subsistence. Studies in Memory of Dennis E. Puleston*. Academic Press, Studies in Archaeology: 327–342.

Gómez-Pompa, A.; y Vázquez-Yañes, C. 1985. Estudio sobre la regeneración de selvas en regiones cálido-húmedas. In: A. Gómez-Pompa y S. Del Amo (eds.). *Investigaciones sobre la regeneración de Selvas Altas en Veracruz, México*. Alhambra Mexicana, SA de CV, Instituto Nacional de Investigaciones sobre Recursos Bióticos (INIREB). México DF; pp. 1–25.

Gómez-Pompa, A.; J.S. Flores y V. Sosa. 1987. The "Pet Kot": a man-made tropical forest of the Maya. *Interciencia* 12: 10–15.

Gómez-Pompa, A.; J.S. Flores and M. Aliphat. 1990. The Sacred Cacao Groves of the Maya. *Latin American Antiquity* 1: 247–257.

Gomez-Pompa, A.; y S. del Amo-Rodríguez (comps.) 1985. *Investigaciones sobre la regeneración de Selvas Altas en Veracruz, México*. Alhambra Mexicana, SA de CV, Instituto Nacional de Investigaciones sobre Recursos Bióticos (INIREB), México, DF.

Gomez-Pompa, A.; y S. del Amo-Rodríguez (comps.) 2001. Retrospectiva histórica del Programa de Acción Forestal Tropical. In: Del Amo, R.S. (ed). *Lecciones del Programa de Acción Forestal Tropical*. México, Secretaría del Medio Ambiente, Recursos Naturales y Pesca; Ed. Plaza y Valdés y Consejo Nacional para la Enseñanza de la Biología.

Gómez-Pompa, A.; M.F. Allen; S.L. Fedick; J.J. Jiménez-Osornio (eds.). 2003. *The Lowland Maya Area. Three Millenia at the Human-wildland Interface*. Harworth Press, New York.

González, J.C. 1992. *Chinampas Prehispánicas*. Consejo Nacional para las Cultura y las Artes (CONACULTA), INAH, Antologías, Serie Arqueología.

González-Jácome, A. 1976. Santa Isabel Xiloxoxtla, Tlaxcala: Un estudio micro- económico. Thesis MA. Sciences (Ethnology), ENAH-UNAM.

González-Jácome, A. 1984a. Ambiente y Caña de Azúcar en Veracruz. Una Historia de la Agricultura Comercial y el Caso del Palmar. *Cuadernos del Departamento de Ciencias Sociales y Políticas.* México, Universidad Iberomericana; pp. 1–31.

González-Jácome, A. 1984b. Agroecosistemas en las tierras altas de México. Economía y Sociedad entre los nahuas contemporáneos. First Nahua Meeting, XX Aniversary, Museo Nacional de Antropología. *Boletín UIA*, México.

González-Jácome, A. 1985. Home gardens in Central Mexico. In: *Prehistoric Intensive Agriculture in the Tropics*; Ian S. Farrington (ed.). England: Manchester, BAR International Series 232, Vol. 2: 521–537.

González-Jácome, A. 1986. Agroecología del Suroeste de Tlaxcala. In: *Historia y Sociedad en Tlaxcala.* México, Universidad Autónoma de Tlaxcala (UAT), Universidad Iberoamericana (UIA), Instituto Tlaxcalteca de Cultura (ITC), Gobierno del estado de Tlaxcala; pp. 201–220.

González-Jácome, A. 1987. *Orígenes del Hombre Americano (Seminario).* México, Secretaría de Educación Pública (SEP), 5° Centenario Collection.

González-Jácome, A. 1988a. La agricultura mesoamericana. In: *La antropología en México. Panorama histórico*, 4. Las cuestiones medulares (Etnología y antropología social); C. García-Mora, and M. Villalobos S. (coords.). México, INAH, Colec. Biblioteca del INAH, CONACULTA, Instituto Nacional de Antropología e Historia (INAH). E-book 2015.

González-Jácome, A. 1988b. Población, Ambiente y Economía en Veracruz Central durante la Colonia. *Documentos de Investigación y Capacitación en Población y Desarrollo.* México, Universidad Iberoamericana (UIA) y Fondo de Naciones Unidas (ONU) para Actividades en Materia de Población; pp. 1–60.

González-Jácome, A. 1988c. Migration as a socioeconomic factor of change in agricultural systems. In: *Global Perspectives on Agroecology and Sustainable Agriculture.* EUA: University of California, Santa Cruz 1988: 253–261.

González-Jácome, A. 1989a [1984]. Agroecosistemas en las Tierras Altas de México. In: *Economía y Sociedad entre los Nahuas Contemporáneos.* First Nahua Meeting: *Los Nahuas de Hoy.* México, Instituto Nacional de Antropología e Historia (INAH): 91–117.

González-Jácome, A. 1989b. Evolución de la Industria Textil en Tlaxcala. Siglo XIX y primera mitad del XX. In: *Historia y Sociedad en Tlaxcala.* México, Universidad Autónoma de Tlaxcala (UAT), UIA, Instituto Tlaxcalteca de la Cultura (ITC), Gob. del Edo. de Tlaxcala, 1989: 86–107.

González-Jácome, A. 1991. *La Economía Desgastada. Historia de la Producción Textil en Tlaxcala.* México: Universidad Autónoma de Tlaxcala (UAT), Universidad Iberoamericana (UIA).

González-Jácome, A. 1992. Manejo del Agua en condiciones de secano. *Terra* Revista del Colegio de Postgraduados (COLPOS), Vol. 10, N° especial. México, Texcoco; pp. 494–502.

González-Jácome, A. 1993. Management of land, water and vegetation in traditional agro-ecosystems in Central Mexico. *Landscape and Urban Planning* 27: 141–150.

González-Jácome, A. 1994. ¿Cuántos eran y cómo vivían? La vida rural de Tlaxcala colonial. In: *Universidad y Sociedad*, Vol. 1, Cuadernos de Investigación, Universidad Autónoma de Tlaxcala (UAT), Tlaxcala; pp. 1–12.

González-Jácome, A. 1995. Antecedentes para la construcción de la agricultura tradicional en México: la agricultura que llega con la conquista. *Agroecología y Desarrollo Sustentable.* México, Universidad Autónoma Chapingo (UACh), Col.1: 15–22.

González-Jácome, A. 1996. ¿Quién pagó la cuenta? La cuenca del Atoyac- Zahuapan y el Distrito de Riego N° 56, en Tlaxcala. Thesis, Doct. History, México, Universidad Iberoamericana.

González-Jácome, A. 1997. Influencia de la antropología estadounidense en México: el caso de la ecología cultural. In: M. Rutsch y C. Serrano (comps.). *Ciencia en los Márgenes. Ensayos de Historia de las Ciencias en México.* México, UNAM, IIA; pp. 167–188.

González-Jácome, A. 1999. El paisaje lacustre y los procesos de desecación en Tlaxcala, México. In: *Estudios sobre historia y ambiente en América. Vol. 1 Argentina, Bolivia, México, Paraguay.* B. García-Martínez, and A. González-Jácome (comps.), El Colegio de México (COLMEX), Instituto Panamericano de Geografía e Historia (IPGH), pp. 191–218.

González-Jácome, A. 2000. Mesoamérica: un desarrollo teórico. *Dimensión Antropológica.* México, INAH, Año 7, Vol. 19, mayo/agosto, 2000: 121–151.

González-Jácome, A. 2001a. Notas sobre algunas concepciones de Ángel Palerm acerca del ambiente y la agricultura. *Historia de la Ciencia en México. La Antropología.* México, Universidad Autónoma del Estado de México (UAEM); pp. 263–278.

González-Jácome, A. 2001b. Legislación y Flora: el traslado de plantas de España a Nueva España. In: B. Barba de Piña Chan; C. Rodríguez L.; L. Berruecos V., and L. Barjau M. (comps.), *Antropología e Historia mexicanas. Homenaje al Maestro Fernando Cámara Barbachano.* México, INAH, Colección Científica No. 408: 315–327.

González-Jácome, A. 2002a. Paisajes del pasado: los calales del suroeste de Tlaxcala. In: *Estudios Sobre Historia y Ambiente en América II.* México, COLMEX, IPGH, pp. 200–220.

González-Jácome, A. 2002b. Cuenca del Alto Lerma: Agricultura y Manejo del Agua. *Memorias de Congreso.* Revista del Departamento de Geografía, University of Varsovia, Polonia; pp. 167–188.

González-Jácome, A. 2002c. La riqueza Natural de Mesoamérica. *Gran Historia de México Ilustrada.* México, Ed. Planeta, 2002: 261–280.

González-Jácome, A. 2003a. Como cambia la agricultura tradicional: ecología cultural en dos comunidades mexicanas. *Reporte de Investigación*, México, Universidad Iberoamericana.

González-Jácome, A. 2003b. *Cultura y Agricultura: Transformaciones en el agro mexicano.* México, Universidad Iberoamericana (UIA).

González-Jácome, A. 2004a. Ambiente y Cultura en la Agricultura Tradicional en México: Casos y Perspectivas. *Ciencia Ergo-Sum* 11(2): 153–163.

González-Jácome, A. 2004b. Dealing with risk: small-scale coffee production in Mexico. *Latin American Perspectives.* Japan: Nanzan, No. 1, 2004: 1–39.

González-Jácome, A. 2005. Reconsiderando a Sauer. Los orígenes de la agricultura en México. *Perspectivas Latinoamericanas*; Japón, Nagoya, University, Nanzan, Centro de Estudios Latinoamericanos, N° 2: 13–28.

González-Jácome, A. 2006. El ambiente y la agricultura en Tlaxcala durante el siglo XVI. *Perspectivas Latinoamericanas*; Japón: Nagoya, University, Nanzan, Centro de Estudios Latinoamericanos, N° 3: 19–46.

González-Jácome, A. 2007a. Agroecosistemas mexicanos: pasado y presente. *Itinerarios*; revista de la Universidad de Varsovia, Polonia; Vol. VI, 2007: 55–80.

González-Jácome, A. 2007b. El Distrito de Riego 056 y los Cambios en el Ambiente y la Agricultura en la Antigua Ciénaga de Tlaxcala. *Memorias del 1er Simposio Internacional sobre Cacaxtla*; México, Gobierno del Estado de Tlaxcala e INAH.

González-Jácome, A. 2008a. *Humedales del Suroeste de Tlaxcala. Agua y Agricultura en el siglo XX en México.* México, UIA, El Colegio de Historia de Tlaxcala.

González-Jácome, A. 2008b. De las Manos Tejedoras a las Fábricas Textiles: el nacimiento de una industria en Tlaxcala. *Tlahcuilo.* Boletín del Archivo Histórico de Tlaxcala. Vol. 2, N° 4, enero-marzo 2008: 9–24.

González-Jácome, A. 2008c. El maíz: planta portentosa. *Iberóforum.* Revista de Ciencias Sociales de la Universidad Iberoamericana, Vol. III, N° 5, enero- junio; pp. 1–17.

González-Jácome, A. 2008d. Del Desarrollo Económico al Sustentable. Pasado, Presente y Posibilidades al Futuro. In: *Integración, Desarrollo, e Interregionalismo en las Relaciones Entre la Unión Europea y América Latina*; Á.M. Casas Gragea y M. Ochman (coords.). México, Miguel Ángel Porrúa y EGAP; pp. 137–172.

González-Jácome, A., and J.J. Velasco Orozco. 2008e. Agua y Agricultura en sociedades rurales: el desarrollo por cuencas hidrológicas en México. In: *Nuevas Rutas para el Desarrollo en América Latina. Experiencias globales y locales*; J. Maestre Alfonso, Á. Casas-Gragea, and A. González-Jácome (comps). México, Universidad Iberoamericana; pp. 215–238.

González-Jácome, A. 2009a. El maíz como producto cultural desde los tiempos antiguos. *Desgranando una mazorca. Orígenes y etnografía de los maíces nativos*; C. Morales V. and C. Rodríguez L. (comps.); *Diario de Campo*, Suplemento 52, enero-febrero, México, Instituto Nacional de Antrop. e Historia (INAH), pp. 40–65.

González-Jácome, A. 2009b.Una Historia de Dependencia en el México Independiente: Productos Agrícolas Para la Exportación. *Perspectivas Latinoamericanas* Japón: Nagoya, Centro de Estudios Latinoamericanos, University Nanzan, N° 6; pp. 1–15.

González-Jácome, A. 2009c. Las faldas de La Malinche: el paisaje de las tierras templado-frías y sus pueblos. In: *Matlalcuéyetl: visiones plurales sobre cultura, ambiente y desarrollo*; F. Castro Pérez y T.M. Tucker (coords.). *México, CONACYT, El Colegio de Tlaxcala A.C. and Mesoamerican Research Foundation*, Vol. 1, pp. 250–281.

González-Jácome, A. 2010a. Natural Resources and Out-Migration in Local Communities of Southern Mexico: Non-NAFTA Issues Impacting NAFTA. *The Impacts of NAFTA on North America. Challenges outside the Box*; I. Hussain (ed.). USA, Palgrave, MacMillan: 141–158.

González-Jácome, A. 2010b. Monarquía e independencia en el mundo Hispano: algunas cuestiones sobre Tlaxcala y su situación política en esos aciagos tiempos. In: *La Guerra de conciencias: monarquía o independencia en los mundos hispánico y lusitano.* M. Kroprivitza-Acuña et al. (eds.). México, Coordinación Nacional de las Conmemoraciones 2010, Comisión Estatal de Conmemoraciones 2010; Tlaxcala, Centro de Estudios CARSO, Fideicomiso Colegio de Historia de Tlaxcala, Fomento Cultural Banamex A.C. 2010: 521–554.

González-Jácome, A. 2010c. Agua para ciudades sedientas: La desecación de la zona lacustre en la cuenca alta del río Lerma (México) y la conformación de nuevos paisajes. In: *Documento/Monumento, Rev. Electrónica.* Núcleo de Documentaçao e Informacao Histórica Reginal (NDIHR), Universidade Federal de Mato Grosso, Brasil, Vol. 3, No-1, Dez 2010: 177–192.

González-Jácome, A. 2010d. Mexico: traditional agriculture as a foundation for sustainability. In: S.R. Gliessman y M. Rosemeyer (eds.); *The Convertion to Sustainable Agriculture. Principles, Processes and Practices.* CRC Press, Taylor & Francis Group, Boca Raton, London, New York; pp. 179–204.

González-Jácome, A., and M.Á. Ramírez-Martínez 2010e. Cacao y chocolate: los senderos de la globalización en un contexto regional. In: *Consumos globales: de México para el mundo*; C. Bueno Castellanos y S. Ayora Díaz (coords.). México, Universidad Iberoamericana, Universidad Autónoma de Yucatán, 2010: 37–65.

González-Jácome, A. 2011a. *Historias Varias. Caminando con los agricultores mexicanos.* México, Universidad Iberoamericana.

González-Jácome, A. 2011b. Mexico's Agriculture under NAFTA: The Passing of Traditional Agroecosystems. *North America at the Crossroad: NAFTA after 15 Years*; I. Hussain (ed.). México, Universidad Iberoamericana: 71–111.

González-Jácome, A. 2012a. Del huerto a los jardines y vecindades: procesos de cambio en un agroecosistema de origen antiguo. In: *El Huerto Familiar del Sureste de México*; R. Mariaca M. (coord.). México, Secretaría de Recursos Naturales y Protección Ambiental del Estado de Tabasco y el Colegio de la Frontera Sur (ECOSUR): 522–534.

González-Jácome, A. 2013a. El Manejo del Agua y Algunos Elementos de la Agricultura en Tlaxcala. *Perspectivas Latinoamericanas.* Japón: Nagoya, Centro de Estudios Latinoamericanos, University Nanzan, N° 10, 2013: 69–82.

González-Jácome, A. 2013b. El maíz en el altiplano central mexicano y en la cuenca *alta del río Lerma. Historia, economía, sociedad y cultura. Zanbatha, Valle de la Luna.* Bastida M., M.C., B.A. Albores Z. y G.A. Patrick E. (coords.). México, Universidad Autónoma Metropolitana (UAM)-Unidad Lerma, Colección Inteligente 2: 95–129.

González-Jácome, A. 2013c. El reparto Agrario en Tlaxcala. Estudio Introductorio. In: *Isidro Candia. El reparto agrario en Tlaxcala*; A. González J. (comp.). Gobierno del estado de Tlaxcala, SEPE/USET; pp. 1–52.

González-Jácome, A. 2013d. El impacto ambiental y social de la construcción de presas en el México de los siglos XX y XXI. En: C.A. Casas Mendoza, J.G. Rivera González y L.E. Márquez-Mireles (coords.); *Sujetos emergentes nuevos y viejos contextos de negociación de las identidades en América Latina.* México, Ed. Eón Sociales, Centro Académico de Estudios Regionales de Frontera Interior en América Latina, Universidad Autónoma de San Luis Potosí, Escuela de Ciencias Sociales y Humanidades; pp. 131–144.

González-Jácome, A. 2016a. Sistemas Agrícolas en Orografías Complejas: Las Terrazas de Tlaxcala. In: *Etnoagroforestería en México*; A.I. Moreno C.; A. Casas; V.M. Toledo; M. Vallejo R. (comps.). México, UNAM: 111–144.

González-Jácome, A. 2016b. *Notas sobre Tlaxcala: Geografía, Historia, Ecología, Arqueología y Antropología.* México, CONACULTA, Gobierno del estado de Tlaxcala, Secretaría de Educación Pública de Tlaxcala, Instituto Tlaxcalteca de Cultura (ITC).

González-Jácome, A. 2016c. Analysys of Tropical Homegardens through and Agroecological and Anthropological Ecology Perspective. In: V.E. Méndez, C.M. Bacon; R. Cohen; and S.R. Gliessman (eds.); *Agroecology. A Transdisciplinary, Participatory and Action Oriented Approach.* CRC Press, S. Francisco, Taylor and Francis; pp. 233–257.

González-Jácome, A. 2016d. Orígenes, domesticación y dispersión del maíz (Zea mays) en México. In: López y Vizcarra (comps.). *El maíz nativo en México. Una aproximación crítica desde los estudios rurales.* México, Universidad Autónoma Metropolitana-Lerma, Ciencias Sociales y Humanidades y Juan Pablos Ed; pp. 25–63.

González-Jácome, A. 2017. Los mercados. Una Introducción Antropológica a su Estudio. In: R. Mariaca-Méndez y J.A. López.Gómez (Comps.). El mercado Tradicional de San Cristóbal de Las Casas: Una Experiencia Etnobiológica. ECOSUR y Red del Patrimonio Biocultural del CONACYT.

González-Jácome, A. 2018a. Historia y orígenes de un agroecosystema: Los huertos en México. In: M.J. Ordoñez-Díaz (comp.); *Atlas biocultural de huertos familiares en México: Chiapas, Hidalgo, Oaxaca, Veracruz y península de Yucatán.* México, CRIM-UNAM; pp. 58–150.

González-Jácome, A. 2018b. Aspectos de la Cultura y la Agricultura en el Noroccidente de México. In: González-Jácome, J.J. Velasco-Orozco; S. Moctezuma-Pérez; y A. Cruz-León (coords); *Sonora: La Sierra, el Desierto y la Costa en el Contexto de los guarijíos.* México, Texcoco, Universidad Autónoma Chapingo (UACH); pp. 21–74.

González-Jácome, A. 2018c. Maíz y alimentación. Historia breve de un largo viaje. Revista de G*eografía Agrícola, estudios regionales de la agricultura mexicana.* N° 60, enero-junio; pp. 9–69.

González-Jácome, A. 2018d. La antropología mexicana y la agricultura tradicional. Breve historia, dilemas y perspectivas. In: L. Reyes Montes; J.M. Pérez.Sánchez y S. Moctezuma-Pérez (coords). *Sistemas agrícolas tradicionales. Biodiversidad y cultura.* Zinacantepec, Estado de México; El Colegio Mexiquense A.C.; pp. 23–50.

González-Jácome, A. y S. Del Amo-Rodríguez. 1999. *Agricultura y Sociedad en México. Diversidad, Enfoques, Estudios de Caso.* México, Universidad Iberoamericana (UIA), PROAFT y Plaza y Valdés.

González-Jácome, A.; S. Del Amo R. y F.G. Gurri G. 2007. *Los Nuevos Caminos de la Agricultura: Procesos de Conversión, Nuevos Enfoques y Paradigmas.* México, Plaza y Valdés, PROAFT y Universidad Iberoamericana.

González-Jácome, A. y L. Reyes-Montes. 2015. El conocimiento agrícola tradicional, la milpa y la alimentación: el caso del Valle de Ixtlahuaca, Estado de México. *Geografía Agrícola* 52–53: 21–42.

González-Jácome, A. y J.J. Velasco-Orozco. 2008. Agua y Agricultura en sociedades rurales: el desarrollo por cuencas hidrológicas en México. In: *Nuevas Rutas para el Desarrollo en América Latina. Experiencias globales y locales*; J. Maestre; Á. Casas-Gragea y A. González-Jácome (comps.) México, Universidad Iberoamericana; pp. 215–238.

González-Jácome, A. y J.J. Velasco Orozco. 2015. *La Muerte de un Humedal. Procesos de desecación en la cuenca del Alto Lerma.* Argentina, Editorial Alpes.

González-Jácome, A.; J.J. Velasco-Orozco; S. Moctezuma-Pérez.; A. Cruz-León (comps.). 2018. *Sonora: La Sierra, el Desierto y la Costa en el Contexto de los Guarijíos.* México, Texcoco, Universidad Autónoma Chapingo.

González-Martínez, L. 1992. *Respuesta Campesina a la Revolución Verde en el Bajío.* México, Universidad Iberoamericana, Programa Institucional de Análisis Regional.

González, M.V.S. 2008. *Agroecología, Saberes Campesinos y Agricultura como forma de vida.* Estado de México, Texcoco, Universidad Autónoma Chapingo (UACH).

González-Pozo, A. (coord.). 2010. *Las chinampas de Xochimilco al despertar el siglo XXI: inicio de su catalogación.* México, UAM-Xochimilco. CDMX, Universidad Autónoma Metropolitana-Xochimilco. 2016. Las Chinampas: Patrimonio Mundial de la *Ciudad de México.* México, UAM-Xochimilco. CDMX, Universidad Autónoma Metropolitana-Xochimilco.

González, R. 2001. *Zapotec Science. Farming and Food in the Northern Sierra of Oaxaca.* University of Texas Press, USA.

González-Solano, J.M.F. 2007. Sistema hidráulico Cerro de Oro en la cuenca del Papaloapan, Veracruz, Oaxaca e indígenas chinantecos y el conflicto por su reubicación, 1972–1989. Tesis MA., History. México, Universidad Autónoma Metropolitana (UAM-I).

González-Torres, Y. 2007. Notas sobre el maíz entre los indígenas mesoamericanos antiguos y modernos. *Dimensión Antropológica*, Vol. 41, sep-diciembre, pp. 45–80. En: http://www.dimensionantropologica.inah.gob.mx/?p=1716.

González-Torres, Y 2008. Etnografía del maíz: variedades, tipos de suelo y rituales en treinta monografías. In: *Centro de Estudios Mexicanos y Centroamericanos*; pp. 179–219. En: http://www.openedition.org/6540.

González-Turmo, I. 2001. Antropología de la alimentación: propuestas metodológicas. In: A. Garrido Aranda (comp.); *Comer cultura. Estudios de cultura alimentaria.* España, Servicio de Publicaciones, Universidad de Córdoba, pp. 13–25.

Goody, J. 1982. *Cooking, cuisine and class. A study in comparative sociology.* Cambridge, Cambridge University Press.

Gorman, M. 1992. Paleoecological evidence for prehistoric agriculture and tropical forest clearence in the Sierra de Los Tuxtlas, Veracruz, Mexico. MA. Thesis, Geography, University of California, Berkeley.

Götz C.H. y K.F. Emery (eds.). 2013. *La arqueología de los animales de Mesoamérica*. Georgia, Atlanta, Lockwood Press, N° 1.

Grobman, A., W. Salhauana y R. Sevilla. 1961. *Races of maize in Peru: their origins, evolution and classification*. Pub. 915. Washington, DC, Natural Academy of Sciences, National Research Council.

Grobman A., D. Bonaviab, T.D. Dillehayc, D.R. Piperno, J. Iriartef; and I. Holst. 2012. Preceramic maize from Paredones and Huaca Prieta, Peru. *Proceedings of the National Academy of Sciences*109(5): 1755–1759.

Grobman, A.T. 2013. Conferencia, Universidad Nacional Agraria La Molina, Lima, Perú (09/2013). In: *Gaceta Molinera*, Jueves, 05/11/2020.

Guerrero-Herrera, M.J. 2015. Informe final del Proyecto FZ016. "Conocimiento de la diversidad y distribución actual del maíz nativo y sus parientes silvestres en México, 2a. etapa 2008–2009. Instituto Nacional de Investigaciones Forestales, Agrícolas y Pecuarias (INIFAP), Centro de Investigación Regional Noroeste, CONABIO.

Guiller, K.E. 2001. *Nitrogen Fixation in Tropical Cropping Systems*. New York, CABI Publishing.

Guzmán, M. y J. Palerm-Viqueira. 2005. Los jagüeyes en la región de los Altos Centrales de morelos. *Boletín, Archivo Histórico del Agua (AHA)* 10(2): 21–26.

Gurri-García, F.D. 2007. Globalización y cambios en la calidad de vida en familias campesinas de Yucatán, México. *Ciencia* 58(2): 60–68.

Gurri-García, F.D. 2011. La doble carga de la transición nutrimental en zonas rurales de la península de Yucatán ¿consecuencia de la alteración de los sistemas agrícolas de subsistencia tradicionales en la segunda mitad del siglo XX? In: J.M. Muñoz (comp.), *Obesidad problema multifactorial*. Villahermosa: Universidad Juárez Autónoma de Tabasco; pp. 65–84.

Gurri-García, F.D.; G. Balam-Pereira and E.F. Morán. 2001. Well being changes in response to 30 years of regional integration in Maya populations from Yucatan, Mexico. *American Journal of Human Biology* 13: 590–602.

Gurri-García, F.D.; and E.F. Moran. 2002. Who is interested in commercial agriculture? Subsistence agriculture and salaried work in the city among Yucatec Maya from the state of Yucatán. *Culture and Agriculture* 24(1): 41–47.

Gurri-García, F.D.; and Molina-Rosales, D. 2011. Agricultura comercial tradicional y vulnerabilidad en Campesinos. *Política y Cultura*, México, Universidad Metropolitana (UAM) Iztapalala; pp. 71–98.

Guzman-Sebastián, K.C.; V.A. Velasco-Velasco; J. Ruiz-Luna; G.V. Campos- Ángeles; G. Rodríguez-Ortiz y J.R. Enríquez-del Valle. 2016. Productos agroalimentarios, comercialización en la "Plaza" de Villa de Zaachila, Oaxaca, México. *Revista Mexicana de Ciencias Agrícolas*; Vol. 7, N° 4: 871–883.

Harper, C.L. and B.F. Le Beau. 2003. *Food, Society and Environment*. New Jersey, Prentice Hall.

Harrison, P.D. 1982. Subsistence and Society in Eastern Yucatán. In: Flannery, K.V. (ed.), *Maya Subsistence. Studies in Memory of Dennis E. Puleston*. New York, London. Academic Press; pp. 119–130.

Harrison, P.D. 2001. La agricultura maya. In: N. Grube (ed.); *Los mayas: una civilización milenaria*. Colonia, Könemann.

Harrison, P.D.; and B.L. Turner II. 1978. *Pre-Hispanic Maya Agriculture*. Studies in Memory of Dennis E. Puleston. New York, Academic Press.

Hastorf, C.A. 2009. Rio Balsas most likely region for maize domestication. *Proceedings of the National Academy of Sciences*106(13): 4957–4958.

Henderson, J.S.; R.A. Joyce; G.R. Half; W.J. Hurst y P.E. McGovern. 2007. Chemical and archaeological evidence for the earliest cacao beverages. *Proceedings of the National Academy of Sciences* 104(48): 18937–18940.

Heine, K. 2003. Paleopedological evidence of human-induced environmental change in the Puebla-Tlaxcala area (Mexico) during the last 3,500 years. En: *Revista Mexicana de Ciencias Geológicas*, 20: 235–244.

Hemond, A., and M. Goloubinoff. 2008. El vía crucis del agua. Clima, calendario agrícola y religioso entre los nahuas de Guerrero. In: *Aires y lluvias. Antropología del clima en México*; A. Lammel, M. Goloubinoff y E. Katz (eds.). México, CIESAS, CEMCA, Publicaciones de la Casa Chata; pp. 133–169.

Hernández-Matamoros, E.R. 2016. "Una propuesta de política pública agrícola para la población rural del Centro de México. La agroecología como modelo de la agricultura de autoconsumo". Thesis, Doct. Política Pública. Tecnológico de Monterrey, Ciudad de México.

Hernández, R.C. 2014. *La Tierra del Maíz*. Gobierno del Estado de Tlaxcala, CONACULTA, ITC.

Hernández-Rodríguez, R. 1982. Epidemias y calamidades en el México prehispánico. In: Florescano E. y E. Malvido (comps). *Ensayos sobre la historia de las epidemias en México*. México, Instituto Mexicano del Seguro Social (IMSS), 2 Vols.; pp. 139–153.

Hernández X.E. 1978. Exploración etnobotánica para la obtención de plasma germinal para México. *Soc. Mexicana de Fitogenética*; UACH; pp. 3–12.
Hernández X.E. 1981 [1977]. *Agroecosistemas de México: contribuciones a la enseñanza, investigación y divulgación agrícola*. Estado de México, Texcoco, Colegio de Postgraduados (COLPOS), Universidad Autónoma Chapingo (UACH).
Hernández X.E. 1982. La dinámica de la milpa. *Agrosistemas* 32: 3–4; Centro de Botánica, Colegio de Postgraduados (COLPOS), Universidad Autónoma Chapingo (UACH), México.
Hernández X.E. 1986. *Biología Agrícola. CECSA*, México.
Hernández X.E. 1987a. Exploración Etnobotánica en Maíz. En: *Xolocotzia, Obras de Efraín Hernández Xolocotzi*, Vol. II. *Revista de Geografía Agrícola*. Universidad Autónoma Chapingo; pp. 751–756.
Hernández X.E. 1987b. Etnobotánica de Tlaxcala. In: Ruiz Figueroa, J.F. (ed.) *Uso y manejo de los tepetates para el desarrollo rural*. México: Universidad Autónoma Chapingo (UACh), Texcoco, Chapingo, pp. 1–8.
Hernández X.E. 1988. La agricultura tradicional en México. In: *Comercio Exterior* Vol. 38, N° 8, agosto; pp. 673–678.
Hernández X.E. 1993. Aspects of plant domestication in México. A personal View. In: T.P. Ramamoorthy, R. Bye, and J. Fa (eds.). *Biological Diversity of Mexico: Origins and distribution*; Oxford University Press. New York; pp. 733–753.
Hernández X.E. 2013a [1949]. Graneros de Maíz en México. *Xolocotzia*, Texcoco, UACH; Vol. 1: 233–265.
Hernández X.E. 2013b. *Xolocotzia*, 2 Vols. Texcoco, UACH.
Hernández X.E.; y G. Alanis F. 1970. Estudio Morfológico de cinco nuevas razas De maíz de la Sierra Madre Occidental de México: Implicaciones filogenéticos y fitogeográficas". *Agrociencia* 1: 3–30.
Hernández X.; S. Levy T. and L.M. Arias-Reyes 1987. Hacia una evaluación de los recursos naturales renovables bajo el sistema de roza-tumba-quema en México. In: *Evaluación de tierras y recursos para la planeación nacional de zonas tropicales*. USDA; pp. 338–340.
Hernández X.; E. Bello; and S. Levy. 1995. *La milpa en Yucatán: Un sistema de producción agrícola tradicional*. Estado de México, Montecillo, Colegio de Postgraduados (COLPOS), 2 Vols.
Hernández-Palomo, J.J. 2004. El Pulque: Usos Indígenas y Abusos Criollos. *Simposio 2001 en Jerez de la la Frontera*; Asociación Española de Americanistas; pp. 239–257.
Heyden, D. 1983. *Mitología y simbolismo de la flora en México prehispánico*. México, UNAM, Inst. de Investig. Antropológicas. 1995. Jardines botánicos prehispánicos. In: *Coloquio Cantos de Mesoamérica*. México, UNAM.
Heyden, D. 2002. Jardines botánicos prehispánicos. Antiguos jardines mexicanos. *Historia de la Arqueología en México* X(57): 18–23.
Hilger, W. 1973. Observaciones respecto a la geología de la región de Puebla– Tlaxcala. *Comunicaciones 7*. Fundación Alemana para la Investigación Científica, Puebla, Puebla, México; pp. 3–6.
Hinton, E.G. 1945. El frijol-soya en Cuernavaca. *Boletín Indigenista*. Instituto Nacional Indigenista (INI), Vol. 2-B: 180–182.
Hirth, K.G. 2012. Producción azteca de maíz y almacenamiento institucional en la cuenca de México: el modelo de descentralización del almacenaje. In: Botot et al. (coords); *Almacenamiento Prehispánico, del Norte de México al Altiplano Central*. CEMCA, Universidad París 1 Pantheon-Sorbone, Laboratorio Arqueológico de las Américas, Universidad de San Luis Potosí (UASLP); pp. 178–186.
Ho, Ping-li. 1956. The introduction of American food plants into China. *American Anthropology* 57: 191–201.
Hodell, D.A.; M. Breener and J.H. Curtis. 2000. "Climate change in the Northern American tropics and subtropics since the last ice age: implications for environment and culture. In: D.L. Lentz (ed.); *Imperfect Balance. Landscape Tramsformations in the Precolumbian Americas*. Columbia University Press; pp. 13–38.
Holtzman, J. 2006. Food and memory. *Annual Review of Anthropology* 35: 361–378.
Hoppe, W.A.; A. Medina and R. Weitlaner. 1975. The Popoloca. In: R. Wachope General Editor; E.Z. Vogt (Vol. ed.); *Handbook of Middle American Indians*, Vol.7; University of Texas Press; pp. 489–505.
Hobsbawn, E.J. 1975a. *The Age of Capital* 1848–1875. Weidenfeld and Nicolson, London.
Housbawn, E.J. 1987 [1975b]. *The Age of Empire* 1875–1914. Weidenfeld and Nicolson, London.
Hobsbawn, E.J. 2003 [1971]. *La era de la revolución*. Editorial Crítica.
Huckell, B.B. 1996. The archaic prehistory of the North American Southwest. *Journal of World Prehistory* 10: 305–373.
Huckell, L.W. 2006. Ancient maize in the American Southwest. What does it look like and what can it tell us? In: Staleer, Tykot and Benz (comps.). *Histories of Maize*. Massachusets, San Diego and London, Academic Press; pp. 97–107.

Hughes C.E.; E. Rajanikanth G.; A. Robertson; D.L. Filer; S.A. Harris and D. Bailey. 2007. Serendipityous backyard hybridization and the origins of crops. *Proceedings of the National Academy of Sciences* 104(36): 14389–14394.

Hurtado M.E., and J.F. Román G. 2012. *The Great Maya Droughts in Cultural Context*. Boulder, University of Colorado, Colorado.

Hurtado M.E. 2013. *Con tinta de agua; historiografía, tecnologías y usos*. México, Universidad Autónoma de Zacatecas, Programa Integral de Fortalecimiento Institucional. Iannone G. (ed.). Ibarra-Crespo, E. 1992. La Vitivinicultura en México. In: L. Hidalgo (coord); *La Viticultura Americana y sus Raíces*. España, Ministerio de Agricultura, Pesca y Alimentación, Secretaría Técnica.

Ibarra Vázquez, C.; R. Tavera y E. Novelo. 2009. Diversidad y estructura de las comunidades de diatomeas del perifiton y el metafiton en un humedal tropical en México. *Revista Mexicana de Biodiversidad* 80: 763–769.

Ignacio-Felipe, E. 2007. *Nahuas de la Montaña*. México, Comisión Nacional para los Pueblos Indios (CDI). En: http:www.cdi.gob.mx

Iltis, H.H. 1983. From teosinte to Maize. The catastrophic sexual transmutation. *Science*, 222, pp. 886–894.

Iltis, H.H. 2000. Homeotic sexual translocations and the origin of maize (*Zea mays* Poaceae): a new look at an old problem. *Economic Botany* 54(1): 7–42.

Iltis, H.H. 2006. Origins of polystichy in maize. En: *Histories of Maize. Multidisciplinary Approaches to the Prehistory, Linguistics, Biogeography, Domestication, and Evolution of Maize*. In: J.E. Staller, R.H. Tykot and B.F. Benz (eds). Elsevier, Academic Press; pp. 3–53.

Instituto Nacional para el Federalismo y el Desarrollo Municipal (INAFED). 2009. Gobierno del Estado de Tlaxcala. En: www.inafed.gob.mx/work/.../regi.htm

Instituto Nacional de Antropología e Historia (INAH). 2018. Boletín N° 1

Instituto Nacional Indigenista (INI). 1943a Resurgimiento agrícola de los yaquis. *Boletín Indigenista* III(2): 93–96.

Instituto Nacional Indigenista (INI). 1943b. Edición para enseñar a construir herramientas e implementos a los agricultores pequeños. *Boletín Indigenista* 3(1-B): 268.

Instituto Nacional Indigenista (INI). 1944. Editorial. *América Indígena* IV(2): 93–96.

Instituto Nacional Indigenista (INI). 1945. Recetas para cocinar con frijol de soya, semillas tostadas, pinole, atoles, tortillas, gordas y leche. *Boletín Indigenista* V(3-B): 254–259.

Instituto Nacional Indigenista (INI). 1946a. Causas de la deforestación en vertientes y costos de la tala inmoderada de vegetación silvestre. *Boletín Indigenista* 6(1-B): 2–5.

Instituto Indigenista Interamericano (III). 1944.Editorial. *América Indígena* IV(2): 93–96.

Instituto Nacional de Estadística, Geografía e Informática (INEGI). 1981. Carta Uso del suelo y vegetación. Escala 1:50,000. México.

Instituto Nacional de Estadística, Geografía e Informática. 1990. Censo General de Población. México.

Instituto Nacional de Estadística, Geografía e Informática (INEGI). 1998. Estadísticas Históricas. México, Aguascalientes.

Instituto Nacional de Estadística, Geografía e Informática (INEGI). 2005.

Instituto Nacional de Estadística, Geografía e Informática (INEGI). 2007. Censo Agrícola y Ganadero. México, Aguascalientes.

Instituto Nacional de Estadística, Geografía e Informática. 2009a. Prontuario (30149), Soteapan, Veracruz, de Ignacio de la Llave. Marco Geoestadístico Municipal; pp. 1–9.

Instituto Nacional de Estadística, Geografía e Informática (INEGI). 2009b. Prontuario (30124), Papantla, Veracruz, de Ignacio de la Llave. Marco Geoestadístico Municipal; pp. 1–8.

Instituto Nacional de Estadística, Geografía e Informática (INEGI). 2010. Censo General de población. México, Aguascalientes.

Instituto Nacional de Estadística, Geografía e Informática (INEGI). 2015. Sistema de Información Municipal, estado de Veracruz, Cuadernillos Municipales. Puente Nacional (Código INEGI 301340001); pp. 1–10.

International Council for Science. 2002. Report of the Scientific and Technological Community to the World Summit on Sustainable Development (Int. Council for Science, Paris). In: www.icsu.org/Library/WSSD-Rep/Vol 1.pdf.

Inomata, T.; J. MacLellana; D. Triadana; J. Munsonb; M. Burhama; K. Aoyama; H. Nasu; F. Pinzóne and H. Yonenobu. 2015. Development of sedentary communities in the Maya lowlands: Coexisting mobile groups and public ceremonias at Ceibal, Guatemala. *Proceedings of the National Academy of Sciences*112(14): 4268–4273.

Iturriaga, J.N. 2013. *La cultura del antojito. De tacos, tamales y tortas*. México, CONACULTA, Direc. Gral. de Publicaciones.

Jácome, Pedro. 2010. *Catálogo de Recursos Gastronómicos de México.* México, Patromonio Cultural y Turismo, Cuadernos 17. CONACULTA.

Jaenike-Després, V.R. and Smith B.D. 2006. Ancient DNA and the integration of archaeological and genetic approache to the study of maize domestication. In: Staller, J.E., R.H. Tykot and B. Benz. *Histories of Maize. Multidisciplinary Approaches to the Prehistory, Linguistics, Biogeography, Domestication, and Evolution of Maize.* New York, London, Oxford, Paris, Elsevier, Academic Press; pp. 83–95.

Jakemann, M. Well. 1952. *The 'Historical Recollections' of Gaspar Antonio Chi. An Early Source account of Ancient Yucatan.* Brigham Young University. Pub. in Archaeology and Early History, N° 3, Provo, Utah.

Jiménez-Osornio, J.J. 1999. La ecología química en la agricultura y su consideración en los agroecosistemas tradicionales. En: A. González-Jácome y S. del Amo; *Agricultura y Sociedad en México: Diversidad, Enfoques, Estudios de Caso.* México, Universidad Iberoamericana, Plaza y Valdés Editores; pp. 103–122.

Jiménez-Osornio, J.J.; y S. del Amo R. 1988. La pérdida de un agroecosistema tradicional intensivo. Las chinampas de Mixquic. In: *Cuatro estudios sobre sistemas tradicionales.* Serie Investigaciones Sociales N° 17. México, Instituto Nacional Indigenista (INI).

Jiménez-Osornio, J.J.; T. Rojas R.; S. del Amo R. y A. Gómez-Pompa. 1995a. Conclusiones y recomendaciones del Taller. In: T. Rojas R. (coord.). *Presente, pasado y futuro de las Chinampas.* México, CIESAS y Patronato del Parque Ecológico de Xochimilco, AC. pp. 19–43.

Jiménez-Osornio, J.J. 1995b. Decálogo para la propuesta del rescate ecológico de Xochimilco. In: *Pasado, presente y futuro de las Chinampas.* México, CIESAS y Patronato del Parque Ecológico de Xochimilco, A.C.; pp. 299–301.

Juan-Pérez J.I. 2003. Tiempo con dinero y tiempo sin dinero. Agricultura tradicional y comercial en una zona de transición ecológica del Estado de México. Thesis Doct. Social Anthropology. México, Universidad Iberoamericana.

Juan-Pérez J.I. 2005. *Los huertos familiares en una provincia del subtrópico mexicano. Análisis espacial. Económico y sociocultural.* España, Fundación Universitaria Andaluza Inca Garcilaso.

Juárez-Choque, Y. 2013. Actividades comparadas de Subsistencia Campesina en Tambillo, Perú y Santa María Nopalucan Tlaxcala, México. Thesis, MA. Social Anthropology. México, Universidad Iberoamericana.

Kaplan, L.N. 1992. Beans of the Americas. *Chilies to Chocolate. Food the Americas Gave the World*; N.F. and L.S. Cordell (eds.). Tucson, The University of Arizona Press.

Kato, T.A.; C. Mapes; l.M. Mera; J.A. Serratos, R.A. Bye. 2009. *Origen y diversificación del maíz: una revisión analítica.* México, UNAM, Instituto de Biología y Comisión Nacional Para el Conocimiento y Uso de la Biodiversidad.

Katz, E. 2008. Emigration, mutations sociales et changements culinaires en pays mixtèque (Oaxaca, Mexique). *Anthropology of Food*, N° S4, Modèles alimentaires et recompositions sociales en Amérique latine. En: http://aof.revues.org/2912

Katz, S.H.; L. Hediger and L.A. Valleroy. 1974. Traitional maize processing techniques in the new world. *Science* 184(4138): 765–773.

Kelly, I. and Á. Palerm. 1952. *The Tajin Totonac: History, subsistence, shelter and technology.* Smithsonian Institution, Institute of Social Anthropology, Washington, D.C.

Kennedy, D. 2014. Las raíces de la comida mexicana. CONABIO. En: www. biodiversidad.gob.mx/usos/dk.

Kennett, J.D.; H.B. Thakarb; A.M. Vanderworker; D.L. Webstera; B.J. Culletona; T.K. Harpera; L. Kistlerd; T.E. Schefflere; y K. Hirtha. 2017. High-precision chronology for Central American maize diversification from El Gigante rockshelter, Honduras. *Proceedings of the National Academy of Sciences of the United States of America* 114(34): 9026–9031.

Kepecs, S. 2007. Chichén Itzá, Tula and the epiclassic early postclassic mesoamerican world system. In: J.K. Kowalsy y C. Kristan-Garham (eds.); *Many Tollans, Chichen Itza, Tula and the Epiclassic-Early Potclassic Mesoamerican World.* Washington DC, Dumbarton Oaks; pp. 129–150.

Kepecs, S. and S. Boucher. 1996. The pre-hispanic cultivation of rejolladas and stone-lands: new evidence from Northeast Yucatán. In: *The Managed Mosaic: Ancient Maya Agriculture and Resource Use* (S.L. Fedick ed.). University of Utah Press, Salt Lake City; pp. 69–91.

Kepecs, S. and M. Masson. 2003. Political organization in Yucatan and belize. In: M.E. Smith, and F.F. Berdan (eds.); *The Posclassic Mesoamerican World.* Utha, Salt Lake City, University of Utah Press; pp. 40–44.

Klee, G.A. 1980. *World Systems of Traditional Resource Management.* Wiley & Sons, New York.

Killion, T.W. 1990a. Cultivation intensity and residential site structure: an ethnoarchaeological examination of peasant agriculture in the Sierra de los Tuxtlas, Veracruz, Mexico. *Latin American Antiquity* 1: 191–215.

Killion, T.W. 1990b. *Gardens of Prehistory. The Archaeology of Settlement Agriculture in Greater Mesoamerica*. University of Alabama Press.

Kimber, C. 1966. Dooryard Gardens of Martinique. *Yearbook of the Pacific Coast Geographers* 28: 97–118.

King, M.E. 1986. Preceramic Cordage and Basketry from Guilá Naquitz. In: Flannery K.V. (ed.) *Guilá Naquitz: Archaic Foraging and Early Agriculture in Oaxaca, Mexico*. Academic Press, New York; pp. 157–161.

Köhler, U. 2007. Los dioses de los cerros entre los tzotziles en su contexto interétnico. In: *Estudios de Cultura Maya*, México, UNAM, Vol. 30.

Konrad, H. 1985. Fallout of the wars of the chacs: the impact ofhurricanes and implications for the pre-Hispanic Quintana Roo Maya processes. In: Marc Thompson et al. (eds.). *States, Structure and Stratification: Current Archaeological Reconstruction*. Canada, Calgary, University of Calgary Press; pp. 321–330.

Konrad, H. 1996. Caribbean tropical storm. Ecological implications for the pre- Hispanic and contemporary Maya Subsistence practices on the Yucatan Peninsula. *Revista Mexicana del Caribe* (I): 98–130.

Krupinsky, J.M.; K.L. Bailey; M.P. McMullen, B.D. Gossen and T.K. Turkington. 2002. Managing plant disease risk in diversified cropping systems. *Agronomy Journal* 94(2): 198–209.

Kumar B.M., P.K.R. Nair (eds.). 2006. *Tropical Homegardens. A Time-Tested Example of Sustainable Agroforestry*. Netherlands, Springer.

Kumate, J. (comp.). 1993. *La investigación científica de la herbolaria medicinal mexicana*. México, Secretaría de Salud.

Laborde, J.D. 2009. "Los Habitantes". In: *Los Paisajes de los Tuxtlas*; S. Guevara; J. Laborde and G. Sánchez-Ríos (comps.). Instituto de Ecología AC (INECOL). pp. 65–78.

Lachat, C.; J.E. Raneri; K. Walker S.; P. Kolsteren; P.V. Damme; K. Verzelen: D. Peñafiel; W. Vanhove; G. Kennedy; D. Hunter; F. Oudoir O.; G. Ntandou- Bouzitou; B. de Baets; D. Ratnasekera; H. The Ky; R. Remans y C. Termote. 2018. Dietary species richness as a measure of food biodiversity and nutritional quality of diets. *Proceedings of the National Academy of Sciences* 115(1): 127–132. www.pnas.org/cgi/doi/10.1073/pnas.1709194115.

Lammel, A.; M. Goloubinoff, and E. Katz. 2008 [2013, Open books]. *Aires y lluvias. Antropología del Clima en México*. CEMCA, CIESAS, IRD.

LaRochelle, S. and F. Berkes. 2003. Traditional ecological knowledge and practice for edible wild plants: biodiversity use by the Rarámuri in the Sierra Tarahumara, Mexico. *International Journal of Sustainable Development & World Ecology* 10: 361–375.

Lascurain, M.; S. Avendaño; S. del Amo; A. Niembro. 2010. *Guía de Frutos Silvestres comestibles en Veracruz*. México, CONACYT, CONAFOR, Instituto de Ecología A.C.

Latournerie-Moreno, L.; J. Tuxill; E. Yupit-Moo; L.M. Arias-Reyes; J.C. Alejo and D.I. Jarvis. 2006. Traditional maize storage methods of Maya farmers in Yucatan, Mexico: Implication for seed selection and crop diversity. *Biodiversity and Conservation* 15(5): 1771–1795.

Lauer W. 1979. Medio ambiente y desarrollo cultural en la región de Puebla–Tlaxcala. In: *Comunicaciones*. Fundación Alemana para la Investigación Científica, Proyecto Puebla- Tlaxcala, 2° Simposio, N° 16: 29–54.

Lazos, E. 1994. La diversification des stratégies agricoles des paysans mayas du sud du Mexique et ses conséquences sur leur système alimentaire. *JATBA*, 36(2): 79–109.

LeCount, L. 2001. Like water for chocolate: feasting and political ritual among the Late Classic Maya at Xunantunich, Belize. *American Anthropologist* 103(4): 935–953.

Lentfer C. and W. Boyd. 2018. *Tracing antiquity of banana cultivation in Papua New Guinea*. Centre for Geoarchaeology and Palaeoenvironmental Research, School of Environmental Science and Management, Southern Cross University, Lismore, NSW.

Lentz, D.L. 2000. Anthropocentric food webs in the Pre-Columbian Americas. In: D.L. Lenz (ed.). *Imperfect Balance: Landscape Transformations in the Pre-Columbian America*. New York; Columbia University Press; pp. 89–120.

Lerner-Martínez, T. 2008. Aporte de alimentos del huerto familiar a la economía campesina Ch'ol, Suclumpá, Chiapas, México. Thesis MA. ECOSUR, San Cristóbal Las Casas, Chiapas.

Lerner-Martínez, T.; R. Mariaca-Méndez; E.B. Salvatierra-Izaba; A. González- Jácome y E. Wahl. 2009. Aporte de alimentos del huerto familiar a la economía campesina Ch'ol, Suclumpá, Chiapas, México. *Etnobiología* 7: 30–44.

Leonti, M.; H. Vibrans; O. Sticher; M. Heinrich. 2001. Ethnopharmacology of the Popoluca, Mexico: an evaluation. *Journal of Pharmacy and Pharmacology (JPP)* 53: 1653–1669.

Leriche, G.L.F. 1983. The transformation of a small peasant community by the state´s proyect for rural development: a case study of a collective ejido in the tropics of Tabasco, Mexico, during the oil boom 1975–1980. MA. Thesis, Univesity of Texas, Austin, Latin American Studies.

Lestage, F. 2008. De la circulation des nourritures, La perpétuation et l'extension des liens sociaux des migrants mexicains via l'approvisionnement en produits alimentaires. *Anthropology of Food*, N° S4, Modèles alimentaires et recompositions sociales en Amérique latine. En: http:// aof.revues.org/2942

Leyva, T.D.A. 2017. Conocimiento tradicional en agroecosistemas de Ocotal Texizapan, Veracruz: Seguridad alimentaria y nutricional. Thesis Doct. COLPOS/Veracruz.

Lewis, O. 1944a. Social and economic changes for a Mexican villages: Tepoztlan 1926–1944. *América Indígena* IV(4): 281–314.

Lewis P.M. and R.T. Carter. 1989. An assessment of the effects of climatic change on agriculture. *Climatic Change* 15: 95–116.

Liebman M. 1999. Sistemas de policultivos. In: Altieri, M.A. *Agroecología. Bases para una agricultura sustentable.* Nordan Comunidad. Montevideo; pp. 191–203.

Linares, O.F. 1976. Garden hunting in the American tropics. *Human Ecology* 4(4): 331–349.

Linares, E. M. and R. Bye. B. 1996. Pericón. The herb of the clouds. *Voices of México* 37: 55–56.

Linares, E. M. and R. Bye. B. 2011. sobre razas de maíz.

Linares, E. M. and R. Bye. B 2013. Jitomate, cebollas y cebollejas, ¿Son plantas del viejo mundo? In: G. Castillo (ed.). *El jitomate y la cebolla, secreto de nuestra sazón.* México Fundación Herdez; pp. 29–41.

Linares, E. M. and R. Bye. B. 2014. Traditional markets in Mesoamerica: a mosaic of history and traditions. In: *Ethnobotany of Mexico.* Interactions of People and Plants in Mesoamerica. Springer; pp. 151–178.

Linares, E. M. and R. Bye. B. 2015. Las especies subutilizadas de la milpa. *Revista Digital Universitaria* 16(5): 22 (1° de mayo).

Linares, E. M. and R. Bye. B 2016. ¡La milpa no solo es maíz! México, UNAM; pp. 9–12.

Lipper, L. et al. 2014. Climate-smart agriculture for food security. *Nature Climate change* 4: 1068–1072.

Lippi, R.D. 1984. Maize Recovetred at La Ponga, An Early Ecuatorian Site. *American Antiquity* 49(1): 118–124.

Long, S.J. 1997. *La cocina mexicana a través de los siglos. IV, La Nueva España.* México, Ed. Clío y Fundación Herdez.

Long, S.J. 2003. *Conquista y comida. Consecuencias del encuentro de dos mundos.* México, UNAM.

Long, S.J. 2009. Tecnología alimentaria prehispánica. *Estudios de Cultura Náhuatl.* México, UNAM; pp. 127–136.

Long, S.J. and L.A. Vargas-Guadarrama. 2005. *Food culture in Mexico.* Serie Food culture around the World. Westpoint, Connecticut and London, Greenwood Press.

López-Acosta, J.C.; G. Zamora-Pedraza; E. Mendoza-Ramírez, M.C. MacSwiney. 2018. Recuperación de la diversidad florística en plantaciones de cítricos abandonadas en el sur de Veracruz, México: implicaciones para la conservación. *Revista Mexicana de Biodiversidad* 89(2018): 1212–1233.

López-Arenas G. 2006. Deidades de la fertilidad agrícola en el panteón mexica. México, UNAM, *Estudios mesoamericanos* 7: 45–52.

López-Austin, A. 1992. Homshuk. Análisis temático del relato. *Anales de Antropología*, Vol. 29: 261–283.

López-Austin, A. 1994. *Tamoanchan y Tlalocan.* México, Fondo de Cultura Económica (FCE).

López-Austin, A. 1996. *Los mitos del tlacuache. Caminos de la mitología mesoamericana.* 3ª edic. México, UNAM, IIA.

López-Austin, A. 1997. La cosmovisión mesoamericana. In: S. Lombardo, and E. Nalda (comps.); *Tierras mesoamericanas.* México, INAH; pp. 471–507.

López-Austin, A.; y L. López-Luján. 2001. *El Pasado Indígena.* México, El Colegio de México.

López-Austin, A. 2003. Cuatro mitos mesoamericanos del maíz. *Sin maíz no hay país*; G. Esteva y C. Marielle (coords.). México, CONACULTA: 17–35.

López-Austin, A. 2012. Cosmovisión y pensamiento indígena. In: Conceptos y fenómenos fundamentales de nuestro tiempo. México, UNAM, Instituto de Investigaciones Sociales.

López-Arenas G. 2006. Platillos suculentos en vajillas elegantes: un acercamiento a la "Alta Cocina" del Clásico maya. *Revista Lakamhá* 20: 3–8.

López-Luján, L. 2016. El capitán Guillermo Dupaix y su álbum arqueológico de 1794. *Revista de Historiografía* 25: 432–435.

López-Moreno, I.; I. Vizcarra-Bordi. 2016. *El maíz nativo en México. Una aproximación crítica desde los estudio rurales.* México, Universidad Autónoma Metropolitana-Lerma, División de Ciencias Sociales y Humanidades y Juan Pablos Ed.

López-Piñero, J.M. y J. Pardo-Tomás. 1994. *Nuevos Materiales y Noticias sobre la Historia de las Plantas de Nueva España de Francisco Hernández*. España, Cuadernos Valencianos de Historia de la Medicina y de la Ciencia, XLIV, Instituto de Estudios Documentales e Históricos Sobre la Ciencia, Universidad de Valencia, CSIC.

López-Piñero, J.M.; and M.L. López-Terrada. 1997. *La influencia española en la introducción en Europa de las plantas americanas (1493–1623).* In: España, Cuadernos Valencianos de Historia de la Medicina y de la Ciencia, XLIV, Inst. de Estudios Documentales e Históricos Sobre la Ciencia, Universidad de Valencia, CSIC. Vol. LIII, Serie A (monografías).

Louette, D. and M. Smale. 2000. Farmers' seed selection practices and traditional varieties in Cuzalapa, Mexico. *Euphytica* 113(1): 25–41.

Lucero, L.J. 1999. Water control and Maya politics in the southern Maya lowlands. In: E.A. Bacus and L.J. Lucero (eds.); *Complex Polities in the Ancient Tropical World.* Archeological Papers of the American Anthropological Association N° 9. Arlington, Va., American Anthropological Association; pp. 34–49.

Lucero, L.J. 2001. *Social Integration in the Ancient Maya Hinterlands: Ceramic Variability in the Belize River Area.* Arizona State University, Anthropological Research Paper 53.

Lucero, L.J. 2008. The collapse of the Classic Maya: a case for the role of water control. *American Anthropologist* 104: 814–826.

Lucero, L. J.; Fletcher, R. and R. Coningham. 2015. "From "collapse" to urban diaspora: the transformation of low-density, dispersed agrarian urbanism". *Antiquity* 89: 1139–1154.

Lucero, L.J. 2017. "Ancient Maya Water Management, Droughts and urban Diaspora Implications for the Present". En: *Tropical Forest Conservation: Long Term Processes of Human Evolution*". UNESCO.

Lohse, J. 2006. Impacts of the ancient Maya soils and soil erosion in the central Maya Lowlands. *Catena* 2006(65): 166–178.

Lohse, J. 2020. Upland agriculture in the Maya Lowlands: ancient Maya soil conservation in Northwestern Belize. *Geographical Review* 92(3): 372–397.

Lumsden, R.D., R. García E., J.A. Lewis, and G.A. Frías T. 1989. Reduction of Damping-Off Disease in Soils from Indigenous Mexican Agroecosystems. *Agroecology. Researching the Ecological Basis for Sustainable Agriculture.* Springer-Verlag: 83–103.

Luna Morales, C. D. 1993. *Cambios en el Aprovechamiento de Los Recursos Naturales de la Antigua Ciénega de Tlaxcala.* México, Texcoco, UACh.

Luna-Morales, C.D.; and Aguirre, R.R. 2001. Clasificación tradicional, aprovechamiento y distribución ecológica de la pitaya mixteca en México. *Interciencia*, pp. 18–24.

Luna-Morales, C.D. 2002. La Mixteca Baja y las cactáceas columnares (Too/Tnu Dichi). *Revista de Geografía Agricola* (32): 25–42.

Luna-Morales, C.D. 2006. Cambio y Persistencia en la agricultura del Suroeste de Tlaxcala. Ponencia, *Memoria* del Simposio Internacional 30 años de Cacaxtla. México: INAH, Tlaxcala, Tlaxcala, 22 pp. (CD version).

Lundell, C. 1937. *The Vegetation of Petén.* Washington DC, The Carnegie Institute of Washington Publication, No. 478.

Luzzadder–Beach, S. 2000. Water resources of the Chunchucmil Maya. *Geographical Review* 90(4): 493–510.

Luzzadder-Beach, S.; T. Beach. 2006. Wetlands as the intersection of soils, water, and indigenous human society in the Americas. In: McNeil, J.R. Windwater, V. (eds.). *Soils and Societies: Perspectives from Environmental History.* White Horse Press, UK; pp. 91–117.

Luzzadder-Beach, S.; and T. Beach. 2009. Arising from the wetlands: Mechanisms and chronology of landscape aggradation in the northern coastal plain of Belize. *Annals of the Association of American Geographers* 99(1): 1–26.

Mabry, T.J. 1998. The Origins of Food production. Francia, UNESCO office in Mexico, Puebla.

Machuca, P. 2013. El arribo de plantas a las Indias Occidentales: el caso del Balsas-Jalisco a través de las Relaciones Geográficas del siglo XVI. In: *Relaciones*, Estudios de historia y sociedad, Vol. 14, N° 136, Zamora sep/nov.

Mack, R.N. 1991. The commercial seed trade: an early disperser of weeds in the United States. *Economic Botany* 45: 257–273.

MacMahon, D.F. 1973. *Antropología de una presa: los mazatecos y el proyecto del Papaloapan.* México, Instituto Nacional Indigenista (INI).

MacNeish, R.S. 1947. A preliminary report on coastal Tamaulipas, Mexico. *American Antiquity* 13(1): 1–15.

MacNeish, S.R.; T.W. Whitaker and H.C. Cutler. 1957. Cucurbit materials from three caves near Ocampo, Tamaulipas. *American Antiquity* 22(4):352–358.

MacNeish, S.R.; and L. Kaplan. 1960. Prehistoric bean remains from caves in the Ocampo region of Tamaulipas, Mexico. *Botanical Museum Leaflets Harvard University* 19(2):33–56.

MacNeish, S.R. 1964. The food-gathering and incipient agriculture stage of prehistoric Middle America. In: *Handbook of Middle American Indians*, Vol. 1, *Natural Environment and Early Cultures*, R. Wauchope and R.C. West (eds.). Austin: University of Texas Press; pp. 413–426.

MacNeish, S.R.; T.W. Whitaker and H.C. Cutler. 1958. Preliminary archaeological investigations in the Sierra de Tamaulipas, Mexico. *Transactions of the American Philosophical Society* 48(6) 210 pp.

McNeish, S.R. 1967a. Introduction. In: Douglas S. Byers (ed.). *The Prehistory of the Tehuacan Valley: Environment and Subsistence* Vol.1. Austin, University of Texas Press.

McNeish, S.R. 1967b. An interdisciplinary approach to an archaeological problem. *The Prehistory of the Tehuacan Valley*, Vol. 1, *Environment and Subsistence*. D.S. Byers (ed.). The University of Texas Press: 14–24.

McNeish, S.R. 1981. Tehuacan. In: *Handbook of Middle American Indians*, Suppl. I, Archaeology. Austin, The University of Texas Press: 31–47.

McNeish, S.R. 1987. La importancia de los primeros doce sitios del Nuevo Mundo. In: *Orígenes del hombre americano (seminario)*. Mexico, SEP; pp. 57–67.

McNeish, S.R. 1992. *The origins of agriculture and settled life*. Oklahoma University Press.

McNeish, S.R. 1997. El Origen de la Civilización Mesoamericana visto desde Tehuacán. In: *Simposio Internacional Tehuacán y su entorno: balance y perspectivas*; E. de la Lama (comp.). México, INAH, Colección Científica N° 313: 80–93.

Maestre-Alfonso, J. 2018. Entrevistas sobre el ambiente y la agricultura en Galicia. 17-27/08/18.

Maestre-Alfonso, J.; Á. Casas-Gragea y A. González-Jácome (comps.). 2008. *Nuevas Rutas para el Desarrollo en América Latina. Experiencias globales y locales*. México, Universidad Iberoamericana.

Maestre-Alfonso, J. 2011. Comunicación personal.

Malinowski, B. y J. de la Fuente. 1952. *La economía de un sistema de mercados en México*. México, ENAH, Acta Antropológica, 2ª época, Vol. 1, N° 2. Hay edición de 2004 por la Univ. Iberoamericana.

Malvido, E. 1982. Cronología de epidemias y crisis agrícolas en la época colonial. In: Florescano y Malvido (Comps.). *Ensayos sobre la historia de las epidemias en México*. 2 Vols. México, IMSS.

Malvido, E. 2003. La Epidemiología, una Propuesta Para Explicar la Despoblación Americana. *Revista de Indias* LXIII(227): 65–78.

Malvido, E.; y S. del Amo R. 1990. "Farmacéuticos mexicanos del siglo XIX". *Arqueología Mexicana* 39(VII): 345–381.

Mandujano, A.; L. Camarillo S. y M.A. Mandujano. 2003. Historia de las epidemias en el México Antiguo. Algunos Aspectos Biológicos y Sociales. México, Universidad Autónoma Metropolitana (UAM), *Revista Tiempo Laberinto*, abril; pp. 9–21.

Manning, W.E. 1949. The genus Carya in Mexico. *Journal Arnold Arbor* 30: 425–432.

Mangelsdorf, P.C. 1974. *Corn: Its Origin, Evolution and Improvement*. Belknap Press, Harvard University, Cambridge, MA.

Mangelsdorf, P.C. 1986. The origin of corn. *Scintific American* 255(2): 72–78.

Mangelsdorf, P.C.; R.S. MacNeish and W.C. Galinat. 1967a. *The Prehistory of the Tehuacan Valley*, Vol. 1, *Environment and Subsistence*. D.S. Byers (ed.). The University of Texas Press.

Mangelsdorf, P.C. 1967b. Prehistoric wild and cultivated maize. In: D.S. Byers (ed.), *The Prehistory of the Tehuacan Valley, Vol. 1, Environment and Subsistence*. University of Texas Press, Austin, TX, pp. 178– 200.

Mangelsdorf, P.C. 1967c. Prehistoric Maize, Teosinte and Tripsacum from Tamaulipas, Mexico. *Botanical Museum Leaflets*; Harvard University, Cambridge Massachusetts, Dec. 27, Vol. 22, N° 2; pp. 33–63.

Mangelsdorf, P.C.; L.M. Roberts y J. S. Rogers. 1981. *The Probable Origin of Annual Teosinites*. Bussey Institute, Harvard Univ., Cambridge, MA.

Manzanero-Medina, G.I.; A. Flores-Martínez y E. Hunn. 2009. Huertos familiares zapotecos de San Miguel Talea de Castro, Sierra Norte de Oaxaca, México. *Etnobiología* 7: 9–29.

Manzanilla, L.R.N. 1993a. (coord.). *Anatomía de un conjunto residencial en Oztoyahualco*. Vol. 1, México, UNAM, Instituto de Investigaciones Antropológicas (IIA).

Manzanilla, L.R.N. 1993b. Ensayo de interpretación. In: L. Manzanilla (coord.); *Anatomía de un conjunto residencial en Oztoyahualco*. Vol. 1, México, UNAM; pp. 548.566.

Manzanilla, L.R.N. 2012. El almacenamiento de Teotihuacán: enfoques metodológicos. In: Botot et al. (coords); *Almacenamiento Prehispánico, del Norte de México al Altiplano Central*. CEMCA, U. París 1 Pantheon-Sorbone, Lab. Arqueología de las Américas y Universidad de S. Luis Potosí; pp. 73–90.

Manzanilla, L.R.N. 2013. La vida doméstica. Dónde y cómo vivía la gente. In: *Arqueología mexicana* N° 121.

Manzanilla, L.R.N. 2016. Los conjuntos departamentales teotihuacanos. *Arqueología Mexicana*, Vol. 140, *La casa mesoamericana*, pp. 53–60.

Mapes-Sánchez, C. 1987. El maíz entre los purépecha de la cuenca del lago de Pátzcuaro, Michoacán, México. *América Indígena* 47(2): 345–379.

Mapes-Sánchez, C. 2010. El amaranto (*Amaranthus* spp.). Planta originaria de México. Jardín Botánico, Instituto de Biología, UNAM, APPAUNAM, Academia, Ciencia y Cultura; pp. 217–222.

Mariaca-Méndez, R. 1988. Análisis estadístico de seis años de cultivo continuo xperimental de una milpa bajo roza-tumba-quema en Yucatán, México (1980–7986). Thesis Ma. Centro de Botánica, Colegio de Postgraduados (COLPOS), México.

Mariaca-Méndez, R. 1992. Características tecnológicas del sistema marceño de cultivo de maíz en las tierras bajas de tabasco. *Geografía Agrícola*, N° 18; pp. 69–76.

Mariaca-Méndez, R. 1997. *¿Qué es la Agricultura*? México, UACh y UAEM.

Mariaca-Méndez, R. 2004a. Marques de Comillas, Chiapas, Mexico: Proceso de Inmigracion y Adaptabilidad en el Trópico Calido Humedo de México. Thesis, Doct. Social Anthropology, México, Universidad Iberoamericana,

Mariaca-Méndez, R. 2004b. Personal communication. 01/08/2004.

Mariaca-Méndez, R. 2011. Personal communication. 2011.

Mariaca-Méndez, R. 2012. *El Huerto Familiar del Suroeste de México*. México, Sría. de Recursos Naturales y Protección Ambiental del Estado de Tabasco, El Colegio de la Frontera Sur (ECOSUR); pp. 323–359.

Mariaca-Méndez, R. 2013. El Conocimiento de la gallina (*Gallus gallus domesticus*) entre los tseltales y tsotsiles de los altos de Chiapas, México. *Etnobiología* 11(1): 29–43.

Mariaca-Méndez, R. 2017. Avances en el estudio de la fauna de los traspatios familiares en el sureste de México. In: *Domesticación en el Continente Americano*, Vol. 2, Investigación para el manejo sustentable de recursos genéticos en el Nuevo Mundo: A. Casas, J. Torres-Guevara y F. Parra (coords.). México UNAM, Universidad Nacional Agraria La Molina (UNALM) del Perú. pp. 475–508.

Mariaca-Méndez, R. 2018. Personal communication 05/08/18.

Mariaca-Méndez, R. 2019. Personal communication 20/03/19.

Mariaca-Méndez, R.; J. Pérez P., N. Samuel-León M. y A. López M.

Mariaca-Méndez, R. 2003. *La Milpa Tzotzil en los Altos De Chiapas y Sus Recursos Genéticos*. Consejo Estatal para la Ciencia y las Artes de Chiapas, ECOSUR, SIBEJ-CONACyT, PRODUCE, Chiapas.

Mariaca-Méndez, R.; A. González-Jácome y T. Lerner-Martínez. 2007a. El huerto familiar en México: Un Agroecosistema Antiguo que Puede ser Sustentable. *Avances en Agroecología y Ambiente*, Vol. 1, J.F. López O. y colegas (Comps.); México, Texcoco, UACh, BUAP, pp. 119–138.

Mariaca-Méndez, R.; and W. Robins. 2007b. *El pepenado: una tradición otomí del bordado en San Juan Ixtenco*. Tlaxcala, Tlaxcala, México, Gobierno del estado de Tlaxcala, Colegio de Historia de Tlaxcala, Instituto Tlaxcalteca de la Cultura.

Mariaca-Méndez, R.; A. González-Jácome, and L.M. Arias-Reyes. 2010a. *El huerto maya yucateco en el siglo* XVI. México, ECOSUR, FOMIX, UIMQROO, CONCYTEY.

Mariaca-Méndez, R. 2010b. Avances en el estudio de los huertos familiares del sur de México. In: Moreno Fuentes, Á.; M.T., Pulido-Silva; R. Mariaca-Méndez; R. Valadéz-Azúa; P. Mejía-Correa; and T. Gutiérrez-Santillán; *Sistemas Biocognitivos Tradicionales. Paradigmas en la Conservación Biológica y el Fortalecimiento Cultural*. México, Asociación Etnobiológica Mexicana, AC.; Global Diversity Foundation, Universidad Autónoma del estado de Hidalgo, El Colegio de la Frontera Sur (ECOSUR), Sociedad Latinoamericana de Etnobiología.

Marica-Méndez, R. (comp.). 2012. *El huerto Familiar del Sureste de México*. México, Secretaría de Recursos naturales y Protección Ambiental del Estado de Tabasco, El Colegio de la Frontera Sur (ECOSUR).

Mariaca-Méndez, R.; J.E. Hernández-Sánchez et al. 2014. La milpa en la región Serrana Chiapas-Tabasco de Huitiupán, Tacotalpa. In: M. González-Espinosa; M.C. Brunel-Mansel (coords.). *Montañas, Pueblos y Agua. Dimensiones y Realidades de la Cuenca Grijalva*. Juan Pablo, Editor, El Colegio de la Frontera Sur (ECOSUR), México; pp. 323–359.

Marcus, J. 1982. The plant world of the sixteenth and seventeenth-century Lowland Maya. In: Flannery, K.V. (ed.), *Maya Subsistence. Studies in Memory of Dennis E. Puleston*. N. York, London: Academic Press: 239–274.

Marcus, J. 1993. Ancient Maya political organization. Lowland Maya Civilization in the Eighth Century A.D. J.A. Sabloff y J.S. Henderson (eds.); Dumbarton Oaks, Washington, DC; pp. 111–172.

Márquez-Morfín, L. 1994. *La desigualdad ante la muerte en la Ciudad de México: el tifo y el cólera (1813 y 1833)*. México, Siglo XXI Editores.

Márquez-Mireles L.E. 2000. Mayas Yucatecos en Quintana Roo. Agricultura de Roza en El Naranjal. Thesis, MA., Social Anthropology, México, Universidad Iberoamericana (UIA).

Márquez-Mireles L.E. 2006. De la agricultura tradicional a la convencional. Ahorro y capital en Cruz de Piedra, Estado de México. Thesis, Doct., Social Anthropology, México, Universidad Iberoamericana (UIA).

Márquez-Mireles L.E. 2007. De la agricultura tradicional a la convencional en Cruz de Piedra, Estado de México. In: A. González et al. *Los nuevos caminos de la agricultura: procesos de conversión y perspectivas.* México, Universidad Iberoamericana, Programa de Acción Forestal Tropical (PROAFT), Plaza y Valdés; pp. 351–373.

Marszewski, T. 1978. The problema of the introduction of primitive into Southeast Asia. *Folio Orient* 19: 127–163.

Martínez C.A. 2006. Ritos al cultivo del maíz en el Valle de Ixtlahuaca. In: L. Reyes Montes (comp.); *Estudio antropológico en una región agrícola al norte del Estado de México. Estado de México.* UAEM, Cuadernos de Investigación, 4ª época N° 46: 77–98.

Martínez-Hernández, J. (ed.) 1929. *Diccionario de Motul: Maya-Español.* México, Mérida, Compañía Tipográfica Yucateca SA.

Martínez, J.L. 1990–1992. *Documentos Cortesianos.* México, Fondo de Cultura Económica (FCE) y Univ. Nacional Autónoma de México (UNAM), 4 Vols.

Martínez-López, L.; E. Zapata-Martelo; M. del R. Ayala-Carrillo; B. Martínez- Corona; G. Vázquez-Carrillo; C. Jacinto-Hernández; A. Espinosa-Calderón. 2018. Conocimiento práctico y teórico de maíz y frijol en la región triqui Alta, Oaxaca. In: *Revista Mexicana de Ciencias Agrícolas*; Vol. 9 N° 1, 01 enero-14 febrero; pp. 111–121.

Martínez, M. 1987. *Catálogo de nombres vulgares y científicos de plantas mexicanas.* México, FCE.

Martínez, M. 1996. *Las plantas medicinales de México.* México, Ed. Botas.

Martínez-Menes M.R., Rubio-Granados E., Palacios-Espinosa C. 1999. *Terrazas.* México: SAGARPA, Colegio de Postgraduados (COLPOS).

Martínez-Saldaña T. 1997 (2ª ed. 1998). *La Diáspora Tlaxcalteca. La Historia Agrícola de la Expansión Mesoamericana al Norte de México.* México, Tlaxcala, Ediciones del Gobierno del Estado.

Martínez-Saldaña, T.; et al. (ed.). 2011. *Riegos Ancestrales en Iberoamérica. Técnicas y organización social del pequeño riego.* Col. de Postgraduados (COLPOS), publicaciones de la bba.

Martínez-Saldaña, T. 2015. La herencia hidraúlica de Nezahualcoyotl: el caso del Acolhuacan. In: M.L. de la Isla de Bauer (comp.); *El Gran Tlahtoani Nezahualcoyotl.* Texcoco, Comité de Acción para el Saneamiento del Ambiente AC (CASA); pp. 95–138.

Masso-Guijarro, E. and S. Santos-Fraile. 2017. Etnografías Multisituadas en la era Global. Propuestas Sobre Epísteme y Métodos. In: XIV Congreso de Antropología, Valencia, España; 5-8/09/2017.

Masson, M.A. 2012. Maya collapse cycles. *Proceedings of the National Academy of Sciences of the United States of America* 109(45): 18237–18238.

Masson, M.A. y C. Peraza-Lope. 2013. "The distribution ans Diversity of Faunal Exploitation at Mayapan: From Temple to Houselot". In: C.M. Götz y K.F. Emery (Eds.); *The Archaeology of Mesoamerican Animals.* Cambridge, MA, Cambridge Scholars Press.

Matheny, R.T. 1978. "Northern Maya Lowland Water Control Systems". In: P. Harrison y B.L. Turner (eds.); Prehispanic Maya Agriculture. Albuquerque, University of New Mexico Press; pp. 185–210.

Matheny, R.T.; and D. Gurr. 1979. Ancient hydraulic techniques in the chiapas highlands. *American Scientist* 67: 441–449.

Matheny, R.T.; and D. Gurr. 1983. *Investigations at Edzná, Campeche, Mexico.* Vol. 1, Part 1, The Hydraulic System. Provo, Utah, New World Archaeology.

McAnany, P.A. 2010. *Ancestral Maya Economies in Archaeological Perspective.* Cambridge, Cambride University Press.

McClung de Tapia, E. 2000. Prehispanic agricultural systems in the basin of Mexico. In: D.L. Lentz (ed.); *Imperfect Balance. Landscape Transformations in the Precolumbian Americas.* Columbia University Press; pp. 121–146.

McClung de Tapia, E. 2010. "Reflexiones en Torno al Paisaje Prehispánico: Reconstrucción del Paleoambiente del Valle de Teotihuacan". In: E. Ortiz-Díaz (ed.); VI C*oloquio Pedro Bosch Gimpera. Lugar, espacio y paisaje en arqueología: Mesoamérica y otras áreas culturales.* México, UNAM-Instituto de Investigaciones Antropológicas; pp. 245–266.

McClung de Tapia, E.; E. Solleiro-Rebolledo; J. Gama-Castro; J.L. Villalpando; S. Sedov. 2003. "Paleosols in the Teotihuacan valley, Mexico: evidence for paleoenvironment and human impact". *Rev. Mexicana de Ciencias Geológicas* 20: 270–282.

McClung de Tapia, E.; D. Martínez-Yrízar; E. Ibarra-Morales y C.C. Adriano Morán. 2014. Los Orígenes Prehispánicos de una Tradición Alimentaria en la Cuenca de México. *Anales de Antropología* 48-I: 97–121.

McC. Netting, R. 1993. *Smallholders, Householders. Farn Families and the Ecology of Intensive, Sustainable Agriculture*. California, Stanford Univ. Press.

McCormack, V.J. 2002. "Sedentism, Site Ocupation, and Settlement Organization at La Joya. A Formative Village in the Sierra de los Tuxtlas, Veracruz, Mexico". PhD Dissertation, Univ. of Pittsburgh, Dept. of Anthropology.

Meiners, M.; C. Sánchez-Garduño y S. de Blois. 2009. "El Ramón: fruto de nuestra cultura y raíz para la conservación". En: *BioDiversitas*, México, CONABIO, N° 87, noviembre/diciembre; pp. 7–10.

Melesio, N.M. 2002. "Legislación, justicia y pueblos indígenas". Ponencia, Congreso: *Impacto de los grandes proyectos de desarrollo sobre los derechos humanos y las libertades fundamentales de las comunidades indígenas*. Buenos Aires.

Melville, E.G.K. 1994. *A plague of sheep. Environmental consequences of the conquest of Mexico*. Cambridge Univ. Press. 1999. *Plaga de ovejas: consecuencias ambientales de la conquista de México*. México, Fondo de Cultura Económica.

Méndez-Mercado, L. 1993. Cambio de dieta entre los migrantes mixtecos. *Antropológicas* 1993, 7: 32–40.

Méndez, G.D. 1999. "El agua entre los mayas de la península de Yucatán". En: *El agua en la cosmovisión y terapéutica de los pueblos indígenas de México*. México, Instituto Nacional Indigenista (INI).

Méndez-Guevara, D.N. 2004. "Nuevo Campechito, Campeche: Ambiente, economía y cultura en una sociedad de pescadores". Tesis Ma. Antrop. Social. México, Univ. Iberoamericana.

Mestries, F. y G. Igartúa. 1988. "El Movimiento Cañero de Veracruz a principios de los setentas". En: L. Paré (coord.); *El Estado, los cañeros y la Industria Azucarera*: 1940–1980. México, UNAM, UAM-Azcapozalco.

Metcalfe, S.E.; Bimpson A.; Courtice A.J.; O'Hara S.L.; Taylor D.M. 1997. Climate change at the monsoon/westerly boundary in northern Mexico. *Journal of Paleolimnology* 17: 155–171.

Mhammed, S.; D. Sharon; D. Kaufman; E. Lahav; J. Hillel; C. Degani; U. Lavi. 1997. Genetic relationships within avocado (*Persea americana* Mill.) cultivars and between Persea species. *Theoretical Applied Genetics* 94: 279–286.

Millar, C.I. 1996. "Tertiary vegetarion History". En: *Sierra Nevada Ecosystem*

Proyect: Final report. Univ. of California; pp. 71–122.

Miller, A. and B. Schaal. 2005. Domestication of a Mesoamerican cultivated fruit tree, Spondias purpurea. *Proceedings of the National Academy of Sciences of the United States of America* 102(36): 12801–12806.

Millon, R. 1957. Irrigation systems in the Valley of Teotihuacan. *American Antiquity* 26: 160–166.

Milton, K. 1993. Environmentalism and anthropology. In: K. Milton (Ed.) *Environmentalism. The View from Antrhopology*. Londres y New York, Routledge; pp. 1–17.

Mintz, S. 1965. *Sweetness and Power. The Place of Sugar in Modern History*. New York, Viking.

Mintz, S. 1996. *Tasting Food and Tasting Freedom. Excursions in culture, eating and the past*. Boston, Beacon Press.

Miracle, M.P. 1966. *Maize in tropical Africa*. Wisconsin, Madison, The Univ. of Wisconsin Press.

Miranda-González, J. 1972. "Notas sobre la introducción de la Mesta en la Nueva España". Rev. *Historia de América*; Inst. Panamericano de Geografía e Historia; N° 17, junio; pp. 1–26.

Moctezuma-Pérez, S. 2008. "Ambiente, cafeticultura y migración". Thesis, Ma. Social Anthropology. México, Universidad Iberoamericana (UIA).

Moctezuma-Pérez, S. 2010. "Aproximación al sistema de cargos y mayordomías en San Francisco Tepeyanco, Tlaxcala. In: *Perspectivas Latinoamericanas*. Japón, University Nanzan, (7) 26–46.

Moctezuma-Pérez, S. 2013a. "San Francisco Tepeyanco: Ambiente, Cultura y Agricultura". Thesis Doct. Social Anthropology. México, Universidad Iberoamericana,

Moctezuma-Pérez, S. 2013b. "Cambios y Continuidades en el Manejo de Huertos Familiares del Suroeste de Tlaxcala, México". *Perspectivas Latinoamericanas*. Japón, University Nanzan, N° 10: 83–101.

Moctezuma-Pérez, S.; y H. Thome Ortiz. 2015a. Leisure Time and Food Memory: An Anthropological Approach to Culinary Tourism in Central Mexico. *Folia Turística*, Antropología del Turismo, Krakovia, Polonia, Vol. 37; pp. 69–84.

Moctezuma-Pérez, S.; J.M. Pérez-Sánchez, M.G. Rivera Herrejón. 2015b. "Aportes alimenticios de los agroecosistemas tradicionales en México rural". México, Castellanos Editores.

Moctezuma-Pérez, S. 2016. "Teoría y Praxis del aprendizaje vicario en las sociedades rurales de México". In: Rebolledo R.N. (comp.); *Aprender a aprender. Nuevas rutas en Pedagogía*. Gobierno del Estado de Tlaxcala, CONACULTA, SEP; pp. 79–94.

Moctezuma-Pérez, S. 2018a. Biodiversidad y alimentación en huertos familiares del suroeste de Tlaxcala. En: L. Reyes Montes; J.M. Pérez-Sánchez y S. Moctezuma-Pérez (coords). *Sistemas agrícolas tradicionales. Biodiversidady cultura*. El Colegio Mexiquense A.C., Zinacantepec, Estado de México; pp. 96–116.

Moctezuma-Pérez, S.; y J.M. Pérez-Sánchez. 2018b. La percepción de los Guarijíos de Sonora ante el Proyecto de la Presa Bicentenario Los Pilares. In: A. González-Jácome, J.J. Velasco-Orozco, S. Moctezuma-Pérez y A. Cruz-León (coords.). *Sonora: La Sierra, el Desierto y la Costa en el contexto de los guarijíos*. Estado de México, Texcoco, Universidad Autónoma Chapingo; pp. 219–234.

Moguel-Pérez, P. y V.M. Toledo. 1999. El café en México. Ecología, cultura indígena y sustentabilidad. En: Rodríguez, H. y J.M. Gutiérrez (Coords.); *El Jarocho Verde*. Red de información y acción ambiental de Veracruz; N° 11.

Moguel-Pérez, P. 2011. Café y biodiversidad en México: ¿por qué deben conservarse los cafetales bajo sombra? In: V.M. Toledo; *La biodiversidad de México. Inventarios, manejos, usos, informática, conservación e importancia cultural*. México, CONACULTA y FCE; pp. 193–219.

Molina-Hampshire, D. 2009. De la laguna de la abundancia a la tierra del buen vestir. El programa Lerma y sus impactos ambientales y socioculturales en Almoloya del Río, Edo. de México. Thesis Doct. Social Anthropology. Mexico, Universidad Iberoamericana.

Molina-Ludy, V. 1976. *San Bartolomé de los Llanos: una urbanización frenada*. México, CISINAH.

Molina-Luna M.G. y G.V. Campos Á. 2016. Historia y Situación Actual de los Mercados Semanales en Valles Centrales de Oaxaca. *Revista Mexicana de Agroecosistemas* 3(2): 272–290.

Molina-Rosales, M.D. 2010. "Colonización en Calakmul, Campeche". Tesis, Doct., Antropología Social. México, Universidad Iberoamericana.

Mollison, B. y D. Holmgren. 1978. *Permaculture One. A Perennial Agriculture for Human Settlements*. Australia, Transworld Publishers.

Montañez, C. (coord.). 1982. *El cultivo de maíz en México, diversidad, limitaciones y alternativas: seis estudios de caso*. México, Centro de Ecodesarrollo.

Montero-García I.A. 2012. *Matlalcueye. El volcán del alma tlaxcalteca*. México: Gobierno del Estado de Tlaxcala, SEPE/USET, Mesoamerican Research Foundation.

Montero-García I.A 2015. Plantas y rituales en las altas montañas del México Antiguo. In: *Flor–flora. Su uso ritual en Mesoamérica*; B.A. Albores.Zárate (Coord.). El Colegio Mexiquense y Fondo Editorial del Gobierno del Estado de México; pp. 85–98.

Montero L.C. y C.M. Varela S. 2017. "¡Tamales para todos! El consumo del venado y perro doméstico en los banquetes de Chinikihá". *Anales de Antropología* 51(2017): 183–191.

Montgomery, David R. 2007. Soil erosion and agricultural sustainability, *Proceedings of the National Academy of Sciences USA* (*PNAS*), August 14, Vol. 104, No. 22: 13268–13272.

Mora-Jarvio, M.A. 2010. Flora del Valle de Tehuacán-Cuicatlán. Fascículo 77. Juglandaceae A.Rich. ex Kunth. UNAM, Instituto de Biología; pp. 1–8.

Mora L., M.I. 2013. *Los caminos de la trashumancia. Territorio, persistencia y representaciones de la ganadería pastoril en el altiplano potosino*. S. Luis Potosí, SLP, El Colegio de San Luis A.C.

Morales, V.R. y P. Blanco-Fernández. 2013. Breve historia de la investigación botánica en España. Chapter IV in R. Morales Valverde (coord.). *Las plantas silvestres en España*. Sevilla, Consejo Superior de Investigaciones Científicas (CSIC); pp. 55–68.

Morales-Folguera J.M. 2004. *Jardines Prehispánicos de Mexico en las Crónicas de Indias*. In: Consejo Superior de Investigaciones Científicas. AEA, LXXVII, 308; pp. 351–373.

Morehart, C.T. 2002 [1998]. Plants of the Underworld: Ritual Use in Ancient Maya Cave Ceremonies. FAMSI Organization 2002; pp. 1–27.

Moreno, T.A. 1987. El siglo de la conquista. In: Cossío Villegas, Daniel (coord.), *Historia general de México*. Tomo 1, 2a Reimpresión (1976). Harla, El Colegio de México (COLMEX); pp. 289–369.

Mountjoy D.C. 1985. "Adaptation and Change in a Local Agroecosystem of Tlaxcala, Mexico". Senior Thesis. EUA: Consejo de Estudios Ambientales. University of California, Santa Cruz.

Morley, S.G. 1992. *La civilización maya*. México, Fondo de Cultura Económica (FCE). Sección Obras de Antropología, 9ª edic.

Morrison, B. and R. Cózatl-Manzano. 2003. Initial evidence for use of periphyton as an agricultural fertilizer by the ancient Maya associated with the El Edén wetland, Northern Quintana Roo, Mexico. In: *Lowland Maya Area: Three Millennia at the Human-Wildland Interface*; Gómez-Pompa, A.; M. Allen, S. Fedick, and J.J. Jiménez- Osornio (eds.). The Haworth Press, Nueva York.

Murra V.J. 2009. *El Mundo Andino. Población, Medio Ambiente y Economía*. Fondo Editorial Pontificia Universidad Católica del Perú e Instituto de Estudios Peruanos IEP, Lima. (Eds. 1975, 2002, 2004, 2009).

Naranjo-Piñera, E.J. and J.L. Rangel-Salazar. 2012. Wild life uses and hunting patterns in rural community of the Yucatan península, Mexico. *Journal of Ethnobiology and Ethnomedicine* 8(1): 38.

Naranjo-Piñera, E.J.; E. Bello.Baltasar; E.I.J. Estrada-Lugo; R. Mariaca-Méndez; and P.A. Macario-Mendoza. 2013. La milpa comedero-trampa como una estrategia de cacería tradicional Maya. *Estudios de Cultura Maya* 42: 87–118.

Navarrete, C. 1968. La relación de Ocozocoautla, Chiapas. *Tlalocan* V(4): 368–373.

Navarrete, C. 1997. Los mitos del maíz entre los mayas de las tierras altas. *Arqueología Mexicana* 5(25): 56–61.

Nazarea, V.D. 2005. A view from a point. Ethnoecology as situated knowledge. In: N. Haenn y R. Wilk (eds.); *The Environmet in Anthropology*. New York University Press; pp. 34–39.

Nazarea, V.D. 2006. Local knowledge and memory in biodiversity conservation. *Annual Review of Anthropology* 35(1): 317–335.

Navarrete, C. Prehistoric Agricultural Intensification in the Tehuacán Valley. The beginnings of Water Management and Agricultural Intensification in Mesoamerica: The Case of the Prehistoric San Marcos Well, the Purron Dam and the Fossilized Canal Systems of the Tehuacan Valley, Puebla, Mexico. In: Centros de Domesticación: Mesoamérica. Researchgate.net, June, 9.

Navarrete, C.; Caran, S.C.; Winsborough, B.M.; Ramírez-Sorensen, F. and S. Valastro Jr. 1995. An early Holocene hand-dug water well in the Tehuacán Valley of Puebla, Mexico. *Current Research in the Pleistocene* 12: 38–40.

Navarrete, C. y Castellón-Huerta, B.R. 2014. Una síntesis del manejo prehispánico del agua en el Valle de Tehuacán, Puebla, México. *Arqueología*, N° 47. Mexico, Instituto Nacional de Antropología e Historia (INAH); pp. 182–198.

Navarrete, C.; M.J. Aiuvalasit; and V.A. Clause. 2015. New light on the prehistoric Purrón dam complex: small corporate group collaboration in the Tehuacán Valley, Puebla, Mexico. *Journal of Field Archaeology* 40(3): 347–364.

Neely, J. A. and Castellón Huerta, B. R. 2014. Una síntesis del manejo prehispánico del agua en el Valle de Tehuacán, Puebla, México. *Arqueología*, No. 47. Mexico City, Instituto Nacional de Antropología e Historia, pp. 182–198.

Neely, J. 2017. "Prehistoric Agricultural Intensification in the Tehuacan Valley. The beggining of water management and agricultural intensification in Mesoamerica: the case of the pre-Historic S. Marcos well, the Purrón dam, and the fosilized canal system of the Tehuacan Valley, Puebla, Mexico". In: Centro de Domesticación: Mesoamerica. Researchgate.net, June, 9.

Nelken-Terner, A. 1968. Los implementos de molienda prehispánicos. Ensayo metodológico. Thesis, Ma., Archaeology. México, Escuela Nacional de Antropología e Historia (ENAH).

Neurath, J. 2008. *Por los caminos del maíz. Mito y ritual en la periferia septentrional de Mesoamérica*. México, Fondo de Cultura Económica (F.C.E.), CONACULTA.

Neusius, S.W. 1996. Game procurement among temperate horticularists: the case for garding hunting by the Dolores Anazasi. In: *Case Studies in Environmental Archaeology*; E.J. Reitz, A. Newson and S.J. Séudder (eds.). New York: Plenum Press; pp. 273–287.

Nickel, H.J. 1996. *Morfología Social de la Hacienda Mexicana*. México, Fondo de Cultura Económica.

Niederberger, B.C. 1976. *Zohapilco. Cinco milenios de ocupación humana en un sitio lacustre de la Cuenca de México, México*, INAH, Colección Científica.

Niederberger, B.C. 1979. "Early Sedentary Economy in the Basin of Mexico". *Science*; American Association for the Advancement of Science, Washington, DC; pp. 132–142.

Niederberger, B.C. 1996. Paisajes, economía de subsistencia y agrosistemas en Mesomérica a principios del siglo XVI. In: S. Lombardo y E. Nalda (eds.); *Temas mesoamericanos*. México, INAH; pp. 11–50.

Nolasco- Armas, M. 1985. *Café y sociedad en México*. México, Centro de Ecodesarrollo.

Nolasco, T.C. 2010. Cactáceas, una alternativa para el rescate económico de pueblos en pobreza extrema. *Semanario de la UAM*, pp. 4–5.

Novelo, E. and R. Tavera. 2003. The role of periphyton in the regulation and supply of nutrients in a wetland at El Eden, Quintana Roo. In: Lowland Maya Area: Three Millennia at the Human-Wildland Interface. A. Gómez- Pompa, M. Allen, S. Fedick, and J.J. Jiménez-Osornio (eds). The Haworth Press, Nueva York.

Ocaranza F. (comp.). 1995. *Historia de la medicina en México*. México, CONACULTA.

Ochoa S.L. 1989. *Comercio, comerciantes y rutas de intercambio en el México Antiguo*. México, Secretaría de Comercio y Fomento Industrial (SECOFI). OET–CGE.

Ochoa S.L. 2002. Programa de Ordenamiento Ecológico General del Estado de Tlaxcala. Coordinación General de Ecología, Gobierno del estado de Tlaxcala. Tlaxcala, México.

Ochoa, S.L. y A. González-Jácome. 2009. El Antes y el Después de los Humedales de la Península de Xicalanco, Campeche. *Itinerarios*, Revista del Instituto de Estudios Ibéricos e Iberoamericanos de la Universidad de Varsovia, Polonia, Vol. 9, 2009a: 145–168.

O'Hara, S.; F.A. Street-Perrot and T.P. Bur. 1993. Accelerated soil erosion around a Mexican highland lake caused by prehispanic agriculture. *Nature* 362: 48–51.

Olivares-Rodríguez, F. 2007. Agricultura Campesina, Cambio y Permanencia: el Caso de Míxquic. Thesis Doct., Social Anthropology, México, Universidad Iberoamericana.

Oliveros-Espinosa, R.G. 2015. Flores sagradas. Ciclo del maíz, rituales y ofrendas. In: Albores (coord.), *Flor-flora. Su uso ritual en Mesoamérica.* El Colegio Mexiquense, Fondo Editorial del Gobierno del Estado de México; pp. 491–514.

Ordoñez-Díaz, M.J. (comp.). 2018. *Atlas Biocultural de Huertos Familiares en México. Chiapas, Hidalgo, Oaxaca, Veracruz y Península de Yucatán.* México, UNAM, Centro Regional de Investig. Interdisciplinarias; Cuernavaca, Morelos.

Ornelas, D.J. 1993. *Ordenamiento territorial y política regional en México.* México, Universidad Autónoma de Tlaxcala.

Orozco-Segovia, A.D.L. 1999. El marceño en las zonas inundables de Tabasco. *Agricultura y Sociedad en México. Diversidad, Enfoques, Estudios de Caso.* México, Plaza y Valdés, PROAFT, UIA: 111–122.

Ortega C.A.; M.J. Guerrero H.; O. Cota A.; L.A. Peinado F. 2007–2008. Informe Final del Proyecto FZ002: Conocimiento de la diversidad y distribución actual del maíz nativo y sus parientes silvestres en México. México, CONABIO, INIFAP, INE.

Ortega-Paczka, R. 1991. Estado Actual de los estudios sobre maíces nativos de México. In: Ortega-Paczka, R.; G. Palomino H.; F. Castillo G.; V.A. González H. y M. Livera M. (eds.); *Avances en el Estudio de los Recursos filogenéticos de México.* Sociedad Mexicana de citogenética; Chapingo, México; pp. 161–185.

Ortega-Paczka, R. 1994. Documentos. Algunos datos introductorios a la vida de N.I. Vavilov y su artículo: 'México y Centroamérica como centro básico de origen de las plantas cultivadas del Nuevo Mundo. *Revista de Geografía agrícola estudios de la agricultura mexicana*; Recursos Filogenéticos de México, Universidad Autónoma Chapingo, N° 20, Dic.; pp. 7–13. (Traducción del ruso al españo); pp. 15–34.

Ortega-Paczka, R. 2002. Los proyectos de conservación in situ y mejoramiento participativo de maíz en México. *Revista de Geografía Agrícola*; Documento; pp. 97–108.

Ortega-Paczka, R. 2003. La diversidad del maíz en México. In: *Sin maíz no hay país*; G. Esteva y C. Marielle (coord.). México, CONACULTA: 123–154.

Ortega-Paczka, R. 2017. Factores e indicios de conservación in situ y evolución bajo la domesticación del maíz en México en el último siglo. In: *Domesticación en el Continente Americano*, Vol. 2, Investigación para el manejo sustentable de recursos genéticos en el Nuevo Mundo: A. Casas, J. Torres-Guevara y F. Parra (coords.). México UNAM, Universidad Nacional Agraria La Molina (UNALM), Perú. pp. 231–272.

Ortiz-Díaz, E. 2010. Los espacios rituales de los antiguos zapotecos de la cuenca del río Caxonos, Oaxaca, durante el posclásico tardío y la época colonial. In: E. Ortiz-Díaz (ed.); VI *Coloquio Pedro Bosch Gimpera. Lugar, espacio y paisaje en arqueología: Mesoamérica y otras áreas culturales.* México, UNAM, Instituto de Investigaciones Antropológicas; pp. 425–446.

Ortiz, F. 1947. *El huracán; su mitología y su símbolos.* México, F.C.E.

Oseguera P.D. y R. Ortega P. 2016. "Gente de Maíz. Historia y Diversidad en la Cocina Mexicana del Maíz". In: López y Vizcarra (comps.). *El maíz nativo en México. Una aproximación crítica desde los estudios rurales.* México, UAM-Lerma, División Ciencias Sociales y Humanidades y Juan Pablos Ed; pp. 113–136.

Osorio-Hernández, C. 1997. El solar en la zona maya de Quintana Roo. Thesis, BA., Biology, México, UNAM.

Otero-Prevost, D.E.; D.F. Gurri-García; R. Mariaca-Méndez y F. Guízar-Vásquez. 2017. La incorporación y el aumento de oferta de alimentos industrializados en las dietas de las unidades domesticas y su relación con el abandono del sistema de subsistencia propio en las comunidades rurales mayas de Yucatán, México. *Cuadernos de Desarrollo Rural* 14(80): 1–16. https://doi.org/10.11144/Javeriana.cdr14-80.iaoa

Pacheco-Olvera, R.M. 2010. El intercambio de plantas en la Nao de China y su impacto en México. In: J. Long-Towell y A. Attolini-Lecón (coords.), *Caminos y mercados de México.* México, UNAM, INAH; pp. 593–607.

Paillés, H.M. 1989. *Cuevas de la región de Ocozocoautla y el río La Venta: El Diario de Campo, 1945, de Matthew W. Stirling con Notas Arqueológicas.* Brigham Young University Notes of the New World Archaeological Foundation 6, Provo.

Palacios, G. 1999. *La pluma y el arado: los intelectuales pedagogos y la construcción social del "problema campesino" en México, 1932–1934*. México, Centro de Estudios Históricos, Colegio de México (COLMEX); División de Estudios Políticos, Centro de Investigación y Docencia Económicas (CIDE).

Palerm, A. 1952. La civilización urbana. *Historia Mexicana* II(2): 184–209.

Palerm, A. 1955a. [1953]. The agricultural basis of urban civilization in Mesoamerica. In: *Irrigation Civilizations. A Comparative Study*. Julian H. Steward (Ed.). Washington DC, Pan American Union, Social Science Monograph N° 1: 28–42.

Palerm, A. 1955b. Sistemas de regadío Prehispánico en Teotihuacán y en el Pedregal de San Ángel. Revista iberoamericana de Ciencias Sociales, Segunda época, Vol. 1, N° 2.

Palerm, A. 1957. "Ecological potential and cultural development in mesoamerica". In: Á. Palerm, *Studies in Human Ecology*; Washington D.C., Unión Panamericana, Secretaría General (Español, 1960).

Palerm, A. y E. Wolf. 1961. La agricultura y el desarrollo de la civilización en Mesoamérica. *Revista Interamericana de Ciencias Sociales*, Washington D.C., Unión Panamericana, Depto. Ciencias Sociales, Vol. 1, N° 1: 223–245.

Palerm, A. 1962. Observaciones sobre la planificación regional. *Cuadernos Informativos de Desarrollo Económico-Social*, 2a época, Madrid, España, Instituto Balmes de Sociología, fascículo 1; pp. 55–69.

Palerm, A. 1965. *Observaciones sobre la planificación regional*, Washington DC, Unión Panamericana, Documentos técnicos de la Organización de los Estados Americanos claves UP/SER. H/VII.37.1 al 5.

Palerm, A. 1967a. Introducción a la Teoría Etnológica. México, Universidad Iberoamericana, Instituto de ciencias sociales. (Reedición en 1987).

Palerm, A. 1967b. Agricultural systems and food patterns. *Handbook of Middle American Indians*. Austin, The University of Texas Press, Vol. VI; pp. 26–52.

Palerm, A. 1968a. *Productividad agrícola: Un estudio sobre México*. México, Centro Nal. de Productividad, Ediciones Productividad, XII.

Palerm, A. 1968b. [1955]. Irrigation in the Old Acolhua Domain, Mexico. *Man in Adaptation: The Biosocial Background*. Cohen Y.A. (ed.). Chicago, Aldine Publishing Co; pp. 343–353.

Palerm, A. 1970. Informe de los Aspectos socioculturales de la población afectada por el Proyecto Angostura. Estudio y recomendaciones. México, Comisión Federal de Electricidad (CFE).

Palerm, A. 1972a. *Agricultura y Sociedad en México*. México, SepSetentas, 55. Reimpresión en Ed. Diana 1980.

Palerm, A. 1972b. *Agricultura y Civilización en Mesoamérica*. México, SepSetentas, N° 32.

Palerm, A. 1976a. *Obras Hidráulicas Prehispánicas en el Sistema Lacustre del Valle de México*. México, INAH, SEP.

Palerm, A. 1976b. *Modos de producción y formaciones sociales*. México, Ed. Gernika.

Palerm, A. 1978. *Sobre la fórmula M-D-M y la articulación del modo campesino de producción al sistema capitalista dominante*. México, Cuadernos de la Casa Chata.

Palerm, A. 1980. *Antropología y marxismo*. México, CIESAS.

Palerm, A. 1992. Sistemas agrícolas en Mesoamerica contemporánea. In: *Guía y lecturas para una primera práctica de campo*. México, Universidad Autónoma de Querétaro; pp. 241–281.

Palerm, A. 1993. *Planificación regional y reforma agraria*. México, Ediciones Gernika y Universidad Iberoamericana.

Palerm, A. 1997 [1967]. *Introducción a la Teoría Etnológica*. México, Univ. Iberoamericana.

Palerm, A. 2007. *Agua y Agricultura. Ángel Palerm, la discusión con Karl Wittfogel sobre el Modo Asiático de Producción y la construcción de un modelo para el estudio de Mesoamérica*. In: A. González-Jácome (comp.). México, Universidad Iberoamericana y España, Centro Superior de Investigaciones Científicas (CSIC).

Palerm-Viqueira, Juan Vicente. 1997. La persistencia y expansion de sistemas agrícolas tradicionales: el caso del huamil en el bajio mexicano. In: *Monografías*, Jardín Botánico de Córdoba 5:121–133.

Palerm-Viqueira, Jacinta. 2007. Técnicas hidráulicas en México, paralelismos con el Viejo Mundo: I. Entarquinamiento en cajas de agua, II. Galerías Filtrantes y III. Bimbaletes. In: Mexican hydraulic technologies, parallelisms with the Old World. Sesión 17, Technological exchange, modes of production and water utilization in Europe and Latin America (from Ancient times to XXth century). Economic History Congress, Buenos Aires.

Palerm-Viqueira, Jacinta. 2004. Las galerías filtrantes o Qanats en México: introducción y tipología de técnicas. In: *Agricultura, Sociedad y Desarrollo*, Vol.1, N° 2 Texcoco, jul./dic.

Palerm-Viqueira, Jacinta; M. Sánchez-Rodríguez; con la colaboración de H. Eling; E. López-Pacheco; J.L. Pimentel-Equihua y J.G. Rodríguez-Meza. 2008. Captación de Agua de Lluvia. *Boletín del Archivo Histórico del Agua* (AHA), Año 13, N° especial.

Palerm-Viqueira, Jacinta y T. Martínez-Saldaña. 2009. *Aventuras con el Agua. La administración del agua de riego: historia y teoría.* México, Texcoco, Colegio de Postgraduados (COLPOS).

Paliwal, R.L. 2009. "El Maíz en los Trópicos. Mejoramiento y Producción". FAO, Depto. de Agricultura.

Paré, L. (coord.). 1988. *El Estado, los Cañeros y la Industria Azucarera.* México, UNAM y UAM-Azcapozalco.

Parsons J.R. 2006. *The last pescadores of Chimalhuacán, Mexico: an archaeological ethnography.* Ann Arbor, University of Michigan,

Parsons J.R. and Parsons M.H. 1990. *Maguey utilization in Highland Central Mexico. An archaeological ethnography.* Anthropological Papers N° 82, Museo de Antropología, USA, University of Michigan, Ann Arbor.

Patrick L.L. 1977. "A Cultural Geography of the Use of Seasonally Dry Sloping Terrain: The Metepantli Crop Terraces of Central Mexico". PhD Thesis in Geography, University of Pittsburgh, USA.

Pearsall, D.M. 1978. Phytolith analysis of archaeological soils: evidence for maize cultivation in formative ecuador. *Science* 199(4325): 177–178.

Pendergast, D.M. 1985. Stability through change: lamanai, belize, from the ninth to the seventeenth century. In: J.A. Sabloff y E.W. Andrews V. (Eds.); *Late Lowland Maya Civilization. Classic to Postclassic.* University of New Mexico Press, Albuquerque; pp. 223–250.

Perales, H.; S.B. Brush y C.O. Qualset. 2003. Landraces of maize in central Mexico: an altitudinal transect. *Economy Botany* 57(1): 7–20.

Pérez-Castro A.B. 1988. *Entre montañas y cafetales. Luchas agrarias en el norte de Chiapas.* México, UNAM, IIA.

Pérez-Castro A.B. 1996. Bajo el símbolo de la ceiba: la lucha de los indígenas cafeticultores de las tierras de Simojovel. In: *Chiapas, los rumbos de otra historia.* México, UNAM, CIESAS, CEMCA, CEMCUG; pp. 301–317.

Pérez-Castro A.B. 2007. Activando el mundo simbólico para enfrentar la migración. In: *Chungará* (Arica), Rev. De Antropología Chilena, Vol. 39, N° 1; pp. 51–68.

Pérez-Castro A.B. 2008. Riego, agricultura y cultura en la historia de San Nicolás Atecoxco, Hidalgo. In: A. Escobar Ohmstede, M. Sánchez Rodríguez y A.M. Gutiérrez Rivas (coords.); *Agua y Tierra en México, siglos XIX y XX* (2 Vols). México, El Colegio de San Luis.

Pérez-Castro A.B. 2010. Actores sociales en un mundo de agua y tierra. La Huaxteca. In: A.O. Ávila Quijas et al (coords.); *Negociaciones acuerdos y conflictos en México, siglos XIX y XX: agua y tierra.* México, UAAG, CIESAS, COLMICH; pp. 377–400.

Pérez-Castro A.B. 2013a. Los cerros sagrados, patrimonio cosmogónico de la Huaxteca. México, UNAM.

Pérez-Castro A.B. 2013b. El sacrificio de aves en los rituales de la Huaxteca. El Colegio de Michoacán (COLMICH).

Pérez, S.E.; L.A. Vega y G. Romero. 2007. "Prácticas alimentarias de mujeres rurales: ¿una nueva percepción del cuerpo? *Salud Pública de México* 49(1): 52–62.

Pérez, J.; E. McClung; J. Gama-Castro; E. Solleiro-Rebolledo; S. Sedov; A. Peralta-Higuera; Y. Rivera. 2006. "Terrazas agrícolas prehispánicas en suelos de origen volcánico, cerro San Lucas, valle de Teotihuacan". In: *Simposio Internacional sobre suelos volcánicos, ISVO 06.* Publicado en Memoria, p. 7.

Pérez-San Vicente, G. 2000. *Repertorio de tamales. Cocina indígena popular.* Tomo XV. México, CONACULTA y Culturas Populares.

Pérez-Sánchez, J.M. 2007. Desarrollo local en el trópico mexicano: los camellones chontales de Tucta, Tabasco. Thesis, MA, Social Anthropology. México, Universidad Iberoamericana.

Pérez-Sánchez, J.M. 2012. Ambiente, Agricultura y Cultura. Las Terrazas de Ixtacuixtla, Tlaxcala. Thesis, Doct. Social Anthropology. México, UIA.

Pérez-Sánchez, J.M. 2014. *Agricultura de Terrazas en Tlaxcala. La Caridad Cuaxonacayo.* México, SEPE/USET, Gobierno del Edo. de Tlaxcala.

Pérez-Sánchez, J.M.; S. Moctezuma P.; J. Sales C. y J.I. Juan P. 2017. Manejo de suelos en la agricultura tradicional de laderas en Tlaxcala y el valle de Toluca. In: F. Carreño, C. Rodríguez S. y J.A. Castellanos (coords.). *Patrimonio Biocultural. Experiencias Integradoras.* Texcoco, Estado de México, UACh; pp. 79–101.

Pérez-Toro, A. 1942. *La milpa.* Publicaciones del Gobierno del Edo. de Yucatán.

Pfeifer, Gottfried. 1966. The Basin of Puebla-Tlaxcala in Mexico. *Revista Geográfica del Instituto Panamericano de Geografía e Historia* (IPGH), Vol. 64, pp. 85–107.

Perry, L. y K.V. Flannery. 2007. Precolumbian use of chili peppers in the Valley of Oaxaca, Mexico. *Proceedings of the National Academy of Sciences of the United States of America* 104(29): 11905–11909.

Pickersgill, B. 2017. Ajies or chile peppers (*Capsicum* spp.): history and biology. In: *Domesticación en el Continente Americano*, Vol. 2, Investigación para el manejo sustentable de recursos genéticos en el Nuevo Mundo: A. Casas, J. Torres-Guevara y F. Parra (coords.). México UNAM, Universidad Nacional Agraria La Molina (UNALM) del Perú. pp. 177–192.

Pilcher, J.M. 1998. *Que vivan los tamales! Food and the making of Mexican identity.* Albuquerque, University of New Mexico Press.

Pilcher, J.M. 2001. *¡Vivan los tamales! La comida y la construcción de la identidad mexicana.* México, CIESAS, CONACULTA.

Pimentel, D. 1997. Pest management in agriculture. In: D. Pimentel (ed.), *Techniques for Reducing Pesticide Use: Environmental and Economic Benefits.* John Wiley, Chichester (UK).

Piña Chan, R. 1990. Mesoamerica. Ensayo histórico cultural. México, Memorias del INAH, Vol. VI; pp. 37–48.

Piperno D.R. and D.M. Pearsall. 1998. *The Origins of Agriculture in the Lowland Neotropics.* Academic Press, S. Diego.

Piperno, D.R., and K.V. Flannery. 2001. The earliest archaeological maize (*Zea mays L.*) from Highland Mexico: new accelerator mass spectrometry date and their implications. Proceedings of the National Academy of Sciences98(4): 2101–2103.

Piperno D.R. 2003. Phytolith evidence for early Holocene Cucurbita domestication in southwest Ecuador. *Science* 299(5609): 1054–1057.

Piperno D.R.; A.J. Ranereb; I. Holstb.; J. Iriarted y R.R. Dickauc. 2009. Starch grain and phytolith evidence for early ninth millennium B.P. maize from the Central Balsas River Valley, Mexico. *Proceedings of the National Academy of Sciences of the United States of America* 106(13): 5019–5024.

Piperno D.R. and K.E. Stothert. 2011. The origins of plant cultivation and domestication in the new world tropics: patterns, process, and new developments. In the beginnings of agriculture: new data, new ideas. *Special Issue of Current Anthropology* 52(S4): 453–470.

Piperno D.R. and B.D. Smith. 2012. The origins of food production in Mesoamerica. In: D.L. Nichols and C.A. Pool (eds.). *The Oxford Handbook of Mesoamerican Archaeology*; Oxford University Press; pp. 153–164.

Plan Puebla. 1972 (1967–1969). Plan Puebla: Análisis de Beneficios y Costos. *México, El Trimestre Económico* 39(156(4)): 783–796.

Pohl, M.D.; K.O. Pope; J.G. Jones; J.S. Jacob; D.R. Piperno; S.D.D. France; D.L. Lentz; J.A. Gifford; M.E. Danforth; and J.K. Josserand. 1996. Early agriculture in the Maya Lowlands. *Latin American Antiquity* 7(4): 355–372.

Pohl, M.D.; D.R. Piperno; K.O. Pope y J.G. Jones. 2007. Microfossil evidence for preColumbian maize dispersals in the neotropics from San Andres, Tabasco, Mexico. *Proceedings of the National Academy of Sciences* 104(16): 6870–6875.

Poleman T.T. 1964. *The Papaloapan Project. Agricultural Development in the Mexican Tropics.* Stanford University Press, California, Food Research Institute.

Pope, K.O.; A.C. Ocampo; G.L. Kingsland y R. Smith. 1996. Surface expression of the Chicxulub crater. *Geology* 24: 527–530.

M.D. Pohl; Pope, K.O.; J.G. Jones; D.L. Lentz; C.V. Nagy; F.J. Vega; and I.R. Quitmeyer. 2001. "Origin and environmental setting of ancient agriculture in the lowlands of Mesoamerica. *Science* 292: 1370–1373.

Plunket, N.P. and L.G.G. Uruñuela. 2016. Las Casas del Preclásico en el Altiplano Central. *Arqueología mexicana* 140: 41–46.

Puleston, D.E. 1968. Brosimum alicastrum as subsistence alternative for the Classic maya of the Central southern lowlands.Tesis Ma, University of Pennsylvania, Philadelphia.

Puleston, D.E. 1971. An experimental approach to the classic Maya Chultuns. *American Antiquity* 36: 322–335.

Problemas Agrícolas N° 2, Vol., 5. 1953. México, Secretaría de Rec. Hidráulicos (SRH). Prontuario de información geográfica municipal de los E.U. Mexicanos, Tenampa, Veracruz de Ignacio de la Llave.

Problemas Agrícolas. In: www.3.inegi.org.mx/contenidos/app/mexicocifras/datos_geográficos/30/30162.pdf.

Quevedo-Pérez, D.C.; J. Cervantes-Herrera; L. Noriero-Escalante y J.M. Zepeda. 2017. Maíz: sustento de vida en la cultura Teenek. Comunidad Tamaletom, Tancahuitz, S.L.P. México. In: *Revista de Geografía Agrícola, estudios regionales de la agricultura mexicana.* Edo. de México, Texcoco, Universidad Autónoma Chapingo (UACH), Dirección de Centros Regionales, N° 58, enero-junio; pp. 5–19.

Quezada, S. 1995. Epidemias, plagas y hambrunas en Yucatán, México (1520–1700). *Revista Biomédica* 6: 238–242.

Quintana, A. 1944. El ingente problema del maíz en su aspecto agrícola y nutritivo. *América Indígena* IV(2): 129–132.

Quiñonez, A.C.T.M. 2005. Chinampas y chinamperos: los horticultores de S. Juan Tezompa. Thesis, Doct., Social Anthropology. México, Universidad Iberoamericana.

Ramírez, F. 2018. Arqueología de Cuicuilco, ayer y hoy. *Arqueología mexicana* 151: 28–33.

Ramírez-González, I. 2004. Chinampas contemporáneas en la cuenca del Alto Lerma. Tesis Ma., Social Anthropology, Universidad Iberoamericana.

Ramírez-García, A.G., and A. Cruz-León. 2018. El municipio de Álamos, Sonora. Algunas características ambientales y climáticas. In: *Sonora: La Sierra, el Desierto y la Costa en el Contexto de los guarijíos.* Texcoco, Universidad Autónoma Chapingo (UACH); pp. 141–173.

Ramírez-Gutiérrez, L.J. y M. Montenegro-Fragoso. 2012. Las centrales hidroeléctricas en México: pasado, presente y futuro. In: *Tecnología y ciencias del agua*, Vol. 3, N° 2, Juitepec abr/junio; versión On-line ISS 2007-2422.

Ramírez-Martínez, M.A. 2007. Ambiente, cultura y sociedad. Los productores de cacao de pequeña escala de José María Pino Suárez, Comalcalco, Tabasco. Thesis, Doct., Social Anthropology, México, Social Anthropology, México, Universidad Iberoamericana.

Ramos-Elorduy, J.B. 2009. ¿Los insectos se comen? México, UNAM y Fundación Telmex.

Ramos-Elorduy, J.B. 2019. Los insectos comestibles. *Arqueología mexicana* 86: 11–14.

Ramos-Galicia, Y. (coord.). 2014. *El maguey y el pulque en la región central de México.* Tlaxcala, Archivo Histórico del estado de Tlaxcala.

Ramos-Galicia, Y. 1994. *Catálogo de Instrumentos Agrícolas Tradicionales de Tlaxcala.* INAH, Museo de Artes y Tradiciones Populares de la Casa de las Artesanías de Tlaxcala.

Ramos-Madrigal, J.; B.D.Smith; J.V. Moreno-Mayar; S. Gopalakrishnan; J. Ross- Ibarra; M.Thomas P.Gilbert; N. Wales. 2016. Genome sequence of a 5,310-year-old maize cob provides insights into the early stages of maize domestication. *Current Biology* 16: 3195–3201.

Randolph, L. F. 1976. Contribution of wild relatives of maize to the evolutionary history of domesticated maize: a synthesis of divergent hypotheses I. *Economic Botany* 30: 321–345.

Rebato, E. 2009. Las nuevas culturas alimentarias: globalización vs. etnicidad. *Osasunaz* 10: 135–147.

Redfield, R. 1950. Chan Kom. *A Village That Chose Progress:* Chicago, University of Chicago Press.

Redfield, R. and A. Villa Rojas. 1934. *Chan Kom Revisited.* Chicago, University of Chicago Press.

Rees P. 1976. *Los transportes y el comercio entre México y Veracruz.* México, SepSetentas.

Reff, D.T. 1990. Indian mortality in northwestern New Spain, 1500–1678. In: Reff, D.T.; *Disease, depopulation and culture change in nothwestwern New Spain. 1518–1764.* Salt Lake City, University of Utah Press; pp. 1–4.

Reilly, J.M. 1995. Climate change on agriculture: recent findings ans issues. *American Journal of Agricultural Economics* 77(3): 727–733.

Renard, D.; J. Iriarte; J.J. Birk: S. Rostain; B. Glaser; and D. McKey. 2012. Ecological engineers ahead of their time: the functioning of pre-Columbian raised-field agriculture and its potential contributions to sustainability today. *Ecological Engineering* 45(2012) 30–44.

Rendón, S. 1941. La alimentación tarasca. *Anales del INAH* 2: 207–228.

Rendón-Aguilar, B.; Bye, B.R. and J. Núñez-Farfán. 2001. Ethnobotany of *Anoda Cristata* (L.) schl. (Malvaceae) in Central Mexico: uses, management and population differentiation in the community of Santiago Mamalhuazuca, Ozumba, State of Mexico. *Economic Botany* 55: 545–555.

Rendón-Aguilar, B.; L.A. Bernal-Ramírez y G.A. Sánchez-Reyes. 2014. Las arvenses: más que hierbas del campo. In: Oikos, web.ecologia.unam.mx/oikos3.0/index.php/dirección.

Rendón-Garcini, R. 1990. *Dos haciendas pulqueras en Tlaxcala.* 1857–1884. México, Gobierno del estado de Tlaxcala, Universidad Iberoamericana.

Rendón-Garcini, R. 1994. *Haciendas de México.* México, Fomento Cultural Banamex A.C.

Rendón-Garcini, R. 1995. *El Prosperato.* México, Ed. Siglo XXI.

Rendón-Garcini, R. 1996. *Breve Historia de Tlaxcala.* Fideicomiso Historia de las Américas, México, FCE.

Rentería-Garita, C.A. 2006. La colonización, adaptación y desarrollo económico de la selva a través de la ganadería: el caso de la frontera Tabasco-Chiapas. Thesis, Ma. Social Anthropology, Mexico, Universidad Iberoamericana.

Reyes-Escutia, F. y S. Barrasa-García (coords.). 2011. *Saberes Ambientales Campesinos. Cultura y Naturaleza en comunidades indígenas y mestizas de México.* México: Universidad de Ciencias y Artes de Chiapas; Agencia Española de Cooperación Internacional (AECID).

Reyes-Montes, L. 2003. Adaptación sociocultural en Santiago Yeché: un estudio de ecología cultural en México; Thesis, Doct., Social Anthropology, México Universidad Iberoamericana.

Reyes-Montes, L. 2017. La fiesta del maíz y cosmovisión mesoamericana en el área de Ixtlahuaca. In: M.P. Iracheta C. y A.J. Enríquez S. (comps.). El Colegio Mexiquense, Cuadernos municipales 26: 219–249.

Revel-Mouroz, J. 1980. *Aprovechamiento y colonización del Trópico Húmedo Mexicano*. México, Fondo de Cultura Económica (FCE).

Revista *Artes de México*. 2006. Rituales del maíz. *Artes de México*, Vol. 78.

Rhoades, R.E. 1984. Breaking New Ground: Agricultural Anthropology. In: International Potato Center, Lima, Peru.

Rice, D.S. 1985. The Peten Postclassic: a settlement perspective. In: *Late Lowland Maya Civilization: Classic to Postclassic*; J.A. Sabloff and E.W. Andrews (eds.). University of New Mexico Press, Albuquerque, NM; pp. 301–344.

Rice, D.S.; P.M. Rice and G.D. Jones. 1993. Geografía política del Petén central, Guatemala en el siglo XVII: la arqueología de las capitales mayas. *Mesoamerica* 26: 281–318.

Riley G.M. 1973. *Fernando Cortés and the Marquesado in Morelos*, 1522–1547. Albuquerque: University of New Mexico Press.

Rincón-Maurer, C.A. 1999. Man and the Environment in the Coixtlahuaca Basin of Northwestern Oaxaca, Mexico: Two Tousand Years of historical Ecology. PhD., University of Texas at Austin.

Rincón-Maurer, C.A. 2017. Seeking the Imprimatur Dei in Coixtlahuaca's 1571– 1575 Boundary Dispute with Zapotitlán. In: *On the Mount of Interwined Serpents*; Viola Köning (ed.). Michael Imhof Verlag GmbH&Co.KG and Museo Etnológico de Berlín; pp. 109–121.

Rivas, M.G.; B. Rodríguez H. y J. Palerm-Viqueira. 2008. El sistema de jollas una técnica de riego no convencional en la Mixteca. *AHA, Boletín* 40: 6–16.

Rivera-Uria, M.Y.; S. Sedov; E. Solleiro-Rebolledo; J. Pérez-Pérez; E. McClung; González and J. Gama-Castro. 2007. Degradación ambiental en el valle Teotihuacan: evidencias geológicas y paleopedológicas. In: *Boletín de la Sociedad Geológica Mexicana*, Tomo LIX, N° 2; pp. 203–217.

Rivera Dommarco, J.A. (coord.); M. Hernández-Ávila; C.A. Aguilar.Salinas; F. Vadillo; O. Murayama Rendón, C. 2012. *Obesidad en México: recomendaciones para una política de Estado*. México, UNAM; pp. 99–123.

Robles-Cervantes, M. 2008. Purificación Tepetitla: Organización Familiar para la Producción de Plantas Medicinales en los Huertos del Somontano. Thesis Ma. Social Anthropology, México, Universidad Iberoamericana.

Robles-Linares G.M.G. 2018. Mawechi y Verano: Sistemas Agrícolas Tradicionales en la Cuenca del Río Mayo. In: A. González, J.J. Velasco, S. Moctezuma P. y A. Cruz-León (comps.). *Sonora: el desierto y la costa en el contexto de los guarijíos*. Texcoco, Estado de México, UACh; pp. 269–296.

Robles-Linares G.M.G. y M. Curiel-Ruiz. 2017. Crisis, comunidad y tiempo: las resistencias, desplazamientos y reasentamientos de Temacapulín, Acasico y Palmarejo. In: J.C. Domínguez-Virgen (coord.). *Megaproyectos, Desplazamiento Forzado y Reasentamiento Involuntario en México: Testimonios y Reflexiones*. Instituto Dr. José Ma. Luis Mora y CONACYT; pp. 163–184.

Rodríguez-Galicia, B. 2017. *La pesca mesoamericana. Las artes de la actividad pesquera del pasado prehispánico y el presente*. México, UNAM-IIA.

Rodríguez-Tapia, S.; C.A. Ortiz-Solorio; C. Hidalgo-Moreno; M.C. Gutiérrez- Castorena. 2004. Los tepetates de la ladera oeste del cerro Tláloc: Saprolita, sin endurecimiento pedológico. In: *Terra Latinoamericana*, Vol. 22, N° 1, enero-marzo. Sociedad Mexicana de la Ciencia del Suelo, AC., Chapingo, México.

Rodríguez-Yc., J.R. 2013. La molienda en Mesoamérica, formas, funciones, usos y manufactura de los instrumentos. Un estudio etnoarqueológico en México. Tesis, Doct., Historia, Universidad de Barcelona, España.

Roldán-Botello, D.P. 1979. Un caso de desarrollo agrícola en Tlaxcala: S. Francisco Tepeyanco. Thesis, BA, Social Anthropology, México, Universidad Iberoamericana.

Rolmarta. 2007. *El Otro Oro traído de América*. Colaboraciones Alumnado, Industrias Alimentarias F.P.

Romero-Contreras, A.T. 1998. *Los temascales de San Isidro Buensuceso, cultura, medicina y tradición en un pueblo tlaxcalteca*. Tlaxcala, Ediciones del Gobierno del Estado.

Romero-Contreras, A.T.; I. Liendo-Vera; G. Rivera-Herrejón; L. González-Díaz. Romero, P.J. et al. 1986. *Diagnóstico de la Producción Agrícola de las Mixtecas Oaxaqueñas Alta y Baja*. CONACYT y *Centro Regional del Sur*, UACh, Tomos II y III.

Romero-Contreras, A.T. 2004. Las estrategias de transporte y adaptación de las especies agrícolas del Viejo Mundo hacia la Nueva España. *Ciencia Ergo Sum* 11(3): 237–245

Rojas de Perdomo, L. 1993. *Aportes Alimenticios del Viejo al Nuevo Mundo*. Colombia: Ed. Voluntad S.A.

Rojas-Pérez, P.C.; A.S. Gijón-Cruz; Y.J. Luis-García. 2017. Producción de pitaya como alternativa de desarrollo local en Acaquizápan, Santiago Chazumba, Oaxaca. 20° Encuentro Nacional sobre Desarrollo Regional en México. Cuernavaca, Morelos, 17–20/11/2015. AMECIDER–CRIM, UNAM.

Rojas-Rabiela, T.; R. Strauss y J. Lameiras, 1974. *Nuevas Noticias sobre las Obras Hidráulicas Prehispánicas y Coloniales en el Valle de México.* México, SEP/INAH.
Rojas-Rabiela, T. 1983. *La Agricultura Chinampera.* México, Universidad Autónoma Chapingo (UACH).
Rojas-Rabiela, T. 1985. *La cosecha del agua: Pesca, caza de aves y recolección de otros productos biológicos acuáticos de la cuenca de México.* CISINAH, Serie Los Pescadores de México.
Rojas-Rabiela, T. 2009. Las obras hidráulicas en las épocas prehispánica y colonial. In: *Semblanza Histórica del Agua en México.* Comisión Nacional del Agua; pp. 9–26.
Rojas-Rabiela, T. 2012. El agua en la antigua Mesoamérica: usos, tecnología y organización social. In: *Cultura hidráulica y simbolismo mesoamericano del agua en el México prehispánico.* Zamora, Michoacán, El Colegio de Michoacán, A.C.
Rojas-Rabiela, T. 2013. Tecnología hidráulica comparada: de Mesoamérica a la Nueva España. In: É. Hurtado M. y J.F. Román G. (coords.); *Con tinta de agua; historiografía, tecnologías y usos.* México, Universidad Autónoma de Zacatecas, Programa Integral de Fortalecimiento Institucional: pp. 65–120.
Rojas-Rabiela, T. 2014. La Tecnología Agrícola. In: L. Manzanilla y López (coords.); *Historia antigua de México. Vol. IV: Aspectos fundamentales de la tradición cultural mesoamericana.* México, CNCA, Inst. Nal. de Antrop. e Historia, UNAM Coord. de Humanidades, Instituto de Investigaciones. Antropológicas; pp. 13–68.
Rosado May, F. 2015. Historia de la Agroecología en México. *Agroecología* 10(2): 9–17.
Rosado May, F. 2018. Comunicación personal, 05/08/2018.
Rosemblat, Á. 1945. *La población indígena de América desde 1942 hasta la actualidad.* B. Aires.
Rosemblat, Á 1967. *La población de América en 1492. Viejos y nuevos cálculos.* México, El Colegio de México (COLMEX).
Roys, R. 1957. *The Political Geography of the Yucatan Maya.* Washington DC, Carnegie Institution.
Rubio, M.T.; N. Gutiérrez y R.Méndez. 2004. Composition and sensory evaluation of lamb carcasses used for the traditional Mexican lamb dish, Barbacoa. *Meat Science* 67: 359–364.
Ruenes-Morales, M.R. 2012. Etnobotánica de los Abales (Ciruela Mexicana) de Yucatán. In: A. Cruz-León et al. *Jocotes, Jobos, Abales o Ciruelas Mexicanas.* Estado de México, Chapingo, UACh; pp. 63–75.
Ruiz-Corral, J.A.; E. Bravo-Mosqueda; G. Ramírez-Ojeda; A.D. Báez G.; M. Álvarez, C.; J.L. Ramos-González; U. Nava C. y K. F. Byerly M. 2013. *Plagas de Importancia Económica En México: Aspectos de su Biología y Ecología.* México, SAGARPA, INIFAP.
Rutsch. M. y A. González-Jácome. 2011. *Culturas y Políticas del Agua en México y un Caso del Mediterráneo.* México, Instituto Nacional de Antropología e Historia (INAH) y Universidad Iberoamericana. Ruttan, V.W. 2001. *Technology, Growth, and Development.* Oxford University Press, New York.
Ruz, M.A. 1998. Los afanes cotidianos mayas: una historia en minúsculas. In: P. Schmidt; M. de la Garza y E. Nalda (coords.); *Los mayas.* Milán-Mexico, CONACULTA, INAH, Américo Arte Editores; pp. 159–177.
Rzedowski, J. 1978. *La Vegetación de México.* México, Ed. Limusa. Hay varias ediciones en papel y una electrónica.
Sabloff, J.A. y W.L. Rathje. 1975. The rise of the Maya merchant class. *Scientific American* 233: 72–82. Secretaría de Agricultura, Ganadería, Desarrollo Rural, Pesca y Alimentación (SAGARPA).
SNITT. 2016. *Agenda Nacional de Investigación, Innovación y Transferencia de Tecnología Agrícola 2016–2022.* México, SNITT. 2017–2030. *Planeación Agrícola Nacional. Aguacate mexicano.*
Sahlins, M.D. 1960. *Evolution and Culture.* Ann Arbor. Michigan, University of Michigan Press.
Salazar-Barrientos, L.L.; Magaña-Magaña, M.Á.; Aguilar-Jiménez, A.N.; Ricalde- Pérez, M.F. 2016. Factores socioeconómicos asociados al aprovechamiento de la agrobiodiversidad de la milpa en Yucatán. *Ecosistemas y Recursos Agropecuarios* 3(9): 391–400.
Salazar-Goroztieta, L.P. 2006. La flor del pericón: tradición y usos. In: *El Tlacuache*, Suplemento Cultural, N° 228; Centro INAH, Morelos, 1° octubre, 1ª p.
Sales-Colín, J. 2010. El proceso de cambio económico, social y cultural y su impacto en el ambiente en el municipio de St. Cruz Atizapan, Estado de México. Thesis, Doct., Social Anthropology, México, Universidad Iberoamericana.
Sales-Colín, J. 2018. La cuenca del Río Mayo y la presa Los Pilares. In: A. González-Jácome, J.J. Velasco-Orozco, J.J.; S. Moctezuma-Pérez, and A. Cruz-León (coords.). *Sonora: La Sierra, el Desierto y la Costa en el contexto de los guarijíos.* Estado de México, Texcoco, Universidad Autónoma Chapingo (UACH); pp. 195–218.
Sánchez, M.A. 2000. El Impacto de la Modernización y el Cambio Tecnológico en la Agricultura de Riego: El Uso Intensivo de las Aguas Subterráneas en la Cuenca del Río Laja, Guanajuato. Thesis MA, Social Anthropology, México, Universidad Iberoamericana.

Sánchez-G., J.J. y M.M. Goodman, 1992. Relationships among the Mexican races of maize. *Economic Botany* 46: 72–85.

Sánchez-G.; C.W. Stuber y M.M. Goodman. 2000. Isozymatic diversity in the races of maize of the Americas. *Maydica* 45: 185–203.

Sanders, W.T. 1956. The Central Mexican symbiotic region. In: G.R. Willey (Ed.); *Prehistoric Settlement Patterns in the New World.* Viking Fund Publications in Anthropology, N° 23.

Sanders, W.T. 1973. The cultural Ecology of the Lowland Maya: A reevaluation. In: Patrick Culbert (ed.), *The Classic Maya Collapse.* Albuquerque: The University of New Mexico Press: 325–365.

Sanders, W.T. 1976. The population of the Central Mexican Symbiotic Region, the Basin of México, and the Teotihuacán Valley in the sixteenth century. In: W. Denevan (ed.); *The native population of the Américas in 1542.* Madison, The University of Wisconsin Press; pp. 85–150.

Sanders, W.T. 1979. Introduction. In: Sanders, W.T.; J. Parsons y R.S. Santley (eds.); *The Basin of Mexico. Ecological Processes in the Evolution of a Civilization.* Academic Press, Studies in Archaeology; pp. 1–10.

Sanders, W.T., and B.J. Price. 1968. *Mesoamerica. The Evolution of a Civilization.* New York, Random House.

Sanders, W.T.; J.R. Parsons, and R.S. Santley. 1979. *The Basin of Mexico. Ecological Process in the Evolution of a Civilization.* Academic Press Inc., Studies in Archaeology.

Sanjur, O.I; D.R. Piperno; T.C. Andrés and L. Wessel-Beaver. 2002. Phylogenetic relationships among domesticated and wild species of Cucurbita *Cucurbitaceae*) inferred from a mitochondrial gene: Implications for crop plant evolution and areas of origin. *Proceedings of the National Academy of Sciences of the United States of America* 99(1): 535–540.

Samartt, J. 1988. Morphological physeological, and biochemical changes in phaseolus beans. In: *Genetic Resources of Phaseolus Beans.* P. Gepts (ed.). Boston, Kluwer Academic Publishers; pp. 143–162.

Sánchez-Izquierdo, M.A. 2000. Innnovación y cambio tecnológico en la cuenca del río Laja, Guanajuato. Thesis Ma., Social Anthropology, México, Universidad Iberoamericana.

Sánchez-Miranda, G. y J.P. Carpenter S. 2018. El resto de maíz más antiguo encontrado hasta la fecha en el noroeste de México, emergió en bandejas de agua. *Boletín* 13, INAH. México, Secretaría de Cultura, 01/15; pp. 1–3.

Santiago-Mejía, B.E.; M.R. Martínez-Menez; E. Rubio-Granados; H. Vaquera- Huerta y J. Sánchez-Escudero. 2018. Variabilidad espacial de propiedades físicas y químicas del suelo en un sistema lama-bordo en la Mixteca Alta de Oaxaca, México. *Agricultura, sociedad y desarrollo,* Vol. 15, N° 2. Texcoco, México, abril/junio.

Santley, R.S.; T.W. Killion; and M.T. Lycett. 1986. On the Maya collapse. *Journal of Anthropological Research* 42: 123–159.

Santley, R.S. 1990. Demographic Archaeology in the Maya Lowlands. In: T.P. Culbert and O.S. Rice (comps.), *Precolumbian Population History in the Maya Lowlands.* Albuquerque: University of New Mexico Press; pp. 325–343.

Santley, R.S. 1996. Prehispanic settlement patterns in the Tuxtla Mountains, Southern Veracruz, Mexico. *Journal of Field Archaeology* 23: 225–249.

Santley, R.S.; P.J. Arnold, and T.P. Barret. 1997. Formative Period Settlement Patterns in the Tuxtla Mountains. In: *Olmec to Aztecs: Settlement Patterns in the Ancient Gulf Lowlands*; B.L. Stark, and P.J. Arnold III (eds.). Tucson, University of Arizona Press: 174–205.

Sauer, C.O. 1952. *Agricultural Origins and Dispersals.* The American Geographical Society, New York.

Scarborough, V.L. and G.G. Gallopin. 1991. A water storage adaptation in the Maya lowlands. *Science* 251: 658–662.

Scarborough, V.L. 1993. Water management in the southern Maya lowlands: an accretive model for the engineered landscape. In: V.L. Scaborough y B.L. Isaac (eds.); *Economic Aspects of Water Management in the Prehispanic New World, Research in Economic Anthropology*, JAI Press, Greenwich, CT; Suppl. 7th; pp. 17–69.

Scarborough, 2012. Water and sustainable land use at the ancient tropical city of Tikal, Guatemala. *Proceedings of the National Academy of Sciences of the United States of America* 109: 12408–12413.

Scheffran, J.; M. Brzoska; J. Kominek; P.M. Link, and J. Schilling. 2012. Climate change and violent conflicto. *Science* 336: 869–871.

Schlesinger, V. 2001. *Animals and Plants of the Ancient Maya. A Guide.* The University of Texas Press.

Scholes F. V. and E. B. Adams. 1936. Documents relating to the Mirones Expedition to the interior of Yucatan. *Maya Research* 3: 153–176, 251–276.

Schroeder, C.A. 1968. Prehistoric avocados in California. In: *Yearbook* 52; California Avocado Society; pp. 29–34.

Secretaría de Economía. 2012. Análisis de la cadena de valor maíz tortilla: Situación Actual y Factores de Competencia Local. México, Dirección General de Industrias Básicas.

Secretaría de Recursos Hidráulicos (SRH). 1972. *Estudio Socioeconómico del Distrito de Riego Nº 56 Atoyac-Zahuapan*. México, SRH.

Secretaría de Agricultura y Recursos Hidráulicos (SARH) 1956. México y Su Política de Irrigación *Memorandum Técnico Nº 115*. México, SARH.

Serafino, G. 2012. Rituales indígenas en Mesoamérica. La fiesta de Petición de Lluvia en La Montaña de Guerrero (México). Thesis, Doct. University of Bologna/ École Pratique des Hautes Études (ephe). París.

Serafino, G. 2014. La comida ritual de los nahuas de S. Pedro Petlacaca por la petición de lluvia. In: *Nuevo mundo. Mundos nuevos*. http://nuevomundo.revues.org/67391.

Serafino, G. 2015. La fiesta de San Miguel en la montaña nahua de Guerrero. *Cuicuilco* 22(63): 207–227.

Servicio Nacional de Sanidad, Inocuidad y Calidad Agroalimentaria (SENASICA). 2018. SENASICA más de una década de evolución 2007–2018. México, SAGARPA, OIRSA, SENASICA.

Serra-Puche M.C. 2017. En Mesoamérica Conocían la Destilación Antes de la Llegada de los Españoles. In: Boletín UNAM-DGCS-004, Ciudad Universitaria. 06:00 hs. 02/01/2017.

Servicio de Información Agroalimentaria y Pesquera. SAGARPA (SIAP). n/d. El plátano. In: siaprendes.siap.gob.mx/contenidos/2/04-platano/index.html

Servín-Segovia, S.J.A. 2000. Sistemas de cultivo y manejo de recursos en una barranca: el caso de Xopilapa, Veracruz. México. Thesis MA., Social Anthropology. Mexico, Universidad Iberoamericana.

Servín-Segovia, S.J.A. 2001. La globalización en una "Economía Enemiga": el Caso de Xopilapa en el Veracruz Central. In: *Actas Latinoamericanas de Varsovia*, Tomo 24; pp. 165–177.

Servín-Segovia, S.J.A. 2002. Banqueteras: un sistema agrícola mesoamericano". *Suplemento Antropológico*, Vol. XXVII, Nº 1, junio, Paraguay, Universidad Católica de Asunción.

Shaw, J.M. 2003. Climate change and deforestation: implications for the Maya collapse. *Ancient Mesoamerica* 14: 157–167.

Sheets, P.D. 1982. Prehistoric agricultural systems in El Salvador. *Maya Subsistence. Studies in Memory of Dennis E. Puleston*. Academic Press, Studies in Archaeology: 99–118.

Siebe, C. 2009. La erupción del volcán Xitle y las lavas del Pedregal hace 1670 +/-35 años AP y sus implicaciones. In: Lota, A.; Cano-Santana; *Biodiversidad del ecosistema del Pedregal de San Ángel*. México, UNAM.

Siemens, A.H. 1978. Karst and the pre-Hispanic Maya in the southern lowlands. In: P.D. Harrison and B.L. Turner (Eds.); *Pre-Hispanic Maya Agriculture*. Austin, University of Texas Press; pp. 117–143.

Siemens, A.H. 1983a. Modelling prehispanic hydroagriculture on leves backslopes in Northern Veracruz, Mexico. Darch, J.P. (ed.), *Drained Field Agriculture in Central And South America*. Oxford, England, BAR International Series.

Siemens, A.H. 1983b. Oriented raised fields in Central Veracruz. *American Antiquity* 48(1): 85–102.

Siemens, A.H. 1989a. *Tierra configurada: investigaciones de los vestigios de agricultura precolombina en tierras inundables costeras desde el norte de Veracruz hasta Belice*. México, Consejo Nacional para la Cultura y las Artes (CONACULTA).

Siemens, A.H. 1989b. Reducing risk: some indications regarding pre-hispanic wetland agricultural intensification from contemporary use of a wetland/terra firma boundary zone in Central Veracruz. *Agroecology. Researching the Ecological Basis for Sustainable Agriculture*. Springer-Verlag: 233–250.

Siemens, A.H. 1990. *Between the Summit and the Sea. Central Veracruz in the Nineteenth Century*. Canadá, Vancouver, University of British Columbia Press.

Siemens, A.H. 1998. *A Favored Place: The San Juan River Wetlands, Central Veracruz, A.D. 500 to the Present*. Austin, The University of Texas Press.

Siemens, A.H. 2002. El 'Rub al-Kahli' prehispánico de Los Tuxtlas: una interpretación del paisaje. In: B. García-Martínez y M. Prieto. (comps.) *Estudios sobre historia y ambiente en América II*. El Colegio de México y el IPGH. México; pp. 161–202.

Siemens, A.H. 2004. Los Paisajes. In: Part 1, *Los Paisajes de los Tuxtlas*; S. Guevara; J. Laborde y G. Sánchez-Ríos (comps.) Inst. de Ecología AC (INECOL); pp. 41–58.

Siemens, A.H. 2009. *Una Manera de ver Los Tuxtlas. Paisaje de Mesoamérica*. México, SEMARNAT, Comisión Nacional para el Conocimiento y Uso de la Biodiversidad Corredor Biológico Mesoamericano México.

Siemens, A.H. 2010. Sucesión de Paisajes: Una Conceptualización de la Relación Entre el Ser Humano y su Ambiente Natural en el Transcurso del Tiempo. In: E.Ortiz-Díaz (Ed.); *VI Coloquio Pedro Bosch Gimpera*. México, UNAM-Instituto de Investigaciones Antropológicas; pp. 183–203.

Siemens, A.H. and D. Puleston. 1972. Ridged fields and associated features in southern Campeche: new perspectives on the lowland Maya. *American Antiquity* 37: 228–239.

Siemens, A.H.; R.J. Hebda; M. Navarrete H.; D. Piperno; J.K. Stein and M.G. Zola B. 1988. Evidence for a cultivar and a chronology from patterned wetlands in Central Veracruz, Mexico. *Science* 242: 105–107.

Sierra-Carrillo, D. 1988. Los nahuas de Morelos: una ceremonia agrícola. *Estudios Nahuas.* México, INAH, Colección de Divulgación.

Sierra-Carrillo, D. 2000. El yauhtli, planta curativa y protectora. Su importancia mágico-religiosa en el presente y en el pasado. Thesis, UNAM, Filosofía y Letras.

Sierra-Carrillo, D. 2008. "El demonio anda suelto. El poder de la cruz de Pericón". In: México, INAH.

Sierra-Carrillo, D. 2015. Anisado aroma y amarillo color. Los poderes de la flor del pericón. In: BA. Albores (coord.), *Flor-flora. Su uso ritual en Mesoamérica.* El Colegio Mexiquense, Fondo Editorial del Gobierno del Estado de México; pp. 413–428.

Sierra Leor, E. 2018. Entrevista (INIFAP). Disminuye producción de mango de Veracruz por cultivo de malanga. Informó J.L. Rogel Montalvo. Televisa News desde Veracruz, 06/04/2018.

Sluyter, A. 1994. Intensive wetland agriculture in Mesoamerica: space, time, and form. *Annals of the American Association of Geographers* 84: 557–584.

Sluyter, A. and G. Domínguez. 2006. Early maize (*Zea mays* L.) cultivation in Mexico: dating sedimentary pollen records and its implications. *Proceedings of the National Academy of Sciences* 103(4): 1147–1151.

Smalley J y M. Blake. 2003. Stalk Sugar and the Domestication of Maize. *Current Anthropology* 44(5): 675–689.

Smith, B.D. 1995a. *The Emergence of Agriculture.* Scientific American Library, New York.

Smith, B.D. 1995b. Crop cultivation and Domestication in Eastern North America. In: *The emergence of Agriculture.* Scientific American Library, New York.

Smith, B.D. 1997. The initial domestication of *Cucurbita pepo* in the Americas 10,000 years ago. *Science* 276: 932–934.

Smith, B.D. 2001. Documenting plant domestication: the consilience of biological and archaeological Approaches. *Proceedings of the National Academy of Sciences* 98:1324–1326.

Smith, B.D. 2005. Reassessing Coxcatlan cave and the early history of domesticated plants in Mesoamerica. *Proceedings of the National Academy of Sciences* 102(27): 9438–9445.

Smith Jr., E.C. 1967. Plant remains. *The Prehistory of the Tehuacan Valley,* Vol. 1, *Environment and Subsistence.* D.S. Byers (Ed.). The University of Texas Press: 220–255.

Smith Jr., E.C. 1969. Catalog of plant remains recovered from open sites in the Valley of Oaxaca. (Unpublished list).

Smith Jr., E.C. 1976. *Modern vegetation and ancient plant remains of the Nochixtlán Valley, Oaxaca.* Nashville, Vanderbilt, University Publications in Anthropology N° 16.

Smith Jr., E.C. 1986. Preceramic plant remains from Guilá Naquitz. In: Flannery, K.V. (ed.); *Guilá Naquitz. Archaic Foraging and early Agriculture in Oaxaca, Mexico.* Academic Press, New York; pp. 265–301.

Sohn, Y.; E. Morán and F.D. Gurri-García. 1999. Deforestation in North-Central Yucatán (1985–1995): Mapping secondary succession of forest and agricultural land use in Sotuta using the cosine of the angle concept. In: *Photogrammetric Engineering and Remote Sensing*; Vol. 65, N° 8; pp. 947–958.

Spores, R. 1969. Settlement, farming technology and environment in the Nochixtlan Valley. *Science* 166(3905): 557–569.

Spores, R. 1986. Asentamiento, tecnología agrícola y medio ambiente en el Valle de Nochistlán. *Suplemento Canícula* 2(13–14): UACh.

Staller, J.E.; R.H. Tykot and B. Benz. 2006. *Histories of Maize. Multidisciplinary Approaches to the Prehistory, Linguistics, Biogeography, Domestication, and Evolution of Maize.* Amsterdam, New York, Oxford, Paris, Elsevier, Academic Press.

Staller, J.E. 2010. *Maize Cobs and Cultures: History of Zea mays L.* Springer Heldelberg Dordrecht, London, New York.

Sturtevant, W.C. 1964. Studies in ethnoscience trnscultural studies. Cognition. *American Anthropologist* 66: 99–113.

Stross, B. 2006. Maize in word and image in Southeastern Mesoamerica. In: *Histories of Maize. Multidisciplinary Approaches to the Prehistory, Linguistics, Biogeography, Domestication, and Evolution of Maize.* Amsterdam, New York, Oxford, Paris, Elsevier, Academic Press: 577–598.

Steward, J.H. 1949. Cultural causality and law: a trial formulation of the development of early civilizations. *American Anthropologits* 51(1): 1–27.

Stewrad, J.H. 1955 [Spanish 2004]. *Theory of Culture Change.* Urbana, University of Illinois Press.

Steward, J.H. 2006. The concept and method of Cultural Ecology. In: N. Haen and R.R. Wilk, *The Environment in Anthropology. A Reader in Ecology, Culture and Sustainable Living.* New York Univ. Press; pp. 5–9.

Steward, J.H. 2014. [1955a]. *Teoría del Cambio Cultural.* México, CIESAS, UAM, Universidad Iberoamericana.

Steward, J.H.; R.M. Adams; D. Collier; Á. Palerm; K.A. Wittfogel and R.L. Beals. 1955b. *Irrigation civilizations: A Comparative Study.* Social Science Monographs, N° 1, Washington D.C., Panamerican Union Press.

Storey, W.B.; B. Bergh; G.A. Zentmyer. 1986. The origin, indigenous range, and dissemination of the avocado. In: *California Avocado Society 1986 Yearbook* 70: 127–133.

Stuart, S.W. 1993. Contribution of Dooryard gardens to contemporary Yucatan Maya subsistence, *Biótica* 1: 53–61.

Suto, T. and Yoshida, Y. 1956. Characteristics of the oriental maize. H. Kihara (ed.). *Land and crops of Nepal Himalaya*, Vol. 2. Japan: Kyoto, Fauna and Flora Res. Soc. Kyoto University; pp. 375–530.

Sutton, D. 2010. Food and the senses. *Annual Review of Anthropology* 39: 209–223.

Tarango-Rivero, S.H.; B.C. Macías-López; A. Alarcón and J. Pérez Moreno. 2004. Micorrizas en nogal pecanero y pistachero. México, INIFAP, Universidad Autónoma de Chihuahua, Folleto Técnico N° 16; 39 pp.

Tavera, R. y E. Novelo. 2011. El perifiton de los humedales de Yucatán y la agricultura maya. *Ciencias* 102 : 39–45.

Taylor, W.B. 1990. El espacio y los pueblos en la Nueva España. *Historia Mexicana* 39(3): 803–811.

Terán, C.S. 2010. Milpa, biodiversidad y diversidad cultural. In: *Biodiversidad y desarrollo humano en Yucatán*; Documentos CICY; Durán R.M. (comp.). CICY, PPD-FMAM, CONABIO, SEDUMA.

Terán, C.S. y C.H. Rasmussen. 1994. *La milpa de los mayas: la agricultura de los mayas prehispánicos y actuales en el noreste de Yucatán*. Mérida, Danida.

Terán, C.S.; C.H. Rasmussen, and O. May Cauich. 1998. *Las plantas de la milpa entre los mayas*. México, Fundación Tun Ben Kin AC, Mérida, Yucatán.

Thome-Ortiz, H. 2015. Agri-food tourism and, new social metabolisms of local products. *Revista Mexicana de Ciencias Agrícolas* 6: 1373–1386.

Thompson, J.E.S. 1992. *Grandeza y decadencia de los mayas*. México, FCE, Obras de Antropología, 4ª edic.

Toledo, Á. and B. Burlingame. 2006. Biodiversity and nutrition: a common path toward global food security and sustainable development. *Food Composition and Analysis* 19(6–7): 477–483.

Toledo, V.M. 1980. La Ecología del modo campesino de producción. *Antropología y Marxismo* Vol. 3; pp. 35–55.

Toledo, V.M. 1985. *Ecología y Autosuficiencia Alimentaria*. México, Siglo XXI Editores.

Toledo, V.M. 1992. La racionalidad ecológica de la producción campesina. In: E. Sevilla- Guzmán y M.L. González de Molina N. (eds.), *Ecología, campesinado e historia*. Madrid, La Piqueta; pp. 197–218.

Toledo, V.M. 2002. Ethnoecology a conceptual framework for the study of indigenous knowledge of nature. In: J.R. Stepp et al. (eds.); *Ethnobiology and Biocultural Diversity*. Internat. Society of Ethnobiology, Georgia; pp. 511–522.

Toledo, V.M. 2003. Ecología, espiritualidad y conocimiento de la sociedad del riesgo a la sociedad sustentable. *Programa de las Naciones Unidas para el Medio Ambiente*; Oficina Regional para América Latina y El Caribe, México D.F.

Toledo, V.M. 2010 [2015]. *La biodiversidad de México. Inventarios, manejos, usos, informática, conservación e importancia cultural*. México, CONACULTA, FCE.

Toledo, V.M.; J. Carabias, C. Mapes, C. Toledo. 1985. *Ecología y Autosuficiencia alimentaria*. México, Siglo XXI eds.

Toledo, V.M. y N. Barrera-Bassols. 2008. *La Memoria Biocultural. La Importancia Ecológica de las Sabidurías Tradicionales*. España, Junta de Andalucía, Icaria Ed. Perspectivas Agroecológicas.

Toledo, V.M. and E. Böege. 2010 [2015]. La biodiversidad, las culturas y los pueblos indígenas. In: V.M. Toledo; *La biodiversidad de México. Inventarios, manejos, usos, informática, conservación e importancia cultural*. México, CONACULTA, FCE; pp. 160–192.

Tomé, P. 2009. Miradas antropológicas a las relaciones entre naturaleza y cultura. A modo de introducción. *Revista de Dialectología y Tradiciones Populares* LXIV(1): 7–22.

Tortolero-Villaseñor, A. 1995. *De la coa a la máquina de vapor : actividad agrícola e innovación tecnológica en las haciendas mexicanas, 1880–1914*. México, El Colegio mexiquense y Siglo XXI Editores.

Tortolero-Villaseñor, A. 2000. *El agua y su historia: México y sus desafíos hacia el siglo XXI*. México, Siglo XXI Editores (Umbrales de México).

Tortolero-Villaseñor, A. 2001. *Empresarios y navegación en la cuenca de México. La importancia de los canales en los siglos* XVIII *y* XIX. México, UAM-Iztapalapa.

Toscano, S. 1970. *Arte precolombino de México y de la América central*. México, UNAM.

Toscano, S. M.; J.M. García C.; M.J. Iglesias P.de L; A. Lacadena G.G.; L.T. Sanza. 1998. *Anatomía de una Civilización: aproximaciones interdisciplinarias a la cultura maya*. España, Sociedad Española de Estudios Mayas; pp. 185–206.

Tourtellot, G.; A. Clarke and N. Hammond. 1993. Mapping La Milpa: a Maya city in Northwestern Belize. *Antiquity* 254: 96–108.

Tourtellot, G. and N. Hammond. 1994. More light on La Milpa. Maya settlement archaeology in northwestern Belize. *Mexicon* 16: 119–124. https://www.jstor.org/stable/23761047

Tovar-Soto, H.M.; G.A. Ballesteros-Patrón; A. Robledo-Paz y A. Cruz-León. 2012. Botánica de *Spondias* L. (Anacardiaceae): Jocotes, Jobos, Abales o Ciruelas Mexicanas en México. In: A. Cruz-León et al. Jocotes, Jobos, Abales o Ciruelas Mexicanas. Edo. de México, Texcoco, Chapingo, UACh; pp. 11–29.

Tozzer, A.M. y M.A. Glover. 1910. *Animal Figures in the Maya Codices.* Papers of The Peabody Museum of American Archaeology and Ethnology. Harverad University, Vol. 18, N° 3, Published by the Peabody Museum, Cambridge.

Trautmann, W. 1981. *Las Transformaciones en el Paisaje Cultural de Tlaxcala Durante la Época Colonial.* Franz Steiner Verlang GMBH, Wiesbaden, BDR.

Trens, M.B. 1956. El imperio en Chiapas 1863–1864. Tuxtla Gutiérrez, Talleres Gráficos.

Trenti-Very, L.C. 2022. De quien son los agostaderos? "Ganadería extensiva, economía familiar campesina y mercado en el Altiplano Potosino". PhD Thesis, Ecology and Sustainable Development. México, S. Cristóbal Las Casas, Chiapas, El Colegio de la Frontera Sur (ECOSUR).

Tuz Chi, H. 2013. *Cosmovisión e identidad en los rituales agrícolas de los mayas peninsulares.* Sría. de Educación del Gob. del Edo. de Yucatán (SEGEY)- Casa de la Historia de la Educación de Yucatán.

Turner B.L. II. 1990a. The rise and fall of Maya population and agriculture, 1000 B.C. to present: The Malthusian perspective reconsidered. In: *Hunger and History: Food Shortages, Poverty and Deprivation.* L. Newman (ed.). Blackwell, Oxford; pp. 178–211.

Turner B.L. II. 1990b. Population reconstruction of the central Maya Lowlands: 1000 B.C. to A.D. 1500. In: T.P. Culbert and D.S. Rice (eds.); *Precolumbian Population History in the Maya Lowlands.* University of New Mexico Press, Albuquerque; pp. 301–324.

Turner B.L. II. 2010. Land change in the southern Yucatán: Case studies in land change science. In: *Regional Environment Change,* 10: 169–174.

Turner B.L. II. and P.D. Harrison. 1983. *Pulltrouser Swamp: Ancient Maya Habitat, Agricultura an Settement in Northern Belize.* Austin, University of Texas Press.

Turner B.L. II.; R.E. Kasperson; P.A. Matson; J.J. McCarthy; R.W. Corell; L. Christensen; N. Eckley; J.X. Kasperson; A.Luers; M.L. Martello; C. Polsky; and A. Schiller. 2003a. A framework for vulnerability analysis in sustainability science. *Proceedings of the National Academy of Sciences*100(14): 8074–8079.

Turner B.L. II.; A. Matson; J.J. McCarthy; R.W. Corell; L. Christensen; N. Eckley; G.K. Hovelsrud-Broda; J.X. Kasperson; R.E. Kasperson; J. Luers, M.L. Martello; S. Mathiesen; R. Naylor; C. Polsky; A. Pulsipher; A. Schiller; H. Selin and N. Tyler. 2003b. Illustrating the coupled human–environment system for vulnerability analysis: three case studies. *Proceedings of the National Academy of Sciences* 14(100): 8080–8085.

Turner B.L. II.; D. Lawrence. 2011. Land architecture in the Maya lowlands: implications for sustainability. In: P. Gepts et al. (eds.). *Biodiversity in Agriculture: Domestication, Ecolution and Sustainability.* Cambridge University Press, Cambridge, U.K.; pp. 445–463.

Turner B.L. II. and J.A. Sabloff. 2012. Classic period collapse of the central Maya Lowlands: insights about human-environmental relationships for sustainability. *Proceedings of the National Academy of Sciences of the United States of America* 109: 13908–13914.

Tutino, J. 1990. *De la Insurrección a la Revolución en México. Las Bases Sociales de la Violencia Agraria,* 1750–1940. México, Ed. Era, Colec. Problemas de México.

Tylor, W.B. 1979. *Drinking, Homicide and Rebellion in Colonial Mexican Villages.* Stanford, University Press.

Tyrtania, Geidt, L. 1992a. *Yagavila. Un ensayo de ecología cultural.* México, Univ. Autónoma Metropolitana, N° 8, Textos y Contextos.

Tyrtania, Geidt, L. 1992b. Lagos artificiales: la evolución de los impactos ecológicos de la Presa Miguel Alemán. *Alteridades* 24: 103–108.

Uruñuela, L.G.G. y P. Plunket N. 2012. De tronco-cónicos a cuexcomates: hacia la exhibición del almacenamiento a finales del Formativo. In: Bortot, S. y D. Michelet V. Darras (eds.). *Almacenamiento Prehispánico del norte de México al Altiplano Central.* Laboratorio Arqueológico de las Américas, Univ. Paris 1, Phanteon-Sorbonne; Universidad Autónoma de S. Luis Potosí y Centro de Estudios Mexicanos y Centroamericanos (CEMCA); pp. 41–60.

Unidad de Informática para la biodiversidad (UNIBIO). 2004–2018. Colecciones Biológicas Nacionales. México, UNAM, Instituto de Biología (Herbario Nacional). In: unibio.unam.mx

USDA. Departament of Agriculture, USA. 2009, 2013. Data Base.

Valadez-Azúa, R. 1992. Impacto del recurso faunístico en la sociedad teotihuacana. Thesis, Doct., Biology, Facultad de Ciencias, México, UNAM.

Valencia, R. 1981. "Persistencia y expansión de sistemas agrícolas tradicionales. El caso del huamil en El Bajío". Tesis Lic. en Antrop. Social, México, Universidad Autónoma Metropolitana Iztapalapa (UAM-Iztapalapa).

Vallejo-Nieto, M.I.; F.D. Gurri-García; D.O. Molina-Rosales. 2011. "Agricultura Comercial, tradicional y vulnerabilidad en campesinos". *Política y Cultura* 36: 71–98.

Van der Ploeg, J.D. 2008. *The new peasantries. Struglers for authonomy and sustainability in an era of Empire and Globalization.* Earthscan Ltd.

Vanderwarker, A.M. 2006. *Farming, Hunting, and Fishing in the Olmec World.* Texas: The University of Texas Press.

Vara-Morán, A. 1980. "La dinámica de la milpa en Yucatán: el solar". En: Hernández X.E. y R. Padilla (Eds.); *Seminario sobre producción agrícola en Yucatán*. Gob. del Estado de Yucatán, SPP, SARH y COLPOS de Chapingo; pp. 305–341.

Vargas-Guadarrama, L.A. 2007a. La historia incompleta del maíz y su nixtamalización. *Cuadernos de Nutrición* 30(3): 98–104.

Vargas-Guadarrama, L. A. 2007b. "El alimento básico en las cocinas de la humanidad. El caso de Mesoamérica". *Itinerarios. Revista de estudios lingüísticos, literarios, históricos y antropológicos.* Polonia, Varsovia, Inst. de Estudios Ibéricos e Iberoamericanos, Vol. VI: 39–53.

Vargas-Guadarrama, L.A. y L.A. Casillas 2010. Comer, beber, cuerpo y cosmovisión, un viaje de ida y vuelta. *Anales de Antropología* 42: 87–115.

Vargas-Guadarrama, L.A y H. Bourges R. 2012. Los fundamentos biológicos y culturales de los cambios de la alimentación conducentes a la obesidad. En: Rivera D.J.; Hernández Á.M.; Aguilar S.C.; Vadillo O.F.; J.A. Vargas-Velázquez, S. (Eds.); *Obesidad en México. Recomendaciones para una política de Estado.* Academia Nacional de Medicina; pp. 99–123.

Vargas-Velázquez, S. 2008. "De la centralización a la gestión integrada del agua en la cuenca Lerma-Chapala". Tesis Doct. en Antrop. Social. México, Univ. Iberoamericana.

Vargas-Velázquez, S. y E. Mollard (Eds.) n/d. "Riego y gestión del agua en la cuenca Lerma-Chapala. Documentos para su historia, 1896–1985". México, SEMARNAT, Inst. Mexicano de Tecnología del Agua (IMTA).

Vásquez-Dávila, M.A. (coord.). 1992. *Cómo predecimos el clima. Augurios animales.* Instituto Agropecuario de Oaxaca. Casa de la Cultura Oaxaqueña. Colección: Como hacemos N° 4.

Vásquez-Dávila, M.A. y D. Lope-Alzina (comps.). 2012. *Aves y Huertos de México.* México, Oaxaca, CONACYT Red de Etnoecología y Patrimonio Biocultural, Instituto Tecnológico del Valle de Oaxaca y Carteles (eds).

Vavilov N.I. 1992 [1926a]. *Origin and Geography of Cultivated Plants.* Cambridge University Press

Vavilov N.I. 1926b. Studies on the Origin of cultivated plants. *Bulletin of Applied Botany* Vol. XVI, N° 2.

Vavilov N.I. 1927. Regularities in the geographical distribution of genes of cultivated plants. *Bulletin of Applied Botany* Vol. XVII.

Vavilov N.I. 1931a. Mexico and Central America, as a fundamental centre of the origin of cultivated plants of the new world. *Bulletin of Applied Botany.*

Vavilov N.I. 1994 [1931b]. México y Centroamérica como centro básico de origen de las plantas cultivadas del Nuevo Mundo. *Boletín de Botánica aplicada, Genética y Fitomejoramiento,* Tomo 26. N° 3 (en Ruso). Traducción al español por E. Gribovskaia y R. Ortega-Paszka. *Revista de Geografía Agrícola* Vol. 20; pp. 15–33.

Vela, E. (ed.) et al. 2014. *El maguey. Arqueología mexicana*, N° especial 57.

Vela, E. 2019. Insectos en Mesoamérica. Usos, simbolismo, Alimento y materia prima. Abejas, Hormigas, Gusanos, Mariposas. *Arqueología mexicana*, N° especial 86.

Velasco-Orozco, J.J. 2002. Subsistencia campesina y desarrollo sustentable en la región monarca. Thesis, BA., Social Anthropology, Universidad Autónoma del Estado de México (UAEM).

Velasco-Orozco, J.J. 2007. El papel de la agricultura en el patrón de subsistencia de una sociedad campesina en transición. In: A. González-Jácome; S. del Amo-Rodríguez; F.D. Gurri-García (coords.); *Los nuevos caminos de la agricultura: procesos de conversión y perspectivas.* México, Universidad Iberoamericana AC, Plaza y Valdés, PROAFT; pp. 331–335.

Velasco-Toro, J. 1990. El Uxpanapa, aproximación histórica a una experiencia de colonización agrícola con población chinanteca relocalizada. *Anuario* VII, Instituto de Investigaciones Históricas, Instituto de Investigaciones Humanas, Xalapa, Veracruz.

Velásquez-Morlet, A. and E. Nalda. 2005. Los mayas de la Península de Yucatán: viejas ideas, nuevas ideas. *Arqueología Mexicana* XIII(75): 30–37.

Vibrans, H. 1997. Lista florística comentada de plantas vasculares silvestres en S. Juan Quetzalcoapan, Tlaxcala. *Acta Botánica Mexicana* 38: 21–67.

Vibrans, H. 1999. Epianthropochory in Mexican weed communities. *American Journal of Botany* 86: 476–481.

Vibrans, H. 2002. Origins of weeds: benefits of clean sedes. In D. Pimentel (ed.); *Encyclopedia of Pest Management*; Marcel Dekker, New York; pp. 558–561.

Vida, C.; A. de Vicente; F.M. Carzola. 2019. The role of organic amendments to soil for crop protection induction of suppression of soilborne pathogens. *Annals of Applied Biology* Vol. 176, Issue 1, 01/2020; pp. 1–15.

Vielle-Calzada J.P.; Vallebueno M.; Padilla-Calzada J.; C. Álvarez-Mejía; J. Martínez; A. García-Cook y R. Montiel. 2014. Paleogenómica en la cueva San Marcos y su contribución al entendimiento del origen del maíz. *Arqueología* 49: 7–20.

Viesca, C. 1982. Hambruna y epidemia en Anáhuac (1450–1454) en la época de Moctezuma Ilhuicamina. In: Florescano E. y E. Malvido (eds.); *Ensayos sobre la historia de las epidemias en México*. México, Instituto Mexicano del Seguro Social (IMSS); pp. 157–165.

Viesca, C. 2010. Epidemias y enfermedades en tiempos de la Independencia. *Revista Médica del Instituto Mexicano del Seguro Social* 48(1): 47–54.

Villalobos-Nájera, H. 1998. *El pan y los panaderos de San Juan Huaczinco*. Gobierno del estado de Tlaxcala.

Villalobos-Nájera, H. 2008. *Artesanos y artesanías de Tlaxcala*. Gob. del Edo. de Tlaxcala, Instituto Tlaxcalteca de la Cultura.

Villalobos-Nájera, H. 2009. *Sobre artesanos y artesanías en Tlaxcala*. CONACULTA, Gob. del estado deTlaxcala.

Villalobos-Nájera, H. 2012. *La vida sobre rieles. Una historia del ferrocarril en Tlaxcala*. México, SEP/USET, Gobierno del Estado de Tlaxcala.

Villalobos-Nájera, H. 2013. El gobierno de Isidro Candia, Hacia un Nuevo Proyecto de Nación: Tlaxcala y el Cardenismo. In: A, González-Jácome (coord.); *Isidro Candia. El reparto agrario en Tlaxcala*. México, Gobierno del estado de Tlaxcala, SEPEUSET; pp. 63–104.

Villa-Rojas, A. 1948a. (ed. 1973). Antropología de una Presa: Los Mazatecos y el Proyecto de Papaloapan. In: D. McMahon; *Antropología de una Presa*: Los Mazatecos y el Proyecto de Papaloapan. México.

Villa-Rojas, A. 1948b. El papel de la antropología en las obras del Papaloapan. *América Indígena* VIII(4): 301–312.

Villa-Rojas, A. 1955. *Los mazatecos y el problema indígena de la cuenca del Papaloapan*. México, INI.

Villanueva, V.C. (comp.). 2017. *Cultivo de Calabazas*. Estado de México, Texcoco, Chapingo, UACH.

Villela S.F. 2016. El huauhtli sagrado. Los tamales *tzoalli* entre los nahuas de Guerrero. *Arqueología Mexicana* 138: 46–53.

Wake, E. 2015. The Maize Theatre: Mexica State Cult and Symbolic Representations of Maize in Prehispanic Religious Ritual. In: B.A. Albores (coord.), *Flor-flora. Su uso ritual en Mesoamérica*. El Colegio Mexiquense, Fondo Editorial del Gobierno del Estado de México; pp. 155–175.

Warde, A. and L. Martens. 2000. *Eating Out. Social Differentiation, Consumption and Pleasure*. Cambridge University Press, United Kingdom.

Wallerstein, I.M. 1979 [English ed. 1974]. *El moderno sistema mundial. La agricultura capitalista y los orígenes de la economía-mundo europea en el siglo XVI*. (Vol.I). Madrid, Siglo XXI.

Wallerstein, I.M A. 2005 [2004]. Análisis de sistemas-mundo. Una introducción. In: *Conocer el mundo, saber el mundo*. México, Siglo XXI.

Weatherwax, P. 1955. History and origin of corn. Early history of corn and theories as to its origin. G.F. Sprague (ed.). *Corn and corn improvement*, 1a Ed. New York, Academic Press; pp. 1–16.

Weirsum, K.F. 1997. From natural forest to tree crops, co-domestication of forests and tree species, an overview. *Netherlands Journal of Agricultural Science* 45: 425–438.

Weitzel, R.L. 2005. *Periphyton: Ecology, Exploitation, and Management*. Wallingford, CABI Publishing.

Wellhausen, E.J., L.M. Roberts y E. Hernández X. 1952. *Races of Maize in Mexico. Their Origin, Characteristics and Distribution*. The Bussey Institution of Harvard University In Spanish 1951 by SAG.

Werner, G. 1986 [1988, 2012]. *Los Suelos en el Estado de Tlaxcala. Altiplano Central Mexicano*. México: Gobierno Estatal de Tlaxcala y Universidad Autónoma de Tlaxcala. Talleres Gráficos del Estado. Ed. SEPE/USET, estado de Tlaxcala.

Werner, G. 1992. Suelos volcánicos endurecidos (tepetates) en el estado de Tlaxcala: distribución, rehabilitación, manejo y conservación. *Terra* 10: 318–331.

West R.C. 1970. Population Densities and Agricultural Practices in Pre-Columbian Mexico, with Emphasis on Semiterracing. Verth, d. 38. International. In: Amerikanistes Kongresses, Stuttgar München, 12–18 August 1968, Vol. 2: 361–369.

West R.C. y P. Armillas. 1950. Las chinampas de México. Poesía y Realidad de los jardines Flotantes. *Cuadernos Americanos* 150: 165–182.

West R.C.; N.P. Psuty B.G. Thron. 1969. *The Tabasco Lowlands of Southeastern Mexico*. Baton Rouge, Lousiana State Universty.

Whitaker T.; H. Cutler; R. MacNeish. 1957. Cucurbit materials from three caves near Ocampo, Tamaulipas. *American Antiquity* 22(4): 352–358.

Whitaker T. 1986 [1965]. Cucurbits and cultures in the Americas. *Economic Botany* 4: 344–349.

White C.D.; F.J. Longstaffe y H.P. Schwarcz. 2006. Social directions in the isotopic anthropology of maize in the Maya region. In: *Histories of Maize. Multidisciplinary Approaches to the Prehistory, Linguistics, Biogeography, Domestication and Evolution of Maize*. New York, Oxford, Paris, Elsevier, Academic Press: 143–160.

Whiteford, S. y K. Enge. 1989. *The Keepers of Water and Earth. Mexican Rural Social Organization and Irrigation.* Texas University Press.

Withmore, T. 1992. *Disease and Death in Early Colonial México. Simulating Amerindian Depopulation.* Boudler, Colorado, Westview Press.

Wilken G.C. 1969. Drained field agriculture: an intensive farming system in Tlaxcala, Mexico. *The Geographical Review* LIX(2): 215–241.

Wilken G.C. 1987. *Good Farmers: Traditional Agricultural Management in Mexico and Central America.* Boulder, Colorado, Westview Press.

Wilkerson, J.K. 1983. So green and like a garden: intensive agriculture in ancient veracruz. Darch, J.P. (ed.). *Drained Field Agriculture in Central and South America.* Oxford, England, BAR International Series 189.

Wilkerson, J.K. 1987. Perspectivas de la Prehistoria de Veracruz y de la costa del golfo de México. In: A. González-Jácome (comp.) *Orígenes del hombre americano* (Seminario). México, SEP; pp. 209–222.

Wilkes, G. 2004. Corn, strange and marvelous: but is a definitive origin known? In: Corn: Origin, History, Technology, and Production. C. Wayne Smith (Ed.); John Wiley & Sons, Inc, pp. 3–63.

Willey, G.R. 1981. Maya Lowland settlement patterns: a summary view. *Lowland Maya Settlement Patterns.* W. Ashmore (ed.). Albuquerque, New Mexico University Press; pp. 385–415.

Willey, G.R. 1983. *Suelos y Sistemas de Producción Agroforestales. Programa Suizo de Coperación para el Desarrollo DDA. INFORAT: Información y Documentación Forestal para América Tropical.* Turrialba. Costa Rica.

Williams, B. 1980. Pictorial representation os soils in the valley of Mexico: evidence from Codex, Vergara. *Geoscience and Man* 21: 51–60.

Williams, B. and C. Ortiz-Solorio. 1981. Middle American folk soil taxonomy. *Annals of the Association of American Geographers* 71: 335–358.

Wilk, R.R. 1997. *Household Ecology. Economic Change and Domestic Life among the Kekchi Maya of Belize.* University of Arizona Press.

Wilk, R.R. 2019. *The Hosehold Economy. Reconsidering the Domestic Mode of Production.* Routlege.

Wittfogel, K.A. 1956. *The Hydraulic Civilizations.* Chicago.

Wittfogel, K.A. 1966. *Despotismo oriental: estudio comparativo del poder totalitario.* Ed Guadarrama.

Wonderly, W.L. 1946. Textos en zoque sobre el concepto del nagual. *Tlalocan* II(2): 97–105.

Wonderly, W.L. 1947. Textos folklóricos en zoque. Tradiciones acerca de los alrededores de Copainalá, Chiapas. *Revista mexicana de estudios antropológicos* 9: 135–163.

Wood, T. and W. H. Bollinger. 1990. Tree improvement from the ground up: the potential for a select microbial inocula in forestry. In: Gliessman, S.R. (ed.), *Agroecology. Researching the Ecological Bases for Sustainable Agriculture.* New York, Berlin, London, Springer-Verlag; pp. 184–204.

Xelhuantzi-López, M.S. 2015. Consideraciones sobre la flora arqueológica. In: B.A. Albores-Zárate (comp.). *Flor-Flora, Su uso ritual en Mesoamérica.* Fondo Editorial del Estado de México, Colegio Mexiquense; pp. 101–111.

Yaeger J.; D.A. Hodell. 2008. The collapse of Maya civilization: assessing the interaction of culture, climate, and environment. El Niño, Catastrophism, and Culture Change in Ancient America. D.H. Sandweiss; J. Quilter (eds.). Dumbarton Oaks, Washington, DC; pp. 197–251.

Zambardino, R. 1980. Mexico's population in the sixteenth century: demographic anomaly or mathematical ilusión? *Journal of Interdisciplinary History*, I: 1–27.

Zárate Z.; R., M. Anaya-G.; C. Ortiz-S and M. E. Ramírez-G. 1991. Diagnóstico y control de la degradación de la tierra en el estado de Tlaxcala. *Agrociencia* 2: 59–81.

Zizumbo, V.D. and H.J. Quero. 1998. Re-evaluation of early observations on coconut in the new world. *Economic Botany* 52(1): 68–77.

Zizumbo, V.D y P. Colunga, G.M. 2008. El origen de la agricultura, la domesticación de plantas y el establecimiento de corredores biológico-culturales en Mesoamérica. *Revista de Geografía Agrícola* 41: 85–113.

Zizumbo, V.D. 2017. La milpa del occidente de Mesoamérica: profundidad histórica, Dinámica evolutiva y rutas de dispersión a Suramérica. *Revista de Geografía Agrícola, estudios regionales de la agricultura mexicana*, Estado de México, Texcoco, Universidad Autónoma Chapingo (UACH), Dirección de Centros Regionales, N° 58, enero-junio; pp. 33–46.

Zolla, C. 2008. Antropología médica, salud y medicina en la obra de Gonzalo Aguirre Beltrán. México, UNAM. In: http://www.nacionmulticultural.unam.mx/portal/cultura_politica/carlos_zolla_20081008.html.

Index

achenes 320
agave (*maguey*) 4, 8, 64–65, 69, 88, 90–93, 99, 102, 108, 110–113, 115, 121–124, 127, 129, 130, 132, 134, 139, 143, 187–188, 191, 215, 217, 225, 229–232, 238–240, 277, 290–292, 295, 297, 300, 301, 302, 308, 329–330, 346–347, 351, 356–358, 389–390, 400–402, 408–409, 414, 416, 420, 436, 443
agave rows (*metepantles*) 113, 127, 139, 143, 215, 217, 231–232, 239–240, 349, 356, 390, 409
agroecology 1, 21, 43, 66, 74–75, 101, 116, 285, 381–382, 385, 386–387, 390, 404
agroecosystems 1–2, 6, 12, 22, 25, 28–33, 35–36, 40, 43, 45, 49–50, 53, 56, 66, 72–73, 75, 78–79, 82, 88–89, 99, 108, 113–116, 119, 124, 125, 127, 141, 147, 153–154, 160, 166–167, 169, 174–175, 178, 187–188, 198, 206, 209, 211, 214–217, 219, 222, 236, 238, 243, 263, 264, 269, 272, 275–276, 287–288, 308, 309, 310, 315, 321, 328, 341–342, 356, 358, 381–383, 385–387, 394, 499, 406, 409, 419
agroforestry 30, 55, 77, 117, 125, 153, 156, 174, 202, 213, 219, 329, 384, 395
agronomy 5, 7–8, 19, 66, 382
alpine grasslands 84, 127
alternative agriculture 1, 21, 25–26, 43, 66, 72, 342, 386
altitude 30–31, 36, 49, 55, 60, 83–84, 86, 90, 92, 99, 121, 141, 179, 187, 193, 215, 245, 250, 253, 255, 267, 269–271, 273, 276, 311, 313, 348, 357, 397, 407–408, 411
altitudinal 17, 30, 84, 215, 219, 233, 243, 247, 274, 276, 295, 315
amaranth 32, 64, 92–93, 96–97, 123, 137, 144, 148, 209, 224, 234, 252, 271, 291, 297–298, 302, 308, 310, 312, 316, 329, 389, 400
aquatic 57, 72, 74, 83–84, 91, 97, 145, 170–171, 225, 228, 234–235, 270, 289, 309, 401, 405, 418
archaeology 1, 5, 23, 60–61, 82, 89, 180, 226, 391, 412
arid 36, 65, 84–85, 90–92, 94, 109, 119, 121–122, 186, 199, 231, 340, 345, 358, 389, 403, 414–415
aromatic flowers/herbs/plants 108, 124, 127, 132, 134, 145–146, 189, 191, 193, 240, 253, 295, 297–298, 300–301, 306, 309, 329, 390, 393, 414
atole (*Atolli*) 69, 133, 136, 149–150, 184–185, 252, 265, 295, 300, 304, 306, 309, 314, 318–319, 334, 336, 370, 391–393, 417

bands 24, 31–32, 34, 81, 84–86, 90, 92–103, 110, 117–118, 120–123, 153, 171, 175–176, 179, 182, 235, 245, 288–292, 295, 336, 391, 411
barn 50–51, 105–106, 266, 398, 405
basin of Balsas River 14, 31, 37, 65, 67, 81, 84–90, 96–97, 99, 103, 169, 224, 233, 288, 290, 297, 300, 336, 355
beans 10, 14, 21–22, 32–33, 64, 76, 88, 90, 96–97, 104–105, 107–108, 114, 119, 124, 129–130, 136, 138, 144, 151, 154, 160, 165, 173, 182–183, 189, 218, 220–222, 227–228, 230, 234, 240–241, 255, 260–261, 264–267, 273, 277–284, 289, 291–292, 294–295, 297–298, 300–301, 303, 307, 309, 312, 314–315, 316–318, 320, 325–326, 329–331, 337, 342, 370, 383, 389, 397, 401, 418
Belize 83, 89, 125, 181, 234, 277–278, 294, 343, 360, 375, 394, 408
biodiversity 33, 48–49, 51, 56, 72–76, 85, 95, 97, 117, 141, 169, 175, 188, 200, 202, 213, 218, 287–288, 309, 315, 330, 386
broad beans 148, 182, 260–261, 312, 316, 322, 325, 329, 331–332

cacao 3, 64–65, 69, 76, 87, 103, 132, 134, 140, 145, 153, 156, 160–161, 181, 191, 225, 252, 255, 271–272, 278–280, 284, 295, 299–301, 303, 304, 309, 379, 392–393, 396, 409
Cacaxtla/Tlaxcala 50, 365–367
Campeche 35–36, 54, 84, 115, 159, 165, 177, 197, 203, 211, 253, 309, 344, 368, 372
canal irrigation 16, 34–38, 41, 57, 59, 68, 78, 93–94, 99, 102, 113, 124, 127, 128, 134, 136–137, 141, 145, 147, 148, 150, 154, 171, 189, 192, 196, 212, 217, 219, 222, 256, 258, 286, 311, 315, 329, 346, 349, 351–353, 354–356, 358, 377, 390, 391–392, 396, 402–403, 409, 410–411, 414, 416
cane planter (*coa*) 10, 113–114, 235, 240, 251, 254, 263, 407
ceremonies 33, 58, 71–72, 76, 138, 142, 191, 235, 241, 247, 281, 295, 362–364, 367, 369–374, 389, 419
chinampa 5, 35, 37, 39, 56–57, 127, 128, 137, 141, 145, 147, 239, 256, 297, 298, 352–353, 355, 390, 392
clay plate (*comal*) 110, 148, 159, 182, 198, 292, 300, 304, 305–306, 316, 323, 332, 401–402, 407, 418
climate 21, 28–30, 32–36, 44, 49, 53–57, 66, 75, 84–85, 88, 90–92, 101, 112, 115, 117, 121–122, 124–125, 127, 140, 163–166, 174, 176, 186–187, 193, 196, 199, 222, 250, 255–256, 267–268, 270–271, 276, 287, 288, 298, 311, 313, 340–341, 342, 361, 363, 368, 381, 383–386, 391, 397, 416
coffee cultivation 4, 10–11, 15, 28, 30, 44, 69, 208, 212, 237–238, 252–253, 268, 272, 274–276, 282–284, 314, 322, 391
Costa Rica 83, 89, 112, 179, 295, 408
cultural ecology 1, 5, 7, 9, 19–21, 23, 381, 395, 397, 400, 411–412
cultural values 22, 29, 75, 395

drained fields/raised fields/ridges (*camellones*) 35–36, 38, 53, 76, 120, 128, 138, 154, 177, 190, 192, 215, 222–223, 226, 236, 255, 258, 261, 329, 342, 346, 355, 392

domestication 1, 21, 30–32, 58, 65, 67, 72, 81–84, 86–90, 92–97, 99–101, 107, 115, 118, 121–125, 127, 147, 154, 160–161, 169, 175–176, 179–181, 185, 200–201, 224–225, 233–234, 285, 288–289, 291–292, 297, 320, 336–337, 343, 369, 374, 389–391, 396, 399, 406, 417, 428
diffusion 1, 21, 24, 52, 63, 65, 72, 82, 87, 89, 97, 102, 109, 115, 133, 135, 179, 209, 244, 295, 297, 302, 314, 377, 383, 400
diseases 10, 22, 37, 45–46, 48–50, 52–54, 68, 75, 114, 116, 124, 146, 158, 160, 178, 188, 228–229, 236, 267, 364, 373, 384, 395

earth banks/earth edges/land edges (*bordos de tierrasetos*) 53, 57, 132, 134, 162, 215, 219, 222–223, 231, 272, 329, 346, 349, 351–352, 356, 358, 396
ecological mosaic 84, 270, 358, 397
edible green weeds (*quelites*) 6, 41, 68, 89, 113, 124, 129, 131, 137, 148–150, 174–175, 182–185, 191, 193, 225, 240, 261, 266, 271, 277, 287, 291, 298, 300–301, 310, 311–312, 310–319, 322, 328–329, 331–336, 402
extended family 71, 76, 213, 251, 275, 406, 411

fallow 48, 79, 154, 216, 235, 250–253, 266, 272–273, 276, 281, 331
fermented Agave mead (*pulque*) 30, 88, 99, 110–112, 139, 143, 229–231, 239, 290, 300, 308, 346–347, 390, 400, 408, 420
foggara (quanatin-line wells) 34–35, 358, 403, 414–415
folk 20, 45, 75, 80, 193, 208, 264, 268, 385–387, 395, 402, 418
food security 74–75, 213, 286

garbage dumps 32, 92, 117–121, 153, 179, 197, 200, 225, 234, 288–289
gourds (*Crescentia alata* Kunth) (*jícaro, jícara, tecomate, cuezcomate, guaje cirial*) 97, 124, 140, 151, 181, 252, 392
gourds (*Lagenaria siceraria*) (*bule, guaje, jícaro, porongo, acocote*) 93, 122–123, 151, 181, 291–292, 298, 389, 406–407
granted lands (*ejido*) 77, 251, 257, 262, 264, 267, 282, 394, 416
Guatemala 11, 37, 65, 79, 82–83, 89, 96, 135, 140, 143, 151, 171, 179, 224, 236, 294–295, 339, 343, 357, 360, 392, 408–409, 420

hardened soils (*tepetate*) 139, 271, 244, 348–349, 352, 358–359
holes/slat-board systems (*jollas, hoyos, ollas*) 34–35, 41, 264, 352, 406
hominy maize (*maíz nixtamalizado*) 120, 176, 253, 277, 292, 295, 298, 300, 303, 306, 313–314, 322–323, 402, 407
Honduras 83, 87, 89, 156, 180–181, 225, 295, 360, 392, 407, 452
hunting garden model 120, 156, 166, 170, 176–177, 182, 191, 199, 204, 225, 251, 288, 294, 313–315, 330, 405

incidental homegarden 32, 93, 116–124, 147, 153, 175–176, 180–182, 200–201, 225, 234, 288, 291–292, 294, 297, 313, 343, 406

intermingled/intermixed crops 113, 215, 217, 221, 230, 255–256, 329, 390

jagüey/jagüeyes (open cisterns) 34–35, 59, 292, 349–352, 356, 406
jaltun (*sarteneja*) 34, 416
jobo fruit/Mexican hog plum (*Spondias mombin* L.) 32, 46, 64–65, 82, 84, 89, 93, 111, 121–124, 132, 140, 154, 179–180, 190, 201, 224, 252, 280, 289, 291, 297, 302, 308–309, 320, 323, 389, 391, 400, 417
jobo fruit/Mexican hog plum (*Spondias purpurea*) 32, 84, 132, 140, 154, 180, 187–188, 190, 201, 224, 234, 252, 271–272, 280, 289, 302, 308–309, 320, 323, 417
jobón (Honeycomb, *panal*) 124–126, 162, 167, 282

kilns 126, 163, 290, 293, 295, 301, 347, 409–411

La Joya 119, 156, 170, 225, 234, 293
La Playa archaeological site in Sonora 31, 100, 101, 290
Los Tuxtlas/Veracruz 30–32, 97, 119–120, 154, 156, 170–171, 176–177, 181, 200, 204, 206, 225, 234, 240, 248, 267–268, 277, 279, 292–295, 309, 368–369, 405

March Maize (*marceño*) 36, 38, 76, 153, 247, 254–255, 276, 390, 407, 413
Mescal/Mexcal/Mezcal 8, 71, 231–232, 290, 297, 408
mestizo 3, 5, 13, 24, 124, 171–172, 214, 230–231, 239, 265, 283, 316, 362, 373, 398, 403, 411
Mexican black cherry (*capulín*) 64, 97, 129–133, 140, 143, 188, 190, 211–212, 222, 225, 259, 297–298, 301–302, 329, 401, 413
muddy water (*agualodo*) 137, 141, 212, 222, 275, 329, 344, 353, 358, 390, 392, 406

nahuas 92, 128–129, 145, 265, 369, 389
nuclear family 59, 70–71, 76, 160, 211, 214, 275, 331, 398, 406, 411

open cisterns (*jagüeyes, sartenejas*) 34–35, 59, 292, 349–352, 356, 405–406
Opuntia (*Nopal*) 64, 67, 88, 90, 92–93, 96, 103, 110, 119, 121–124, 129–130, 136–137, 139–140, 143, 149, 183, 188, 194, 206, 209, 215, 225, 230, 261, 272, 289–292, 297–298, 301–308, 308, 309, 310, 314, 317, 330, 333, 349, 355, 358, 390
orchard 43, 79, 119, 124–125, 127–131, 166, 169, 186–188, 191–192, 208–211, 213, 217–219, 255–256, 271, 283–284, 311, 390, 410, 413

pests/pesticides 6, 22, 50, 52, 55, 72, 75, 105, 114, 120, 124, 128, 170, 182, 215–216, 222, 243–244, 267, 276, 286, 310, 328, 402, 404
petenes (*akalché*) 36, 38, 341, 352, 355, 377, 413
Puebla (Purrón Dam) 35, 37, 93–95, 121, 123, 292, 297, 358, 403, 414
Puebla (Tehuacán) 20, 31, 35, 37, 40, 83, 88–98, 101, 121–123, 153–154, 177, 181, 224–225, 291–292, 297, 355, 358, 364, 391, 403, 414–415, 417

Quintana Roo 39, 154, 194, 197, 253, 352, 364–365, 368–369, 372

resilience 17, 20, 33, 39, 49, 60, 66, 74–75, 113, 116, 174, 186, 198, 202, 205, 213, 243, 316, 341, 381–382, 387, 415
ridges/drained fields/raised fields (camellones, chinampas de tierra adentro, Campos elevados, Campos drenados) 16, 35–36, 38, 53, 76, 128, 138, 177, 190, 226, 236, 240, 245, 255, 258, 261, 342, 346, 355
risk 6, 43, 49, 54, 60, 66, 120, 139, 142, 170, 177, 266–267, 364, 368, 373, 381, 383, 385, 415

semi-terraces (*bancales*) 29, 41, 78, 110, 115, 230–231, 239, 266, 311, 346, 349, 390, 409, 416
stirrups (*estriboscajetes*) 35, 68, 349, 352–354, 356–358, 377
swamp agriculture 38, 127, 153, 254–255, 277, 341–342, 407
sustainable 1, 20–21, 49, 51–52, 56, 58, 72, 74–75, 78, 116–118, 125, 160, 169, 174, 198, 200–202, 205–206, 209–210, 213, 243, 253, 287, 339, 341, 343, 360, 382, 366, 387, 390, 414, 419
soil erosion 4, 6–7, 30, 53, 55, 68, 116, 128, 138, 142, 201, 215, 263–264, 272, 310, 312, 331, 341–349, 353, 356–360, 363, 377, 384, 390, 409, 416
slash and burn (*roza y quema*) 4, 6–7, 11, 20, 28–30, 32, 53, 60, 76, 79, 116, 154–156, 176, 219–220, 225, 227, 234, 236, 244, 250–251, 253–254, 263–264, 272–274, 278, 282–284, 298, 330, 339–342, 377

TAK (traditional agricultural knowledge) 44, 60, 66, 72, 75–76, 247, 287–288, 385, 387, 415–416, 418–419
TEK (traditional ecological knowledge) 72, 75, 247, 288, 385, 386, 413, 418
teosinte (parviglumis) 64, 82–87, 89–90, 93, 95–100, 115, 124, 127, 181, 187, 224, 233, 244, 285, 288–289, 295, 315
Teotihuacan/State of Mexico 2, 4, 21, 35, 129, 167, 229–230, 298–299, 352, 367
Tlatelolco 21, 84, 98, 132, 134, 137, 151, 163, 174, 191, 229, 297, 300–302, 353
Tlaxcala/S. Francisco Tepeyanco (Topoyanco) 119, 130, 136, 140, 140, 173, 198, 201, 204, 209, 211–212, 218, 249–250, 257–258, 261–262, 392, 403, 414
Tlaxcala/St. Isabel Xiloxoxtla 250, 257–259, 261, 263, 328, 347, 363, 370, 403, 414, 415

underground cistern (*chultun*) 34, 36, 351, 356, 393
United States (USA) 1, 5, 9–10, 12, 15–16, 25, 27, 46, 77, 79, 90, 97, 100–101, 195, 290, 395, 403

Vavilov 64–65, 83, 392, 419–420
Veracruz Bezuapan archaeological site 119, 170, 225, 256, 293
vicarious observation 73, 80, 384, 420
vulnerability framework analysis model 6, 20, 23, 49, 51, 54, 420

wetlands 36, 38, 54, 57, 76, 83, 90, 154, 225, 234, 254–255, 341–342, 352, 355, 412–413

Xinantecatl volcano/State of Mexico 57, 255, 371
Xochimilco 37, 57, 84, 147, 329, 377, 418

Zahuapan River/Tlaxcala 16–17, 38, 53, 84, 134, 172, 189–190, 223, 231, 239, 260, 286, 244, 345–346, 348, 356

Taylor & Francis Group
an **informa** business

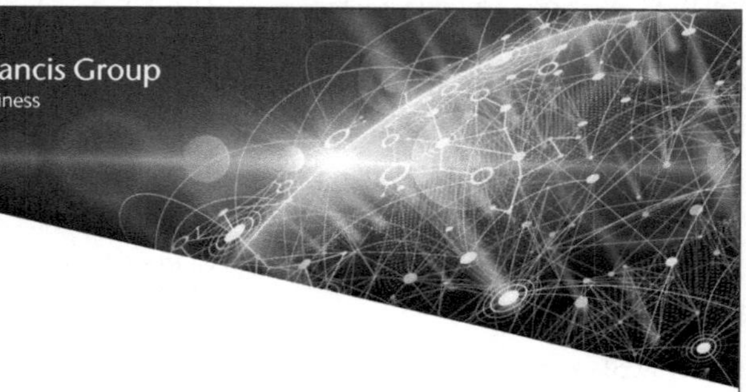

Taylor & Francis eBooks

www.taylorfrancis.com

A single destination for eBooks from Taylor & Francis with increased functionality and an improved user experience to meet the needs of our customers.

90,000+ eBooks of award-winning academic content in Humanities, Social Science, Science, Technology, Engineering, and Medical written by a global network of editors and authors.

TAYLOR & FRANCIS EBOOKS OFFERS:

- A streamlined experience for our library customers
- A single point of discovery for all of our eBook content
- Improved search and discovery of content at both book and chapter level

REQUEST A FREE TRIAL

support@taylorfrancis.com

Routledge
Taylor & Francis Group

CRC Press
Taylor & Francis Group

FSC
www.fsc.org
MIX
Papier | Fördert
gute Waldnutzung
FSC® C083411

Kolibri 360 GmbH
Ferdinand-Jühlke-Straße 7
99095 Erfurt, Deutschland
produktsicherheit@kolibri360.de